Progress in Mathematics
Volume 222

Juan-Pablo Ortega
Tudor S. Ratiu

Momentum Maps and Hamiltonian Reduction

Birkhäuser
Boston • Basel • Berlin

Juan-Pablo Ortega
CNRS-Laboratoire de Mathématiques
de Bensançon
Université de Franche-Comté
UFR des Sciences et Techniques
F-25030 Bensançon Cedex
France

Tudor S. Ratiu
Départment de Mathématiques
Lausanne 1015
Switzerland

Library of Congress Cataloging-in-Publication Data

Ortega, Juan-Pablo.
Momentum maps and Hamiltonian reduction / Juan-Pablo Ortega and Tudor S. Ratiu.
p. cm. — (Progress in mathematics ; v. 222)
Includes bibliographical references and index.
ISBN 0-8176-4307-9 — ISBN 3-7643-4307-9
1. Hamiltonian systems. 2. Global analysis (Mathematics) 3. Global differential geometry. 4. Lie groups. I. Ortega, Juan-Pablo, 1969- II. Ratiu, Tudor S. III. Progress in mathematics (Boston, Mass.) ; v. 222.

QA614.83.H337 2003
515'.39—dc22 2003063012
CIP

AMS Subject Classifications: 70H33, 58F05, 53C80, 22EXX, 81S10

ISBN 0-8176-4307-9 Printed on acid-free paper.

Printed in the United States of America. (HP)

9 8 7 6 5 4 3 2 1 SPIN 10888094

Birkhäuser is a part of *Springer Science+Business Media*

www.birkhauser.com

Ferran Sunyer i Balaguer (1912–1967) was a self-taught Catalan mathematician who, in spite of a serious physical disability, was very active in research in classical mathematical analysis, an area in which he acquired international recognition. His heirs created the Fundació Ferran Sunyer i Balaguer inside the Institut d'Estudis Catalans to honor the memory of Ferran Sunyer i Balaguer and to promote mathematical research.
Each year, the Fundació Ferran Sunyer i Balaguer and the Institut d'Estudis Catalans award an international research prize for a mathematical monograph of expository nature. The prize-winning monographs are published in this series. Details about the prize and the Fundació Ferran Sunyer i Balaguer can be found at

`http://www.crm.es/info/ffsb.htm`

This book has been awarded the Ferran Sunyer i Balaguer 2000 prize.

INSTITVT D'ESTVDIS CATALANS MCMVII

The members of the scientific commitee of the 2000 prize were:

Hyman Bass
University of Michigan

Pilar Bayer
Universitat de Barcelona

Antonio Córdoba
Universidad Autónoma de Madrid

Paul Malliavin
Université de Paris VI

Joseph Oesterlé
Université de Paris VI

Ferran Sunyer i Balaguer Prize winners:

1992	Alexander Lubotzky *Discrete Groups, Expanding Graphs and Invariant Measures*, PM 125
1993	Klaus Schmidt *Dynamical Systems of Algebraic Origin*, PM 128
1994	The scientific committee decided not to award the prize
1995	As of this year, the prizes bear the year in which they are awarded, rather than the previous year in which they were announced
1996	V. Kumar Murty and M. Ram Murty *Non-vanishing of L-Functions and Applications*, PM 157
1997	A. Böttcher and Y.I. Karlovich *Carleson Curves, Muckenhoupt Weights, and Toeplitz Operators*, PM 154
1998	Juan J. Morales-Ruiz *Differential Galois Theory and Non-integrability of Hamiltonian Systems*, PM 179
1999	Patrick Dehornoy *Braids and Self-Distributivity*, PM 192
2000	Juan-Pablo Ortega and Tudor Ratiu *Hamiltonian Singular Reduction*, PM 222
2001	Martin Golubitsky and Ian Stewart *The Symmetry Perspective*, PM 200
2002	André Unterberger *Automorphic Pseudodifferential Analysis and Higher Level Weyl Calculi*, PM 209
	Alexander Lubotzky and Dan Segal *Subgroup Growth*, PM 212

To our wives María Pilar and Lilian

for their love, patience, and understanding

and

to our parents

Joaquín, María Teresa, Mircea, and Rodica

for supporting our decision to

become mathematicians

Contents

Introduction

The use of the symmetries of a physical system in the study of its dynamics has a long history that goes back to the founders of classical mechanics. Symmetry-based techniques are often implemented by using the ***integrals of motion*** that one can sometimes associate to these symmetries. The integrals of motion of a dynamical system are quantities that are conserved along the flow of that system. In classical mechanics symmetries are usually induced by point transformations, that is, they come exclusively from symmetries of the configuration space; the intimate connection between integrals of motion and symmetries was formalized in this context by NOETHER (1918). This idea can be generalized to many symmetries of the entire phase space of a given system, by associating to the Lie algebra action encoding the symmetry, a function from the phase space to the dual of the Lie algebra. This map, whose level sets are preserved by the dynamics of any symmetric system, is referred to in modern terms as a ***momentum map*** of the symmetry, a construction already present in the work of LIE (1890). Its remarkable properties were rediscovered by KOSTANT (1965) and SOURIAU (1966, 1969) in the general case and by SMALE (1970) for the lifted action to the cotangent bundle of a configuration space. For the history of the momentum map we refer to WEINSTEIN (1983b) and MARSDEN AND RATIU (1999), §11.2. It turns out that this "standard" momentum map has various generalizations that are very useful in mathematics as well as in physical and engineering applications. *The first goal of this book is to present the theory of these momentum maps.*

The integrals of motion, also called ***conserved quantities***, have been used by the founders of classical mechanics to eliminate degrees of freedom in the particular systems under investigation. These procedures are precursors of what is today called ***reduction theory***. Jacobi's elimination of the node is a well-known example of this technique. This purely geometric procedure has evolved in parallel with the development of theoretical physics. Its application yields dynamics on a lower-dimensional space as well as a procedure that permits the reconstruction of the solutions of the original system from those of the reduced equations. Reduction is a very general procedure that is applied to arbitrary symmetric dynamical systems (see GOLUBITSKY et al. (1985); CHOSSAT AND LAUTERBACH (1999); GOLUBITSKY AND STEWART (2002)). However, it is particularly powerful for conservative systems when the symmetries induce a momentum map. In this case, more degrees of freedom than one would expect can be eliminated by using jointly the symmetry properties of the flow and the associated conservation laws.

The modern geometric formulation of reduction for Hamiltonian systems, as long as there are no singularities, is due to MARSDEN AND WEINSTEIN (1974) and to MEYER (1973). It associates to a symmetric Hamiltonian system another Hamiltonian system on a smaller-dimensional phase space. If singularities are present, this procedure can be modified (SJAMAAR (1990); SJAMAAR AND LERMAN (1991); ARMS et al. (1991); BATES AND LERMAN (1997)), but there are significant technical complications. *The second goal of this book is to present a self-contained exposition of reduction theory for arbitrary phase spaces. We shall combine the various notions of momentum maps and construct a general reduction method for any phase space without any regularity assumptions. Reduction theory for general symplectic and Poisson manifolds as presented here is now a fairly complete theory.* However, this does not mean that there are not important special cases that need development, especially in view of their potential applications. They will be listed later on when we present a family of topics that should be tied to the present theory.

This book adopts the Hamiltonian point of view in which the phase space is either a symplectic or a Poisson manifold. Recall that a ***symplectic manifold*** is a pair (M, ω), where M is a smooth manifold and ω is a closed nondegenerate two-form on it. The isomorphism between vector fields and one-forms on M defined by ω allows one to associate to any smooth function $h : M \to \mathbb{R}$ a vector field X_h, called a ***Hamiltonian vector field***, by the relation $\omega(X_h, \cdot) = \mathbf{d}h$, where $\mathbf{d}$ denotes the exterior differential on M. ***Hamilton's equations*** are then $\dot{m}(t) = X_h(m(t))$. In addition, the prescription $\{f, h\} := \omega(X_f, X_h)$ endows the space of smooth functions on M with a ***Poisson bracket***, that is, $\{\cdot, \cdot\}$ is a Lie bracket that satisfies also the Leibniz identity in each factor. Examples of such symplectic phase spaces are cotangent bundles of arbitrary manifolds and coadjoint orbits of Lie groups. If there is no symplectic form and one only has a Poisson bracket on the space of smooth functions, then the pair $(M, \{\cdot, \cdot\})$ is called a ***Poisson manifold***. The Hamiltonian vector field of $h : M \to \mathbb{R}$ is defined in that situation by the fundamental property characterizing vector fields as derivations on smooth functions, that is, $X_h = \{\cdot, h\}$. Hamilton's equations for the function h then take the form $\dot{f} = \{f, h\}$ for any smooth function f. It turns out that Poisson manifolds are foliated (in a generalized sense) by injectively immersed submanifolds that are symplectic and whose symplectic structure induces the given Poisson bracket; the leaves of this foliation are called the ***symplectic leaves*** of the Poisson manifold. The tangent space to a symplectic leaf consists of the values of all Hamiltonian vector fields at that point. In particular, any symplectic manifold is Poisson. The standard example of a Poisson manifold that is, in general, not symplectic, is the dual of any Lie algebra; its symplectic leaves are the coadjoint orbits.

In this abstract formulation, a smooth Lie group action of the group G on the manifold M, preserving either the symplectic or Poisson structure, often gives rise to a ***momentum map*** $\mathbf{J} : M \to \mathfrak{g}^*$, where $\mathfrak{g}$ is the Lie algebra of G and $\mathfrak{g}^*$ is its dual. Its definition is given by the requirement that the infinitesimal generators of the action given by each element $\xi \in \mathfrak{g}$ equal the Hamiltonian vector field of the function $\mathbf{J}^\xi : M \to \mathbb{R}$, where $\mathbf{J}^\xi(m) := \langle \mathbf{J}(m), \xi \rangle$, for any $m \in M$ and where $\langle \cdot, \cdot \rangle : \mathfrak{g}^* \times \mathfrak{g} \to \mathbb{R}$ is the natural duality pairing. The ***Marsden–Weinstein*** or ***point-reduced space*** at the value μ of $\mathbf{J}$ is, by definition, $M_\mu := \mathbf{J}^{-1}(\mu)/G_\mu$, where $\mu \in \mathfrak{g}^*$ and $G_\mu := \{g \in G \mid \mathrm{Ad}^*_g \mu = \mu\}$ is the coadjoint isotropy subgroup at μ; for any $g \in G$, the linear isomorphisms $\mathrm{Ad}_g : \mathfrak{g} \to \mathfrak{g}$ denote the adjoint and $\mathrm{Ad}^*_{g^{-1}} : \mathfrak{g}^* \to \mathfrak{g}^*$ the coadjoint representations. If M is symplectic, μ is a regular value of $\mathbf{J}$, and G_μ acts freely and properly on the level manifold $\mathbf{J}^{-1}(\mu)$, then the *Marsden–Weinstein Reduction Theorem* states that M_μ inherits a natural symplectic structure and that the Hamiltonian dynamics of a G-invariant Hamiltonian drops to Hamiltonian dynamics on M_μ, relative to the reduced symplectic form. It is the family of these spaces M_μ, their geometric structure, and their various generalizations that will be studied in this book, even if all the regularity assumptions above are dropped, in which case M_μ is not even a manifold; in this case one talks about ***singular reduction***. For a history of reduction we refer to MARSDEN AND WEINSTEIN (2001) and references therein.

We shall give below a detailed description of the contents of the book.

Roughly, one third of the book reproduces results available in the literature, one third consists of new material, and another third consists of nontrivial reformulations and streamlining of existing work.

What this book contains. There are three parts to this book: background, the core material on momentum maps and reduction, and more advanced topics. These parts are not linearly ordered, and we shall give at the end of this presentation of the contents of the book a diagram of the interdependence of the chapters.

The purpose of the first three chapters is to recall various notions, to establish notation and conventions, and to give precise definitions and statements of theorems used in the rest of this book. All of this is standard material and an expert can skip these chapters, returning to them only when a specific reference in a later proposition or proof requires it. The point of view taken in these chapters is that if the material is easily found in the literature, then only definitions and results are presented, whereas if certain notions and statements are either hard to find or simply "folklore", then a proof will be given. It should be mentioned that there are parts of this material that may not be considered elementary or well known by everyone. For example, Chapter 1 introduces initial manifolds and their properties that are essential in later chapters. The key results on stratified spaces and sheaves of functions to be used later on in the text are also found here. Chapter 2 presents the full theory of proper Lie group actions, including slices, the properties of invariant functions, and a review of type and local type submanifolds. Chapter 3 introduces pseudogroups of transformations and groupoids. It provides a detailed account of various results on generalized distributions, the problem of integrability, and discusses integral submanifolds and accessible sets. Special attention is given to the study of equivariant vector fields and the distributions generated by them.

The core material of the book begins in Chapter 4. After quickly providing an account of the basic results on symplectic and Poisson geometry needed in the rest of the book, the notion of ***Noether momentum map*** is introduced. These are maps defined on a Poisson manifold with values in some topological space with the property that they are constant on the dynamics of any symmetric Hamiltonian vector field, that is, Noether's theorem holds. After giving several examples of such momentum maps, a thorough discussion of the "standard" momentum map, whose values are in the dual of the Lie algebra of symmetries, is presented. This is the map discussed previously. Its properties are studied in detail, not all of them being easily retrievable from the existing literature. For example, it is shown that there exists a canonical action of a suitably defined Lie subgroup on each connected component of an isotropy type manifold; this action admits a momentum map which can be computed directly from the momentum map on the ambient manifold. Throughout this chapter no equivariance assumptions are imposed on the momentum map and the theory is developed in this general framework. The results on the momentum map presented here are essential in the rest of the book.

Chapter 5 is mainly dedicated to two generalizations of the standard momentum map, namely ***cylinder valued momentum maps*** and the ***optimal momentum map***. The cylinder valued momentum map is a construction due to CONDEVAUX, DAZORD, AND MOLINO (1988) which, in the context of symplectic manifolds, generalizes the standard momentum map and has the important property of being always defined, unlike the standard momentum map. Cylinder valued momentum maps are genuine generalizations of the standard ones in the sense that whenever a Lie algebra action admits a standard momentum map, there is a cylinder valued momentum map that coincides with it. For Abelian symmetries, cylinder valued momentum maps are closely related to the so-called ***Lie group valued*** momentum maps. This relationship is discussed in detail. The study of the cylinder valued momentum maps is followed by a section on ***Hamiltonian covering spaces***. A Hamiltonian covering of a symplectic manifold, acted canonically upon by a Lie algebra, is a usual covering of that manifold by another symplectic manifold acted upon by the same Lie algebra but, this time, the action is required to have an associated standard momentum map. The strategy that led to the introduction of the cylinder valued momentum map can be used to construct a Hamiltonian covering space that satisfies the same universality properties as the covering space of a manifold but with the arrows reversed. We refer to this universal object as the ***universal covered space*** of a symplectic Lie algebra action. A significant part of the chapter is dedicated to the optimal momentum map, defined as the projection from the phase space to its quotient by the generalized distribution defined by the Hamiltonian vector fields of symmetric functions. This Noether momentum map has remarkable reduction and universality properties that are crucial in the chapters on singular reduction and dual pairs. The optimal momentum map is in most cases not a genuine generalization of the standard momentum map. The relation between these two objects is by now well understood in the symplectic case. This point is explained in detail in the last section of the chapter where it is also shown that the optimal momentum map can sometimes be seen as the ***moment map*** associated to a natural groupoid action on the manifold in question. The analog of this result in the Poisson category is still the subject of ongoing research.

Chapter 6 gives the full reduction theory, the various approaches to it, and their interconnections, if all regularity assumptions are made and the existence of a standard momentum map is supposed. The reconstruction of the original dynamics from the reduced one is also presented in detail. This material can be considered standard for anyone working in symplectic geometry or geometric mechanics. Nevertheless, there are several points worth mentioning. First, even though the proofs may seem standard, they are all anchored in Poisson and not in symplectic geometry. This is a subtle difference that is necessary for later chapters, because in the singular and optimal reduction case this point of view eases the exposition in a significant way. Second, orbit reduction, which considers instead of the point reduced space M_μ, the quotient $\mathbf{J}^{-1}(\mathcal{O}_\mu)/G$, called the ***orbit reduced space***, could be applied only if the coadjoint orbit $\mathcal{O}_\mu \subset \mathfrak{g}^*$ containing μ was an embedded submanifold, or equivalently, a locally closed subset of $\mathfrak{g}^*$. In this chapter it is shown how to work out the general case when no assumptions on $\mathcal{O}_\mu$ are made. It turns out that this is not that easy since there are several natural manifold structures available on the orbit reduced space and only one of them renders satisfactory results. Third, regular reduction by stages is discussed in detail. This also turns out to be a relatively involved subject that answers in the symplectic category a question that is trivial in the Poisson category: if M is a Poisson manifold on which a Lie group G acts freely properly and canonically and if $N \subset G$ is a normal subgroup, then it is clear that the quotient Poisson manifolds M/G and $(M/N)/(G/N)$ are isomorphic. Can one obtain the leaves of M/G as a two stage process, first reducing by the N-action and then by the G/N-action? It is immediately clear that this question, as just posed, makes no sense because the group G/N does not act on the N-point reduced spaces. How this works is precisely the theory of reduction by stages presented in this chapter. Fourth, there is also an extensive section on cotangent bundle reduction but no proofs are given. The reason for this is that the theory in the singular case is still under development and the elaboration of all the statements in this section would have required an additional lengthy chapter. The statements in this section are scattered throughout the literature and some are difficult to find. These results provide powerful tools in the stability and bifurcation analysis of symmetric Hamiltonian systems and they deserve to be more widely known.

Chapter 7 presents one of the main technical tools in singular reduction and, in general, in the study of stability, persistence, and bifurcation phenomena in symmetric Hamiltonian systems on symplectic manifolds. The *Symplectic Slice Theorem* provides a privileged system of semiglobal coordinates in which all the ingredients related to a canonical group action on a symplectic manifold take a particularly simple form. This is the case of the symplectic form itself, or of the standard, or the cylinder valued momentum maps associated to this action. Important byproducts of this tool are the so-called ***reconstruction equations***; this is the name used for the expression in these coordinates of the equations that determine the Hamiltonian vector field corresponding to an invariant Hamiltonian function.

If one drops all regularity assumptions in Chapter 6 with the exception of the properness of the action, the reduced space M_μ is not a manifold anymore but a symplectically stratified space. Chapter 8 presents, for the first time, this theory in full generality. If $\mu = 0$ this was shown by SJAMAAR (1990); SJAMAAR AND LERMAN (1991). From the point of view of orbit reduction, certain results were obtained by BATES AND LERMAN (1997), whereas an algebraic approach to singular reduction was proposed by ARMS et al. (1991). This chapter fills in all the gaps of this theory, bases it on point reduction, and shows that all these approaches are equivalent. The stratification theorem is also proved in detail, both for point as well as orbit reduction. It should be mentioned that our approach to singular reduction is not historically faithful, but closely follows the spirit of the original paper by MARSDEN AND WEINSTEIN (1974). We begin by constructing the singular reduced spaces or symplectic strata, we discuss their properties, and then we prove how they fit together into a stratified space or, more specifically, into a cone space. A major difference between the approach taken here and others found in the literature is the limited use of the Symplectic Slice Theorem. More specifically, the structure of the symplectic point and orbit strata is globally studied; this allows, for instance, for the writing of an actual formula that characterizes their symplectic form. The Symplectic Slice Theorem is used only in the proof of the local structure of the stratification of the quotients $\mathbf{J}^{-1}(\mu)/G_\mu$ by the symplectic strata presented in the chapter. The global characterization of the symplectic strata is based on an idea – we call it ***Sjamaar's Principle*** – that is used to look at these strata as regular Marsden–Weinstein reduced spaces relative to a natural action available on certain isotropy type manifolds. We were motivated to proceed using this approach by considerations from theoretical mechanics and bifurcation theory. The use of this strategy allowed us to recover the other singular reduction procedures in the literature, to correct several errors present in some of the original papers, and also to generalize the reduction scheme to various extensions of momentum maps, all inspired by physical applications.

Chapter 9 is the beginning of a group of three chapters dealing with more advanced topics. It addresses reduction theory when no momentum map is present. In this situation, the optimal momentum map introduced in Chapter 5 becomes essential and reveals itself as an extremely powerful tool. The entire reduction theory, including reduction by stages, is redone here from this point of view. Surprisingly, the problems related to singularities disappear in this context because reduction with the optimal momentum map yields directly the connected components of the strata of the (point or orbit) reduced spaces if a standard momentum map is already present. This serves indirectly as a check that the choices involved in the construction of the strata in Chapter 8 were correctly made. Optimal reduction is an extremely general procedure that is always available; in the presence of an arbitrary canonical action on any Poisson manifold it yields a symplectic manifold, provided a certain properness condition is satisfied. In particular, if the symmetry group is trivial, one recovers the singular foliation of the Poisson manifold into its symplectic leaves, that is, the classical symplectic foliation theorem for Poisson manifolds into symplectic leaves appears as the trivial case of the reduction procedure by the optimal momentum map. The reduction procedures linked to the optimal momentum map are powerful tools that are well adapted to deal with singularities.

There remain many open problems in this recently introduced theory that are discussed in this chapter. A particularly interesting and intriguing one is the study of the presymplectic homogeneous spaces that show up in the context of ***polar reduction***; these manifolds appear naturally when carrying out orbit reduction with the optimal momentum map. In Chapter 6 it is shown how the symplectic structure of the regular orbit reduced spaces, in the presence of a standard momentum map, is characterized by a beautiful formula that intertwines the symplectic forms of the original manifold and the Kostant–Kirillov–Souriau symplectic form of the specific coadjoint orbit used to perform the reduction. In the optimal context, the coadjoint orbit is replaced by an orbit in the momentum space, called the ***polar reduced space***, and the analog in this situation of the orbit reduction formula defines on it a closed two-form that, in general, is degenerate. The circumstances under which this natural presymplectic form on the polar reduced space is nondegenerete can be fully characterized in the case of the optimal reduction of a symplectic manifold. In the Poisson case, a sufficient condition is provided. We believe that this generalization of the Kostant–Kirillov–Souriau construction could be of much use not only in reduction theory but also in other domains such as representation theory.

Poisson reduction is the subject of Chapter 10. The main goal is to obtain new Poisson manifolds or Poisson algebras by combining the operations of restriction and passage to a quotient. In this context, the first point that needs to be specified are the functions or, more specifically, the sheaf of functions to which the newly obtained bracket will be applied. This has been carried out with great care in the text, filling in some gaps in the literature on this subject. The reduction results in this chapter are divided mainly into two parts: reduction with respect to a symmetry defined on the entire manifold (Section 10.2) and reduction with respect to a symmetry defined only on the subspace to which the Poisson structure is restricted (Section 10.4). Various known reduction schemes in the Poisson category are shown to be corollaries of these results. This is the case, for instance, for the reduction of the coisotropic submanifolds of a Poisson submanifold by their characteristic distributions (also known as first class constraints) or of the Poisson structure of the cosymplectic manifolds given by WEINSTEIN (1983a) (also known as second class constraints). Remarkably, it is shown that the Dirac formula for the constrained Poisson bracket on a symplectic submanifold of a given symplectic manifold can be generalized to the purely Poisson context if the constraint submanifolds are cosymplectic.

Chapter 11 is dedicated to the notion of duality and polarity in Poisson geometry. The chapter starts with a review of the various notions of dual pair available in the literature, studies the relationships between them, examines the properties of Poisson manifolds in duality, and points out some common misconceptions that are part of the "folklore" in this area. Special attention is given to the theory of bifoliations, which is widely used in the theory of noncommutative integrable systems. The rest of the chapter presents a generalization of the standard notion of dual pair designed to accommodate the concept of polarity which was used to define the optimal momentum map. It is shown that this generalization reproduces the correspondence between the symplectic leaves of the dual pair formed by the projection onto orbit space and the standard momentum map, that is, the correspondence between symplectic orbit reduced spaces and coadjoint orbits. We introduce the notions of singular Howe pair as well as the dual pair induced by two pseudogroups of local Poisson diffeomorphisms of a Poisson manifold. The last section in this chapter identifies a variety of sufficient conditions that, when imposed on a canonical group action on a Poisson manifold, ensure that the projection onto the orbit space of the group action and the associated optimal momentum map are in duality in our sense. The *Leaf Correspondence Theorem* available in this case establishes a relation between the optimal orbit reduced spaces and the (in general only presymplectic) polar reduced spaces discussed before.

We hope that the interesting links established in this book between several apparently disjoint results and that the beautiful and coherent picture of reduction theory described here will be useful in applications to both pure mathematics as well as physical and engineering problems.

The following diagram gives a rough picture of the plan of the book. One cannot, of course, just blindly follow the flow chart since the book tends to be linearly ordered, even though we have tried to keep it as modular as possible to facilitate its use as a reference work. We hope that the constant cross-referencing of results and the detailed index will aid the reader navigate towards the specific topic of interest. Also, it should not be assumed that it is always the case that an entire chapter is needed for the comprehension of the one following it, since the chapters are organized to have internal coherence and to collect material of the same type. A glance on the part of the reader interested in a specific subject will easily identify sections in the chapters sitting above the one of interest that could be omitted. In the diagram, the chapters marked with a dagger consist mainly of introductory material, while those with a star contain more advanced topics that can be read separately.

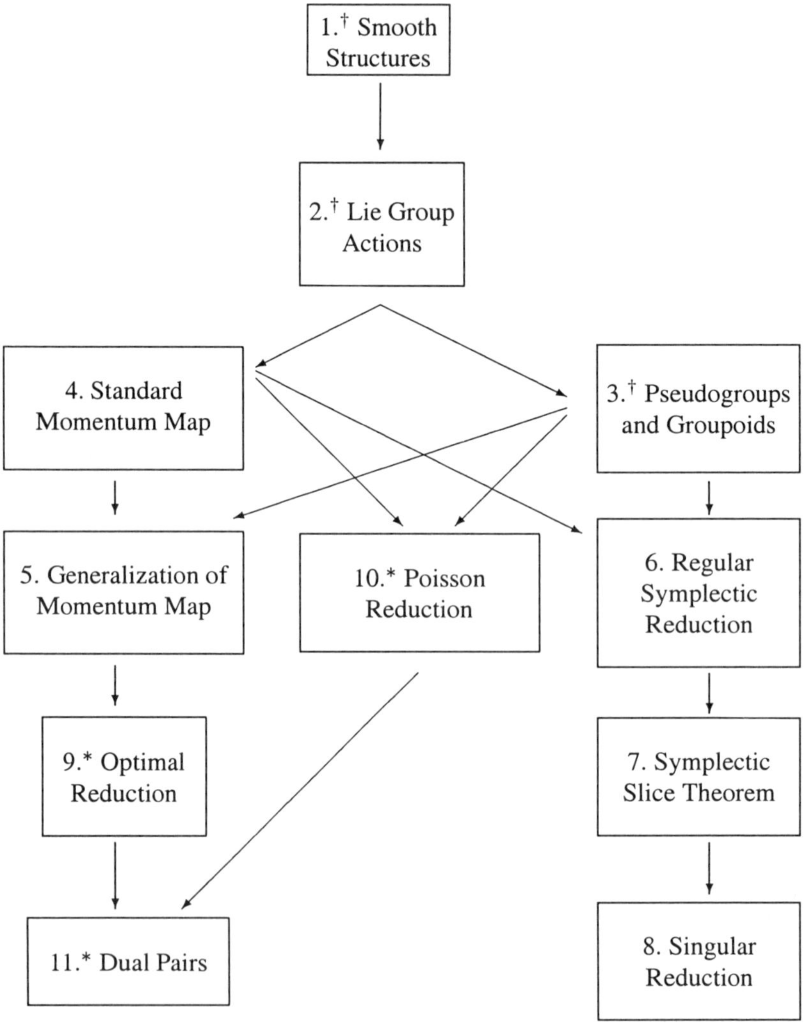

What this book does not contain. In spite of its length and completeness for general Hamiltonian systems, there are several important areas in the theory of momentum maps and reduction that are not covered in this book. Space considerations and the fact that most of these topics are still under development were the main reasons to omit them. They all represent fertile areas of current research that continue to pose extremely interesting questions from the purely geometrical as well as the applied points of view.

There are two groups of subjects: topics that directly involve momentum maps and reduction, and topics for which momentum maps and reduction are an essential tool in their development, without actually belonging to the general theory.

TOPICS IN THE THEORY OF MOMENTUM MAPS AND REDUCTION

There are several subjects that belong to the core theory of momentum maps and reduction that will not be discussed in the main text of the book. Below is a short presentation of these topics.

1. *Symplectic convexity.* There is an obvious connection between the convexity properties of the momentum map and symmetric Hamiltonian bifurcation theory. This is a subject of great interest that is in the process of being developed. The momentum polytope plays the role of an a priori bifurcation diagram for any symmetric Hamiltonian system. Closely related to this subject is the study of the Duistermaat–Heckman measures (see DUISTERMAAT AND HECKMAN (1982); GUILLEMIN et al. (1996)).

The *Symplectic Convexity Theorem* states that for a globally Hamiltonian action of a compact connected Lie group on a compact connected symplectic manifold the intersection of the image of the associated momentum map with a Weyl chamber of the Lie algebra of a maximal torus of the symmetry group is a compact convex polytope. This result is an extension of the *Linear Convexity Theorem of* KOSTANT (1974) that states that the orthogonal projection of the adjoint orbit of a compact connected Lie group on a maximal toral algebra is the convex hull of the corresponding Weyl group orbit. This theorem in turn generalizes the classical result of SCHUR (1923) and HORN (1954a), which represents the $U(n)$ case of Kostant's theorem. The Schur–Horn Theorem states that if $\mathcal{O}_\lambda$ is the set of all $n \times n$ Hermitian matrices with fixed eigenvalues $\lambda_1, \ldots, \lambda_n$ (so $\mathcal{O}_\lambda$ is a $U(n)$-coadjoint orbit and hence a symplectic manifold) and if π denotes the projection that sends a matrix to its diagonal, thought of as a vector in $\mathbb{R}^n$ (so π is the momentum map of the maximal torus action on $\mathcal{O}_\lambda$), then $\pi(\mathcal{O}_\lambda)$ is the convex compact polytope in $\mathbb{R}^n$ obtained as the convex hull of $\{\lambda_{\sigma(1)}, \ldots, \lambda_{\sigma(n)} \mid \sigma \text{ a permutation of } \{1, \ldots, n\}\}$.

In §4.7 a very brief presentation of this topic is given but the general theory is missing; it can be found in in the original papers of ATIYAH (1982) and GUILLEMIN AND STERNBERG (1982, 1984a), for Abelian actions, and in KIRWAN (1984b) for the general case. This subject has been further developed in, for instance, DUISTERMAAT (1983, 1984); DELZANT (1988); CONDEVAUX et al. (1988); AUDIN (1991); HILGERT et al. (1994); PRATO (1994); LERMAN (1995); GUILLEMIN et al. (1996); LERMAN AND TOLMAN (1997); SJAMAAR (1998); KNOP (1997, 2002); KNUTSON AND TAO (1999, 2001). See BLOCH et al. (1992) for an interesting application of the convexity theorem to a dynamic sorting algorithm.

2. *Poisson–Lie convexity.* There is an analogue of the above mentioned convexity theorem for actions of Poisson–Lie groups on compact connected symplectic manifolds. In this setting, the momentum map has values in the dual group. The standard situation, that is, the momentum map has values in the dual of the Lie algebra of the symmetry Lie group, is recovered when the Poisson structure on the Poisson–Lie group is trivial. This general convexity result is motivated by the *Nonlinear Convexity Theorem of* KOSTANT (1974), which in turn generalizes a classical result of HORN (1954b) and WEYL (1949). This result is nonlinear because it involves the logarithm and projections onto group factors rather than Lie algebra factors. Further developments in this direction can be found in LU AND RATIU (1991) and GINZBURG AND WEINSTEIN (1992).

To give an idea of what kind of nonlinear phenomena are involved, we briefly recall the Horn–Weyl theorem. Let P be the set of positive definite Hermitian matrices, Σ_λ the subset of matrices with fixed eigenvalues $\lambda_1, \ldots, \lambda_n$, and let π be the projection onto the vector $(\log(\det p_1), \ldots, \log(\det p_n)) \in \mathbb{R}^n$, where $p_k = (p_{ij})$, $i, j = 1, \ldots, k$ is the $k \times k$ principal submatrix of $p \in P$. Then the image $\pi(\Sigma_\lambda)$ is a convex compact polytope. In Lie group language, Σ_λ is an orbit (by conjugation) of $K = SU(n)$ on its Cartan complement P in $SL(n, \mathbb{C})$, and $\pi(p)$ is (essentially) the logarithm of the diagonal factor in the decomposition $p = kan$, where k is unitary, a is positive diagonal, and n is upper triangular with ones on the diagonal. Kostant's Nonlinear Convexity Theorem represents the general Lie-theoretic version of the above statement. As opposed to the linear convexity theorem discussed in the previous point, the symplectic structure on Σ_λ is not that of a coadjoint orbit; it comes from a Poisson structure on the Lie group $\mathrm{SL}(n, \mathbb{C})$ (having certain compatibility properties relative to multiplication) and, as before, π is an associated momentum map. This is how Poisson–Lie groups and the momentum maps attached to their actions enter the field of convex geometry.

The *Poisson–Lie Convexity Theorem*, generalizing in turn Kostant's nonlinear convexity theorem to arbitrary Poisson–Lie group actions on compact connected symplectic manifolds, is itself a corollary of a more general convexity statement due to FLASCHKA AND RATIU (1996). ALEKSEEV (1997) has shown how to recover the Poisson–Lie Convexity Theorem from the Symplectic Convexity Theorem; it is still unclear if his method can be extended to recover the more general convexity theorem of FLASCHKA AND RATIU (1996).

Finally, it seems that all these convexity theorems, mentioned in this and the previous point, may have natural formulations in the context of sub-Riemannian geometry; the rewriting of these results in this framework should also yield new nonlinear convexity results.

3. *Infinite-dimensional convexity.* There are convexity theorems in infinite dimensions due to ATIYAH AND PRESSLEY (1983), BLOCH et al. (1993), and NEUMANN (1999, 2002). All of them are proved by specialized techniques that do not, unfortunately, rely on the finite-dimensional convexity methods. BIRTEA (2003) and BIRTEA et al. (2003) give a proof that yields both the classical Symplectic Convexity Theorem and the infinite-dimensional Atiyah–Pressley result. It seems that this proof may also yield the Neumann theorems as well as the Poisson–Lie convexity result.

4. *Momentum maps in infinite dimensions.* Another topic missing in this book is the general theory of momentum maps in infinite dimensions and its applications to various geometrical and physical problems, reduction theory included. Such results can be found, for instance, in CHERNOFF AND MARSDEN (1974); ARMS et al. (1981); ATIYAH AND BOTT (1982); MARSDEN AND WEINSTEIN (1982, 1983); ISENBERG AND MARSDEN (1982); HOLM et al. (1998); MEINRENKEN AND WOODWARD (1998); DONALDSON (1999); CATTANEO AND FELDER (2001); WURZBACHER (2001); ODZIJEWICZ AND RATIU (2003), and references therein.

5. *Lie group valued momentum maps.* In point **2** above, momentum maps whose values are in Lie groups have been mentioned. There is a general theory of such momentum maps. Circle valued momentum maps were introduced by MCDUFF (1988) and torus valued ones by GINZBURG (1992). The definition for general group valued momentum maps in the non-Abelian case is due to ALEKSEEV et al. (1998), which is based on prior work by HUEBSCHMANN AND JEFFREY (1994); HUEBSCHMANN (1995a). The motivation behind this construction lies more in the understanding of certain moduli spaces rather than in applications to mechanics and in the study of Hamiltonian conservation laws. The definition of these objects is actually incompatible with the symplectic or Poisson category whenever the symmetry is given by a non-Abelian Lie group. This has led to the introduction of the so-called ***quasi-Hamiltonian*** category in which these momentum maps allow the formulation of conservation laws and reduction schemes. In the Abelian case, the quasi-Hamiltonian spaces constructed on a symplectic manifold and the associated momentum maps are much related to the cylinder valued momentum maps that we shall discuss in Section 5.2. This relationship is spelled out in detail in Section 5.4. Further developments of this theory in various directions (such as the Duistermaat–Heckman formula, the "quantization commutes with reduction" statement, or Kirwan's surjectivity theorem for group valued momentum maps) can be found in ALEKSEEV et al. (2000, 2001); BOTT et al. (2003) and references therein.

6. *Jacobi manifolds.* The theory of Jacobi manifolds is not developed in this book. It turns out that these manifolds have an intimate relationship with other geometric structures, also not discussed in the main text and briefly described below, that are closely related to symplectic forms. Roughly speaking, Jacobi manifolds are to Poisson manifolds what contact manifolds are to symplectic manifolds. A ***Jacobi manifold*** (see LICHNEROWICZ (1978); GUEDIRA AND LICHNEROWICZ (1984); VAISMAN (1985)) is a smooth manifold endowed with a bivector field Λ and a vector field E satisfying $[\Lambda, \Lambda] = 2E \wedge \Lambda$ and $[\Lambda, E] = 0$, where the bracket in these relations is the Schouten bracket. Note that the case $E = 0$ corresponds to Poisson manifolds. Equivalently, a Jacobi manifold is a smooth manifold such that the space of smooth functions $C^\infty(M)$ carries a local Lie algebra structure, that is, $\operatorname{supp}(\{f, g\}) \subseteq \operatorname{supp}(f) \cap \operatorname{supp}(g)$, for any $f, g \in C^\infty(M)$. The relationship between the bracket $\{\cdot, \cdot\}$ and the pair (Λ, E) is given by the relation $\{f, g\} = \Lambda(\mathbf{d}f, \mathbf{d}g) + f\langle \mathbf{d}g, E\rangle - g\langle \mathbf{d}f, E\rangle$. The Hamiltonian vector field X_h of $h \in C^\infty(M)$ is defined by $X_h := \Lambda(\cdot, \mathbf{d}h) - hE$ and with this definition the usual relation $[X_f, X_g] = -X_{\{f,g\}}$, familiar from Poisson geometry, holds. Thus, the Hamiltonian vector fields span a smooth generalized distribution that is involutive and it can be shown that it is integrable. Each leaf carries a unique induced Jacobi structure. The Jacobi manifold is called ***transitive*** if it has only one leaf.

There is an intimate relation between Jacobi structures and two other classes of manifolds that will be described below: the locally conformal symplectic and exact contact manifolds. Namely, *there is a natural bijective correspondence between even-dimensional transitive Jacobi manifolds and locally conformal symplectic manifolds and between odd-dimensional transitive Jacobi manifolds and exact contact manifolds.* In other words, the leaves of a Jacobi manifold are either locally conformal symplectic or exact contact manifolds.

If (M, Λ, E) is a Jacobi manifold, then the space $\Omega^1(M) \times C^\infty(M)$ is a Lie algebra relative to the bracket

$$\begin{aligned}&\{(\alpha, f), (\beta, g)\} := \\ &\quad \big(-\pounds_{\Lambda^\sharp(\alpha)}\beta + \pounds_{\Lambda^\sharp(\beta)}\alpha - \mathbf{d}(\Lambda(\alpha, \beta)) - f\pounds_E\beta + g\pounds_E\alpha + \mathbf{i}_E(\alpha \wedge \beta), \\ &\qquad -\Lambda(\alpha, \beta) + \Lambda(\alpha, \mathbf{d}g) - \Lambda(\beta, \mathbf{d}f) - f\langle \mathbf{d}g, E\rangle + g\langle \mathbf{d}f, E\rangle \big),\end{aligned}$$

where $\alpha, \beta \in \Omega^1(M)$, $f, g \in C^\infty(M)$, and $\Lambda^\sharp$ denotes the map from one-forms to vector fields induced by the tensor Λ.

Regular reduction of Jacobi manifolds by submanifolds and subbundles satisfying certain compatibility conditions in the spirit of MARSDEN AND RATIU (1986) was carried out in NUNES DA COSTA (1989, 1990); MIKAMI (1989); IBORT et al. (1997). The reduction scheme proposed in these papers also works for group actions. The singular case is open.

7. *Poisson–Nijenhuis and Jacobi–Nijenhuis manifolds.* Closely related to Poisson and Jacobi manifolds and motivated by the notion of compatible Poisson structures appearing in integrable systems (MAGRI (1978): two Poisson structures are compatible if their sum is again a Poisson structure) are the Poisson–Nijenhuis (KOSMANN-SCHWARZBACH AND MAGRI (1990)) and the Jacobi–Nijenhuis (MARRERO et al. (1999); PETALIDOU AND NUNES DA COSTA (2003)) manifolds. Their definition is the following. A ***Nijenhuis operator*** is a $(1, 1)$-tensor N with vanishing ***Nijenhuis torsion***, that is,

$$T(N)(X, Y) := [NX, NY] - N[NX, Y] - N[X, NY] + (N \circ N)[X, Y] = 0$$

for all $X, Y \in \mathfrak{X}(M)$, where N is identified with the operator $N : \mathfrak{X}(M) \to \mathfrak{X}(M)$ given by $NX := N(X, \cdot) : \Omega^1(M) \to \mathbb{R}$.

A ***Poisson–Nijenhuis manifold*** is a triple (M, Λ, N), where (M, Λ) is a Poisson manifold, N is a Nijenhuis operator, and the following identities hold: $N \circ \Lambda^\sharp = \Lambda^\sharp \circ N^t$ and $\Lambda^\sharp \circ C(\Lambda, N) : \Omega^1(M) \times \Omega^1(M) \to \mathfrak{X}(M)$ vanishes. In the first identity, $N^t : \Omega^1(M) \to \Omega^1(M)$ is the transpose of N defined by $\langle \alpha, NX\rangle = \langle N^t\alpha, X\rangle$, for any $X \in \mathfrak{X}(M)$ and $\alpha \in \Omega^1(M)$. In the second identity, $C(\Lambda, N) : \Omega^1(M) \times \Omega^1(M) \to \Omega^1(M)$ is defined by

$$C(\Lambda, N)(\alpha, \beta) := \{\alpha, \beta\}_1 - \{N^t\alpha, \beta\}_0 - \{\alpha, N^t\beta\}_0 + N^t\{\alpha, \beta\}_0,$$

where $\{\cdot, \cdot\}_i$ are the Lie brackets on $\Omega^1(M)$ defined by $N^i\Lambda^\sharp$, $i = 0, 1$. The Nijenhuis operator is usually called in this setting the ***recursion operator***. Regular reduction of

Poisson–Nijenhuis manifolds by a subbundle and a submanifold compatible with Λ and N is given in VAISMAN (1996b). The singular case is open.

Jacobi–Nijenhuis manifolds were introduced in MARRERO et al. (1999) and their definition tightened in PETALIDOU AND NUNES DA COSTA (2003). The notion of Nijenhuis operator can be extended in the following manner. Given a $(1,1)$-tensor N on the manifold M, a vector field $Z \in \mathfrak{X}(M)$, a one-form $\gamma \in \Omega^1(M)$, and a smooth function $h \in C^\infty(M)$, define $\mathcal{N} := (N, Z, \gamma, h) : \mathfrak{X}(M) \times C^\infty(M) \to \mathfrak{X}(M) \times C^\infty(M)$ by $\mathcal{N}(X, f) := (NX + fZ, \langle \gamma, X \rangle + hf)$. The space $\mathfrak{X}(M) \times \Omega^1(M)$ is a Lie algebra relative to the bracket $[(X, f), (Y, g)] := ([X, Y], X[g] - Y[f])$ so one can define the ***Nijenhuis torsion*** of $\mathcal{N}$ by

$$\begin{aligned}\mathcal{T}(\mathcal{N})((X, f), (Y, g)) :=& [\mathcal{N}(X, f), \mathcal{N}(Y, g)] - \mathcal{N}[\mathcal{N}(X, f), (Y, g)] \\ &- \mathcal{N}[(X, f), \mathcal{N}(Y, g)] + (\mathcal{N} \circ \mathcal{N})[(X, f), (Y, g)],\end{aligned}$$

where $X, Y \in \mathfrak{X}(M)$ and $f, g \in C^\infty(M)$. A $C^\infty(M)$-linear map $\mathcal{N} : \mathfrak{X}(M) \times C^\infty(M) \to \mathfrak{X}(M) \times C^\infty(M)$ is a ***Nijenhuis operator*** if its torsion vanishes.

Two Jacobi structures (Λ_0, E_0) and (Λ_1, E_1) are ***compatible*** if their sum $(\Lambda_0 + \Lambda_1, E_0 + E_1)$ is again a Jacobi structure.

A ***Jacobi–Nijenhuis manifold*** is a Jacobi manifold (M, Λ, E) together with a Nijenhuis operator $\mathcal{N}$ such that $\mathcal{N} \circ (\Lambda_0^\sharp, E_0) = (\Lambda_0^\sharp, E_0) \circ \mathcal{N}^t$ and $(\Lambda^\sharp, E) \circ \mathcal{C}(\Lambda, E, \mathcal{N}) : \Omega^1(M) \times C^\infty(M) \times \Omega^1(M) \times C^\infty(M) \to \mathfrak{X}(M) \times C^\infty(M)$ vanishes, where

- $\mathcal{N}^t : \Omega^1(M) \times C^\infty(M) \to \Omega^1(M) \times C^\infty(M)$ is the transpose of $\mathcal{N}$, that is, $\langle (\alpha, g), \mathcal{N}(X, f) \rangle = \langle \mathcal{N}^t(\alpha, g), (X, f) \rangle$, for any $X \in \mathfrak{X}(M)$, $\alpha \in \Omega^1(M)$, and $f, g \in C^\infty(M)$, the pairing being defined componentwise: $\langle (\alpha, g), (X, f) \rangle := \langle \alpha, X \rangle + fg$;

- $\mathcal{C}(\Lambda, E, \mathcal{N}) : \Omega^1(M) \times C^\infty(M) \times \Omega^1(M) \times C^\infty(M) \to \Omega^1 \times C^\infty(M)$ is defined by

$$\begin{aligned}\mathcal{C}(\Lambda, E, \mathcal{N})((\alpha, f), (\beta, g)) :=& \{(\alpha, f), (\beta, g)\}_1 - \{\mathcal{N}^t(\alpha, f), (\beta, g)\}_0 \\ &- \{(\alpha, f), \mathcal{N}^t(\beta, g)\}_0 + \mathcal{N}^t\{(\alpha, f), (\beta, g)\}_0\end{aligned}$$

  for any $\alpha, \beta \in \Omega^1(M)$, $f, g \in C^\infty(M)$;

- $\{\cdot, \cdot\}_i$ is the Lie bracket on $\Omega^1(M)$ defined by the Jacobi structure (Λ, E) and $\mathcal{N} \circ (\Lambda^\sharp, E)$, respectively.

Regular Jacobi–Nijenhuis reduction is due to NUNES DA COSTA AND PETALIDOU (2002). Singular Jacobi–Nijenhuis reduction is open.

8. *Reduction for manifolds with other geometric structures.* There are several classes of symplectic manifolds or closely related geometric structures that warrant special attention and for which reduction theory, even in the regular case, is considerably richer and more varied than that for general symplectic manifolds, with important consequences for applications.

Group I: Structures not involving a metric

a. The case of cotangent bundle reduction is a well-developed area in the regular case. We review this theory without any proofs in §6.6. There are two main theorems in this field: an *Embedding Theorem* in the cotangent bundle of the quotient of configuration space by the coadjoint isotropy subgroup and a *Fibration Theorem* over the cotangent bundle of shape space, that is, the quotient of configuration space by the full group of symmetries. In both theorems a central role is played by the curvature of a certain connection that provides in the case of the first theorem the correct symplectic form on the cotangent bundle containing the reduced space (one adds to the canonical symplectic form a "magnetic term") and in the case of the second theorem it provides a specific term in the Poisson bracket of the reduced space. The statements in this section are scattered throughout the literature and a systematic theory can be found in MARSDEN AND RATIU (2003). The singular case is just beginning to be developed; see RODRÍGUEZ-OLMOS et al. (2003); SCHMAH (2001).

Cotangent bundle reduction, even in the regular case, has proved to be an extremely powerful tool in various mechanical applications. The block diagonalization method of SIMO, LEWIS, AND MARSDEN (1991) extending the energy-Casimir method (see, for example, ARNOLD (1965, 1966b); HOLM et al. (1985)) has proved to be crucial in the study of the nonlinear stability of concrete mechanical systems as well as in the study of various instability phenomena (see BLOCH et al. (1994, 1996); DERKS et al. (1995)). Extensions of these methods to the singular case should provide deep insights in the bifurcation theory of symmetric Hamiltonian systems.

b. A ***locally conformal symplectic manifold*** (GUEDIRA AND LICHNEROWICZ (1984); VAISMAN (1985)) is a triple (M, ω, η), where M is a $2n$-dimensional smooth manifold, ω is a nondegenerate two-form on M, η is a closed one-form on M, and $\mathbf{d}\omega + \eta \wedge \omega = 0$. The standard example of a locally conformal symplectic manifold that is not symplectic is $S^3 \times S^1$. This manifold is not symplectic for topological reasons: $H^2(S^3 \times S^1) = 0$, which implies that if there were a symplectic form on $S^3 \times S^1$, then it would have to be exact and so the symplectic volume would also be exact, contradicting $H^4(S^3 \times S^1) \neq 0$ by compactness of $S^3 \times S^1$. The locally conformal symplectic structure on $S^3 \times S^1$ is constructed in the following way: S^3 is the Lie group of unit quaternions whose Lie algebra is $(\mathbb{R}^3, \times)$. Let α be the left invariant one-form on S^3 whose value at the identity $(1, 0, 0, 0)$ is the element in the dual basis of $(\mathbb{R}^3)^*$ corresponding to $\mathbf{i} = (1, 0, 0) \in \mathbb{R}^3$. If one defines $\eta := \mathbf{d}t \in \Omega^1(S^3 \times S^1)$, where $\mathbf{d}t$ is the standard length element on the circle S^1, and $\omega := \mathbf{d}\alpha + \mathbf{d}t \wedge \alpha \in \Omega^2(S^3 \times S^1)$, then $(S^3 \times S^1, \omega, \eta)$ is a locally conformal symplectic manifold.

Regular locally conformal symplectic reduction at zero is due to HALLER AND RYBICKI (2001) and MCLACHLAN AND PERLMUTTER (2001). Nothing seems to be known in the singular case and no relationship of this reduction procedure, even in the regular case by a Lie group action, has been established with the reduction of Jacobi manifolds.

c. A ***contact structure*** on a smooth $(2n + 1)$-dimensional manifold N is a codimension one vector subbundle $H \subset TN$ locally given by the kernel of a one-form η satisfying $\eta \wedge (\mathbf{d}\eta)^n \neq 0$. Such an η is called a (local) ***contact form***.

Any two proportional contact forms define the same contact structure. A contact structure that is the kernel of a global contact form is called ***exact*** or ***co-orientable***. If

η is a one-form of an exact contact structure, the pair (N, η) is called an ***exact contact manifold***.

Regular contact reduction of exact contact manifolds by a Lie group action at the zero value of the contact momentum map was developed by ALBERT (1989); GEIGES (1997); LOOSE (2001). WILLET (2002) introduced another reduction method that works at any value of the contact momentum map. Reduction for general contact manifolds follows easily from that for exact contact manifolds. By a classical theorem of Palais one can choose a global one-form, whose kernel is the given contact structure, that is invariant by the given proper Lie group action. Then one performs reduction for the resulting exact contact manifold. GEIGES (1997) has shown that the reduced contact structure does not depend on the choice of the contact form on the original manifold. Singular reduction for exact contact manifolds can be found in LERMAN AND WILLETT (2001).

Regular contact reduction for the cosphere bundle was treated in DRĂGULETE et al. (2003), providing the analogue for the Embedding Cotangent Bundle Reduction Theorem. There is also a slight generalization of regular contact reduction to CR manifolds due to LOOSE (2000).

The main reason for not presenting this material in the current book, space considerations aside, is the lack of applications in mechanics. While the theory is complete in both the regular and the singular cases (with the exception of singular reduction of the cosphere bundle), the definition of the Hamiltonian vector field diverges from what is used in theoretical mechanics: for a product of a symplectic manifold with the real axis, the Hamiltonian vector field defined by those working in this area *does not coincide* with the Hamiltonian vector field of a time dependent system (see, for instance, ALBERT (1989); ARNOLD (1989)). Thus, reduction theory for contact manifolds is useful in the construction of new contact manifolds or contact stratified spaces, and therefore is of considerable mathematical interest, but so far, links to classical time dependent mechanics have not been made.

d. An ***almost cosymplectic manifold*** is a triple (N, ω, η), where N is a $2n + 1$-dimensional smooth manifold, ω is a two-form and η is a one-form on N such that $\eta \wedge \omega^n$ is a volume form on N. If both ω and η are closed, (N, ω, η) is called a ***cosymplectic manifold***. The standard example of a cosymplectic manifold is motivated by the theory of time dependent Hamiltonian systems: the manifold $M \times \mathbb{R}$ is cosymplectic, where (M, ω) is a symplectic manifold, the two-form on $M \times \mathbb{R}$ is the pull back of ω by the projection on the first factor, and η is the pull back of $\mathbf{d}t$ by the projection on the second factor, with t is the coordinate on $\mathbb{R}$. Since for any almost cosymplectic manifold N, the map $X \in \mathfrak{X}(N) \mapsto \mathbf{i}_X\omega + \eta(X)\eta \in \Omega^1(N)$ defines an isomorphism of $C^\infty(N)$-modules, each function h determines a vector field. In the case $N = M \times \mathbb{R}$, this vector field is the time dependent vector field defined by the time dependent Hamiltonian function $h \in C^\infty(M \times \mathbb{R})$.

Regular reduction of cosymplectic manifolds was carried out in ALBERT (1988, 1989); DE LEÓN AND SARALEGI (1993); DE LEÓN AND TUYNMAN (1996). The singular case is open.

Group II: Structures involving a metric

The reduction procedure has also been extended to structures compatible with a fixed Riemannian metric, the motivation coming not just from mechanics, but from the need to search for new examples of (often compact) manifolds carrying a given additional geometric structure. The viewpoint that has been used for classifying Riemannian manifolds is via the holonomy of the Levi-Civita connection (see BESSE (1987)). The ***holonomy group*** of a Riemannian n-manifold is the set of all elements of $O(n)$ that map a given orthogonal frame in the tangent space to a point to the orthogonal frame obtained by parallel transporting the first one about a closed loop (the holonomy is a subgroup of $O(n)$ because parallel transport by the Levi-Civita connection is orthogonal). It is a remarkable fact that a Riemannian manifold whose holonomy group is not $SO(n)$ or $U(n)$ is an Einstein manifold.

e. ***Kähler manifolds*** are Riemannian manifolds with holonomy $U(n)$, the case $SU(n)$ corresponding to the class of ***Kähler–Einstein*** manifolds. The general theory of regular Kähler reduction and its implications for the computation of the cohomology of reduced spaces can be found in KIRWAN (1984a). Singular reduction for Kähler manifolds is a recent development due to HUEBSCHMANN (2003). While these works formulate a general theory of Kähler reduction, much remains to be done, and this whole area is a very active field of current research. Applications of this theory include, for example, the motion of point vortices on the sphere (see KIRWAN (1988)) and the resonant three-wave interaction (see PEKARSKY AND MARSDEN (2001)). Note that the Einstein property is not automatically inherited by the quotient. Sufficient conditions for the quotient to be Kähler–Einstein are given, for instance, in FUTAKI (1988).

f. A ***hyper-Kähler manifold*** is a Riemannian manifold with holonomy $Sp(n)$, the compact symplectic group ($Sp(n)$ is defined in, for example, KNAPP (1996) or the internet supplement for MARSDEN AND RATIU (1999)). Regular hyper-Kähler reduction by a group action, a subject also missing in this book, has been introduced in LINDSTRÖM AND ROČEK (1983); HITCHIN et al. (1987) and developed in, for example, BIELAWSKI (1999a,b,c); BIELAWSKI AND DANCER (2000); BIQUARD AND GAUDUCHON (1997); DANCER (1999); HITCHIN (2000); KONNO (2002); LINDSTRÖM et al. (2000); IKEDA (2000). The singular case (with the exception of orbifold quotients) appears not to have been worked out. More generally, a reduction for hypercomplex structures (a generalization of hyper-Kähler manifolds) has been defined in JOYCE (1991) and applied to hyper-Kähler manifolds with torsion in GRANTCHAROV et al. (2002).

g. A ***quaternion-Kähler manifold*** is a Riemannian manifold with holonomy $Sp(n) \cdot Sp(1) := (Sp(n) \times Sp(1))/\mathbb{Z}_2$. Regular reduction of quaternionic manifolds (always by a group action) has been defined in GALICKI (1987) then developed in, for example, GALICKI AND LAWSON (1988); BOYER et al. (1993); KOBAK AND SWANN (1996, 2001); BATTAGLIA (1999); NITTA (1990). The difficulty here is the lack of a symplectic form. One uses instead a closed four-form.

There exist other structures, not having special holonomy, but related to special holonomy by some natural constructions, as the cone over them. For some of these structures a reduction scheme is already available.

h. The Hermitian version of locally conformal symplectic manifolds are the locally conformal Kähler manifolds (see DRAGOMIR AND ORNEA (1997)). A ***locally conformal Kähler manifold*** is a locally conformal symplectic manifold (M, ω, η) carrying, in addition, a complex structure $\mathbb{J}$ and a Hermitian metric compatible with ω, that is,

$\omega(\mathbb{J}\cdot, \cdot)$ is a positive definite scalar product. Their regular reduction has been developed by BIQUARD AND GAUDUCHON (1997); GINI et al. (2003).

i. ***Sasakian*** (respectively 3-***Sasakian***) manifolds are Riemannian manifolds (M, g) whose Riemannian cones $(M \times \mathbb{R}_+, t^2 g + dt^2)$ carry a Kähler (respectively hyper-Kähler) structure. Their regular reduction (under a group action) has been developed in BOYER AND GALICKI (1999, 2000); BOYER et al. (1994, 1998a,b); BOYER, GALICKI, AND PICCINNI (2002); GRANTCHAROV AND ORNEA (2001).

The relation between symplectic and contact structures has an analog for all these structures presented above. As in the case of contact and symplectic reduction, this relation is compatible with reduction.

9. *Symplectic and contact cuts.* The theory of symplectic and contact cuts (see LERMAN (1995, 2001), LERMAN et al. (1998)) and its application to the theory of singular symplectic, Kähler, and contact reduction is also not covered in the present book. This is a powerful tool that links the phenomena of blow-up to Morse theoretical ideas in the context of reduction in these categories.

10. *Poisson–Lie reduction.* The theory of regular reduction can be extended to the case of Poisson actions of Poisson–Lie groups (see LU (1990, 1991); LU AND WEINSTEIN (1990)). A theorem of ALEKSEEV (1997) links these constructions to those for usual canonical actions of Lie groups on symplectic manifolds. So far, nothing has been done to extend this theory to the singular case.

11. *Lagrangian and discrete reduction.* Also lacking in this book is the theory of Lagrangian reduction, in which one reduces the variational principle rather than symplectic or Poisson structures. The regular case has been developed in MARSDEN AND SCHEURLE (1993b); CENDRA et al. (2001a,b); HOLM et al. (1998, 2002). Regular Routhian reduction can be found in ROUTH (1977); ARNOLD et al. (1988); MARSDEN AND SCHEURLE (1993a); JALNAPURKAR AND MARSDEN (2000); MARSDEN, RATIU, AND SCHEURLE (2000). The singular case of Lagrangian and Routhian reduction is, as of today, an untouched area of research.

Variational methods, geometric mechanics, Lagrangian and Routh reduction as well as multisymplectic variational methods have already had a significant impact on the theory of structure preserving numerical algorithms; see, e.g., WENDLANDT AND MARSDEN (1997); MARSDEN AND WEST (2001); LEW et al. (2003) and references therein. These methods are closely allied with the general theory of discrete mechanics and discrete reduction; in that subject, the general discrete mechanical and discrete integrable systems have been extensively developed in MOSER AND VESELOV (1991); MARSDEN et al. (1998); BOBENKO AND SURIS (1999); MARSDEN et al. (1999); SURIS (2003); JALNAPURKAR et al. (2003). The discrete analog of singular reduction is also an untouched area of research.

12. *Nonholonomic reduction.* Reduction of nonholonomically constrained mechanical systems is also not presented. Certain aspects, including the Lagrangian and Hamiltonian points of view, can be found in BLOCH (2003) along with a vast literature on the subject. The momentum defined in this setting is not conserved; in fact, there is a very interesting equation satisfied by the momentum map which is ultimately responsible for locomotion generation. On the other hand, the fact that the momentum map is not conserved raises fundamental questions about the reduction process. From the Lagrange–d'Alembert point of view, the theory is relatively complete and was developed by KOILLER (1992); BLOCH et al. (1996); CENDRA et al. (2001b). The theory from the point of view of almost Poisson and Leibniz reduction is also in reasonable shape; see, for example, BATES AND SNIATYCKI (1993); VAN DER SCHAFT AND MASCHKE (1994); MARLE (1995); KOON AND MARSDEN (1997, 1998); MARLE (1998); CUSHMAN AND SNIATYCKI (2002); ORTEGA AND PLANAS (2003b).

Progress on singular nonholonomic reduction is still underway; a start in this direction can be found , for instance, in BATES (1998, 2002); SNIATYCKI (2001). Nonholonomic reduction, even in the regular case, taking into account the momentum equation is still lacking.

13. *Dirac reduction.* The theory of Dirac structures (COURANT AND WEINSTEIN (1986); COURANT (1990); DORFMAN (1993); VAN DER SCHAFT (1998a); DALSMO AND VAN DER SCHAFT (1999); BLANKENSTEIN AND VAN DER SCHAFT (2001)) and the associated reduction are also are missing in this book. Regular Dirac reduction from a Hamiltonian point of view has been developed by LIU et al. (1998); BLANKENSTEIN (2000); BLANKENSTEIN AND VAN DER SCHAFT (2001) and the singular case by BLANKENSTEIN AND RATIU (2003). The Lagrangian point of view as well as its links to nonholonomic systems and circuits has been carried out in YOSHIMURA AND MARSDEN (2003). This whole subject needs further development, including its connection to reduction theory (for instance, for nonholonomic mechanics) and control theory (see, for instance, NIJMEIJER AND VAN DER SCHAFT (1990); VAN DER SCHAFT (1998b); VAN DER SCHAFT AND MASCHKE (1994, 1995a,b, 1997)).

14. *Contact convexity.* The convexity theorem for toral momentum maps for contact manifolds due to LERMAN (2002) is also not presented in this book. This subject needs further development and a link of the convexity with contact singular reduction should be established.

15. *Intrinsic convexity.* The Poisson convexity conjecture of WEINSTEIN (2000) states that the moment map of a proper symplectic groupoid action has intrinsic convexity properties; this is a map from the symplectic manifold to the base of the groupoid. The fundamental idea behind this conjecture is that proper groupoid actions generalize both Hamiltonian proper actions and Poisson actions, for which convexity theorems are already available. For compact Poisson–Lie group actions, this was discussed in point **2**. For noncompact real semisimple proper Lie group actions such that the momentum map has its image contained in the subset of the dual of the Lie algebra consisting of points having a compact coadjoint isotropy, such convexity theorems were given by WEINSTEIN (2001). Further evidence for this conjecture is discussed in WEINSTEIN (1987b, 2002b) and ZUNG (2003) sketches its proof. All of these results use a momentum map whose values are in some Poisson manifold with a proper symplectic groupoid. This suggests that the optimal momentum map should have a tight relationship with this circle of ideas and that, moreover, it should also possess convexity properties.

TOPICS LINKED TO REDUCTION

While the topics above are directly concerned with momentum maps and reduction, there are subjects that are closely linked to them and can be seen as direct applications of this theory. Below is a short biased list of such areas, all in dynamics, that we believe could have had their place in a more extensive book. Most of these topics are still unsatisfactorily tied to the theory of singular reduction. We hope that this book will provide the necessary foundation on which these links can be further developed.

1. *Geometric phases.* While the reconstruction method of the dynamics is discussed in several places, the subject of geometric phases is absent. A short presentation of reconstruction phases is given in §6.6, but the general theory of geometric phases for more complicated situations as well as its link to averaging connections, adiabatic invariants, the long time behavior of the dynamics, and applications to mechanics, control, integrable systems, and optics is lacking. The literature on this subject is vast; we mention for illustration FROHLICH (1979); SIMON (1983); BERRY (1984, 1985, 1988); GUICHARDET (1984); HANNAY (1985); GOZZI AND THACKER (1987a,b); KRITSIS (1987); SHAPERE AND WILCZEK (1987, 1988); ANANDAN (1988); BERRY AND HANNAY (1988); LITTLEJOHN (1988); MONTGOMERY (1988, 1990, 1991a,b, 1996, 2002); WILCZEK AND SHAPERE (1989); AVRON et al. (1989); GOLIN et al. (1989); WEINSTEIN (1990); KRISHNAPRASAD (1990); KRISHNAPRASAD AND YANG (1991); MARSDEN et al. (1989, 1990); YANG AND KRISHNAPRASAD (1994); MARSDEN AND OSTROWSKI (1996); MARSDEN, RATIU, AND SCHEURLE (2000); BLAOM (2000), and references therein.

2. *Relative equilibria and relative periodic orbits.* The stability of relative dynamical elements has been extensively treated; see, for example, MADDOCKS (1991); PATRICK (1992); LERMAN AND SINGER (1998); ORTEGA AND RATIU (1999a,b,c); PATRICK et al. (2003), and references therein. Space also did not permit to present another application to dynamics, namely the theory of persistence and bifurcation of relative equilibria and relative periodic orbits. A persistence theorem is a result that predicts the existence of certain dynamical elements around a given one that satisfies a nondegeneracy hypothesis. Examples and applications of this kind of results can be found in, for instance, MONTALDI et al. (1988); OH et al. (1989); LEWIS (1993); BARTSCH (1994); LEWIS AND RATIU (1996); MONTALDI (1997a,b); ORTEGA AND RATIU (1997); ROBERTS AND DE SOUSA DIAS (1997); LEWIS (1998); MONTALDI AND ROBERTS (1999); LERMAN AND TOKIEDA (1999); ROBERTS et al. (2002); WULFF AND ROBERTS (2002); WULFF (2002, 2003); ORTEGA (2003a). When nondegeneracy hypotheses are not available then a bifurcation theorem is obtained. Bifurcation problems in the symmetric conservative context have been treated in, for instance, CHOSSAT (1986); VAN DER MEER (1986, 1990, 1996); DELLNITZ et al. (1992); CHOSSAT AND IOOSS (1994); VANDERBAUWHEDE AND VAN DER MEER

(1995); GOLUBITSKY et al. (1995); CHOSSAT AND DIAS (1995); CHOSSAT, KOENIG, AND MONTALDI (1995); CHOSSAT, ORTEGA, AND RATIU (2002); CHOSSAT et al. (2003); BIRTEA et al. (2003); HERNÁNDEZ AND MARSDEN (2003); ORTEGA AND RATIU (2003b).

3. *Integrable systems.* The theory of integrable systems is so vast and there are already so many excellent books and review articles available, that it could not possibly find its way into this work. Nevertheless, there are purely symplectic aspects of the theory of integrable systems that are intimately connected to reduction. For example, a symmetric Hamiltonian system is said to be ***noncommutative*** or ***superintegrable*** if, for generic values of the momentum map, the reduced space be discrete. One could strengthen this notion by requiring that at *each* value of the momentum map the reduced space is discrete. Both notions are used in the literature and these systems are *integrable* because one can integrate them by the reconstruction method in §6.1.10. The words *non-Abelian* appear in order to distinguish this notion from the classical *Abelian* one due to Liouville, in which a cylinder (mostly a torus) is the symmetry group in question whose generators are given by the flows of n independent integrals in involution, where n is the number of degrees of freedom of the system. In §11.2.10 we shall briefly discuss aspects of integrable systems related to dual pairs and bifoliations, but the full theory is not presented. For example DUISTERMAAT (1980) analyzed the obstructions to the existence of global action–angle variables. The relation between Abelian and non-Abelian integrability has been discussed in MISHCHENKO AND FOMENKO (1978) (see also the book by FOMENKO AND TROFIMOV (1988)). For the symplectic geometry of integrable systems see, for example, BOGOYAVLENSKIJ (1996b,c, 1997, 1998a,b,c); DAZORD AND DELZANT (1987); FASSÒ AND RATIU (1998); FASSÒ (1999) and references therein. The topological classification of various integrable systems is due to Fomenko and his school (see, for instance, FOMENKO (1991a,b); ZUNG (1996); BOLSINOV AND FOMENKO (2000) and references therein). This work also describes the bifurcation of the Liouville tori. No tight relationship between singular reduction, singular action–angle variables (ELIASSON (1990); ITO (1991)), convexity of the momentum map, and Fomenko's bifurcation theory of Liouville tori seems to have been established. ZUNG (2003) contains an up-to-date bibliography on this subject and is an excellent introduction to this circle of ideas.

BOGOYAVLENSKIJ (1996a) introduced a notion of integrability that goes beyond Hamiltonian systems; see BOGOYAVLENSKIJ (1996d); FASSÒ AND GIACOBBE (2002) and references therein for the development of this theory. Section 11.2 contains some paragraphs on the interplay between bifoliations and noncommutative integrability.

4. *Multisymplectic reduction.* For PDEs, the multisymplectic (as opposed to symplectic) framework seems appropriate, both for relativistic and nonrelativistic systems. In fact, this approach has experienced somewhat of a revival since it has been realized that it is rather useful for numerical computation (see MARSDEN et al. (1998)). Only a few instances and examples of multisymplectic reduction, in Hamiltonian or Lagrangian formulation, are really well understood (see MARSDEN et al. (1986); CASTRILLÓN-LÓPEZ et al. (2000); CASTRILLÓN-LÓPEZ, GARCÍA PÉREZ, AND RATIU

(2001); CASTRILLÓN-LÓPEZ AND RATIU (2003); CASTRILLÓN-LÓPEZ AND MARSDEN (2003)), so one can expect to see more activity in this domain. As with finite-dimensional systems mentioned before, the whole subject of discrete multisymplectic systems and its link to numerical algorithms is a rich area of current research. Little is known about reduction for discrete multisymplectic systems.

Conventions. Unfortunately, there are several possible conventions in differential geometry and Hamiltonian dynamics that are incompatible. We shall adopt here those in ABRAHAM AND MARSDEN (1978); ABRAHAM et al. (1988); BOURBAKI (1971); CHOQUET-BRUHAT AND DEWITT-MORETTE (1982); KOLÁŘ et al. (1993); KNAPP (1996); LANG (1999); MARSDEN AND RATIU (1999); SPIVAK (1979), which differ from those used in other widely quoted classical texts such as ARNOLD (1989); GREUB et al. (1970a,b); KOBAYASHI AND NOMIZU (1963); MATSUSHIMA (1972); YANO AND KON (1984), for example. Throughout the book, all manifolds are assumed to be finite-dimensional and smooth. No assumption on paracompactness will be made and if partitions of unity are used, this will be clearly mentioned. If some other topological property (such as Hausdorff, second countability, connectedness, or compactness) is needed, it will be explicitly specified.

All Lie groups are Hausdorff and their actions on various manifolds are assumed to be smooth. Quotients are not automatically Hausdorff and we shall comment on this when appropriate. Also, we will consider several submanifolds, usually associated to group actions, that turn out not to be connected, and it may even happen that their connected components have different dimensions; we shall warn the reader whenever this is the case.

Prerequisites. The present book is self-contained and, in principle, very little outside standard knowledge of calculus on manifolds is needed to be able to read it. However, prior acquaintance with symplectic geometry and geometric mechanics (as found, for example, in standard books such as ARNOLD (1989); GUILLEMIN AND STERNBERG (1984c); LIBERMANN AND MARLE (1987); MARSDEN AND RATIU (1999); MCDUFF AND SALAMON (1998)) will significantly speed up the reading. We have tried to keep referral to outside references to a minimum, in the sense that we review, without proofs, the required material each time this is necessary and at the same time provide the bibliographical sources.

The book is addressed to graduate students and researchers in the field and can be used as a reference or as material for various courses in symplectic or Poisson geometry as well as geometric mechanics. Space did not permit us to include many concrete examples because, in general, these are involved, especially in the singular situation. We have often given bibliographical references where various examples are carried out in detail.

Acknowledgments. We thank Anton Alekseev, Judith Arms, Larry Bates, María-Pilar Bernal, Petre Birtea, Anthony Blaom, Anthony Bloch, Dan Burghelea, Ana Cannas da Silva, Pascal Chossat, Richard Cushman, Francesco Fassò, Viktor Ginzburg, Martin Golubistky, Jean-Claude Hausmann, Johannes Huebschmann, Alan Knutson, Ann Kostant, Bertram Kostant, Eugene Lerman, Rui Loja Fernandes, Charles-Michel

Marle, Jerry Marsden, Peter Michor, James Montaldi, Richard Montgomery, Liviu Ornea, George Patrick, Víctor Planas-Bielsa, Mark Roberts, Miguel Rodríguez-Olmos, Tanya Schmah, Reyer Sjamaar, Jędrzej Sniatycki, Ian Stewart, Tadashi Tokieda, Gijs Tuynman, Alan Weinstein, Claudia Wulff, and the anonymous referees for their input and comments that influenced our presentation of this material.

This book was written in a time interval when the authors were at the University of California at Santa Cruz, at the École Polytechnique Fédérale de Lausanne, at the Institut Non Linéaire de Nice, at the Technische Unversität München, and at the Centre National de la Recherche Scientifique. We want to thank all these institutions for providing the necessary infrastructure such that this research could be performed and for the friendly and collegial atmosphere during our stay there.

Partial financial support during the writing period of this book was provided by the US and Swiss National Science Foundations, by the US Department of Energy and the US International Agency, by the Fulbright Commission and the Banco Central Hispano, by the Humboldt and Rotary Foundations, by the CNRS, and by the European Commission through the Fifth Framework Research Training Network *Mechanics and Symmetry in Europe*. We hereby thank all these agencies for their funding that made our collaboration possible.

A first version of this book was awarded the 2000 Ferran Sunyer i Balaguer Prize. We hereby thank the Board of the Institut d'Estudis Catalans for the patience it has shown and for the freedom it has granted us by giving us the necessary time to considerably enlarge that work into this final published version. In the process, we have greatly benefited from the input given to us by our editor, Ann Kostant, whose excellent advice has had a significant impact on the exposition of the material in this book.

Chapter 1

Manifolds and Smooth Structures

The purpose of this chapter is to have a handy glossary of definitions, notation, conventions, and theorems in elementary manifold theory. The reader should use it only as a source of information whenever the need arises. The style used in the exposition will be merely descriptive and the statements will be framed under the form of a result only when either a short proof is given because it is not readily available in the literature or when they will be needed for future reference. The results quoted here can be found, for example, in ABRAHAM AND MARSDEN (1978); ABRAHAM, MARSDEN, AND RATIU (1988); ABRAHAM AND ROBBIN (1967); BOURBAKI (1971); BRICKELL AND CLARK (1970); PFLAUM (2001a); KOBAYASHI AND NOMIZU (1963); KOLÁŘ, MICHOR, AND SLOVÁK (1993); LIBERMANN AND MARLE (1987); SPANIER (1966); TONDEUR (1997); WARNER (1983).

1.1 Manifolds and smooth maps

1.1.1 The definitions of standard notions such as manifolds, smooth maps, locally trivial fiber bundles, vector bundles, algebraic operations with vector bundles (such as the direct or Whitney sum, or tensor and exterior products of vector bundles), and tensor calculus on manifolds will not be reviewed. All objects are assumed to be smooth, that is, are of class C^∞, unless otherwise stated. All manifolds and bundles are finite dimensional. Occasionally, additional topological assumptions on the manifold are needed, such as Hausdorff, paracompactness (so that there exist partitions of unity), second countability (to use the topological version of Sard's theorem) or connectedness (to avoid working on manifolds whose connected components have different dimensions); in such cases, these extra topological hypotheses will be explicitly stated. Charts on a manifold M are denoted by (U, φ), where U is an open subset of M and $\varphi : U \to U' \subset \mathbf{E}$, $\mathbf{E} = \mathbb{R}^n$, for some $n \in \mathbb{N}$, is a homeomorphism onto an open set U' in the Euclidean space $\mathbf{E}$. The number n, called the ***(local) dimension*** of M, is constant on each connected component of M. The vector space $\mathbf{E}$ is called the ***model space*** of the manifold M and one says that M ***is modeled on*** $\mathbf{E}$. The ring of smooth functions on M will be denoted by $C^\infty(M)$. Every finite-dimensional manifold is a ***locally compact topological space***, that is, it is locally Hausdorff and each point admits an open neighborhood whose closure is compact.

1.1.2 Smooth maps. If $f : M \to N$ is a map, $m \in M$, (U, φ) is a chart at m in M, and (V, ψ) is a chart at $f(m)$ in N such that $f(U) \subset V$, we shall denote by $f_{\varphi\psi} := \psi \circ f \circ \varphi^{-1} : U' \to V'$ its ***local representative*** at m. Often, we shall simply ignore the indices specifying the chart maps and say that the local representative of f has a certain form, using the same notation for the map and its local representative. The map $f : M \to N$ is ***smooth*** at the point $m \in M$ if there is a local smooth representative of f at m. If f is smooth at any point of M we will simply say that f is smooth.

1.1.3 The submanifold property. Throughout this book various subsets of a given manifold will be studied. These will have different regularity properties. In order to avoid possible confusion, we recall the classical definition of a submanifold. The subset $N \subset M$ is called a ***submanifold*** if it satisfies the following condition: for each $n \in N$ there is a chart (U, φ) such that $n \in U$, $\varphi : U \to \mathbf{E} = \mathbf{E}_1 \times \mathbf{E}_2$, and $\varphi(U \cap N) = \varphi(U) \cap (\mathbf{E}_1 \times \{0\})$. Such charts of M are said to have the ***submanifold property***. The intersection of N with the charts on M having the submanifold property endows N with a smooth manifold structure whose underlying topology coincides with the induced topology on N. It is important to note here that a submanifold is necessarily a locally closed subset of M. Recall that a subset N of a topological space M is said to be ***locally closed*** if each point $n \in N$ has an open neighborhood U in M such that $U \cap N$ is closed in U, where U is endowed with the relative topology inherited from M. Equivalently, N is locally closed if it is the intersection of an open and a closed set in M, which happens if and only if N is open in its closure in M.

1.1.4 Tangent vectors and the tangent bundle. For a smooth manifold M modeled on $\mathbf{E}$, TM denotes its ***tangent bundle***. Recall that two C^1 curves $c_1, c_2 :]-\epsilon, \epsilon[\to M$, $\epsilon > 0$, $c_1(0) = c_2(0) = m$ are said to be ***tangent at*** m if there is some chart (U, φ) at m such that $\frac{d}{dt}(\varphi \circ c_1)|_{t=0} = \frac{d}{dt}(\varphi \circ c_2)|_{t=0}$. This is an equivalence relation independent of the chart used to state it. A ***tangent vector*** at $m \in M$ to the manifold M is defined as an equivalence class $[c]_m$ of C^1 tangent curves at m. The ***tangent space*** $T_m M$ is the vector space of all such classes $[c]_m$. Then one defines $TM := \bigcup_{m \in M} T_m M$.

If $f : M \to N$ is a smooth map between the smooth manifolds M and N (modeled on $\mathbf{E}$ and $\mathbf{F}$ respectively), $Tf : TM \to TN$ is its ***derivative*** or ***tangent map*** defined by

$$Tf(v_m) := \left.\frac{d}{dt}\right|_{t=0} f(c(t)), \tag{1.1.1}$$

where $c(t)$ is an arbitrary smooth curve satisfying $c(0) = m$ and $\dot{c}(0) = v_m \in T_m M$. In terms of equivalence classes of curves, this definition reads $T_m f([c]_m) = [f \circ c]_{f(m)}$, where $v_m = [c]_m$.

If $U' \subset \mathbf{E}$ is open in $\mathbf{E}$, there is a natural identification

$$(u, \mathbf{e}) \in U' \times \mathbf{E} \leftrightarrow [c_{u,\mathbf{e}}]_u \in TU',$$

where $c_{u,\mathbf{e}}(t) := u + t\mathbf{e}$. We shall not distinguish in what follows between TU' and $U' \times \mathbf{E}$. Thus, any chart (U, φ) on M naturally induces a chart $(TU, T\varphi : TU \to \varphi(U) \times \mathbf{E})$, called the ***induced vector bundle chart*** on TM. The atlas formed by all these induced charts defines a vector bundle structure on TM.

Let (U, φ) be a chart at m on M and (V, ψ) a chart at $f(m)$ on N such that $f(U) \subset V$. The local representative $(Tf)_{T\varphi T\psi}$ of Tf in the naturally induced tangent bundle charts $(TU, T\varphi : TU \to \varphi(U) \times \mathbf{E})$ and $(TV, T\psi : TV \to \psi(V) \times \mathbf{F})$ is given by

$$(u, \mathbf{e}) \in \varphi(U) \times \mathbf{E} \mapsto (f_{\varphi\psi}(u), \mathbf{D}f_{\varphi\psi}(u) \cdot \mathbf{e}) \in \psi(V) \times \mathbf{F}, \tag{1.1.2}$$

where $\mathbf{D}f_{\varphi\psi}(u) : \mathbf{E} \to \mathbf{F}$ is the linear map given by the Fréchet derivative of $f_{\varphi\psi}$ at u. This formula shows that if $f : M \to N$ is smooth, then $Tf : TM \to TN$ is also a smooth map.

1.1.5 Smooth functions and tangent vectors. We recall that a ***sheaf*** $\mathcal{F}$ of functions on a topological space P is a map that assigns to any open set U a set of real valued functions $\mathcal{F}(U)$ defined on U which is an algebra under multiplication. In the definition it is also required that for every inclusion $V \subset U$ of open sets there is a given homomorphism $\operatorname{res}_V^U : \mathcal{F}(U) \to \mathcal{F}(V)$ called the ***restriction*** from U to V that satisfies the following conditions:

(SH1) $\mathcal{F}(\emptyset) = \{0\}$ and $\operatorname{res}_U^U : \mathcal{F}(U) \to \mathcal{F}(U)$ is the identity map.

(SH2) If $W \subset V \subset U$ are open sets, then $\operatorname{res}_W^V \circ \operatorname{res}_V^U = \operatorname{res}_W^U$.

(SH3) Let U be an open set and $\{V_i\}_{i \in I}$ an open covering of U. If $f \in \mathcal{F}(U)$ is such that the restriction $\operatorname{res}_{V_i}^U(f)$ of f to each V_i is 0, then $f = 0$.

(SH4) Let U be an open set, $\{V_i\}_{i \in I}$ an open covering of U, and let $f_i \in \mathcal{F}(V_i)$ be given for each $i \in I$. Suppose that the restrictions of f_i and f_j to $V_i \cap V_j$ are equal for all $i, j \in I$. Then there exists a unique $f \in \mathcal{F}(U)$ whose restriction to each V_i is f_i for all $i \in I$.

When the map $\mathcal{F}$ satisfies only properties **(SH1)** and **(SH2)** we say that $\mathcal{F}$ is a ***presheaf***. The elements in $\mathcal{F}(U)$ are called the ***sections*** of $\mathcal{F}$ over U. The elements in $\mathcal{F}(P)$ are called ***global sections***. The direct limit (see LANG (2002) for a definition)

$$\mathcal{F}_z := \varinjlim_{U} \mathcal{F}(U),$$

where U represents the directed system by inclusion of all the open neighborhoods of a point $z \in P$, is called the ***stalk*** of $\mathcal{F}$ at the point z. The stalk $\mathcal{F}_z$ has a natural algebra structure and it is isomorphic to the set of germs at z. Recall that two functions f and g defined on open sets containing z are said to have the same ***germ*** at z if they agree on some neighborhood of z. This introduces an equivalence relation in the set $\{\mathcal{F}(U) \mid U \text{ neighborhood of } z\}$, two functions being equivalent if and only if they have the same germ at z. The equivalence classes are called ***germs*** at z.

When M is a smooth manifold we can naturally define a presheaf of functions on M, denoted by C_M^∞, or C^∞ for short, that will be used extensively in this book. Let U

be any open subset of M. As U has the submanifold property, it is by itself a manifold. Let $C^\infty(U)$ be the set of smooth real valued functions on U according to the definition in §1.1.2. The assignment $U \to C^\infty(U)$ defines a presheaf of functions on M that we will call the ***presheaf of smooth functions*** on M. Under certain topological conditions this presheaf is actually a sheaf (see Proposition 1.1.22).

Let C_z^∞ be the stalk of C^∞ at z that we will identify with the set of germs at z. As an algebra, C_z^∞ contains as an ideal the subspace $\mathfrak{m}_z$ of germs in C_z^∞ that vanish at z. Let $\mathfrak{m}_z^2$ be its square.

It can be proved (see WARNER (1983)) that the tangent space T_zM to the manifold M at the point z, as defined in §1.1.4, is naturally isomorphic to the space of linear derivations of the algebra C_z^∞ or, equivalently, to $(\mathfrak{m}_z/\mathfrak{m}_z^2)^*$. (The symbol $*$ denotes dual vector space.)

1.1.6 Smooth maps between spaces with a (pre)sheaf of functions. Let $\mathcal{F}$ be a (pre)sheaf of functions defined on the topological space P. The preceding discussion on the description of the tangent space of a manifold at a point in terms of its presheaf of smooth functions motivates the following definitions: for any point $z \in P$ we define the ***tangent space*** T_zP of P at z as the vector space $(\mathfrak{m}_z/\mathfrak{m}_z^2)^*$. The ***tangent bundle*** TP of P is the union of all the tangent spaces, that is

$$TP = \bigcup_{z \in P} T_zP.$$

The notion of smooth map and its derivative can also be adapted to this context. Let P_1 and P_2 be two topological spaces where two sheaves of functions $\mathcal{F}_1$ and $\mathcal{F}_2$, respectively, have been defined. We say that the continuous map $f : (P_1, \mathcal{F}_1) \to (P_2, \mathcal{F}_2)$ is ***smooth*** if for any open set $U \subset P_2$ we have that

$$f^*\mathcal{F}_2(U) \subset \mathcal{F}_1(f^{-1}(U)),$$

where $f^*s := s \circ f$ for any $s \in \mathcal{F}_2(U)$. The ***derivative*** or ***tangent map*** $T_zf : T_zP_1 \to T_{f(z)}P_2$ of f at z is the dual of the map $[g] \in \mathfrak{m}_{f(z)}/\mathfrak{m}_{f(z)}^2 \mapsto [g \circ f] \in \mathfrak{m}_z/\mathfrak{m}_z^2$.

1.1.7 The cotangent bundle. The dual T^*M of the vector bundle TM (obtained by taking the dual space of every fiber) is called the ***cotangent bundle***. It is a smooth vector bundle whose charts are induced by those on M in the following way. Every chart (U, φ) on the manifold M modeled on $\mathbf{E}$ naturally induces a chart $(T^*U, \varphi_* : T^*U \to \varphi(U) \times \mathbf{E}^*)$ on T^*M by

$$\varphi_*(\alpha_u) := \left(\varphi(u), \alpha_u \circ (T_u\varphi)^{-1}\right).$$

Given a smooth map $f : M \to N$ and $m \in M$, $T_m^*f : T_{f(m)}^*N \to T_m^*M$ is the dual map to T_mM, that is,

$$\left\langle T_m^*f(\alpha_{f(m)}), v_m \right\rangle := \left\langle \alpha_{f(m)}, T_mf(v_m) \right\rangle$$

for $v_m \in T_mM$ and $\alpha_{f(m)} \in T_{f(m)}^*N$. Unlike the case of tangent bundles, there is no globally defined map T^*f in general.

Any smooth map $f : M \to N$ has a convenient local representation at a given point $m_0 \in M$. Using the notation above, let $k = \operatorname{rank} T_{m_0} f$, and let $\mathbf{F}_1 := \mathbf{D} f_{\varphi\psi}(\varphi(m_0))(\mathbf{E}) \cong \mathbb{R}^k$ have complement $\mathbf{F}_2$, that is, $\mathbf{F}_1 \oplus \mathbf{F}_2 = \mathbf{F}$. Let $\mathbf{E}_1 \cong \mathbb{R}^k$ be the complement of $\mathbf{E}_2 := \ker \mathbf{D} f_{\varphi\psi}(\varphi(m_0))$ in $\mathbf{E}$. Then there are open sets $\tilde{U} \subset \varphi(U) \subset \mathbf{E}$, $\varphi(m_0) \in \tilde{U}$, and $\tilde{V} \subset \mathbf{F}_1 \times \mathbf{E}_2$, as well as a diffeomorphism $\rho : \tilde{V} \to \tilde{U}$ such that $(f_{\varphi\psi} \circ \rho)(u, v) = (u, \eta(u, v))$, where $\eta : \tilde{V} \to \mathbf{F}_2$ is a smooth map satisfying $\mathbf{D}\eta(\rho^{-1}(\varphi(m_0))) = 0$. Informally stated, f is locally of the form $f(u, v) = (u, \eta(u, v))$ with $\mathbf{D}\eta(u_0, v_0) = 0$, where m_0 is locally (u_0, v_0).

1.1.8 Smooth maps and submanifolds. Several classes of smooth maps will be important later. The smooth map $f : M \to N$ is called

- a ***diffeomorphism*** if f is bijective and its inverse $f^{-1} : N \to M$ is also smooth;
- an ***immersion*** if $T_m f : T_m M \to T_{f(m)} N$ is injective for all $m \in M$;
- a ***submersion*** if $T_m f : T_m M \to T_{f(m)} N$ is surjective for all $m \in M$;
- a ***subimmersion***, or ***constant rank map***, if $T_m f : T_m M \to T_{f(m)} N$ has the same rank for all m in each connected component of M;
- an ***injective immersion*** if f is an injective map that is also an immersion;
- a ***regular immersion*** if it is an injective immersion satisfying the following condition: for any smooth manifold P, an arbitrary map $g : P \to M$ is smooth if and only if $f \circ g : P \to N$ is smooth;
- an ***embedding*** if it is an injective immersion that is a homeomorphism onto its image $f(M)$, where $f(M)$ is endowed with the relative topology induced by N. In this case, $f(M)$ is a submanifold of N.

We can use these definitions to introduce various notions of submanifolds that will be used throughout the book. Let M be a manifold and $N \subset M$ a subset of M endowed with its own smooth manifold structure. We say that N is

- an ***immersed submanifold*** when the inclusion $i : N \hookrightarrow M$ is an immersion;
- an ***initial submanifold*** when the inclusion $i : N \hookrightarrow M$ is a regular immersion;
- an ***embedded submanifold*** when the inclusion $i : N \hookrightarrow M$ is an embedding.

If N is an immersed submanifold of M, then its manifold topology is stronger than the relative topology of N with respect to M. Every embedded submanifold is initial and any initial submanifold is immersed. These inclusions are strict (see §1.1.12). Embedded submanifolds are submanifolds in the sense that around any point in N we can find a chart with the submanifold property. Conversely, if N is a submanifold of M, the inclusion $N \hookrightarrow M$ is an embedding.

1.1.9 Lemma *Let M be a smooth manifold and $S \subset M$ an embedded submanifold. Then for any $s \in S$*

$$(T_s S)^\circ = \{\mathbf{d}f(s) \mid f \in C^\infty(M),\ f|_S \equiv 0\}. \tag{1.1.3}$$

The symbol $(T_sS)^\circ := \{\alpha_s \in T_s^*M \mid \alpha_s|_{T_sS} \equiv 0\}$ *denotes the* ***annihilator*** *of the vector subspace* T_sS *of* T_sM *in* T_s^*M.

Proof. We shall show that for any $\alpha_s \in (T_sS)^\circ$ there exists a function $f \in C^\infty(M)$ satisfying $f|_S \equiv 0$ and $\mathbf{d}f(s) = \alpha_s$. This is done in the following way. Since S is a submanifold, there is chart around s in M whose image is the product $U \times V$, where $U \subset \mathbf{E}_1$, $V \subset \mathbf{E}_2$ with $\mathbf{E}_1$, $\mathbf{E}_2$ Euclidean spaces, and s has image $(0, 0)$ by this chart map. Then α_s is identified with an element $\alpha \in \mathbf{E}_2^*$. Let ψ be a bump function around the origin on the open set $V \subset \mathbf{E}_2$, φ a bump function around the origin on the open set $U \subset \mathbf{E}_1$, and $\bar{f} : V \to \mathbb{R}$ a function such that $\bar{f}(0) = 0$ and $\mathbf{d}\bar{f}(0) = \alpha$. Define $f : U \times V \to \mathbb{R}$ by $f(u, v) := \varphi(u)\psi(v)\bar{f}(v)$. Then $f(u, 0) = \varphi(u)\psi(0)\bar{f}(0) = 0$ since $\psi(0) = 1$ and $\bar{f}(0) = 0$. Also, $\mathbf{d}f(0, 0) = \psi(0)\bar{f}(0)\mathbf{d}\varphi(0) + \varphi(0)\psi(0)\mathbf{d}\bar{f}(0) + \varphi(0)\bar{f}(0)\mathbf{d}\psi(0) = \mathbf{d}\bar{f}(0) = \alpha$. Thus, f is a smooth function on $U \times V$, compactly supported, such that $f|_{U\times\{0\}} \equiv 0$ and $\mathbf{d}f(0, 0) = \alpha$. Use the chart map to bring this function f back to the manifold and extend it by zero outside the chart domain; call the extension also f. We obtain thus a function $f \in C^\infty(M)$ such that $f|_S \equiv 0$ and $\mathbf{d}f(s) = \alpha_s$. ■

1.1.10 Initial submanifolds. The properties stated in the following lemma will be used later on in the book.

1.1.11 Lemma *Let M be a smooth manifold.*

(i) *Let N be an initial submanifold of M. The smooth manifold structure that makes the subset $N \subset M$ into an initial submanifold is unique.*

(ii) *Let K and N be two initial submanifolds of M such that $K \subset N \subset M$. Then K is an initial submanifold of N.*

(iii) *Let N be an injectively immersed submanifold of the smooth manifold M. Suppose that N can be written as the disjoint union of the elements of a family $\{S_\alpha\}_{\alpha\in I}$ of open subsets of N such that each S_α is an initial submanifold of M. Then, N is initial.*

Proof. **(i)** Suppose that N admits another smooth manifold structure, call it N', that makes it into an initial submanifold of M. Then, the identity map $id_N : N \to N'$ is a diffeomorphism. Indeed, as the injection $N \hookrightarrow M$ is smooth and N' is by hypothesis initial, then the identity map $id_N : N \to N'$ is smooth. Since the same argument can be made for $id_{N'} : N' \to N$, the result follows.
(ii) This is a straightforward consequence of the definition.
(iii) Let $i_N : N \hookrightarrow M$ and $i_\alpha : S_\alpha \hookrightarrow N$ be the injections. Let Z be an arbitrary smooth manifold and $f : Z \to M$ a smooth map such that $f(Z) \subset N$. As the sets S_α are open and partition N, the manifold Z can be written as a disjoint union of open sets $Z_\alpha := f^{-1}(S_\alpha)$, that is

$$Z = \coprod_{\alpha\in I} f^{-1}(S_\alpha).$$

Given that for each index α the map $f_\alpha : Z_\alpha \to M$ obtained by restriction of f to Z_α is smooth, the corresponding map $\bar{f}_\alpha : Z_\alpha \to S_\alpha$ defined by the identity $i_\alpha \circ \bar{f}_\alpha = f_\alpha$

is also smooth by the initial character of S_α. Let $\bar{f} : Z \to N$ be the map obtained by union of the mappings $\bar{f}_\alpha$. This map is smooth and satisfies that $i_N \circ \bar{f} = f$ which proves that N is initial. ■

1.1.12 The local structure of initial submanifolds. Let N be an injectively immersed submanifold of the smooth manifold M. Let $\dim M = m$ and $\dim N = n$. Given an arbitrary point $z \in N$, let $C_z(N)$ be the set of all $z' \in N$ that can be joined to z by a piecewise smooth curve in M lying in N. It can be proved MICHOR (2001) that N is an initial submanifold of M if and only if for each $z \in N$ there exists a chart $(U, \varphi : U \to \varphi(U) \subset \mathbb{R}^m)$ at z on M such that

$$\varphi(C_z(U \cap N)) = \varphi(U) \cap (\mathbb{R}^m \times \{0\}). \tag{1.1.4}$$

This characterization of the initial submanifolds allows one to easily prove that the open "figure eight" $N \subset \mathbb{R}^2$ in the plane given by the parametric equation $t \in (0, 2\pi) \longmapsto (\sin^3(t), \sin(t)\cos(t)) \in \mathbb{R}^2$ is an injectively immersed submanifold in $\mathbb{R}^2$ which is not initial. This is indeed so because for any open disk U centered at the origin in the plane (1.1.4) is violated because $C_0(U \cap N) = U \cap N$, which is the figure $\times$ and is therefore incompatible with (1.1.4).

The irrational (or Kronecker) flow on the torus is an initial submanifold of $\mathbb{T}^2$ (because it is the leaf of a foliation; see §1.4.2) that is not embedded.

1.1.13 Submersions, immersions, subimmersions, and transversality. Let $f : M \to N$ be a smooth map and $m \in M$. We say that f is a ***local diffeomorphism*** around m if the linear map $T_m f : T_m M \to T_{f(m)} N$ is an isomorphism. If f is a local diffeomorphism around m, then f is a diffeomorphism from some neighborhood of m onto some neighborhood of $f(m)$.

Let $f : M \to N$ be smooth. A point $m \in M$ is said to be ***regular*** if $T_m f$ is a surjective linear map. If not, that is, the rank of $T_m f$ does not coincide with the dimension of N, then m is said to be a ***singular*** or a ***critical point*** of f. If C is the set of critical points of f, then $f(C)$ is called the set of ***critical values*** of f. Its complement, $\mathcal{R}(f) := N \backslash f(C)$, is called the set of ***regular values*** of f. Thus, a point $n \in N$ is a regular value of f, if either n does not lie in the range of f, or if it does, then for any $m \in f^{-1}(n)$ the linear map $T_m f$ is surjective. The set of regular points of f is open in M; thus the set of critical points of f is a closed subset of M.

If $T_m f$ is surjective, then it is locally a projection $(u, v) \mapsto v$ (m is locally $(0, 0)$). In particular, f is locally onto. This implies that if f is a submersion, then it is an ***open map*** (that is, the image under f of any open set in M is an open set in N). Additionally, this local characterization implies that if $f : M \to N$ is a surjective submersion, then

$$(\ker T_m f)^\circ = \operatorname{span}\{\mathbf{d}(g \circ f)(m) \mid g \in C^\infty(N)\}, \quad \text{for any } m \in M, \tag{1.1.5}$$

and where $(\ker T_m f)^\circ$ denotes the annihilator of the vector subspace $\ker T_m f$ of $T_m M$ in $T_m^* M$.

The implicit function theorem implies that if $n \in N$ is a regular value, then $f^{-1}(n)$ is a submanifold of M whose tangent spaces are given by $T_m f^{-1}(n) = \ker T_m f$ for all $m \in f^{-1}(n)$. If $f : M \to N$ is a smooth surjective map it is possible that the derivative vanishes on an open set, even if $M = N = \mathbb{R}$. However, if f is an *open* map, then the set of its regular points is open *and dense* in M. In particular, if $f : M \to N$ is a smooth open map, then $\dim M \geq \dim N$. If $f : M \to N$ is a submersion of M onto N and $g : N \to K$ is a map such that $g \circ f : M \to K$ is smooth, then g is also smooth. Submersions have local sections; moreover, we have the following characterization (see Proposition 6.1.4 in BRICKELL AND CLARK (1970) for a proof):

1.1.14 Proposition *A smooth map $f : M \to N$ is a submersion of M onto N if and only if for each point $m \in M$ there is a neighborhood U of $f(m)$ and a smooth map $g : U \to M$ such that $f \circ g = \mathrm{id}|_U$.*

A smooth map $f : M \to N$ is an immersion if and only if it is locally an injection $u \mapsto (u, 0)$ (m is locally 0), which in turn is equivalent to the statement that m has an open neighborhood U such that $f(U)$ is a submanifold of N and $f|U : U \to f(U)$ is a diffeomorphism. If $f : M \to N$ is an immersion and $g : Z \to M$ is a continuous function such that $f \circ g$ is smooth, then g is also smooth. For a smooth map $f : M \to N$, the set $\{m \in M \mid T_m f \text{ is injective}\}$ is open in M. An injective immersion which is an open or closed map onto its image is an embedding. Recall that $f : M \to N$ is a ***closed map*** if the image of any closed set in M is closed in N. The map f is called ***proper*** if each sequence $\{m_p \mid p \in \mathbb{N}\} \subset M$ for which the sequence $\{f(m_p) \mid p \in \mathbb{N}\} \subset N$ is convergent in N admits a convergent subsequence in M. Since N is a locally compact space, a map is proper if and only if the inverse image of any compact set is also compact. The composition of two proper maps is proper. Any proper map is necessarily a closed map. Thus, a proper injective immersion is an embedding.

If $f : M \to N$ is a smooth *injective* map, then the set $\{m \in M \mid T_m f \text{ is injective}\}$ is open *and dense* in M. In particular, $\dim M \leq \dim N$ in this case.

Using the local representation theorem for maps given before, it follows that $f : M \to N$ is a subimmersion, or a constant rank map, if and only if it is locally of the form $(u, v) \mapsto (u, 0)$, where both m and $f(m)$ are locally the origin in their respective model spaces. This statement is in turn equivalent to the fact that $\ker Tf$ is a subbundle of TM. Another equivalent statement is the following: for each $m \in M$ there are open neighborhoods U of m in M and V of $f(m)$ in N, $f(U) \subset V$, and a submanifold Z of M, $m \in Z$, such that $f(U)$ is a submanifold of N and the restriction $f : f^{-1}(V) \cap Z \cap U \to f(U) \cap V$ is a diffeomorphism. These statements imply that if $f : M \to N$ has constant rank in a neighborhood of a point $n \in N$, then $f^{-1}(n)$ is a submanifold of M whose tangent spaces are given by $T_m f^{-1}(n) = \ker T_m f$ for all $m \in f^{-1}(n)$.

The notion of subimmersivity can be localized as follows. A smooth map $f : M \to N$ is a subimmersion at $m \in M$ if there is an open neighborhood U of m such that the rank of $T_m f$ is constant in U. Obviously, the preceding statements have local versions by simply stating them in an open neighborhood of a point. Equivalently, f is subimmersive at $m \in M$ if there is an open neighborhood U of m in M, a manifold P, a submersion $s : U \to P$, and an immersion $j : P \to N$ such that $f|U = j \circ s$. The set $\{m \in M \mid f \text{ is a subimmersion at } m\} = \{m \in M \mid \mathrm{rank}(T_m f) \text{ is locally constant}\}$ is open and dense in M (density follows by the lower semicontinuity of the rank).

The smooth map $f : M \to N$ is said to be ***transversal*** at $m \in M$ to the immersed submanifold P of N if $(T_m f)(T_m M) + T_{f(m)} P = T_{f(m)} N$ (the sum need not be direct). The map f is said to be transversal to the immersed submanifold P

of N, denoted $f \pitchfork P$, if either $f^{-1}(P) = \emptyset$ or if for every $m \in f^{-1}(P)$ we have $(T_m f)(T_m M) + T_{f(m)}P = T_{f(m)}N$ (the sum does not need to be direct). The set $\{m \in M \mid f \text{ is transversal to } P \text{ at } m\}$ is open in M. The proof of the following theorem can be found in BLAOM (2001) for immersed submanifolds, in MICHOR (2001) for initial submanifolds, and in ABRAHAM et al. (1988) for embedded submanifolds.

1.1.15 Theorem (Transversal Mapping Theorem) *Let $f : M \to N$ be a smooth map transversal to the immersed submanifold P of N. Then:*

(i) *There is a smooth manifold structure on $f^{-1}(P)$ with respect to which the inclusion $f^{-1}(P) \hookrightarrow M$ is an immersion and such that the map $f^{-1}(P) \to P$ obtained from f by restriction is a submersion.*

(ii) *If P is an initial submanifold of N, then so is $f^{-1}(P)$.*

(iii) *If P is an embedded submanifold of N, then so is $f^{-1}(P)$.*

In the three cases we have that $T_m(f^{-1}(P)) = (T_m f)^{-1}(T_{f(m)}P)$ for all $m \in f^{-1}(P)$. In particular, $\dim M - \dim f^{-1}(P) = \dim N - \dim P$.

Two submanifolds M_1 and M_2 of M are said to be ***transversal***, denoted $M_1 \pitchfork M_2$, if for every $m \in M_1 \cap M_2$ we have $T_m M_1 + T_m M_2 = T_m M$, that is, if the inclusion $M_i \hookrightarrow M$ is transversal to M_j, $i \neq j$. Thus, if $M_1 \pitchfork M_2$, then $M_1 \cap M_2$ is a submanifold of M whose tangent spaces are given by $T_m(M_1 \cap M_2) = T_m M_1 \cap T_m M_2$ for all $m \in M_1 \cap M_2$. If the latter condition holds (but the two manifolds are not necessarily transversal), then they are said to intersect ***cleanly***.

1.1.16 Sard's Theorem. A fundamental theorem in differential topology is the ***Sard Theorem*** that states that if $f : U \to \mathbb{R}^m$ is of class C^k, U open in $\mathbb{R}^n$, $k > \max(0, m - n)$, then the set of critical values of f has (Lebesgue) measure zero in $\mathbb{R}^n$. To extend this theorem to manifolds and draw density conclusions, one needs additional topological assumptions. Recall that a closed measure zero set in $\mathbb{R}^n$ has dense complement. A subset of a topological space is called ***residual*** if it is a countable intersection of open dense sets. The ***Baire category theorem*** states that a residual subset of a locally compact topological space (hence of a finite-dimensional manifold) is dense. A topological space is called ***Lindelöf*** if every covering of the topological space by open sets admits a countable subcovering. Every closed subspace of a Lindelöf space is Lindelöf. Let $f : X \to Y$ be a continuous map with X Lindelöf. Then the subspace $f(X)$ of Y is also Lindelöf.

The extension of Sard's theorem to manifolds states that if $f : M \to N$ is a C^k map between C^k manifolds, $k > \max(0, \dim M - \dim N)$ and M is Lindelöf, then the set $\mathcal{R}(f)$ of regular values of f is residual and hence dense in N. In particular, if M is second countable, then it is Lindelöf. Recall that a topological space is ***second countable*** if there is a countable family of open subsets such that any other open set is an arbitrary union of the sets in this family. The following lemma is a direct application of Sard's theorem.

1.1.17 Lemma *Let M and N be smooth manifolds. If $f : M \to N$ is a smooth bijective immersion and M is either Lindelöf or paracompact, then f is a diffeomorphism.*

Proof. WARNER (1983) To show this, it suffices to prove that f is a local diffeomorphism, which in turn is implied by the fact that $\dim M = \dim N$. The immersivity hypotheses implies that $\dim M \leq \dim N$. Assume, by contradiction, that $\dim M < \dim N$. By Sard's theorem, the set of critical values has dense complement in N. But if $\dim M < \dim N$, every point is critical, that is, $f(M)$ has measure zero in N. In particular, there are points in N that are not in $f(M)$, which contradicts the bijectivity of f. If M is paracompact, the same proof works by replacing the density argument by the fact that the set of critical values has measure zero in N. ■

1.1.18 A first approach to quotient spaces. In this book we shall use many quotient constructions. These present topological and analytical problems that we shall address at the corresponding places. Here we only give a brief review of the general theory of quotient manifolds. We begin by recalling the basic definitions and results of quotient topological spaces. Let M be a topological space, let $\mathfrak{R}$ be an equivalence relation on it, and denote by $\pi : M \to M/\mathfrak{R}$ the canonical projection that associates to each element $m \in M$ its equivalence class $[m] \in M/\mathfrak{R}$. The ***quotient topology*** on $M/\mathfrak{R}$ is given by the following condition: a subset U of $M/\mathfrak{R}$ is open if, by definition, $\pi^{-1}(U)$ is open in M. The quotient topology on $M/\mathfrak{R}$ is equivalently defined by the condition that it be the strongest topology for which the projection is continuous. This immediately implies that a map $g : M/\mathfrak{R} \to N$, where N is another topological space, is continuous if and only if $g \circ \pi : M \to N$ is continuous. The ***graph*** of the equivalence relation $\mathfrak{R}$ is defined by $\operatorname{graph}(\mathfrak{R}) := \{(m_1, m_2) \in M \times M \mid m_1 \mathfrak{R} m_2\}$. If the topological space $M/\mathfrak{R}$ is Hausdorff, then $\operatorname{graph}(\mathfrak{R})$ is a closed subset of $M \times M$. Conversely, if $\operatorname{graph}(\mathfrak{R})$ is a closed subset of $M \times M$ and the projection π is an open map, then $M/\mathfrak{R}$ is Hausdorff.

An equivalence relation $\mathfrak{R}$ on the manifold M is called ***regular*** if the (possibly non-Hausdorff) quotient topological space $M/\mathfrak{R}$ carries a manifold structure whose underlying topological structure is the quotient topology, such that the canonical projection $\pi : M \to M/\mathfrak{R}$ is a submersion. Since submersions are open maps, the quotient topology on the manifold $M/\mathfrak{R}$ is Hausdorff if and only if $\operatorname{graph}(\mathfrak{R})$ is closed in $M \times M$. If there is a manifold structure on $M/\mathfrak{R}$ for which the projection π is a submersion, then this manifold structure is unique. A map $g : M/\mathfrak{R} \to N$, where N is another manifold, is smooth if and only if $g \circ \pi : M \to N$ is smooth. Let $f : M \to N$ be a submersion and define the equivalence relation $m_1 \mathfrak{R} m_2$ if and only if $f(m_1) = f(m_2)$; then $f(M)$ is open in N, $\mathfrak{R}$ is regular, and $M/\mathfrak{R}$ is diffeomorphic to $f(M)$. ***Godement's Theorem*** states that $\mathfrak{R}$ is regular if and only if $\operatorname{graph}(\mathfrak{R})$ is a submanifold of $M \times M$ and the first (or second) projection $p_1 : \operatorname{graph}(\mathfrak{R}) \to M$, $p_1(m_1, m_2) = m_1$ (or $p_2(m_1, m_2) = m_2$), is a submersion. The main technical point in the proof of this theorem is the construction of a ***local slice*** at $m \in M$ which is a submanifold S contained in an open neighborhood U of m together with a smooth map $s : U \to S$ such that $m \in S$, the restriction $s|S$ of s to S is the identity map on S, and $[u] \cap S = \{s(u)\}$ for all $u \in U$.

1.1.19 Presheaves of functions on subspaces, quotients, and products. The objects that we introduce in this section will be of much use in our treatment of Poisson reduction and dual pairs that we will cover in chapters 10 and 11. Let M be a topological space and $\mathcal{F}_M$ a presheaf of functions on M. Let $S \subset M$ be a subset of M endowed with a given topology $\mathcal{T}$ that does not necessarily coincide with the subspace topology. The presheaf $\mathcal{F}_M$ induces a natural presheaf of functions $\mathcal{F}_{S,M}$ on S that consists of assigning to each open subset V of S the set of functions $\mathcal{F}_{S,M}(V)$ consisting of the functions on V such that for any $z \in V$ there is a open neighborhood U_z of z in M and a function $F \in \mathcal{F}_M(U_z)$ such that

$$f|_{U_z \cap V} = F|_{U_z \cap V}. \tag{1.1.6}$$

We will refer to this construction as the presheaf of ***Whitney smooth functions*** on $(S, \mathcal{T})$ induced by $(M, \mathcal{F}_M)$. In (1.1.6) we will say that F is a ***local extension*** of f at z. Let $f : (M, \mathcal{F}_M) \to (N, \mathcal{F}_N)$ be a smooth function in the sense of §1.1.6 and S and T two topological subspaces of M and N, respectively, such that $f(S) \subset T$. Then the map $\bar{f} : (S, \mathcal{F}_{S,M}) \to (T, \mathcal{F}_{T,N})$ constructed by restricting the domain and range of f to S and T, respectively, is also smooth.

Consider now an equivalence relation $\mathfrak{R}$ on $(M, \mathcal{F}_M)$ and let $\pi : M \to M/\mathfrak{R}$ the canonical projection. The presheaf of functions $\mathcal{F}_M$ on M naturally induces a presheaf of functions $\mathcal{F}_{M/\mathfrak{R}}$ on $M/\mathfrak{R}$ defined by

$$\mathcal{F}_{M/\mathfrak{R}}(U) := \{f \text{ function on } U \mid f \circ \pi|_{\pi^{-1}(U)} \in \mathcal{F}_M(\pi^{-1}(U))\}.$$

We will refer to $\mathcal{F}_{M/\mathfrak{R}}$ as the ***quotient presheaf*** of functions on $M/\mathfrak{R}$. If $\mathcal{F}_M$ is actually a sheaf, then so is $\mathcal{F}_{M/\mathfrak{R}}$.

Consider for instance a regular equivalence relation $\mathfrak{R}$ on the smooth manifold M and let $\pi : M \to M/\mathfrak{R}$ be the corresponding surjective submersion. In this case the quotient $M/\mathfrak{R}$ is a smooth manifold by itself and hence it has a natural presheaf of smooth functions associated that we will denote by $C^\infty_{M/\mathfrak{R}}$. At the same time, the presheaf C^∞_M of smooth functions on M induces a quotient presheaf of functions on $M/\mathfrak{R}$ that we will denote by $C^\infty_{M/\mathfrak{R},\pi}$. The fact that π is a submersion implies that

$$C^\infty_{M/\mathfrak{R}} = C^\infty_{M/\mathfrak{R},\pi}. \tag{1.1.7}$$

An equivalence relation $\mathfrak{R}$ on the topological space M with a presheaf of functions $\mathcal{F}_M$ can be used to define another presheaf on the topological space of saturated open sets. We say that an open subset of M is $\mathfrak{R}$-***invariant*** or $\mathfrak{R}$-***saturated*** when it can be written as the union of equivalence classes of $\mathfrak{R}$. The $\mathfrak{R}$-saturated sets of M form a topology for M, in general strictly weaker than the original topology. The presheaf $\mathcal{F}^{\mathfrak{R}}_M$ of $\mathfrak{R}$-***invariant*** or $\mathfrak{R}$-***saturated functions*** associates to each $\mathfrak{R}$-invariant open subset U the set

$$\mathcal{F}^{\mathfrak{R}}_M(U) := \{f \in \mathcal{F}_M(U) \mid f \text{ is constant in the equivalence classes of } \mathfrak{R}\}.$$

Let $S \subset M$ be a subset of M endowed with a given topology $\mathcal{T}$ and consider the restriction of the equivalence relation $\mathfrak{R}$ to S. Consider now the presheaf $(\mathcal{F}_{S,M})^{\mathfrak{R}}$ of $\mathfrak{R}$-invariant functions on S and the restriction $(\mathcal{F}^{\mathfrak{R}}_M)_{S,M}$ to S of the presheaf $\mathcal{F}^{\mathfrak{R}}_M$ of $\mathfrak{R}$-invariant functions of M. A presheaf of much importance in our future discussion is the intersection of these two, which we will note by $\mathcal{F}^{\mathfrak{R}}_{S,M}$, that is,

$$\mathcal{F}^{\mathfrak{R}}_{S,M} := (\mathcal{F}_{S,M})^{\mathfrak{R}} \cap (\mathcal{F}^{\mathfrak{R}}_M)_{S,M}.$$

The presheaf $(\mathcal{F}_{S,M})^{\mathfrak{R}}$ limits the domain of $\mathcal{F}^{\mathfrak{R}}_{S,M}$ to $\mathfrak{R}$-invariant open sets of $(S, \mathcal{T})$. To be more explicit, for any such set V, $\mathcal{F}^{\mathfrak{R}}_{S,M}(V)$ is made of $\mathfrak{R}$-invariant functions f defined on V such that for any $z \in V$ there exists an open $\mathfrak{R}$-saturated neighborhood U_z of z in M and a function $F \in \mathcal{F}^{\mathfrak{R}}_M(U_z)$ such that

$$f|_{U_z \cap V} = F|_{U_z \cap V}.$$

We will refer to $\mathcal{F}^{\mathfrak{R}}_{S,M}$ as the ***presheaf of Whitney invariant functions*** on S induced by $\mathcal{F}^{\mathfrak{R}}_M$.

1.1.20 Proposition *Let M be a topological space with a presheaf $\mathcal{F}_M$ of functions on it. Let $\mathfrak{R}$ be an equivalence relation on M and S an $\mathfrak{R}$-invariant subset of M endowed with a given topology $\mathcal{T}$. Let $\mathcal{F}_{S/\mathfrak{R},M/\mathfrak{R}}$ be the presheaf of Whitney smooth functions on $S/\mathfrak{R}$ considered as a subset of $M/\mathfrak{R}$; then*

$$\mathcal{F}_{S/\mathfrak{R},M/\mathfrak{R}} \subset \mathcal{F}^{\mathfrak{R}}_{S/\mathfrak{R}},$$

where $\mathcal{F}^{\mathfrak{R}}_{S/\mathfrak{R}}$ is the quotient presheaf on $S/\mathfrak{R}$ corresponding to $\mathcal{F}^{\mathfrak{R}}_{S,M}$. If the projection $\pi : M \to M/\mathfrak{R}$ is an open map, then

$$\mathcal{F}_{S/\mathfrak{R},M/\mathfrak{R}} = \mathcal{F}^{\mathfrak{R}}_{S/\mathfrak{R}}.$$

Proof. Let $\pi : M \to M/\mathfrak{R}$ and $\pi_S : S \to S/\mathfrak{R}$ be the projections and V an arbitrary open subset of $S/\mathfrak{R}$. Let $f \in \mathcal{F}_{S/\mathfrak{R},M/\mathfrak{R}}(V)$. We will now show that $f \in \mathcal{F}^{\mathfrak{R}}_{S/\mathfrak{R}}(V)$ by proving that $f \circ \pi_S \in \mathcal{F}^{\mathfrak{R}}_{S,M}(\pi_S^{-1}(V))$. This amounts to showing that for any $m \in \pi_S^{-1}(V)$ there exists an open $\mathfrak{R}$-invariant neighborhood U_m of m in M and $F \in \mathcal{F}^{\mathfrak{R}}_M(U_m)$ such that

$$f \circ \pi_S|_{\pi_S^{-1}(V)\cap U_m} = F|_{\pi_S^{-1}(V)\cap U_m}. \tag{1.1.8}$$

Let $z = \pi_S(m) \in V$. Since $f \in \mathcal{F}_{S/\mathfrak{R},M/\mathfrak{R}}(V)$ there exists an open neighborhood U_z of z in $M/\mathfrak{R}$ and $H \in \mathcal{F}_{M/\mathfrak{R}}(U_z)$ such that $f|_{V\cap U_z} = H|_{V\cap U_z}$. Define now $U_m := \pi^{-1}(U_z)$ and $F := H \circ \pi|_{U_m} \in \mathcal{F}^{\mathfrak{R}}_M(U_m)$. We now check that (1.1.8) holds. Firstly, since the map f is defined on $V \cap U_z$, then $f \circ \pi_S$ is defined on

$$\pi_S^{-1}(V \cap U_z) = \pi_S^{-1}(\pi_S(\pi_S^{-1}(V)) \cap U_z) = \pi_S^{-1}\left(\pi_S^{-1}(V)/\mathfrak{R} \cap U_m/\mathfrak{R}\right).$$

The $\mathfrak{R}$-invariance of S and $\pi_S^{-1}(V) \cap U_m$ implies that the last expression equals

$$\pi_S^{-1}\left(\left(\pi_S^{-1}(V) \cap U_m\right)/\mathfrak{R}\right) = \pi_S^{-1}(\pi_S(\pi_S^{-1}(V) \cap U_m)) = \pi_S^{-1}(V) \cap U_m.$$

Now, for any point $p \in \pi_S^{-1}(V) \cap U_m$ we have that $f \circ \pi_S(p) = H \circ \pi_S(p) = F(p)$, as required.

Suppose now that π is an open map and let $f \in \mathcal{F}^{\mathfrak{R}}_{S/\mathfrak{R}}(V)$, for some open subset V of $S/\mathfrak{R}$. We have to show that for any $z \in V$ there exists an open neighborhood U_z of z in $M/\mathfrak{R}$ and $F \in \mathcal{F}_{M/\mathfrak{R}}(U_z)$ such that

$$f|_{V\cap U_z} = F|_{V\cap U_z}. \tag{1.1.9}$$

Let $m \in \pi_S^{-1}(z) \subset \pi_S^{-1}(V)$. Since by hypothesis $f \in \mathcal{F}^{\mathfrak{R}}_{S/\mathfrak{R}}(V)$, there exist an open $\mathfrak{R}$-invariant neighborhood U_m of m in M and a function $H \in \mathcal{F}^{\mathfrak{R}}_M(U_m)$ such that

$$f \circ \pi_S|_{\pi_S^{-1}(V)\cap U_m} = H|_{\pi_S^{-1}(V)\cap U_m}. \tag{1.1.10}$$

Additionally, since the map π is open we have that $U_z := \pi(U_m)$ is an open neighborhood of $z \in V$. Let $F \in \mathcal{F}_{M/\mathfrak{R}}(U_z)$ be the map uniquely determined by the relation $F \circ \pi|_{U_m} = H$. We now check that this choice of neighborhood U_z and the map $F \in \mathcal{F}_{M/\mathfrak{R}}(U_z)$ satisfy (1.1.9). First, by (1.1.10) f is well defined in $\pi_S(\pi_S^{-1}(V) \cap U_m)$. Second, due to the $\mathfrak{R}$-invariance of U_m and S we can write, by considering $S/\mathfrak{R}$ as a subset of $M/\mathfrak{R}$, that

$$\pi_S(\pi_S^{-1}(V) \cap U_m) = \pi_S(\pi_S^{-1}(V)) \cap \pi(U_m) = V \cap U_z.$$

Due to this equality we have that for any $p \in V \cap U_z$, there exists an element $q \in \pi_S^{-1}(V)\cap U_m$ such that $p = \pi_S(q)$ and hence $f(p) = f\circ\pi_S(q) = H(q) = F\circ\pi(q) = F(p)$, as required. ■

Let $(M, \mathcal{F}_M)$ and $(N, \mathcal{F}_N)$ be two presheaves of functions on the topological spaces M and N, respectively. Consider now the product topological space $M \times N$. The presheaves of functions $\mathcal{F}_M$ and $\mathcal{F}_N$ naturally induce a presheaf $\mathcal{F}_{M\times N}$ on the product $M \times N$ that we will call the ***product presheaf*** and is defined by

$$\begin{aligned}\mathcal{F}_{M\times N}(U) := \{&f \text{ function on } U \mid f(\cdot, y) \in \mathcal{F}_M(\pi_M(U)),\\ &f(x, \cdot) \in \mathcal{F}_N(\pi_N(U)), \text{ for all } x \in \pi_M(U) \text{ and } y \in \pi_N(U)\},\end{aligned}$$

where $\pi_M : M \times N \to M$ and $\pi_N : M \times N \to N$ are the projections and $x \in \pi_M(U)$ and $y \in \pi_N(U)$ are arbitrary.

Finally, we say that $f : (M, \mathcal{F}_M) \to (N, \mathcal{F}_N)$ is ***compatible*** with two equivalent relations $\mathfrak{R}_M$ and $\mathfrak{R}_N$ defined on M and N, respectively, if for any $x, y \in M$ such that $x\mathfrak{R}_M y$, then $f(x)\mathfrak{R}_N f(y)$. In this situation and if f is smooth there exists a unique smooth function $\hat{f} : (M/\mathfrak{R}_M, \mathcal{F}_{M/\mathfrak{R}_M}) \to (N/\mathfrak{R}_N, \mathcal{F}_{N/\mathfrak{R}_N})$ such that $\hat{f} \circ \pi_{\mathfrak{R}_M} = \pi_{\mathfrak{R}_N} \circ f$.

1.1.21 Paracompactness, partitions of unity and extensions of functions. A standard tool in differential topology is the smooth ***partition of unity***. where $\{U_i \mid i \in I\}$ is a locally finite covering of M by open sets (that is, every point has a neighborhood that intersects only finitely many sets U_i), ρ_i are smooth positive functions (that is, $\rho_i(m) \geq 0$ for all $m \in M$), and for every $m \in M$, $\sum_i \rho_i(m) = 1$ (which is a finite sum at every point). A partition of unity is ***subordinate to an atlas*** If any atlas on M admits a subordinate partition of unity, we say that M admits partitions of unity. If M is a smooth finite-dimensional manifold, then it admits partitions of unity if and only if M is paracompact. Recall that a topological space is ***paracompact*** if it is Hausdorff and every open covering admits a locally finite refinement; that is, there is another open covering of the space which is locally finite and is such that each of its elements is a subset of some open set in the given covering. Every closed subspace of a paracompact space is paracompact. An open subset of a paracompact topological space need not be paracompact. Every open submanifold of a paracompact manifold is paracompact. Second countable locally compact Hausdorff spaces are paracompact. For (Hausdorff finite-dimensional) manifolds, paracompactness and metrizability are equivalent. A connected (Hausdorff finite-dimensional) manifold admits a Riemannian metric if and only if it is second countable.

1.1.22 Proposition *Let M be a smooth paracompact manifold. The presheaf of smooth functions on M is a sheaf.*

Proof. The only property of a sheaf that is not obviously satisfied by the presheaf of smooth functions on M is **(SH4)**. We show that the paracompactness of M implies that this condition holds. Let U be an open subset of M and $\{V_i\}_{i\in I}$ an open covering of U. Let $f_i \in \mathcal{F}(V_i)$ be given for each $i \in I$. Suppose that the restrictions of f_i and f_j to $V_i \cap V_j$ are equal for all $i, j \in I$. Now, since U is an open submanifold of a paracompact manifold it is paracompact, and hence there exists a locally finite refinement $\{U_\alpha\}_{\alpha\in J}$ of the covering $\{V_i\}_{i\in I}$ and a partition of unity $\{(U_\alpha, \rho_\alpha) \mid \alpha \in J\}$ subordinate to it. Let $F \in C^\infty(U)$ be the function defined by $F = \Sigma_{\alpha\in J}\rho_\alpha f_\alpha$. Let now $v \in U$ be a fixed element such that $v \in U_{\alpha_1} \cap \ldots \cap U_{\alpha_n} \subset V_{\alpha_1} \cap \ldots \cap V_{\alpha_n}$. Then $F(v) = \rho_{\alpha_1}(v) f_{\alpha_1}(v) + \cdots + \rho_{\alpha_n}(v) f_{\alpha_n}(v) = (\rho_{\alpha_1}(v) + \cdots + \rho_{\alpha_n}(v)) f_\alpha(v) = f_\alpha(v)$, with α any element in the list $\{\alpha_1, \ldots, \alpha_n\}$. In particular, this proves that $F|_{V_i} = f_i$, for any $i \in I$. ■

1.1.23 Proposition *Let M be a smooth manifold and S a smooth embedded submanifold of M. Let C^∞_M and C^∞_S be the presheaves of smooth functions of M and S, respectively. Let $C^\infty_{S,M}$ be the presheaf of Whitney smooth functions of S considered as a topological subspace of M. Then*

$$C^\infty_S \subset C^\infty_{S,M}. \tag{1.1.11}$$

If S is paracompact, then

$$C^\infty_S = C^\infty_{S,M}. \tag{1.1.12}$$

Proof. Let U be an arbitrary open subset of M and $f \in C^\infty_S(U \cap S)$. Let $z \in U \cap S$ and choose an open neighborhood U_z of z in M such that $U_z \cap U$ is the domain of a submanifold chart for S around z. In this chart one can find in a straightforward manner a function $F \in C^\infty_M(U_z \cap U)$ such that

$$f|_{U_z\cap U\cap S} = F|_{U_z\cap U\cap S}.$$

This argument proves (1.1.11). Suppose now that S is paracompact and let $f \in C^\infty_{S,M}(U \cap S)$. Then for any $z \in U \cap S$ there exists an open neighborhood U_z of z in M and a function $F \in C^\infty_M(U_z)$ such that $f|_{U_z\cap U\cap S} = F|_{U_z\cap U\cap S}$. The paracompactness of S implies that $U \cap S$ is paracompact, which allows us to construct a locally finite covering $\{U_i \cap S\}_{i\in I}$ of $U \cap S$ such that there are functions $F_i \in C^\infty_M(U_i)$ satisfying $F_i|_{U_i\cap S} = f|_{U_i\cap S}$. Let $\{(U_i \cap S, \rho_i) \mid i \in I\}$ be a partition of unity subordinate to this covering and let $\overline{f} := \sum_{i\in I} \rho_i F_i|_{U_i\cap S}$. The map $\overline{f}$ belongs to $C^\infty_S(U \cap S)$ and $\overline{f} = f$, which proves the claim. ■

As we can see in the following result, the choice of topologies in a subset has a great impact in the definition of its presheaf of Whitney smooth functions.

1.1.24 Proposition *Let S be an immersed submanifold of the smooth manifold M and let C_S^∞ and C_M^∞ be the corresponding presheaves of smooth functions. Consider now the relative topology $\mathcal{R}$ on S and the topology $\mathcal{I}$ on S that comes from its immersed submanifold structure. Let $C^\infty_{(S,\mathcal{R}),M}$ and $C^\infty_{(S,\mathcal{I}),M}$ be the presheaf of Whitney smooth functions associated to $(S, \mathcal{R})$ and $(S, \mathcal{I})$, respectively. Then for any open subset U of M we have that*

(i) $C^\infty_{(S,\mathcal{R}),M}(U \cap S) \subset C^\infty_{(S,\mathcal{I}),M}(U \cap S)$.

(ii) $C_S^\infty \subset C^\infty_{(S,\mathcal{I}),M}$. *If $(S, \mathcal{I})$ is paracompact we then have equality.*

Proof. **(i)** It is a straightforward consequence of the fact that the topology $\mathcal{I}$ is stronger than $\mathcal{R}$.

(ii) Let $V \in \mathcal{I}$ and $f \in C_S^\infty(V)$. Then, for any $z \in V$ there exists an open neighborhood V_z of z in S such that $V_z \subset V$ and V_z is an embedded submanifold of M. Shrinking V_z if necessary and using the expression (1.1.11) we can guarantee the existence of a neighborhood U_z of z in M and of a function $F \in C_M^\infty(U_z)$ such that $f|_{U_z \cap V_z} = F|_{U_z \cap V_z}$ which proves that $f \in C^\infty_{(S,\mathcal{I}),M}(V)$. Suppose now that S is paracompact and let $f \in C^\infty_{(S,\mathcal{I}),M}(V)$ for some $V \in \mathcal{I}$. By definition, for any $z \in V$ there exists an open neighborhood U_z of z in M and a function $F \in C_M^\infty(U_z)$ such that $F|_{U_z \cap V} = f|_{U_z \cap V}$. Notice that $U_z \cap V = U_z \cap V \cap S$ and hence $U_z \cap V$ can be written as the intersection of $U_z \cap S \in \mathcal{R} \subset \mathcal{I}$ and $V \in \mathcal{I}$ which implies that $U_z \cap V \in \mathcal{I}$. If S is paracompact, then so is V, and then we can proceed by mimicking what we did in the proof of the expression (1.1.12). ∎

One important reason for the interest in partitions of unity lies in the fact that they enable the separation of closed sets by smooth functions and the extension of various objects from a closed set to the entire manifold. Thus, if M admits smooth partitions of unity and if A and B are closed subsets of M, then there exists a smooth function $f : M \to [0, 1]$ such that $f(A) = 0$ and $f(B) = 1$. Let $\pi : E \to M$ be a vector bundle, A a closed subset of M, and $\sigma : A \to E$ a smooth section (that is, $\pi \circ \sigma$ is the identity on A and σ is a smooth map defined on a neighborhood of A). If M admits smooth partitions of unity, then σ can be extended to a smooth global section $\bar{\sigma} : M \to E$. In particular, we have the following result:

1.1.25 Lemma *Let M be a paracompact manifold and $A \subset M$ a closed subset. Let $f : A \to \mathbf{F}$ be a smooth function defined on the closed set A (which means that f is defined and smooth on a neighborhood of A) with values in a normed vector space $\mathbf{F}$. Then there exists a smooth extension $\bar{f} : M \to \mathbf{E}$ of f; moreover, if there is a bound $C > 0$ for f on A, that is, $\|f(a)\| \leq C$ for all $a \in A$, then $\bar{f}$ has the same bound, that is, $\|\bar{f}(m)\| \leq C$ for all $m \in M$.*

A useful and often quoted application of partitions of unity is the ***Ehresmann Fibration Theorem*** which states that if $f : M \to N$ is a smooth submersion between finite-dimensional manifolds with M paracompact such that $f^{-1}(n)$ is compact for every $n \in N$, then f is a ***locally trivial fibration*** or a ***locally trivial fiber bundle***, that is, for each $n \in N$ there is an open neighborhood V of n in N and a diffeomorphism $\varphi : V \times f^{-1}(n) \to f^{-1}(V)$ such that $f(\varphi(y, u)) = y$ for all $y \in V$ and all $u \in f^{-1}(n)$.

Thus, a locally trivial fiber bundle $f : M \to N$ is locally the product of an open set in the ***base*** manifold N with a typical ***fiber*** $f^{-1}(n)$; the diffeomorphism φ is called a ***bundle chart***. Since properness of the map f ensures that its fibers are compact, a convenient way to state the Ehresmann Fibration Theorem is the following: a smooth proper submersion $f : M \to N$ with M paracompact is a locally trivial fibration.

1.1.26 Homotopy and covering spaces. Given a connected manifold M, the space of C^k paths $\gamma : [0, 1] \to M$ starting at $\gamma(0) = m_0 \in M$ endowed with the C^k topology (that is, the topology of uniform convergence of the first k derivatives) is denoted by $\mathcal{P}^k_{m_0}M$. Two paths $\gamma_0, \gamma_1 \in \mathcal{P}^k_{m_0}M$ such that $\gamma_0(0) = \gamma_1(0) = m_0$ and $\gamma_0(1) = \gamma_1(1)$ are called ***equivalent*** if there is a C^k ***homotopy*** with fixed endpoints between them, that is, a C^k map $\chi : [0, 1] \times [0, 1] \to M$ such that $\chi(0, t) = \gamma_0(t)$, $\chi(1, t) = \gamma_1(t)$, $\chi(s, 0) = \gamma_0(0) = \gamma_1(0) = m_0$, and $\chi(s, 1) = \gamma_0(1) = \gamma_1(1)$, for all $s, t \in [0, 1]$. We first recall the topological situation, that is when $k = 0$. The continuity of χ is equivalent to the continuity of $s \in [0, 1] \longmapsto \chi(s, \cdot) \in \mathcal{P}^0_{m_0}M$. The quotient topological space of equivalence classes $[\gamma]$ of paths in $\mathcal{P}^0_{m_0}M$, called ***homotopy classes***, is denoted by $\tilde{M}$. Since M is a connected manifold it is a locally Euclidean space and therefore pathwise connected; therefore the map $\tilde{\pi} : \tilde{M} \to M$ given by $\tilde{\pi}([\gamma]) = \gamma(1)$ is well defined, continuous, and surjective. Given two paths $\lambda_0 \in \mathcal{P}^0_{m_0}M$ and $\lambda_1 \in \mathcal{P}^0_{\lambda_0(1)}M$, their composition $\lambda_0 * \lambda_1 \in \mathcal{P}^0_{m_0}M$ is defined by $(\lambda_0 * \lambda_1)(t) = \lambda_0(2t)$ if $t \in [0, 1/2]$ and $(\lambda_0 * \lambda_1)(t) = \lambda_1(2t - 1)$ if $t \in [1/2, 1]$. This concatenation of paths is compatible with the equivalence relation defined above so it induces an operation on homotopy classes of closed continuous loops $\tilde{\pi}^{-1}(m_0)$ based at m_0 endowing it with the structure of a (in general noncommutative) group, called the ***fundamental*** or ***first homotopy group*** of M and denoted by $\pi_1(m_0, M)$. The terminology need not mention the base point m_0 since if m_1 is another point, letting λ be a continuous path satisfying $\lambda(0) = m_0$ and $\lambda(1) = m_1$, the map $[\gamma] \in \pi_1(m_0, M) \mapsto [\lambda]*[\gamma]*[\lambda]^{-1} \in \pi_1(m_1, M)$ is an isomorphism of groups. In general, this isomorphism depends on the path λ unless the fundamental group is commutative. The manifold M is called ***simply connected*** if the fundamental group is trivial, which means that every closed loop can be shrunk to its base point or, equivalently, that $\tilde{\pi}$ is bijective. The topological space $\tilde{M}$ is the universal covering space of M defined below.

If X and Y are topological spaces, a continuous surjective map $p : X \to Y$ is called a ***covering map*** and Y is called a ***covering space*** of X if every point in Y has a connected open neighborhood V such that $p^{-1}(V)$ is a disjoint union of open connected sets in X, each of them homeomorphic to V by p. The covering is said to be ***finite*** if the set $p^{-1}(x)$ has finite cardinality, for all $x \in X$. If X is connected, then the number of points in $p^{-1}(x)$ is independent of x. The space Y is called a ***universal covering space*** of X if for any other covering space $p' : Y' \to X$ there is a covering map $q : Y \to Y'$ (in general not unique) such that $p = p' \circ q$. In spite of the absence of uniqueness for the morphism $q : Y \to Y'$, the use of the term universal is well justified because one can prove (see for instance SPANIER (1966, Corollary 6, page 80)) that all universal covering spaces are homeomorphic. Incidentally, this is why one refers to all of them collectively as *the universal covering space* of X. A simply connected covering space is the universal covering space.

The prior considerations used only the fact that M was a connected locally pathwise connected topological space. To have the same statements for manifolds one proves first that two C^k paths are C^k equivalent if and only if they are C^0 equivalent (this is a standard uniform approximation argument). Thus, the inclusions $\mathcal{P}^k_{m_0}M \hookrightarrow \mathcal{P}^0_{m_0}M$ induce homeomorphisms of their quotients (relative to C^k and C^0 homotopy equivalence, respectively). Therefore, one can work with continuous objects, even though one needs ultimately smooth ones. Next, one shows that $\tilde{M}$ itself is a smooth manifold. The charts of $\tilde{M}$ are naturally induced by those of M in the following way: let γ be a continuous path connecting m_0 to m. Let (U, φ) be a chart on M at the point m such that $\varphi(U)$ is the unit ball in the model space of M. Modify continuously the path γ (by changing its value close to 1 independently of the point $x \in U$) in order to get a continuous path connecting m_0 to an arbitrary point $x \in U$. This defines a section $\sigma : U \subset M \to \tilde{M}$ of $\tilde{\pi}$ on U by mapping x to the homotopy class of this path that ends at x. This section σ sends m to $[\gamma]$. Now note that $(\sigma(U), (\varphi \circ \tilde{\pi})|_{\sigma(U)})$ is a chart at $[\gamma]$. The collection of these charts forms an atlas on $\tilde{M}$.

Given a smooth map $f : M \to N$ between two smooth connected manifolds M and N with universal covers $\tilde{\pi}_M : \tilde{M} \to M$, $\tilde{\pi}_N : \tilde{N} \to N$, and two points $\tilde{m}_0 \in \tilde{M}$, $\tilde{n}_0 \in \tilde{N}$ such that $f(\tilde{\pi}_M(\tilde{m}_0)) = \tilde{\pi}_N(\tilde{n}_0)$, there is a unique smooth map $\tilde{f}_{\tilde{m}_0\tilde{n}_0} : \tilde{M} \to \tilde{N}$ satisfying $\tilde{\pi}_N \circ \tilde{f}_{\tilde{m}_0\tilde{n}_0} = f \circ \tilde{\pi}_M$ and $\tilde{f}_{\tilde{m}_0\tilde{n}_0}(\tilde{m}_0) = \tilde{n}_0$. The map $\tilde{f}_{\tilde{m}_0\tilde{n}_0}$ is called a ***covering map*** of f; it depends on the two chosen points $\tilde{m}_0 \in \tilde{M}$ and $\tilde{n}_0 \in \tilde{N}$.

In particular, given the universal covering $\tilde{\pi} : \tilde{M} \to M$ of the manifold M, the set $\mathcal{G}_{\tilde{\pi}}$ of smooth diffeomorphisms of $\tilde{M}$ covering the identity map on M is a group that has the following remarkable property. Given two arbitrary points $\tilde{m}_1, \tilde{m}_2 \in \tilde{M}$ there is a unique diffeomorphism of $\tilde{M}$ covering the identity of M that maps $\tilde{m}_1$ to $\tilde{m}_2$. This immediately implies that the fundamental group $\pi_1(M)$ is isomorphic to $\mathcal{G}_{\tilde{\pi}}$ in the following way. Take a point $m_0 \in M$ relative to which we shall construct the fundamental group $\pi_1(M, m_0)$ and let $\tilde{m}_0 \in \tilde{\pi}^{-1}(m_0) \subset \tilde{M}$ be arbitrary. Given $\tilde{f} \in \mathcal{G}_{\tilde{\pi}}$, consider all the paths connecting $\tilde{m}_0$ to $\tilde{f}(\tilde{m}_0)$ and project them to M, thereby obtaining a set of smooth closed paths based at m_0 in M which are all homotopic. Now associate to $\tilde{f}$ the class of these closed paths in M; this is the isomorphism between $\mathcal{G}_{\tilde{\pi}}$ and $\pi_1(M, m_0)$.

In Section 5.3 we will discuss an analog of this and other related objects in the context of symplectic manifolds acted canonically upon by a Lie algebra.

1.2 Vector fields and Lie derivatives

1.2.1 Vector fields and flows. A ***vector field*** on M is a smooth section of TM, that is, a smooth map $X : M \to TM$ satisfying $X(m) \in T_mM$. The set of all vector fields on M is denoted by $\mathfrak{X}(M)$; it is a real vector space and a $C^\infty(M)$ module. Each vector field $X \in \mathfrak{X}(M)$ determines uniquely a ***flow*** F_t by the identity

$$\frac{d}{dt}F_t(m) = X(F_t(m)) \tag{1.2.1}$$

for all $m \in M$. Thus, $t \mapsto F_t(m)$ is the integral curve of X passing through $m \in M$ at $t = 0$. The set of pairs $(m, t) \in M \times \mathbb{R}$ for which $F_t(m)$ is well defined is denoted by $\mathcal{D}_X$. It can be proved that $\mathcal{D}_X$ is an open subset of $M \times \mathbb{R}$ such that $\mathcal{D}_X \supset M \times \{0\}$.

The family of local diffeomorphisms $\{F_t\}$ on M (for t in some open interval around the origin in $\mathbb{R}$) satisfies $F_0 = identity$ and, whenever it makes sense, $F_{t+s} = F_t \circ F_s$. The vector field and its flow are called ***complete*** if $\mathcal{D}_X = M \times \mathbb{R}$. If the vector field has compact support, then it is complete.

Conversely, let us consider a smooth ***path of diffeomorphisms*** $\{\varphi_t\}$, for t in some open interval around the origin in $\mathbb{R}$, that is, each φ_t is a smooth diffeomorphism of M, $\varphi_0 = identity$, and the map $(t, m) \mapsto \varphi_t(m)$ is smooth. Any such path of diffeomorphisms defines a ***time-dependent vector field*** X_t on M (that is, $X : \mathbb{R}\times M \to TM$ is a smooth map satisfying $X(t, m) \in T_mM$ for all $m \in M$ and $t \in \mathbb{R}$), by the formula

$$X(t, m) := \left.\frac{d}{dt}\right|_{t=0} \varphi_t(m). \tag{1.2.2}$$

The time-dependent vector field X_t is characterized through its ***time-dependent flow*** $\psi_{t,s}$ by the requirement that $t \mapsto \psi_{t,s}(m)$ be the integral curve of X_t with initial condition m at $t = s$. Thus

$$\frac{d}{dt}\psi_{t,s}(m) = X(t, \psi_{t,s}(m)) \qquad \text{and} \qquad \psi_{s,s}(m) = m \tag{1.2.3}$$

for every $m \in M$ and $t, s \in \mathbb{R}$. Comparing (1.2.2) and (1.2.3), it follows that $\varphi_t = \psi_{t,0}$. The group property of the flow of a usual vector field is replaced by the time-dependent flow property $\psi_{t,s} \circ \psi_{s,r} = \psi_{t,r}$ for $t, s, r \in \mathbb{R}$. In particular, a usual (time independent) vector field $X \in \mathfrak{X}(M)$ defines both a flow F_t and a time-dependent flow $\psi_{t,s}$; they are related by $F_{t-s} = \psi_{t,s}$.

1.2.2 Lie derivatives and Lie brackets. A vector field $X \in \mathfrak{X}(M)$ with flow F_t is completely determined by the ***directional*** or ***Lie derivative*** on functions

$$X[f] \equiv \langle \mathbf{d}f, X\rangle \equiv \pounds_X f := \left.\frac{d}{dt}\right|_{t=0} F_t^* f, \tag{1.2.4}$$

for any $f \in C^\infty(U)$, U open in M, where $F_t^* f := f \circ F_t$ is the ***pull-back*** of f by F_t. The notation $\langle\cdot\,, \cdot\rangle$ will denote throughout this book a natural ***duality pairing*** between a vector space or vector bundle and its dual. In this case, $\langle\cdot\,, \cdot\rangle : T^*M \times TM \to \mathbb{R}$ is the natural evaluation of a one-form in the cotangent space T_m^*M on a tangent vector in T_mM. The Lie derivative is a ***local operator***, that is, if $U \subset V \subset M$ are open sets, and $f \in C^\infty(V)$ is arbitrary, then $(\pounds_X f)|U = \pounds_X(f|U)$. The operation $f \mapsto \mathbf{d}f$ is the ***derivative*** of f.

The ***Lie bracket*** of $X, Y \in \mathfrak{X}(M)$ is denoted by $[X, Y] \in \mathfrak{X}(M)$. It is completely characterized by the requirement

$$[X, Y][f] = X[Y[f]] - Y[X[f]] \tag{1.2.5}$$

for any $f \in C^\infty(M)$. The ***Lie derivative*** on vector fields is defined by $\pounds_X Y := [X, Y]$. The Lie bracket is $\mathbb{R}$-bilinear, skew symmetric, and satisfies the ***Jacobi identity***

$$[[X, Y], Z] + [[Y, Z], X] + [[Z, X], Y] = 0. \tag{1.2.6}$$

Thus, $\mathfrak{X}(M)$ is a real ***Lie algebra*** relative to the bracket operation defined by (1.2.5). The bracket is also a derivation over $C^\infty(M)$ in every factor, that is, it satisfies the ***Leibniz identity***

$$\pounds_X(fY) = [X, fY] = f[X, Y] + X[f]Y = f\pounds_X Y + (\pounds_X f)Y \tag{1.2.7}$$

for any $X, Y, Z \in \mathfrak{X}(M)$ and $f \in C^\infty(M)$. The Lie bracket on vector fields is also local in the sense that if $U \subset V \subset M$ are open sets and $X, Y \in \mathfrak{X}(V)$, then $[X, Y]|U = [X|U, Y|U]$.

If $\psi : M \to N$ is a diffeomorphism and $Y \in \mathfrak{X}(N)$, then the ***pull-back*** ψ^*Y is a vector field on M defined by $\psi^*Y := T\psi^{-1} \circ Y \circ \psi$; its flow is $\psi^{-1} \circ G_t \circ \psi$, where G_t is the flow of Y.

With this notation, the Lie bracket of two vector fields $X, Y \in \mathfrak{X}(M)$ is related to the flow F_t of X by the ***Lie derivative formula***

$$F_t^*[X, Y] = \frac{d}{dt} F_t^* Y. \tag{1.2.8}$$

This also shows that for any diffeomorphism $\psi : M \to M$, we have

$$\psi_*[X, Y] = [\psi_* X, \psi_* Y], \tag{1.2.9}$$

where $\psi_* := (\psi^{-1})^*$ is the ***push-forward*** operation on vector fields. Another consequence of (1.2.8) that will often be used in later chapters is that if $X, Y \in \mathfrak{X}(M)$ have flows F_t and G_t respectively, then $[X, Y] = 0$ if and only if $F_t \circ G_s = G_s \circ F_t$ for all t, s for which the flows are defined; this is in turn equivalent to $F_t^*Y = Y$ or to $G_s^*X = X$. These consequences can also be drawn from the following formula of the Lie bracket:

$$[X, Y] = \left.\frac{d}{dt}\frac{d}{ds}\right|_{t,s=0} F_{-t} \circ G_s \circ F_t. \tag{1.2.10}$$

There are two useful expressions that provide explicit formulas for the flow of the Lie bracket of two vector fields and for the flow of the sum. For any $m \in M$, the flow H_t of $[X, Y]$ is given by

$$H_t(m) = \lim_{n\to\infty} \left(G_{-\sqrt{t/n}} \circ F_{-\sqrt{t/n}} \circ G_{\sqrt{t/n}} \circ F_{\sqrt{t/n}} \right)^n (m), \ t \geq 0. \tag{1.2.11}$$

This expression is historically know by the name of the ***exponential formula***. The flow S_t of the sum $X + Y$ can be expressed as

$$S_t(m) = \lim_{n\to\infty} \left(F_{t/n} \circ G_{t/n} \right)^n (m). \tag{1.2.12}$$

This expression is referred to as the ***Trotter product formula***.

1.2.3 Derivations and vector fields. Let $(V, \cdot)$ be a real algebra. A linear map $D : V \to V$ is called a ***derivation*** of the algebra $(V, \cdot)$ when the ***Leibniz identity*** holds, that is,

$$D(u \cdot v) = D(u) \cdot v + u \cdot D(v), \text{ for any } u, v \in V.$$

Any smooth vector field $X \in \mathfrak{X}(M)$ defines a derivation D_X on $C^\infty(M)$ when we consider this set as a real, associative (infinite-dimensional) algebra with respect to function multiplication. More explicitly, D_X is defined via the directional or Lie derivative, that is,

$$\begin{aligned} D_X : C^\infty(M) &\longrightarrow \quad C^\infty(M) \\ f \quad &\longmapsto X[f] \equiv \pounds_X f. \end{aligned}$$

A fact that will be very important later on in the book is that there is a bijective correspondence between derivations of $C^\infty(M)$ and vector fields, that is, the map that assigns to each vector field $X \in \mathfrak{X}(M)$ the derivation D_X is one-to-one and onto the space of derivations of $C^\infty(M)$ (see, for instance MATSUSHIMA (1972, page 73)).

1.2.4 Vector fields and immersed submanifolds. Let N be an immersed submanifold of M and denote by $i : N \hookrightarrow M$ the inclusion. We say that the vector field $X \in \mathfrak{X}(M)$ is ***tangent*** to N if for any $n \in N$ we have that $X(n) \in T_n i(T_n N)$. If $X \in \mathfrak{X}(M)$ is tangent to N, there exists a unique vector field $X' \in \mathfrak{X}(N)$ such that $T_n i \cdot X'(n) = X(n)$, for any $n \in N$. The flows F_t and F'_t of X and X', respectively, are such that $i \circ F'_t = F_t \circ i$. As a corollary of this property we have that if $X, Y \in \mathfrak{X}(M)$ are tangent to the immersed submanifold N, so is their Lie bracket $[X, Y]$.

1.3 Calculus on manifolds

1.3.1 The considerations in Section 1.2 have analogues for tensor fields. The space of smooth sections of T^*M, called the ***space of one-forms***, is denoted by $\Omega^1(M)$ or by $\mathfrak{X}^*(M)$, since it is the dual of $\mathfrak{X}(M)$ thought of as a real vector space or a $C^\infty(M)$ module. Recall that a tensor field of type (p, r) is a $C^\infty(M)$ multilinear map $\Upsilon : \mathfrak{X}^*(M) \times \cdots \times \mathfrak{X}(M) \to C^\infty(M)$ (p copies of $\mathfrak{X}^*(M)$ and r copies of $\mathfrak{X}(M)$). This is equivalent to saying that Υ is a section of the ***tensor bundle*** $T^p_r(M)$ on M obtained by attaching to every point $m \in M$ the vector space of (p, r) tensors on $T_m M$; $T^p_r(M)$ is a vector bundle relative to the naturally induced atlas from M and its space of sections is denoted by $\mathfrak{T}^p_r(M)$. Elements of $\mathfrak{T}^p_r(M)$ are said to be p times ***contravariant*** and r times ***covariant*** tensor fields. An important covariant symmetric tensor field is a ***Riemannian metric*** $g \equiv \langle \cdot, \cdot \rangle \in \mathfrak{T}^0_2(M)$ which defines on each tangent space $T_m M$ an inner product that depends smoothly on m. It can be proved that a connected Hausdorff manifold admits a Riemannian metric if and only if it is second countable.

1.3.2 The ***Lie derivative*** on tensor fields is defined (and uniquely determined) by the requirements that it is a local operator, that it commutes with contractions, and that it coincides with the already defined Lie derivative on functions and on vector fields. Thus, if $\Upsilon \in \mathfrak{T}^p_r(M)$, the condition that $\pounds_X$ commutes with contractions reads as follows:

$$\begin{aligned}X[\Upsilon(\alpha_1,\dots,\alpha_p,X_1,\dots,X_r)] &= \pounds_X(\Upsilon(\alpha_1,\dots,\alpha_p,X_1,\dots,X_r))\\ &= (\pounds_X\Upsilon)(\alpha_1,\dots,\alpha_p,X_1,\dots,X_r)\\ &\quad+\sum_{j=1}^{p}\Upsilon(\alpha_1,\dots,\pounds_X\alpha_j,\dots,\alpha_p,X_1,\dots,X_r)\\ &\quad+\sum_{k=1}^{r}\Upsilon(\alpha_1,\dots,\alpha_p,X_1,\dots,\pounds_X X_k,\dots,X_r),\end{aligned} \tag{1.3.1}$$

for any $\alpha_1,\dots,\alpha_p \in \mathfrak{X}^*(M)$ and any $X_1,\dots,X_r \in \mathfrak{X}(M)$. These conditions imply that the Lie derivative also satisfies the Leibniz rule relative to the tensor product, that is,

$$\pounds_X(\Upsilon_1\otimes\Upsilon_2) = \pounds_X\Upsilon_1\otimes\Upsilon_2+\Upsilon_1\otimes\pounds_X\Upsilon_2 \tag{1.3.2}$$

for any $X \in \mathfrak{X}(M)$ and any two tensor fields Υ_1 and Υ_2 on M.

The fundamental link between flows and Lie derivatives is given by the analogue of formulas (1.2.4) and (1.2.8), namely

$$\frac{d}{dt}F_t^*\Upsilon = F_t^*\pounds_X\Upsilon, \tag{1.3.3}$$

where F_t is the flow of X. This formula immediately implies that the Lie derivative commutes with pull-back, that is, if $\psi : M \to M$ is a diffeomorphism, then

$$\psi^*\pounds_X\Upsilon = \pounds_{\psi^*X}\psi^*\Upsilon, \tag{1.3.4}$$

for any vector field X and any tensor field Υ; the pull-back on tensor fields is defined pointwise as the standard pull-back operation on multilinear maps. The previous two-formulas immediately imply that for any $X, Y \in \mathfrak{X}(M)$ and any $\Upsilon \in \mathfrak{T}^p_r(M)$

$$\pounds_{[X,Y]}\Upsilon = \pounds_X\pounds_Y\Upsilon - \pounds_Y\pounds_X\Upsilon = [\pounds_X,\pounds_Y]\Upsilon. \tag{1.3.5}$$

There are two generalizations of (1.2.8) and (1.2.9) that we shall use later on. First, if $\psi_{t,s}$ is the evolution operator of the time-dependent vector field X_t and Υ_t is a time-dependent tensor field, then

$$\frac{d}{dt}\psi_{t,s}^*\Upsilon_t = \psi_{t,s}^*\left(\frac{\partial\Upsilon_t}{\partial t}+\pounds_{X_t}\Upsilon_t\right). \tag{1.3.6}$$

It should be noted here that the commutation relation (1.3.4) cannot be applied to the second term since the diffeomorphisms and vector fields are time-dependent.

Second, if $\rho : M \to N$ is a smooth map between the manifolds M and N, the vector fields $X \in \mathfrak{X}(M)$ and $Y \in \mathfrak{X}(N)$ are called ***ρ-related*** if $T\rho\circ X = Y\circ\rho$. In particular, if ρ is a diffeomorphism, then X and ρ_*X are ρ-related. A basic result generalizing (1.2.9) states that if X_i is ρ-related to Y_i, $i = 1, 2$, $X_1, X_2 \in \mathfrak{X}(M)$, $Y_1, Y_2 \in \mathfrak{X}(N)$, then the vector fields $[X_1, X_2] \in \mathfrak{X}(M)$ and $[Y_1, Y_2] \in \mathfrak{X}(N)$ are also ρ-related:

$$T\rho\circ X_i = Y_i\circ\rho,\ i=1,2 \quad\Longrightarrow\quad T\rho\circ[X_1,X_2] = [Y_1,Y_2]\circ\rho. \tag{1.3.7}$$

In addition, if $X \in \mathfrak{X}(M)$ and $Y \in \mathfrak{X}(N)$ are ρ-related and Γ is a *covariant* tensor field on N (which means that it is of type $(0, r)$), then

$$\rho^* \pounds_Y \Gamma = \pounds_X \rho^* \Gamma. \tag{1.3.8}$$

Formula (1.3.3) shows that if the covariant tensor field Υ is symmetric or skew symmetric, then so is $\pounds_X \Upsilon$.

1.3.3 For the case of ***exterior differential forms***, that is, skew symmetric covariant tensor fields, the above remark defines the Lie derivative. Antisymmetrization of (1.3.2) shows that the Lie derivative satisfies the Leibniz rule relative to the wedge product (whose definition we shall recall below), that is,

$$\pounds_X(\alpha, \wedge\beta) = \pounds_X \alpha \wedge \beta + \alpha \wedge \pounds_X \beta \tag{1.3.9}$$

for any $X \in \mathfrak{X}(M)$ and any $\alpha, \beta \in \Omega(M)$. There are two more operations that need to be quickly reviewed: the exterior differential and the interior product.

We begin by recalling some standard notations and definitions. Let M be a smooth n-dimensional manifold. The ***bundle*** $\Lambda^k(M)$ ***of exterior differential forms*** is a vector bundle over M whose $n!/k!(n-k)!$-dimensional fiber consists of all k-linear real skew symmetric maps on $T_m M$. Its space of sections, which is the $C^\infty(M)$ module of exterior differential k-forms on M, is denoted by $\Omega^k(M)$ and the space of all forms on M is the direct sum $\Omega(M) := \oplus_{k \in \mathbb{N}} \Omega^k(M)$. By convention, $\Lambda^0_m(M) = \mathbb{R}$, for all $m \in M$, so that $\Omega^0(M) := C^\infty(M)$. Note that if $k > n$, then $\Lambda^k_m(M) = \{0\}$ for all $m \in M$, which implies that also $\Omega^k(M) = \{0\}$ for all $k > n$. In dealing with the ***exterior*** or ***wedge product*** (the skew symmetrization of the usual tensor product), we need to specify the conventions we shall adopt since there are more than one in standard use. In this book, the Bourbaki (BOURBAKI (1971)) conventions will be used, that is, if $\alpha \in \Omega^p(M)$ and $\beta \in \Omega^r(M)$, then

$$\begin{aligned}(\alpha \wedge \beta)&(X_1, \dots, X_{p+r}) \\ &= \sum (\operatorname{sign} \sigma) \alpha(X_{\sigma(1)}, \dots, X_{\sigma(p)}) \beta(X_{\sigma(p+1)}, \dots, X_{\sigma(p+r)}),\end{aligned} \tag{1.3.10}$$

where $X_1, \dots, X_{p+r} \in \mathfrak{X}(M)$ and the sum is taken over all (p, r)-shuffles σ, that is, σ is a permutation of the set $\{1, \dots, p+r\}$ such that $\sigma(1) < \cdots < \sigma(p)$ and $\sigma(p+1) < \cdots < \sigma(p+r)$, and $\operatorname{sign}\sigma$ is the signature of σ. In particular, if $\alpha^1, \dots, \alpha^k \in \Omega^1(M) \equiv \mathfrak{X}^*(M)$ and $X_1, \dots, X_k \in \mathfrak{X}(M)$, then

$$(\alpha^1 \wedge \alpha^2)(X_1, X_2) = \alpha^1(X_1)\alpha^2(X_2) - \alpha^1(X_2)\alpha^2(X_1),$$

and, more generally,

$$\begin{aligned}(\alpha^1 \wedge \cdots \wedge \alpha^k)(X_1, \dots, X_k) &= \sum_\sigma (\operatorname{sign} \sigma) \alpha^1(X_{\sigma(1)}, \dots, \alpha^k(X_{\sigma(k)}) \\ &= \det\left[\alpha^i(X_j)\right],\end{aligned}$$

where the sum is taken over all permutations σ of the set $\{1, \dots, k\}$. The wedge product of two-forms is again a form; it is bilinear over $C^\infty(M)$, associative, distributive relative to the addition of forms, and ***anticommutative***, that is, $\alpha \wedge \beta = (-1)^{pr}\beta \wedge \alpha$ if $\alpha \in \Omega^p(M)$ and $\beta \in \Omega^r(M)$. The convention adopted for the wedge product influences various coefficients in the definitions of the exterior derivative and the interior product.

If $\rho : M \to N$ is a smooth map (not necessarily a diffeomorphism), then it induces pull-back operations $\rho^* : \Omega^k(N) \to \Omega^k(M)$, $k \in \mathbb{N}$, by

$$(\rho^*\beta)(m)(v_1, \dots, v_k) = \beta(\rho(m))(T_m\rho(v_1), \dots, T_m\rho(v_k)), \tag{1.3.11}$$

for any $\beta \in \Omega^k(N)$, $v_1, \dots, v_k \in T_mM$. The push-forward operation is defined only for diffeomorphisms $\psi : M \to N$ by $\psi_* := (\psi^{-1})^*$. The key property of pull-back is that it is an exterior algebra homomorphism, that is, $\rho^*(\alpha \wedge \beta) = \rho^*\alpha \wedge \rho^*\beta$ for all $\alpha, \beta \in \Omega(M)$.

1.3.4 The ***exterior derivative*** on $\Omega(M)$ is a family of local operators $\mathbf{d} : \Omega^k(U) \to \Omega^{k+1}(U)$ (that is, they commute with restrictions to open sets) defined for each $k \in \mathbb{N}$ and each open subset U in M, which coincides with the derivative on functions defined in (1.2.4) for $k = 0$, is a $\wedge$-derivation, that is,

$$\mathbf{d}(\alpha \wedge \beta) = \mathbf{d}\alpha \wedge \beta + (-1)^p \alpha \wedge \mathbf{d}\beta \tag{1.3.12}$$

for all $\alpha \in \Omega^p(M)$, $\beta \in \Omega^r(M)$, and satisfies $\mathbf{d} \circ \mathbf{d} \equiv \mathbf{d}^2 = 0$. The exterior derivative commutes with pull-back, that is, if $\rho : M \to N$ is a smooth map and $\beta \in \Omega(N)$, then $\rho^*\mathbf{d}\alpha = \mathbf{d}\rho^*\beta$. In view of (1.3.3) this implies that the exterior and the Lie derivatives commute, that is, $\mathbf{d} \circ \pounds_X = \pounds_X \circ \mathbf{d}$ for any $X \in \mathfrak{X}(M)$. The following formula due to Cartan gives an intrinsic expression of the exterior derivative:

$$\begin{aligned} \mathbf{d}\alpha(X_0, X_1, \dots, X_k) &= \sum_{r=0}^{k} X_r[\alpha(X_0, \dots, \hat{X}_r, \dots, X_k)] \\ &+ \sum_{0 \le i < j \le k} (-1)^{i+j} \alpha([X_i, X_j], X_0, \dots, \hat{X}_i, \dots, \hat{X}_j, \dots, X_k), \end{aligned} \tag{1.3.13}$$

where $\alpha \in \Omega^k(M)$ and the symbol $\hat{X}_r$ means that X_r is deleted. In particular, if $\alpha \in \Omega^1(M)$ and $X, Y \in \mathfrak{X}(M)$ this formula becomes

$$\mathbf{d}\alpha(X, Y) = X[\alpha(Y)] - Y[\beta(X)] - \alpha([X, Y]).$$

A form $\alpha \in \Omega(M)$ is said to be ***closed*** if $\mathbf{d}\alpha = 0$. A form $\beta \in \Omega(M)$ is called ***exact*** if there is another form $\gamma \in \Omega(M)$ such that $\beta = \mathbf{d}\gamma$. Since $\mathbf{d}^2 = 0$ it follows that every exact form is also closed. The converse is true only locally (that is, in small enough neighborhoods of any point), in general, and globally if M is simply connected; this statement is known as the ***Poincaré Lemma***. A generalization of this result is the so-called ***Relative Poincaré Lemma***.

1.3.5 Lemma *Let N be a submanifold of a smooth manifold M and let α be a k-form on M. Assume that α is closed and that the form induced by α on N vanishes identically, that is,*

$$\mathbf{d}\alpha = 0, \qquad i_N^*\alpha = 0,$$

where $i_N : N \to M$ denotes the inclusion map. Then there is a $(k-1)$-form β defined on an open neighborhood U of N in M, which vanishes on N, and is such that for any $z \in U$

$$\alpha(z) = \mathbf{d}\beta(z).$$

1.3.6 Cohomology. The space of closed k-forms on M is denoted by $Z^k(M, \mathbb{R})$ and the exact k-forms by $B^k(M, \mathbb{R})$. The quotient space $H^k(M, \mathbb{R}) := Z^k(M, \mathbb{R})/B^k(M, \mathbb{R})$ of closed k-forms by the exact k-forms is called the kth-***de Rham cohomology group*** of the manifold M.

1.3.7 The ***interior product*** $\mathbf{i}_X : \Omega^{k+1}(M) \to \Omega^k(M)$ is defined as the contraction on the first variable, that is,

$$\mathbf{i}_X\alpha(X_1, \ldots, X_k) := \alpha(X, X_1, \ldots, X_k), \tag{1.3.14}$$

for any $X, X_1, \ldots, X_k \in \mathfrak{X}(M)$. By convention, $\mathbf{i}_X f = 0$ for any $f \in \Omega^0(M) = C^\infty(M)$. The interior product is bilinear over $C^\infty(M)$ in both variables, that is,

$$\mathbf{i}_{fX}\alpha = f\mathbf{i}_X\alpha = \mathbf{i}_X f\alpha, \tag{1.3.15}$$

for any $X \in \mathfrak{X}(M)$, $\alpha \in \Omega(M)$, and $f \in C^\infty(M)$; it satisfies $\mathbf{i}_X \circ \mathbf{i}_X = 0$, and is a $\wedge$-derivation, that is,

$$\mathbf{i}_X(\alpha \wedge \beta) = \mathbf{i}_X\alpha \wedge \beta + (-1)^p \alpha \wedge \mathbf{i}_X\beta, \tag{1.3.16}$$

for any $X \in \mathfrak{X}(M)$ and any $\alpha \in \Omega^p(M)$, $\beta \in \Omega^r(M)$. In addition, if $\rho : M \to N$ is a smooth map and $X \in \mathfrak{X}(M)$ is ρ-related to $Y \in \mathfrak{X}(N)$, then $\mathbf{i}_X\rho^*\beta = \rho^*\mathbf{i}_Y\beta$ for any $\beta \in \Omega(N)$.

The fundamental relation between $\mathbf{d}$, $\pounds_X$, and $\mathbf{i}_X$ is given by the following formula of Cartan

$$\pounds_X = \mathbf{i}_X \circ \mathbf{d} + \mathbf{d} \circ \mathbf{i}_X. \tag{1.3.17}$$

This immediately implies that $\mathbf{d}$ and $\pounds_X$, as well as $\mathbf{i}_X$ and $\pounds_X$ commute, that is, $\mathbf{d} \circ \pounds_X = \pounds_X \circ \mathbf{d}$ and $\mathbf{i}_X \circ \pounds_X = \pounds_X \circ \mathbf{i}_X$, for any $X \in \mathfrak{X}(M)$. More generally, we have

$$\mathbf{i}_{[X,Y]} = \pounds_X \circ \mathbf{i}_Y - \mathbf{i}_Y \circ \pounds_X = [\pounds_X, \mathbf{i}_Y] \tag{1.3.18}$$

for any $X, Y \in \mathfrak{X}(M)$. Formulas (1.3.15) and (1.3.17) also imply that

$$\pounds_{fX}\alpha = f\pounds_X\alpha + \mathbf{d}f \wedge \mathbf{i}_X\alpha \tag{1.3.19}$$

for any $X \in \mathfrak{X}(M)$, $f \in C^\infty(M)$, and $\alpha \in \Omega(M)$.

1.3.8 Volume and orientation. For an n-dimensional manifold, a ***volume form*** is an element $\mu \in \Omega^n(M)$ that is nowhere zero. If one can define on M a volume form, the manifold is called ***orientable***. If the volume form μ on M is specified, M is said to be ***oriented*** by μ, and the pair (M, μ) is called a ***volume manifold***. If M is a connected n-dimensional manifold, then M is orientable if and only if $\Omega^n(M)$ is a one-dimensional

$C^\infty(M)$ module or, equivalently, if there is an atlas such that the Jacobian determinants of all chart transition maps are positive; such an atlas is called an ***oriented atlas***. Two volume forms μ_1 and μ_2 are ***equivalent*** if $\mu_2 = g\mu_1$, for some $g \in C^\infty(M)$ satisfying $g(m) > 0$, for all $m \in M$. An equivalence class of volume forms is called an ***orientation***. An orientable manifold is connected if and only if it has precisely two orientations. A smooth map $f : (M, \mu_M) \to (N, \mu_N)$ between oriented manifolds is called ***volume preserving*** if $f^*\mu_N = \mu_M$, ***orientation preserving*** if $f^*\mu_N$ equals a strictly positive function on M times μ_M, and ***orientation reversing*** if $f^*\mu_N$ equals a strictly negative function on M times μ_M. If the manifold M is connected, then $f : M \to N$ is a local diffeomorphism if and only if it is an orientation preserving or orientation reversing map.

The orientation on M induces an orientation on a submanifold $P \subset M$ if certain conditions are satisfied. Let $p = \dim P$, $n = \dim M$, and suppose that P has trivial normal bundle, that is, there are smooth maps $N_i : P \to TM$, $i = 1, \ldots, p$, such that $T_m M = T_m P \oplus \operatorname{span}\{N_1(x), \ldots, N_p(x)\}$, for all $m \in P$. Then, P is orientable. In particular, if M is paracompact, if $h \in C^\infty(M)$, and if $c \in \mathbb{R}$ is a regular value of h, then $h^{-1}(c)$ (if nonempty) is an orientable submanifold.

Let $f : M \to N$ be a smooth map between the volume manifolds (M, μ_M) and (N, μ_N). The ***Jacobian determinant*** of f is the unique function $J_{\mu_M,\mu_N}(f) \in C^\infty(M)$ satisfying $f^*\mu_N = J_{\mu_M,\mu_N}(f)\mu_M$. If $M = N$ we shall simply write $J(f) := J_{\mu_M,\mu_M}(f)$. Thus, f is a local diffeomorphism (volume preserving, orientation preserving, orientation reversing) if and only if $J_{\mu_M,\mu_N}(f)$ never vanishes ($J_{\mu_M,\mu_N}(f)) = 1$, $J_{\mu_M,\mu_N}(f) > 0$, $J_{\mu_M,\mu_N}(f) < 0$). The Jacobian determinant of the identity map is one and if f is a diffeomorphism of M, then $J(f^{-1}) = 1/[J(f) \circ f^{-1}]$.

Let $X \in \mathfrak{X}(M)$ where (M, μ) is a volume manifold. The unique function $\operatorname{div}_\mu X \in C^\infty(M)$ satisfying $\pounds_X \mu = (\operatorname{div}_\mu X)\mu$ is called the ***divergence*** of the vector field X.

1.3.9 Integration. The ***integral*** of a compactly supported n-form α on an n-dimensional manifold is defined in the following way. One starts by defining the integral of the n-form α which has compact support in a chart domain as being the usual $\mathbb{R}^n$-integral of the push-forward of this form by the chart map. The change of variables formula shows that on an oriented manifold for two positively oriented charts, this definition of the integral gives the same value. Next, one assumes that the oriented manifold M is paracompact and uses smooth partitions of unity to patch integrals defined in charts to a global integral on the manifold. The change of variables formula and paracompactness show that this definition is independent of the positively oriented atlas and the partition of unity one has initially chosen to define the integral. Thus, the integral $\int_M \alpha$ is a well-defined object.

Let (M, μ) be a paracompact n-dimensional volume manifold. Each connected component of M is hence second countable (since connected paracompact manifolds are second countable). The reasoning below will be carried out on each connected component separately. The ***integral of a function*** $f \in C^\infty(M)$ with compact support is defined as the integral $\int_M f\mu$ of the compactly supported n-form $f\mu$ on M. Uniformly approximating any compactly supported continuous function f by smooth compactly supported functions f_n, one defines the integral $\int_M f\mu$ as the limit of the sequence $\int_M f_n\mu$. Showing that this number $\int_M f\mu$ is independent of the approximating sequence, it follows that $\int_M f\mu$ is well defined for any continuous compactly supported function f. Since M is locally compact, the σ-algebras generated by open, closed, or compact sets coincide (on each connected component which is second countable) and the ***Riesz Representation Theorem*** holds, which states that there is a unique measure m_μ on this σ-algebra such that the integral we just defined is identical to the integral associated to this measure, that is,

$$\int_M f\mu = \int_M f\, dm_\mu$$

for all continuous compactly supported functions f. An immediate consequence is that $X \in \mathfrak{X}(M)$ has zero divergence if and only if

$$\int_M f\, dm_\mu = \int_M (f \circ F_t)\, dm_\mu$$

for all $f \in L^1(M, m_\mu)$, where F_t is the flow of X. Another useful statement is the ***Transport Theorem***: if $f \in C^\infty(\mathbb{R} \times M)$, then

$$\frac{d}{dt}\int_{F_t(U)} f\, dm_\mu = \int_{F_t(U)} \left(\frac{\partial f}{\partial t} + \operatorname{div}_\mu(fX)\right) dm_\mu$$

for any open set $U \subset M$ with compact closure.

1.3.10 Integration and manifolds with boundary. The same considerations on the integral hold if M has a boundary. Recall that a manifold with boundary partitions as $\operatorname{Int} M \cup \partial M$ where both $\operatorname{Int} M$ and ∂M are boundaryless manifolds, called respectively the ***interior*** and the ***boundary*** of M; ∂M is the topological boundary of $\operatorname{Int} M$ in M but $\operatorname{Int} M$ is *not* the topological interior of M which, of course, coincides with M itself. The atlases on these manifolds are obtained by intersecting them with the atlas of M (in the sense of an atlas for manifolds with boundary). If M is n-dimensional, then $\dim \operatorname{Int} M = n$ and $\dim \partial M = n-1$. The tangent space $T_m M$, even at points $m \in \partial M$, is always $\dim M$-dimensional; it does not coincide with $T_m \partial M$ which has codimension one. Any diffeomorphism $f : M \to N$ between two manifolds with boundary induces two diffeomorphisms $\operatorname{Int} f : \operatorname{Int} M \to \operatorname{Int} N$ and $\partial f : \partial M \to \partial N$.

For an n-dimensional manifold with boundary we choose a positively oriented atlas and let $(U, \varphi : U \to \mathbb{R}^n_+)$ be a positively oriented chart, that is, $T_u\varphi : T_u U \to \mathbb{R}^n$ is orientation preserving. Here $\mathbb{R}^n_+ := \{(x_1, \ldots, x_n) \mid x_n \geq 0\}$ is the closed upper half space. A basis $\{v_1, \ldots, v_{n-1}\}$ of $T_m \partial M$ is, by definition, ***positively oriented*** if $\{(T_m\varphi)^{-1}(-\mathbf{e}_n), v_1, \ldots, v_{n-1}\}$ is positively oriented in the given orientation of M; $\mathbf{e}_n$ is the standard basis vector of $\mathbb{R}^n$ along the x_n axis in the usual positively oriented orthonormal basis of $\mathbb{R}^n$. This defines the ***induced boundary orientation*** on the manifold ∂M. Explicitly, if μ_M denotes a volume form defining the positive orientation of M, the induced boundary volume form $\mu_{\partial M}$ defining the induced positive orientation on ∂M is given by

$$\mu_{\partial M}(m)(v_1, \ldots, v_{n-1}) := \mu_M(m)((T_m\varphi)^{-1}(-\mathbf{e}_n), v_1, \ldots, v_{n-1}).$$

This definition does not depend on the chosen positively oriented chart, since the property of a tangent vector to M at the boundary point m of pointing inwards or outwards at the boundary (the inner product with $\mathbf{e}_n$ of the local representative is strictly positive, respectively strictly negative) is chart independent. Therefore, one can integrate continuous functions and smooth $n-1$-forms on the boundary.

Let M be a (second countable, smooth) n-dimensional manifold and suppose that $\alpha \in \Omega^{n-1}(M)$ and $X \in \mathfrak{X}(M)$ has compact support. ***Stokes' Theorem*** states that

$$\int_M \mathbf{d}\alpha = \int_{\partial M} \alpha.$$

If μ_M is a volume form on M and $X \in \mathfrak{X}(M)$ has a compact support, taking in the above formula $\alpha = \mathbf{i}_X \mu_M$ one gets the ***Divergence Theorem***

$$\int_M (\operatorname{div}_{\mu_M} X)\mu_M = \int_{\partial M} \mathbf{i}_X \mu_M.$$

If $\langle \cdot, \cdot \rangle$ is a Riemannian metric on M, there is a unique outward pointing unit normal $n_{\partial M}$ along ∂M. The right hand side on the previous expression can be further calculated in terms of the induced boundary volume $\mu_{\partial M}$ to get ***Gauss' Theorem***

$$\int_M (\operatorname{div}_{\mu_M} X)\mu_M = \int_{\partial M} \langle X, n_{\partial M} \rangle \mu_{\partial M}.$$

1.4 Foliated spaces and distributions

One of the topics at the core of this book is the use of generalized distributions and foliations to describe the conservation laws associated to the symmetries of Hamiltonian dynamical system. This will be described in detail in Chapters 5 and 9. In the meantime we briefly describe these standard mathematical tools, whose connection with the theory of pseudogroups and groupoids is spelled out in Chapter 3.

1.4.1 Distributions. For a manifold M, a smooth vector subbundle $E \subset TM$ is also called a ***distribution*** on M. The distribution E is ***involutive*** if for any two vector fields X, Y defined on some open set of M and which take values in E, their Lie bracket $[X, Y]$ also takes values in E. The distribution E is called ***integrable*** if for any $m \in M$ there is a local submanifold N containing m whose tangent bundle TN coincides with the restriction $E|_N$ of E to N. Such a local manifold is called a ***local integral manifold*** of E at m. By the ***restriction*** of the subbundle E to N we mean the vector bundle with base is N and whose fiber at $n \in N$ equals the fiber E_n of E. The ***local Frobenius Theorem*** states that, for a distribution E on M, involutivity and integrability are equivalent. In particular, if X is a nowhere vanishing vector field and E is the distribution given as the span of X at every point, then the Frobenius Theorem in this case is just the classical theorem of the existence and uniqueness of integral curves of X.

1.4.2 Foliations. To understand the global version of the Frobenius Theorem one needs to introduce the concept of a foliation. Let M be a manifold and let $\Phi = \{\mathcal{L}_\alpha\}_{\alpha \in A}$ be a partition of M into disjoint connected sets, called ***leaves***. This partition is called a ***foliation*** if each point $m \in M$ has a chart $(U, \varphi : U \to U' \times V' \subset \mathbf{E} \times \mathbf{F})$ such that for each $\mathcal{L}_\alpha$ the connected components $(U \cap \mathcal{L}_\alpha)^i$ of $U \cap \mathcal{L}_\alpha$ are given by

$\varphi((U \cap \mathcal{L}_\alpha)^i) = U' \times \{c_\alpha^i\}$, where $c_\alpha^i \in V' \subset \mathbf{F}$ are constants for each $\alpha \in A$ and i. Such charts are called ***foliated*** and the manifold M is said to be foliated by Φ. The ***dimension*** of the foliation is by definition $\dim \mathbf{E}$ and its ***codimension*** is $\dim \mathbf{F}$. The atlas of foliated charts on M defines on each $\mathcal{L}_\alpha$ the structure of a smooth manifold in the following manner. If $(U, \varphi : U \to U' \times V' \subset \mathbf{E} \times \mathbf{F})$ is a foliated chart and $p_{\mathbf{E}} : \mathbf{E} \times \mathbf{F} \to \mathbf{E}$ is the projection, then $p_{\mathbf{E}} \circ \varphi$ restricted to $(U \cap \mathcal{L}_\alpha)^i$ defines a chart on $\mathcal{L}_\alpha$. Each $\mathcal{L}_\alpha$ is an ***initial connected submanifold*** of M. In general, the inclusion $\mathcal{L}_\alpha \hookrightarrow M$ is not an embedding since the manifold topology on $\mathcal{L}_\alpha$ does not coincide with the relative topology induced from M on $\mathcal{L}_\alpha$. It is possible that the leaf $\mathcal{L}_\alpha$ accumulates on itself. The set

$$T(M, \Phi) := \bigcup_{\alpha \in A} \bigcup_{m \in \mathcal{L}_\alpha} T_m \mathcal{L}_\alpha$$

is a vector subbundle of TM called the ***tangent bundle of the foliation***. The elements of $T(M, \Phi)$ are also called ***vectors tangent to the foliation***. The quotient vector bundle $TM/T(M, \Phi)$ is called the ***normal bundle to the foliation*** Φ. In terms of the tangent bundle to a foliation Φ, its leaves are characterized by the following property: $x, y \in M$ lie in the same leaf if and only if x can be joined to y by a finite concatenation of integral curves of vector fields tangent to the foliation Φ.

The ***global Frobenius Theorem*** states that a distribution $E \subset TM$ is involutive if and only if it is integrable, which in turn is equivalent to the existence of a foliation Φ on M such that $E = T(M, \Phi)$.

A ***first integral*** of the foliation Φ on M is a mapping that has M as domain and that is constant on the leaves of Φ. A ***local first integral*** of Φ is a locally defined smooth function $f \in C^\infty(U)$, with U an open subset of M, such that $\mathbf{d}f\,(T(M, \Phi)|_U) \equiv 0$. Using foliated charts it is easy to prove that for any point $m \in M$ there exists an open neighborhood U of m and $k := \operatorname{codim}(\Phi)$ local first integrals $f_1, \ldots, f_k \in C^\infty(U)$ of Φ such that their differentials are linearly independent on U and

$$T_m(M, \Phi) = (\operatorname{span}\{\mathbf{d}f_1(m), \ldots, \mathbf{d}f_k(m)\})^\circ, \quad \text{for any } m \in U. \tag{1.4.1}$$

A similar argument using foliated charts guarantees that for any $m \in M$,

$$T_m(M, \Phi) = (\{\mathbf{d}f(m) \mid f \text{ local first integral of } D\})^\circ. \tag{1.4.2}$$

1.4.3 The space of leaves. Since the leaves of a foliation form a partition of M, they define an equivalence relation $\mathfrak{F}$ on M. The ***space of leaves*** is the quotient topological space $M/\mathfrak{F}$. Foliations for which $\mathfrak{F}$ is a regular equivalence relation, that is, $M/\mathfrak{F}$ is a smooth manifold and the projection $M \to M/\mathfrak{F}$ is a submersion, are called ***regular foliations***. It turns out that a foliation Φ is regular if and only if it admits a ***local slice*** at every point (see, for instance, PALAIS (1957)). A local slice of a foliation at m is a local submanifold S of M such that S intersects every leaf in at most one point and $T_m S \oplus T_m(M, \Phi) = T_m M$.

1.4.4 Generalized foliations. The leaves of the foliations presented in the previous paragraph have all the same dimension as immersed submanifolds of M which, by definition, is the dimension of the foliation. This condition is often too restrictive.

This has motivated the introduction of the so-called generalized foliations and distributions in STEFAN (1974a,b); SUSSMANN (1973); see also DAZORD (1985), LIBERMANN AND MARLE (1987), and references therein. By definition a ***generalized foliation*** of M is a partition $\Phi = \{\mathcal{L}_\alpha\}_{\alpha \in A}$ of this manifold into disjoint connected sets such that each point z has a ***generalized foliated chart*** that is defined as the pair $(U, \varphi : U \to W \subset \mathbb{R}^m)$ with $z \in U$ and such that for each leaf $\mathcal{L}_\alpha$ there is a natural number $n \leq m$, called the ***dimension*** of $\mathcal{L}_\alpha$, and a subset $A_\alpha \subset \mathbb{R}^{m-n}$ such that $\varphi(U \cap \mathcal{L}_\alpha) = \{(z_1, \ldots, z_m) \in W \mid (z_{n+1}, \ldots, z_m) \in A_\alpha\}$. Each element $(z^i_{n+1}, \ldots, z^i_m) \in A_\alpha$ determines a connected component $(U \cap \mathcal{L}_\alpha)^i$ of $U \cap \mathcal{L}_\alpha$, that is, $\varphi((U \cap \mathcal{L}_\alpha)^i) = \{(z_1, \ldots, z_n, z^i_{n+1}, \ldots, z^i_m) \in W\}$. Notice that, unlike in the case of standard foliations, the number n may change from leaf to leaf. The generalized foliated charts induce on the leaves the smooth structure that makes them into initial submanifolds of M.

The possibility of having leaves of different dimensions motivates the distinction between regular and singular leaves. A leaf $\mathcal{L}_\alpha$ of a foliation $\Phi = \{\mathcal{L}_\alpha\}_{\alpha \in A}$ is ***regular*** when it has an open neighborhood that intersects leaves of only the same dimension as $\mathcal{L}_\alpha$. The leaf is called ***singular*** when such a neighborhood is not available. The points $m \in M$ are said to be regular or singular with respect to Φ depending on whether they lie on a regular or a singular leaf of Φ.

1.4.5 Lemma *Let $\Phi = \{\mathcal{L}_\alpha\}_{\alpha \in A}$ be a generalized foliation of the second countable manifold M. Then*

(i) *for each leaf $\mathcal{L}_\alpha$ of dimension n and any foliated chart $(U, \varphi : U \to W \subset \mathbb{R}^m)$ that intersects it, the corresponding set $A_\alpha \subset \mathbb{R}^{m-n}$ is countable;*

(ii) *the set of regular points of the foliation is open and dense in M;*

(iii) *any closed leaf is automatically an embedded submanifold of M.*

The statement in part **(iii)** of the previous lemma is specific to the leaves of a foliation. More specifically, an injectively immersed submanifold that is a closed subset in the relative topology is not necessarily embedded (consider, for instance, the open "figure eight" in the plane described in §1.1.12).

1.4.6 Generalized Distributions. The notion of distribution can be generalized so that the relation between standard foliations and integrable distributions will hold between generalized foliations and generalized distributions, to be discussed in what follows. A ***generalized distribution*** D on M is a subset of the tangent bundle TM such that, for any point $m \in M$, the fiber $D(m) := D \cap T_m M$ is a vector subspace of $T_m M$. The dimension of $D(m)$ is called the ***rank*** or the ***dimension*** of the distribution D at the point m. A ***differentiable section*** of D is a differentiable vector field X defined on an open subset U of M, such that for any point $z \in U$, $X(z) \in D(z)$. An immersed connected submanifold N of M is said to be an ***integral manifold*** of the distribution D if, for every $z \in N$, $T_z i(T_z N) \subset D(z)$, where $i : N \to M$ is the injection. The integral submanifold N is said to be of ***maximal dimension*** at a point $z \in N$ if $T_z i(T_z N) = D(z)$.

The generalized distribution D is ***differentiable*** if, for every point $z \in M$, and for every vector $v \in D(z)$, there exists a differentiable section X of D, defined on an open neighborhood U of z, such that $X(z) = v$. The generalized distribution D is ***completely integrable*** if, for every point $z \in M$, there exists an integral manifold of D everywhere of maximal dimension which contains z. The generalized distribution D is ***involutive*** if it is invariant under the (local) flows associated to differentiable sections of D. This definition of involutivity is more general than the one we introduced before and it only coincides with it when the dimension of $D(m)$ is the same for any $m \in M$. Various characterizations of the complete integrability of a distribution can be found in the literature:

- STEFAN (1974b) and SUSSMANN (1973): D is completely integrable if and only if it is involutive (see also Theorem 3.2.1).

- VIFLYANTSEV (1980): D is completely integrable if and only if the following two conditions hold:

  (i) there exists a Lie subalgebra $\mathfrak{X}_D(M)$ of $\mathfrak{X}(M)$ such that $\{X(m) \mid X \in \mathfrak{X}_D(M)\} = D(m)$, for any $m \in M$;

  (ii) for any $m \in M$ and any $v \in D(m)$ there exists a smooth path $m(t)$, $t \in (-\epsilon, \epsilon)$, such that $m(0) = m$, $\dot{m}(0) = v$, and $\dim\big(D(m(t))\big) = \dim\big(D(m(0))\big)$, for any $t \in (-\epsilon, \epsilon)$.

- KOLÁŘ, MICHOR, AND SLOVÁK (1993): D is completely integrable if and only if the following two conditions hold:

  (i) there exists a Lie subalgebra $\mathfrak{X}_D(M)$ of $\mathfrak{X}(M)$ such that $\{X(m) \mid X \in \mathfrak{X}_D(M)\} = D(m)$, for any $m \in M$;

  (ii) for any $m \in M$ and any $X \in \mathfrak{X}_D(M)$, $\dim(D(F_t(m))) = \dim(D(m))$, where F_t is the flow of X.

Let D be an integrable generalized distribution. Then, for every point $m \in M$, there exists a unique connected integral manifold $\mathcal{L}_m$ of D that contains m and which is maximal in the following sense: it is everywhere of maximal dimension, and if there is any other connected integral manifold $\mathcal{L}'$ of maximal dimension that intersects $\mathcal{L}_m$, then $\mathcal{L}'$ is an open submanifold of $\mathcal{L}_m$. The submanifold $\mathcal{L}_m$ is called the ***maximal integral manifold*** or the ***accessible set*** of D going through m. The maximal integral manifolds of D are always initial submanifolds of M and constitute a generalized foliation Φ_D of M. We denote by $M/D := M/\Phi_D$ the leaf space of Φ_D.

The term accessible set is justified by the fact that the maximal integral manifold $\mathcal{L}_m$ of D going through the point m coincides with the set of points that can be reached by applying to m finite compositions of flows of the (locally defined) differentiable sections that span D. We will study these concepts in more detail in the context of pseudogroups of transformations (see Section 3.2).

1.5 Stratified spaces

One of the main results that we will cover in this book is the Symplectic Stratification Theorem (see Chapter 8). The statement and proof of this result is somewhat technical and requires a certain familiarity of the reader with the concepts that we introduce in the next three sections.

We will follow the conventions introduced in PFLAUM (2001a,b). The reader is encouraged to check with these excellent references for the proofs of the results that we now recall. See also GORESKY AND MACPHERSON (1988), SJAMAAR AND LERMAN (1991), and references therein.

1.5.1 Decomposed spaces. Let $\mathcal{Z}$ be a locally finite partition of the topological space P into smooth manifolds $S_i \subset P$, $i \in I$. We assume that the manifolds $S_i \subset P$, $i \in I$, with their manifold topology are locally closed topological subspaces of P. We say that the pair $(P, \mathcal{Z})$ is a ***decomposition*** of P with ***pieces*** in $\mathcal{Z}$ when the following condition is satisfied:

(DS) If $R, S \in \mathcal{Z}$ are such that $R \cap \bar{S} \neq \emptyset$, then $R \subset \bar{S}$. In this case we write $R \preceq S$. If, in addition, $R \neq S$, we say that R is ***incident*** to S or that it is a ***boundary piece*** of S and write $R \prec S$.

Recall that a ***locally closed*** subset A of a topological space P is a subset with the property that each of its points has an open neighborhood U in P such that $U \cap A$ is closed in U.

An injectively immersed submanifold is embedded if and only if its image is locally closed in the ambient manifold.

Condition **(DS)** is usually referred to as the ***frontier condition*** and the pair $(P, \mathcal{Z})$ is called a ***decomposed space***. The ***dimension*** of P is defined as $\dim P = \sup\{\dim S_i \mid S_i \in \mathcal{Z}\}$. If $k \in \mathbb{N}$, the ***k-skeleton*** P^k of P is the union of all the pieces of dimension smaller than or equal to k; its topology is the relative topology induced by P. The ***depth*** $\mathrm{dp}(z)$ of any point $z \in P$ relative to the decomposition $\mathcal{Z}$ is defined by

$$\mathrm{dp}(z) := \sup\{k \in \mathbb{N} \mid \exists\, S_0, S_1, \ldots, S_k \in \mathcal{Z} \text{ with } z \in S_0 \prec S_1 \prec \ldots \prec S_k\}.$$

Note that $\mathrm{dp}(x) = \mathrm{dp}(y)$ for any $x, y \in S$, $S \in \mathcal{Z}$. Thus the depth $\mathrm{dp}(S)$ of the piece $S \in \mathcal{Z}$ is well defined by $\mathrm{dp}(S) := \mathrm{dp}(x)$, $x \in S$. Finally, the depth $\mathrm{dp}(P)$ of $(P, \mathcal{Z})$ is defined by $\mathrm{dp}(P) := \sup\{\mathrm{dp}(S) \mid S \in \mathcal{Z}\}$.

A continuous mapping $f : P \to Q$ between the decomposed spaces $(P, \mathcal{Z})$ and $(Q, \mathcal{Y})$ is a ***morphism of decomposed spaces*** if for every piece $S \in \mathcal{Z}$, there is a piece $T \in \mathcal{Y}$ such that $f(S) \subset T$ and the restriction $f|_S : S \to T$ is smooth. If $(P, \mathcal{Z})$ and $(P, \mathcal{T})$ are two decompositions of the same topological space we say that $\mathcal{Z}$ is ***coarser*** than $\mathcal{T}$ or that $\mathcal{T}$ is ***finer*** than $\mathcal{Z}$ if the identity mapping $(P, \mathcal{T}) \to (P, \mathcal{Z})$ is a morphism of decomposed spaces. A topological subspace $Q \subset P$ is a ***decomposed subspace*** of $(P, \mathcal{Z})$ if for all pieces $S \in \mathcal{Z}$, the intersection $S \cap Q$ is a submanifold of S, and the corresponding partition $\mathcal{Z} \cap Q$ forms a decomposition of Q.

1.5.2 Set germs. Let P be a topological space and $z \in P$. Two subsets A and B of P are said to be ***equivalent*** at z if there is an open neighborhood U of z such that $A \cap U = B \cap U$. This relation constitutes an equivalence relation on the power set of P. The class of all sets equivalent to a given subset A at z will be denoted by $[A]_z$ and called the ***set germ*** of A at z. If $A \subset B \subset P$ we say that $[A]_z$ is a ***subgerm*** of $[B]_z$, and denote $[A]_z \subset [B]_z$.

1.5.3 Stratifications. A ***stratification*** of the topological space P is a map $\mathcal{S}$ that associates to any $z \in P$ the set germ $\mathcal{S}(z)$ of a closed subset of P such that the following condition is satisfied:

(ST) For every $z \in P$ there is a neighborhood U of z and a decomposition $\mathcal{Z}$ of U such that for all $y \in U$ the germ $\mathcal{S}(y)$ coincides with the set germ of the piece of $\mathcal{Z}$ that contains y.

The pair $(P, \mathcal{S})$ is called a ***stratified space***. Any decomposition of P defines a stratification of P by associating to each of its points the set germ of the piece on which it is sitting. The converse is, by definition, locally true. A continuous map $f : P \to Q$ between two stratified spaces $(P, \mathcal{S})$ and $(Q, \mathcal{T})$ is called a ***morphism of stratified spaces*** or a ***stratified map***, if for every $z \in P$ there exist neighborhoods V of $f(z)$ and $U \subset f^{-1}(V)$ of z, together with decompositions $\mathcal{Z}$ of U and $\mathcal{Y}$ of V that induce $\mathcal{S}|_U$ and $\mathcal{T}|_V$, respectively, and such that the following condition holds: for every $y \in U$ there exists an open neighborhood $W \subset U$ such that the map $f|_{S \cap W}$ restricted to the open subset $S \cap W$ of the piece $S \in \mathcal{Z}$ containing y has its range in the piece $R \in \mathcal{Y}$ containing $f(y)$ and is such that $f|_{S \cap W}$ is a smooth map from $S \cap W$ to R.

A ***stratified subspace*** of $(P, \mathcal{S})$ is a topological subspace Q of P such that for every $z \in Q$ there is an open neighborhood U of z in P and a decomposition $\mathcal{Z}$ inducing $\mathcal{S}|_U$ such that $(Q \cap U, \mathcal{Z} \cap Q)$ is a decomposed subspace of $(U, \mathcal{Z})$. Under such circumstances the pair $(Q, \mathcal{S}|_Q \cap Q)$ is a stratified space.

1.5.4 The strata. Two decompositions $\mathcal{Z}_1$ and $\mathcal{Z}_2$ of P are said to be ***equivalent*** if they induce the same stratification of P. If $\mathcal{Z}_1$ and $\mathcal{Z}_2$ are equivalent decompositions of P, then for all $z \in P$, we have that $\mathrm{dp}_{\mathcal{Z}_1}(z) = \mathrm{dp}_{\mathcal{Z}_2}(z)$ (see MATHER (1973)). Any stratified space $(P, \mathcal{S})$ has a unique decomposition $\mathcal{Z}_{\mathcal{S}}$ associated with the following maximal property: for any open subset $U \subset P$ and any decomposition $\mathcal{Z}$ of P inducing $\mathcal{S}$ over U, the restriction of $\mathcal{Z}_{\mathcal{S}}$ to U is coarser than the restriction of $\mathcal{Z}$ to U. The decomposition $\mathcal{Z}_{\mathcal{S}}$ is called the ***canonical decomposition*** associated to the stratification $(P, \mathcal{S})$. It is often denoted by $\mathcal{S}$ and its pieces are called the ***strata*** of P. The local finiteness of the decomposition $\mathcal{Z}_{\mathcal{S}}$ implies that for any stratum S of $(P, \mathcal{S})$ there are only finitely many strata R with $S \prec R$.

In the sequel the symbol $\mathcal{S}$ in the stratification $(P, \mathcal{S})$ will denote both the map that associates to each point a set germ and the set of pieces associated to the canonical decomposition induced by the stratification of P.

Let $f : P \to Q$ be a morphism between two stratified spaces $(P, \mathcal{S})$ and $(Q, \mathcal{T})$. For every connected component S_0 of a stratum S of P there exists a stratum T of Q such that $f(S_0) \subset T$ and $f|_{S_0} : S_0 \to T$ is smooth. If all the restrictions $f|_{S_0}$ are immersions (resp. submersions, subimmersions) we say that f is a ***stratified immersion*** (resp. ***stratified submersion, subimmersion***).

1.6 Stratified spaces with smooth structure

1.6.1 Let $(P, \mathcal{S})$ be a stratified space. A ***singular*** or ***stratified chart*** of P is a homeomorphism $\phi : U \to \phi(U) \subset \mathbb{R}^n$ from an open set $U \subset P$ to a subset of $\mathbb{R}^n$ such that for every stratum $S \in \mathcal{S}$ the image $\phi(U \cap S)$ is a submanifold of $\mathbb{R}^n$ and the restriction $\phi|_{U\cap S} : U \cap S \to \phi(U \cap S)$ is a diffeomorphism. Two singular charts $\phi : U \to \phi(U) \subset \mathbb{R}^n$ and $\varphi : V \to \varphi(V) \subset \mathbb{R}^m$ are ***compatible*** if for any $z \in U \cap V$ there exist an open neighborhood $W \subset U \cap V$ of z, a natural number $N \geq \max\{n, m\}$, open neighborhoods $O, O' \subset \mathbb{R}^N$ of $\phi(U) \times \{0\}$ and $\varphi(V) \times \{0\}$, respectively, and a C^k-diffeomorphism $\psi : O \to O'$ such that $i_m \circ \varphi|_W = \psi \circ i_n \circ \phi|_W$, where i_n and i_m denote the natural embeddings of $\mathbb{R}^n$ and $\mathbb{R}^m$ into $\mathbb{R}^N$ by using the first n and m coordinates, respectively. The notion of ***singular*** or ***stratified atlas*** is the natural generalization for stratifications of the concept of atlas existing for smooth manifolds. Analogously, we can talk of compatible and maximal stratified atlases. If the stratified space $(P, \mathcal{S})$ has a well defined maximal atlas, then we say that this atlas determines a ***C^k-smooth*** or ***C^k-differentiable structure*** on P. We will refer to $(P, \mathcal{S})$ as a ***C^k-smooth stratified space***. If $k = \infty$, only the word "smooth" will be used.

1.6.2 The smooth functions of a smooth stratified space. The procedure of construction of the sheaf of smooth functions on a manifold out of its smooth structure can be mimicked to define a sheaf of smooth functions on a smooth stratified space. Suppose now that $(P, \mathcal{S})$ is a stratified space with a smooth structure corresponding to a maximal atlas $\mathcal{A}$. We now use this atlas to construct a ***presheaf C^∞_P of smooth functions*** on P. Let $U \subset P$ be an open set. We define the section $C^\infty_P(U)$ of C^∞_P over U as the set of all continuous functions $f : U \to \mathbb{R}$ such that for all $z \in U$ and any chart $\phi : V \to \mathbb{R}^n$ such that $z \in V$, there exists an open neighborhood W of z such that $W \subset U \cap V$ and a smooth function $\bar{f} : \mathbb{R}^n \to \mathbb{R}$ such that $f|_W = \bar{f} \circ \phi|_W$. A straightforward verification shows that C^∞_P is indeed a presheaf.

In this case, for any point $z \in P$, the stalk $(C^\infty_P)_z$ of C^∞_P at z also contains as an ideal the subspace $\mathfrak{m}_z$ of germs that vanish at z. We define the ***rank*** $\mathrm{rk}(z)$ of the point z as the dimension $\dim(\mathfrak{m}_z/\mathfrak{m}_z^2)$. It can be shown (see PFLAUM (2001a, Proposition 1.3.10)) that there exists a chart (U, ϕ) around z of the form $\phi : U \to \mathbb{R}^{\mathrm{rk}(z)}$. Moreover, there exists an open neighborhood V of z such that $\mathrm{rk}(y) \leq \mathrm{rk}(z)$, for every $y \in V$.

A continuous map $f : P \to Q$ between two stratified spaces with smooth structure is ***smooth*** (respectively of ***class*** C^m) if $f^* C^\infty_Q \subset C^\infty_P$ (respectively $f^* C^m_Q \subset C^m_P$), that is, for any open subset V of Q and any $g \in C^\infty_Q(V)$, the composition $g \circ f \in C^\infty_P(f^{-1}(V))$. It can be proved that $f : P \to Q$ is smooth if and only if for every $z \in P$, there exist singular charts $\varphi : U \to \mathbb{R}^n$, with $z \in U$, and $\psi : V \to \mathbb{R}^m$, with $f(z) \in V$, as well as an open neighborhood $W \subset U$ with $f(W) \subset V$, and a smooth mapping $\bar{f} : \mathbb{R}^n \to \mathbb{R}^m$ such that $\psi \circ f|_W = \bar{f} \circ \varphi|_W$.

1.6.3 Remark We caution that a smooth map between two stratified spaces is not, in general, a stratified map and, conversely, a stratified map needs not be smooth.

1.7 Whitney stratifications

In general, stratifications may be rather pathological. Nevertheless, most of the stratifications that we will encounter in our work correspond to a particular class of stratifications with a smooth structure whose behavior is much better. We follow the presentation of PFLAUM (2001a).

1.7.1 The Whitney conditions. Let M be a smooth stratified space and $R, S \subset M$ two strata. We say that the pair (R, S) satisfies the ***Whitney condition* (A)** at the point $z \in R$, or that (R, S) is **(A)*-regular*** at z, if the following condition is satisfied:

(A) For any sequence of points $\{z_n\}_{n\in\mathbb{N}}$ in S converging to $z \in R$ for which the sequence of tangent spaces $\{T_{z_n}S\}_{n\in\mathbb{N}}$ converges in the Grassmann bundle of $\dim S$-dimensional subspaces of TM to $\tau \subset T_zM$, we have that $T_zR \subset \tau$.

Let $\phi : U \to \mathbb{R}^n$ be a smooth chart of M around the point z. The ***Whitney condition* (B)** at the point $z \in R$ with respect to the chart (U, ϕ) is given by the following statement:

(B) Let $\{x_n\}_{n\in\mathbb{N}} \subset R \cap U$ and $\{y_n\}_{n\in\mathbb{N}} \subset S \cap U$ be two sequences with the same limit $z = \lim_{n\to\infty} x_n = \lim_{n\to\infty} y_n$ and such that $x_n \neq y_n$, for all $n \in \mathbb{N}$. Suppose that the set of connecting lines $\overline{\phi(x_n)\phi(y_n)} \subset \mathbb{R}^n$ converges in projective space to a line L and that the sequence of tangent spaces $\{T_{y_n}S\}_{n\in\mathbb{N}}$ converges in the Grassmann bundle of $\dim S$-dimensional subspaces of TM to $\tau \subset T_zM$. Then $(T_z\phi)^{-1}(L) \subset \tau$.

If the condition (A) (respectively (B)) is verified for every point $z \in R$, the pair (R, S) is said to satisfy the ***Whitney condition* (A)** (respectively **(B)**) or that S is **(A)-** (respectively **(B)-**) ***regular*** over R.

It can be verified that Whitney's condition (B) does not depend on the chart used to formulate it and that (B)-regularity implies (A)-regularity.

A stratified space with smooth structure such that for every pair of strata Whitney's condition (A) is satisfied is called a ***Whitney* (A)*-space*** or an **(A)*-stratified space***. A stratified space with smooth structure such that for every pair of strata Whitney's condition (B) is satisfied is called a ***Whitney space*** or a **(B)*-stratified space***.

1.7.2 Cone spaces and local triviality. Let P be a topological space. Consider the equivalence relation $\sim$ in the product $P \times [0, \infty)$ given by $(z, a) \sim (z', a')$ if and only if $a = a' = 0$. We define the ***cone*** CP on P as the quotient topological space $P \times [0, \infty)/\sim$. If P is a smooth manifold, then the cone CP is a decomposed space with two pieces, namely, $P \times (0, \infty)$ and the ***vertex*** which is the class corresponding to any element of the form $(z, 0)$, $z \in P$, that is, $P \times \{0\}$. Analogously, if $(P, \mathcal{Z})$ is a decomposed (stratified) space, then the associated cone CP is also a decomposed (stratified) space whose pieces (strata) are the vertex and the sets of the form $S \times (0, \infty)$, with $S \in \mathcal{Z}$. This implies, in particular, that

$$\dim CP = \dim P + 1,$$
$$\mathrm{dp}(CP) = \mathrm{dp}(P) + 1.$$

A stratified space $(P, \mathcal{S})$ is said to be ***locally trivial*** if for any $z \in P$ there exist a neighborhood U of z, a stratified space $(F, \mathcal{S}^F)$, a distinguished point $\mathbf{0} \in F$, and an isomorphism of stratified spaces

$$\psi : U \to (S \cap U) \times F,$$

where S is the stratum that contains z and ψ satisfies that $\psi^{-1}(y, \mathbf{0}) = y$, for all $y \in S \cap U$. When F is given by a cone CL over a compact stratified space L, then L is called the ***link*** of z.

An important corollary of the so-called ***Thom's first isotopy lemma*** guarantees that every Whitney stratified space is locally trivial (see THOM (1969); MATHER (1970)). A converse to this implication needs the introduction of the so-called cone spaces. Their definition is given by recursion on the depth of the space.

1.7.3 Definition *Let $m \in \mathbb{N} \cup \{\infty, \omega\}$. A **cone space** of class C^m and depth 0 is the union of countably many C^m manifolds together with the stratification whose strata are the unions of the connected components of equal dimension. A cone space of class C^m and depth $d+1$, $d \in \mathbb{N}$, is a stratified space $(P, \mathcal{S})$ with a C^m differentiable structure such that for any $z \in P$ there exists a connected neighborhood U of z, a compact cone space L of class C^m and depth d called the **link**, and a stratified isomorphism*

$$\psi : U \to (S \cap U) \times CL,$$

where S is the stratum that contains the point z, the map ψ satisfies that $\psi^{-1}(y, \mathbf{0}) = y$, for all $y \in S \cap U$, and $\mathbf{0}$ is the vertex of the cone CL.

*If $m \neq 0$, then L is required to be embedded into a sphere via a fixed smooth global singular chart $\varphi : L \to S^l$ that determines the smooth structure of CL. More specifically, the smooth structure of CL is generated by the global chart $\tau : [z, t] \in CL \longmapsto t \cdot \varphi(z) \in \mathbb{R}^{l+1}$. The maps $\psi : U \to (S \cap U) \times CL$ and $\varphi : L \to S^l$ are referred to as a **cone chart** and a **link chart** respectively. Moreover, if $m \neq 0$, then ψ and ψ^{-1} are required to be differentiable of class C^m as maps between stratified spaces with a smooth structure.*

The cone charts and the link charts in the definition of a cone space imply that these structures are stratified spaces with smooth structure. The following theorem proved in PFLAUM (2001a) shows that cone spaces have additional structure.

1.7.4 Theorem *Any cone space of class C^m with $m \geq 2$ is a Whitney stratified space.*

1.7.5 Whitney stratified spaces are in general *not* cone spaces. A counterexample is given by ***Neil's parabola*** (see PFLAUM (2001a)). However, Mather's theory of control data (see MATHER (1970) and page 410 of SJAMAAR AND LERMAN (1991) for an outline of the construction of the link) implies that Whitney stratified subsets of Euclidean space are cone spaces. The reader should be aware that some authors (for instance SJAMAAR AND LERMAN (1991)) take cone spaces as the definition of stratified space.

Chapter 2

Lie Group Actions

One of the main themes of this book is the notion of symmetry. The main goal in the chapters at the core of this book is explaining to the reader how the symmetries of a Hamiltonian dynamical system can be used to simplify or *reduce* the study of that system. From the mathematical point of view the description of symmetries is implemented via the use of Lie group actions and, more generally, pseudogroups and groupoids. In the following two chapters we review all the material concerning these topics that will be needed in the rest of the book.

This chapter presents a brief review of the theory of Lie group actions on a manifold. Experts can safely skip this chapter and consult it only for reference. However, as opposed to Chapter 1 the presentation here is more detailed and there are even some proofs of harder to find standard facts that will be used later. The results quoted in this chapter can be found in ABRAHAM AND MARSDEN (1978), BOURBAKI (1989b), BREDON (1972), BRÖCKER AND TOM DIECK (1985), BURGHELEA et al. (1975), CUSHMAN AND BATES (1997), DUISTERMAAT AND KOLK (1999), GUILLEMIN AND STERNBERG (1984c), HOCHSCHILD (1965), IWASAWA (1949), KAWAKUBO (1991), KNAPP (1996), KOBAYASHI (1995), KOBAYASHI AND NOMIZU (1963), KOLÁŘ et al. (1993), PALAIS (1961), and TOM DIECK (1987).

2.1 Lie groups

2.1.1 Lie groups. A ***Lie group*** G is a smooth manifold which is a group such that multiplication $(g, h) \in G \times G \mapsto gh \in G$ is a smooth map. The implicit function theorem implies that the inversion operation is also smooth. It turns out that only the differentiability class C^2 is needed for everything that follows. We shall denote by $L_g, R_g : G \to G$ the left and right translations defined by $g \in G$:

$$L_g(h) := gh, \qquad R_g(h) := hg$$

for any $h \in G$. These maps are diffeomorphisms of G.

Any open subgroup of a Lie group G is closed. A subgroup of G is open if and only if it contains a neighborhood of the identity element. A Lie group is connected if and only if it is generated by any neighborhood of the identity. The connected component of the identity G^0 of the Lie group G is a normal open and closed subgroup of G. Since G^0 is generated by a neighborhood of the identity and one can always take such a neighborhood to be compact, it immediately follows that G^0 is the countable union of all the positive integer powers of this compact neighborhood; each such power is clearly a compact set. An easy point set topological argument shows then that G^0, and hence G, is paracompact.

2.1.2 Lie algebras. A vector field $X \in \mathfrak{X}(G)$ is ***left (right) invariant*** if $L_g^* X = X$ ($R_g^* X = X$) for every $g \in G$. The real vector subspace $\mathfrak{X}_L(G) \subset \mathfrak{X}(G)$ of left invariant vector fields on G is isomorphic to the tangent space T_eG at the identity via the restriction map that sends a vector field $X \in \mathfrak{X}_L(G)$ to its value $X(e) \in T_eG$ at the identity e. The inverse of this map is given by $\xi \in T_eG \mapsto \xi_L \in \mathfrak{X}_L(G)$, where ξ_L is the left invariant vector field on G whose value at the identity is ξ, that is, $\xi_L(g) = T_eL_g\xi$ for all $g \in G$. Via this isomorphism T_eG becomes a Lie algebra whose Lie bracket operation is also denoted by $[\cdot, \cdot]$, that is, $[\xi_L, \eta_L] = [\xi, \eta]_L$ for all $\xi, \eta \in \mathfrak{g}$. The vector space T_eG together with this Lie algebra structure is called the ***Lie algebra of*** G and is denoted by $\mathfrak{g}$ or by $\operatorname{Lie}(G)$. Its dual space, which is an object that plays an important role in this book, is denoted by $\mathfrak{g}^*$.

One could define the Lie algebra structure on T_eG starting with the right invariant vector fields $\mathfrak{X}_R(G)$. In this case, the inverse of the isomorphism $X \in \mathfrak{X}_R(G) \mapsto X(e) \in T_eG$ is given by $\xi \in T_eG \mapsto \xi_R \in \mathfrak{X}_R(G)$, where ξ_R is the right invariant vector field on G whose value at the identity is ξ, that is, $\xi_R(g) = T_eR_g\xi$ for all $g \in G$. It should be noted, however, that if the Lie bracket on T_eG is defined using left invariant vector fields, which is the convention that we shall use in this book, then $[\xi_R, \eta_R] = -[\xi, \eta]_R$ for all $\xi, \eta \in \mathfrak{g}$, that is, the two Lie algebra structures on T_eG defined by left and right invariant vector fields on G are *anti*-isomorphic.

2.1.3 The exponential. Any left (right) invariant vector field $X \in \mathfrak{X}_L(G)$ is complete. If $c(t)$ is the integral curve through e of $\xi_L \in \mathfrak{X}_L(G)$, $\xi \in \mathfrak{g}$, then the ***exponential map*** $\exp : \mathfrak{g} \to G$ is defined by $\exp\xi := c(1)$. The exponential map is smooth. If one defines the exponential map of ξ as the value at 1 of the integral curve of ξ_R through the origin, it coincides with the one above. This means that the definition of the exponential map is independent of the definition of the Lie algebra structure on T_eG using left or right invariant vector fields. Thus, the flow of ξ_L is given by $(t, g) \mapsto g\exp t\xi$ and that of ξ_R by $(t, g) \mapsto \exp t\xi\, g$. The derivative of the exponential map at the origin is the identity, and so it defines a diffeomorphism between an open neighborhood of the origin in $\mathfrak{g}$ and an open neighborhood of the identity element in G. Thus, the inverse of the exponential map in an open neighborhood of the identity element, that is the ***logarithm***, can be used as a chart around the identity; by left translations this generates an atlas defining the manifold structure of G. This proves that the manifold structure of G is uniquely determined by its Lie algebra via the exponential map, both in the real and the complex cases.

If $\xi, \eta \in \mathfrak{g}$ and $[\xi, \eta] = 0$, then $\exp\xi \exp\eta = \exp(\xi + \eta)$. This is a straightforward consequence of the Baker–Campbell–Hausdorff formula that we review in §2.1.6. Also, $[\xi, \eta] = 0$ for all $\xi, \eta \in \mathfrak{g}$ if and only if the connected component of the identity G^0 is Abelian. If G^0 is Abelian, then $\exp : \mathfrak{g} \to G^0$ is a group homomorphism and its kernel is a discrete subgroup of $\mathfrak{g}$.

2.1.4 The ***Structure theorem for Abelian Lie groups*** says that every connected Abelian n-dimensional Lie group is isomorphic to a cylinder, that is, to $\mathbb{T}^k \times \mathbb{R}^{n-k}$ for some $k = 1, \ldots, n$. We denoted here and in what follows by $\mathbb{T}^k$ the k-dimensional torus, that is, the product of k (flat) circles or, equivalently, $\mathbb{T}^k \equiv \mathbb{R}^k/\mathbb{Z}^k$. Thus, every connected compact Abelian Lie group is a torus. This theorem is a straightforward consequence of the fact that in the Abelian case the exponential map $\exp : (\mathfrak{g}, +) \to G$ is a Lie group homomorphism whose kernel ker(exp) is a discrete subgroup of $(\mathfrak{g}, +)$. The result follows from the chain of isomorphisms $G \simeq \mathfrak{g}/\ker(\exp) \simeq \mathbb{T}^k \times \mathbb{R}^{n-k}$.

2.1.5 The following standard notation will be used. If $\xi, \eta \in \mathfrak{g}$, the ***adjoint representation*** of $\mathfrak{g}$ on itself will be denoted by $\mathrm{ad}_\xi \eta := [\xi, \eta]$. The dual of the linear map $\mathrm{ad}_\xi : \mathfrak{g} \to \mathfrak{g}$ for a fixed $\xi \in \mathfrak{g}$ is denoted by $\mathrm{ad}_\xi^* : \mathfrak{g}^* \to \mathfrak{g}^*$. The operation $(\xi, \mu) \in \mathfrak{g} \times \mathfrak{g}^* \mapsto -\mathrm{ad}_\xi^* \mu \in \mathfrak{g}^*$ is called the ***coadjoint representation*** of $\mathfrak{g}$ on $\mathfrak{g}^*$.

With this notation, the derivative of the exponential map at any $\xi \in \mathfrak{g}$ is given by

$$
\begin{aligned}
T_\xi \exp &= T_e R_{\exp \xi} \circ \sum_{n=0}^{\infty} \frac{1}{(n+1)!} \mathrm{ad}_\xi^n \\
&= T_e R_{\exp \xi} \circ \frac{e^{\mathrm{ad}_\xi} - I}{\mathrm{ad}_\xi} \\
&= T_e R_{\exp \xi} \circ \int_0^1 e^{t\, \mathrm{ad}_\xi}\, dt
\end{aligned}
\tag{2.1.1}
$$

or, using left translations

$$
\begin{aligned}
T_\xi \exp &= T_e L_{\exp \xi} \circ \sum_{n=0}^{\infty} \frac{(-1)^n}{(n+1)!} \mathrm{ad}_\xi^n \\
&= T_e L_{\exp \xi} \circ \frac{I - e^{-\mathrm{ad}_\xi}}{\mathrm{ad}_\xi} \\
&= T_e L_{\exp \xi} \circ \int_0^1 e^{-t\, \mathrm{ad}_\xi}\, dt.
\end{aligned}
\tag{2.1.2}
$$

The second set of equalities in the above formulas is a convenient shorthand notation for the series appearing in the first set of equalities. Using the spectral mapping theorem on the complexification of $\mathfrak{g}$ it follows that if $\lambda \neq 0$ is an eigenvalue of ad_ξ, then the eigenvalues of

$$
\frac{e^{\mathrm{ad}_\xi} - I}{\mathrm{ad}_\xi} \quad \text{are} \quad \frac{e^\lambda - 1}{\lambda}
$$

and of

$$
\frac{I - e^{-\mathrm{ad}_\xi}}{\mathrm{ad}_\xi} \quad \text{are} \quad \frac{1 - e^{-\lambda}}{\lambda},
$$

whereas if $\lambda = 0$, then 1 is an eigenvalue of these operators. Thus, (2.1.1) and (2.1.2) show that $T_\xi \exp$ is not invertible for those $\xi \in \mathfrak{g}$ for which ad_ξ has an eigenvalue of the form $2\pi i k$ with $k \in \mathbb{Z}$, $k \neq 0$.

In particular, the subset $\mathfrak{E} \subset \mathfrak{g}$ where the derivative of the exponential is invertible is an open set containing the origin. Its complement is a countable union of algebraic varieties indexed by $k \in \mathbb{Z}\backslash\{0\}$, where the variety for $k = 1$ is given by those $\xi \in \mathfrak{g}$ for which the complexification of the operator ad_ξ in the complexification of $\mathfrak{g}$ has $2\pi i$ as an eigenvalue and the variety for an arbitrary k is simply its homothetic k-image. According to (2.1.1), for $\xi \in \mathfrak{E}$,

$$\left(T_\xi \exp\right)^{-1} = \frac{\mathrm{ad}_\xi}{e^{\mathrm{ad}_\xi} - I} \circ T_e R_{\exp(-\xi)},$$

where, as usual, the fraction is the shorthand notation for the linear operator which is the sum of the convergent series expansion of $s/(e^s - 1)$ with s replaced by ad_ξ.

2.1.6 The Baker–Campbell–Hausdorff formula. The ***Baker–Campbell–Hausdorff*** or ***Dynkin formula*** (see DUISTERMAAT AND KOLK (1999) for a historical account) states that:

1. The subset of all $(\xi, \eta) \in \mathfrak{g} \times \mathfrak{E}$ for which the solution of the differential equation in $\mathfrak{E}$

$$\frac{d}{dt}\zeta(t) = \frac{\mathrm{ad}_{\zeta(t)}}{e^{\mathrm{ad}_{\zeta(t)}} - I}(\xi), \qquad \zeta(0) = \eta$$

is defined for all $t \in [0, 1]$, is open in $\mathfrak{g} \times \mathfrak{g}$ and contains $(0, 0)$;

2. If for (ξ, η) in the subset above one defines $\zeta(1) := \mathcal{M}(\xi, \eta)$, then

$$\exp\xi \exp\eta = \exp\mathcal{M}(\xi, \eta) \tag{2.1.3}$$

for all pairs (ξ, η) in this subset;

3. The map $\mathcal{M}$ defined on the open subset given in the first item above is analytic (real or complex, depending on whether $\mathfrak{g}$ was a real or a complex Lie algebra) and the following explicit formula holds, where the function f in the first equality is the series

$$f(z) = \frac{\log z}{z - 1} = \sum_{k=0}^{\infty} \frac{(-1)^k}{k+1}(z - 1)^k$$

and where the second sum in the third equality is taken over all p-tuples of integers $(n_1, \ldots, n_p)$, $(m_1, \ldots, m_p)$ satisfying $n_j \geq 0$, $m_j \geq 0$, and $n_j + m_j \geq 1$ for all $j = 1, \ldots, p$:

$$
\begin{aligned}
\mathcal{M}(\xi, \eta) &= \eta + \int_0^1 f\left(e^{t\,\mathrm{ad}_\xi} \circ e^{t\,\mathrm{ad}_\eta}\right)(\xi)dt \\
&= \xi + \eta + \sum_{p=1}^{\infty} \frac{(-1)^p}{p+1} \int_0^1 \left(\sum_{n,m \geq 0,\, n+m \geq 1} t^n \frac{\mathrm{ad}_\xi^n}{n!} \circ \frac{\mathrm{ad}_\eta^m}{m!} \right)^p (\xi)dt \\
&= \xi + \eta + \\
&\sum_{p=1}^{\infty} \frac{(-1)^p}{p+1} \sum \frac{1}{n_1 + \cdots + n_p + 1} \left(\frac{\mathrm{ad}_\xi^{n_1}}{n_1!} \circ \frac{\mathrm{ad}_\eta^{m_1}}{m_1!} \circ \cdots \frac{\mathrm{ad}_\xi^{n_p}}{n_p!} \circ \frac{\mathrm{ad}_\eta^{m_p}}{m_p!} \right)(\xi) \\
&= \xi + \eta + \frac{1}{2}[\xi, \eta] + \frac{1}{12}[\xi, [\xi, \eta]] + \frac{1}{12}[\eta, [\eta, \xi]] \\
&\qquad - \frac{1}{24}[\xi, [\eta, [\xi, \eta]]] + o(\|\xi\|^5 + \|\eta\|^5).
\end{aligned}
$$

Using the Baker–Campbell–Hausdorff formula and the logarithm in a neighborhood of the identity, one concludes, when $\mathfrak{g}$ is a real Lie algebra, that the differentiable structure on G, which, as we remarked at the beginning, has to be only of class C^2, is that of a real analytic manifold and that the group operations are also analytic; thus G is a real analytic Lie group. If $\mathfrak{g}$ is a complex Lie algebra, this argument only shows that the manifold structure on G is complex analytic. To conclude that G is a complex analytic Lie group one needs to add the condition that the adjoint operators $\mathrm{Ad}_g : \mathfrak{g} \to \mathfrak{g}$ are complex linear for every $g \in G$ (which automatically holds for all $g \in G^0$, thereby establishing the uniqueness of the complex analytic structure of G^0 if $\mathfrak{g}$ is a complex Lie algebra). This result is only one of many *automatic smoothness* results for Lie groups; we shall encounter others below.

2.1.7 The Haar measure. For an arbitrary Lie group G there is a volume form μ, unique up to nonzero multiplicative constants, which is left (right) invariant. If the group G is, in addition, compact, then this volume form is right (left) invariant as well. The Lebesgue measure m_μ associated to μ is called the ***Haar measure*** on G. For compact Lie groups the Haar measure is usually normalized by the condition that $m_\mu(G) = 1$. The Haar measure is usually denoted by dg, integration relative to the Haar measure is written as $\int_G f(g)dg$, and left invariance means that

$$\int_G f(hg)dg = \int_G f(g)dg$$

for all $h \in G$ and all $f \in L^1(G)$.

2.1.8 Homomorphisms. A ***Lie group homomorphism*** $f : G \to H$ is a smooth map that is at the same time a group homomorphism. Then $T_e f : \mathfrak{g} \to \mathfrak{h}$ is a ***Lie algebra homomorphism***, that is, $T_e f \cdot [\xi, \eta] = [T_e f \cdot \xi, T_e f \cdot \eta]$ for any $\xi, \eta \in \mathfrak{g}$. Here $\mathfrak{g}$ is the Lie algebra of G and $\mathfrak{h}$ is the Lie algebra of H. In addition,

$$f \circ \exp_G = \exp_H \circ T_e f, \tag{2.1.4}$$

where $\exp_G$ and $\exp_H$ are the exponential maps of G and H, respectively.

Formula (2.1.4) implies that if $f_1, f_2 : G \to H$ are Lie group homomorphisms and $T_e f_1 = T_e f_2$, then f_1 and f_2 coincide on the connected component of the identity G^0.

Another interesting consequence of (2.1.4) is that if $f : G \to H$ is a Lie group homomorphism, then f is injective if and only if $Tf : TG \to TH$ is injective. Thus, if f is a bijective Lie group homomorphism, then it is an isomorphism of Lie groups. Indeed, the previous statement implies that Tf is injective and thus f is an immersion. Therefore f is a bijective immersion and since G is always paracompact, as we remarked before, it follows from Lemma 1.1.17 that f is a diffeomorphism.

A ***one parameter subgroup*** of a Lie group G is a Lie group homomorphism from $\mathbb{R}$ (endowed with the additive structure) to G. Any such parameter subgroup is of the form $\exp t\xi$ for some $\xi \in \mathfrak{g}$, the Lie algebra of G.

Each $g \in G$ defines the ***inner automorphism*** or ***conjugations*** $\mathrm{AD}_g \equiv I_g : G \to G$ defined for every $g \in G$ by $I_g(h) = ghg^{-1}$. The induced Lie algebra isomorphism $\mathrm{Ad}_g : \mathfrak{g} \to \mathfrak{g}$ is called the ***adjoint operator***. The key relationship between the adjoint operator and the Lie bracket is given by

$$\left.\frac{d}{dt}\right|_{t=0} \mathrm{Ad}_{\exp t\xi}\, \eta = [\xi, \eta] = \mathrm{ad}_\xi\, \eta, \tag{2.1.5}$$

for any $\xi, \eta \in \mathfrak{g}$. Equivalently, this formula can be written as

$$[\xi, \eta] = \left.\frac{d}{dt}\frac{d}{ds} g(t)h(s)g(t)^{-1}\right|_{s=0, t=0}, \tag{2.1.6}$$

where $g(t)$ and $h(s)$ are curves in G satisfying $g(0) = h(0) = e$ and $\dot{g}(0) = \xi$, $\dot{h}(0) = \eta$.

Since I_g is a Lie group homomorphism, formula (2.1.4) gives

$$\exp(\mathrm{Ad}_g\, \xi) = g(\exp \xi)g^{-1}. \tag{2.1.7}$$

Moreover, for any $g \in G$ the linear map Ad_g is a Lie algebra isomorphism of $\mathfrak{g}$, that is,

$$[\mathrm{Ad}_g \xi, \mathrm{Ad}_g \eta] = \mathrm{Ad}_g[\xi, \eta], \text{ for any } \xi, \eta \in \mathfrak{g}.$$

This expression implies that for any $g \in G$ and $\xi \in \mathfrak{g}$ we have

$$\mathrm{ad}_{\mathrm{Ad}_g \xi} \circ \mathrm{Ad}_g = \mathrm{Ad}_g \circ \mathrm{ad}_\xi$$

as well as its dual

$$\mathrm{Ad}_g^* \circ \mathrm{ad}^*_{\mathrm{Ad}_g \xi} = \mathrm{ad}_\xi^* \circ \mathrm{Ad}_g^*.$$

Let $\mathrm{GL}(\mathfrak{g})$ be the Lie group of linear isomorphisms of $\mathfrak{g}$ and $\mathrm{Aut}(\mathfrak{g})$ the closed Lie subgroup of $\mathrm{GL}(\mathfrak{g})$ consisting of Lie algebra isomorphisms of $\mathfrak{g}$. Since $g \in G \mapsto \mathrm{Ad}_g \in \mathrm{Aut}(\mathfrak{g}) \subset \mathrm{GL}(\mathfrak{g})$ is a Lie group homomorphism, the two formulas (2.1.4) and (2.1.5) give

$$\mathrm{Ad}_{\exp \xi} = e^{\mathrm{ad}_\xi} \tag{2.1.8}$$

for any $\xi \in \mathfrak{g}$, since the usual exponential e^{ad_ξ} of the linear operator ad_ξ is the exponential map of the Lie group $\mathrm{GL}(\mathfrak{g})$, whose Lie algebra $\mathfrak{gl}(\mathfrak{g})$ consists of all linear operators on $\mathfrak{g}$.

As we have remarked before, in Lie group theory there are a number of smoothness results. For homomorphisms this is quite striking: every continuous group homomorphism between Lie groups is analytic. This result has an interesting slight weakening: any group homomorphism from a Lie group to another Lie group that has only countably many components and whose graph is closed in the product is automatically analytic. If the hypothesis on the connected components of the target Lie group is dropped, the result is, in general, false. An immediate corollary of this second statement is the following: any continuous group isomorphism between two Lie groups such that the source Lie group has only countably many components is an analytic Lie group isomorphism, that is, both the isomorphism and its inverse are analytic maps.

Since every continuous group homomorphism between Lie groups is analytic it follows that on a topological group there is at most one (analytic) compatible Lie group structure, that is, the underlying topological group structure of this Lie group is the given one. Moreover, the solution to ***Hilbert's Fifth Problem*** due to Gleason, Montgomery, and Zippin, states that a structure exists if and only if the group is ***locally Euclidean*** (that is, each point has a neighborhood homeomorphic to some open set in $\mathbb{R}^n$ for some $n \in \mathbb{N}$). A locally compact topological group admits a compatible Lie group structure if and only if there is a neighborhood of the identity element e that does not contain any other subgroup other than $\{e\}$ itself.

2.1.9 Subgroups and subalgebras. A ***Lie subgroup*** H of a Lie group G is a subgroup of G that is a Lie group in its own right and such that the inclusion map $H \hookrightarrow G$ is an immersion. If H is in addition a submanifold of G, then H is called a ***regular*** Lie subgroup of G. A basic theorem states that H is a closed subgroup of G if and only if it is a regular Lie subgroup of G. For example, if $f : G \to H$ is a Lie group homomorphism, then its ***kernel***

$$\ker f := \{g \in G \mid f(g) = e\}$$

is a regular Lie normal subgroup of G. We shall show later on that the ***range***

$$\operatorname{range} f := \{f(g) \mid g \in G\}$$

is a Lie subgroup of H, which is, in general (see §2.3.11), not regular. Any open Lie subgroup is also closed. A Lie subgroup is closed if and only if it contains the closure of any one parameter subgroup contained in it. A Lie group has no subgroups that are included in all neighborhoods of the identity element. An arbitrary intersection (even infinite) of Lie subgroups is again a Lie subgroup whose Lie algebra is the intersection of the Lie algebras of the corresponding Lie subgroups. The group generated by a family of connected Lie subgroups (possibly infinite) is a connected Lie subgroup whose Lie algebra coincides with the Lie algebra generated by the Lie algebras of these subgroups.

Let G be a Lie group with Lie algebra $\mathfrak{g}$. If H is a Lie subgroup of G, then its Lie algebra $\mathfrak{h}$ is a Lie subalgebra of $\mathfrak{g}$ and is characterized by the condition

$$\mathfrak{h} = \{\xi \in \mathfrak{g} \mid \exp t\xi \in H \text{ for all } t \in \mathbb{R}\}. \tag{2.1.9}$$

Conversely, given a Lie subalgebra $\mathfrak{h} \subset \mathfrak{g}$, there exists a unique connected regularly immersed (initial) Lie subgroup H of G whose Lie algebra is $\mathfrak{h}$; H is generated by $\exp_G \mathfrak{h}$. A stronger version of this result can be found in Chapter III, §4.5 of BOURBAKI (1989b):

2.1.10 Theorem *Let G be a Lie group and $H \subset G$ a subgroup of G. Then there exists a unique smooth structure on H that makes it an initial submanifold of G and thereby a Lie subgroup of G. The Lie algebra $\mathfrak{h}$ of H with respect to this smooth structure is given by*

$$\mathfrak{h} = \{c'(0) \mid c : I \to G \text{ smooth such that } c(0) = e \text{ and } c(I) \subset H\},$$

where $I \subset \mathbb{R}$ is an open interval in $\mathbb{R}$ that contains zero.

The Lie group structure on H defined in the previous theorem is called the structure induced on H by the Lie group structure on G. If H is a Lie subgroup of G, its Lie group structure is induced by that on G. We therefore have the following corollary:

2.1.11 Corollary *Let H be a Lie subgroup of the Lie group G. Then H is an initial submanifold of G.*

The ***Freudenthal–Kuranishi–Yamabe theorem*** states that given a subgroup H of a Lie group G, the smooth structure induced on H by the Lie group structure on G makes it into a connected Lie subgroup of G if and only if H is arcwise connected as a topological subspace of G (see YAMABE (1950)).

2.1.12 Centralizers and normalizers. We shall now introduce several examples of Lie subgroups that will be used later on. For A a subset of G and $\mathfrak{a}$ a subset of $\mathfrak{g}$ there are four notions of ***centralizer*** of A and $\mathfrak{a}$ in G or $\mathfrak{g}$, as the case may be:

$$Z(A) := \{g \in G \mid ga = ag \text{ for all } a \in A\} \tag{2.1.10}$$
$$Z(\mathfrak{a}) := \{g \in G \mid \mathrm{Ad}_g \zeta = \zeta \text{ for all } \zeta \in \mathfrak{a}\} \tag{2.1.11}$$
$$\mathfrak{z}(A) := \{\xi \in \mathfrak{g} \mid \mathrm{Ad}_a \xi = \xi \text{ for all } a \in A\} \tag{2.1.12}$$
$$\mathfrak{z}(\mathfrak{a}) := \{\xi \in \mathfrak{g} \mid [\xi, \zeta] = 0 \text{ for all } \zeta \in \mathfrak{a}\}. \tag{2.1.13}$$

If $A = G$ or $\mathfrak{a} = \mathfrak{g}$ they are called ***centers***. The centralizers $Z(A)$ and $Z(\mathfrak{a})$ are both closed Lie subgroups of G, $\mathfrak{z}(A)$ and $\mathfrak{z}(\mathfrak{a})$ are both Lie subalgebras of $\mathfrak{g}$, the Lie algebra of $Z(\mathfrak{a})$ is $\mathfrak{z}(\mathfrak{a})$, and the Lie algebra of $Z(A)$ is $\mathfrak{z}(A)$. If, in addition, A is a Lie subgroup of G whose Lie algebra is $\mathfrak{a}$, then we also have the inclusions $Z(A) \subset Z(\mathfrak{a})$ and $\mathfrak{z}(A) \subset \mathfrak{z}(\mathfrak{a})$. Moreover, if A is a connected Lie subgroup of G with Lie algebra $\mathfrak{a}$, then $Z(A) = Z(\mathfrak{a})$, $\mathfrak{z}(A) = \mathfrak{z}(\mathfrak{a})$, and the Lie algebra of $Z(A)$ is $\mathfrak{z}(\mathfrak{a})$.

Let again A be a subset of G and $\mathfrak{a}$ a subset of $\mathfrak{g}$. There are three notions of ***normalizer***

$$N(A) := \{g \in G \mid gA = Ag\} \tag{2.1.14}$$
$$N(\mathfrak{a}) := \{g \in G \mid \mathrm{Ad}_g \mathfrak{a} = \mathfrak{a}\} \tag{2.1.15}$$
$$\mathfrak{n}(\mathfrak{a}) := \{\xi \in \mathfrak{g} \mid [\xi, \mathfrak{a}] \subset \mathfrak{a}\}. \tag{2.1.16}$$

If A is a closed subset of G, then $N(A)$ is a closed Lie subgroup of G. If $\mathfrak{a}$ is a vector subspace of $\mathfrak{g}$, then $N(\mathfrak{a})$ is a closed Lie subgroup of G whose Lie algebra is $\mathfrak{n}(\mathfrak{a})$. If A is a connected Lie subgroup (not necessarily closed) of G whose Lie algebra is $\mathfrak{a}$, then $N(A) = N(\mathfrak{a})$ is a closed Lie subgroup of G containing the closure of A whose Lie algebra is $\mathfrak{n}(\mathfrak{a})$. In the absence of connectedness all that can be said is that Lie($N(A)$) is a subalgebra of $\mathfrak{n}(\mathfrak{a})$. In general Lie($N(A)$) and $\mathfrak{n}(\mathfrak{a})$ do not coincide. If $\mathfrak{a} = \mathfrak{n}(\mathfrak{a})$ and if A is the topological subgroup of G generated by $\exp \mathfrak{a}$, then A is a closed Lie subgroup of G which coincides with the connected component of the identity of $N(A)$.

The Lie subgroup H of G is ***normal*** if $N(H) = G$. Thus, the normalizer of a subgroup K of G is the largest subgroup in which K is normal. The kernel of any Lie group homomorphism is a normal subgroup. The connected component of the identity element G^0 is normal in G. The connected components of G are the cosets $gG^0 = G^0 g$, $g \in G$. The vector subspace $\mathfrak{h} \subset \mathfrak{g}$ is an ***ideal*** of $\mathfrak{g}$ is $\mathfrak{n}(\mathfrak{h}) = \mathfrak{g}$. Any ideal is automatically a Lie subalgebra. Thus, the normalizer of a subalgebra $\mathfrak{k}$ is the largest subalgebra of $\mathfrak{g}$ that contains $\mathfrak{k}$ as an ideal. The bijective correspondence between Lie subalgebras and connected Lie subgroups induces a similar one between ideals and connected normal subgroups. More precisely, if H is a Lie subgroup of G with Lie algebra $\mathfrak{h}$, then the connected component of the identity element H^0 is normal in G^0 if and only if $\mathfrak{h}$ is an ideal in $\mathfrak{g}$ which in turn is equivalent to $N(\mathfrak{h}) = G^0$. If $\mathfrak{h}$ is an ideal in $\mathfrak{g}$, then the connected subgroup generated by $\exp \mathfrak{h}$ is normal in G^0 and its Lie algebra is $\mathfrak{h}$. If $H \subset G$ is a normal Lie subgroup, then H^0 is also a normal Lie subgroup of G. Note also that the center $Z(G)$ of a connected Lie group G is a normal subgroup whose Lie algebra is the center $\mathfrak{z}(\mathfrak{g})$ of $\mathfrak{g}$.

If H and K are subgroups of G such that $H \subset K \subset G$, we will denote by

$$\begin{aligned} N(H) &= \{n \in G \mid nHn^{-1} = H\} \quad \text{and} \\ N_K(H) &= \{n \in K \mid nHn^{-1} = H\} = N(H) \cap K \end{aligned}$$

the ***normalizers*** of H in G and K, respectively. The following lemmas will be of importance in what follows.

2.1.13 Lemma *Let K be a Lie group and let N a normal Lie subgroup with corresponding Lie algebras $\mathfrak{k}$ and $\mathfrak{n}$. Then for all $\xi \in \mathfrak{k}$ and all $n \in N$, we have*

$$\mathrm{Ad}_n \xi - \xi \in \mathfrak{n}\,.$$

Conversely, let K be a Lie group and H a Lie subgroup with corresponding Lie algebras $\mathfrak{k}$ and $\mathfrak{h}$. If $\xi \in \mathfrak{k}$ satisfies

$$\mathrm{Ad}_h \xi - \xi \in \mathfrak{h}$$

for all $h \in H$, then ξ lies in the Lie algebra of the normalizer $N(H)$ of H.

Proof. Let $\mathrm{AD}_n : K \to K$ denote the inner automorphism for $n \in N$, defined by $\mathrm{AD}_n(m) = nmn^{-1}$. Since Ad_n is the derivative of AD_n at the identity, we get for any $\xi \in \mathfrak{k}$,

$$\begin{aligned}\mathrm{Ad}_n \xi - \xi &= \left.\frac{d}{dt}\right|_{t=0} \mathrm{AD}_n(\exp t\xi)\exp(-t\xi)\\ &= \left.\frac{d}{dt}\right|_{t=0} (n(\exp t\xi)n^{-1})\exp(-t\xi)\\ &= \left.\frac{d}{dt}\right|_{t=0} n[\exp(t\xi)n^{-1}\exp(-t\xi)].\end{aligned}$$

Since N is a normal subgroup of K, $\exp(t\xi)n^{-1}\exp(-t\xi)$ is a curve in N (passing through the point n^{-1} at $t = 0$), so the result is some element in $\mathfrak{n}$.

Conversely, let H be a closed subgroup of K and assume that $\mathrm{Ad}_h \xi - \xi \in \mathfrak{h}$ for all $h \in H$. By taking the derivative relative to h at the identity, it follows that $[\eta, \xi] \in \mathfrak{h}$ for all $\eta \in \mathfrak{h}$. By the Baker–Campbell–Hausdorff formula (2.1.3), $\exp(t\mathrm{Ad}_h\xi)\exp(-t\xi) = \exp(t(\mathrm{Ad}_h \xi - \xi) + O(t^2))$, where $O(t^2)$ is a convergent series each of whose terms is some iterated bracket of $\mathrm{Ad}_h \xi$ and ξ in some order, but always applied to $[\mathrm{Ad}_h \xi, \xi] = [\mathrm{Ad}_h \xi - \xi, \xi] \in \mathfrak{h}$ since $\mathrm{Ad}_h \xi - \xi \in \mathfrak{h}$. Thus each time one takes a bracket with ξ the result is in $\mathfrak{h}$, and each time one takes the bracket with $\mathrm{Ad}_h \xi$, one adds and subtracts a ξ to get two terms: the first, a bracket with ξ which lies in $\mathfrak{h}$, the second, a bracket with $\mathrm{Ad}_h \xi - \xi \in \mathfrak{h}$, which again lies in $\mathfrak{h}$, because both elements in the bracket are in $\mathfrak{h}$. The conclusion is that each term in this series lies in $\mathfrak{h}$ and hence the sum of the series is in $\mathfrak{h}$. Therefore, $h\exp(t\xi)h^{-1}\exp(-t\xi) = \exp(t\,\mathrm{Ad}_h \xi)\exp(-t\xi) = \exp(t(\mathrm{Ad}_h\xi - \xi) + O(t^2)) \in H$ for all $h \in H$ and all $t \in \mathbb{R}$. Therefore $\exp(t\xi)H\exp(-t\xi) \subset H$ for all $t \in \mathbb{R}$ which says that $\exp(t\xi) \in N(H)$ for all $t \in \mathbb{R}$, that is, ξ is in the Lie algebra of $N(H)$. ■

2.1.14 Lemma *Let H be a compact subgroup of a Lie group G. Then any of the inclusions $gHg^{-1} \subset H$ or $H \subset gHg^{-1}$ with $g \in G$ is equivalent to the identity $gHg^{-1} = H$.*

Proof. Clearly, it is enough to prove the result for just one of the inclusions. If $gHg^{-1} \subset H$, we can construct a descending sequence

$$H \supset gHg^{-1} \supset g^2Hg^{-2} \supset g^3Hg^{-3} \ldots .$$

If this sequence stops, let's say at the nth step, that is, $g^{n+1}Hg^{-n-1} = g^nHg^{-n}$, then the lemma follows. The latter is true because of the compactness of H. Indeed, $g^{i+1}Hg^{-i-1}$ is obtained from g^iHg^{-i} by conjugation by g. Conjugation is a group isomorphism and so $\dim(g^{i+1}Hg^{-i-1}) = \dim(g^iHg^{-i}) = \dim H$. Thus, $g^{i+1}Hg^{-i-1}$ is an open Lie subgroup of H and therefore also closed, which in turn implies that $g^{i+1}Hg^{-i-1}$ is made of connected components of H. At the same time, the inclusion $g^{i+1}Hg^{-i-1} \subset g^iHg^{-i}$ implies that g^iHg^{-i} has strictly more connected components of H than $g^{i+1}Hg^{-i-1}$. However, since H is compact, it has only a finite number of connected components, so the strict inclusions have to stop after a finite number of steps. ■

2.1.15 Commutators. The ***commutator subgroup*** is the subgroup (G, G) generated by the subset $\{ghg^{-1}h^{-1} \mid g, h \in G\} \subset G$. Equivalently, (G, G) is the smallest normal subgroup H of G such that G/H is commutative. Given a Lie algebra $\mathfrak{g}$, the ***commutator Lie subalgebra*** $[\mathfrak{g}, \mathfrak{g}]$ is the Lie subalgebra of $\mathfrak{g}$ generated by the subset $\{[\xi, \eta] \mid \xi, \eta \in \mathfrak{g}\}$. It is the smallest ideal $\mathfrak{h}$ of $\mathfrak{g}$ such that $\mathfrak{g}/\mathfrak{h}$ is Abelian. For a connected Lie group G with Lie algebra $\mathfrak{g}$, the commutator subgroup (G, G) is a connected Lie subgroup whose Lie algebra is the commutator Lie algebra $[\mathfrak{g}, \mathfrak{g}]$. If G is, in addition, simply connected, then (G, G) is closed. This implies that for a connected Lie group G, $\mathfrak{g} = [\mathfrak{g}, \mathfrak{g}]$ if and only if $(G, G) = G$.

2.1.16 Automorphisms. Another family of examples are Lie groups and Lie algebras naturally associated to a given Lie group G and its Lie algebra $\mathfrak{g}$. We have already encountered the first such object before: ignoring the Lie algebra structure on $\mathfrak{g}$, the set of all linear isomorphisms $\mathrm{GL}(\mathfrak{g})$ of $\mathfrak{g}$ is an open subset of the vector space of all linear maps $\mathfrak{gl}(\mathfrak{g})$ on $\mathfrak{g}$. Thus, $\mathrm{GL}(\mathfrak{g})$ is a Lie group with Lie algebra $\mathfrak{gl}(\mathfrak{g})$ and whose exponential map is the usual exponential of operators. The set of all ***automorphisms*** $\mathrm{Aut}(\mathfrak{g})$ of the Lie algebra $\mathfrak{g}$ (that is, all Lie algebra isomorphisms $\mathfrak{g} \to \mathfrak{g}$) is a closed subgroup, and thus a Lie subgroup of $\mathrm{GL}(\mathfrak{g})$; it is not connected in general. Its Lie algebra $\mathfrak{aut}(\mathfrak{g})$ consists of all ***derivations*** on $\mathfrak{g}$, that is, of all linear maps $\delta \in \mathfrak{gl}(\mathfrak{g})$ satisfying

$$\delta[\xi, \eta] = [\delta\xi, \eta] + [\xi, \delta\eta]$$

for all $\xi, \eta \in \mathfrak{g}$. Since $\mathrm{ad} : \xi \in \mathfrak{g} \mapsto \mathrm{ad}_\xi \in \mathfrak{aut}(\mathfrak{g})$ is a Lie algebra homomorphism (that is, $\mathrm{ad}_{[\xi,\eta]} = [\mathrm{ad}_\xi, \mathrm{ad}_\eta]$, which is a reformulation of the Jacobi identity), its range

$$\mathrm{ad}\, \mathfrak{g} = \{\mathrm{ad}_\xi \mid \xi \in \mathfrak{g}\}$$

is a Lie subalgebra of $\mathfrak{aut}(\mathfrak{g})$. The identity $[\delta, \mathrm{ad}_\xi] = \mathrm{ad}_{\delta(\xi)}$ for any derivation δ, shows that $\mathrm{ad}\, \mathfrak{g}$ is an ideal of $\mathfrak{aut}(\mathfrak{g})$. Elements of $\mathrm{ad}\, \mathfrak{g}$ are called ***inner derivations*** of $\mathfrak{g}$. Thus, there exists a unique connected normal Lie subgroup, the ***adjoint group***, $\mathrm{Ad}(\mathfrak{g}) \subset \mathrm{Aut}(\mathfrak{g})$ whose Lie algebra is $\mathrm{ad}\, \mathfrak{g}$; $\mathrm{Ad}(\mathfrak{g})$ is in general not closed in $\mathrm{Aut}(\mathfrak{g})$ or in $\mathrm{GL}(\mathfrak{g})$. The Lie group $\mathrm{Ad}(\mathfrak{g})$ is generated by the Lie algebra automorphisms e^{ad_ξ} for all $\xi \in \mathfrak{g}$ and its elements are called ***inner automorphisms*** of $\mathfrak{g}$.

Given a Lie group G, the map $\mathrm{Ad} : g \in G \mapsto \mathrm{Ad}_g \in \mathrm{Aut}(\mathfrak{g})$ is a Lie group homomorphism and $T_e \mathrm{Ad} = \mathrm{ad}$ by (2.1.5). Thus $T_e \mathrm{Ad} : \mathfrak{g} \to \mathrm{ad}\, \mathfrak{g}$ is surjective. Since $\mathrm{Ad} : G \to \mathrm{Ad}\, \mathfrak{g}$ is a Lie group homomorphism, its derivative at every point is also surjective, that is, Ad is a submersion and hence an open map. Therefore $\mathrm{Ad}(G^0)$ is a connected, open, and thus closed, subgroup of the connected Lie group $\mathrm{Ad}\, \mathfrak{g}$, hence equal to it: $\mathrm{Ad}(G^0) = \mathrm{Ad}\, \mathfrak{g}$. Using (2.1.7) in a neighborhood of the origin where the exponential map is a diffeomorphism, it follows that $\ker(\mathrm{Ad}|_{G^0}) = \ker \mathrm{Ad} \cap G^0$ is the center $Z_{G^0}(G^0) := \{g \in G^0 \mid gh = hg \text{ for all } h \in G^0\}$ of G^0, which is closed and is hence a normal Lie subgroup of G^0. It follows that $G^0/Z_{G^0}(G^0)$ is isomorphic to $\mathrm{Ad}\, \mathfrak{g}$, the isomorphism being induced by Ad. The Lie group structure of the quotient will be described in the next section.

For a Lie group G let $\mathrm{Aut}(G)$ denote the group of ***Lie group automorphisms*** of G (that is, all Lie group isomorphisms $G \to G$) and let $\mathrm{AD}(G) := \{g \longmapsto hgh^{-1} \mid h \in G\}$ be the normal subgroup of ***inner automorphisms*** of G. If G is connected, the map that associates to each element of $\mathrm{Aut}(G)$ its tangent map at the identity, which is an element of $\mathrm{Aut}(\mathfrak{g})$, is an injective group homomorphism whose image is a closed subgroup of $\mathrm{Aut}(\mathfrak{g})$. Thus, this image is naturally a Lie subgroup of $\mathrm{Aut}(\mathfrak{g})$ and therefore $\mathrm{Aut}(G)$ becomes a Lie group isomorphic, by definition, with its image in $\mathrm{Aut}(\mathfrak{g})$. Since for a connected Lie group G, $\mathrm{AD}(G)$ is isomorphic to $G/Z(G)$, it follows that $\mathrm{AD}(G)$ is also isomorphic to the adjoint group $\mathrm{Ad}(\mathfrak{g})$.

2.1.17 The universal covering group. If G is a connected Lie group, the universal covering manifold $\tilde{G}$ discussed in §1.1.26 is also a Lie group relative to the group operation induced by pointwise group multiplication of paths on the quotient; it is called the ***universal covering group*** of G. Let $\tilde{\pi} : \tilde{G} \to G$ be the covering projection that sends a class $[\gamma] \in \tilde{G}$ to $\gamma(1)$. There is an explicit construction of this Lie group that is quite interesting and will be recalled below. The space $\mathcal{P}^k_e G$ of C^k paths starting at the identity element is a Banach Lie group, called the ***path group*** of G, relative to pointwise multiplication $(\gamma_1\gamma_2)(t) = \gamma_1(t)\gamma_2(t)$. The ***loop group*** $\mathcal{L}^k G := \{\gamma \in \mathcal{P}^k_e G \mid \gamma(e) = e\}$ of G is a closed normal Banach Lie subgroup of $\mathcal{P}^k_e G$ whose connected component $(\mathcal{L}^k G)^0$ containing the identity (that is, the constant path equal to the identity element e) consists of loops homotopic to the constant path e. Then $\tilde{G}$ is isomorphic to $\mathcal{P}^k_e G/(\mathcal{L}^k G)^0$, so one usually takes it to be the universal covering group of G. The projection $\tilde{\pi} : \tilde{G} \to G$ is a Lie group homomorphism. Of course, as in the case of manifolds, one can set $k = 0$ in the above considerations, since the quotient on the right-hand side is independent of k. In addition, as groups, $\ker \tilde{\pi} = \pi_1(G, e)$, the fundamental group of G, and $\pi_1(G, e) \subset Z(G)$; in particular the fundamental group is a discrete commutative subgroup of $\tilde{G}$. The Lie group G is isomorphic to $\tilde{G}/\pi_1(G, e)$ (the Lie group structure on the quotient will be discussed later on).

If K is a compact connected Lie group with discrete center, then its universal covering group is also compact, a result due to Weyl. A compact connected Lie group admits a finite covering group that is isomorphic to the product of a torus with a compact simply connected semisimple Lie group.

If G is a connected simply connected Lie group, then not only does the first de Rham cohomology group $H^1(G, \mathbb{R})$ vanish but also its second one: $H^2(G, \mathbb{R}) = 0$.

Lie's Third Fundamental Theorem states that for any finite-dimensional real (complex) Lie algebra $\mathfrak{g}$ there exists a unique (up to isomorphisms) connected simply connected real (complex) Lie group G whose Lie algebra equals $\mathfrak{g}$. In addition, the exponential map induces an isomorphism between the center of $\mathfrak{g}$ and the component of the identity of the center of $\tilde{G}$. This connected and simply connected Lie group $\tilde{G}$ is the universal covering Lie group of a matrix group. Indeed, ***Ado's Theorem*** states that any finite-dimensional Lie algebra is isomorphic to a subalgebra of $\mathfrak{gl}(n, \mathbb{R})$ for some $n \in \mathbb{N}$. Thus there is a connected Lie subgroup of $\mathrm{GL}(N, \mathbb{R})$ with Lie algebra $\mathfrak{g}$ and its universal covering group $\tilde{G}$ is the connected simply connected Lie group whose Lie algebra is $\mathfrak{g}$. It should be noted that there is no analogue of Ado's Theorem for Lie groups. For example, the universal covering group of $\mathrm{SL}(2, \mathbb{R})$ is not isomorphic to a matrix group. If G and H are Lie groups with Lie algebras $\mathfrak{g}$ and $\mathfrak{h}$ and G is connected and simply connected, then for any Lie algebra homomorphism $\phi : \mathfrak{g} \to \mathfrak{h}$ there is a unique Lie group homomorphism $f : G \to H$ such that $T_e f = \phi$. Thus, a connected and simply connected Lie group is uniquely determined by its Lie algebra.

If G is a connected simply connected Lie group the map that assigns to each element of $\mathrm{Aut}(G)$ its tangent map at the identity, which is an element of $\mathrm{Aut}(\mathfrak{g})$, is an isomorphism of groups. This induces a natural isomorphism of Lie groups $\mathrm{Aut}(G)/\mathrm{AD}(G) \simeq \mathrm{Aut}(\mathfrak{g})/\mathrm{Ad}(\mathfrak{g})$.

The topology of general connected Lie groups turns out to be that of their maximal compact subgroups. More precisely, any compact subgroup is contained in a maximal compact subgroup that is necessarily connected. All maximal compact subgroups of a connected Lie group are conjugate. In addition, any connected Lie group is diffeomorphic to the product of a maximal compact subgroup with a vector space (which is obtained by exponentiating Abelian subalgebras, but not just one).

2.1.18 Lie algebra cohomology. We say that a differential k-form $\omega \in \Omega^k(G)$ is left invariant when $L_g^*\omega = \omega$, for any $g \in G$. Analogously, a vector field $X \in \mathfrak{X}(G)$ is left invariant when $L_g^*X = X$. As we already pointed out, the set of left invariant vector fields on G can be identified with the Lie algebra $\mathfrak{g}$ of G. Moreover, left invariant k-forms on G are completely determined by its values on the left invariant vector fields which allows us to identify them with the set $\Lambda^k(\mathfrak{g})$ of k-forms on $\mathfrak{g}$. The exterior derivative $\mathbf{d}$ on $\Omega^k(G)$ induces a derivative on $\Lambda^k(\mathfrak{g})$ given by the formula

$$\begin{aligned}\mathbf{d}\omega(\xi_0, \xi_1, \ldots, \xi_k) = \\ \sum_{0\leq i<j\leq k} (-1)^{i+j}\omega([\xi_i, \xi_j], \xi_0, \ldots, \hat{\xi}_i, \ldots, \hat{\xi}_j, \ldots, \xi_k),\end{aligned} \tag{2.1.17}$$

where $\omega \in \Lambda^k(\mathfrak{g})$ and the symbol $\hat{\xi}_r$ means that ξ_r is deleted. In particular, if

(i) $\omega \in \Lambda^0(\mathfrak{g}) \simeq \mathbb{R}$, then $\mathbf{d}\omega = 0$.

(ii) $\omega \in \Lambda^1(\mathfrak{g}) \simeq \mathfrak{g}^*$, then $\mathbf{d}\omega(\xi, \eta) = -\omega([\xi, \eta])$, for any $\xi, \eta \in \mathfrak{g}$.

(iii) $\omega \in \Lambda^2(\mathfrak{g})$, then $\mathbf{d}\omega(\xi, \eta, \zeta) = -(\omega([\xi, \eta], \zeta) + \omega([\zeta, \xi], \eta) + \omega([\eta, \zeta], \xi))$, for any $\xi, \eta, \zeta \in \mathfrak{g}$.

The space of closed k-forms on $\mathfrak{g}$ is denoted by $Z^k(\mathfrak{g}, \mathbb{R})$ and the exact k-forms by $B^k(\mathfrak{g}, \mathbb{R})$. The elements in $Z^k(\mathfrak{g}, \mathbb{R})$ are referred to as ***k-algebra cocycles*** and those in $B^k(\mathfrak{g}, \mathbb{R})$ as ***k-algebra coboundaries***. The quotient space $H^k(\mathfrak{g}, \mathbb{R}) := Z^k(\mathfrak{g}, \mathbb{R})/B^k(\mathfrak{g}, \mathbb{R})$ of closed k-forms by the exact k-forms is called the kth-***cohomology*** of the Lie algebra $\mathfrak{g}$. For example, the first cohomology of $\mathfrak{g}$ is very easy to compute in view of **(i)** and **(ii)** above. Indeed,

$$H^1(\mathfrak{g}, \mathbb{R}) = [\mathfrak{g}, \mathfrak{g}]^\circ \simeq \mathfrak{g}/[\mathfrak{g}, \mathfrak{g}]. \tag{2.1.18}$$

If $\mathfrak{g}$ is Abelian, a quick inspection of (2.1.17) reveals that $\mathbf{d} = 0$ and hence $H^k(\mathfrak{g}, \mathbb{R}) = \Lambda^k(\mathfrak{g})$. Also, if $\mathfrak{g}$ is semisimple the ***First and Second Whitehead lemmas*** guarantee that $H^1(\mathfrak{g}, \mathbb{R}) = 0$ and $H^2(\mathfrak{g}, \mathbb{R}) = 0$, respectively. Finally, if G is a compact group with Lie algebra $\mathfrak{g}$, the Lie algebra cohomology of $\mathfrak{g}$ coincides with the de Rham cohomology of G.

2.1.19 Lie group cohomology. Let G be a Lie group and $\rho : G \to \mathrm{GL}(V)$ a group homomorphism, where V is some vector space (that is, ρ is a ***representation*** of G on V). A ***V-valued one-cocycle*** of G is a map $\sigma : G \longrightarrow V$ such that the ***cocycle identity***

$$\sigma(gh) = \sigma(g) + \rho(g)\sigma(h)$$

holds. Taking in this identity $g = h = e$ implies

$$\sigma(e) = 0. \tag{2.1.19}$$

A V-cocycle Δ is said to be a ***one-coboundary*** if there is a $v \in V$ such that

$$\Delta(g) = v - \rho(g)v.$$

The set of one-cocycles of a group forms a vector space and the one-coboundaries are a vector subspace of it. The quotient space is called the ***first V-valued cohomology*** of G. Even though we will not use it in our work, the reader should know that a kth-Lie group cohomology is available, for any natural number k.

Whitehead's Lemma for groups (see for instance JACOBSON (1979)) proves that if G is a semisimple Lie group, then every V-cocycle is a coboundary.

The following result is due to SOURIAU (1969)[1] and provides a fairly large category of situations in which a Lie algebra two-cocycle can be seen as the infinitesimal version of a $\mathfrak{g}^*$-valued group one-cocycle.

2.1.20 Proposition *Let $\sigma : G \to \mathfrak{g}^*$ be a $\mathfrak{g}^*$-valued differentiable group one-cocycle relative to the coadjoint representation of G on $\mathfrak{g}^*$. Define the map $\Sigma : \mathfrak{g} \times \mathfrak{g} \to \mathbb{R}$ by $\Sigma(\xi, \eta) := \langle T_e\sigma(\xi), \eta\rangle$, for any $\xi, \eta \in \mathfrak{g}$. If the bilinear map Σ is skew symmetric, then it is a Lie algebra two-cocycle, that is, it satisfies the satisfies the* ***two-cocycle identity***

$$\Sigma([\xi, \eta], \zeta) + \Sigma([\eta, \zeta], \xi) + \Sigma([\zeta, \xi], \eta) = 0. \tag{2.1.20}$$

Proof. It can be easily verified that the skew symmetry of Σ is equivalent to

$$T_e\sigma = -T_e^*\sigma. \tag{2.1.21}$$

Note that by definition and by (2.1.21)

$$\begin{aligned}
&\Sigma([\xi, \eta], \zeta) + \Sigma([\eta, \zeta], \xi) + \Sigma([\zeta, \xi], \eta) \\
&\qquad = \langle \mathrm{ad}_\eta^* T_e\sigma(\xi), \zeta\rangle - \langle \mathrm{ad}_\xi^* T_e\sigma(\eta), \zeta\rangle + \langle T_e\sigma(\zeta), [\xi, \eta]\rangle \\
&\qquad = \langle \mathrm{ad}_\eta^* T_e\sigma(\xi), \zeta\rangle - \langle \mathrm{ad}_\xi^* T_e\sigma(\eta), \zeta\rangle - \langle T_e\sigma([\xi, \eta]), \zeta\rangle,
\end{aligned}$$

for any $\xi, \eta, \zeta \in \mathfrak{g}$. Therefore the two-cocycle identity holds if and only if

$$T_e\sigma([\xi, \eta]) = \mathrm{ad}_\eta^* T_e\sigma(\xi) - \mathrm{ad}_\xi^* T_e\sigma(\eta), \qquad \text{for any } \xi, \eta \in \mathfrak{g}. \tag{2.1.22}$$

In order to prove this we will use the fact that σ is a $\mathfrak{g}^*$-valued group one-cocycle

[1] We thank J. Huebschmann for letting us know about this result.

relative to the coadjoint representation of G on $\mathfrak{g}^*$. Indeed,

$$\begin{aligned}
T_e\sigma([\xi,\eta]) = T_e\sigma(\mathrm{ad}_\xi\eta) &= \left.\frac{d}{dt}\right|_{t=0}\left.\frac{d}{ds}\right|_{s=0}\sigma(\exp t\xi \exp s\eta \exp(-t\xi)) \\
&= \left.\frac{d}{dt}\right|_{t=0}\left.\frac{d}{ds}\right|_{s=0}\sigma(\exp t\xi) + \mathrm{Ad}^*_{\exp(-t\xi)}\sigma(\exp s\eta \exp(-t\xi)) \\
&= \left.\frac{d}{dt}\right|_{t=0}\left.\frac{d}{ds}\right|_{s=0}\mathrm{Ad}^*_{\exp(-t\xi)}\sigma(\exp s\eta) + \mathrm{Ad}^*_{\exp(-t\xi)}\mathrm{Ad}^*_{\exp(-s\eta)}\sigma(\exp(-t\xi)) \\
&= -\mathrm{ad}^*_\xi T_e\sigma(\eta) - \left.\frac{d}{dt}\right|_{t=0}\mathrm{Ad}^*_{\exp(-t\xi)}\mathrm{ad}^*_\eta\sigma(\exp(-t\xi)) \\
&= -\mathrm{ad}^*_\xi T_e\sigma(\eta) + \mathrm{ad}^*_\xi\mathrm{ad}^*_\eta\sigma(e) + \mathrm{ad}^*_\eta T_e\sigma(\xi) \\
&= -\mathrm{ad}^*_\xi T_e\sigma(\eta) + \mathrm{ad}^*_\eta T_e\sigma(\xi),
\end{aligned}$$

where the last equality follows from (2.1.19). This chain of equalities proves (2.1.22), as required. ■

2.2 Actions of Lie groups

This section presents the elementary facts about smooth Lie group actions and quotients that will be needed later in the text.

2.2.1 Definition *Let M be a manifold and G a Lie group. A* ***left action*** *of G on M is a smooth mapping $\Phi : G \times M \to M$ such that*

(i) *$\Phi(e, z) = z$, for all $z \in M$ and*

(ii) *$\Phi(g, \Phi(h, z)) = \Phi(gh, z)$ for all $g, h \in G$ and $z \in M$.*

To simplify notation, we will often write in what follows $g \cdot z := \Phi(g, z) := \Phi_g(z) := \Phi^z(g)$. For any $g \in G$, the map $\Phi_g := \Phi(g, \cdot) : M \to M$ is a diffeomorphism of M with inverse $\Phi_{g^{-1}}$. We will denote by A_G the subgroup of the diffeomorphism group $\mathrm{Diff}(M)$ of M associated to the G-action on M, that is,

$$A_G := \{\Phi_g \mid g \in G\}.$$

The triple (M, G, Φ) is called a ***G-space*** or a ***G-manifold***.

A ***right action*** is a map $\Psi : M \times G \to M$, such that $\Psi(z, e) = z$, for all $z \in M$, and $\Psi(\Psi(z, g), h) = \Psi(z, gh)$ for all $g, h \in G$ and $z \in M$. In this case, the shorthand notation for the action will be $z \cdot g := \Psi(z, g)$. Most of the theoretical developments in the coming pages will be carried out for left actions; the analogues for right actions are left most of the time as exercises for the reader.

Given a G-space, we can obtain additional group actions by restriction. This procedure is spelled out in the following statement which is a trivial consequence of the definition of initial submanifold.

2.2.2 Lemma *Let (M, G, Φ) be a G-space, H a Lie subgroup of G, and N an H-invariant initial submanifold of M. The map $\phi : H \times N \to N$ given by $\phi(h, n) = \Phi(h, n)$ makes (N, H, ϕ) into an H-space.*

2.2.3 Examples of group actions.

(i) Translation and conjugation. The ***left translation*** on G defined by $g \in G$, that is, the map $L_g : G \to G$, given by $h \mapsto gh$, induces a left action of G on itself. ***Right translation***, $R_g : G \to G$, $h \mapsto hg$, defines a right action. The ***inner automorphism*** $\mathrm{AD}_g \equiv I_g : G \to G$, given by $I_g := R_{g^{-1}} \circ L_g$ defines a left action of G on itself called ***conjugation***.

(ii) Adjoint and coadjoint action. The differential at the identity of the conjugation mapping defines a linear left action of G on $\mathfrak{g}$ called the ***adjoint representation*** of G on $\mathfrak{g}$

$$\mathrm{Ad}_g := T_e I_g : \mathfrak{g} \longrightarrow \mathfrak{g}.$$

If $\mathrm{Ad}_g^* : \mathfrak{g}^* \to \mathfrak{g}^*$ is the dual of Ad_g, then the map

$$\begin{aligned} \Phi : G \times \mathfrak{g}^* &\longrightarrow \quad \mathfrak{g}^* \\ (g, \nu) &\longmapsto \mathrm{Ad}_{g^{-1}}^* \nu, \end{aligned}$$

defines also a linear left action of G on $\mathfrak{g}^*$ called the ***coadjoint representation*** of G on $\mathfrak{g}^*$.

(iii) Group representation. If in Definition 2.2.1 the manifold M is a vector space V and G acts linearly on V, that is, $\Phi_g \in \mathrm{GL}(V)$ for all $g \in G$, where $\mathrm{GL}(V)$ denotes the group of all linear automorphisms of V, then the action is said to be a ***representation*** of G on V. For example, the adjoint and coadjoint actions of G defined above are representations.

For every representation of G on V, there exists a natural representation $\Phi^* : G \times V^* \to V^*$ of G on the dual vector space V^* of V defined by $(\Phi^*)_g := \Phi_{g^{-1}}^*$, that is,

$$\langle (\Phi^*)_g(\eta), v \rangle := \langle \eta, \Phi_{g^{-1}}(v) \rangle, \qquad \text{for any} \quad g \in G,\ \eta \in V^*,\ v \in V.$$

This action is usually referred to as the ***contragredient representation*** of G on V^*, associated to the linear representation Φ. For example, the coadjoint representation of G on $\mathfrak{g}^*$ is the contragredient representation induced by the adjoint representation of G on $\mathfrak{g}$.

(iv) Tangent lifts of group actions. Let $\Phi : G \times M \to M$ be a smooth Lie group action on the manifold M. The map Φ induces a natural action on the tangent bundle TM of M by

$$g \cdot v_m := T_m \Phi_g \cdot v_m,$$

where $g \in G$ and $v_m \in T_m M$.

(iv) Cotangent lifts of group actions. Let $\Phi : G \times M \to M$ be a smooth Lie group action on the manifold M. The map Φ induces a natural action on the cotangent bundle T^*M of M by

$$g \cdot \alpha_m := T_{g \cdot m}^* \Phi_{g^{-1}} \cdot \alpha_m$$

where $g \in G$ and $\alpha_m \in T_m^* M$.

2.2.4 Semidirect products. Group actions are used to define the notion of semidirect product of two groups. Since there are many conventions in the literature regarding the multiplication we shall specify here those that are used in this book. Let G and H be groups and assume that there is a left action $\alpha : G \times H \to H$ of G on H by group homomorphisms, that is, $\alpha_g : H \to H$ is an endomorphism of H for any $g \in G$. Since α is an action, it follows that α_g is an isomorphism of H for any $g \in G$. As usual, we shall denote $\alpha(g, h) = g \cdot h$.

The ***semidirect product group*** $G{\circledS}H$ is the set $G \times H$ relative to the multiplication

$$(g_1, h_1)(g_2, h_2) := (g_1 g_2, h_1(g_1 \cdot h_2)). \tag{2.2.1}$$

The neutral element is (e_G, e_H), where e_G and e_H are the neutral elements of G and H respectively, and $(g, h)^{-1} = (g^{-1}, g^{-1} \cdot h^{-1})$. The subgroup $\{e_G\} \times H$, which is isomorphic to H, is a normal subgroup of $G{\circledS}H$.

If G and H are Lie groups with Lie algebras $\mathfrak{g}$ and $\mathfrak{h}$, respectively, and the action α is by Lie group isomorphisms of H and is smooth, then $G{\circledS}H$ is a Lie group whose Lie algebra is the semidirect product $\mathfrak{g}{\circledS}\mathfrak{h}$. The underlying vector space of $\mathfrak{g}{\circledS}\mathfrak{h}$ is $\mathfrak{g} \times \mathfrak{h}$ and its bracket is given by

$$[(\xi_1, \eta_1), (\xi_2, \eta_2)] = ([\xi_1, \xi_2], \xi_1 \cdot \eta_2 - \xi_2 \cdot \eta_2 + [\eta_1, \eta_2]), \tag{2.2.2}$$

where $\xi_1, \xi_2 \in \mathfrak{g}$, $\eta_1, \eta_2 \in \mathfrak{h}$, and $\xi_1 \cdot \eta_2, \xi_2 \cdot \eta_1 \in \mathfrak{h}$ denotes the Lie algebra action of $\xi_1, \xi_2 \in \mathfrak{g}$ on $\eta_2, \eta_1 \in \mathfrak{h}$, respectively. This Lie algebra action is naturally induced by α in the following manner (Hochschild (1965)). For every $g \in G$, $\alpha_g : H \to H$ is a Lie group automorphism of H whose derivative $\tilde{\alpha}_g := T_{e_H}\alpha_g : \mathfrak{h} \to \mathfrak{h}$ is a Lie algebra automorphism. In this way one obtains a Lie group homomorphism $\tilde{\alpha} : g \in G \mapsto \tilde{\alpha}_g \in \mathrm{Aut}(\mathfrak{h})$ from G to the Lie group $\mathrm{Aut}(\mathfrak{h})$ of Lie algebra automorphisms of $\mathfrak{h}$, which induces a Lie algebra homomorphism $\phi := T_{e_G}\tilde{\alpha} : \mathfrak{g} \to \mathrm{aut}(\mathfrak{h})$ from $\mathfrak{g}$ to the Lie algebra $\mathrm{aut}(\mathfrak{h})$ of derivations of $\mathfrak{h}$. Now set $\xi \cdot \eta := \phi(\xi)(\eta)$ if $\xi \in \mathfrak{g}$ and $\eta \in \mathfrak{h}$.

Let $\mathrm{Aut}(H)$ denote the Lie group of automorphisms of H and let $\mathrm{Lie}(\mathrm{Aut}(H))$ be its Lie algebra. Then $\mathrm{Lie}(\mathrm{Aut}(H)) \subset \mathfrak{X}(H)$ and every element of $\mathrm{Lie}(\mathrm{Aut}(H))$ vanishes at e_H. Since $\bar{\alpha} : g \in G \mapsto \alpha_g \in \mathrm{Aut}(H)$ is a Lie group homomorphism, the induced Lie algebra homomorphism $\bar{\alpha}' := T_{e_G}\bar{\alpha} : \mathfrak{g} \to \mathrm{Lie}(\mathrm{Aut}(H))$ satisfies $\bar{\alpha}'(\xi) \in \mathfrak{X}(H)$ for every $\xi \in \mathfrak{g}$. With these notations the adjoint action of $G{\circledS}H$ on $\mathfrak{g}{\circledS}\mathfrak{h}$ is given by

$$\mathrm{Ad}_{(g,h)}(\xi, \eta) = \left(\mathrm{Ad}_g \xi,\ \left(\mathrm{Ad}_h \circ \tilde{\alpha}_g\right)(\eta) + T_{h^{-1}}L_h\left(\bar{\alpha}'(\mathrm{Ad}_g \xi)(h^{-1})\right)\right),$$

where $\xi \in \mathfrak{g}$, $\eta \in \mathfrak{h}$, $g \in G$, $h \in H$. To compute the coadjoint action one has to introduce one more notation. Given a Lie algebra homomorphism $F : \mathfrak{g} \to \mathrm{Lie}(\mathrm{Aut}(H))$ and $h \in H$, denote $F^{\#}(h) : \mathfrak{g} \to T_hH$ the linear map given by $F^{\#}(h)(\xi) := F(\xi)(h)$, for any $\xi \in \mathfrak{g}$. Let $F^{\#}(h)^* : T_h^*H \to \mathfrak{g}^*$ denote its dual map. With this notation the coadjoint action of $G{\circledS}H$ on $(\mathfrak{g}{\circledS}\mathfrak{h})^* = \mathfrak{g}^* \times \mathfrak{h}^*$ is given by

$$\begin{aligned}
&\mathrm{Ad}^*_{(g,h)^{-1}}(\mu, \nu) \\
&\quad = \left(\mathrm{Ad}^*_{g^{-1}} \mu + \left(\bar{\alpha}' \circ \mathrm{Ad}_{g^{-1}}\right)^{\#}(g^{-1} \cdot h)^*(T^*_{g^{-1}\cdot h} L_{g^{-1}\cdot h^{-1}} \nu)\right., \\
&\qquad\qquad\qquad\qquad\qquad \left.\tilde{\alpha}^*_{g^{-1}}\, \mathrm{Ad}^*_{g^{-1}\cdot h^{-1}}\, \nu\right),
\end{aligned}$$

where $\mu \in \mathfrak{g}^*$, $\nu \in \mathfrak{h}^*$, $g \in G$, and $h \in H$ (KUPERSHMIDT AND RATIU (1983)).

If the Lie group H is a vector space V regarded as an Abelian Lie group under addition and $\alpha : G \times V \to V$ is a representation, then the Lie algebra of V is V itself, $\tilde{\alpha}_g = \alpha_g$ for all $g \in G$, and $\phi = \bar{\alpha}' : \mathfrak{g} \to \mathfrak{gl}(V)$. In this case, the action of $\mathfrak{g}$ on V is the induced Lie algebra representation, denoted by $\xi \cdot v$ for $\xi \in \mathfrak{g}$ and $v \in V$. The multiplication in $G Ⓢ V$ is

$$(g_1, v_1)(g_2, v_2) = (g_1 g_2, v_1 + g_1 \cdot v_2),$$

$(g, v)^{-1} = (g^{-1}, -g^{-1} \cdot v)$, the Lie bracket in $\mathfrak{g} Ⓢ V$ is

$$[(\xi_1, v_1), (\xi_2, v_2)] = ([\xi_1, \xi_2], \xi_1 \cdot v_2 - \xi_2 \cdot v_1),$$

the adjoint action is given by

$$\mathrm{Ad}_{(g,u)}(\xi, v) = (\mathrm{Ad}_g \xi,\ g \cdot v - (\mathrm{Ad}_g \xi) \cdot u),$$

and the coadjoint action has the expression

$$\mathrm{Ad}^*_{(g,u)^{-1}}(\mu, a) = (\mathrm{Ad}^*_{g^{-1}} \mu + \phi^*_u(g \cdot a), g \cdot a),$$

where $g_1, g_2, g \in G$, $v_1, v_2, u, v \in V$, $\xi_1, \xi_2, \xi \in \mathfrak{g}$, $\mu \in \mathfrak{g}^*$, $a \in V^*$, and where $g \cdot a$ is defined by $\langle g \cdot a, w \rangle = \langle a, g^{-1} \cdot w \rangle$ for any $w \in V$. In addition, the linear map $\phi^*_u : V^* \to \mathfrak{g}^*$ is the dual of the linear map $\phi_u : \mathfrak{g} \to V$ given by $\phi_u(\zeta) := \zeta \cdot u$, for $\zeta \in \mathfrak{g}$.

In general, if $\mathfrak{g}$ is a Lie algebra and $\phi : \mathfrak{g} \to \mathrm{aut}(\mathfrak{h})$ is a Lie algebra homomorphism from $\mathfrak{g}$ to the Lie algebra $\mathrm{aut}(\mathfrak{h})$ of derivations of $\mathfrak{h}$, the ***semidirect product Lie algebra*** is defined by (2.2.2) with $\xi \cdot \eta := \phi(\xi)(\eta)$, for $\xi \in \mathfrak{g}$ and $\eta \in \mathfrak{h}$. The Lie subalgebra $\{0\} \times \mathfrak{h}$, which is isomorphic to $\mathfrak{h}$, is an ideal in $\mathfrak{g} Ⓢ \mathfrak{h}$.

2.2.5 The tangent bundle Lie group and the tangent bundle action. Let G be a Lie group with multiplication given by the map $m : G \times G \to G$. The tangent bundle TG of G is a Lie group whose multiplication law is given by the tangent map $Tm : TG \times TG \to TG$ of m. More explicitly, if $u_g \in T_gG$ and $v_h \in T_hG$, then $u_g \cdot v_h := T_{(g,h)}m(u_g, v_h) = T_g R_h(u_g) + T_h L_g(v_h)$. The identity element is $0_e \in T_eG$ and the inverse u_g^{-1} of $u_g \in T_gG$ is given by $u_g^{-1} = -T_e L_{g^{-1}} T_g R_{g^{-1}}(u_g)$.

The ***right trivialization*** $u_g \in TG \longmapsto (g, T_g R_{g^{-1}}(u_g)) \in G \times \mathfrak{g}$ of TG is a Lie group isomorphism between TG and the semidirect product group $G Ⓢ \mathfrak{g}$, where $\mathfrak{g}$ is the representation space of G by the adjoint action. The ***left trivialization*** $u_g \in TG \longmapsto (g, T_g L_{g^{-1}}(u_g)) \in G \times \mathfrak{g}$ of TG induces another multiplication law on $G \times \mathfrak{g}$ that makes this space isomorphic as a Lie group to $G Ⓢ \mathfrak{g}$.

If $\Phi : G \times M \to M$ is a smooth action of the Lie group G on the manifold M, then the tangent bundle Lie group TG also acts smoothly on M by the ***tangent bundle action***

$$v_g \cdot m := (T_g R_{g^{-1}}(v_g))_M(g \cdot m) = T_m \Phi_g (T_g L_{g^{-1}}(v_g))_M(m),$$

where $g \in G$, $v_g \in T_gG$, and $m \in M$.

2.2.6 Infinitesimal generators. Given a (left) action $\Phi : G \times M \to M$, the ***infinitesimal generator*** $\xi_M \in \mathfrak{X}(M)$ associated to $\xi \in \mathfrak{g}$ is the vector field on Mdefined by

$$\xi_M(m) := \left.\frac{d}{dt}\right|_{t=0} \Phi_{\exp t\xi}(m) = T_e\Phi^m \cdot \xi.$$

The infinitesimal generators are complete vector fields. Indeed, the flow of ξ_M equals $(t, m) \mapsto \exp t\xi \cdot m$. Moreover, the map $\xi \in \mathfrak{g} \mapsto \xi_M \in \mathfrak{X}(M)$ is a ***Lie algebra antihomomorphism***, that is,

(i) $(a\xi + b\eta)_M = a\xi_M + b\eta_M$,

(ii) $[\xi, \eta]_M = -[\xi_M, \eta_M]$,

for all $a, b \in \mathbb{R}$ and $\xi, \eta \in \mathfrak{g}$. The second formula is obtained by taking the derivative at $g = e$ of the identity

$$(\mathrm{Ad}_{g^{-1}}\xi)_M = \Phi_g^*\xi_M, \text{ for any } g \in G,\ \xi \in \mathfrak{g}. \tag{2.2.3}$$

Motivated by these properties one introduces the following definition.

2.2.7 Definition *Let* $\mathfrak{g}$ *be a Lie algebra and* M *a smooth manifold. A* ***right (left) Lie algebra action*** *of* $\mathfrak{g}$ *on* M *is a Lie algebra (anti)homomorphism* $\xi \in \mathfrak{g} \longmapsto \xi_M \in \mathfrak{X}(M)$ *such that the mapping* $(m, \xi) \in M \times \mathfrak{g} \longmapsto \xi_M(m) \in TM$ *is smooth.*

Given a Lie group action, we will refer to the Lie algebra action induced by its infinitesimal generators as the ***associated Lie algebra action***.

2.2.8 Stabilizers and orbits. The ***isotropy subgroup*** or ***stabilizer*** of an element m in the manifold M acted upon by the Lie group G is the closed subgroup

$$G_m := \{g \in G \mid \Phi_g(m) = m\} \subset G$$

whose Lie algebra $\mathfrak{g}_m$ equals

$$\mathfrak{g}_m = \{\xi \in \mathfrak{g} \mid \xi_M(m) = 0\}. \tag{2.2.4}$$

In particular, if one considers the coadjoint representation, the stabilizer G_μ of $\mu \in \mathfrak{g}^*$ is called the ***coadjoint isotropy subgroup*** of G. Its Lie algebra is the ***coadjoint isotropy subalgebra***

$$\mathfrak{g}_\mu := \{\xi \in \mathfrak{g} \mid \mathrm{ad}_\xi^* \mu = 0\}.$$

The coadjoint isotropy subgroups corresponding to connected groups that, at the same time, are either compact or semisimple are connected (see Theorem 3.3.1 in DUISTERMAAT AND KOLK (1999)).

When $\mathfrak{g}$ is a Lie algebra acting on the manifold M we define, in view of (2.2.4), $\mathfrak{g}_m := \{\xi \in \mathfrak{g} \mid \xi_M(m) = 0\}$, for any $m \in M$. We will refer to $\mathfrak{g}_m$ in this context as the ***isotropy subalgebra*** of the point m.

The ***orbit*** $\mathcal{O}_m$ of the element $m \in M$ under the group action Φ is the set

$$\mathcal{O}_m \equiv G \cdot m := \{\Phi_g(m) \mid g \in G\}.$$

In particular, in the case of the adjoint and coadjoint representations, the orbits will be called ***adjoint*** and, respectively, ***coadjoint orbits***. We will study the natural smooth structure of the group orbits in §2.3.11. The isotropy subgroups of the elements in a group orbit are related by the expression

$$G_{g \cdot m} = gG_m g^{-1} \text{ for all } g \in G.$$

The notion of orbit allows the introduction of an equivalence relation in the manifold M, namely, two elements $x, y \in M$ are equivalent if and only if they are in the same G-orbit, that is, if there exists an element $g \in G$ such that $\Phi_g(x) = y$. The space of classes with respect to this equivalence relation is usually referred to as the ***space of orbits*** and, depending on the context, it is denoted by the symbols M/G or M/A_G. If we endow the space of orbits with the quotient topology, the projection $\pi_G : M \to M/G$ is a continuous (see §1.1.18) and open map. The openness of π_G is a consequence of the following argument: if U is an open set in M, $\pi_G(U)$ is open if and only if $\pi_G^{-1}(\pi_G(U))$ is open. Since $\pi_G^{-1}(\pi_G(U)) = G \cdot U = \cup_{g \in G} \Phi_g(U)$, $\pi_G^{-1}(\pi_G(U))$ is a union of open sets and therefore open.

In the next section we will provide sufficient conditions that ensure that the orbit spaces of a group action are regular quotient manifolds.

The group action on M is said to be ***transitive*** if there is only one orbit, ***free*** if the isotropy of every element in M consists only of the identity element, and ***effective*** or ***faithful*** if $\Phi_g = \mathrm{id}_M$ implies that $g = e$ (equivalently, the map $g \mapsto \Phi_g$ is injective). A Lie algebra action on M is ***locally free*** when $\mathfrak{g}_m = \{0\}$, for any $m \in M$.

2.2.9 Example Let G be a Lie group and $H \subset G$ a Lie subgroup. We can use Lemma 2.2.2 to restrict the actions of G on itself by left and right translations to left and right actions, respectively, of H on G. These actions are free and their orbits are given by the cosets Hg, $g \in G$, for the left action, and gH, $g \in G$, for the right action. The respective orbit spaces are usually denoted by the symbols $G\backslash H$ and G/H. In the next section we will show that when H is a closed subgroup of G, then G/H and $G\backslash H$ are regular quotient manifolds, usually referred to as ***homogeneous manifolds***.

2.2.10 Lemma (Leibniz rule for group actions) *Let $m(t)$ be a curve in M and $\xi \in \mathfrak{g}$. Then*

(i) $\frac{d}{dt}\big|_{t=0} \exp t\xi \cdot m(t) = \xi_M(m) + \frac{d}{dt}\big|_{t=0} m(t)$.

(ii) *If $g(t)$ is a curve in G then*

$$\frac{d}{dt} g(t) \cdot m(t) = g(t) \cdot \left(\left(T_{g(t)} L_{g(t)^{-1}} \dot{g}(t)\right)_M (m(t)) + \frac{d}{dt} m(t) \right), \tag{2.2.5}$$

where the dot in the left hand side stands for the lifted G-action on TM.

Proof. The first part of the lemma is a trivial consequence of the decomposition of the differential of a mapping in terms of its partial derivatives and the definition of the infinitesimal generator. The second part follows from the first and the relation

$$T_g \Phi^m \left(T_e L_g \cdot \xi \right) = T_m \Phi_g \left(\xi_M(m) \right),$$

for any $m \in M$, $g \in G$, and $\xi \in \mathfrak{g}$. ■

2.2.11 Invariant and equivariant maps. A subset $S \subset M$ of the G-space (M, G, Φ) is called G-***invariant*** or G-***saturated*** if $\Phi_g(S) = S$, for all $g \in G$ or, equivalently, if it is a union of G-orbits. The ***saturation*** $\mathrm{Sat}(N) \equiv G \cdot N$ of an arbitrary subset N of M is defined as

$$\mathrm{Sat}(N) \equiv G \cdot N := \bigcup_{g \in G} \Phi_g(N).$$

A set is G-saturated if and only if it coincides with its saturation. Lemma 2.2.2 implies in particular that any open G-invariant subset U of (M, G, Φ) is a G-manifold. This fact allows us to define the ***presheaf of smooth G-invariant functions*** $C^\infty_M(\cdot)^G$ on (M, G, Φ) as the map that assigns to each open G-invariant subset U the set of smooth functions $C^\infty_M(U)^G$ defined by

$$C^\infty_M(U)^G := \{f \in C^\infty_M(U) \mid f \circ \Phi_g|_U = f \text{ for all } g \in G\}.$$

The set of global sections $C^\infty_M(M)^G = \{f \in C^\infty_M(M) \mid f \circ \Phi_g = f, \text{ for all } g \in G\}$ of this presheaf will be referred to as the set of smooth G-***invariant functions*** on M. When there is no danger of confusion we will just write $C^\infty(\cdot)^G$.

We can use the remarks on quotient presheaves made in §1.1.19 and the presheaf of smooth G-invariant functions on M to define a presheaf of functions $C^\infty_{M/G}$ on the quotient M/G by assigning, to each open subset $U/G \subset M/G$, U an open G-invariant subset of M, the family of functions $C^\infty_{M/G}(U/G)$ defined by

$$C^\infty_{M/G}(U/G) := \{f \in C^0(U/G) \mid f \circ \pi|_U \in C^\infty(U)^G\},$$

where $\pi : M \to M/G$ is the projection onto the orbit space.

If S is a G-invariant subset of M endowed with a given topology $\mathcal{T}$, then the ***presheaf of Whitney invariant functions*** $C^\infty_{S,M}(\cdot)^G$ on S, induced by $C^\infty_M(\cdot)^G$, assigns to each open G-invariant subset V of S the functions f on V, such that for any $z \in V$, there exists an open G-invariant open neighborhood U_z of z in M and a function $F \in C^\infty_M(U_z)^G$ such that $f|_{U_z \cap V} = F|_{U_z \cap V}$. Given that the projection $\pi : M \to M/G$ is always open we have by Proposition 1.1.20 that

$$C^\infty_{S/G,M/G} = C^\infty_{S/G}(\cdot)^G,$$

where $C^\infty_{S/G}(\cdot)^G$ denotes the quotient presheaf associated to $C^\infty_{S,M}(\cdot)^G$.

A mapping $\varphi : M_1 \to M_2$, between two G-spaces M_1 and M_2 is said to be G-***equivariant*** provided that for any $g \in G$ and $z \in M_1$, the mapping φ satisfies the identity

$$\varphi(g \cdot z) = g \cdot \varphi(z).$$

Using again the statements in §1.1.19 we can easily conclude that if the G-equivariant map $\varphi : M_1 \to M_2$ is smooth, then it drops to a unique smooth map $\hat{\varphi} : (M_1/G, C^\infty_{M_1/G}) \to (M_2/G, C^\infty_{M_2/G})$.

We say that the global section $C^\infty(M)^G$ ***separates the G-orbits*** when the following condition is satisfied: if two orbits $G \cdot x$, $G \cdot y$ are such that $f(G \cdot x) = f(G \cdot y)$ for all $f \in C^\infty(M)^G$, then $G \cdot x = G \cdot y$ necessarily.

We say that the G-action on M has the ***extension property*** when any G-invariant function $f \in C^\infty(U)^G$ defined on any G-invariant open subset U has the following feature: for any $z \in U$, there is a G-invariant open neighborhood $V \subset U$ of z and a G-invariant smooth function $F \in C^\infty(M)^G$ such that $f|_V = F|_V$.

Let S be a G-invariant immersed, initial, or embedded submanifold of M. We say that S has the ***local extension property*** with respect to the G-action when, for any open G-invariant subset V of S and any $f \in C^\infty_S(V)^G$, we have that for any $z \in V$ there exists an open G-invariant neighborhood U_z of z in M and $F \in C^\infty_M(U_z)^G$ such that $f|_{V\cap U_z} = F|_{V\cap U_z}$.

2.2.12 Proposition *Let M be a smooth manifold acted upon by a Lie group G. Let S be an embedded G-invariant submanifold of M such that the orbit space S/G is a regular quotient manifold. Then if any open G-invariant subset of S admits G-invariant partitions of unity subordinate to any cover by G-invariant open sets, we then have that*

$$C^\infty_{S/G,M/G} \subset C^\infty_{S/G}. \tag{2.2.6}$$

Conversely, if S has the local extension property with respect to the G-action, then

$$C^\infty_{S/G} \subset C^\infty_{S/G,M/G}. \tag{2.2.7}$$

Proof. Let $\pi_S : S \to S/G$ be the projection onto the orbit space that, by hypothesis, is a surjective submersion and $f \in C^\infty_{S/G,M/G}(V)$, for some open subset V of S/G. Given that by Proposition 1.1.20 the equality $C^\infty_{S/G,M/G} = C^\infty_{S/G}(\cdot)^G$ holds, we then have that, for any $m \in \pi_S^{-1}(V)$, there exists an open G-invariant neighborhood U_m of m in M and $F \in C^\infty_M(U_m)^G$ such that $f \circ \pi_S|_{\pi_S^{-1}(V)\cap U_m} = F|_{\pi_S^{-1}(V)\cap U_m}$. The subsets of the form $\pi_S^{-1}(V) \cap U_m$ form an open G-invariant covering of $\pi_S^{-1}(V)$. By the hypothesis on the existence of partitions of unity, there exists a locally finite G-invariant open subcover $\{\pi_S^{-1}(V) \cap U_i\}_{i\in I}$ of $\pi_S^{-1}(V)$ and functions $F_i \in C^\infty_M(U_i)^G$ such that $f \circ \pi_S|_{\pi_S^{-1}(V)\cap U_i} = F_i|_{\pi_S^{-1}(V)\cap U_i}$, for any $i \in I$. Let $\{(\rho_i, \pi_S^{-1}(V) \cap U_i)\}_{i\in I}$ be a G-invariant partition of unity subordinate to this covering. The function $F \in C^\infty_S(\pi_S^{-1}(V))^G$ defined by $F := \sum_{i\in I} \rho_i F_i$ is such that $F = f \circ \pi_S|_{\pi_S^{-1}(V)}$ and since π_S is a submersion we have that $f \in C^\infty_{S/G}(V)$, as required.

Suppose now that S has the local extension property with respect to the G-action and let $f \in C^\infty_{S/G}(V)$, for some open set $V \subset S/G$. Given that π_S is a surjective submersion we have that $f \circ \pi_S|_{\pi_S^{-1}(V)} \in C^\infty_S(\pi_S^{-1}(V))^G$. Using the local extension property we can guarantee that, for any $m \in \pi_S^{-1}(V)$, there exist an open G-invariant neighborhood U_m of m in M and a function $F \in C^\infty_M(U_m)^G$ such that $f \circ \pi_S|_{\pi_S^{-1}(V)\cap U_m} = F|_{\pi_S^{-1}(V)\cap U_m}$. This implies that $f \in C^\infty_{S/G}(V)^G = C^\infty_{S/G,M/G}(V)$, as required. ■

2.2.13 Infinitesimally invariant and equivariant functions. Let $\mathfrak{g}$ be a Lie algebra acting on the manifold M. We say that a function $f \in C^\infty(M)$ is $\mathfrak{g}$-***invariant*** if

$$\xi_M[f] = 0, \quad \text{for any} \quad \xi \in \mathfrak{g}.$$

We will denote by $C^\infty(M)^{\mathfrak{g}}$ the set of $\mathfrak{g}$-invariant functions on M.

Let M and N be two manifolds acted upon by the same Lie algebra $\mathfrak{g}$. A smooth map $F : M \to N$ is said to be $\mathfrak{g}$-***equivariant*** if

$$\xi_N(F(m)) = T_m F \cdot \xi_M(m),$$

for any $m \in M$ and any $\xi \in \mathfrak{g}$.

2.3 Proper Lie group actions

Most of the Lie group actions considered in this book satisfy the properness condition. The theory of proper Lie group actions was introduced in PALAIS (1961), who proved that this hypothesis is enough to ensure that the main properties of compact Lie group actions are available. These properties had been obtained previously by GLEASON (1950), KOSZUL (1953), MONTGOMERY AND YANG (1957), and MOSTOW (1957).

2.3.1 Proper maps. Let X and Y be two topological spaces with Y first countable. A continuous map $f : X \to Y$ is called ***proper*** if for any sequence $\{x_n\}_{n \in \mathbb{N}}$ such that $f(x_n) \to y$ there exists a convergent subsequence $\{x_{n_k}\}$ such that $x_{n_k} \to x$ and $f(x) = y$. A map $f : X \to Y$ is proper if and only if it is closed and $f^{-1}(y)$ is compact, for any $y \in Y$ (see Theorem 1 in §10.2 of BOURBAKI (1989a) for a proof). If Y is locally compact, then the continuous map $f : X \to Y$ is proper if and only if the inverse image by f of compact sets in Y are compact sets in X. Moreover, if Y is locally compact and $f : X \to Y$ is proper, then X is also locally compact. It is easy to show that any continuous map defined on a compact space is proper. If $f : X \to Y$ and $g : Y \to Z$ are continuous maps, then the following hold: if f and g are proper, then the composition $g \circ f$ is proper; if $g \circ f$ is proper and f (respectively g) is surjective (respectively injective), then g (respectively f) is proper. Finally, if $f : X \to Y$ is proper and $T \subset Y$, then the induced map $f^{-1}(T) \to T$ is proper.

2.3.2 Definition *Let G be a Lie group acting on the manifold M via the map $\Phi : G \times M \to M$. We say that Φ is a* ***proper action*** *whenever the map $\Theta : G \times M \to M \times M$ defined by $\Theta(g, z) = (z, \Phi(g, z))$ is proper. The properness of the action is equivalent to the following condition: for any two convergent sequences $\{m_n\}$ and $\{g_n \cdot m_n\}$ in M, there exists a convergent subsequence $\{g_{n_k}\}$ in G. We say that the action Φ is* ***proper at the point*** *$z \in M$ when for for any two convergent sequences $\{m_n\}$ and $\{g_n \cdot m_n\}$ in M such that $m_n \to z$ and $g_n \cdot m_n \to z$, there exists a convergent subsequence $\{g_{n_k}\}$ in G.*

2.3.3 Example Compact Group Actions. Let G be a compact Lie group acting on the manifold M. The compactness of G implies that the requirement in Definition 2.3.2 is automatically satisfied. Consequently, compact group actions are always proper.

There are cases where the condition of properness in a Lie group action automatically implies the compactness of the group in question. For instance, if the manifold M is Hausdorff and compact, the properness of the map $\Theta : G \times M \to M \times M$ implies that the inverse image of $M \times M$, that is, $G \times M$ is compact. Consequently, the group G is necessarily compact. In particular, this argument implies that any discrete group acting properly on a compact manifold is necessarily finite.

2.3.4 Example The Euclidean Group. The ***special Euclidean group*** SE(n) is defined as

$$\mathrm{SE}(n) = \{(A, a) \mid A \in \mathrm{SO}(n) \text{ and } a \in \mathbb{R}^n\},$$

with the composition law given by

$$(A, a) \cdot (B, b) = (AB, Ab + a).$$

The group SE(n) is not compact but it acts properly on $\mathbb{R}^n$ by

$$(A, a) \cdot z = Az + b.$$

Indeed, let $z_n \to z_1$ and $(A_n, a_n) \cdot z_n \to z_2$ be two convergent sequences. Since $\{A_n\} \subset \mathrm{SO}(n)$, and SO($n$) is a compact Lie group, the sequence $\{A_n\}$ admits a convergent subsequence $A_{n_k} \to A$. Since clearly $A_{n_k} z_{n_k} \to Az_1$ and $A_{n_k} z_{n_k} + a_{n_k} \to z_2$, it follows that $a_{n_k} \to z_2 - Az_1$ and, therefore, $(A_{n_k}, a_{n_k}) \to (A, z_2 - Az_1)$ is a convergent subsequence of (A_n, a_n), as required.

2.3.5 Example Lie groups acting on themselves. Let G be a Lie group acting on itself by left translations (for right translations the situation is completely analogous). This action is proper. Indeed, let $g_n \cdot h_n \to l$ and $h_n \to h$ be two convergent sequences. The continuity of the inversion mapping on G guarantees that $h_n^{-1} \to h^{-1}$. Moreover, the continuity of the group operation implies that $g_n \cdot h_n \cdot h_n^{-1} \to l \cdot h^{-1}$ and hence $g_n \to l \cdot h^{-1}$.

2.3.6 Example Lifted group actions on its own tangent and cotangent bundles. The lifts of the left and right translations of a Lie group G on itself to its tangent and cotangent bundles, TG and T^*G respectively (see §2.2.3), are proper actions. Indeed, let $\{(g_n, v_{g_n})\}_{n \in \mathbb{N}}$ and $\{(h_n g_n, T_{g_n}\Phi_{h_n} \cdot v_{g_n})\}_{n \in \mathbb{N}}$ be two convergent sequences in TG. This implies in particular that $\{g_n\}_{n \in \mathbb{N}}$ and $\{h_n g_n\}_{n \in \mathbb{N}}$ are convergent. Since the G action on itself by left translations is proper, as proved in the previous example, this implies that $\{h_n\}_{n \in \mathbb{N}}$ has a convergent subsequence, as required. Analogously, it is easy to check that the lifts of the left and right translations of a Lie group G on itself to its cotangent bundle T^*G, given respectively by $\alpha_g \mapsto T^*_{hg} L_{h^{-1}} \alpha_g$ and $\alpha_g \mapsto T^*_{gh} R_{h^{-1}} \alpha_g$, are a proper actions.

2.3.7 The restriction of a proper action. Let G be a Lie group acting properly on the manifold M. Let H be a Lie subgroup of G that leaves the initial submanifold N of M invariant. By Lemma 2.2.2, the action of H on N by restriction is smooth. If H is closed in G this action is also proper.

Properness is a very powerful assumption for a group action because it guarantees that some of the technically important properties of compact group actions are still valid, as it can be seen in the following proposition.

2.3.8 Proposition *Let $\Phi : G \times M \to M$ be a proper action of the Lie group G on the manifold M. Then:*

(i) *For any $m \in M$, the isotropy subgroup G_m is compact.*

(ii) *The orbit space M/G is a Hausdorff topological space. (Even when M and G are not Hausdorff.)*

(iii) *If the action is free, M/G is a smooth manifold, and the canonical projection $\pi : M \to M/G$ defines on M the structure of a smooth left principal G-bundle.*

(iv) *If all the isotropy subgroups of the elements of M under the G-action are conjugate to a given one, say $H \subset G$, then M/G is a smooth manifold and the canonical projection $\pi : M \to M/G$ defines on M the structure of a smooth locally trivial fiber bundle with structure group $N(H)/H$ and fiber G/H.*

(v) *If the manifold M is paracompact, then there exists a G-invariant Riemannian metric on it. Also, there exists a G-invariant partition of unity subordinate to any cover of M by open G-invariant sets. Additionally, if the orbit space M/G is a regular quotient smooth manifold, the presheaf $C^\infty_{M/G}$ of smooth functions on it is actually a sheaf.*

(vi) *If the manifold M is paracompact, then smooth G-invariant functions separate the G-orbits.*

(vii) *Suppose that M is paracompact. Let N be any G-invariant subset of M and let $f \in C^\infty(M)$ be such that the restriction $f|_N$ is constant on each G-orbit. Then there is a smooth G-invariant function $F \in C^\infty(M)^G$ such that $F|_N = f|_N$.*

Proof. **(i)** Let $\Theta : G \times M \to M \times M$ be the proper map given by $\Theta(g, z) = (z, \Phi(g, z))$. Notice that $\Theta^{-1}(\{m\} \times M) = G \times \{m\}$. Thus the map $G \times \{m\} \to \{m\} \times M$ given by $(g, m) \longmapsto (m, g \cdot m)$ is proper and, consequently, so is the map $\theta_m : G \to M$ defined by $\theta_m(g) := g \cdot m$. Therefore $\theta_m^{-1}(m) = G_m$, is compact. **(ii)** The image set $\Theta(G \times M)$ is the graph of the equivalence relation that defines the quotient M/G. Since the action is proper, Θ is a closed map and therefore $\Theta(G \times M)$ is closed in $M \times M$. According to the remarks made in §1.1.18 we just need to show that the projection $\pi : M \to M/G$ onto orbit space is an open map. This is so because if $U \subset M$ is open, then $\pi(U)$ is open if and only if $\pi^{-1}(\pi(U))$ is open; but $\pi^{-1}(\pi(U)) = G \cdot U = \cup_{g \in G} \Phi_g(U)$ is open. The proofs of the remaining important but standard facts are longer so we just indicate where to find them. For **(iii)** and **(iv)** see Proposition 10 in Chapter 3, §1.5 of BOURBAKI (1989b), as well as Proposition 4.1.23 and Exercise 4.1M of ABRAHAM AND MARSDEN (1978). Point **(v)** is Theorem 4.3.1 in PALAIS (1961). A different proof is provided in Proposition 2.5.2 of DUISTERMAAT AND KOLK (1999). The statement **(vii)** is proved in Proposition 2 of ARMS et al. (1991), whose Corollary 1 coincides with **(vi)**. For more information, see also BATES AND LERMAN (1997); CUSHMAN AND BATES (1997); ABRAHAM et al. (1988). ■

2.3.9 A straightforward corollary to part **(v)** in Proposition 2.3.8 is that the presheaf $C^\infty(\cdot)^G$ of G-invariant functions on a paracompact manifold acted properly upon by G is actually a sheaf.

2.3.10 Let G be a Lie group acting linearly and properly on the vector space V. The isotropy subgroup of zero is the entire group G. Hence the properness of the action and the previous proposition imply that G is necessarily compact. We have thus proved that linear proper G-actions require the compactness of G.

2.3.11 Homogeneous manifolds and the smooth structure of the orbits of a group action. The previous proposition gives a natural way to endow the quotients introduced in Example 2.2.9 with a smooth manifold structure. Recall that in that setup we had a Lie group G and a Lie subgroup $H \subset G$ acting on G by left and right translations. As we have already seen, these actions are smooth, free, and their orbits are given by the cosets Hg, $g \in G$, for the left action, and gH, $g \in G$, for the right action. The respective orbit spaces are denoted by the symbols $G\backslash H$ and G/H. It is easy to check that these H-actions are proper if and only if H is closed in G. Therefore, Proposition 2.3.8 guarantees that if H is closed, the quotients G/H and $G \setminus H$ are smooth manifolds such that the canonical projections $G \rightarrow G/H$ and $G \rightarrow G\backslash H$ are surjective submersions. These quotients are referred to as ***homogeneous manifolds***. The group G acts transitively on the right on $G\backslash H$ and on the left on G/H. If, additionally, H is a normal subgroup of G, the quotient $G/H = G\backslash H$ is a Lie group with this smooth structure and the projection $G \rightarrow G/H$ is a Lie group homomorphism. An example of this construction is the ***group of connected components*** of a given group G: the connected component of the identity G^0 is an open and closed normal subgroup of G and therefore the quotient G/G^0 is a Lie group whose cardinality coincides with the number of connected components of G.

This construction allows us to naturally endow the orbits of a group action with a smooth structure. Let $\Phi : G \times M \rightarrow M$ be a smooth action and let $m \in M$ be arbitrary. As the isotropy subgroup G_m of m is always a closed subgroup of G, the previous construction guarantees that the quotient G/G_m is always a homogeneous smooth manifold. Let $\mathcal{O}_m := G \cdot m$ be the G-orbit through the point m. The map $G/G_m \rightarrow \mathcal{O}_m$ given by $gG_m \longmapsto g \cdot m$, $g \in G$, is a well-defined bijection. We will consider the orbit $\mathcal{O}_m$ as a manifold endowed with the unique smooth structure that makes this bijection into a diffeomorphism.

2.3.12 Proposition *Let $\Phi : G \times M \rightarrow M$ be a smooth Lie group action on the manifold M. Let $\mathcal{O}_m := G \cdot m$ be the orbit through m endowed with the smooth structure that makes it diffeomorphic to the homogeneous manifold G/G_m. Then:*

(i) *The orbit $\mathcal{O}_m$ is an initial submanifold of M of dimension* $\dim G - \dim G_m$ *whose tangent space at the point $z \in \mathcal{O}_m$ is given by*

$$\mathfrak{g} \cdot z := T_z\mathcal{O}_m = \{\xi_M(z) \mid \xi \in \mathfrak{g}\}. \tag{2.3.1}$$

(ii) *The connected components of $\mathcal{O}_m$ are given by the sets $gG^0 \cdot m$, with $g \in G$ and G^0 the connected component of the identity of G. Moreover,*

$$\mathcal{O}_m = \coprod_{[g] \in G/G^0 G_m} gG^0 \cdot m. \tag{2.3.2}$$

In particular, the orbit $\mathcal{O}_m$ has $\operatorname{Card}(G/G^0G_m)$ *connected components.*

(iii) *If the G-action is proper the orbit $\mathcal{O}_m$ is closed. If, additionally, the manifold M is second countable, then $\mathcal{O}_m$ is an embedded submanifold of M.*

2.3.13 Remark Corollary 2.3.33 will show how the existence of slices guaranteed by the properness of the action implies that the orbits are embedded submanifolds of M, without assuming that M is second countable.

2.3.14 Notation. Let $\mathfrak{g}$ be a Lie algebra acting on the manifold M. In order to be consistent with the notation introduced if (2.3.1) we define, for any $z \in M$,

$$\mathfrak{g} \cdot z := \{\xi_M(z) \mid \xi \in \mathfrak{g}\}.$$

Proof of the proposition. Let D be the generalized distribution on M given by

$$D(m) := \{\xi_M(m) \mid \xi \in \mathfrak{g}\}, \text{ for all } m \in M.$$

This distribution is smooth and involutive in the sense of §1.4.6. Indeed, let F_t be the flow of the infinitesimal generator vector field η_M, $\eta \in \mathfrak{g}$; its flow is given by $F_t(z) = \exp t\eta \cdot z$. Now, for any $\xi \in \mathfrak{g}$, it is easy to check that

$$T_z F_t \cdot \xi_M(z) = (\mathrm{Ad}_{\exp t\eta}\xi)_M(\exp t\eta \cdot z) \in D(\exp t\eta \cdot z),$$

which proves that the distribution is involutive and therefore, by the Stefan–Sussmann Theorem, also integrable.

We now show that the maximal integral manifolds of this generalized distribution are the connected components of the G-orbits. Indeed, notice first that any orbit $\mathcal{O}_m$ is an injectively immersed submanifold of M because the map $j : G/G_m \to M$ given by $j(gG_m) = g \cdot m$ is an injective immersion. Moreover, a simple diagram inspection shows that $T_{gG_m} j \cdot T_{gG_m}(G/G_m) = \{\xi_M(g \cdot m) \mid \xi \in \mathfrak{g}\} = D(g \cdot m)$, which proves (2.3.1) and shows that the orbits are integral manifolds of D of maximal dimension. At the same time, the Stefan–Sussmann Theorem claims that the maximal integral manifolds of D are the accessible sets in M obtained by finite compositions of the flows of the infinitesimal generator vector fields that define this distribution. It is then clear that for any $g \cdot m \in \mathcal{O}_m$, the accessible set of D going through that point is the set $gG^0 \cdot m$. This implies that the sets of the form $gG^0 \cdot m$ are the connected components of $\mathcal{O}_m$. Indeed, let S be the connected component of $\mathcal{O}_m$ that contains the connected set $gG^0 \cdot m$, that is, $gG^0 \cdot m \subset S$. As $\mathcal{O}_m$ is a manifold, it is locally connected and therefore its connected components are open and closed. In particular S is an open connected subset of $\mathcal{O}_m$ and therefore a connected integral submanifold of the distribution D. By the maximality of $gG^0 \cdot m$ as an integral submanifold of D it follows that $gG^0 \cdot m = S$, which proves that the connected components of $\mathcal{O}_m$ are given by the sets $gG^0 \cdot m$, with $g \in G$. Also, as the sets $gG^0 \cdot m$ are the integral manifolds of a generalized distribution, they are necessarily initial submanifolds of M. Lemma 1.1.11 guarantees that $\mathcal{O}_m$ is an initial submanifold of M.

We now prove the identity (2.3.2). Notice first that as G^0 is normal in G, the product G^0G_m is a subgroup of G. Now, in order to show that the equality (2.3.2) makes sense we have to prove that if $[g] = [g'] \in G/G^0G_m$, then $gG^0 \cdot m = g'G^0 \cdot m$. Indeed, if $[g] = [g']$, there exists $h \in G^0$ and $k \in G_m$ such that $g' = ghk$. Consequently, $g'G^0 \cdot m = ghkG^0 \cdot m = ghG^0k \cdot m = gG^0 \cdot m$, as required. Since we obviously have that $\mathcal{O}_m = \bigcup_{[g] \in G/G^0G_m} gG^0 \cdot m$ it only remains to be shown that this union is disjoint. We proceed by contradiction; suppose that $[g]$ and $[g'] \in G/G^0G_m$ are such that $[g] \neq [g']$ and $gG^0 \cdot m \cap g'G^0 \cdot m \neq \emptyset$. Let $z \in \mathcal{O}_m$ be an element in this intersection that can be written as $z = gh \cdot m = g'h' \cdot m$, with $h, h' \in G^0$. The last equality implies that $h^{-1}g^{-1}g'h' \in G_m$ and therefore $g^{-1}g' \in hG_m(h')^{-1} \subset G^0G_m$. Hence $[g] = [g']$ which contradicts our hypothesis.

We now show that if the G-action is proper, then the orbit $\mathcal{O}_m$ is closed. Recall the proper map $\theta_m : G \to M$ given by $\theta_m(g) = g \cdot m$ introduced in the proof of Proposition 2.3.8. As θ_m is proper it is also a closed map and therefore $\theta_m(G) = \mathcal{O}_m$ is closed in M. The same argument works for each connected component $gG^0 \cdot m = \theta_m(gG^0)$, $[g] \in G/G^0G_m$, of $\mathcal{O}_m$. Finally, if M is second countable, Lemma 1.4.5 implies that the connected components of $\mathcal{O}_m$, and hence $\mathcal{O}_m$, are embedded submanifolds of M. ■

Homogeneous manifolds can also be used to study the range of a Lie group homomorphism $f : G \to H$. As we already mentioned, the range of f is a subgroup of H and it is isomorphic to the quotient Lie group $G/\ker f$ (recall that $\ker f$ is a closed normal subgroup of G) by the map $G/\ker f \to \operatorname{range} f$ defined by $g \ker f \longmapsto f(g)$, $g \in G$. It can be checked that if we take on range f the smooth structure that makes this bijection a diffeomorphism, range f becomes a Lie subgroup (in general not closed) of H.

2.3.15 Twisted products. The properness condition in a group action guarantees the existence of a very convenient semilocal model that is obtained via the Slice Theorem. This will be reviewed later on in this section. The mathematical formulation of this model uses the constructions that we introduce in the following paragraphs.

2.3.16 Definition *Let G be a Lie group and $H \subset G$ a subgroup. Suppose that H acts on the left on the manifold A. The* **twisted action** *of H on the product $G \times A$ is defined by*

$$h \cdot (g, a) = (gh, h^{-1} \cdot a), \quad h \in H, \ g \in G, \text{ and } a \in A.$$

Note that this action is free and proper by the freeness and properness of the action on the G-factor (see Example 2.3.5). The **twisted product** *$G \times_H A$ is defined as the orbit space $(G \times A)/H$ corresponding to the twisted action. The elements of $G \times_H A$ will be denoted by $[g, a]$, $g \in G$, $a \in A$.*

2.3.17 Proposition *The twisted product $G \times_H A$ is a G-space relative to the left action defined by $g' \cdot [g, a] := [g'g, a]$. Also, the action of H on A is proper if and only if the G-action on $G \times_H A$ just defined is proper.*

Proof. The fact that the operation $g' \cdot [g, a] := [g'g, a]$ defines a smooth left action of G on $G \times_H A$ is left to the reader as a straightforward exercise.

Suppose now that the action of H on A is proper. We will show that the G-action on $G \times_H A$ is proper. Indeed, let $g_n \cdot [k_n, a_n] = [g_n k_n, a_n] \to [l, b]$ and $[k_n, a_n] \to [k, a]$ be two convergent sequences. Since the natural projection $\pi : G \times A \to G \times_H A$ is a surjective submersion, it admits local smooth sections (Proposition 1.1.14). Let σ_1 and σ_2 be two smooth sections for π around $[l, b]$ and $[k, a]$; then $\sigma_1([g_n k_n, a_n]) \to \sigma_1([l, b])$ and $\sigma_2([k_n, a_n]) \to \sigma_2([k, a])$. Hence there exist two sequences $\{h_n\}$, $\{l_n\} \subset H$ and elements $h, j \in H$ such that

$$(g_n k_n h_n, h_n^{-1} \cdot a_n) \to (lh, h^{-1} \cdot b) \quad \text{and} \quad (k_n l_n, l_n^{-1} \cdot a_n) \to (kj, j^{-1} \cdot a).$$

The fact that $h_n^{-1} \cdot a_n \to h^{-1} \cdot b$ can be rewritten as $(h_n^{-1} l_n) \cdot (l_n^{-1} \cdot a_n) \to h^{-1} \cdot b$. Since at the same time $l_n^{-1} \cdot a_n \to j^{-1} \cdot a$, the hypothesis on the properness of the H-action on

A implies the existence of a subsequence $\{h_{n_k}^{-1} l_{n_k}\}$ and of an element $f \in H$ such that $h_{n_k}^{-1} l_{n_k} \to f$. Given that $g_n k_n h_n \to lh$, we have that $g_{n_k} k_{n_k} h_{n_k} h_{n_k}^{-1} l_{n_k} = g_{n_k} k_{n_k} l_{n_k} \to lhf$. At the same time $k_{n_k} l_{n_k} \to kj$. Consequently, we have the convergence of the subsequence $g_{n_k} \to lhfj^{-1}k^{-1}$, as required.

Conversely, suppose that the G-action on $G \times_H A$ is proper. Let $a_n \to a$ and $h_n \cdot a_n \to b$ be two convergent sequences in A with $a_n \in A$ and $h_n \in H$, for any $n \in \mathbb{N}$. This implies that the sequences $[e,\, a_n] \to [e,\, a]$ and $h_n \cdot [e,\, a_n] = [h_n,\, a_n] = [e,\, h_n \cdot a_n] \to [e,\, b]$ are convergent. The properness of the G-action on $G \times_H A$ guarantees that the sequence $\{h_n\}_{n\in\mathbb{N}}$ admits a convergent subsequence, as required. ■

2.3.18 Proposition *The isotropy subgroups of the G-action on the twisted product $G \times_H A$, defined in the previous proposition, satisfy*

$$G_{[g,\, a]} = gH_a g^{-1}, \quad \text{for any} \quad g \in G,\, a \in A.$$

Proof. Let $k \in G_{[g,\, a]}$. By definition, $k \cdot [g,\, a] = [g,\, a]$ or, equivalently, $[kg,\, a] = [g,\, a]$, which implies that there exists an element $h \in H$ such that $kg = gh$ and $a = h^{-1} \cdot a$. The last equality guarantees that $h \in H_a$ and, since $k = ghg^{-1}$, we have that $k \in gH_a g^{-1}$. Conversely, let $ghg^{-1} \in gH_a g^{-1}$. By definition,

$$ghg^{-1} \cdot [g,\, a] = [gh,\, a] = [g,\, h \cdot a] = [g,\, a], \quad \text{since } h \in H_a. \; ■$$

2.3.19 Proposition *Let $G \times_H A$ be the twisted product introduced in Definition* 2.3.16. *Let K be a subgroup of G such that $H \subset K \subset G$. Then the map*

$$\begin{array}{rccc} \phi : & G \times_H A & \longrightarrow & G \times_K (K \times_H A) \\ & [g, a] & \longmapsto & [g,\, [e,\, a]], \end{array}$$

is a G-equivariant diffeomorphism.

Proof. Using the action introduced in Proposition 2.3.17, the twisted product $K \times_H A$ is a K-space and therefore, the twisted product $G \times_K (K \times_H A)$ is well defined. Also, the map ϕ is smooth, well defined, and clearly G-equivariant since it is the projection onto $G \times_H A$ of the H-invariant and G-equivariant smooth map

$$\begin{array}{ccc} G \times A & \longrightarrow & G \times_K (K \times_H A) \\ (g, a) & \longmapsto & [g,\, [e,\, a]]. \end{array}$$

Let now be the map $G \times K \times A \to G \times A$ given by $(g, k, a) \longmapsto (gk, a)$. By H-equivariance, this map drops to a smooth map $G \times (K \times_H A) \to G \times_H A$ and, by K-invariance, to another smooth map $G \times_K (K \times_H A) \to G \times_H A$ given by $[g,\, [k,\, a]] \longmapsto [gk,\, a]$ that constitutes the inverse of ϕ, which consequently is a diffeomorphism. ■

2.3.20 Associated bundles. Twisted actions and products are closely related to the concept of ***associated bundle*** that we briefly recall in what follows (see GREUB et al. (1970b) or BREDON (1972) for more details).

Let $\mathcal{P} = (P,\, \pi,\, B,\, G)$ be a smooth principal bundle with principal action

$$T : P \times G \longrightarrow P.$$

Recall that the very definition of principal bundle implies that this action is free and that the orbit through an arbitrary point $z \in P$ coincides with the fiber containing the point $z \in P$. Consider now

$$S : G \times F \longrightarrow F,$$

a fixed left action of the Lie group G on the manifold F and define

$$Q : (P \times F) \times G \longrightarrow P \times F$$

the right action of G on the product manifold $P \times F$ by

$$Q_g(z, y) := (z \cdot g, g^{-1} \cdot y), \qquad g \in G, \; z \in P, \; y \in F.$$

Being consistent with the definition of the twisted product we will denote the orbit space of this action by $P \times_G F$. Let

$$\pi_G : P \times G \longrightarrow P \times_G F$$

be the canonical projection and let $\rho : P \times_G F \to B$ be the mapping uniquely determined by the commutative diagram

$$\begin{array}{ccc} P \times F & \xrightarrow{\pi_G} & P \times_G F \\ {\scriptstyle \pi_P}\downarrow & & \downarrow{\scriptstyle \rho} \\ P & \xrightarrow{\pi} & B, \end{array}$$

where π_P is the projection onto the P factor. The interest of all these constructions lies in the following standard result, whose proof can be found, for instance, in GREUB et al. (1970b).

2.3.21 Theorem *There is a unique smooth structure on $P \times_G F$ such that:*

(i) $\mathcal{A} := (P \times_G F, \rho, B, F)$ *is a smooth fiber bundle;*

(ii) *the projection $\pi_G : P \times G \to P \times_G F$ is a smooth fiber preserving map, restricting to diffeomorphisms*

$$(\pi_G)_z : \{z\} \times F \longrightarrow F_{\pi(z)}, \qquad z \in P,$$

on each fiber $F_{\pi(z)}$;

(iii) $(P \times F, \pi_G, P \times_G F, G)$ *is a smooth principal bundle with principal action Q;*

(iv) π_P *is a homomorphism of principal bundles covering ρ.*

2.3.22 Definition *With the notation introduced in the preceding theorem, $\mathcal{A}$ is called the fiber bundle with fiber F and structure group G* ***associated*** *to $\mathcal{P}$. The projection π is called the* ***principal map****.*

2.3.23 Slices. The main feature that makes proper actions extremely manageable is the existence of slices. The Slice Theorem that we will present in the following paragraphs is due to KOSZUL (1953) in the case of compact group actions and was generalized to proper group actions by PALAIS (1961).

2.3.24 Definition *Let M be a manifold and G a Lie group acting properly on M. Let $m \in M$ and denote $H := G_m$. A* ***tube*** *around the orbit $G \cdot m$ is a G-equivariant diffeomorphism*

$$\varphi : G \times_H A \longrightarrow U,$$

where U is a G-invariant neighborhood of $G \cdot m$ and A is some manifold on which H acts. Note that the G-action on the twisted product $G \times_H A$ is proper since by Proposition 2.3.8 *the isotropy subgroup H is compact and, consequently, its action on A is proper. Hence, by Proposition* 2.3.17, *the G-action on $G \times_H A$ is proper.*

2.3.25 Definition *Let M be a manifold and G a Lie group acting properly on M. Let $m \in M$ and denote $G_m := H$. Let S be a submanifold of M such that $m \in S$ and $H \cdot S = S$. We say that S is a* ***slice*** *at m if the G-equivariant map*

$$\begin{aligned} \varphi : G \times_H S &\longrightarrow U \\ [g, s] &\longmapsto g \cdot s \end{aligned}$$

is a tube about $G \cdot m$ for some G-invariant open neighborhood U of $G \cdot m$. Notice that if S is a slice at m, then $\Phi_g(S)$ is a slice at the point $\Phi_g(m)$.

The following theorem provides several equivalent characterizations of the concept of slice that are available in the literature. Recall that given a set X and a subset $A \subset X$ a map $r : X \to A$ is called a ***retraction*** of the inclusion $\iota : A \hookrightarrow X$ if $r(a) = a$ for all $a \in A$, or, equivalently, if $r \circ \iota = id_A$.

2.3.26 Theorem *Let M be a manifold and G a Lie group acting properly on M. Let $m \in M$, denote $H := G_m$, $\mathfrak{h}$ the Lie algebra of H, and let S be a submanifold of M containing m. Then the following statements are equivalent:*

(i) *There is a tube $\varphi : G \times_H A \longrightarrow U$ about $G \cdot m$ such that $\varphi[e, A] = S$.*

(ii) *S is a slice at m.*

(iii) *The submanifold S satisfies the following properties:*

- **(a)** *The set $G \cdot S$ is an open neighborhood of the orbit $G \cdot m$ and S is closed in $G \cdot S$.*
- **(b)** *For any $z \in S$ we have that $T_z M = \mathfrak{g} \cdot z + T_z S$. Moreover, $\mathfrak{g} \cdot z \cap T_z S = \mathfrak{h} \cdot z$. In particular, for $z = m$ the sum $\mathfrak{g} \cdot z + T_z S$ is direct.*
- **(c)** *S is H-invariant. Moreover, if $z \in S$ and $g \in G$ are such that $g \cdot z \in S$, then $g \in H$.*
- **(d)** *Let $\sigma : U \subset G/H \to G$ be a local section of the submersion $G \to G/H$. Then, the map $F : U \times S \to M$ given by $F(u, z) := \sigma(u) \cdot z$ is a diffeomorphism onto an open subset of M.*

(iv) *$G \cdot S$ is an open neighborhood of $G \cdot m$ and there is an equivariant smooth retraction*

$$r : G \cdot S \longrightarrow G \cdot m$$

of the injection $G \cdot m \hookrightarrow G \cdot S$ such that $r^{-1}(m) = S$.

Proof. **(i) ⇒ (ii)** is straightforward since the map $\psi : G \times_H S \to U$ given by $[g, s] \longmapsto g \cdot s$ is clearly a tube about the orbit $G \cdot m$.
(ii) ⇒ (iii): **(a)**: $G \cdot S = \psi(G \times_H S) = U$ is an open neighborhood of $G \cdot m$ in M. Let now $z \in G \cdot S$ be an element in the closure of S in $G \cdot S$ and $\{s_n\}_{n \in \mathbb{N}} \subset S$ a sequence such that $s_n \to z$. Let $[g, t] := \psi^{-1}(z)$. The continuity of ψ implies that $[e, s_n] \to [g, t]$. Let now τ be a local section around $[g, t]$ of the projection $\pi_H : G \times S \to G \times_H S$ and consider now only the elements in the sequence $\{[e, s_n]\}_{n \in \mathbb{N}}$ with n big enough so that they belong to the domain of τ. The convergence $[e, s_n] \to [g, t]$ implies that $\tau([e, s_n]) \to \tau([g, t])$; hence, there exists a sequence $\{h_n\}_{n \in \mathbb{N}} \subset H$ and an element $h \in H$ such that $(h_n, h_n^{-1} \cdot s_n) \to (gh, h^{-1} \cdot t)$. This implies, in particular, that $h_n \to gh$, and as H is compact and therefore closed, it follows that $gh \in H$. Consequently, $g \in H$ and thus $[g, t] = [e, g \cdot t] = \psi^{-1}(z)$. Hence, $z = g \cdot t \in S$, which proves the closedness of S in $G \cdot S$. **(b)**: Let $z \in S$. Then, $T_z M = T_{[e,z]}\psi \cdot T_{[e,z]}(G \times_H S) = T_{[e,z]}\psi \cdot T_{(e,z)}\pi_H(\mathfrak{g} \times T_z S) = \mathfrak{g} \cdot z + T_z S$. We now check that $\mathfrak{g} \cdot z \cap T_z S = \mathfrak{h} \cdot z$. The inclusion $\mathfrak{h} \cdot z \subset \mathfrak{g} \cdot z \cap T_z S$ is straightforward since S is H-invariant. If $\xi_M(z) \in \mathfrak{g} \cdot z \cap T_z S$ there is a curve $z(t) \subset S$ such that $z(0) = z$ and $T_z \psi^{-1} \cdot \xi_M(z) = \frac{d}{dt}\big|_{t=0} [\exp t\xi, z] = \frac{d}{dt}\big|_{t=0} [e, z(t)]$. The last equality implies that $\frac{d}{dt}\big|_{t=0} [\exp -t\xi, z(t)] = 0$ and therefore $\frac{d}{dt}\big|_{t=0} (\exp -t\xi, z(t)) = (-\xi, z'(0)) \in \ker T_{(e,z)}\pi_H = \{(\eta, (-\eta)_M(z)) \mid \eta \in \mathfrak{h}\}$. Consequently, $\xi \in \mathfrak{h}$, as required. This obviously implies that if $z = m$ we have $\mathfrak{g} \cdot m \cap T_m S = \{0\}$. **(c)**: Let $z \in S$ and $g \in G$ be such that $g \cdot z \in S$. Now, as the map ψ^{-1} is G-equivariant we have that $\psi^{-1}(g \cdot z) = g \cdot \psi^{-1}(z)$ or, equivalently, $[e, g \cdot z] = g \cdot [e, z] = [g, z]$. Hence there exists an element $h \in H$ such that $(h, h^{-1} g \cdot z) = (g, z)$ and therefore $g = h \in H$, as required. **(d)**: The map F is clearly smooth. It is also injective because if $(u, s), (u', s') \in U \times S$ are such that $F(u, s) = F(u', s')$, then $\sigma(u) \cdot s = \sigma(u') \cdot s'$ and hence $\sigma(u')^{-1}\sigma(u) \cdot s = s'$. Property **(c)** implies that $\sigma(u')^{-1}\sigma(u) \in H$. Hence there exists an element $h \in H$ such that $\sigma(u) = \sigma(u')h$ and therefore $u = \pi(\sigma(u')h) = u'$. Consequently, $\sigma(u) = \sigma(u')$ and $s = s'$.

We now show that F is a diffeomorphism of $U \times S$ onto its image by proving that for any element $(u, s) \in U \times S$ the linear map $T_{(u,s)}F$ is an isomorphism. Suppose that $\sigma(u) = g \in G$ and hence $u = gH$. Let $\xi \in \mathfrak{g}$ and $s(t) \subset S$ be any curve in S such that $s(0) = s$. Then there exists a smooth curve $h(t) \subset H$ such that $h(0) = e$ and $h'(0) = \eta \in \mathfrak{h}$ for which

$$\begin{aligned} T_{(u,s)}F \cdot \frac{d}{dt}\bigg|_{t=0} (g \exp t\xi H, s(t)) &= \frac{d}{dt}\bigg|_{t=0} g \exp t\xi h(t) \cdot s(t) \\ &= T_s \Phi_g(\xi_M(s) + \eta_S(s) + s'(0)). \end{aligned}$$

Given that $\xi \in \mathfrak{g}$ and $s'(0) \in T_s S$ in the previous expression are arbitrary and that $\eta_S(s) \in T_s S$ we have that

$$\operatorname{Im}(T_{(u,s)}F) = T_s \Phi_g(\mathfrak{g} \cdot s + T_s S) = T_s \Phi_g(T_s M) = T_{g \cdot s} M,$$

which proves that $T_{(u,s)}F$ is onto. Suppose now that $T_{(u,s)}F \cdot \frac{d}{dt}\big|_{t=0}(g \exp t\xi H, s(t)) = T_s\Phi_g(\xi_M(s) + \eta_S(s) + s'(0)) = 0$. This implies that $\xi_M(s) + \eta_S(s) + s'(0) = 0$ and therefore $\xi_M(s) \in \mathfrak{g} \cdot s \cap T_sS$. Hence, by part **(b)**, $\xi_M(s) \in \mathfrak{h} \cdot s$, and then our hypothesis reduces to $T_{(u,s)}F \cdot \frac{d}{dt}\big|_{t=0}(g \exp t\xi H, s(t)) = T_s\Phi_g(s'(0)) = 0$, which in turn implies that $s'(0) = 0$. Consequently, $\frac{d}{dt}\big|_{t=0}(g \exp t\xi H, s(t)) = (0, 0)$, which proves that $T_{(u,s)}F$ is injective, as required.

(iii) ⇒ (iv): Define $r : G \cdot S \longrightarrow G \cdot m$ by $r(g \cdot s) := g \cdot m$, for all $g \in G$ and $s \in S$. This is a good definition because if $g' \in G$ and $s' \in S$ are such that $g \cdot s = g' \cdot s'$, then $g^{-1}g' \cdot s' = s$ and by property **(c)** in **(iii)** we have that $(g')^{-1}g \in H$. Hence, $r(g' \cdot s') = g' \cdot m = g'(g')^{-1}g \cdot m = g \cdot m = r(g \cdot s)$. The map r is clearly a retraction of the inclusion $G \cdot m \hookrightarrow G \cdot S$ and it is G-equivariant. Also, the H-invariance of S assumed in **(b)** implies that $r^{-1}(m) = H \cdot S = S$. It only remains to be shown the r is smooth, which is guaranteed by **(d)**. Indeed, let V be an open subset of $G \cdot S$ small enough so that there exists a local section $\sigma : U \subset G/H \to G$ such that the associated map $F : U \times S \to M$ introduced in **(d)** maps onto V, that is, $F(U \times S) = V$. Let $\pi_1 : U \times S \to U$ be the projection onto the first factor. The map r is smooth because its restriction to V can be written as $r(z) = \sigma(\pi_1(F^{-1}(z))) \cdot m$, for all $z \in V$: if $g \cdot s \in V$, then $\sigma(\pi_1(F^{-1}(g \cdot s))) \cdot m = \sigma(gH) \cdot m = g \cdot m$, as required.

(iv) ⇒ (i): We first show that S is H-invariant. By hypothesis, r is G-equivariant and $r^{-1}(m) = S$. Then, by definition,

$$\begin{aligned} S = r^{-1}(m) &= \{g \cdot s \in G \cdot S \mid r(g \cdot s) = m\} = \{g \cdot s \in G \cdot S \mid g \cdot r(s) = m\} \\ &= \{g \cdot s \in G \cdot S \mid g \cdot m = m\} = \{g \cdot s \in G \cdot S \mid g \in H, s \in S\} = H \cdot S. \end{aligned}$$

The H-invariance of S allows us to construct the twisted product $G \times_H S$. The map $\varphi : G \times_H S \to G \times S$ given by $\varphi[g, s] := g \cdot s$ provides us with the tube that we need. Indeed, φ is a smooth map onto an open G-invariant neighborhood of the orbit $G \cdot m$. It is also injective because if $[g, s], [g', s'] \in G \times_H S$ are such that $g \cdot s = g' \cdot s'$, then $r(g \cdot s) = r(g' \cdot s')$ or, equivalently, $g \cdot m = g' \cdot m$. Hence $g^{-1}g' \in H$ and $[g, s] = [gg^{-1}g', (g')^{-1}g \cdot s] = [g', s']$, as required. The map φ is a diffeomorphism onto its image because the inverse map $\varphi^{-1} : G \times S \to G \times_H S$ given by $\varphi^{-1}(g \cdot s) := [g, s]$ is also smooth. This is so because for any $g \cdot s \in G \cdot S$, there is a local section of the projection $G \to G/H$ around the point $r(g \cdot s) = g \cdot m \in G \cdot m \simeq G/H$ that allows us to locally write φ^{-1} as $\varphi^{-1}(z) = [\sigma(r(z)), (\sigma(r(z)))^{-1} \cdot z]$. ■

2.3.27 The Slice Theorem. In this section we prove that slices do exist for proper smooth actions on manifolds.

2.3.28 Theorem (Tube Theorem) *Let M be a manifold and G a Lie group acting properly on M at the point $m \in M$, $H := G_m$. There exists a tube $\varphi : G \times_H B \longrightarrow U$ about $G \cdot m$. B is an open H-invariant neighborhood of 0 in a vector space H-equivariantly isomorphic to $T_mM/T_m(G \cdot m)$ on which H acts linearly by $h \cdot (v + T_m(G \cdot m)) := T_m\Phi_h \cdot v + T_m(G \cdot m)$.*

Proof. The proof below is due to SCHMAH (2002). We start with a couple of lemmas.

2.3.29 Lemma *Let H be a compact Lie group that acts on the manifold M. Assume that $m \in M$ is a fixed point of the action, that is, $H \cdot m = \{m\}$. Then any open neighborhood of m contains an H-invariant open neighborhood of m.*

Proof of Lemma 2.3.29. Let $\Phi : H \times M \to M$ be the group action and U an arbitrary neighborhood of m. The set $\Phi^{-1}(U)$ is clearly open and contains $H \times \{m\}$. Now, for any $g \in H$, there exist neighborhoods W_g of g in H and V_g of m in M such that $W_g \times V_g \subset \Phi^{-1}(U)$. Since by hypothesis H is compact, the open cover $\{W_g \mid g \in H\}$ of H has a finite subcover $\{W_{g_1}, \ldots, W_{g_n}\}$. Let $V := \bigcap_{i=1}^n V_{g_i}$. Then the H-invariant set $W := \Phi(H, V) \subset U$ is clearly open since $W = \bigcup_{g \in H} \Phi_g(V)$ and, by construction, contains the point m. ▼

2.3.30 Lemma *Let H be a compact Lie group that acts on the manifold M and let $m \in M$ be such that $H \cdot m = \{m\}$. Then there exists an H-invariant Riemannian metric defined on some H-invariant neighborhood of m.*

Proof of Lemma 2.3.30. Let $\varphi : U \to \varphi(U) \subset \mathbb{R}^n$ be a local chart of M around the point m. By the previous lemma, the open neighborhood U can be chosen to be H-invariant. The pull-back of the Euclidean metric on $\varphi(U) \subset \mathbb{R}^n$ gives a Riemannian metric g on U. Let now g' be the ***averaged metric*** on U given by

$$g'(z)(u, v) := \int_H g(h \cdot z)(T_z\Phi_h \cdot u, T_z\Phi_h \cdot u)dh, \tag{2.3.3}$$

where the integral is defined using the normalized Haar measure on H. By construction, g' is an H-invariant Riemannian metric on U. ▼

We now proceed to the proof of the Tube Theorem. Since the G-action is proper, the isotropy subgroup $H := G_m$ is compact. Let g be an H-invariant metric defined on some H-invariant neighborhood of m, whose existence is guaranteed by the previous lemmas. Let Exp be the corresponding Riemannian exponential and N_m the orthogonal complement to $\mathfrak{g} \cdot m$ in T_mM with respect to the inner product induced by $g(m)$. The vector subspace N_m is clearly H-invariant and equivariantly isomorphic to $T_mM/T_m(G \cdot m)$ via the map $v \longmapsto v + \mathfrak{g} \cdot m$; recall that $\mathfrak{g} \cdot m = T_m(G \cdot m)$. Now, an elementary fact in Riemannian geometry (see for instance PERDIGÃO DO CARMO (1993, Theorem 3.7)) guarantees the existence of a neighborhood W of the origin in T_mM such that the restriction of (the Riemannian exponential map) Exp_m to W is a diffeomorphism onto its image. By Lemma 2.3.29, W can be chosen to be H-invariant. Let $V := W \cap N_m$. The open set V is H-invariant by construction and hence the twisted product $G \times_H V$ is well defined. Let now τ be the map defined by

$$\begin{aligned} \tau : G \times_H V &\longrightarrow M \\ [g, v] &\longmapsto g \cdot \mathrm{Exp}_m v. \end{aligned} \tag{2.3.4}$$

The map τ is well defined due to the H-equivariance of Exp and is obviously G-equivariant.

We now show that $T_{[e,0]}\tau$ is an isomorphism. Let $\pi_H : G \times V \to G \times_H V$ be the projection. It is easy to see that for any $(\xi, u) \in T_{(e,0)}(G \times V) = \mathfrak{g} \times N_m$ we have

$$T_{(e,0)}(\tau \circ \pi_H) \cdot (\xi, u) = \xi_M(m) + u.$$

Since, by construction, N_m is a complement to $\mathfrak{g} \cdot m$ in T_mM, $T_{[e,0]}\tau$ is necessarily surjective. For injectivity, if $\xi_M(m) + u = 0$, then we must have that $u = 0$ and that $\xi \in \mathfrak{h}$. Consequently, $T_{(e,0)}\pi_H \cdot (\xi, u) = 0$ and thus $T_{[e,0]}\tau$ is an isomorphism.

The Local Diffeomorphism Theorem implies the existence of a neighborhood W' of $[e, 0]$ in $G \times_H V$ such that the restriction of τ to W' is a diffeomorphism onto its image. In particular, there exists an open neighborhood V' of 0 in V such that for any $v \in V'$, the point $[e, v]$ is contained in W'. Since $\tau|_{W'}$ is a diffeomorphism, it follows thus that its tangent map $T_{[e,v]}\tau$ is an isomorphism. Next, the equivariance of τ implies that $T_{[g,v]}\tau$ is an isomorphism for any $g \in G$ and $n \in V'$. Hence, τ restricted to $G \times_H V'$ is a local diffeomorphism.

We now use the properness condition on the G-action to show that there exists an open H-invariant subset B of V' containing 0 such that the restriction of τ to $G \times_H B$ is injective. Suppose, by contradiction, that such a subset does not exist. Then there exist two sequences $\{[g_i, v_i]\}_{i\in\mathbb{N}}$ and $\{[g_i', v_i']\}_{i\in\mathbb{N}}$ such that both $\{v_i\}_{n\in\mathbb{N}}$ and $\{v_i'\}_{n\in\mathbb{N}}$ tend to zero, and for every i, $[g_i, v_i] \neq [g_i', v_i']$ but $g_i \cdot \operatorname{Exp}_m v_i = g_i' \cdot \operatorname{Exp}_m v_i'$. We assume without loss of generality that $g_i' = e$ for every i. Since $\{v_i\}_{n\in\mathbb{N}}$ and $\{v_i'\}_{n\in\mathbb{N}}$ tend to zero, the sequences $\operatorname{Exp}_m v_i$ and $\operatorname{Exp}_m v_i'$ tend to m and the equality $g_i \cdot \operatorname{Exp}_m v_i = \operatorname{Exp}_m v_i'$ implies that $\{g_i \cdot \operatorname{Exp}_m v_i\}_{i\in\mathbb{N}}$ does as well. The properness of the G-action implies that $\{g_i\}_{i\in\mathbb{N}}$ has a convergent subsequence which we assume, without loss of generality, to be the original sequence. Let $g = \lim_{i\to\infty} g_i$. The continuity of the group action and the equalities $g_i \cdot \operatorname{Exp}_m v_i = \operatorname{Exp}_m v_i'$ show that $g \cdot m = m$. In other words, $g \in H$ and hence $[g, 0] = [e, 0]$.

Let W' be the neighborhood of $[e, 0]$ in $G \times_H V$ on which τ is a diffeomorphism onto its image; the existence of such a neighborhood was just proved a few lines above. Since $[g_i, v_i] \to [g, 0] = [e, 0] \in W'$ and $[e, v_i'] \to [e, 0] \in W'$ it follows that for i big enough both $[g_i, v_i]$ and $[e, v_i']$ are in W'. By hypothesis $[g_i, v_i] \neq [e, v_i']$ and $g_i \cdot \operatorname{Exp}_m v_i = \operatorname{Exp}_m v_i'$, which contradicts the bijectivity of τ on W'. Hence, there exists an H-invariant neighborhood B of zero in $V \subset N_m$ such that the restriction of τ to $G \times_H B$ is injective.

The restriction of τ to $G \times_H B$ is an injective local G-equivariant diffeomorphism onto its image and therefore a diffeomorphism onto an open G-invariant neighborhood of $m \in M$. This is the map φ needed in the statement of our theorem. ∎

2.3.31 Theorem (Slice Theorem) *Let M be a manifold and G a Lie group acting properly on M at the point $m \in M$. Then there exists a slice for the G-action at m.*

Proof. This is a straightforward consequence of the Tube Theorem: take $S := \varphi[e, B]$ as the slice at m. ∎

2.3.32 Some elementary consequences of the Slice Theorem. The Slice Theorem has as a consequence that the orbits of a proper action are always closed and embedded submanifolds, that is, one can drop the second countability hypothesis on the manifold needed in Proposition 2.3.12.

2.3.33 Corollary *Let M be a manifold and G a Lie group acting properly on M at the point $m \in M$. The orbit $G \cdot m$ is a closed and embedded submanifold of M.*

Proof. The Tube Theorem implies that in a G-invariant neighborhood U of the orbit $G \cdot m$, the G-manifold M can be identified with the twisted product $G \times_H B$ via a G-equivariant map $\varphi : G \times_H B \to U$, where B is an open H-invariant neighborhood of 0 in the vector space $T_m M / T_m(G \cdot m)$, on which $H := G_m$ acts linearly. In this model, the orbit $G \cdot m$ is given by the subset $G \times_H \{0\} \simeq G/H$ which is a closed and embedded submanifold of $G \times_H B$. The orbit $G \cdot m = \varphi(G \times_H \{0\})$ is therefore a closed and embedded submanifold of U, and consequently of M. ■

The Slice Theorem and the Tube Lemma in point set topology imply the following improvement of Lemma 2.3.29, which we state for future reference.

2.3.34 Corollary *Let G be a compact Lie group that acts on the manifold M and let $m \in M$. Any open neighborhood of the orbit $G \cdot m$ contains a G-invariant open neighborhood of $G \cdot m$.*

Proof. Let H be a compact Lie group acting linearly on the finite-dimensional vector space E. Averaging an arbitrary inner product on E (see (2.3.3)), one obtains an H-invariant inner product on E. Since all norms on E are equivalent, it follows that any open set containing the origin also contains an open ball in this invariant metric, that is, an H-invariant neighborhood.

We now prove the corollary. Let V be an open neighborhood of the orbit $G \cdot m$. Since G is compact the action is proper. Let $\varphi : G \times_H B \to U$ be a tube around $G \cdot m$ where B is an open H-invariant neighborhood of 0 in the vector space $T_m M / T_m(G \cdot m)$, on which $H := G_m$ acts linearly. As U and V contain the orbit $G \cdot m$ so does $W := U \cap V$. Hence, $\varphi^{-1}(W) \subset G \times_H B$ is an open neighborhood of $G \times_H \{0\}$ and $\pi_H^{-1}(\varphi^{-1}(W)) \subset G \times B$ is an open neighborhood of $G \times \{0\}$, where $\pi_H : G \times B \to G \times_H B$ is the projection. The Tube Lemma (MUNKRES (1975), Lemma 5.8, page 169) implies the existence of an open neighborhood T of 0 in B such that $G \times T \subset \pi_H^{-1}(\varphi^{-1}(W))$. By the argument at the beginning of the proof, we can choose T to be H-invariant. Hence, $\varphi(G \times_H T) \subset W \subset V$ is the open G-invariant neighborhood whose existence we wanted to prove. ■

2.3.35 Corollary *Let G be a Lie group acting freely and properly on the manifold M. Let S be a slice at the point $m \in M$. Then there exists a smooth surjective map $\chi : G \cdot S \to S$ such that for any $z \in G \cdot S$ we have $G \cdot \chi(z) = G \cdot z$ and $\chi|_S = \mathrm{id}_S$.*

Proof. Let $\varphi : G \times S \to G \cdot S$ be a tube around the orbit $G \cdot m$ and $\pi_S : G \times S \to S$ the projection onto the S factor. The map $\chi : G \cdot S \to S$ is given by $\chi := \pi_S \circ \varphi^{-1}$. ■

2.3.36 Equivariant sections of homogeneous manifolds. In part **(iii)–(d)** of Theorem 2.3.26 we saw how a slice at the point m in a proper G-manifold can be combined with a local section of the projection $G \to G/G_m$ to produce a particularly convenient set of coordinates around m. We will see in Section 3.3 that in order for this construction to be of more use, the local section of $G \to G/G_m$ should have certain equivariance properties that we will develop below. The following technical results have been introduced in FIELD (1982, 1983, 1991) in the context of the study of equivariant vector fields.

Let G be a Lie group and $H \subset G$ a compact Lie subgroup. We define $T(H)$ as the product $H \times H$ endowed with the Lie group structure induced by the multiplication defined by

$$(a, b)(c, d) := (ac, bada^{-1}), \quad a, b, c, d \in H.$$

The neutral element is (e, e) and the inverse is given by $(a, b)^{-1} = (a^{-1}, a^{-1}b^{-1}a)$. $T(H)$ acts on G by

$$(a, b) \cdot g = aga^{-1}b^{-1}, \quad (a, b) \in T(H),\ g \in G.$$

The subsets $H_c := H \times \{e\}$ and $H_r := \{e\} \times H$ are subgroups of $T(H)$ isomorphic to H. H_r is normal in $T(H)$ and $T(H)$ can be written as the semidirect product of H_c and H_r (see 2.2.4 for the definitions and conventions on semidirect products). The notation used for these two subgroups is justified by the fact that when we restrict the $T(H)$-action on G to an H_r and an H_c-action on G we obtain right translation by the inverse and conjugation, respectively. We emphasize that both actions are left actions.

The hypothesis on the compactness of H implies that the $T(H)$-action on G is automatically proper. Let S_e be a slice at $e \in G$ for this action.

2.3.37 Lemma *In the setup that we just introduced we have that:*

(i) $H = T(H) \cdot e = H_r \cdot e$.

(ii) $T(H)_e = H_c$.

(iii) $H_c \cdot S_e = S_e$. *Equivalently,* $hS_eh^{-1} = S_e$ *for all* $h \in H$.

(iv) $T(H) \cdot S_e = H_r \cdot S_e$.

(v) *The* H_r*-action on* G *is free and* S_e *is a slice at* e *for this action.*

(vi) *There exists a smooth map* $\chi : H_r \cdot S_e \to S_e$ *such that*

  (a) $H_r \cdot \chi(g) = H_r \cdot g$, *for all* $g \in H_r \cdot S_e$. *Equivalently,* $\chi(g)H = gH$, *for all* $g \in H_r \cdot S_e$.

  (b) $\chi(hgh^{-1}t) = h\chi(g)h^{-1}$, *for all* $t, h \in H$ *and all* $g \in H_r \cdot S_e$.

Proof. **(i)** and **(ii)** are straightforward.

(iii) is a consequence of **(ii)** and of the fact that the slice is invariant with respect to the action of the isotropy subgroup of the point at which it is constructed.

(iv) The inclusion $H_r \cdot S_e \subset T(H) \cdot S_e$ is obvious. Conversely, let $g \in S_e$ and $(a, b) \in T(H)$. By **(iii)** $aga^{-1} \in S_e$ and hence $(a, b) \cdot g = aga^{-1}b^{-1} \in S_eb^{-1} \subset H_r \cdot S_e$.

(v) The H_r-action on G is clearly free. We will prove that S_e is a slice at e for this action using the characterization for the slices provided in point **(iv)** of Theorem 2.3.26. Since S_e is a slice at e for the $T(H)$-action on G, there exists a $T(H)$-equivariant smooth retraction $r : T(H) \cdot S_e \to T(H) \cdot e$ of the inclusion $T(H) \cdot e \hookrightarrow T(H) \cdot S_e$ such that $r^{-1}(e) = S_e$. However, as $T(H) \cdot S_e = H_r \cdot S_e$, $T(H) \cdot e = H = H_r \cdot e$, and r is also H_r-equivariant, then $r : H_r \cdot S_e \to H_r \cdot e$ can also be seen as an H_r-equivariant retraction of the inclusion $H_r \cdot e \hookrightarrow H_r \cdot S_e$, whose existence guarantees, by Theorem 2.3.26, that S_e is a slice at e for the H_r-action on G.

(vi) Part **(a)** is a consequence of Corollary 2.3.35 applied to the free H_r-action on G and using S_e as a slice for it at the point e. Regarding **(b)** note first that for any $t, h \in H$ and $g \in H_r \cdot S_e$ we have by **(iv)** that $hgh^{-1}t = (h, t^{-1}) \cdot g \in T(H) \cdot (H_r \cdot S_e) = T(H) \cdot S_e = H_r \cdot S_e$. Now, by part **(a)** we have that $\chi(hgh^{-1}t)H = hgh^{-1}tH = hgh^{-1}H =$

$hgHh^{-1} = h\chi(g)Hh^{-1} = h\chi(g)h^{-1}H$, which is equivalent to $H_r \cdot \chi(hgh^{-1}t) = H_r \cdot h\chi(g)h^{-1}$. This equality implies the existence of an element $k \in H_r$ such that $k \cdot \chi(hgh^{-1}t) = h\chi(g)h^{-1}$. Given that both $\chi(hgh^{-1}t)$ and $h\chi(g)h^{-1}$ are in S_e (by the construction of χ and **(iii)**, respectively) and the H_r-action on G is free we have that $k = e$ necessarily. ■

Let $\pi_H : G \to G/H$ be the projection onto the orbit space of the left H_r-action on G. Recall that G (and consequently H) acts on G/H on the left by $k \cdot gH := kgH$. The projection π_H is equivariant with respect to this action, that is, $\pi_H(gk) = g \cdot \pi_H(k)$.

2.3.38 Proposition *Let G be a Lie group and $H \subset G$ a compact Lie subgroup. There exists an H-invariant open neighborhood U of H in G/H and a smooth local section $\sigma : U \subset G/H \to G$ of the projection $\pi_H : G \to G/H$ such that*

$$\sigma(h \cdot x) = h\sigma(x)h^{-1}, \text{ for all } h \in H \text{ and } x \in U. \tag{2.3.5}$$

Proof. Take S_e as in the previous lemma and define $U := \pi_H(H_r \cdot S_e)$. U is an open H-invariant neighborhood of H in G/H (recall that π_H is open and part **(iii)** in the previous lemma). Consider now the map $\chi : H_r \cdot S_e \to S_e$ introduced in Lemma 2.3.37. Since $\chi(gt) = \chi(g)$ for all $g \in H_r \cdot S_e$ and $t \in H$, the map χ induces a map $\sigma : U \to G$ on the quotient, that is, $\sigma(gH) =: \chi(g)$, for all $gH \in U$. However, $\chi(hgh^{-1}) = h\chi(g)h^{-1}$ for all $h \in H$ and $g \in H_r \cdot S_e$, so the equivariance property (2.3.5) required for σ follows. ■

Any local section of the homogeneous manifold G/H that satisfies (2.3.5) will be called an ***equivariant*** or ***admissible section***. As a straightforward corollary of the equivariance relation (2.3.5), we obtain the following observation that we will use later on in the book.

2.3.39 Corollary *Let $\sigma : U \subset G/H \to G$ be an equivariant section of the projection $\pi_H : G \to G/H$. Let T be a subset of H and take $g \in N(T)$ such that $gH \in U$. Then $\sigma(gH) \in Z(T)$. In particular, $\sigma(H) \in Z(H)$.*

If in part **(iii)–(d)** of Theorem 2.3.26 we use the local section that we just constructed we obtain the following result:

2.3.40 Corollary *Let G be a Lie group acting properly on the manifold M. Let $m \in M$ be a point with isotropy subgroup $H := G_m$ and S a slice at m. Then there exists an H-invariant open neighborhood U of H in G/H and a smooth local section $\sigma : U \subset G/H \to G$ of the projection $\pi_H : G \to G/H$ such that the map*

$$\begin{aligned} F : U \times S &\longrightarrow M \\ (u, s) &\longmapsto \sigma(u) \cdot s \end{aligned}$$

is an H-equivariant diffeomorphism onto an H-invariant open subset of M, with respect to the H-action $h \cdot (gH, s) := (hgH, h \cdot s)$ on $U \times S$ and the H-action on M given by (2.3.5).

2.4 The structure of proper G-manifolds

A group action on a manifold naturally induces on it and on the corresponding orbit space a partition by means of the so-called orbit type submanifolds. When the group action is proper these partitions have outstanding geometrical properties that we will review in this section. To be more specific, we will see that proper G-manifolds and their corresponding orbit spaces are Whitney (B) stratified spaces. In the process we will introduce several notions that will be heavily used throughout the book.

2.4.1 Type submanifolds and fixed point subspaces. The subsets defined in the following definition will be of extreme importance in our future discussions.

2.4.2 Definition *Let G be a Lie group acting on the manifold M. Let H, J, and K be closed subgroups of G such that $H \subset K \subset G$. We define the following conjugacy classes:*

$$\begin{aligned}
(H) &= \{L \subset G \mid L = gHg^{-1},\ g \in G\},\\
(H)^K &= \{L \subset G \mid L = gHg^{-1},\ g \in K\},\\
(H)_K &= \{L \subset G \mid L = gHg^{-1} \subset K,\ g \in G\},\\
(H)_K^J &= (H)_K \cap (H)^J = \{L \subset G \mid L = gHg^{-1} \subset K,\ g \in J\}.
\end{aligned}$$

We use these conjugacy classes to define the following subsets of M:

$$\begin{aligned}
M_{(H)} &= \{z \in M \mid G_z \in (H)\},\\
M_{(H)^K} &= \{z \in M \mid G_z \in (H)^K\},\\
M_{(H)_K} &= \{z \in M \mid G_z \in (H)_K\},\\
M_{(H)_K^J} &= \{z \in M \mid G_z \in (H)_K^J\},\\
M^H &= \{z \in M \mid H \subset G_z\},\\
M_H &= \{z \in M \mid H = G_z\}.
\end{aligned}$$

*The set $M_{(H)}$ is referred to as the (H)-**orbit type submanifold**. M_H is the H-**isotropy type submanifold** and M^H is the H-**fixed point submanifold**. We will collectively call these subsets **type submanifolds**. As we will see later on, the connected components of these sets will have special relevance: we will denote by $M^z_{(H)}$, $M^z_{(H)^K}$, $M^z_{(H)_K}$, $M^z_{(H)_K^J}$, M^H_z, and M^z_H the connected components of $M_{(H)}$, $M_{(H)^K}$, $M_{(H)_K}$, $M_{(H)_K^J}$, M^H, and M_H that respectively contain a given point $z \in M$.*

The previous definitions admit an infinitesimal version. For instance, if $\mathfrak{h} \subset \mathfrak{g}$ is a Lie subalgebra of $\mathfrak{g}$, we define

$$(\mathfrak{h}) = \{\mathfrak{k} \subset \mathfrak{g} \mid \mathfrak{k} = \mathrm{Ad}_g(\mathfrak{h}),\, g \in G\}$$

and

$$\begin{aligned}
M_{(\mathfrak{h})} &= \{z \in M \mid \mathfrak{g}_z \in (\mathfrak{h})\},\\
M^{\mathfrak{h}} &= \{z \in M \mid \mathfrak{h} \subset \mathfrak{g}_z\},\\
M_{\mathfrak{h}} &= \{z \in M \mid \mathfrak{g}_z = \mathfrak{h}\}.
\end{aligned}$$

*The set $M_{(\mathfrak{h})}$ is referred to as the **infinitesimal** $(\mathfrak{h})$**-orbit type submanifold**. $M_{\mathfrak{h}}$ is the* $\mathfrak{h}$***-infinitesimal isotropy type submanifold** and $M^{\mathfrak{h}}$ is the* $\mathfrak{h}$***-infinitesimal fixed point submanifold**. We will collectively call these subsets **infinitesimal type submanifolds**.*

2.4.3 Remark If M is a vector space on which H acts linearly, the set M^H is found in the physics literature under the names of ***space of singlets*** or ***space of invariant vectors***. See ABUD AND SARTORI (1981); SARTORI (1983); ABUD AND SARTORI (1983).

2.4.4 Proposition *The sets defined in the previous definition satisfy:*

(i) *M^H and $M^{\mathfrak{h}}$ are closed in M.*

(ii) *$M_{(H)} = G \cdot M_H$ and $M_{(\mathfrak{h})} = G \cdot M_{\mathfrak{h}}$.*

(iii) *$M_{(H)^K} = K \cdot M_H$.*

(iv) *$M_{\mathfrak{h}} = M^{\mathfrak{h}} \cap M_{(\mathfrak{h})}$ and $M_{\mathfrak{h}}$ is closed in $M_{(\mathfrak{h})}$.*

(v) *If H is compact, then $M_H = M^H \cap M_{(H)}$ and M_H is closed in $M_{(H)}$.*

Proof **(i)** Let z be an element in the closure of M^H in M and $\{z_n\}_{n\in\mathbb{N}} \subset M^H$ a sequence such that $z_n \to z$. By the continuity of the action we have that $h \cdot z_n = z_n \to h \cdot z$, for any $h \in H$. The uniqueness of the limit implies that $h \cdot z = z$, for all $h \in H$ and hence $z \in M^H$. Analogously, let z be an element in the closure of $M^{\mathfrak{h}}$ in M and $\{z_n\}_{n\in\mathbb{N}} \subset M^{\mathfrak{h}}$ a sequence such that $z_n \to z$. By definition of $M^{\mathfrak{h}}$ we have that $\eta_M(z_n) = 0$ for all $\eta \in \mathfrak{h}$. Taking the limit we get $\eta_M(z) = 0$ and hence $z \in M^{\mathfrak{h}}$.

(ii) Let $z \in M_{(H)}$ be arbitrary. By definition, there exists an element $g \in G$, such that $G_z = gHg^{-1}$. Hence $G_{g^{-1}\cdot z} = g^{-1}gHgg^{-1} = H$, and therefore $z = g \cdot (g^{-1} \cdot z) \in G \cdot M_H$. The converse inclusion follows from $M_H \subset M_{(H)}$, and the G-invariance of $M_{(H)}$. The second equality is proved in the same way.

(iii) Obvious.

(iv) The only nontrivial inclusion is $M^{\mathfrak{h}} \cap M_{(\mathfrak{h})} \subset M_{\mathfrak{h}}$. If $z \in M^{\mathfrak{h}} \cap M_{(\mathfrak{h})}$, then there exists $g \in G$ such that $\mathfrak{g}_z = \mathrm{Ad}_g\mathfrak{h}$ and $\mathfrak{h} \subset \mathfrak{g}_z$. Therefore, $\mathfrak{g}_z = \mathrm{Ad}_g\mathfrak{h} \subset \mathrm{Ad}_g\mathfrak{g}_z$. Since Ad_g is an isomorphism the vector spaces $\mathfrak{g}_z$ and $\mathrm{Ad}_g\mathfrak{g}_z$ have the same dimension. Since $\mathfrak{g}_z \subset \mathrm{Ad}_g\mathfrak{g}_z$ we have that $\mathfrak{g}_z = \mathrm{Ad}_g\mathfrak{g}_z$. Therefore, $\mathfrak{g}_z = \mathrm{Ad}_g\mathfrak{h} = \mathrm{Ad}_g\mathfrak{g}_z$ and hence $\mathfrak{h} = \mathfrak{g}_z$, that is, $z \in M_{\mathfrak{h}}$. We now prove that $M_{\mathfrak{h}}$ is closed in $M_{(\mathfrak{h})}$. Let $\overline{M_{\mathfrak{h}}}$ be the closure of $M_{\mathfrak{h}}$ in $M_{(\mathfrak{h})}$. For any $z \in \overline{M_{\mathfrak{h}}}$ there exists an element $g \in G$ such that $\mathfrak{g}_z = \mathrm{Ad}_g\mathfrak{h}$ and a sequence $\{z_n\}_{n\in\mathbb{N}} \subset M_{\mathfrak{h}}$ such that $\lim_{n\to\infty} z_n = z$. Since $\{z_n\}_{n\in\mathbb{N}} \subset M_{\mathfrak{h}}$ we have $\xi_M(z_n) = 0$, for all $\xi \in \mathfrak{h}$, $n \in \mathbb{N}$, and hence $\xi_M(z) = 0$, for all $\xi \in \mathfrak{h}$. Consequently $z \in M^{\mathfrak{h}}$, that is, $\mathfrak{h} \subset \mathfrak{g}_z$. As $\mathfrak{g}_z = \mathrm{Ad}_g\mathfrak{h} \subset \mathrm{Ad}_g\mathfrak{g}_z$ and Ad_g is an isomorphism we have that $\mathfrak{g}_z = \mathrm{Ad}_g\mathfrak{h} = \mathrm{Ad}_g\mathfrak{g}_z$ and hence $\mathfrak{h} = \mathfrak{g}_z$.

(v) The only nontrivial inclusion is $M^H \cap M_{(H)} \subset M_H$. If $m \in M^H$, then $H \subset G_m$. If also $m \in M_{(H)}$, there is a $g \in G$ such that $G_m = gHg^{-1}$, and hence $H \subset gHg^{-1}$ which, by the compactness of H and Lemma 2.1.14, implies that $G_m = H$. We now prove that M_H is closed in $M_{(H)}$ following BIRTEA (2002). Let $\overline{M_H}$ be the closure of M_H in $M_{(H)}$. For any $z \in \overline{M_H}$ there exists an element $g \in G$ such that $G_z = gHg^{-1}$

and a sequence $\{z_n\}_{n\in\mathbb{N}} \subset M_H$ such that $\lim_{n\to\infty} z_n = z$. Since $\{z_n\}_{n\in\mathbb{N}} \subset M_H$ we have $h\cdot z_n = z_n$, for all $h \in H$, $n \in \mathbb{N}$, and hence $z_n = h\cdot z_n \to h\cdot z$. Uniqueness of the limit implies then that $h \cdot z = z$ for all $h \in H$, that is, $H \subset G_z = gHg^{-1}$. Lemma 2.1.14, implies that $H = gHg^{-1}$ and hence $G_z = H$. ■

2.4.5 Proposition *Let G be a Lie group acting smoothly on the manifold M. If $m \in M$ and $H := G_m$, then*

$$T_m(G\cdot m) \cap (T_mM)^H = T_m(N(H)\cdot m). \tag{2.4.1}$$

Proof. We clearly have $T_m(N(H)\cdot m) \subset T_m(G\cdot m)$. If ξ is in the Lie algebra of $N(H)$ and $h \in H$, by Lemma 2.1.13 we conclude that $\mathrm{Ad}_h\,\xi - \xi \in \mathfrak{h}$ so that $(\mathrm{Ad}_h\,\xi - \xi)_M(m) = 0$ and hence

$$\begin{aligned}\xi_M(m) &= (\mathrm{Ad}_h\,\xi - \xi)_M(m) + \xi_M(m) = (\mathrm{Ad}_h\,\xi)_M(m)\\ &= (\mathrm{Ad}_h\,\xi)_M(h\cdot m) = (\Phi^*_{h^{-1}}\xi_M)(h\cdot m)\\ &= (T_m\Phi_h \circ \xi_M \circ \Phi_{h^{-1}})(h\cdot m) = T_m\Phi_h(\xi_M(m))\end{aligned}$$

which shows that $\xi_M(m) \in T_mM$ is fixed by all $h \in H$, that is, $\xi_M(m) \in (T_mM)^H$. We have hence $T_m(N(H)\cdot m) \subset (T_mM)^H$ and therefore $T_m(N(H)\cdot m) \subset T_m(G\cdot m) \cap (T_mM)^H$.

To prove the converse inclusion, note that if $\xi_M(m) \in T_m(G\cdot m) \cap (T_mM)^H$, that is, $\xi_M(m)$ is fixed by all $h \in H$, then

$$\begin{aligned}\xi_M(m) &= T_m\Phi_h(\xi_M(m)) = (T_m\Phi_h \circ \xi_M \circ \Phi_{h^{-1}})(h\cdot m)\\ &= (\Phi^*_{h^{-1}}\xi_M)(h\cdot m) = (\mathrm{Ad}_h\,\xi)_M(m)\end{aligned}$$

which states that $\mathrm{Ad}_h\,\xi - \xi \in \mathfrak{g}_m = \mathfrak{h}$ for all $h \in H$. By Lemma 2.1.13 it follows that ξ is in the Lie algebra of $N(H)$ which proves the inclusion $T_m(G\cdot m) \cap (T_mM)^H \subset T_m(N(H)\cdot m)$ and hence the proposition. ■

2.4.6 Proposition *Let A be an H-manifold and $G\times_H A$ the twisted product introduced in Definition 2.3.16, with $H \subset G$ a compact subgroup of G. Then, relative to the left action of G on $G\times_H A$ given by $g'\cdot[g,a] = [g'g, a]$ for $g', g \in G$ and $a \in A$, we have*

(i) $(G\times_H A)_H = N(H)\times_H A^H = N(H)\times_H A_H = (G\times_H A)^H$,

(ii) $(G\times_H A)_{(H)} = G\times_H A^H = G\times_H A_H$.

(iii) *If $K \subset G$ is a closed subgroup of G such that $H \subset K \subset G$, then* $(G\times_H A)_{(H)^K} = K\,N(H)\times_H A^H = K\,N(H)\times_H A_H$.

(iv) *If $K \subset G$ is an isotropy subgroup of the G-action on $G\times_H A$, then we have that* $(G\times_H A)_{(K)} = G\times_H A_{(K)^G_H}$.

(v) $(G\times_H A)_{\mathfrak{h}} = N(\mathfrak{h})\times_H A^{\mathfrak{h}} = N(\mathfrak{h})\times_H A_{\mathfrak{h}} = (G\times_H A)^{\mathfrak{h}}$,

(vi) $(G\times_H A)_{(\mathfrak{h})} = G\times_H A^{\mathfrak{h}} = G\times_H A_{\mathfrak{h}}$.

(vii) *Let S be an (immersed, initial, or embedded) G-invariant submanifold of $G \times_H A$. Then there exists an H-invariant (immersed, initial, or embedded, respectively) submanifold B of A such that $S = G \times_H B$.*

Proof. Since H is compact, its action on A is proper and therefore, by Proposition 2.3.17 the G-action on $G \times_H A$ is necessarily proper.
(i) If $n = [g, a] \in (G \times_H A)_H$, by Proposition 2.3.18 we have

$$H = G_n = gH_a g^{-1}.$$

Since $H_a \subset H$, we have that $H = gH_a g^{-1} \subset gHg^{-1}$. The compactness of H and Lemma 2.1.14 imply that $H = gHg^{-1} = gH_a g^{-1}$, which guarantees that $g \in N(H)$ and that $H = H_a$. Hence $a \in A_H \subset A^H$, thereby showing that $(G \times_H A)_H \subset N(H) \times_H A_H \subset N(H) \times_H A^H$.

Conversely, if $n = [g, a] \in N(H) \times_H A^H$, then $H \subset H_a$ and $g \in N(H)$. However, since we always have $H_a \subset H$, it follows that $H_a = H$. Hence, by Proposition 2.3.18, $G_n = gH_a g^{-1} = gHg^{-1} = H$. This shows that $N(H) \times_H A^H \subset (G \times_H A)_H$, thus proving the equalities

$$(G \times_H A)_H = N(H) \times_H A_H = N(H) \times_H A^H.$$

We now prove that $(G \times_H A)^H = N(H) \times_H A^H$. One inclusion is clear since $N(H) \times_H A^H = (G \times_H A)_H \subset (G \times_H A)^H$. Conversely, if $n = [g, a] \in (G \times_H A)^H$ we have that for any $h \in H$, $h \cdot n = n$ or $[hg, a] = [g, a]$. This implies the existence of some $h' \in H$ such that $hg = gh'$, that is, $g^{-1}hg \in H$. Since h is arbitrary we have that $g^{-1}Hg \subset H$ and, by Lemma 2.1.14 we obtain that $g^{-1}Hg = H$ and hence $g \in N(H)$. Let now $h \in H$ be arbitrary and consider the element $ghg^{-1} \in H$ (since $g \in N(H)$). Since, by hypothesis, $ghg^{-1} \cdot n = n$ we have that

$$[g, a] = ghg^{-1} \cdot [g, a] = [ghg^{-1}g, a] = [gh, a] = [g, ha],$$

which implies that $h \cdot a = a$ for all $h \in H$, that is, $a \in A^H$.

(ii) Follows from part **(i)**, by taking into account that (see Proposition 2.4.4) $(G \times_H A)_{(H)} = G \cdot (G \times_H A)_H = G \cdot (N(H) \times_H A_H) = G \cdot (N(H) \times_H A^H) = G \times_H A_H = G \times_H A^H$. An analogous argument proves **(iii)**: $(G \times_H A)_{(H)^K} = K \cdot (G \times_H A)_H = K \cdot (N(H) \times_H A_H) = K\, N(H) \times_H A_H$.

(iv) Let $[g, a] \in (G \times_H A)_{(K)}$. By definition $G_{[g,a]} = gH_a g^{-1} = lKl^{-1}$ for some $l \in G$. This implies that $H_a = g^{-1}lKl^{-1}g \in (K)_H^G$ and consequently $[g, a] \in G \times_H A_{(K)_H^G}$. Conversely, let $[g, a] \in G \times_H A_{(K)_H^G}$. By definition there exists an element $l \in G$ such that $H_a = lKl^{-1} \subset H$ and hence $G_{[g,a]} = gH_a g^{-1} = glKl^{-1}g^{-1} \in (K)$, which implies that $[g, a] \in (G \times_H A)_{(K)}$.

(v) Let $[g, a] \in (G \times_H A)_{\mathfrak{h}}$. By definition, $\mathfrak{g}_{[g,a]} = \mathfrak{h}$. Equivalently, $\mathrm{Ad}_g \mathfrak{h}_a = \mathfrak{h}$. Now, since $\mathfrak{h}_a \subset \mathfrak{h}$, we have that $\mathfrak{h} = \mathrm{Ad}_g \mathfrak{h}_a \subset \mathrm{Ad}_g \mathfrak{h}$. Since Ad_g is a linear isomorphism we have that $\mathfrak{h} = \mathrm{Ad}_g \mathfrak{h}$ and hence $\mathfrak{h}_a = \mathfrak{h}$ and hence $g \in N(\mathfrak{h})$ which proves that

$$(G \times_H A)_{\mathfrak{h}} \subset N(\mathfrak{h}) \times_H A_{\mathfrak{h}} \subset N(\mathfrak{h}) \times_H A^{\mathfrak{h}}. \tag{2.4.2}$$

We now show that

$$N(\mathfrak{h}) \times_H A^{\mathfrak{h}} \subset (G \times_H A)^{\mathfrak{h}}. \tag{2.4.3}$$

Let $[n, a] \in N(\mathfrak{h}) \times_H A^{\mathfrak{h}}$ and $\xi \in \mathfrak{h}$ arbitrary. Then, if $\pi : G \times A \to G \times_H A$ is the projection, we have that

$$\xi_{G\times_H A}([n, a]) = T_{(n,a)}\pi(T_e R_n \cdot \xi, 0).$$

Given that $n \in N(\mathfrak{h})$ we have that $\mathrm{Ad}_{n^{-1}}\xi \in \mathfrak{h}$, that is, $T_e(L_{n^{-1}} \circ R_n) \cdot \xi \in \mathfrak{h}$, which implies that $T_e R_n \cdot \xi = T_e L_n \cdot \eta$, for some $\eta \in \mathfrak{h}$. Therefore, $\xi_{G\times_H A}([n, a]) = T_{(n,a)}\pi(T_e L_n \cdot \eta, 0) = T_{(n,a)}\pi(0, -\eta_A(a)) = 0$, because $a \in A^{\mathfrak{h}}$. This proves (2.4.3). We conclude by showing that

$$(G \times_H A)^{\mathfrak{h}} \subset (G \times_H A)_{\mathfrak{h}}. \tag{2.4.4}$$

Indeed, if $[g, a] \in (G \times_H A)^{\mathfrak{h}}$ we have that $\mathfrak{h} \subset \mathfrak{g}_{[g,a]} = \mathrm{Ad}_g \mathfrak{h}_a \subset \mathrm{Ad}_g \mathfrak{h}$. Since Ad_g is a linear isomorphism $\mathfrak{h} = \mathrm{Ad}_g \mathfrak{h}$ necessarily and hence $\mathfrak{h} = \mathfrak{g}_{[g,a]}$ which implies that $[g, a] \in (G \times_H A)_{\mathfrak{h}}$. The expressions (2.4.2), (2.4.3), and (2.4.4) prove the equalities in the statement.

(vi) By part **(ii)** in Proposition 2.4.4 we have that

$$(G \times_H A)_{(\mathfrak{h})} = G \cdot (G \times_H A)_{\mathfrak{h}} = G \cdot (N(\mathfrak{h}) \times_H A_{\mathfrak{h}}) = G \times_H A_{\mathfrak{h}}.$$

Analogously, we have that $(G \times_H A)_{(\mathfrak{h})} = G \times_H A^{\mathfrak{h}}$.

(vii) Let $B := \{a \in A \mid [g, a] \in S \text{ for some } g \in G\}$. We first show that the set B is H-invariant. Indeed, for any $a \in B$ and any $h \in H$ there exists by definition an element $g \in G$ such that $[g, a] \in S$. Now, since S is G-invariant, the element $[g, h \cdot a] = [gh, a] = ghg^{-1} \cdot [g, a]$ belongs to S because $[g, a] \in S$ and hence $h \cdot a \in B$. The H-invariance of B allows us to form the twisted product $G \times_H B$.

The importance of the set B is given by the fact that $S = G \times_H B$. Indeed, if $[g, a] \in S$, then $a \in B$ and hence $[g, a] \in G \times_H B$. Conversely, if $[g, a] \in G \times_H B$ there exists $g' \in G$ such that $[g', a] \in S$, consequently, by the G-invariance of S we have that $[g, a] = g(g')^{-1} \cdot [g', a] \in S$ by the G-invariance of S.

It only remains to be shown that B is an (immersed, initial, or embedded) submanifold of A. Consider the smooth map $i : A \to G \times_H A$ given by $a \longmapsto [e, a]$, for any $a \in A$. Notice that the G-invariance of S implies that an element $a \in i^{-1}(S)$ if and only if $[e, a] \in S$, which is equivalent to $a \in B$. This argument guarantees that $i^{-1}(S) = B$. Additionally, the map i is transversal to the submanifold S. Indeed, if $\pi : G \times A \to G \times_H A$ is the projection and $a \in B = i^{-1}(S)$, then the G-invariance of S implies that

$$T_a i \cdot T_a A + T_{[e,a]} S \supset T_a i \cdot T_a A + \mathfrak{g} \cdot [e, a] = T_{(e,a)}\pi \cdot (\mathfrak{g}, T_a A) = T_{[e,a]}(G \times_H A)$$

which guarantees that

$$T_{[e,a]}(G \times_H A) = T_a i \cdot T_a A + T_{[e,a]} S.$$

Theorem 1.1.15 proves the claim. ■

The next proposition shows that when the G-action is proper, the connected components of some of the subsets introduced in Definition 2.4.2 are actual submanifolds of M.

2.4.7 Proposition *Let G be a Lie group acting properly on the manifold M. Let H be an isotropy subgroup of this action and K a closed subgroup of G such that $H \subset K \subset G$. The connected components of the sets $M_{(H)}$, $M_{(H)^K}$, M^H, and M_H are locally closed embedded submanifolds of M. Moreover, M_H is open in M^H. For any $m \in M_H$, the tangent space to the connected component M_H^m of M_H containing m is given by*

$$\begin{aligned} T_m M_H^m &= \{v_m \in T_m M \mid T_m \Phi_h \cdot v_m = v_m, \ \forall h \in H\} \\ &= (T_m M)^H = T_m M_m^H. \end{aligned} \tag{2.4.5}$$

Proof. This statement is a straightforward consequence of the Slice Theorem and the preceding propositions. Indeed, let $m \in M$ be such that $G_m = H$. The Slice Theorem implies the existence of a tube $\varphi : G \times_H B \to U$ around the orbit $G \cdot m$. The G-equivariance of the map φ yields, by Proposition 2.4.6, that $\varphi^{-1}(U \cap M_H) = \varphi^{-1}(U \cap M^H) = (G \times_H B)_H = N(H) \times_H B^H$, $\varphi^{-1}(U \cap M_{(H)}) = (G \times_H B)_{(H)} = G \times_H B^H$, $\varphi^{-1}\left(U \cap M_{(H)^K}\right) = (G \times_H B)_{(H)^K} = G \times_H B^H = K\, N(H) \times_H B^H$. Since the subsets $N(H) \times_H B^H$, $G \times_H B^H$, and $K\, N(H) \times_H B^H$ are closed submanifolds of $G \times_H B$, the connected components of the subsets M_H, M^H, $M_{(H)}$, and $M_{(H)^K}$ are necessarily locally closed embedded submanifolds of M.

We now prove the equality $T_m M_H^m = (T_m M)^H$ in (2.4.5) by showing that

$$\begin{aligned} T_m M_H^m &= T_{[e,0]}\varphi \cdot T_{[e,0]}(G \times_H B)_H = T_{[e,0]}\varphi \cdot T_{[e,0]}(N(H) \times_H B^H) \\ &= T_{[e,0]}\varphi \cdot (T_{[e,0]}(G \times_H B))^H = (T_{[e,0]}\varphi \cdot T_{[e,0]}(G \times_H B))^H \\ &= (T_m M)^H. \end{aligned} \tag{2.4.6}$$

All the equalities in this chain follow from either Proposition 2.4.6 or from the G-equivariance of φ, except for $T_{[e,0]}\varphi \cdot T_{[e,0]}(N(H) \times_H B^H) = T_{[e,0]}\varphi \cdot (T_{[e,0]}(G \times_H B))^H$ or, equivalently, $T_{[e,0]}(N(H) \times_H B^H) = (T_{[e,0]}(G \times_H B))^H$, that we now prove: let $w \in (T_{[e,0]}(G \times_H B))^H$. If we denote by $\pi_H : G \times B \to G \times_H B$ the canonical projection, there exists $\xi \in \mathfrak{g}$ and $v \in T_0 B$ such that $w = T_{(e,0)}\pi_H(\xi, v)$. The condition $w \in (T_{[e,0]}(G \times_H B))^H$ is equivalent to having $T_{[e,0]}\Phi_h \cdot T_{(e,0)}\pi_H(\xi, v) = T_{(e,0)}\pi_H(\xi, v)$, for all $h \in H$. If $v(t) \subset B$ is a smooth curve in B such that $v(0) = 0$ and $v'(0) = v$, we can rewrite this as

$$\left.\frac{d}{dt}\right|_{t=0} \pi_H(h \exp t\xi, v(t)) = \left.\frac{d}{dt}\right|_{t=0} \pi_H(\exp t\xi, v(t)), \quad \text{for all } h \in H,$$

or, equivalently,

$$\left.\frac{d}{dt}\right|_{t=0} \pi_H(h \exp t\xi h^{-1}, h \cdot v(t)) = \left.\frac{d}{dt}\right|_{t=0} \pi_H(\exp t\xi, v(t)), \quad \text{for all } h \in H,$$

which in turn amounts to

$$T_{(e,0)}\pi_H \cdot (\mathrm{Ad}_h \xi - \xi, T_0\psi_h \cdot v - v) = 0,$$

where ψ denotes the H-action on B. Given that $\ker T_{(e,0)}\pi_H = \mathfrak{h} \times \{0\}$, the previous expression is equivalent to having $\mathrm{Ad}_h\xi - \xi \in \mathfrak{h}$ and $T_0\psi_h \cdot v = v$, for all $h \in H$. By Lemma 2.1.13 the relation on ξ holds if and only if $\xi \in \mathrm{Lie}(N(H))$, which proves the desired equality.

Notice that the remaining equality in 2.4.5, that is, $T_m M_m^H = (T_m M)^H$ follows from (2.4.6) and Proposition 2.4.6. Indeed,

$$\begin{aligned} T_m M_m^H &= T_{[e,0]}\varphi \cdot T_{[e,0]}(G \times_H B)^H \\ &= T_{[e,0]}\varphi \cdot T_{[e,0]}(N(H) \times_H B^H) = (T_m M)^H. \end{aligned}$$

This equality and (2.4.6) guarantee that for any $m \in M_H \subset M^H$, $T_m M_H^m = T_m M_m^H$ and, therefore, the connected components of M_H are necessarily open submanifolds of the connected components of M^H. Thus M_H is open in M^H. ■

2.4.8 Remark If the G-action is proper, the equality in the statement of Proposition 2.4.5 reads:

$$T_m(G \cdot m) \cap T_m M_H^m = T_m(N(H) \cdot m).$$

In this case, an alternative proof of this proposition can be given by using (2.4.5). Indeed,

$$\begin{aligned} T_m(G \cdot m) \cap T_m M_H^m &= T_m(G \cdot m) \cap (T_m M)^H \quad \text{(by (2.4.5))} \\ &= (T_m(G \cdot m))^H \\ &= T_m(G \cdot m)_H \quad \text{(again, by (2.4.5))}. \end{aligned}$$

However,

$$(G \cdot m)_H = \{g \cdot m \mid H = G_{g \cdot m} = gHg^{-1}\} = \{g \cdot m \mid g \in N(H)\} = N(H) \cdot m.$$

So,

$$T_m(G \cdot m) \cap T_m M_H^m = T_m(G \cdot m)_H = T_m(N(H) \cdot m),$$

as required.

2.4.9 Connected components of the type submanifolds and local type submanifolds. The requirement in the previous proposition on the restriction to the connected components of the type manifolds is not vacuous. In order to convince the reader that these connected components may have different dimensions we now consider the following example due to SJAMAAR AND LERMAN (1991). Let $S^5 \subset \mathbb{C}^3$ be given by $S^5 := \{(z_0, z_1, z_2) \in \mathbb{C}^3 \mid |z_0|^2 + |z_1|^2 + |z_2|^2 = 1\}$. The circle S^1 acts freely and properly on S^5 by $e^{i\theta} \cdot (z_0, z_1, z_2) := (e^{i\theta} z_0, e^{i\theta} z_1, e^{i\theta} z_2)$ and therefore the corresponding orbit space is, by Proposition 2.3.8, a smooth Hausdorff manifold usually referred to as the ***complex projective plane*** and denoted by $\mathbb{CP}^2$. We will denote the elements of $\mathbb{CP}^2$ by $[z_0, z_1, z_2]$. It is easy to see that the equality $e^{i\varphi} \cdot [z_0, z_1, z_2] := [e^{i\varphi} z_0, z_1, z_2]$ defines a smooth S^1-action on $\mathbb{CP}^2$ and is such that $(\mathbb{CP}^2)_{S^1} = (\mathbb{CP}^2)^{S^1}$ consists of the point $[1, 0, 0]$ and, at the same time, of all the points of the form $[0, z_1, z_2]$.

A particularly convenient way to deal with components of different dimensions is the use of ***local type manifolds***: given M a proper G-space and $z \in M$ such that $H := G_z$, the ***local orbit type manifold through the point*** z is the subset $M^{l_z}_{(H)}$ of M containing all the points $x \in M$ for which there exists a G-invariant open neighborhood U_x around x and a G-equivariant diffeomorphism from U_x onto a G-invariant neighborhood of z. Using the Tube Theorem, the local orbit type manifold through the point z can be visualized as the subset of $M_{(H)}$ consisting of the points $x \in M_{(H)}$ for which the linear representation of G_x on $T_xM/\mathfrak{g} \cdot x$ is isomorphic to the linear representation of G_z on $T_zM/\mathfrak{g} \cdot z$. It can be shown (see DUISTERMAAT AND KOLK (1999)) that the local orbit type manifolds are locally closed G-invariant submanifolds of M and that they are open and closed subsets of the orbit type manifold that contains them.

A related concept is the ***local isotropy type manifold*** $M^{l_z}_H$ through the point z with isotropy H that is defined as

$$M^{l_z}_H := M^{l_z}_{(H)} \cap M^H.$$

The local isotropy type manifolds are locally closed submanifolds of M. Moreover, $M^{l_z}_H$ is open in M^H and $M^{l_z}_{(H)} = G \cdot M^{l_z}_H$.

2.4.10 The invariance properties of the isotropy type manifolds. Let M be a proper G-space and M_H the isotropy type manifold corresponding to a fixed isotropy subgroup $H \subset G$. It is easy to see that the biggest subset of G that leaves M_H invariant coincides with the normalizer subgroup $N(H)$ of H in G, that is, $N(H)$ acts on M_H. Moreover, this action induces a free action of the quotient group $N(H)/H$ on M_H.

The local isotropy type manifold $M^{l_z}_H$ that contains a point $z \in M_H$ is $N(H)$-invariant (see DUISTERMAAT AND KOLK (1999) for a proof) and the $N(H)$-action on $M^{l_z}_H$ is smooth.

Given $z \in M_H$ we will denote by M^z_H the connected component of M_H that contains the point z and by $N(H)^z$ the subset of $N(H)$ consisting of the elements that preserve M^z_H, that is,

$$N(H)^z := \{n \in N(H) \mid n \cdot z \in M^z_H, \text{ for all } z \in M^z_H\}. \tag{2.4.7}$$

2.4.11 Proposition *$N(H)^z$ is a closed subgroup of $N(H)$ and contains $N(H)^0$, the connected component of the identity of $N(H)$. Consequently,*

$$\operatorname{Lie}(N(H)^z) = \operatorname{Lie}(N(H)), \tag{2.4.8}$$

and

$$\operatorname{Lie}(N(H)^z/H) = \operatorname{Lie}(N(H)/H). \tag{2.4.9}$$

Proof. We start by showing that $N(H)^z$ is a subgroup of $N(H)$. The definition (2.4.7) clearly shows that if $g, h \in N(H)^z$, then their product $gh \in N(H)^z$. We now show that $N(H)^z$ is invariant with respect to the inverse operation. If $g \in N(H)^z$, the action map $\Phi_g : M_H \to M_H$ is a diffeomorphism of M_H. As M^z_H is a connected component of M_H it is a closed and open subset of M_H and consequently so is $\Phi_g(M^z_H)$. This implies that $\Phi_g(M_H^z)$ is a union of connected components of M_H. Now, since $g \in N(H)^z$ we have that $\Phi_g(M_H^z) \subset M_H^z$ and as M_H^z is a connected component of M_H we have that

$$\Phi_g(M_H^z) = M_H^z \tag{2.4.10}$$

necessarily. Let now $m \in M_H^z$ arbitrary. By (2.4.10) there exists an element $m' \in M_H^z$ such that $m = g \cdot m'$. Now $g^{-1} \cdot m = g^{-1} \cdot (g \cdot m') = m' \in M_H^z$, which proves that $g^{-1} \in N(H)^z$ and hence that $N(H)^z$ is a subgroup of $N(H)$.

In order to show that $N(H)^z$ is a closed subgroup of $N(H)$ consider an element g in the closure $\overline{N(H)^z}$ of $N(H)^z$. Let $\{g_n\}_{n\in\mathbb{N}} \subset N(H)^z$ be a sequence such that $g_n \to g$. The continuity of the group action implies that for any $m \in M_H^z$ we have that $g_n \cdot m \to g \cdot m$, that is, $g \cdot m$ belongs to the closure $\overline{M_H^z}$ of M_H^z in M_H. As M_H^z is a connected component of M_H it is necessarily closed, which implies that $g \cdot m \in M_H^z$ and hence that $g \in N(H)^z$, due to the arbitrary character of the point $m \in M_H^z$.

We now show that $N(H)^0 \subset N(H)^z$. Recall that any element $g \in N(H)^0$ can be written as $g = \exp \xi_1 \cdots \exp \xi_n$ with $\{\xi_1, \ldots, \xi_n\} \in \mathrm{Lie}(N(H))$. Consequently, for any $m \in M_H^z$, the element $g \cdot m$ is connected to m via the path $c(t) := \exp t\xi_1 \cdots \exp t\xi_n \cdot m$ that satisfies that $c(0) = m$, $c(1) = g \cdot m$, and that lies in M_H^z for any $t \in [0, 1]$. This shows that $g \in N(H)^z$. The inclusion $N(H)^0 \subset N(H)^z$ implies that $\mathrm{Lie}(N(H)) = \mathrm{Lie}(N(H)^0) \subset \mathrm{Lie}(N(H)^z) \subset \mathrm{Lie}(N(H))$, which guarantees that $\mathrm{Lie}(N(H)^z) = \mathrm{Lie}(N(H))$. Since we also have that $(N(H)/H)^z = N(H)^z/H$, the equality (2.4.8) implies that

$$\mathrm{Lie}(N(H)^z/H) = \mathrm{Lie}(N(H)/H). \quad \blacksquare$$

2.4.12 Theorem (Structure Theorem of G-manifolds) *Let G be a Lie group acting properly on the manifold M and let H be a fixed isotropy subgroup. If $z \in M$ denote $G_z = H$. The local (H)-orbit type manifold $M_{(H)}^{l_z}$ containing z is a G-invariant submanifold of M such that the orbit space $M_{(H)}^{l_z}/G$ is a smooth manifold and the projection $\pi_{(H)}^{l_z} : M_{(H)}^{l_z} \to M_{(H)}^{l_z}/G$ is a smooth fiber bundle with fiber G/H and structure group $N(H)/H$. Furthermore, the projection $\pi_H^{l_z} : M_H^{l_z} \to M_H^{l_z}/(N(H)/H)$ defines a smooth principal $N(H)/H$-bundle. Regarding G/H as a right $N(H)/H$-space and $M_H^{l_z}$ as a left $N(H)/H$-space, we consider the bundle with fiber G/H and structure group G, associated with $\pi_H^{l_z}$:*

$$G/H \times_{N(H)/H} M_H^{l_z} \longrightarrow M_H^{l_z}/(N(H)/H).$$

This bundle is G-isomorphic to $\pi_{(H)} : M_{(H)}^{l_z} \to M_{(H)}^{l_z}/G$. In particular, $M_{(H)}^{l_z}$ is G-diffeomorphic to $G/H \times_{N(H)/H} M_H^{l_z} \simeq G \times_{N(H)} M_H^{l_z}$.

The proof of this theorem can be found in Kawakubo (1991) for compact group actions and in Duistermaat and Kolk (1999) for proper actions. Even though we do not present this proof, we state the explicit form of the G-bundle isomorphism in its statement. Firstly, the diffeomorphism $G \times_{N(H)} M_H^{l_z} \to M_{(H)}^{l_z}$ is given by $[g, z] \mapsto g \cdot z$, $g \in G$, $z \in M_H^{l_z}$. Taking its derivative at $z \in M_{(H)}^{l_z}$ shows that any vector v in

$T_{g\cdot z}M^{l_z}_{(H)}$, with $g \in G$ and $z \in M^{l_z}_H$ can be written as

$$v = T_z\Phi_g(\xi_M(z) + v_z), \qquad \text{with} \quad \xi \in \mathfrak{g},\ v_z \in T_z M^{l_z}_H.$$

Secondly, the diffeomorphism between $M^{l_z}_H/(N(H)/H)$ and $M^{l_z}_{(H)}/G$ is given by the projection into the respective quotients of the $(N(H)/H,\ G)$-equivariant injection $M^{l_z}_H \hookrightarrow M^{l_z}_{(H)}$.

2.4.13 The Stratification Theorem. The partition of a G-proper manifold M into orbit type manifolds has remarkable topological properties that have as a corollary the ***Stratification Theorem***:

Let M be a smooth manifold and G a Lie group acting properly on it. The connected components of the orbit type manifolds $M_{(H)}$ and their projections onto orbit space $M_{(H)}/G$ constitute a Whitney stratification of M and M/G, respectively. This stratification of M/G is minimal among all Whitney stratifications of M/G.

The proof of this result, which can be found in DUISTERMAAT AND KOLK (1999) or PFLAUM (2001a), is based on the Slice Theorem and on a series of extremely important properties of the orbit type manifolds decomposition that we enumerate in the next section. The smooth structure in the quotient space M/G needed to formulate the Whitney condition is spelled out in Section 2.5.4.

2.4.14 The set of conjugacy classes of subgroups of a Lie group G admits a partial order by defining $(K) \preceq (H)$ if and only if H is conjugate to a subgroup of K. Also, a point $z \in M$ in a proper G-space M (or its corresponding G-orbit, $G \cdot z$) is called ***principal*** if its corresponding local orbit type manifold $M^{l_x}_{(G_x)}$ is open in M. The orbit $G \cdot z$ is called ***regular*** if the dimension of the orbits nearby coincides with the dimension of $G \cdot z$. The set of principal and regular orbits will be denoted by M^{princ}/G and M^{reg}/G, respectively. Using this notation we have that:

(i) For any $z \in M$ there exists an open neighborhood U of z that intersects only finitely many connected components of finitely many orbit type manifolds. If M is compact or a linear space where G acts linearly, then the G-action on M has only finitely many distinct connected components of orbit type manifolds.

(ii) For any $z \in M$ there exists an open neighborhood U of z such that $(G_z) \preceq (G_x)$, for all $x \in U$. In particular, this implies that $\dim G \cdot z \leq \dim G \cdot x$, for all $x \in U$.

(iii) Principal Orbit Theorem: For every connected component M^0 of M the subset $M^{reg} \cap M^0$ is connected, open, and dense in M^0. Each connected component $(M/G)^0$ of M/G contains only one principal orbit type, which is a connected open and dense subset of it.

(iv) If the orbit space M/G is connected, then $M^{reg} = M_{(\mathfrak{g}_x)}$ for any $x \in M^{reg}$. Also, for any $z \in M^{princ}$ we have that $M^{princ} = M_{(G_z)} = M^{l_z}_{(G_z)}$.

(v) If the group G is compact and the manifold M is connected, there is a unique maximal conjugacy class (H) such that $(G_x) \preceq (H)$ for all $x \in M$ and whose associated orbit type manifold $M_{(H)}$ is an open and dense set in M. The orbit space $M_{(H)}/G$ is connected.

Most of these results have their origin in the works of SAMELSON (1952), MONTGOMERY AND YANG (1957), MONTGOMERY (1960), and YANG (1957) dealing with mostly continuous compact group actions. The statements above dealing with smooth proper group actions are quoted from DUISTERMAAT AND KOLK (1999). The reader can find in this reference a variety of interesting related results.

2.5 The invariant functions of a proper G-space

2.5.1 The theorems of Hilbert, Schwarz, and Mather. We start by quoting two standard results that we will use later on in this section. The proofs can be found in WEYL (1946), SCHWARZ (1974), POÈNARU (1976), MATHER (1977), BIERSTONE (1980), KEMPF (1987), and PFLAUM (2001a).

2.5.2 Theorem (Hilbert–Weyl) *Let H be a compact Lie group acting linearly on the real vector space V. Then the algebra of H-invariant polynomials on V is finitely generated.*

This theorem guarantees that when a compact Lie group H acts linearly on a vector space V one can always find a finite number of H-invariant polynomials $\{\pi_1, \ldots, \pi_k\}$ on V, such that every H-invariant polynomial P can be written as a polynomial function of them. More specifically, there exists certain $\widehat{P} \in \mathbb{R}[X_1, \ldots, X_k]$ such that $P = \widehat{P}(\pi_1, \ldots, \pi_k)$. Note that the generating family $\{\pi_1, \ldots, \pi_k\}$ can be chosen to be minimal. In that situation we say that $\{\pi_1, \ldots, \pi_k\}$ is a ***Hilbert basis*** of $\mathcal{P}(V)^H$. In applications it is very practical to choose the Hilbert basis made out of homogeneous polynomials. Note also that the Hilbert basis is not necessarily free and that therefore there are in general relations between its elements.

The Hilbert–Weyl Theorem can is extremely useful at the time of providing a model for the orbit space V/H of the linear representation of the compact group H on V. This is a corollary of the fact that the elements of a Hilbert basis separate the H-orbits, that is, if $\{\pi_1, \ldots, \pi_k\}$ is a Hilbert basis of the algebra of H-invariant polynomials on V, then two vectors $v, v' \in V$ are such that

$$(\pi_1(v), \ldots, \pi_k(v)) = (\pi_1(v'), \ldots, \pi_k(v'))$$

if and only if v and v' lie in the same H-orbit. This property, whose proof can be found in POÈNARU (1976); ABUD AND SARTORI (1983), provides us with a parameterization of the quotient space since we can identify the quotient space V/H with the image of the ***Hilbert map***, defined by

$$\begin{aligned} \pi : V &\longrightarrow \mathbb{R}^k \\ v &\longmapsto (\pi_1(v), \ldots, \pi_k(v)). \end{aligned}$$

The generalization of the Hilbert–Weyl Theorem from polynomials to smooth functions has been carried out in SCHWARZ (1974).

2.5.3 Theorem (Schwarz–Mather) *Let H be a compact Lie group acting linearly on the real vector space V and $\mathcal{B} = \{\pi_1, \ldots, \pi_k\}$ a Hilbert basis of the algebra of H-invariant polynomials on V. Then the map*

$$p : C^\infty(\mathbb{R}^k) \longrightarrow C^\infty(V)^H, \quad f \longmapsto f \circ (\pi_1, \ldots, \pi_k)$$

is surjective.

A refinement of this theorem in MATHER (1977) guarantees that the quotient presheaf of smooth functions on V/H is isomorphic to the presheaf of Whitney smooth functions on $\pi(V) \subset \mathbb{R}^k$ induced by the sheaf of smooth functions on $\mathbb{R}^k$. Additionally since π is a polynomial map, the Tarski–Seidenberg theorem guarantees that V/H can be considered as a semi-algebraic subset of $\mathbb{R}^k$ COLLINS (1974); ABUD AND SARTORI (1983). By a well-known result (see for instance GIBSON et al. (1976)) every semi-algebraic set admits a canonical Whitney stratification into a finite number of semi-algebraic subsets. It turns out that this canonical stratification coincides with the stratification of V/H via orbit types BIERSTONE (1975).

2.5.4 A smooth structure for the orbit space M/G**.** The combination of the Slice Theorem with Theorems 2.5.2 and 2.5.3 allows the construction of a smooth structure for the quotient space M/G compatible with its stratification by orbit types. Indeed, given any point $m \in M$, the Slice Theorem guarantees the existence of an open neighborhood U of the orbit $G \cdot m$ and of an equivariant diffeomorphism $\varphi : U \to G \times_{G_m} B$, with B an open G_m-invariant neighborhood of the vector space $V := T_m M / T_m(G \cdot m)$. The equivariance of φ implies the existence of a projected map $\phi : U/G \to (G \times_{G_m} B)/G \simeq B/G_m$. Let now $\mathcal{B} = \{\pi_1, \ldots, \pi_k\}$ be a Hilbert basis of the algebra of G_m-invariant polynomials on V and $p : V \to \mathbb{R}^k$ be the map given by $v \longmapsto (\pi_1(v), \ldots, \pi_k(v))$, for any $v \in V$. The G_m-invariance of the polynomials in $\mathcal{B}$ allows us to drop the map p to a map $q : V/G_m \to \mathbb{R}^k$.

The pairs of the form $(U/G, \psi := q \circ \phi)$ constitute an atlas of M/G compatible with its stratification in orbit types. For a proof of this fact check with PFLAUM (2001a).

2.5.5 Proper actions and extensions. The properness condition on an action makes available certain extensions of functions. For instance, any proper action has the extension property introduced in §2.2.11:

2.5.6 Proposition *Any proper action of a Lie group G on a manifold M satisfies the extension property. That is, for any open G-invariant subset U of M, and $f \in C^\infty(U)^G$ we have that for each $z \in U$ there exist a G-invariant open neighborhood $V \subset U$ of z and a G-invariant smooth function $F \in C^\infty(M)^G$ such that $f|_V = F|_V$.*

Proof. Choose $U_1 \subset U$ an open G-invariant neighborhood of z that by the Slice Theorem can be modeled by the tube $G \times_{G_z} B_r$, where B_r is the open ball of radius r in the vector space $T_z M / T_z(G \cdot z)$ on which G_z acts orthogonally. Let $\varphi : G \times_{G_z} B_r \to U_1$ be the G-equivariant diffeomorphism provided by the Slice Theorem. Define $g : B_r \to \mathbb{R}$ as the smooth G_z-invariant function given by $g(v) := (f \circ \varphi)([e, v])$. Since G_z is compact, there exists a G_z-invariant bump function $\phi : B_r \to [0, 1]$ such that

$$\phi|_{B_{r/2}} = 1 \qquad \text{and} \qquad \phi|_{B_r \setminus B_{3r/4}} = 0.$$

Define $f' \in C^\infty(U_1)^G$ by $f'(\varphi^{-1}([h, v])) := \phi(v) g(v)$, for any $h \in G$ and $v \in B_r$. Since f' and all its derivatives vanish on the boundary of U_1, its extension off U_1 by the identically zero function yields $F \in C^\infty(M)^G$. Take $V = \varphi^{-1}(G \times_{G_z} B_{r/2})$. It is clear that $F|_V = f'|_V = f|_V$. ■

2.5.7 Proposition *Let G be a Lie group acting properly on the smooth manifold M and S a G-invariant embedded submanifold of M. Then S has the local extension property with respect to the G-action.*

Proof. Let $f \in C_S^\infty(V)^G$, for some open G-invariant subset V of S and $z \in S$. By the Tube Theorem 2.3.28 there exists an open G-invariant neighborhood U of z in M and a tube $\varphi : U \to G \times_{G_z} B$ about the orbit $G \cdot z$, where B is an open G_z-invariant neighborhood of the origin in certain G_z-representation space. Shrinking U if necessary we can arrange things so that $U \cap S \subset V$. Moreover, by Property **(vii)** in Proposition 2.4.6 there exists an embedded G_z-invariant submanifold A of B containing the origin such that $\varphi(U \cap S) = G \times_{G_z} A$. The G-invariance of f implies that the map $f \circ \varphi^{-1}|_{G \times_{G_z} A}$ can be seen as a G_z-invariant map on A. More specifically, if $\pi : G \times B \to G \times_{G_z} B$ is the projection, then there exists a unique map $\overline{f} \in C^\infty(A)^{G_z}$ such that $f \circ \varphi^{-1}\pi(g, a) = \overline{f}(a)$, for any $(g, a) \in G \times A$. Let now $W \subset B$ be the domain of a chart (W, ψ) in B that has the submanifold property with respect to A at 0. By Corollary 2.3.34 the set W can be chosen G_z-invariant. By construction, there exist vector spaces E_1 and E_2 such that $\psi : W \to E_1 \times E_2$ and $\psi(A \cap W) = \psi(W) \cap (E_1 \times \{0\})$.

Let $\overline{F} \in C^\infty(W)$ be the smooth map defined by $\overline{F}(\Psi^{-1}(v_1, v_2)) = \overline{f}(\psi^{-1}(v_1, 0))$. The map $\overline{F}$ is clearly a smooth extension of $\overline{f}|_{W \cap A}$. By Part **(vii)** of Proposition 2.3.8 there exists another such extension $\widehat{F}$ that can be chosen G_z-invariant, that is, $\widehat{F} \in C^\infty(W)^{G_z}$. Consider now the open G-invariant neighborhood U_z of z in M defined by $U_z := \varphi^{-1}(G \times_{G_z} W)$ and the map $F \in C^\infty(U_z)^G$ given by $F(m) = \widehat{F}(a)$, for any $m = \varphi^{-1}([g, a]) \in U_z$. The pair (U_z, F) provides the local G-invariant extension of f at z needed in the statement of the proposition. ■

2.5.8 Corollary *Let G be a Lie group acting properly on the smooth manifold M and let S be a G-invariant embedded submanifold such that the orbit space S/G is a regular quotient manifold. Then*

$$C_{S/G}^\infty \subset C_{S/G,M/G}^\infty. \tag{2.5.1}$$

If S is paracompact we, then have an equality.

Proof. The inclusion (2.5.1) is a consequence of Proposition 2.5.7 and (2.2.7). The second claim follows from Part **(v)** in Proposition 2.3.8 and (2.2.6).

We will also need the following property dealing with the existence of local G-invariant extensions of invariant functions defined on the isotropy type manifolds.

2.5.9 Proposition *Let G be a Lie group acting properly on the manifold M. Let $m \in M$ be an arbitrary point and denote $H := G_m$. Then every $N(H)$-invariant function $f \in C^\infty(M_H^{l_m})^{N(H)}$ defined on the local isotropy type manifold containing the point m admits a local extension to a G-invariant function $F \in C^\infty(M)^G$ on M, that is, there exists a G-invariant neighborhood U of m in M and a G-invariant function $F \in C^\infty(M)^G$, such that $F|_{U \cap M_H^{l_m}} = f|_{U \cap M_H^{l_m}}$. This claim is still valid if we replace $M_H^{l_m}$ by M_H^m and $N(H)$ by $N(H)^m$.*

Proof. Let $\varphi : G \times_H B_r \to V$ be a tube about the orbit $G \cdot m$. Recall that by Proposition 2.4.6 we can write

$$\varphi^{-1}(V \cap M_H^{l_m}) = (G \times_H B_r)_H = N(H) \times_H B_r^H.$$

Let now $g : B_r \to \mathbb{R}$ be the smooth function defined by $g(v) := f(\varphi([e, v]))$ for any $v \in B_r^H$. Using a bump function similar to the one that was introduced in the proof of Proposition 2.5.6 we can construct another function $g_1 \in C^\infty(B_r^H)$ such that $g_1|_{B^H_{r/2}} = g|_{B^H_{r/2}}$ and $g_1|_{B_r^H \setminus B^H_{3r/4}} = 0$. Due to the compactness of H, the vector space B can be decomposed as the direct sum $B = B^H \oplus W$ of two H-invariant subspaces B^H and W. Define $g_2 \in C^\infty(B_r)^H$ by $g_2(v + w) = g_1(v)$, for any $v \in B_r^H$ and $w \in W$. We now let $g_3 \in C^\infty(V)^H$ be given by $g_3(\varphi^{-1}([h, v])) = g_2(v)$, for any $\varphi([h, v]) \in V = \varphi(G \times_H B_r)$. Finally, let $F \in C^\infty(M)^G$ be the function given by

$$F(z) = \begin{cases} g_3(z) & \text{if} \quad z \in V \\ 0 & \text{if} \quad z \notin V. \end{cases}$$

By taking the function F above and $U = \varphi^{-1}(G \times_H B_{r/2})$, the lemma follows. ■

We now use the Schwarz–Mather Theorem to prove a result that will be of great importance for later developments.

We start by recalling that given a real vector space V, its dual V^*, the natural pairing $\langle \cdot, \cdot \rangle : V^* \times V \to \mathbb{R}$, and a subspace $U \subset V$, the ***annihilator*** $U^\circ \subset V^*$ of U in V^* is defined by

$$U^\circ := \{\nu \in V^* \mid \langle \nu, u \rangle = 0, \ \forall u \in U\}.$$

2.5.10 Theorem (ORTEGA (1998)) *Let G be a Lie group acting properly on the smooth manifold M and $m \in M$ a point with isotropy subgroup $H := G_m$. Then*

$$\left((T_m(G \cdot m))^\circ\right)^H = \{\mathbf{d}f(m) \mid f \in C^\infty(M)^G\} \tag{2.5.2}$$

$$= \{\mathbf{d}f(m) \mid f \in C^\infty(U)^G, \ U \subset M \textit{ open } G\textit{-invariant set}\} \tag{2.5.3}$$

Proof. The identity

$$\{\mathbf{d}f(m) \mid f \in C^\infty(M)^G\} = \{\mathbf{d}f(m) \mid f \in C^\infty(U)^G, \ U \subset M \text{ open } G\text{-invariant}\}$$

is a straightforward consequence of Proposition 2.5.6.

We now establish (2.5.2). We start by showing that if $f \in C^\infty(M)^G$, then $\mathbf{d}f(m) \in ((T_m(G \cdot m))^\circ)^H$. It is clear that for any $\xi \in \mathfrak{g}$,

$$\langle \mathbf{d}f(m), \xi_M(m) \rangle = \left.\frac{d}{dt}\right|_{t=0} f(\exp t\xi \cdot m) = \left.\frac{d}{dt}\right|_{t=0} f(m) = 0.$$

Hence, $\mathbf{d}f(m) \in T_m(G \cdot m)^\circ$. Now, $\mathbf{d}f(m)$ is also H-fixed since for any $h \in H$ and any $v = \left.\frac{d}{dt}\right|_{t=0} m(t) \in T_m M$ with $m(0) = m$,

$$\begin{aligned}\langle h \cdot \mathbf{d}f(m), v \rangle &= \langle \mathbf{d}f(m), h^{-1} \cdot v \rangle \\ &= \left.\frac{d}{dt}\right|_{t=0} f(h^{-1} \cdot m(t)) = \left.\frac{d}{dt}\right|_{t=0} f(m(t)) = \langle \mathbf{d}f(m), v \rangle.\end{aligned}$$

Since the vector v is arbitrary, $h \cdot \mathbf{d}f(m) = \mathbf{d}f(m)$, as required.

In order to show the converse inclusion take $\varphi : G \times_H V_r \to U$ a tube around the orbit $G \cdot m$, where V_r is a ball of radius r around the origin in the vector space $V = T_mM/T_m(G \cdot m)$. Notice that the G-equivariance of φ and Proposition 2.5.6 guarantee that it suffices to show that

$$((T_{[e,0]}(G \cdot [e,0]))^\circ)^H \subset \{\mathbf{d}f([e,0]) \mid f \in C^\infty(G \times_H V_r)^G\}. \tag{2.5.4}$$

Indeed, recall that if $A : T \to S$ is an isomorphism between the vector spaces T and S, and R is a vector subspace of S, then $A^*(R^\circ) = (A^{-1}(R))^\circ$. If we use this relation on $(T_m\varphi^{-1})^*((T_{[e,0]}(G \cdot [e,0]))^\circ)^H$ and use the G-equivariance of φ, we obtain

$$\begin{aligned}(T_m\varphi^{-1})^*&((T_{[e,0]}(G \cdot [e,0]))^\circ)^H \\ &= ((T_m\varphi^{-1})^*(T_{[e,0]}(G \cdot [e,0]))^\circ)^H = ((T_m(G \cdot m))^\circ)^H.\end{aligned}$$

Hence, if we apply $(T_m\varphi^{-1})^*$ on both sides of (2.5.4) and we use this relation, we obtain

$$((T_m(G \cdot m))^\circ)^H \subset \{\mathbf{d}f(m) \mid f \in C^\infty(U)^G\} \subset \{\mathbf{d}f(m) \mid f \in C^\infty(M)^G\},$$

where the last inclusion is a consequence of Proposition 2.5.6.

In order to prove (2.5.4) we first notice that

$$T_{[e,0]}(G \cdot [e,0]) = \{T_{(e,0)}\pi(\xi, 0) \in T_{[e,0]}(G \times_H V_r) \mid \xi \in \mathfrak{g}\} \cong \mathfrak{g}/\mathfrak{h} \times \{0\},$$

where $\pi : G \times V_r \to G \times_H V_r$ is the canonical projection. Clearly,

$$(T_{[e,0]}(G \cdot [e,0]))^\circ \cong \{0\} \times V^* \cong V^*. \tag{2.5.5}$$

At this point we introduce the following lemma (ORTEGA (1998), SCHMAH (2001)):

2.5.11 Lemma *Let H be a compact Lie group acting linearly on the vector space V, as well as on its dual V^* via the associated contragredient representation. Then, the restriction to $(V^*)^H$ of the dual map associated to the inclusion $i_{V^H} : V^H \hookrightarrow V$ is an H-equivariant isomorphism from $(V^*)^H$ to $(V^H)^*$. The H-fixed vector subspace V^H has a unique H-invariant complement in V.*

Proof. Note that for any $\beta \in V^*$, $i^*_{V^H}(\beta) = \beta|_{V^H}$. Take an H-invariant inner product $\langle\langle\cdot,\cdot\rangle\rangle$ on V, always available by the compactness of H. Let W be the H-invariant orthogonal complement to V^H with respect to $\langle\langle\cdot,\cdot\rangle\rangle$.

Any element $\alpha \in \left(V^H\right)^*$ can be extended to $\beta \in V^*$ by setting $\beta|_W = 0$. Moreover, note that $i^*_{V^H}(\beta) = \beta|_{V^H} = \alpha$ and also, for any $v \in V^H$, $w \in W$, and $h \in H$, we have

$$\langle h \cdot \beta, v + w\rangle = \langle \beta, h^{-1} \cdot (v + w)\rangle = \langle \beta, v + h^{-1} \cdot w\rangle = \langle \beta, v + w\rangle,$$

since both w and $h^{-1} \cdot w$ are in W and $\beta|_W = 0$. This implies that $\beta \in (V^*)^H$, and hence $i^*_{V^H}|_{(V^*)^H}$ is surjective.

For injectivity, suppose $\beta|_{V^H} = 0$, for some $\beta \in (V^*)^H$. Let $v \in V$ be such that $\langle \beta, w \rangle = \langle\langle v, w \rangle\rangle$, for all $w \in W$. Then, for any $h \in H$ and any $w \in V$ we have that

$$\langle\langle h \cdot v, w \rangle\rangle = \langle\langle v, h^{-1} \cdot w \rangle\rangle = \langle \beta, h^{-1} \cdot w \rangle = \langle h \cdot \beta, w \rangle = \langle \beta, w \rangle = \langle\langle v, w \rangle\rangle,$$

which, by the nondegeneracy of the inner product, implies that $h \cdot v = v$ and hence $v \in V^H$. But $\beta|_{V^H} = 0$, which implies that $v = 0$, which in turn implies $\beta = 0$. Hence $i^*_{V^H}|_{(V^*)^H}$ is an isomorphism.

The H-equivariance of $i^*_{V^H}|_{(V^*)^H}$ follows trivially from the following chain of equalities that are satisfied for any $h \in H$, $\beta \in (V^*)^H$, and $v \in V^H$

$$\langle h \cdot i^*_{V^H}(\beta), v \rangle = \langle i^*_{V^H}(\beta), h^{-1} \cdot v \rangle = \langle \beta, h^{-1} \cdot v \rangle = \langle \beta, v \rangle = \langle i^*_{V^H}(\beta), v \rangle.$$

We conclude by showing the uniqueness of the H-invariant complement to V^H in V. which is a consequence of the injectivity of $i^*_{V^H}|_{(V^*)^H}$. Suppose that W_1 and W_2 are two distinct H-invariant complements to V^H in V. Let $w \in W_2 \setminus W_1$. The vector w admits a unique decomposition as $w = w_1 + w^H$ with $w_1 \in W_1$ and $w^H \in V^H$. Let now $\langle\langle \cdot, \cdot \rangle\rangle$ be the H-invariant inner product on V that we used before and define $\alpha \in (V^H)^*$ by $\alpha = \langle\langle w^H, \cdot \rangle\rangle$. The functional α admits two distinct H-invariant extensions, say $\beta_1, \beta_2 \in V^*$, constructed by setting $\beta_i|_{W_i} = 0$, $i \in \{1, 2\}$. Indeed, as $\beta_1(w) = \langle\langle w^H, w^H \rangle\rangle \neq 0$ and $\beta_2(w) = 0$, the two extensions are different, which contradicts the injectivity of $i^*_{V^H}|_{(V^*)^H}$. ▼

Using this lemma in the expression (2.5.5) we get

$$((T_{[e,0]}(G \cdot [e, 0]))^\circ)^H \simeq (V^*)^H \cong (V^H)^*.$$

Notice now that in the tubular model, the G-invariant functions $f \in C^\infty(G \times_H V_r)^G$ are characterized by the condition $f \circ \pi \in C^\infty(V_r)^H$. The claim (2.5.4) then follows if we show that

$$(V^*)^H = \{\mathbf{d}g(0) \in V^* \mid g \in C^\infty(V_r)^H\}.$$

Let $g \in C^\infty(V_r)^H$ and $h \in H$ be arbitrary. Then, for any $v = \frac{d}{dt}\big|_{t=0} c(t) \in V$ with $c(0) = 0$, we have

$$\begin{aligned}\langle h \cdot \mathbf{d}g(0), v \rangle &= \langle \mathbf{d}g(0), h^{-1} \cdot v \rangle \\ &= \left.\frac{d}{dt}\right|_{t=0} g(h^{-1} \cdot c(t)) = \left.\frac{d}{dt}\right|_{t=0} g(c(t)) = \langle \mathbf{d}g(0), v \rangle.\end{aligned}$$

Since $v \in V$ is arbitrary, it follows that $h \cdot \mathbf{d}g(0) = \mathbf{d}g(0)$.

To prove the converse we will use the H-invariant inner product $\langle\langle \cdot, \cdot \rangle\rangle$ on V introduced in Lemma 2.5.11 to split

$$V = V^H \oplus W,$$

where W is the H-invariant orthogonal complement to V^H with respect to $\langle\langle \cdot, \cdot \rangle\rangle$. Let $\{v_1, \ldots, v_k, w_1, \ldots, w_r\}$ be an orthonormal basis of V adapted to this splitting, that is, $\{v_1, \ldots, v_k\} \subset V^H$ and $\{w_1, \ldots, w_r\} \subset W$. Define $\pi_1, \ldots, \pi_k \in V^*$ by

$\pi_i := \langle\langle v_i, \cdot\rangle\rangle$, $i \in \{1, \ldots, k\}$. By construction, the subspace $\operatorname{span}\{\pi_1, \ldots, \pi_k\} \subset V^*$ consists of linear invariants of the H-action on V. Moreover, it contains all of them. Indeed, if $\alpha : V \to \mathbb{R}$ is an arbitrary linear H-invariant, there exists an element $u \in V$ such that $\alpha = \langle\langle u, \cdot\rangle\rangle$ and that, additionally, satisfies $h \cdot \alpha = \alpha$, for all $h \in H$. This implies that for any $v \in V$ and $h \in H$ we have that $\langle h \cdot \langle\langle u, \cdot\rangle\rangle, v\rangle = \langle\, \langle\langle u, \cdot\rangle\rangle, v\rangle$, or equivalently $\langle\, \langle\langle u, \cdot\rangle\rangle, h^{-1}\cdot v\rangle = \langle\, \langle\langle u, \cdot\rangle\rangle, v\rangle$, which can be rewritten as $\langle\langle u, h^{-1}\cdot v\rangle\rangle = \langle\langle u, v\rangle\rangle$. The H-invariance of $\langle\langle \cdot, \cdot\rangle\rangle$ implies that $\langle\langle h \cdot u, v\rangle\rangle = \langle\langle u, v\rangle\rangle$ for all $v \in V$ and hence $h \cdot u = u$ for any $h \in H$. This implies that $\alpha \in \operatorname{span}\{\pi_1, \ldots, \pi_k\}$.

We have thus shown that $\pi_1, \ldots, \pi_k \in V^*$, or, in general, that any basis of $(V^H)^*$ spans the set of all independent linear invariants of the H-action on V. By Hilbert's Theorem 2.5.2, the ring of H-invariant polynomials on V is finitely generated. We complete the set $\{\pi_1, \ldots, \pi_k\}$ to a generating system $\{\pi_1, \ldots, \pi_k, \pi_{k+1}, \ldots, \pi_q\}$ of the this ring. The Schwarz–Mather Theorem 2.5.3 guarantees that every H-invariant function $f \in C^\infty(V)^H$ can be written as $f = g(\pi_1, \ldots, \pi_q)$, with $g \in C^\infty(\mathbb{R}^q)$. Let now $\alpha \in (V^*)^H \cong (V^H)^*$ be arbitrary. The form $\alpha \in (V^H)^*$ can be expanded as $\alpha = \alpha_1\pi_1 + \ldots + \alpha_k\pi_k$, with $\alpha_1, \ldots, \alpha_k \in \mathbb{R}$. Let $g \in C^\infty(\mathbb{R}^q)$ be such that

$$\frac{\partial g(0)}{\partial \pi_i} = \alpha_i, \quad i \in \{1, \ldots, k\}.$$

With this choice, the function $f := g(\pi_1, \ldots, \pi_q)$ belongs to $C^\infty(V)^H$ and satisfies

$$\mathbf{d}f(0) = \frac{\partial g(0)}{\partial \pi_1}\pi_1 + \ldots + \frac{\partial g(0)}{\partial \pi_k}\pi_k = \alpha_1\pi_1 + \ldots + \alpha_k\pi_k = \alpha,$$

where we used that $\mathbf{d}\pi_j(0) = 0$ for $j \in \{k+1, \ldots, q\}$ because the invariants π_j in this range of the indices are at least quadratic. Since α is arbitrary, the result follows. ■

2.5.12 Remark The properness condition in the statement of the previous proposition is essential (and is not tied to the technical need of slices in the proof) since there are examples of nonproper actions where this result does not hold. Indeed, consider the irrational flow on the torus. Since the orbits of this action fill densely the torus, the only invariant functions in this particular case are the constant functions. Hence the right-hand side of the equality in Theorem 2.5.10 is trivial. However, if the torus in question is larger than one-dimensional, the vector space $(T_m(G \cdot m))^\circ$ is nontrivial.

Chapter 3

Pseudogroups and Groupoids

The action $\Phi : G \times M \to M$ of Lie group G on a manifold M can be seen as the choice of a subgroup $A_G := \{\Phi_g \mid g \in G\}$ of Diff(M), that is, the globally defined diffeomorphisms of M. There are mathematical structures, such as distributions and foliations, where the transformations of the manifold M that naturally appear in the problem are only locally defined. It is in the study of those structures that the objects constituting the subject of this chapter become relevant.

Pseudogroups of transformations and the theory of groupoids constitute, as it is the case with Lie theory, subjects on their own that we will not review here in detail. In the following pages we will just recall some rudiments that we will need in what follows when dealing with singular foliations and distributions. The style in the presentation will be similar to the one in the previous chapter; proofs will be provided only when they are not readily available in the literature.

The theory of pseudogroups has its origin in the works of CARTAN (1904, 1905, 1908, 1909). Other references where the reader can explore in detail the interplay between pseudogroups, groupoids, and other branches of mathematics, as well as more in-depth presentations of the material treated in this chapter are: CANNAS AND WEINSTEIN (1999), KURANISHI (1959, 1961), LANDSMAN (1998), LIBERMANN AND MARLE (1987), MACKENZIE (1987), MIKAMI AND WEINSTEIN (1988), PATERSON (1999), STEFAN (1974a,b), SUSSMANN (1973), and references therein.

3.1 Pseudogroups of transformations

3.1.1 Local diffeomorphisms. Let M be a smooth manifold. The symbol $\mathrm{Diff}_L(M)$ will denote the set of ***local diffeomorphisms*** of M. More explicitly, the elements of $\mathrm{Diff}_L(M)$ are diffeomorphisms $F : \mathrm{Dom}(F) \subset M \to F(\mathrm{Dom}(F))$ of an open subset $\mathrm{Dom}(F) \subset M$ onto its image $F(\mathrm{Dom}(F)) \subset M$. We will denote the elements of $\mathrm{Diff}_L(M)$ as pairs $(F, \mathrm{Dom}(F))$. The local diffeomorphisms can be composed using the binary operation defined as

$$(G, \mathrm{Dom}(G)) \cdot (F, \mathrm{Dom}(F)) := (G \circ F, F^{-1}(\mathrm{Dom}(G)) \cap \mathrm{Dom}(F)), \qquad (3.1.1)$$

for all $(G, \mathrm{Dom}(G)), (F, \mathrm{Dom}(F)) \in \mathrm{Diff}_L(M)$. It is easy to see that this operation is associative and has $(\mathbb{I}, M)$, the identity map of M, as a (unique) two-sided identity element, which makes $\mathrm{Diff}_L(M)$ into a monoid (set with an associative operation which contains a two-sided identity element). Notice that only the elements in $\mathrm{Diff}(M) \subset \mathrm{Diff}_L(M)$ have an inverse since, in general, for any $(F, \mathrm{Dom}(F)) \in \mathrm{Diff}_L(M)$, we have that

$$(F^{-1}, F(\mathrm{Dom}(F))) \cdot (F, \mathrm{Dom}(F)) = (\mathbb{I}|_{\mathrm{Dom}(F)}, \mathrm{Dom}(F)) \tag{3.1.2}$$

$$(F, \mathrm{Dom}(F)) \cdot (F^{-1}, F(\mathrm{Dom}(F))) = (\mathbb{I}|_{F(\mathrm{Dom}(F))}, F(\mathrm{Dom}(F))). \tag{3.1.3}$$

Consequently, the only way to obtain the identity element $(\mathbb{I}, M)$ out of the composition of F with its inverse is having $\mathrm{Dom}(F) = M$. It follows from this argument that $\mathrm{Diff}(M)$ is the biggest subgroup contained in the monoid $\mathrm{Diff}_L(M)$ with respect to the composition law (3.1.1).

3.1.2 Pseudogroups. In the sequel we will frequently encounter submonoids A of $\mathrm{Diff}_L(M)$ that satisfy the following property:

> **(PS)** For any $F : \mathrm{Dom}(F) \to F(\mathrm{Dom}(F))$ in A there exists another element $F^{-1} : F(\mathrm{Dom}(F)) \to \mathrm{Dom}(F)$ also in A that satisfies the identities (3.1.2) and (3.1.3).

Such submonoids will be referred to as ***pseudogroups*** of $\mathrm{Diff}_L(M)$. Recall that A being a submonoid implies that it is closed under composition and $(\mathbb{I}, M) \in A$.

One of the important features of pseudogroups is that they have an associated orbit space. Indeed, if A is a pseudogroup we define the ***orbit*** $A \cdot m$ under A of any element $m \in M$ as the set $A \cdot m := \{F(m) \mid F \in A, \text{ such that } m \in \mathrm{Dom}(F)\}$. A being a pseudogroup implies that the relation *being in the same A-orbit* is an equivalence relation and induces a partition of M into A-orbits. The ***space of A-orbits*** will be denoted by M/A. If we endow the space of orbits M/A with the quotient topology, the projection $\pi_A : M \to M/A$ is a continuous (see §1.1.18) and open map. To see that it is an open map, recall that if U is an open set in M, $\pi_A(U)$ is open in M/A if and only if $\pi_A^{-1}(\pi_A(U))$ is open in M. Since $\pi_A^{-1}(\pi_A(U)) = A \cdot U = \bigcup_{F \in A} F(U \cap \mathrm{Dom}(F))$ is a union of open sets, it follows that it is open.

Using this notion of orbit we can generalize to the pseudogroup context in a straightforward manner the concepts of ***saturated sets***, invariant functions, and the properties related to the existence of ***extensions***, ***local extensions*** and ***separation of orbits*** introduced in §2.2.11 in the context of group actions. The following result is the generalization to the framework of pseudogroups of Proposition 2.2.12. The proof of that result can be adapted in a straightforward manner to this context.

3.1.3 Proposition *Let M be a smooth manifold and A a pseudogroup of* $\mathrm{Diff}_L(M)$. *Let S be an embedded A-invariant submanifold of M such that the orbit space S/A is a regular quotient manifold. Then if any open A-invariant subset of S admits A-invariant partitions of unity subordinate to any cover by A-invariant open sets, we then have that*

$$C^\infty_{S/A, M/A} \subset C^\infty_{S/A}.$$

Conversely, if S has the local extension property with respect to the A-action, then

$$C^\infty_{S/A} \subset C^\infty_{S/A, M/A}.$$

3.1.4 Integrable pseudogroups. We say that a pseudogroup $A \subset \mathrm{Diff}_L(M)$ is ***integrable*** when its orbits form a generalized foliation of M. In particular, the orbits of an integrable pseudogroup are initial submanifolds of M.

A significant number of integrable pseudogroups are generated by collections of arrows (see STEFAN (1974a,b)). An ***arrow*** is a differentiable mapping $a : U \subset \mathbb{R} \times M \to M$ whose domain U is an open subset of $\mathbb{R} \times M$ and that, additionally, satisfies:

(i) For every $t \in \mathbb{R}$, the map $a_t := a(t, \cdot)$ is a local diffeomorphism of M (possibly with empty domain).

(ii) If the point (t, x) belongs to the domain of a, then so does (s, x) for every $s \in [0, t]$. Moreover, $a(0, x) = x$

An example of an arrow is the flow of any differentiable vector field on M.

Let R be a collection of arrows on M. We associate to R a set $\mathcal{A}_R \subset \mathrm{Diff}_L(M)$ of local diffeomorphisms defined by $\mathcal{A}_R := \{a_t \mid a \in R\}$ which, at the same time, generates a pseudogroup

$$A_R := (\mathbb{I}, M) \bigcup \{a^1_{t_1} \circ \cdots \circ a^n_{t_n} \mid n \in \mathbb{N} \text{ and } a^n_{t_n} \in \mathcal{A}_R \text{ or } (a^n_{t_n})^{-1} \in \mathcal{A}_R\}.$$

We also define, for any $z \in M$, the following vector subspaces of $T_z M$:

$$\mathcal{D}_R(z) := \operatorname{span}\left\{ \left.\frac{d}{dt}\right|_{t=t_0} a(t, y) \,\middle|\, a \in R,\ a(t, y) = z \right\}, \text{ and}$$
$$D_R(z) := \operatorname{span}\{T_y \mathcal{F}_T \cdot \mathcal{D}_R(y) \mid \mathcal{F}_T \in A_R,\ \mathcal{F}_T(y) = z\}.$$

3.1.5 Theorem (Stefan) *Let R be a collection of arrows on the smooth manifold M. Then the pseudogroup A_R is integrable. If we denote by Ψ_R the associated generalized foliation, then for any $z \in M$ we have that*

$$T_z(M, \Psi_R) = D_R(z).$$

The leaves of Ψ_R are usually referred to as the ***accessible sets*** of the collection of arrows R.

Another example of integrable pseudogroup (actually it is a group in this case) is the isotopic component of the identity of any subgroup of $\mathrm{Diff}(M)$. Let $G \subset \mathrm{Diff}(M)$ be an arbitrary subgroup of M. Two elements $g, h \in G$ are ***G-isotopic*** if there exists a differentiable mapping $a : \mathbb{R} \times M \to M$ such that $a_t \in G$ for every $t \in \mathbb{R}$, $a_t = g$ for $t \leq 0$, and $a_t = h$ for $t \geq 1$. This is an equivalence relation and the isotopic component G^0 containing the identity is a normal subgroup of G. The subgroup G^0 is integrable, that is, the partition of M into G^0-orbits is a generalized foliation (STEFAN (1974a)).

3.2 Pseudogroups generated by families of vector fields

Let M be a manifold and F an everywhere defined family of local vector fields. By *everywhere defined* we mean that for every $m \in M$ there exists $X \in F$ such that $m \in \mathrm{Dom}(X)$. The domains $\mathrm{Dom}(X) \subset M$, $X \in F$, will be taken open in M. The flows of the vector fields in F constitute a collection of arrows to which we can associate, following the scheme in §3.1.4, a set of local diffeomorphisms $\mathcal{A}_F := \{F_t \mid F_t \text{ flow of } X \in F\}$ of M and a pseudogroup of transformations generated by it,

$$A_F := (\mathbb{I}, M) \bigcup \{F^1_{t_1} \circ \cdots \circ F^n_{t_n} \mid n \in \mathbb{N} \text{ and } F^n_{t_n} \in \mathcal{A}_F \text{ or } (F^n_{t_n})^{-1} \in \mathcal{A}_F\}.$$

Analogously, we also define, for any $z \in M$, the following vector subspaces of T_zM:

$$\begin{aligned} \mathcal{D}_F(z) &:= \operatorname{span}\left\{ \left.\frac{d}{dt}\right|_{t=t_0} F_t(y) \middle| F_t \text{ flow of } X \in F,\ F_t(y) = z \right\} \\ &= \operatorname{span}\{X(z) \in T_zM \mid X \in F \text{ and } z \in \operatorname{Dom}(X)\}, \\ D_F(z) &:= \operatorname{span}\{T_y\mathcal{F}_T \cdot \mathcal{D}_F(y) \mid \mathcal{F}_T \in A_F, \mathcal{F}_T(y) = z\}. \end{aligned}$$

Note that, by construction, $\mathcal{D}_F$ is a differentiable generalized distribution in the sense of §1.4.6. We will say that $\mathcal{D}_F$ is the smooth generalized distribution ***spanned*** by F. Conversely, any differentiable generalized distribution is spanned by its differentiable sections.

Theorem 3.1.5 guarantees that the A_F-orbits, also called the ***accessible sets***, of the family F) form a generalized foliation whose leaves have as tangent spaces the values of D_F. An important question is determining when the smooth distribution $\mathcal{D}_F$ spanned by F is integrable:

3.2.1 Theorem (STEFAN (1974b) and SUSSMANN (1973)). *Let $\mathcal{D}_F$ be a differentiable generalized distribution on the smooth manifold M spanned by an everywhere defined family of vector fields F. The following properties are equivalent:*

(i) *The distribution '$\mathcal{D}_F$ is invariant under the pseudogroup of transformations generated by F, that is, for each $\mathcal{F}_T \in A_F$ and for each $z \in M$ in the domain of $\mathcal{F}_T$,*

$$T_z\mathcal{F}_T(\mathcal{D}_F(z)) = \mathcal{D}_F(\mathcal{F}_T(z)).$$

(ii) $\mathcal{D}_F = D_F$.

(iii) *For any $X \in F$ with flow F_t and any $x \in \operatorname{Dom}(X)$, there exist:*

- **(a)** *A finite set $\{X_1, \ldots, X_p\} \subset F$ such that $\mathcal{D}_F(x) = \operatorname{span}\{X_1(x), \ldots, X_p(x)\}$.*
- **(b)** *A constant $\epsilon > 0$ and Lebesgue integrable functions $\lambda_{ij} : (-\epsilon, \epsilon) \to \mathbb{R}$ $(1 \le i, j \le p)$ such that for every $t \in (-\epsilon, \epsilon)$ and $j \in \{1, \ldots, p\}$:*

$$[X, X_j](F_t(x)) = \sum_{i=1}^{p} \lambda_{ij}(t) X_i(F_t(x))$$

and $\mathcal{D}_F(F_t(x)) = \operatorname{span}\{X_1(F_t(x)), \ldots, X_p(F_t(x))\}$.

(iv) *The distribution $\mathcal{D}_F$ is integrable and its maximal integral manifolds are the A_F-orbits.*

3.2.2 The topology of the accessible sets of a family of vector fields. Given that the accessible sets of an everywhere defined family of local vector fields F on the manifold M form a generalized foliation of M, they are in particular initial submanifolds of it. The associated pseudogroup of transformations A_F provides an alternative characterization of the topology of the accessible sets or, on other words, of the A_F-orbits. Consider the strongest topology on M such that for any $k \in \mathbb{N}$, any $\mathcal{F}_T = F^1_{t_1} \circ \cdots \circ F^k_{t_k}$, and any $z \in M$, the map

$$T \longmapsto \mathcal{F}_T(z)$$

defined on an open neighborhood of the origin in $\mathbb{R}^k$ and with values in M is continuous.

We will refer to this construction as the ***F-topology*** on M. This topology is in general strictly stronger than the manifold topology of M and, in particular, it is Hausdorff. The connected components of M for the F-topology coincide with the accessible sets of F. Given an accessible set of F, its topology as an initial submanifold of M coincides with the relative topology induced by the F-topology on M.

3.2.3 Sufficient conditions for integrability and Chow's Theorem. The Stefan–Sussmann Theorem that we just presented provides a full characterization of the integrability of a generalized distribution spanned by a family of vector fields. We now recall a few sufficient conditions for the integrability of such distributions that prove useful in specific applications and can be readily obtained as a corollary of Theorem 3.2.1. We refer the reader to STEFAN (1974b) for proofs and references.

3.2.4 Theorem *Let $\mathcal{D}_F$ be a differentiable generalized distribution on the smooth manifold M spanned by an everywhere defined family of vector fields F. If any of the following three conditions is satisfied, then the distribution $\mathcal{D}_F$ is integrable:*

(i) Frobenius Theorem: *the family F is closed under formation of the Lie bracket and the dimension of $\mathcal{D}_F$ is locally constant on M.*

(ii) *The manifold M is analytic, the family F is closed under formation of the Lie bracket, and it consists of analytic vector fields defined on the entire M.*

(iii) *For any $m \in M$, $\dim \mathcal{D}_F(m) \leq 1$.*

We emphasize that the conditions just introduced for integrability are sufficient but not necessary. For instance, the distribution on $\mathbb{R}^2 \setminus \{(0,0)\}$ generated by the vector fields $X = x^2 \frac{\partial}{\partial x} + y^2 \frac{\partial}{\partial y}$ and $Y = y \frac{\partial}{\partial x} + x \frac{\partial}{\partial y}$ is clearly integrable (the entire $\mathbb{R}^2 \setminus \{(0,0)\}$ integrates this distribution) however this family is not closed under the Lie bracket since $[X, Y] = (y^2 - 2xy)\frac{\partial}{\partial x} + (x^2 - 2xy)\frac{\partial}{\partial y}$.

The following sufficient condition due to CHOW (1939) is of great use in control theory.

3.2.5 Theorem (Chow) *Let M be a smooth manifold and F an everywhere defined family of locally defined vector fields. Let $\bar{F}$ be the minimal set of vector fields on M that contains F and is closed under the formation of the Lie bracket. Let $\mathcal{L}_z$ be the accessible set through $z \in M$ of the distribution $\mathcal{D}_F$ spanned by F. If $\dim \mathcal{D}_{\bar{F}} = \dim M$, then L is an open subset of M.*

3.2.6 The completion of a generalized distribution. An everywhere defined family F of local vector fields on a manifold M has a unique pseudogroup of transformations A_F and a generalized distribution $\mathcal{D}_F$ associated but not the other way around, that is, a variety of families of locally defined vector fields on M can be chosen to define the same distribution $\mathcal{D}_F$. Nevertheless, if $\mathcal{D}_F$ is integrable and F' is another generating family of vector fields for $\mathcal{D}_F$, the uniqueness of the maximal integral leaves of an integrable distribution and the fact that, by Theorem 3.2.1, these are given by the pseudogroup orbits, we have that for any $z \in M$, $A_F \cdot z = A_{F'} \cdot z$. Consequently, the corresponding leaf spaces $M/\mathcal{D}_F = M/A_F = M/A_{F'} = M/\mathcal{D}_{F'}$ coincide even though the pseudogroups A_F and $A_{F'}$ themselves may be different. This uniqueness property makes integrability a feature of much interest in many situations that we will encounter.

Under some circumstances, the freedom in the choice of the generating family of a generalized smooth distribution can be used to pick a family of vector fields whose associated pseudogroup is actually a subgroup of the diffeomorphism group of the manifold and hence the maximal integral manifolds appear as group orbits. This remark motivates the introduction of the following definition.

3.2.7 Definition *Let D be an integrable generalized distribution on the smooth manifold M. We will say that D is* ***complete*** *when we can choose a generating family $F \subset \mathfrak{X}(M)$ of D made out of complete vector fields. In this case, the associated pseudogroup A_F forms a subgroup of* Diff(M). *If F is a generating family of D that contains a subfamily that still generates D and is made of complete vector fields, then we say that F is* ***completable****; any such subfamily will be called a* ***completion*** *of F.*

The relation between generalized distributions and pseudogroups of transformations allows us to talk in the integrable distribution context of concepts that in principle seem to be related only to group actions and pseudogroups of transformations:

3.2.8 Definition *Let D be an integrable generalized distribution on the smooth manifold M spanned by an everywhere defined family of vector fields F.*

- *A subset $U \subset M$ is called D-**invariant** or D-**saturated** if it is A_F-invariant.*
- *Let U be an open D-saturated subset of M. A smooth function $f \in C^\infty(U)$ is D-**invariant** if it belongs to $C^\infty(U)^{A_F}$. We will denote the set of D-invariant functions by $C^\infty(U)^D$.*
- *The distribution D has the* ***extension property*** *when the pseudogroup A_F has it.*
- *Finally, we say that $C^\infty(M)^D$* ***separates*** *the integral leaves of D when $C^\infty(M)^{A_F}$ separates the A_F-orbits.*

Recall that the integrability of D guarantees that the previous definitions DO NOT depend on the generating family F of D chosen to formulate them.

3.2.9 Proposition *Let D be a smooth integrable regular distribution on the manifold M, that is, D has constant rank and the projection $\pi_D : M \to M/D$ onto the leaf space is assumed to be a surjective submersion. Then D has the extension property.*

Proof. Let U be a D-saturated open subset of M, z a point in U, and $f \in C^\infty(U)^D$. Since π_D is a submersion, there are charts (V, φ) and (W, ψ) around z and $\pi_D(z)$, respectively, such that $\pi_D(V) = W$, $\varphi : V \to V' \times W'$, $\psi : W \to W'$, $\varphi(z) = (0, 0)$, and the local representative of π_D, that is, $\psi \circ \pi_D \circ \varphi^{-1} : V' \times W' \to W'$ is the projection onto the second factor. We will shrink V if necessary so that $V \subset U$.

Let $B_\epsilon(0) \subset W'$ be a ball of radius $\epsilon > 0$ and $\phi : B_\epsilon(0) \to [0, 1]$ a bump function such that $\phi|_{B_{\epsilon/2}(0)} = 1$ and $\phi|_{B_\epsilon(0)\setminus B_{3\epsilon/4}(0)} = 0$. Let $F'_D : \psi^{-1}(B_\epsilon(0)) \subset W \to \mathbb{R}$ be the smooth function given by $F'_D(l) = f_D(l)\phi(\psi(l))$, $l \in \psi^{-1}(B_\epsilon(0))$, where $f_D : \pi_D(U) \to \mathbb{R}$ is the unique smooth function determined by the relation $f = f_D \circ \pi_D|_U$. As the function F'_D and all its derivatives are zero in the boundary of $\psi^{-1}(B_\epsilon(0))$, F'_D can be extended to a function $F_D \in C^\infty(M/D)$.

Let $F \in C^\infty(M)^D$ be the function given by $F := F_D \circ \pi_D$ and Σ the submanifold of M through z defined as $\Sigma := \varphi^{-1}(\{0\} \times B_{\epsilon/2}(0))$. Notice that $\pi_D(\Sigma)$ is an open subset of $\pi_D(U)$ since $\psi(\pi_D(\Sigma)) = B_{\epsilon/2}(0)$ is an open subset of $\psi(\pi_D(U))$. Consequently, $T = \pi_D^{-1}(\pi_D(\Sigma)) \subset U$ is an open D-invariant subset of U.

We will prove the lemma by showing that $F|_T = f|_T$. Indeed, let $m \in T$ arbitrary. By definition $m \in T$ if and only if $\pi_D(m) \in \pi_D(\Sigma)$ or, equivalently, there exists an element $z \in \varphi^{-1}(\{0\} \times B_{\epsilon/2}(0))$ such that $\pi_D(m) = \pi_D(z)$. Due to the local expression of π_D in the charts (V, φ) and (W, ψ) we have that $\psi(\pi_D(z)) \in B_{\epsilon/2}(0)$ or, equivalently, $\pi_D(z) \in \psi^{-1}(B_{\epsilon/2}(0))$. With this in mind, we have that

$$F(m) = F_D \circ \pi_D(m) = F_D \circ \pi_D(z) = F'_D \circ \pi_D(z),$$

where the previous equality follows from the fact that $\pi_D(z) \in \psi^{-1}(B_{\epsilon/2}(0))$. We now use the definition of F'_D and $F'_D \circ \pi_D(z) = f_D(\pi_D(z))\phi(\psi(\pi_D(z))) = f_D(\pi_D(z)) = f_D(\pi_D(m)) = f(m)$, which proves that $F(m) = f(m)$, as required. ■

In this section we have explored an approach to generalized distributions based on vector fields. The reader interested in the dual point of view constructed on the use of one-forms or ***Pfaffian systems*** is encouraged to check, for instance, with FREEMAN (1984) and LIBERMANN AND MARLE (1987).

3.3 Equivariant vector fields and invariant distributions

In this section we explore the properties of vector fields that are equivariant with respect to a transformation group as well of distributions that are left invariant by such operations.

3.3.1 Definition *Let M be a smooth manifold and A a subgroup of its group of diffeomorphisms* Diff(M). *Let U be an A-equivariant open subset of M and $X \in \mathfrak{X}(U)$ a vector field defined on it. We say that X is A-**equivariant** when*

$$X \circ \phi = T\phi \circ X \quad \textit{for all} \quad \phi \in A.$$

We will denote the set of A-equivariant vector fields on U by $\mathfrak{X}(U)^A$.

*Let F be an everywhere defined family of local vector fields on M and $\mathcal{D}_F$ the generalized distribution spanned by it. We say that the distribution $\mathcal{D}_F$ is A-**invariant** when for any $z \in M$ and any $\phi \in A$ we have that*

$$T_z\phi \cdot \mathcal{D}_F(z) = \mathcal{D}_F(\phi(z)).$$

A distribution spanned by a family of A-equivariant vector fields is always A-invariant but the reverse implication is not necessarily true; consider, for instance, a Lie group G acting on the manifold M via the map $\Phi : G \times M \to M$. Let $A_G = \{\Phi_g \mid g \in G\}$ be the group of associated diffeomorphisms and F the family of vector fields on M consisting of all the infinitesimal generators, that is, $F = \{\xi_M \mid \xi \in \mathfrak{g}\}$. The elements in F are not A_G-equivariant since $T\Phi_g \cdot \xi_M = (\mathrm{Ad}_g \xi)_M \circ \Phi_g$, for all $g \in G$ and $\xi \in \mathfrak{g}$ nevertheless, the associated distribution $\mathcal{D}_F(z) = \mathfrak{g} \cdot z$ is A-invariant. We will discuss more in detail these relations in §3.5.3.

3.3.2 Proposition *Let M be a smooth manifold, A a subgroup of its group of diffeomorphisms* $\mathrm{Diff}(M)$, *and $\pi_A : M \to M/A$ the projection onto the corresponding orbit space.*

(i) *Let U be an A-equivariant open subset of M and $X \in \mathfrak{X}(U)^A$ an A-equivariant vector field defined on it. Then, the domain of definition* $\mathrm{Dom}(F_t) \subset U$ *of the flow F_t of X is A-invariant and F_t itself is A-equivariant.*

(ii) *Let F be an everywhere defined family of local A-equivariant vector fields and A_F the pseudogroup of transformations generated by it. Then, A commutes with A_F, that is, for any $(\mathcal{F}_T, \mathrm{Dom}(\mathcal{F}_T)) \in A_F$ the domain $\mathrm{Dom}(\mathcal{F}_T)$ is an open A-invariant set and, for any $(\phi, M) \in A$ we have that $(\mathcal{F}_T \circ \phi, \mathrm{Dom}(\mathcal{F}_T)) = (\phi \circ \mathcal{F}_T, \mathrm{Dom}(\mathcal{F}_T))$.*

(iii) *Any $(\mathcal{F}_T, \mathrm{Dom}(\mathcal{F}_T)) \in A_F$ induces a local diffeomorphism $(\bar{\mathcal{F}}_T, \pi_A(\mathrm{Dom}(\mathcal{F}_T)))$ of the quotient space $(M/A, C^\infty_{M/A})$, uniquely determined by the relation $\bar{\mathcal{F}}_T \circ \pi_A = \pi_A \circ \mathcal{F}_T$.*

Proof. **(i)** Let $\phi \in A$ be arbitrary and $(F_t, \mathrm{Dom}(F_t))$ the flow of X. The A-equivariance of X implies that

$$X = T\phi \circ X \circ \phi^{-1}. \tag{3.3.1}$$

Let now $G_t : \phi(\mathrm{Dom}(F_t)) \to \phi(F_t(\mathrm{Dom}(F_t)))$ be the local diffeomorphism defined by $G_t := \phi \circ F_t \circ \phi^{-1}$. The chain rule and expression (3.3.1) show that for any $z \in \mathrm{Dom}(F_t)$:

$$\begin{aligned}\frac{d}{dt}G_t(\phi(z)) &= (T\phi \circ X)(F_t(\phi^{-1}(\phi(z)))) \\ &= (T\phi \circ X \circ \phi^{-1})(G_t(\phi(z)) = X(G_t(\phi(z)).\end{aligned} \tag{3.3.2}$$

The uniqueness of the flow of a vector field implies that $\phi(\mathrm{Dom}(F_t)) \subset \mathrm{Dom}(F_t)$. Since $\phi \in A$ is arbitrary, we also have that $\phi^{-1}(\mathrm{Dom}(F_t)) \subset \mathrm{Dom}(F_t)$ and, consequently $\phi(\mathrm{Dom}(F_t)) = \mathrm{Dom}(F_t)$. Expression (3.3.2) also implies that $F_t = G_t = \phi \circ F_t \circ \phi^{-1}$ which guarantees the A-equivariance of F_t.

(ii) Just notice that the elements in A_F are finite compositions of flows as those in **(i)** together with their inverses. **(iii)** See §1.1.19. ■

3.3.3 The case of proper Lie group actions. The previous paragraphs described a few properties of the vector fields and distributions that are invariant by a group of diffeomorphisms $A \subset \text{Diff}(M)$. We now consider the case in which A consists of the diffeomorphisms associated to a proper action $\Phi : G \times M \to M$ of the Lie group G on M. We will denote, as customary, $A_G = \{\Phi_g \mid g \in G\} \subset \text{Diff}(M)$. The following proposition is a corollary of Proposition 3.3.2.

3.3.4 Proposition *Let G be a Lie group acting properly on the manifold M via the map $\Phi : G \times M \to M$ and $X \in \mathfrak{X}(M)^G$, a G-equivariant vector field with flow F_t. Then,*

(i) *The domain of definition* $\text{Dom}(F_t)$ *of F_t is an invariant open subset of M and F_t is equivariant.*

(ii) *Let $m \in M$ be an arbitrary point and $M^m_{G_m}$ the connected component of the isotropy type manifold containing it. Then, F_t leaves this manifold invariant and consequently X is tangent to it.*

Proof. **(i)** It is a straightforward consequence of Proposition 3.3.2.
(ii) The equivariance of F_t implies that $G_{F_t(m)} = G_m$ and hence $F_t(m) \in M^m_{G_m}$, for any t where $F_t(m)$ is defined. In particular, this implies that $X(m) \in T_m M^m_{G_m}$. ■

The existence of slices in the proper Lie group actions case implies various important properties that we analyze in what follows. Our next result was originally proved in Krupa (1990), Field (1991), and Chossat and Golubitsky (1988) in various degrees of generality, and it is a statement that locally allows the splitting of an equivariant vector field as the sum of a vector field tangent to the group orbits and another one in the slice. This result is of much use in the study of G-equivariant dynamics, in particular in the description of G-equivariant flows around a relative equilibrium. In Chapter 7 we will present an analogue of this result, namely the so-called *reconstruction equations*, in the framework of symmetric symplectic manifolds.

3.3.5 Theorem (The tangent-normal decomposition) *Let G be a Lie group acting properly on the manifold M via the map $\Phi : G \times M \to M$ and X a G-equivariant vector field defined on a G-invariant open subset of M. Let m be a point in the domain of X and S a slice at m. Then there exist two vector fields X_T and X_N such that:*

(i) *$X_T \in \mathfrak{X}(G \cdot S)^G$ and $X_T(z) = (\xi(z))_M(z)$, $z \in G \cdot S$, where $\xi : G \cdot S \to \mathfrak{g}$ is a smooth G-equivariant map such that $\xi(z) \in \text{Lie}\, N(G_z) \cdot z$, for all $z \in G \cdot S$. Moreover, the flow F_t of X_T is given by $F_t(z) = \exp t\xi(z) \cdot z$. In particular, X_T is a complete vector field.*

(ii) $X_N \in \mathfrak{X}(S)^{G_m}$.

(iii) *If $z = g \cdot s \in G \cdot S$ with $g \in G$ and $s \in S$, then*

$$X(z) = X_T(z) + T_s\Phi_g \cdot X_N(s) = T_s\Phi_g \cdot (X_T(s) + X_N(s)). \tag{3.3.3}$$

(iv) *If G_t is the flow of X_N, then the integral curve of X through the point $g \cdot s \in G \cdot S$ is*

$$F_t(g \cdot s) = g(t) \cdot G_t(s) \tag{3.3.4}$$

where $g(t)$ is the solution of the first order differential equation

$$\dot{g}(t) = T_e L_{g(t)} \cdot \xi\left(G_t(s)\right), \quad g(0) = g. \tag{3.3.5}$$

Proof. The properness of the G-action guarantees the existence of a G_m-invariant open neighborhood U of G_m in G/G_m and a smooth local G_m-equivariant section $\sigma : U \subset G/G_m \to G$ of the projection $\pi_{G_m} : G \to G/G_m$ such that the map

$$\begin{aligned} F : U \times S &\longrightarrow Z \subset M \\ (u, s) &\longmapsto \sigma(u) \cdot s \end{aligned} \tag{3.3.6}$$

is a G_m-equivariant diffeomorphism onto a G_m-invariant open subset Z of M, with respect to the G_m-action $h \cdot (gG_m, s) := (hgG_m, h \cdot s)$ on $U \times S$ and the G_m-action on M (see Corollary 2.3.40). Recall that equivariance of σ means that $\sigma(hgG_m) = h\sigma(gG_m)h^{-1}$, for all $h \in G_m$ and $gG_m \in U$. The G_m-equivariance of F implies that of F^{-1}.

Also, notice that for any $s \in S$ we have that $F^{-1}(s) = (G_m, \sigma(G_m)^{-1} \cdot s)$. Hence,

$$T_s F^{-1} \cdot X(s) =: (X_U(s), X_S(s)) \in T_{G_m} U \times T_{\sigma(G_m)^{-1} \cdot s} S.$$

Let

$$X_N(s) := T_{\sigma(G_m)^{-1} \cdot s} \Phi_{\sigma(G_m)} \cdot X_S(s), \tag{3.3.7}$$

$$X_T(g \cdot s) := T_s \Phi_g \cdot \left(T_{\sigma(G_m)} R_{\sigma(G_m)^{-1}} \cdot T_{G_m} \sigma \cdot X_U(s)\right)_M (s) \tag{3.3.8}$$

$$= \left(\mathrm{Ad}_g \left(T_{\sigma(G_m)} R_{\sigma(G_m)^{-1}} \cdot T_{G_m} \sigma \cdot X_U(s)\right)\right)_M (g \cdot s), \tag{3.3.9}$$

where the last equality follows from (2.2.3). Before we get into the study of X_N and X_T notice that as X and F^{-1} are G_m-equivariant we have that X_U and X_S satisfy

$$X_U(h \cdot s) = T_{G_m} \Psi_h \cdot X_U(s) \in T_{G_m} U \tag{3.3.10}$$

$$X_S(h \cdot s) = T_{\sigma(G_m)^{-1} \cdot s} \Phi_h \cdot X_S(s) \in T_{h\sigma(G_m)^{-1} \cdot s} S, \tag{3.3.11}$$

where $h \in G_m$, $s \in S$, and $\Psi : G_m \times G/G_m \to G/G_m$ is the G_m-action on G/G_m defined by $\Psi(h, gG_m) := hgG_m$.

We now check that X_N and X_T, as given in (3.3.7) and (3.3.8), are actually well-defined and satisfy the requirements **(i)** through **(iv)** in the statement of the theorem.

(i) We start by defining the function $\xi : G \cdot S \to \mathfrak{g}$ by

$$\xi(z) := \mathrm{Ad}_g \left(T_{\sigma(G_m)} R_{\sigma(G_m)^{-1}} \cdot T_{G_m} \sigma \cdot X_U(s)\right), \tag{3.3.12}$$

where $g \in G$ and $s \in S$ are such that $z = g \cdot s$. We first check that (3.3.12) is a good definition. Let $z \in G \cdot S$ and $g, g' \in G$, $s, s' \in S$ be such that $z = g \cdot s = g' \cdot s'$. The last equality and the properties of the slices imply the existence of an element $h \in G_m$ such that $g' = gh$ and $s' = h^{-1} \cdot s$. Note that G_m-equivariance of σ implies that $h\sigma(G_m)h^{-1} = \sigma(hG_m) = \sigma(G_m)$ for any $h \in G_m$. Thus $h\sigma(G_m)^{-1}h^{-1} = (h\sigma(G_m)h^{-1})^{-1} = \sigma(G_m)^{-1}$ and hence $R_{h^{-1}} \circ R_{\sigma(G_m)^{-1}} \circ R_h = R_{\sigma(G_m)^{-1}}$, for any

$h \in G_m$. Using (3.3.10) in the third, equivariance of σ in the fourth, and this identity in the sixth equality below we get

$$\begin{aligned}
\xi(g' \cdot s') &= \xi(gh \cdot (h^{-1} \cdot s)) \\
&= \mathrm{Ad}_{gh} \left(T_{\sigma(G_m)} R_{\sigma(G_m)^{-1}} \cdot T_{G_m}\sigma \cdot X_U(h^{-1} \cdot s) \right) \\
&= \mathrm{Ad}_{gh} \cdot \left(T_{\sigma(G_m)} R_{\sigma(G_m)^{-1}} \cdot T_{G_m}\sigma \cdot T_{G_m}\Psi_{h^{-1}} \cdot X_U(s) \right) \\
&= \mathrm{Ad}_{gh} \left(T_{\sigma(G_m)} R_{\sigma(G_m)^{-1}} \cdot T_{G_m}(L_{h^{-1}} \circ R_h \circ \sigma) \cdot X_U(s) \right) \\
&= \mathrm{Ad}_{gh} \left(T_{G_m}(L_{h^{-1}} \circ R_h \circ R_{h^{-1}} \circ R_{\sigma(G_m)^{-1}} \circ R_h \circ \sigma) \cdot X_U(s) \right) \\
&= \mathrm{Ad}_{gh} \left(\mathrm{Ad}_{h^{-1}} \left(T_{G_m}(R_{\sigma(G_m)^{-1}} \circ \sigma) \cdot X_U(s) \right) \right) \\
&= \mathrm{Ad}_g \left(T_{G_m}(R_{\sigma(G_m)^{-1}} \circ \sigma) \cdot X_U(s) \right) = \xi(g \cdot s),
\end{aligned} \tag{3.3.13}$$

which proves that $\xi : G \cdot S \to \mathfrak{g}$ is a well-defined function on $G \cdot S$.

Notice that by its very definition the map ξ is G-equivariant and $X_T(z) = (\xi(z))_M(z)$ for any $z \in G \cdot S$. The string of equalities leading to (3.3.13) implies that the smooth map $(g, s) \in G \times S \longmapsto \mathrm{Ad}_g \left(T_{\sigma(G_m)} R_{\sigma(G_m)^{-1}} \cdot T_{G_m}\sigma \cdot X_U(s) \right) \in \mathfrak{g}$ is invariant with respect to the twisted action of G_m on $G \times S$ (see Definition 2.3.16) and therefore it induces a smooth map $\psi : G \times_{G_m} S \to \mathfrak{g}$. Since S is a slice the map $\varphi : [g, s] \in G \times_{G_m} S \mapsto g \cdot s \in G \cdot S$ is a G-equivariant diffeomorphism. Therefore, the map $\xi = \psi \circ \varphi^{-1}$ is smooth and therefore so is X_T. Also, the definition (3.3.8) shows that X_T is a G-equivariant vector field on $G \cdot S$.

We will show now that the integral curve of X_T through z equals $F_t(z) := \exp t\xi(z) \cdot z$. To see this it suffices to observe that the curve $t \mapsto \exp t\xi(z) \cdot z$ passes through z at $t = 0$ and that it is an integral curve of X_T:

$$\begin{aligned}
\frac{d}{dt} \exp t\xi(z) \cdot z &= \left.\frac{d}{dr}\right|_{r=0} \exp(t + r)\xi(z) \cdot z = T_z\Phi_{\exp t\xi(z)}((\xi(z))_M(z)) \\
&= T_z\Phi_{\exp t\xi(z)}(X_T(z)) = X_T(\exp t\xi(z) \cdot z),
\end{aligned}$$

the last equality being valid because of the G-equivariance of X_T.

Finally, we show that $\xi(z) \in \mathrm{Lie}(N(G_z))$, for any $z \in G \cdot S$. Since X_T is G-equivariant, by Proposition 3.3.4 $X_T(z) \in T_z M_{G_z} = (T_z M)^{G_z}$, for any $z \in G \cdot S$. Hence $X_T(z) = (\xi(z))_M(z) \in (T_z M)^{G_z} \cap \mathfrak{g} \cdot z = T_z(N(G_z) \cdot z)$ by Proposition 2.4.5. Consequently, $\xi(z) \in \mathrm{Lie}(N(G_z))$.

(ii) Expression (3.3.7) defines the vector field X_N on S, so we just have to show that X_N is G_m-equivariant; this will be a consequence of (3.3.11). Indeed, for any $h \in G_m$ and $s \in S$ we have that

$$\begin{aligned}
X_N(h \cdot s) &= T_{\sigma(G_m)^{-1}h \cdot s}\Phi_{\sigma(G_m)} \cdot X_S(h \cdot s) \\
&= T_{\sigma(G_m)^{-1}h \cdot s}\Phi_{\sigma(G_m)} \cdot T_{\sigma(G_m)^{-1} \cdot s}\Phi_h \cdot X_S(s) \\
&= T_{\sigma(G_m)^{-1} \cdot s}\Phi_{\sigma(G_m)h} \cdot X_S(s) \\
&= T_{\sigma(G_m)^{-1} \cdot s}\Phi_{hh^{-1}\sigma(G_m)h} \cdot X_S(s) \\
&= T_{\sigma(G_m)^{-1} \cdot s}(\Phi_h \circ \Phi_{\sigma(h^{-1}G_m)}) \cdot X_S(s) \\
&= T_{\sigma(G_m)^{-1} \cdot s}(\Phi_h \circ \Phi_{\sigma(G_m)}) \cdot X_S(s) = T_s\Phi_h \cdot X_N(s).
\end{aligned}$$

(iii) Let $s \in S$. Using just the definitions we can write

$$\begin{aligned} X(s) &= T_{(G_m, \sigma(G_m)^{-1} \cdot s)} F \cdot (X_U(s), X_S(s)) \\ &= T_{(G_m, \sigma(G_m)^{-1} \cdot s)} F \cdot (X_U(s), 0) + T_{(G_m, \sigma(G_m)^{-1} \cdot s)} F \cdot (0, X_S(s)). \end{aligned}$$

Now using the definition (3.3.6) of F this equals

$$\left(T_{\sigma(G_m)} R_{\sigma(G_m)^{-1}} \cdot T_{G_m} \sigma \cdot X_U(s)\right)_M (s) + T_{\sigma(G_m)^{-1} \cdot s} \Phi_{\sigma(G_m)} \cdot X_S(s) \\ = X_T(s) + X_N(s).$$

In order to recover (3.3.3) we just use this equality and the G-equivariance of X and X_T.

(iv) We start by proving that if $g(t)$ is a solution of (3.3.5), then $k(t) := g(t)h, h \in G_m$, solves

$$\dot{k}(t) = T_e L_{g(t)h} \cdot \xi(G_t(h^{-1} \cdot s)), \quad k(0) = gh. \tag{3.3.14}$$

Indeed, we have

$$\begin{aligned} \dot{k}(t) &= T_{g(t)} R_h \cdot \dot{g}(t) = T_e \left(R_h \circ L_{g(t)}\right) \cdot \xi(G_t(s)) = T_{g(t)} R_h \cdot \dot{g}(t) \\ &= T_e \left(L_{g(t)h} \circ L_{h^{-1}} \circ R_h\right) \cdot \xi(G_t(s)) = T_e L_{g(t)h} \cdot \mathrm{Ad}_{h^{-1}} \xi(G_t(s)) \\ &= T_e L_{g(t)h} \cdot \xi(h^{-1} \cdot G_t(s)) = T_e L_{g(t)h} \cdot \xi(G_t(h^{-1} \cdot s)). \end{aligned}$$

We now use (3.3.14) to show that (3.3.4) is well defined. Let $g, g' \in G$, $s, s' \in S$ be such that $g \cdot s = g' \cdot s'$. The properties of the slices imply the existence of an element $h \in G_m$ such that $g' = gh$ and $s' = h^{-1} \cdot s$. Then

$$F_t(g' \cdot s') = F_t(gh \cdot (h^{-1} \cdot s)) = k(t) \cdot G_t(h^{-1} \cdot s),$$

where $k(t) = g(t)h$ solves (3.3.14). Consequently, this equals $g(t)h \cdot G_t(h^{-1} \cdot s) = g(t) \cdot G_t(s) = F_t(g \cdot s)$, as required.

We conclude by showing that (3.3.4) is the integral curve of X through the point $g \cdot s$. It is clear that $g(0) \cdot G_0(s) = g \cdot s$. The Leibniz rule (2.2.5) implies that

$$\begin{aligned} &\frac{d}{dt} g(t) \cdot G_t(s) \\ &\quad = T_{G_t(s)} \Phi_{g(t)} \left(T_{g(t)} L_{g(t)^{-1}} \dot{g}(t)\right)_M (G_t(s)) + T_{G_t(s)} \Phi_{g(t)} \cdot \frac{d}{dt} G_t(s) \\ &\quad = T_{G_t(s)} \Phi_{g(t)} \left[(\xi(G_t(s)))_M (G_t(s)) + X_N(G_t(s))\right] \\ &\quad = T_{G_t(s)} \Phi_{g(t)} \left[X_T(G_t(s)) + X_N(G_t(s))\right] = T_{G_t(s)} \Phi_{g(t)} \cdot X(G_t(s)) \\ &\quad = X(g(t) \cdot G_t(s)). \quad \blacksquare \end{aligned}$$

3.3.6 Corollary *Let M be a proper G-manifold and F an everywhere defined family of local G-equivariant vector fields. Then the generalized G-invariant distribution $\mathcal{D}_F$ spanned by F is complete.*

Proof. We will proceed by using the tangent-normal decomposition in order to find a completion of F. Let $m \in M$ be arbitrary and let $F_m := \{X_1, \ldots, X_k\} \subset F$ be such that $\mathcal{D}_F(m) = \operatorname{span}\{X_1(m), \ldots, X_k(m)\}$. Now pick a slice S of the G-action at the point m such that $G \cdot S \subset \operatorname{Dom}(X_i)$, $i \in \{1, \ldots, k\}$. Let U_1 and U_2 be two open G_m-invariant neighborhoods of m (available by Corollary 2.3.34) in S such that $U_1 \subset U_2$ and choose a G_m-invariant bump function ϕ_m such that $\phi_m|_{U_1} = 1$ and $\phi_m|_{S\setminus U_2} = 0$ (available by Lemma 1.1.25 and Proposition 2.3.8(**vii**)). If we denote by $X_T^i \in \mathfrak{X}(G \cdot S)^G$ and $X_N^i \in \mathfrak{X}(S)^{G_m}$ the tangent and normal parts, respectively, of X_i, we define $\tilde{X}_N^i := \phi X_N^i \in \mathfrak{X}(S)^{G_m}$ and

$$\tilde{X}_i(g \cdot s) = X_T^i(g \cdot s) + T_s\Phi_g \cdot \tilde{X}_N^i(s), \quad i \in \{1, \ldots, k\},\ g \in G,\ s \in S.$$

We will denote $\tilde{F}_m := \{\tilde{X}_1, \ldots, \tilde{X}_k\}$. Notice that since $\tilde{X}_N^i$ is compactly supported, it is a complete vector field with flow G_t^i. Since the flow F_t^i of $\tilde{X}_i$ is given by $F_t^i(g \cdot s) = g \exp t\xi \cdot G_t^i(s)$ with $\xi \in \mathfrak{g}$ such that $X_T^i(s) = \xi_M(s)$, the vector field $\tilde{X}_i$ is also complete. By construction, $\tilde{X}_i$ is zero outside $G \cdot U_2$, so it can be trivially extended to a complete vector defined on the entire M that will be denoted by the same symbol. At the same time, we have that $\mathcal{D}_F(m) = \operatorname{span}\{\tilde{X}_1(m), \ldots, \tilde{X}_k(m)\}$. Therefore, the family $\tilde{F} := \bigcup_{m\in M} \tilde{F}_m$ is a completion of F. ■

3.4 The saturated sets of an invariant distribution

Later on in our work we will discuss orbit reduction techniques that will involve the use of distributions generated by vector fields that are equivariant with respect to a proper group action and, more specifically, the saturation of its accessible sets by that group action. The study of these saturated sets is the subject of this section.

3.4.1 The leaves of an integrable generalized distribution spanned by equivariant vector fields. Let M be a smooth manifold and G a Lie group acting properly on it via the mapping $\Phi : G \times M \to M$. Let F be an everywhere defined family of local G-equivariant vector fields on M that span an integrable generalized distribution $\mathcal{D}_F$. The integrability hypothesis implies, by Theorem 3.2.1, that $\mathcal{D}_F = D_F$. Denote by A_F the pseudogroup of local diffeomorphisms associated to the flows of the family F. By Corollary 3.3.6 D_F is complete. Also, the integral leaves of D_F coincide with the A_F orbits and they are initial submanifolds of M. We will write

$$\mathcal{J}_F : M \longrightarrow M/D_F$$

for the projection onto the space of leaves. The choice of the symbol $\mathcal{J}_F$ for this projection will be justified later on. As customary, the space of leaves will be considered as a quotient topological space endowed with its quotient presheaf of functions denoted by C^∞_{M/D_F}.

The G-equivariance properties of the local diffeomorphisms in A_F imply that all the elements in a leaf of D_F share the same G-isotropy subgroup. More specifically, if z is an arbitrary element of the integral leaf $\mathcal{L}_m$ of D_F containing m, then $G_z = G_m$. Indeed, for any $z \in \mathcal{L}_m$ there exists $\mathcal{F}_T \in A_F$ such that $z = \mathcal{F}_T(m)$. As $\mathcal{F}_T$ is G-equivariant (see Proposition 3.3.2) the claim follows.

3.4.2 The G-action on leaf space. The G-equivariance of the elements in A_F can also be used to introduce a natural G-action $\Psi : G \times M/D_F \to M/D_F$ of G on M/D_F defined by $\Psi(g, \mathcal{J}_F(m)) := \mathcal{J}_F(g \cdot m)$, for any $g \in G$ and any $m \in M$. This is indeed a good definition because if m and m' are such that $\mathcal{J}_F(m) = \mathcal{J}_F(m')$, then there exists $\mathcal{F}_T \in A_F$ such that $m' = \mathcal{F}_T(m)$. Consequently, $\mathcal{J}_F(g \cdot m') = \mathcal{J}_F(g \cdot \mathcal{F}_T(m)) = \mathcal{J}_F(\mathcal{F}_T(g \cdot m)) = \mathcal{J}_F(g \cdot m)$, as required.

Notice that, by construction, $\mathcal{J}_F$ is G-equivariant with respect to the G-action on M/D_F introduced above. It is straightforward to check that Ψ is a smooth map in the sense of maps between topological spaces with a presheaf of functions, that is,

$$\Psi^* C^{\infty}_{M/D_F}(U) \subset C^{\infty}_{G \times M/D_F}(\Psi^{-1}(U)),$$

for any open subset $U \subset M/A_F$.

For any $\rho \in M/D_F$ we will write $\mathcal{O}_\rho := G \cdot \rho$. The G-equivariance of $\mathcal{J}_F$ implies that $\mathcal{J}_F^{-1}(\mathcal{O}_\rho) = G \cdot \mathcal{J}_F^{-1}(\rho)$.

3.4.3 The isotropy subgroups of the G-action on leaf space. The fact that the G-action on the leaf space M/D_F is not smooth in the standard sense implies that most of the usual facts about smooth Lie group actions used so far are not valid. The first place where this difficulty appears is in the study of the isotropy subgroups of this action. Since the leaf space M/D_F is, in general, not Hausdorff, one cannot guarantee that the isotropy subgroups are closed in G and, therefore, that they are Lie subgroups of G. However, there is still something that can be said.

3.4.4 Proposition *Let M be a smooth manifold and G a Lie group acting properly on it via the mapping $\Phi : G \times M \to M$. Let F be an everywhere defined family of local G-equivariant vector fields on M that span an integrable generalized distribution $\mathcal{D}_F = D_F$. Let G_ρ be the isotropy subgroup of the element $\rho \in M/D_F$ with respect to the G-action on M/D_F defined in §3.4.2. Then:*

(i) *There is a unique smooth structure on G_ρ with respect to which this subgroup is an initial Lie subgroup of G with Lie algebra $\mathfrak{g}_\rho$ given by*

$$\mathfrak{g}_\rho = \{\xi \in \mathfrak{g} \mid \xi_M(m) \in T_m \mathcal{J}_F^{-1}(\rho), \text{ for all } m \in \mathcal{J}_F^{-1}(\rho)\} \tag{3.4.1}$$

or, equivalently

$$\mathfrak{g}_\rho = \{\xi \in \mathfrak{g} \mid \exp t\xi \in G_\rho, \text{ for all } t \in \mathbb{R}\}. \tag{3.4.2}$$

(ii) *With this smooth structure for G_ρ, the left action $\Phi^\rho : G_\rho \times \mathcal{J}_F^{-1}(\rho) \to \mathcal{J}_F^{-1}(\rho)$ defined by $\Phi^\rho(g, z) := \Phi(g, z)$ is smooth. Moreover, if G_ρ is closed in G, then this action is proper.*

(iii) *This action has fixed isotropies, that is, if $z \in \mathcal{J}_F^{-1}(\rho)$, then $(G_\rho)_z = G_z$, and $G_m = G_z$ for all $m, z \in \mathcal{J}_F^{-1}(\rho)$.*

(iv) *Let $z \in \mathcal{J}_F^{-1}(\rho)$ be arbitrary. Then*

$$\mathfrak{g}_\rho \cdot z = A_F(z) \cap \mathfrak{g} \cdot z = T_z \mathcal{J}_F^{-1}(\rho) \cap \mathfrak{g} \cdot z. \tag{3.4.3}$$

Proof. **(i)** It suffices to take on G_ρ the Lie group structure induced by G that we introduced in Theorem 2.1.10. We now show that with that smooth structure the Lie algebra $\mathfrak{g}_\rho$ of G_ρ is given by (3.4.1) or (3.4.2). We start by proving the equivalence between the expressions (3.4.1) and (3.4.2). First, it is clear that $\{\xi \in \mathfrak{g} \mid \exp t\xi \in G_\rho, \text{ for all } t \in \mathbb{R}\} \subset \{\xi \in \mathfrak{g} \mid \xi_M(m) \in T_m\mathcal{J}_F^{-1}(\rho), \text{ for all } m \in \mathcal{J}_F^{-1}(\rho)\}$. Conversely, let $\xi \in \mathfrak{g}$ such that $\xi_M(m) \in T_m\mathcal{J}_F^{-1}(\rho)$, for all $m \in \mathcal{J}_F^{-1}(\rho)$; given that $\mathcal{J}_F^{-1}(\rho)$ is an immersed submanifold of M there exists a vector field $\xi^\rho_{\mathcal{J}_F^{-1}(\rho)}$ on $\mathcal{J}_F^{-1}(\rho)$ (see §1.2.4) such that $Ti_\rho \cdot \xi^\rho_{\mathcal{J}_F^{-1}(\rho)} = \xi_M \circ i_\rho$, where $i_\rho : \mathcal{J}_F^{-1}(\rho) \hookrightarrow M$ is the inclusion. The flow F_t^ρ of $\xi^\rho_{\mathcal{J}_F^{-1}(\rho)}$ is given by $F_t^\rho(m) = \exp t\xi \cdot m$, for all $m \in \mathcal{J}_F^{-1}(\rho)$. Since $F_t^\rho(m) \in \mathcal{J}_F^{-1}(\rho)$, for all $t \in \mathbb{R}$ and all $m \in \mathcal{J}_F^{-1}(\rho)$ we have that $\exp t\xi \cdot \rho = \rho$, and hence $\exp t\xi \in G_\rho$, as required.

Now, by Theorem 2.1.10 we have that $\mathfrak{g}_\rho = \left\{\xi \in \mathfrak{g} \mid \text{there exists a smooth curve } c : \mathbb{R} \to G_\rho \text{ such that } c(0) = e \text{ and } c'(0) = \xi\right\}$. It is clear that the set $\{\xi \in \mathfrak{g} \mid \exp t\xi \in G_\rho, \text{ for all } t \in \mathbb{R}\}$ in (3.4.2) is included in $\mathfrak{g}_\rho$. Conversely, if $\xi \in \mathfrak{g}_\rho$ it is clear that $\xi_M(m) \in T_m\mathcal{J}_F^{-1}(\rho)$, for all $m \in \mathcal{J}_F^{-1}(\rho)$ and hence the Lie algebra $\mathfrak{g}_\rho$ equals the expressions in (3.4.1) or (3.4.2).

(ii) As $\mathcal{J}_F^{-1}(\rho)$ is an initial submanifold of M, the map Φ^ρ is smooth if and only if the map $i^\rho \circ \Phi^\rho$ is smooth. This is so because $i^\rho \circ \Phi^\rho = \Phi \circ (i^{G_\rho} \times i^\rho) : G_\rho \times \mathcal{J}_F^{-1}(\rho) \to M$ is smooth, where $i^{G_\rho} : G_\rho \hookrightarrow G$ is the inclusion. The properness of this action when G_ρ is closed in G follows from the properness of the G-action on M.

(iii) This is a straightforward consequence of the fact that any two elements $z, m \in \mathcal{J}_F^{-1}(\rho)$ can be written as $z = \mathcal{F}_T(m)$ for some $\mathcal{F}_T \in A_F$. The G-equivariance of $\mathcal{F}_T$ (see Proposition 3.3.2) implies that $G_z = G_m$ for all $z, m \in \mathcal{J}_F^{-1}(\rho)$. Regarding the G_ρ isotropy subgroups notice first that clearly $(G_\rho)_z \subset G_z$. Conversely, if $g \in G_z$, then $g \cdot z = z$. Applying $\mathcal{J}_F$ to both sides of this expression we obtain that $\rho = \mathcal{J}_F(z) = \mathcal{J}_F(g \cdot z) = g \cdot \rho$ and hence $g \in G_\rho$.

(iv) The inclusion $\mathfrak{g}_\rho \cdot z \subset A_F(z) \cap \mathfrak{g} \cdot z$ is a consequence of (3.4.1). Conversely, let $X(z) = \xi_M(z) \in A_F(z) \cap \mathfrak{g} \cdot z$, with X a G-equivariant vector field and $\xi \in \mathfrak{g}$. The G-equivariance of X implies that $[X, \xi_M] = 0$, and hence, if F_t is the flow of X and G_t is the flow of ξ_M (more explicitly $G_t(m) = \exp t\xi \cdot m$ for any $m \in M$), then $F_t \circ G_s = G_s \circ F_t$. By the Trotter product formula (see (1.2.12)), the flow H_t of $X - \xi_M$ is given by

$$
\begin{aligned}
H_t(m) &= \lim_{n\to\infty} \left(F_{t/n} \circ G_{-t/n}\right)^n (m) = \lim_{n\to\infty} \left(F^n_{t/n} \circ G^n_{-t/n}\right)(m) \\
&= (F_t \circ G_{-t})(m) = F_t(\exp(-t\xi) \cdot m),
\end{aligned}
$$

for any $m \in M$. Consequently, since $X(z) = \xi_M(z)$, the point $z \in M$ is an equilibrium of $X - \xi_M$, and hence $F_t(\exp(-t\xi) \cdot z) = z$ or, equivalently, $\exp t\xi \cdot z = F_t(z)$. Applying $\mathcal{J}_F$ on both sides of this equality and taking into account that F_t is the flow of a G-equivariant vector field, it follows that $\exp t\xi \cdot \rho = \exp t\xi \cdot \mathcal{J}_F(z) = \mathcal{J}_F(\exp t\xi \cdot z) = \mathcal{J}_F(F_t(z)) = \mathcal{J}_F(z) = \rho$, and hence $\xi \in \mathfrak{g}_\rho$ by (3.4.2). Thus $\xi_M(z) \in \mathfrak{g}_\rho \cdot z$, as required. ■

3.4.5 The saturating distribution. Our goal is showing that the saturation of the integral leaves of D_F are initial submanifolds of M. We will do this by showing that the connected components of these saturations are the integral leaves of the integrable generalized distribution that we introduce in the following proposition.

3.4.6 Proposition *Let $\Phi : G \times M \to M$ be a smooth proper action of the Lie group G on the manifold M. Let F be an everywhere defined family of local G-equivariant vector fields on M that span an integrable generalized distribution $\mathcal{D}_F = D_F$. Then:*

(i) *The generalized distribution D on M defined by $D(m) := \mathfrak{g} \cdot m + D_F(m)$, for all $m \in M$, is integrable.*

(ii) *If $m \in M$ and $\mathcal{L}_m$ is the maximal integral leaf of D containing m, then $\mathcal{L}_m = G^0 \cdot \mathcal{J}_F^{-1}(\rho)$. The symbol G^0 denotes the connected component of of the identity element in G. The sets $G^0 \cdot \mathcal{J}_F^{-1}(\rho)$ are therefore initial submanifolds of M.*

Proof. **(i)** The distribution D can be written as the span of the vector fields:

$$D = \operatorname{span}\{\xi_M, \ X \mid \xi \in \mathfrak{g} \text{ and } X \in F\}. \tag{3.4.4}$$

By the Stefan–Sussman Theorem, the integrability of D is equivalent to its invariance with respect to the flows of the vector fields in (3.4.4) that generate it. Let $X, Y \in F$, $\xi, \eta \in \mathfrak{g}$, let F_t be the flow of Y, and H_t the flow of η_M. Recall that η_M is a complete vector field with flow $H_t(m) = \exp t\eta \cdot m$, for all $t \in \mathbb{R}$ and $m \in M$. Now, the integrability of D_F guarantees that $T_m F_t \cdot X(m) \in D_F(F_t(m)) \subset D(F_t(m))$. Also, the G-equivariance of F_t and X imply that $T_m F_t \cdot \xi_M(m) = \xi_M(F_t(m))$ and $T_m H_t \cdot X(m) = X(H_t(m))$. Finally, we have that

$$\begin{aligned} T_m H_t \cdot \xi_M(m) &= \left.\frac{d}{ds}\right|_{s=0} \exp t\eta \exp s\xi \cdot m \\ &= \left.\frac{d}{ds}\right|_{s=0} \exp t\eta \exp s\xi \exp(-t\eta) \exp t\eta \cdot m = (\mathrm{Ad}_{\exp t\eta}\xi)_M(\exp t\eta \cdot m), \end{aligned}$$

which proves that D is integrable.

(ii) As D is integrable and is generated by the vector fields (3.4.4), its maximal integral submanifolds coincide with the orbits of the action of the pseudogroup constructed by finite composition of flows of the vector fields in (3.4.4), that is, for any $m \in M$, the integral leaf $\mathcal{L}_m$ of D that contains m is

$$\mathcal{L}_m = \left\{(F_{t_1} \circ \cdots \circ F_{t_n})(m) \mid \text{with } F_{t_i} \text{ the flow of a vector field in (3.4.4)}\right\}.$$

Since $[X, \xi_M] = 0$ for all $X \in F$ and $\xi \in \mathfrak{g}$, the previous expression can be rewritten as

$$\begin{aligned} \mathcal{L}_m = \big\{&(H_{t_1} \circ \cdots \circ H_{t_j} \circ G_{s_1} \circ \cdots \circ G_{s_k})(m) \\ &\mid G_{s_i} \text{ flow of } X_i \in F \text{ and } H_{t_i} \text{ flow of } \xi^i_M,\ \xi^i \in \mathfrak{g}\big\}. \end{aligned}$$

Therefore, $\mathcal{L}_m = G^0 \cdot A_F(m) = G^0 \cdot \mathcal{J}_F^{-1}(\rho)$, as required. ■

3.4.7 The initial smooth structure of the saturated set $\mathcal{J}_F^{-1}(\mathcal{O}_\rho)$.

3.4.8 Proposition *Assume the same hypotheses as in Proposition* 3.4.6. *Let* $\rho \in M/D_F$. *If either* G_ρ *is closed in* G *or, more generally,* G_ρ *acts properly on* $\mathcal{J}_F^{-1}(\rho)$, *then:*

(i) *The* G_ρ *action on the product* $G \times \mathcal{J}_F^{-1}(\rho)$ *defined by* $h \cdot (g, z) := (gh, h^{-1} \cdot z)$ *is free and proper, and therefore the corresponding orbit space* $G \times \mathcal{J}_F^{-1}(\rho)/G_\rho =: G \times_{G_\rho} \mathcal{J}_F^{-1}(\rho)$ *is a smooth regular quotient manifold. We will denote by* $\pi_{G_\rho} : G \times \mathcal{J}_F^{-1}(\rho) \to G \times_{G_\rho} \mathcal{J}_F^{-1}(\rho)$ *the canonical surjective submersion.*

(ii) *The mapping* $i : G \times_{G_\rho} \mathcal{J}_F^{-1}(\rho) \to M$ *defined by* $i([g, z]) := g \cdot z$ *is an injective immersion onto* $\mathcal{J}_F^{-1}(\mathcal{O}_\rho)$ *such that, for any* $[g, z] \in G \times_{G_\rho} \mathcal{J}_F^{-1}(\rho)$, $T_{[g,z]}i \cdot T_{[g,z]}(G \times_{G_\rho} \mathcal{J}_F^{-1}(\rho)) = D(g \cdot z)$. *In other words,* $i(G \times_{G_\rho} \mathcal{J}_F^{-1}(\rho)) = \mathcal{J}_F^{-1}(\mathcal{O}_\rho)$ *is an integral submanifold of* D.

Proof. **(i)** As we saw in §2.3.11 G_ρ is closed in G iff the action of G_ρ on G by right translations is proper. Additionally, if G_ρ is closed in G, then the G_ρ-action on $\mathcal{J}_F^{-1}(\rho)$ is proper (Proposition 3.4.4). In any case, if the action of G_ρ on either G, or on $\mathcal{J}_F^{-1}(\rho)$, or on both, is proper, so is the action on the product $G \times \mathcal{J}_F^{-1}(\rho)$ in the statement of the proposition. As to the freeness, it is inherited from the freeness of the G_ρ-action on G.
(ii) First, the map i is clearly well defined and smooth since it is the projection onto the orbit space $G \times_{G_\rho} \mathcal{J}_F^{-1}(\rho)$ of the G_ρ-invariant smooth map $G \times \mathcal{J}_F^{-1}(\rho) \to M$ given by $(g, z) \longmapsto g \cdot z$. It is also injective because if $[g, z], [g', z'] \in G \times_{G_\rho} \mathcal{J}_F^{-1}(\rho)$ are such that $i([g, z]) = i([g', z'])$, then $g \cdot z = g' \cdot z'$ or, equivalently, $g^{-1}g' \cdot z' = z$, which implies that $g^{-1}g' \in G_\rho$ (by applying $\mathcal{J}_F$ to both sides). Consequently, $[g, z] = [gg^{-1}g', (g')^{-1}g \cdot z] = [g', z']$, as required.

Finally, we check that i is an immersion. Let $[g, z] \in G \times_{G_\rho} \mathcal{J}_F^{-1}(\rho)$ be arbitrary and let $\xi \in \mathfrak{g}$, $X \in F$ be such that $T_{[g,z]}i \cdot T_{(g,z)}\pi_{G_\rho} \cdot (T_eL_g(\xi), X(z)) = 0$. If we denote by F_t the flow of X we can rewrite this equality as

$$\left.\frac{d}{dt}\right|_{t=0} g \exp t\xi \cdot F_t(z) = 0 \quad \text{or equivalently,} \quad T_z\Phi_g(X(z) + \xi_M(z)) = 0.$$

Hence $X(z) = -\xi_M(z)$ which by (3.4.3) implies that $\xi \in \mathfrak{g}_\rho$ and therefore $T_{(g,z)}\pi_{G_\rho} \cdot (T_eL_g(\xi), X(z)) = T_{(g,z)}\pi_{G_\rho} \cdot (T_eL_g(\xi), -\xi_M(z)) = 0$, as required.

Given that for any $\xi \in \mathfrak{g}$, $X \in F$, and $[g, z] \in G \times_{G_\rho} \mathcal{J}_F^{-1}(\rho)$ we have that $T_{[g,z]}i \cdot T_{(g,z)}\pi_{G_\rho} \cdot (T_eL_g(\xi), X(z)) = (\mathrm{Ad}_g\xi)_M(g \cdot z) + X(g \cdot z)$, it is clear that $T_{[g,z]}i \cdot T_{[g,z]}(G \times_{G_\rho} \mathcal{J}_F^{-1}(\rho)) = D(g \cdot z)$ and thereby $i(G \times_{G_\rho} \mathcal{J}_F^{-1}(\rho)) = \mathcal{J}_F^{-1}(\mathcal{O}_\rho)$ is an integral submanifold of D. ■

3.4.9 Definition *Suppose that we are in the setup of Proposition* 3.4.8. *Take on the saturated set* $\mathcal{J}_F^{-1}(\mathcal{O}_\rho)$, *the smooth structure that makes the bijection* $G \times_{G_\rho} \mathcal{J}_F^{-1}(\rho) \to \mathcal{J}_F^{-1}(\mathcal{O}_\rho)$ *given by* $(g, z) \mapsto g \cdot z$ *into a diffeomorphism. We will refer to this structure as the* ***initial smooth structure*** *of* $\mathcal{J}_F^{-1}(\mathcal{O}_\rho)$.

The following theorem justifies the choice of terminology in the previous definition and why we will be able to refer to the smooth structure introduced there as *the* initial smooth structure of $\mathcal{J}_F^{-1}(\mathcal{O}_\rho)$.

3.4.10 Theorem *Assume the hypotheses of Definition* 3.4.9. *Then the set* $\mathcal{J}_F^{-1}(\mathcal{O}_\rho)$ *endowed with the initial smooth structure is an initial submanifold of* M *that can be decomposed as a disjoint union of connected components as*

$$\mathcal{J}_F^{-1}(\mathcal{O}_\rho) = \coprod_{[g]\in G/(G^0 G_\rho)} gG^0 \cdot \mathcal{J}_F^{-1}(\rho). \tag{3.4.5}$$

Each connected component of $\mathcal{J}_F^{-1}(\mathcal{O}_\rho)$ *is a maximal integral submanifold of the saturating distribution* D *defined in Proposition* 3.4.6. *Notice that* (3.4.5) *implies that* $\mathcal{J}_F^{-1}(\mathcal{O}_\rho)$ *has as many connected components as the cardinality of the homogeneous manifold* $G/(G^0 G_\rho)$.

If, additionally, the subgroup G_ρ *is closed in* G, *the topology on* $\mathcal{J}_F^{-1}(\mathcal{O}_\rho)$ *induced by its initial smooth structure coincides with the initial topology induced by the map* $(\mathcal{J}_F)_{\mathcal{J}_F^{-1}(\mathcal{O}_\rho)} : \mathcal{J}_F^{-1}(\mathcal{O}_\rho) \to \mathcal{O}_\rho$ *given by* $z \longmapsto \mathcal{J}_F(z)$, *where the orbit* $\mathcal{O}_\rho$ *is endowed with the smooth structure of the homogeneous manifold* G/G_ρ.

Proof. First, notice that the sets $gG^0 \cdot \mathcal{J}_F^{-1}(\rho)$ are clearly maximal integral submanifolds of D by part **(ii)** in Proposition 3.4.6. As a corollary of this, they are the connected components of $\mathcal{J}_F^{-1}(\mathcal{O}_\rho)$ endowed with the smooth structure in Definition 3.4.10. Indeed, let S be the connected component of $\mathcal{J}_F^{-1}(\mathcal{O}_\rho)$ that contains $gG^0 \cdot \mathcal{J}_F^{-1}(\rho)$, that is, $gG^0 \cdot \mathcal{J}_F^{-1}(\rho) \subset S \subset \mathcal{J}_F^{-1}(\mathcal{O}_\rho)$. As $\mathcal{J}_F^{-1}(\mathcal{O}_\rho)$ is a manifold, it is locally connected, and therefore its connected components are open and closed. In particular, since S is an open connected subset of $\mathcal{J}_F^{-1}(\mathcal{O}_\rho)$, part **(ii)** in Proposition 3.4.8 shows that S is a connected integral submanifold of D. By the maximality of $gG^0 \cdot \mathcal{J}_F^{-1}(\rho)$ as an integral submanifold of D, $gG^0 \cdot \mathcal{J}_F^{-1}(\rho) = S$, necessarily. The set $gG^0 \cdot \mathcal{J}_F^{-1}(\rho)$ is therefore a connected component of $\mathcal{J}_F^{-1}(\mathcal{O}_\rho)$. As it is a leaf of a smooth integrable distribution on M, it is also an initial submanifold of M of dimension $d = \dim \mathcal{J}_F^{-1}(\mathcal{O}_\rho) = \dim G + \dim \mathcal{J}_F^{-1}(\rho) - \dim G_\rho$.

We now show that $\mathcal{J}_F^{-1}(\mathcal{O}_\rho)$ with the smooth structure in Definition 3.4.10 is an initial submanifold of M. Indeed, part **(ii)** in Proposition 3.4.8 shows that $\mathcal{J}_F^{-1}(\mathcal{O}_\rho)$ is an injectively immersed submanifold of M. The initial character can be obtained as a consequence of the fact that its connected components are initial manifolds together with Lemma 1.1.11 **(iii)**.

To prove Expression (3.4.5), note that since G^0 is normal in G, the set $G^0 G_\rho$ is a subgroup (in principle not closed) of G. We obviously have that

$$\mathcal{J}_F^{-1}(\mathcal{O}_\rho) = \bigcup_{g\in G} gG^0 \cdot \mathcal{J}_F^{-1}(\rho). \tag{3.4.6}$$

Moreover, if g and $g' \in G$ are such that $[g] = [g'] \in G/(G^0 G_\rho)$, then we can write $g' = ghk$ with $h \in G^0$ and $k \in G_\rho$. Consequently, $g' G^0 \mathcal{J}_F^{-1}(\rho) = ghkG^0 \mathcal{J}_F^{-1}(\rho) =$

$gh(G^0k)\mathcal{J}_F^{-1}(\rho) = g(hG^0)(k\mathcal{J}_F^{-1}(\rho)) = gG^0\mathcal{J}_F^{-1}(\rho)$, which implies that (3.4.6) can be refined to

$$\mathcal{J}_F^{-1}(\mathcal{O}_\rho) = \bigcup_{[g]\in G/(G^0 G_\rho)} gG^0 \cdot \mathcal{J}_F^{-1}(\rho). \tag{3.4.7}$$

It only remains to be shown that this union is disjoint. Let $gh \cdot z = lh' \cdot z'$ with $h, h' \in G^0$ and $z, z' \in \mathcal{J}_F^{-1}(\rho)$. If we apply $\mathcal{J}_F$ to both sides of this equality we obtain $gh \cdot \rho = lh' \cdot \rho$. Hence, $(h')^{-1}l^{-1}gh \in G_\rho$ and $l^{-1}g \in h'G_\rho h^{-1} \subset G^0G_\rho$. This implies that $[l] = [g] \in G/(G^0G_\rho)$ and $gG^0 \cdot \mathcal{J}_F^{-1}(\rho) = lG^0 \cdot \mathcal{J}_F^{-1}(\rho)$, as required.

We finally show that if G_ρ is closed in G, the topology on $\mathcal{J}_F^{-1}(\mathcal{O}_\rho)$ induced by its initial smooth structure coincides with the initial topology induced by the map $(\mathcal{J}_F)_{\mathcal{J}_F^{-1}(\mathcal{O}_\rho)} : \mathcal{J}_F^{-1}(\mathcal{O}_\rho) \to \mathcal{O}_\rho$ on $\mathcal{J}_F^{-1}(\mathcal{O}_\rho)$. Recall first that this topology is characterized by the fact that for any topological space Z and any map $\phi : Z \to \mathcal{J}_F^{-1}(\mathcal{O}_\rho)$ we have that $\phi : Z \to \mathcal{J}_F^{-1}(\mathcal{O}_\rho)$ is continuous if and only if $(\mathcal{J}_F)_{\mathcal{J}_F^{-1}(\mathcal{O}_\rho)} \circ \phi$ is continuous. Moreover, as the family $\{(\mathcal{J}_F)^{-1}_{\mathcal{J}_F^{-1}(\mathcal{O}_\rho)}(U) \mid U \text{ open subset of } \mathcal{O}_\rho\}$ is a subbase of this topology, the initial topology on $\mathcal{J}_F^{-1}(\mathcal{O}_\rho)$ induced by the map $(\mathcal{J}_F)_{\mathcal{J}_F^{-1}(\mathcal{O}_\rho)}$ is first countable. We prove that this topology coincides with the topology induced by the initial smooth structure on $\mathcal{J}_F^{-1}(\mathcal{O}_\rho)$ by showing that the map $f : G \times_{G_\rho} \mathcal{J}_F^{-1}(\rho) \to \mathcal{J}_F^{-1}(\mathcal{O}_\rho)$, $f([g, z]) := g \cdot z$ is a homeomorphism by considering $\mathcal{J}_F^{-1}(\mathcal{O}_\rho)$ as a topological space with the initial topology induced by $(\mathcal{J}_F)_{\mathcal{J}_F^{-1}(\mathcal{O}_\rho)}$. Indeed, f is continuous if and only if the map $G \times_{G_\rho} \mathcal{J}_F^{-1}(\rho) \to \mathcal{O}_\rho$ given by $[g, z] \mapsto g \cdot \rho$ is continuous, which in turn is equivalent to the continuity of the map $G \times \mathcal{J}_F^{-1}(\rho) \to G/G_\rho$ defined by $(g, z) \longmapsto gG_\rho$, which is true. We now show that the inverse $f^{-1} : \mathcal{J}_F^{-1}(\mathcal{O}_\rho) \to G \times_{G_\rho} \mathcal{J}_F^{-1}(\rho)$ of f given by $g \cdot z \mapsto [g, z]$ is continuous. Since the initial topology on $\mathcal{J}_F^{-1}(\mathcal{O}_\rho)$ induced by $(\mathcal{J}_F)_{\mathcal{J}_F^{-1}(\mathcal{O}_\rho)}$ is first countable it suffices to show that for any convergent sequence $\{z_n\} \subset \mathcal{J}_F^{-1}(\mathcal{O}_\rho) \to z \in \mathcal{J}_F^{-1}(\mathcal{O}_\rho)$, we have that $\lim_{n\to\infty} f^{-1}(z_n) = f^{-1}(\lim_{n\to\infty} z_n) = f^{-1}(z)$. Indeed, as $(\mathcal{J}_F)_{\mathcal{J}_F^{-1}(\mathcal{O}_\rho)}$ is continuous, the sequence $\{\mathcal{J}_F(z_n) = g_n \cdot \rho\} \subset \mathcal{O}_\rho$ converges in $\mathcal{O}_\rho$ to $\mathcal{J}_F(z) = g \cdot \rho$, for some $g \in G$. Let $j : \mathcal{O}_\rho \to G/G_\rho$ be the standard diffeomorphism and let $\sigma : U_{gG_\rho} \subset G/G_\rho \to G$ be a local smooth section of the submersion $G \to G/G_\rho$ in a neighborhood U_{gG_ρ} of $gG_\rho \in G/G_\rho$. Let $V := (\mathcal{J}_F)^{-1}_{\mathcal{J}_F^{-1}(\mathcal{O}_\rho)}(j^{-1}(U_{gG_\rho}))$. V is an open neighborhood of z in $\mathcal{J}_F^{-1}(\mathcal{O}_\rho)$ because $j \circ (\mathcal{J}_F)_{\mathcal{J}_F^{-1}(\mathcal{O}_\rho)}(z) = j(g \cdot \rho) = gG_\rho \in U_{gG_\rho}$. Note that for any $m \in V$ we have

$$f^{-1}(m) = \left[\left(\sigma \circ j \circ (\mathcal{J}_F)_{\mathcal{J}_F^{-1}(\mathcal{O}_\rho)}\right)(m), \left(\left(\sigma \circ j \circ (\mathcal{J}_F)_{\mathcal{J}_F^{-1}(\mathcal{O}_\rho)}\right)(m)\right)^{-1} \cdot m\right].$$

Consequently, since

$$\begin{aligned}
&\lim_{n\to\infty} f^{-1}(z_n) \\
&= \lim_{n\to\infty}\left[\left(\sigma\circ j\circ(\mathcal{J}_F)_{\mathcal{J}_F^{-1}(\mathcal{O}_\rho)}\right)(z_n), \left(\left(\sigma\circ j\circ(\mathcal{J}_F)_{\mathcal{J}_F^{-1}(\mathcal{O}_\rho)}\right)(z_n)\right)^{-1}\cdot z_n\right] \\
&= \left[\left(\sigma\circ j\circ(\mathcal{J}_F)_{\mathcal{J}_F^{-1}(\mathcal{O}_\rho)}\right)(z), \left(\left(\sigma\circ j\circ(\mathcal{J}_F)_{\mathcal{J}_F^{-1}(\mathcal{O}_\rho)}\right)(z)\right)^{-1}\cdot z\right] = f^{-1}(z),
\end{aligned}$$

the continuity of f^{-1} is guaranteed. ■

3.4.11 Remark Notice that Proposition 2.3.12 on the initial submanifold character of the group orbits of a proper group action is a corollary of the previous theorem. Indeed, if we take in Theorem 3.4.10 the zero distribution as D_F, then $M/D_F = M$, $\mathcal{J}_F : M \to M$ is the identity map, the leaves $\mathcal{J}_F^{-1}(\rho)$ are the points of M, and the isotropy subgroups G_ρ are the G-isotropies of the elements of M. With these facts in mind Expression (2.3.2) clearly appears as a particular case of (3.4.5).

3.5 Equivariant vector fields and isotropy type submanifolds

Let M be a smooth manifold and let $\Phi : G \times M \to M$ be a smooth proper action of the Lie group G on it. In this section we will study the distribution spanned by the set of all the G-equivariant vector fields defined on all the open G-invariant neighborhoods of M, that is, we will be looking at the distribution $\mathcal{D}_F$ generated by the family of vector fields

$$F = \bigcup_U \mathfrak{X}(U)^G, \quad U \text{ open } G\text{-invariant subset of } M. \tag{3.5.1}$$

We will show in what follows that

$$\mathcal{D}_F(z) = T_z M_{G_z}^z, \text{ for any } z \in M,$$

which constitutes the dual version of the relation (2.5.3) in Theorem 2.5.10. Additionally, we will show that $\mathcal{D}_F$ is integrable and that its integral leaves coincide with the connected components of the isotropy type submanifolds of M with respect to its G-symmetry.

3.5.1 Theorem (Ortega **(2002b)**) *Let M be a smooth manifold and $\Phi : G \times M \to M$ a smooth proper action of the Lie group G on it. Let $\mathcal{D}_F$ be the generalized distribution spanned by the family F in* (3.5.1). *Then $\mathcal{D}_F$ is integrable.*

Let $\mathcal{J}_F : M \to M/A_F$ be the projection onto the leaf space and let $z \in M$ be an arbitrary point in M; denote $\rho := \mathcal{J}_F(z)$. Then $\mathcal{J}_F^{-1}(\rho)$ coincides with the connected component $M_{G_z}^z$ of the G_z-isotropy type submanifold M_{G_z} containing z.

The isotropy subgroup G_ρ of the element $\rho \in M/A_F$ with respect to the G-action on M/A_F defined in §3.4.2 *equals the closed subgroup $N(G_z)^z$ of the normalizer $N(G_z)$ consisting of the elements in $N(G_z)$ that leave $M_{G_z}^z$ invariant.*

Proof. **$\mathcal{D}_F$ is integrable.** By Corollary 3.3.6, the distribution $\mathcal{D}_F$ is complete. Moreover, the method used in the proof of that result implemented to the family F gives a completion $\bar{F}$ of F:

$$\bar{F} = \{X \in \mathfrak{X}(M)^G \mid X \text{ is a complete vector field}\}.$$

As $\mathcal{D}_F = \mathcal{D}_{\bar{F}}$ we can prove the integrability of $\mathcal{D}_F$ by showing that $\mathcal{D}_{\bar{F}}$ is integrable. Let $X, Y \in \mathfrak{X}(M)^G$ be a pair of arbitrary complete G-equivariant vector fields with flows F_t and G_t, respectively. We are supposed to show that for any $z \in M$ we have that $T_z F_t \cdot Y(z) \in \mathcal{D}_{\bar{F}}(F_t(z))$. Consider the pull-back vector field $Z = TF_t \circ Y \circ F_{-t}$ whose flow is $H_s = F_t \circ G_s \circ F_{-t}$. The G-equivariance and completeness of the flows F_t and G_s imply that Z is a G-equivariant complete vector field and thus it belongs to $\bar{F}$. Consequently, $T_z F_t \cdot Y(z) = Z(F_t(z)) \in \mathcal{D}_{\bar{F}}(F_t(z))$, as required, which shows that $\mathcal{D}_F = \mathcal{D}_{\bar{F}}$ is an integrable distribution by the Stefan–Sussmann Theorem 3.2.1.
The integral leaves of $\mathcal{D}_F$. We now prove that the maximal integral leaves of $\mathcal{D}_F$ are the connected components of the isotropy type manifolds, that is, for any $z \in M$, the maximal integral leaf $\mathcal{L}_z$ of $\mathcal{D}_F$ through z coincides with the connected component $M^z_{G_z}$ of the G_z-isotropy type submanifold M_{G_z} containing z. We start by proving that for any $z \in M$ we have

$$\mathcal{D}_F(z) = T_z M^z_{G_z}. \tag{3.5.2}$$

Let F_t be the flow of an arbitrary element X in the family F. The equivariance of X implies by Proposition 3.3.4 that $\operatorname{Dom}(F_t)$ is an open G-invariant subset of M, that F_t is equivariant, and that $X(z) \in T_z M^z_{G_z}$. As $X \in \mathcal{D}_F$ is arbitrary, this shows that $\mathcal{D}_F(z) \subset T_z M^z_{G_z}$.

In order to prove the converse inclusion we use the Tube Theorem. Let $w \in T_z M^z_{G_z}$ and $\varphi : G \times_{G_z} B \to U$ a tube about the orbit $G \cdot z$ such that $\varphi([e, 0]) = z$. By Proposition 2.4.6 we have that $\varphi^{-1}(U_{G_z}) = N(G_z) \times_{G_z} B^{G_z}$ and hence $T_z\varphi^{-1} \cdot w = T_{(e,0)}\pi_{G_z}(\xi, v)$, where $\pi_{G_z} : G \times B \to G \times_{G_z} B$ is the projection and $\xi \in \operatorname{Lie}(N(G_z))$, $v \in B^{G_z}$. We now construct a G-equivariant vector field X on $G \times_{G_z} B$ such that $X([e, 0]) = T_{(e,0)}\pi_{G_z}(\xi, v)$. Let $\pi : G \to G/G_z$ be the projection and $\sigma : W \subset G/G_z \to G$ a local equivariant section of π. Recall that the action on G relative to which σ is equivariant is conjugation. Let $k := \sigma(G_z)$ and recall that by Corollary 2.3.39, the element k belongs in $Z(G_z)$. Define $\zeta := T_e\pi \cdot \xi$, $\tau := T_k R_{k^{-1}} \cdot T_{G_z}\sigma \cdot \zeta$, and

$$X([g, u]) := T_{(g,u)}\pi_{G_z}(T_e L_g \cdot \tau, v), \quad \text{for any } [g, u] \in G \times_{G_z} B. \tag{3.5.3}$$

We now check that this is a good definition for the vector field X on $G \times_{G_z} B$ that we are looking for. If $h \in G_z$ arbitrary, then, by definition,

$$\begin{aligned} X([gh, h^{-1} \cdot u]) &= T_{(gh,h^{-1}\cdot u)}\pi_{G_z}(T_e L_{gh} \cdot \tau, v) \\ &= \left.\frac{d}{dt}\right|_{t=0} \pi_{G_z}(gh\sigma(\pi(\exp t\xi))k^{-1}, h^{-1} \cdot u + tv) \\ &= \left.\frac{d}{dt}\right|_{t=0} \pi_{G_z}(gh\sigma(\pi(\exp t\xi))h^{-1}hk^{-1}, h^{-1} \cdot u + tv) \\ &= \left.\frac{d}{dt}\right|_{t=0} \pi_{G_z}(g\sigma(h \cdot \pi(\exp t\xi))hk^{-1}, h^{-1} \cdot u + tv). \end{aligned}$$

Since $\xi \in \mathrm{Lie}(N(G_z))$, we have $\sigma(h\cdot\pi(\exp t\xi)) = \sigma(h \exp t\xi\, G_z) = \sigma(\exp t\xi\, G_z) = \sigma(\pi(\exp t\xi))$. Additionally, as $k \in Z(G_z)$ and $v \in B^{G_z}$, we have $hk^{-1} = k^{-1}h$, $h^{-1} \cdot v = v$ and hence

$$\begin{aligned} X([gh, h^{-1} \cdot u]) &= \left.\frac{d}{dt}\right|_{t=0} \pi_{G_z}(g\sigma(\pi(\exp t\xi))hk^{-1}, h^{-1} \cdot u + tv) \\ &= \left.\frac{d}{dt}\right|_{t=0} \pi_{G_z}(g\sigma(\pi(\exp t\xi))k^{-1}h, h^{-1} \cdot u + th^{-1} \cdot v) \\ &= \left.\frac{d}{dt}\right|_{t=0} \pi_{G_z}(g\sigma(\pi(\exp t\xi))k^{-1}, u + tv) = X([g, u]). \end{aligned}$$

Consequently, (3.5.3) is a good definition for a vector field on $G \times_{G_z} B$. This vector field is clearly G-equivariant. We now check that $X([e, 0]) = T_{(e,0)}\pi_{G_z}(\xi, v)$. Define the curve $k : (-\epsilon, \epsilon) \to G_z$ by $k(t) = \exp(-t\xi)\sigma(\pi(\exp t\xi))$. Clearly, $k(0) = k$ and therefore

$$\begin{aligned} X([e, 0]) &= \left.\frac{d}{dt}\right|_{t=0} \pi_{G_z}(\sigma(\pi(\exp t\xi))k^{-1}, tv) \\ &= \left.\frac{d}{dt}\right|_{t=0} \pi_{G_z}(\exp t\xi k(t)k^{-1}, tv) \\ &= \left.\frac{d}{dt}\right|_{t=0} \pi_{G_z}(\exp t\xi, tk(t)^{-1}k \cdot v) \\ &= \left.\frac{d}{dt}\right|_{t=0} \pi_{G_z}(\exp t\xi, tv) = T_{(e,0)}\pi_{G_z}(\xi, v), \end{aligned}$$

as required. Finally, the vector field $Y = T\varphi \cdot X$ belongs to $\mathfrak{X}(U)^G$ and is such that $Y(z) = w$. Given that $z \in M$ and $w \in T_z M^z_{G_z}$ are arbitrary we have proved that $\mathcal{D}_F(z) \supset T_z M^z_{G_z}$ and, consequently, $\mathcal{D}_F(z) = T_z M^z_{G_z}$.

Let now $\mathcal{L}_z$ be the maximal integral leaf of $\mathcal{D}_F$ through z. Since $\mathcal{D}_F$ is integrable, the Stefan–Sussmann Theorem guarantees that $\mathcal{L}_z = A_F \cdot z$, with A_F the pseudogroup of transformations generated by the flows of the vector fields in F. As these flows are G-equivariant $\mathcal{L}_z = A_F \cdot z \subset M^z_{G_z}$. At the same time (3.5.2) shows that $M^z_{G_z}$ is a connected integral manifold of $\mathcal{D}_F$ and, consequently, by the maximality of $\mathcal{L}_z$, we have that $\mathcal{L}_z = M^z_{G_z}$, as required.

Isotropy subgroups. Suppose that $\mathcal{J}_F(z) = \rho$. By definition, $g \in G_\rho$ if and only if $g \cdot \rho = \rho$, which in turn is equivalent to $g \cdot \mathcal{J}_F(z) = \mathcal{J}_F(z)$ and hence to $\mathcal{J}_F(g \cdot z) = \mathcal{J}_F(z)$. In view of the characterization of the fibers of $\mathcal{J}_F$ that we just proved, this equality amounts to $g \cdot z \in M^z_{G_z}$ which is equivalent to $g \in N(G_z)^z$. ■

3.5.2 Corollary *Let M be a smooth manifold and $\Phi : G \times M \to M$ a smooth proper action of the Lie group G on it. Let $H \subset G$ be an isotropy subgroup of this action and M^z_H the connected component of the isotropy type manifold M_H that contains a given point $z \in M$. If $g \in N(H)$ is such that $g \cdot m = m'$ for two given points $m, m' \in M^z_H$, then $g \in N(H)^z$.*

Proof. Given any point $s \in M^z_H$ there exists by Theorem 3.5.1 an element $\mathcal{F}_T$ in the pseudogroup of transformations A_F associated to the flows of the vector fields in the family F defined in (3.5.1) such that $s = \mathcal{F}_T(m)$. Then by Proposition 3.3.4 we have that

$$g \cdot s = g \cdot \mathcal{F}_T(m) = \mathcal{F}_T(g \cdot m) = \mathcal{F}_T(m') \in M_H^z.$$

Since $s \in M_H^z$ is arbitrary we have that $g \in N(H)^z$. ■

3.5.3 Equivariant vector fields, invariant distributions, and distributions with invariant leaves. The distribution just studied can be used to illustrate the relation between the following three concepts whose interplay is not as functorially straightforward as could be expected at first sight: invariant integrable distributions, integrable distributions spanned by equivariant vector fields, and generalized foliations whose leaves are invariant. The following diagram presents in a schematic way what the relations between these objects are:

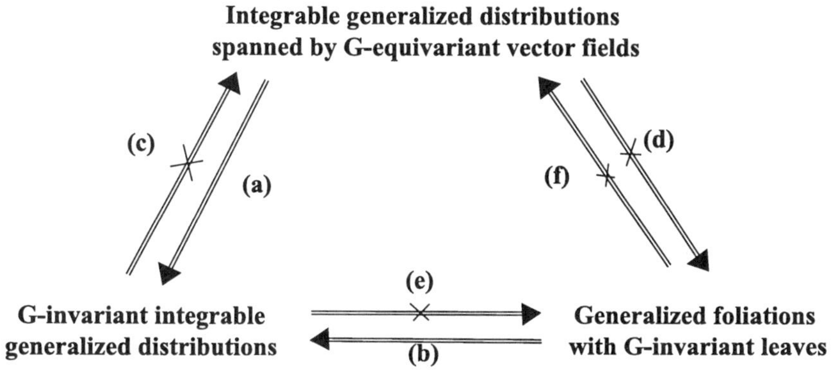

The implications **(a)** and **(b)** are obvious. The crossed arrow **(c)** means that there are G-invariant distributions that are not spanned by equivariant vector fields, as we already illustrated with an example below Definition 3.3.1. The crossed arrows **(d)** and **(e)** mean that the leaves of neither an invariant integrable distribution nor an integrable distribution generated by equivariant vector fields are guaranteed to be invariant, as can be easily seen by looking at the distribution introduced in Theorem 3.5.1. Finally, **(f)** indicates that generalized foliations with invariant leaves can be generated by vector fields that are not necessarily G-equivariant; an example of this is the saturating distribution introduced in Proposition 3.4.6 in the case in which G is connected and non-Abelian.

3.6 Groupoids

When we introduced the pseudogroups of transformations in Section 3.1 we emphasized that the main difference of these structures with respect to the groups of transformations is given by the fact that the composition law is partially defined. Indeed, not any two given local diffeomorphisms can be composed since certain relations between the domain of one and the range of the other have to be satisfied. The category of groupoids is an abstraction of this idea.

Groupoids, whose introduction is usually attributed to BRANDT (1926), appear in many contexts (see PATERSON (1999) and references therein for interesting historical remarks). In connection with some of the parts of mathematics that we have touched so far in this book we can say that they seem to be the right way to approach local symmetries (see WEINSTEIN (1996a)) or that they can be used to model leaf spaces of regular integrable foliations (see CRAINIC AND MOERDIJK (2001); MOERDIJK AND MRČUN (2003)).

Several equivalent definitions for groupoids can be found in the literature. We will adopt the one in CANNAS AND WEINSTEIN (1999).

3.6.1 Definition *A **groupoid** $G \rightrightarrows X$ over the set X, called the **base**, is a set G, called the **total space**, together with the following structure maps:*

(i) $\alpha, \beta : G \to X$. *We will refer to α as the **target** map and to β as the **source** map. An element $g \in G$ is thought of as an arrow from $\beta(g)$ to $\alpha(g)$ in X.*

(ii) *The **set of composable pairs** is defined as:*

$$G^{(2)} := \{(g, h) \in G \times G \mid \beta(g) = \alpha(h)\}.$$

*There is a **product map** $m : G^{(2)} \to G$ that satisfies $\alpha(m(g,h)) = \alpha(g)$, $\beta(m(g,h)) = \beta(h)$, and $m(m(g,h),k) = m(g, m(h,k))$, for any $g, h, k \in G$. We will usually write gh for $m(g,h)$.*

(iii) *An injection $\epsilon : X \to G$, called the **identity section**, such that $\epsilon(\alpha(g))g = g = g\epsilon(\beta(g))$. In particular, $\alpha \circ \epsilon = \beta \circ \epsilon$ is the identity map on X.*

(iv) *An **inversion map** $i : G \to G$, also denoted by $i(g) = g^{-1}$, $g \in G$, such that $g^{-1}g = \epsilon(\beta(g))$ and $gg^{-1} = \epsilon(\alpha(g))$.*

When the total space and the base of a groupoid $G \rightrightarrows X$ are smooth manifolds, the target and source maps are surjective submersions, and the multiplication, inversion, and identity section are smooth maps, we say that $G \rightrightarrows X$ is a ***Lie groupoid***.

In analogy with the Lie group theoretical context there exists a notion of ***Lie algebroid***, which we will not review here. The relation between Lie algebroids and Lie groupoids is more convoluted than that between Lie algebras and Lie groups, and it has only recently been completely settled in CRAINIC AND LOJA FERNANDES (2000).

Notice that any group is a groupoid over a set with just one element. Any set X can be endowed with a ***trivial groupoid*** structure over itself by taking for the source and target maps only the identity. The cartesian product $X \times X$ of any set is a groupoid over X if we take as target and source maps the projection on the first and second factors, respectively. The product is given by $(x,y)(y,z) = (x,z)$, $x, y, z \in X$, the identity section is $\epsilon(x) = (x,x)$, and $(x,y)^{-1} = (y,x)$. This is usually called the ***pair*** or ***coarse groupoid***.

3.6.2 Products, subgroupoids, groupoid homomorphisms, orbits, and isotropies. Given two groupoids G_1 and G_2 over the sets X_1 and X_2, respectively, there is a naturally defined groupoid $G_1 \times G_2 \rightrightarrows X_1 \times X_2$ usually referred to as the ***product groupoid***.

Given the groupoid $G \rightrightarrows X$, we say that a subset $H \subset G$ is a ***subgroupoid*** of G when it is closed under multiplication and inversion. Under those circumstances H is a groupoid over $\alpha(H) = \beta(H) \subset X$. If $\alpha(H) = \beta(H) = X$, $H \rightrightarrows X$ is called a ***wide subgroupoid*** of G.

Given two groupoids G_1 and G_2 over the sets X_1 and X_2, respectively, we say that a pair of maps $T : G_1 \to G_2$ and $b : X_1 \to X_2$ is a ***homomorphism of groupoids*** when T and b commute with all the structural maps of $G_1 \rightrightarrows X_1$ and $G_2 \rightrightarrows X_2$.

For any groupoid $G \rightrightarrows X$, the maps $(\alpha, \beta) : G \to X \times X$ and the identity of X define a groupoid homomorphism between $G \rightrightarrows X$ and the pair groupoid $X \times X \rightrightarrows X$ over X. Its image is a wide subgroupoid of $X \times X$ that defines an equivalence relation on X whose equivalence classes are called the ***orbits*** of G in X. Notice that there is a correspondence between symmetric and transitive relations on X and subgroupoids of $X \times X \rightrightarrows X$, as well as between equivalence relations on X and wide subgroupoids of $X \times X \rightrightarrows X$. A groupoid is called ***transitive*** if it has just one orbit. A groupoid is called ***integrable*** when its orbits form a generalized foliation on M.

Given $x \in X$, the set $G_x := \{g \in G \mid \alpha(g) = \beta(g) = x\}$ is a subgroup of G called the ***isotropy subgroup*** of x.

3.6.3 The groupoid associated to a pseudogroup of transformations. Let M be a smooth manifold and $A \subset \mathrm{Diff}_L(M)$ a pseudogroup of transformations of M. Recall that since A is a pseudogroup, it contains the identity map on M. Let $\bar{A}$ be defined by

$$\bar{A} = \left\{ \bar{\varphi} : M \to M \mid \varphi \in A \text{ and } \bar{\varphi}(x) := \begin{cases} \varphi(x) & \text{if } x \in \mathrm{Dom}(\varphi) \\ x & \text{if } x \notin \mathrm{Dom}(\varphi) \end{cases} \right\}. \tag{3.6.1}$$

The product $M \times \bar{A}$ is a groupoid over M with the following choice of structure maps. Let $x, y \in M$ and $\bar{\varphi}, \bar{\psi} \in \bar{A}$; define $\alpha, \beta : M \times \bar{A} \to M$ by $\beta(x, \bar{\varphi}) = x$ and $\alpha(x, \bar{\varphi}) = \bar{\varphi}(x)$. Also, $m((x, \bar{\varphi}), (y, \bar{\psi})) := (y, \overline{\varphi \circ \psi})$, $(x, \bar{\varphi})^{-1} = (\bar{\varphi}(x), \overline{\varphi^{-1}})$, and $\epsilon(x) = (x, \mathbb{I}_M)$. We will refer to $M \times \bar{A} \rightrightarrows M$ as the ***transformation groupoid*** associated to the pseudogroup A. Note that if $A \subset \mathrm{Diff}(M)$, then $\bar{A} = A$.

The orbits of $M \times \bar{A}$ on M coincide with the orbits of A. This feature can be used to generalize to pseudogroups Stefan's Theorem on the integrability of the isotopic component of the identity of any group of diffeomorphism, quoted at the end of §3.1.4. To this end we first introduce some terminology. Let $g = (x, \bar{\varphi})$, $h = (y, \bar{\psi}) \in M \times \bar{A}$; we say that g and h are ***A-isotopic*** whenever $x = y$ and there exists an open neighborhood U of x and a differentiable mapping $a : \mathbb{R} \times U \to M$ satisfying: $(x, \overline{a_t}) \in M \times \bar{A}$ for all $t \in \mathbb{R}$, $(x, \overline{a_t}) = g = (x, \bar{\varphi})$ for $t \leq 0$, and $(x, \overline{a_t}) = h = (x, \bar{\psi})$ for $t \geq 1$. Let $(M \times \bar{A})^0$ be the subset of $M \times \bar{A}$ defined by

$$(M \times \bar{A})^0 := \{g \in M \times \bar{A} \mid g \text{ is } A\text{-isotopic to } \epsilon(\beta(g))\};$$

$(M \times \bar{A})^0$ is a wide subgroupoid of $M \times \bar{A}$ usually referred to as the ***isotopic component of the identity*** of $M \times \bar{A}$. Moreover (see STEFAN (1974b) for a proof):

3.6.4 Theorem (Stefan) *$(M \times \bar{A})^0$ is integrable.*

3.6.5 Example: The action groupoid. An important particular case of the groupoid in §3.6.3 concerns the situation in which the pseudogroup of transformations in question is the group of diffeomorphisms $A_G = \{\Phi_g \mid g \in G\}$ associated to the action of the Lie group G on the manifold M via a map $\Phi : G \times M \to M$. Definitions similar to those in §3.6.3 produce a groupoid $G \times M \rightrightarrows M$ that is usually referred to as the ***action groupoid*** induced by Φ. To be more specific, the structure maps of this groupoid are given by $\alpha(g, m) := g \cdot m$, $\beta(g, m) := m$, $\epsilon(m) := (e, m)$, $m((g, h \cdot n), (h, n)) := (gh, n)$, and $(g, m)^{-1} := (g^{-1}, g \cdot m)$, for any $g, h \in G$ and $m, n \in M$. The orbits and isotropy subgroups of this groupoid coincide with those of the group action Φ.

3.6.6 Example: The cotangent bundle of a Lie group. Let G be a Lie group, T^*G its cotangent bundle, and $\mathfrak{g}$ its Lie algebra. If we identify T^*G with $G \times \mathfrak{g}^*$ using right translations we can use the previous paragraph to endow T^*G with a groupoid structure over $\mathfrak{g}^*$ by taking the following structure maps: for any $(g, \mu) \in G \times \mathfrak{g}^*$ let $\alpha(g, \mu) := \mathrm{Ad}^*_{g^{-1}}\mu$, $\beta(g, \mu) := \mu$, $\epsilon(\mu) = (e, \mu)$, $m((g, \mathrm{Ad}^*_{h^{-1}}\mu), (h, \mu)) = (gh, \mu)$, and $(g, \mu)^{-1} = (g^{-1}, \mathrm{Ad}^*_{g^{-1}}\mu)$. $T^*G \rightrightarrows \mathfrak{g}^*$ is a ***symplectic groupoid*** in the sense of WEINSTEIN (1987a); COSTÉ et al. (1987); ALBERT AND DAZORD (1990).

3.6.7 The Baer groupoid. We now study a groupoid introduced in BAER (1929) that will be used later in the book. Let G be a group and $\mathfrak{S}(G)$ the set of subgroups of G. Let $\mathfrak{B}(G)$ be the set of cosets of elements in $\mathfrak{S}(G)$. We do not specify if it is the set of right or left cosets since they coincide; indeed, for any $g \in G$ and any $H \in \mathfrak{S}(G)$ we have that $gH = (gHg^{-1})g$.

The set $\mathfrak{B}(G)$ is a groupoid over $\mathfrak{S}(G)$ with the following choice of structural maps. The target and source maps $\alpha, \beta : \mathfrak{B}(G) \to \mathfrak{S}(G)$ are defined by $\alpha(D) = Dg^{-1}$, $\beta(D) = g^{-1}D$ for some $g \in D$. Elementary group theory guarantees that these definitions do not depend on the choice of $g \in D$. The set of composable pairs $(\mathfrak{B}(G))^{(2)}$ is given by

$$(\mathfrak{B}(G))^{(2)} = \{(D_1, D_2) \in \mathfrak{B}(G)^2 \mid g_1^{-1}D_1 = D_2 g_2^{-1}, \text{ for any } g_1 \in D_1, g_2 \in D_2\}.$$

The groupoid product defined on $(\mathfrak{B}(G))^{(2)}$ is given by $m(D_1, D_2) := D_1D_2 = \{gh \mid g \in D_1, h \in D_2\}$. Given $D \in \mathfrak{B}(G)$ define $D^{-1} = g^{-1}Dg^{-1}$, for any $g \in D$. The identity section is given by simple inclusion.

Notice that the orbits of $\mathfrak{B}(G) \rightrightarrows \mathfrak{S}(G)$ are given by the conjugacy classes of subgroups of G.

3.6.8 Groupoid actions. In this paragraph we adopt the terminology introduced in MIKAMI AND WEINSTEIN (1988). Let $G \rightrightarrows X$ be a groupoid over X and M a set. We will use the same notation as in the preceding paragraphs to denote the structure maps of $G \rightrightarrows X$. Let $J : M \to X$ be a map from M into X and

$$G \times_J M := \{(g, m) \in G \times M \mid \beta(g) = J(m)\}.$$

A (left) ***groupoid action*** of G on M with ***moment map*** $J : M \to X$ is a mapping

$$\begin{aligned} \Psi : G \times_J M &\longrightarrow M \\ (g, m) &\longmapsto g \cdot m := \Psi(g, m), \end{aligned}$$

that satisfies the following properties:

(i) $J(g \cdot m) = \alpha(g)$,

(ii) $gh \cdot m = g \cdot (h \cdot m)$,

(iii) $(\epsilon(J(m))) \cdot m = m$,

for any $g, h \in G$ and $m \in M$. Notice that **(i)** guarantees that in **(ii)** each side of the equality is defined if the other is too. Right actions are defined similarly by looking at the set $M \times_J G := \{(m, g) \in M \times G \mid J(m) = \alpha(g)\}$.

3.6.9 Examples of groupoid actions.

(i) A groupoid acts on its total space and on its base. A groupoid $G \rightrightarrows X$ acts on G by multiplication with moment map α. G also acts on X with moment map the identity I_X via the mapping $g \cdot \beta(g) := \alpha(g)$.

(ii) The G-action groupoid acts on G-spaces. Let G be a Lie group acting on two sets M and N and let $J : M \to N$ be any equivariant map with respect to those actions. The map J naturally induces an action of the product groupoid $G \times N \rightrightarrows N$ on the set M. Indeed,

$$(G \times N) \times_J M = \{((g, J(m)), m) \mid g \in G, m \in M\} \subset (G \times N) \times M.$$

The action is defined by

$$\begin{aligned} \Psi : (G \times N) \times_J M &\longrightarrow M \\ (((g, J(m)), m) &\longmapsto g \cdot m. \end{aligned}$$

(iii) The Baer groupoid acts on G-spaces. (WEINSTEIN (2002a)). Let G be a Lie group, M a G-space, and $B : M \to \mathfrak{S}(G)$ the map that assigns to each point $m \in M$ its isotropy subgroup $G_m \in \mathfrak{S}(G)$. Define $\mathfrak{B}(G) \times_B M := \{(gG_m, m) \in \mathfrak{B}(G) \times M \mid m \in M\}$. The map $\mathfrak{B}(G) \times_B M \to M$ given by $(gG_m, m) \mapsto g \cdot m$ defines an action of the Baer groupoid $\mathfrak{B}(G) \rightrightarrows \mathfrak{S}(G)$ on the G-space M with moment map B. Notice that the level sets of the moment map are the isotropy type subsets of M that, when M is a manifold and the G-action is proper, coincide up to connected components with the leaves of the invariant distribution introduced in Section 3.5. The significance of this fact, as well as the pertinence of the terminology utilized in the last few paragraphs, will be better understood in the context of the next chapter.

Chapter 4

The Standard Momentum Map

The term ***momentum map*** (also found in the literature as ***moment map***, a different translation of the French term *application moment*) is mainly used to design a mathematical construction introduced in LIE (1890), KOSTANT (1965), and SOURIAU (1965, 1966) that encodes the conservation laws associated to the symmetries of a Hamiltonian dynamical system. The choice in terminology is justified by the fact that when one deals with the translational and rotational symmetries of a simple mechanical system, the momentum map recovers the standard linear and angular momentum, respectively. The reader interested in the history of this construction is referred to WEINSTEIN (1983b) and the remarks in §11.2 of MARSDEN AND RATIU (1999).

The momentum map is the basic ingredient for symplectic reduction, one of the main themes of this book. In our context, its importance is given by the fact that it is able to describe some of the conservation laws associated to a symmetry of the system. This explains why we review it in detail in the following pages. In the next chapter we will introduce other notions of momentum map that were considered during the last decades in order to deal with various situations at different degrees of generality and technical complication.

We will start this chapter with a brief review of symplectic and Poisson manifolds with symmetry as well as of invariant Hamiltonian dynamics. This summary is intended to recall the necessary background and to introduce the key concepts and results that will be used in the next chapters. The treatment of these classical topics can be found in many books, such as ABRAHAM AND MARSDEN (1978), ARNOLD (1989), ARNOLD et al. (1988), CUSHMAN AND BATES (1997), GUILLEMIN AND STERNBERG (1984c), LIBERMANN AND MARLE (1987), MARSDEN (1992), MARSDEN AND RATIU (1999), and SOURIAU (1969).

4.1 Hamiltonian systems

4.1.1 Symplectic manifolds and Hamilton equations. A ***symplectic manifold*** is a pair (M, ω), where M is a manifold and $\omega \in \Omega^2(M)$ is a closed nondegenerate two-form on M, that is, $\mathbf{d}\omega = 0$ and, for every $m \in M$, the map $v \in T_mM \mapsto \omega(m)(v, \cdot) \in T_m^*M$ is a linear isomorphism between the tangent space T_mM to M at m and the cotangent space T_m^*M. In particular, the dimension of M is always even. If ω is allowed to be degenerate, (M, ω) is called a ***presymplectic manifold***. A ***Hamiltonian dynamical system*** is a triple (M, ω, h), where (M, ω) is a symplectic manifold and $h \in C^\infty(M)$ is the ***Hamiltonian function*** of the system. By nondegeneracy of the symplectic form ω, to each Hamiltonian system one can associate a ***Hamiltonian vector field*** $X_h \in \mathfrak{X}(M)$, defined by the equality

$$\mathbf{i}_{X_h}\omega = \mathbf{d}h. \tag{4.1.1}$$

4.1.2 Example Let V be a vector space and V^* its dual. Let $Z = V \times V^*$. The **canonical symplectic form** Ω on Z is defined by

$$\Omega((v_1, \alpha_1), (v_2, \alpha_2)) := \langle \alpha_2, v_1 \rangle - \langle \alpha_1, v_2 \rangle, \tag{4.1.2}$$

where $v_1, v_2 \in V$ and $\alpha_1, \alpha_2 \in V^*$.

4.1.3 Example Let Q be a smooth manifold and T^*Q its cotangent bundle. Let $\pi_Q : T^*Q \to Q$ be the projection and Θ the one-form on T^*Q defined by

$$\Theta(\beta) \cdot v_\beta := \langle \beta, T_\beta \pi_Q \cdot v_\beta \rangle,$$

where $\beta \in T^*Q$ and $v_\beta \in T_\beta(T^*Q)$. The **canonical symplectic form** Ω on the cotangent bundle T^*Q is defined by $\Omega = -\mathbf{d}\Theta$.

4.1.4 Theorem (Darboux) *Let (M, ω) be a $2n$-dimensional symplectic manifold and let $m \in M$ be an arbitrary point. Then there is a chart (U, φ), containing m such that in the coordinates $\varphi(z) = (q^1, \dots, q^n, p_1, \dots, p_n)$, $z \in U$, the symplectic form ω can be written as*

$$\omega|_U = \sum_{i=1}^n \mathbf{d}q^i \wedge \mathbf{d}p_i.$$

Coordinates in which ω takes the above form are called ***canonical*** *or* ***Darboux coordinates***.

In canonical coordinates, the integral curves of the Hamiltonian vector field X_h are determined by the well-known ***Hamilton equations***,

$$\frac{dq^i}{dt} = \frac{\partial h}{\partial p_i}, \qquad \frac{dp_i}{dt} = -\frac{\partial h}{\partial q^i}. \tag{4.1.3}$$

4.1.5 Symplectic linear algebra. A linear map $A : (V_1, \omega_1) \to (V_2, \omega_2)$ between symplectic vector spaces is called a ***linear symplectic map*** if $\omega_2(Au, Av) = \omega_1(u, v)$ for all $u, v \in V_1$. The linear symplectic map A is called a ***symplectic isomorphism*** if it is an isomorphism between the vector spaces V_1 and V_2. Note that if A is a symplectic isomorphism, then its inverse is also a linear symplectic map.

Let (V, ω) be a symplectic vector space and W a vector subspace of V. The ***symplectic orthogonal complement*** W^ω of W in V is defined by

$$W^\omega := \{v \in V \mid \omega(v, w) = 0 \text{ for all } w \in W\}. \tag{4.1.4}$$

The subspace W is ***isotropic*** if $W \subset W^\omega$. W is ***coisotropic*** when $W^\omega \subset W$. It is ***Lagrangian*** when it is isotropic and it has an isotropic complement, that is, $V = W \oplus W'$, with W' isotropic. W is a ***symplectic*** subspace when ω restricted to $W \times W$ is nondegenerate. Some standard properties related to the definition in (4.1.4) that will be used profusely are the following (see, for instance, WEINSTEIN (1977a) or ABRAHAM AND MARSDEN (1978) for proofs): if $U, W \subset V$ are two vector subspaces of the symplectic vector space (V, ω), then

1. if $U \subset W$, then $W^\omega \subset U^\omega$;
2. $U^\omega \cap W^\omega = (U + W)^\omega$ and $(U \cap W)^\omega = U^\omega + W^\omega$;
3. $U = (U^\omega)^\omega$ and $\dim V = \dim U + \dim U^\omega$;
4. the following are equivalent:
   (a) U is Lagrangian,
   (b) $U = U^\omega$,
   (c) U is isotropic and $\dim U = \frac{1}{2} \dim V$,
   (d) U is coisotropic and $\dim U = \frac{1}{2} \dim V$,
   (e) U is a maximal isotropic subspace,
   (f) U is a minimal coisotropic subspace;
5. if U is Lagrangian, then there is a symplectic isomorphism $A : (V, \omega) \to (U \times U^*, \Omega)$ mapping U to $U \times \{0\}$, where Ω is the canonical symplectic form (4.1.2).

An immersed submanifold S of a symplectic manifold M is called ***isotropic, coisotropic, Lagrangian***, or ***symplectic*** if for each $s \in S$ the tangent space $T_s S$ is an isotropic, coisotropic, Lagrangian, or symplectic subspace of $T_s M$.

4.1.6 Hamiltonian and locally Hamiltonian vector fields. The set of Hamiltonian vector fields on a symplectic manifold (M, ω) will be denoted by $\mathfrak{X}_H(M, \omega)$, or simply, $\mathfrak{X}_H(M)$. In view of (4.1.1) it is clear that $X \in \mathfrak{X}(M)$ is a Hamiltonian vector field if and only if the one-form $\mathbf{i}_X \omega$ is exact. If this form is only closed we say that X is a ***locally Hamiltonian vector field***. The set of locally Hamiltonian vector fields on (M, ω) will be denoted by $\mathfrak{X}_{LH}(M, \omega)$, or simply, $\mathfrak{X}_{LH}(M)$. This terminology is justified by the fact that, by Poincaré's Lemma, for any point $m \in M$ there exists an open neighborhood U and a function $f \in C^\infty(U)$ such that $\mathbf{i}_{X|_U} \omega|_U = \mathbf{d} f$. Using Cartan's formula (1.3.17) and the closedness of the symplectic form ω it is easy to see that $X \in \mathfrak{X}_{LH}(M)$ if and only if

$$\pounds_X \omega = 0. \tag{4.1.5}$$

The Lie bracket $[X, Y]$ of two locally Hamiltonian vector fields $X, Y \in \mathfrak{X}_{LH}(M)$ is Hamiltonian with $\omega(Y, X)$ as associated Hamiltonian function. This proves that $[\mathfrak{X}_{LH}(M), \mathfrak{X}_{LH}(M)] \subset \mathfrak{X}_H(M)$ and that $\mathfrak{X}_H(M)$ is an ideal in $\mathfrak{X}_{LH}(M)$. CALABI (1970) and LICHNEROWICZ (1973) have shown that $[\mathfrak{X}_{LH}(M), \mathfrak{X}_{LH}(M)] = \mathfrak{X}_H(M)$.

4.1.7 Definition *Let $f, g \in C^\infty(M)$. The* ***Poisson bracket*** *of these two functions is the function $\{f, g\} \in C^\infty(M)$ defined by*

$$\{f, g\}(z) = \omega(z)(X_f(z), X_g(z)) = X_g[f](z) = -X_f[g](z).$$

Note that in canonical coordinates, the Poisson bracket takes the traditional form

$$\{f, g\} = \sum_{i=1}^{n} \left(\frac{\partial f}{\partial q^i} \frac{\partial g}{\partial p_i} - \frac{\partial g}{\partial q^i} \frac{\partial f}{\partial p_i} \right).$$

4.1.8 Theorem (Jacobi–Lie–Caratheodory) *Let (M, ω) be a $2n$-dimensional symplectic manifold and let $m \in M$ be an arbitrary point. Assume that there exist $k \leq n$ smooth functions $f_1, \ldots, f_k \in C^\infty(U)$ defined on some open neighborhood U of m such that $\{f_i, f_j\} = 0$ for any $i, j \in \{1, \ldots, k\}$ and $\mathbf{d} f_1, \ldots, \mathbf{d} f_k$ are linearly independent one-forms on U. Then there exist $2n - k$ smooth functions $f_{k+1}, \ldots, f_n, g_1, \ldots, g_n \in C^\infty(V)$, $V \subset U$, such that $\omega = \Sigma_{i=1}^n \mathbf{d} f_i \wedge \mathbf{d} g_i$.*

4.1.9 Theorem (Cartan) *Let (M, ω) be a $2n$-dimensional symplectic manifold and let $m \in M$ be an arbitrary point. Assume that there exist $k + l$, $0 \leq k, l \leq n$ smooth functions $f_1, \ldots, f_k, g_1, \ldots g_l \in C^\infty(U)$ defined on some open neighborhood U of m such that $\{f_i, f_j\} = 0$, $\{g_r, g_s\} = 0$, and $\{f_i, g_r\} = \delta_{ir}$, for any $i, j \in \{1, \ldots, k\}$ and $r, s \in \{1, \ldots, l\}$, and $\mathbf{d} f_1, \ldots, \mathbf{d} f_k, \mathbf{d} g_1, \ldots, \mathbf{d} g_l$ are linearly independent one-forms on U. Then there exist $2n - k - l$ smooth functions $f_{k+1}, \ldots, f_n, g_{l+1}, \ldots, g_n \in C^\infty(V)$, $V \subset U$, such that $\omega = \Sigma_{i=1}^n \mathbf{d} f_i \wedge \mathbf{d} g_i$.*

4.1.10 Poisson manifolds. Given a symplectic manifold (M, ω), the set $C^\infty(M)$ is endowed not only with a commutative ring structure relative to pointwise multiplication, but also with a real Lie algebra structure relative to the Poisson bracket. These two algebraic structures on $C^\infty(M)$ are linked by the property that the Poisson bracket is a ring derivation in each of its arguments, that is, the Leibniz rule holds. It turns out that these properties are just enough to formulate Hamiltonian dynamics on a manifold and lead to the following generalization introduced by DIRAC (1964); BAYENet al. (1978), and LICHNEROWICZ (1977).

4.1.11 Definition *A* ***Poisson manifold*** *is a pair $(M, \{\cdot, \cdot\})$, where M is a manifold and $\{\cdot, \cdot\}$ is a bilinear operation on $C^\infty(M)$ such that $(C^\infty(M), \{\cdot, \cdot\})$ is a Lie algebra and $\{\cdot, \cdot\}$ is a derivation (that is, the Leibniz identity holds) in each argument. The pair $(C^\infty(M), \{\cdot, \cdot\})$ is also called a* ***Poisson algebra****. The functions in the center $\mathcal{C}(M)$ of the Lie algebra $(C^\infty(M), \{\cdot, \cdot\})$ are called* ***Casimir functions****.*

From the natural isomorphism between derivations on $C^\infty(M)$ and vector fields on M (see §1.2.3), it follows that each $h \in C^\infty(M)$ induces a vector field on M via the expression

$$X_h = \{\cdot, h\},$$

called the ***Hamiltonian vector field*** associated to the ***Hamiltonian function*** h. Hamilton's equations $\dot{z} = X_h(z)$ can be equivalently written in Poisson bracket form as

$$\dot{f} = \{f, h\},$$

for any $f \in C^\infty(M)$. The triplet $(M, \{\cdot, \cdot\}, h)$ is called a ***Poisson dynamical system***. The Lie algebra mapping $(C^\infty(M), \{\cdot, \cdot\}) \to (\mathfrak{X}(M), [\cdot, \cdot])$ that assigns to each function $f \in C^\infty(M)$ the associated Hamiltonian vector field $X_f \in \mathfrak{X}(M)$ is a Lie algebra antihomomorphism, that is,

$$X_{\{f,g\}} = -[X_f, X_g], \text{ for all } f, g \in C^\infty(M). \tag{4.1.6}$$

Note that $f \in C^\infty(M)$ is a Casimir function if and only if $X_f \equiv 0$.

Obviously, any Hamiltonian system on a symplectic manifold is a Poisson dynamical system relative to the Poisson bracket induced by the symplectic structure. Given a Poisson dynamical system $(M, \{\cdot, \cdot\}, h)$, its ***integrals of motion*** or ***conserved quantities*** are defined as the centralizer of h in $(C^\infty(M), \{\cdot, \cdot\})$ that is, the subalgebra of $(C^\infty(M), \{\cdot, \cdot\})$ consisting of the functions $f \in C^\infty(M)$ such that $\{f, h\} = 0$. Note that the terminology is justified since by Hamilton's equations in Poisson bracket form we have

$$\dot{f} = X_h[f] = \{f, h\} = 0,$$

that is, f is constant on the flow of X_h.

4.1.12 Example Let $(M, \{\cdot, \cdot\}^M)$ be a Poisson manifold and $U \subset M$ any open subset of M. The pair $(U, \{\cdot, \cdot\}^M_U)$ is also a Poisson manifold where the ***restricted Poisson bracket*** is defined by

$$\{\cdot, \cdot\}^M_U := \{\cdot, \cdot\}^M|_{C^\infty_M(U)\times C^\infty_M(U)}.$$

4.1.13 Example The dual $\mathfrak{g}^*$ of a Lie algebra $\mathfrak{g}$ is a Poisson manifold with respect to the ***±-Lie–Poisson*** brackets $\{\cdot, \cdot\}_\pm$ defined by

$$\{f, g\}_\pm(\mu) := \pm\left\langle \mu, \left[\frac{\delta f}{\delta \mu}, \frac{\delta g}{\delta \mu}\right]\right\rangle, \quad f, g \in C^\infty(\mathfrak{g}^*), \ \mu \in \mathfrak{g}^*,$$

where the element $\frac{\delta f}{\delta \mu} \in \mathfrak{g}$ is defined by the equality $\langle \nu, \frac{\delta f}{\delta \mu}\rangle := Df(\mu) \cdot \nu$, for any $\nu \in \mathfrak{g}^*$; see, for example, MARSDEN AND RATIU (1999) for a direct proof. Given any function $h \in C^\infty(\mathfrak{g}^*)$, the associated Hamiltonian vector fields X_h with respect to the ±-Lie–Poisson brackets on $\mathfrak{g}^*$ are given by the expressions

$$X_h(\mu) = \mp \mathrm{ad}^*_{\delta h/\delta \mu}\mu, \quad \mu \in \mathfrak{g}^*.$$

4.1.14 The Poisson tensor. The derivation property of the Poisson bracket implies that for any two functions $f, g \in C^\infty(M)$, the value of the bracket $\{f, g\}(z)$ at an arbitrary point $z \in M$ (and therefore $X_f(z)$ as well), depends on f only through $\mathbf{d}f(z)$ which allows us to define a contravariant antisymmetric two-tensor $B \in \Lambda^2(T^*M)$ by

$$B(z)(\alpha_z, \beta_z) = \{f, g\}(z),$$

where $\mathbf{d}f(z) = \alpha_z \in T^*_zM$ and $\mathbf{d}g(z) = \beta_z \in T^*_zM$. This tensor is called the ***Poisson tensor*** of M. The vector bundle map $B^\sharp : T^*M \to TM$ naturally associated to B is defined by

$$B(z)(\alpha_z, \beta_z) = \langle \alpha_z, B^\sharp(\beta_z)\rangle.$$

Its range $D := B^\sharp(T^*M) \subset TM$ is called the ***characteristic distribution***. We will justify this terminology in §4.1.27. For any point $m \in M$, the dimension of $D(m)$ as a vector subspace of $T_m M$ is called the ***rank*** of the Poisson manifold $(M, \{\cdot, \cdot\})$ at the point m.

4.1.15 The Weinstein coordinates of a Poisson manifold. The coordinates that generalize the Darboux charts of a symplectic manifold to the Poisson category have been found by WEINSTEIN (1983a). The reader is encouraged to check with the original paper or with LIBERMANN AND MARLE (1987), VAISMAN (1996a), for a proof of the following statement.

4.1.16 Theorem (Weinstein) *Let $(M, \{\cdot, \cdot\})$ be an m-dimensional Poisson manifold and $z_0 \in M$ a point where the rank of $(M, \{\cdot, \cdot\})$ equals $2n$, $0 \leq 2n \leq m$. There exists a chart (U, φ) of M whose domain contains the point z_0 and such that the associated local coordinates, denoted by $(q^1, \ldots, q^n, p_1, \ldots, p_n, z_1, \ldots, z_{m-2n})$, satisfy*

$$\{q^i, q^j\} = \{p_i, p_j\} = \{q^i, z_k\} = \{p_i, z_k\} = 0, \qquad \textit{and} \qquad \{q^i, p_j\} = \delta^i_j,$$

for all i, j, k, $1 \leq i, j \leq n$, $1 \leq k \leq m - 2n$.

*For all k, l, $1 \leq k, l \leq m-2n$, the Poisson bracket $\{z_k, z_l\}$ is a function of the local coordinates $z^1, \ldots, z^{m-2n}$ exclusively, and vanishes at z_0. Hence, the restriction of the bracket $\{\cdot, \cdot\}$ to the coordinates $z^1, \ldots, z^{m-2n}$ induces a Poisson structure that is usually referred to as the **transverse Poisson structure** of $(M, \{\cdot, \cdot\})$ at m. This structure is unique up to isomorphisms.*

If the rank of $(M, \{\cdot, \cdot\})$ is constant and equal to $2n$ on a neighborhood W of z_0, then, by choosing the domain U of the chart contained in W, the coordinates z satisfy

$$\{z_k, z_l\} = 0,$$

for all k, l, $1 \leq k, l \leq m - 2n$.

4.1.17 The last paragraph of the Weinstein Theorem, above, implies that if the rank of $(M, \{\cdot, \cdot\})$ is constant and equal to $2n$ on a neighborhood W of a point $m \in M$, then there exists an open neighborhood $U \subset W$ of this point on which one can define $k := \dim M - 2n$ independent ***locally defined Casimir functions*** of $(M, \{\cdot, \cdot\})$ around m. More specifically, there exist k functions $f_1, \ldots, f_k \in C^\infty(U)$ whose differentials are linearly independent at any point in U and such that $\{f_i, g\}_U \equiv 0$, for any $i \in \{1, \ldots, k\}$ and any $g \in C^\infty(U)$. The differential of any other local Casimir of $(M, \{\cdot, \cdot\})$ on U is a linear combination of $\{\mathbf{d}f_1, \ldots, \mathbf{d}f_k\}$.

4.1.18 Canonical transformations. A smooth mapping $\varphi : M_1 \to M_2$, between the two Poisson manifolds $(M_1, \{\cdot, \cdot\}_1)$ and $(M_2, \{\cdot, \cdot\}_2)$ is called ***canonical*** or ***Poisson*** if for all $g, h \in C^\infty(M_2)$ we have

$$\varphi^*\{g, h\}_2 = \{\varphi^* g, \varphi^* g\}_1 . \tag{4.1.7}$$

This definition can be reformulated in terms of the Poisson tensors B_1 and B_2 of $(M_1, \{\cdot, \cdot\}_1)$ and $(M_2, \{\cdot, \cdot\}_2)$, respectively. Indeed, it is easy to check that $\varphi : M_1 \to M_2$ is canonical if and only if for any $z \in M_1$ and $\alpha_{\varphi(z)}, \beta_{\varphi(z)} \in T^*_{\varphi(z)} M_2$ we have that

$$B_1(z)((T_z\varphi)^* \cdot \alpha_{\varphi(z)}, (T_z\varphi)^* \cdot \beta_{\varphi(z)}) = B_2(\varphi(z))(\alpha_{\varphi(z)}, \beta_{\varphi(z)}), \tag{4.1.8}$$

or, equivalently,

$$B_2^\sharp(\varphi(z)) = T_z\varphi \circ B_1^\sharp(z) \circ T_z^*\varphi. \tag{4.1.9}$$

A particularly important situation takes place when the map $\varphi : M \to M$ is a canonical diffeomorphism of $(M, \{\cdot, \cdot\})$. Such transformations form a group that will be referred to as the ***group of Poisson diffeomorphisms*** of M and will be denoted by $\mathcal{P}(M)$. The local counterpart of this group is the ***pseudogroup of local Poisson diffeomorphisms*** of M, denoted by $\mathcal{P}_L(M)$, and consisting of all the local diffeomorphisms $F \in \mathrm{Diff}_L(M)$ such that if for all $g,\ h \in C^\infty(F(\mathrm{Dom}(F)))$ we have

$$\varphi^*\{g,\ h\}|_{F(\mathrm{Dom}(F))} = \{F^*g,\ F^*h\}|_{\mathrm{Dom}(F)}\,. \tag{4.1.10}$$

For future reference we state the following standard result whose proof can be found, for instance, in MARSDEN AND RATIU (1999, Proposition 10.5.2).

4.1.19 Proposition *Let $\varphi : M_1 \to M_2$ be a smooth map between two Poisson manifolds $(M_1, \{\cdot, \cdot\}_1)$ and $(M_2, \{\cdot, \cdot\}_2)$. Then φ is a Poisson map if and only if $T\varphi \circ X_{h\circ\varphi} = X_h \circ \varphi$ for any $h \in C^\infty(M_2)$. Moreover, if φ is Poisson, $h \in C^\infty(M_2)$, F_t^2 is the flow of X_h, F_t^1 is the flow of $X_{h\circ\varphi}$, and* $\mathrm{Dom}(F_t^1)$ *and* $\mathrm{Dom}(F_t^2)$ *are the domains of definition of F_t^1 and F_t^2, respectively, then* $\mathrm{Dom}(F_t^1) \subset \varphi^{-1}(\mathrm{Dom}(F_t^2))$ *and*

$$(F_t^2 \circ \varphi)(z) = (\varphi \circ F_t^1)(z), \text{ for any } z \text{ in the domain } \mathrm{Dom}(F_t^1) \text{ of } F_t^1. \tag{4.1.11}$$

In the symplectic category, a map $\varphi : M_1 \to M_2$ between the symplectic manifolds (M_1, ω_1) and (M_2, ω_2) is called ***canonical*** or ***symplectic*** if

$$\varphi^*\omega_2 = \omega_1. \tag{4.1.12}$$

Note that symplectic maps are, by the nondegeneracy assumption on the symplectic two-form, necessarily immersions. Canonical transformations between symplectic manifolds of the same dimension are local diffeomorphisms. The relation between this definition and the one given for Poisson manifolds in (4.1.7) is given by the fact that a diffeomorphism $\varphi : M_1 \to M_2$ between two symplectic manifolds (M_1, ω_1) and (M_2, ω_2) is symplectic if and only if it is Poisson with respect to the Poisson structures induced by the symplectic forms ω_1 and ω_2. Symplectic diffeomorphisms are usually referred to as ***symplectomorphisms***. The equivalence between symplectic and Poisson maps holds only in the case of diffeomorphisms: if the symplectic map $\varphi : M_1 \to M_2$ is not a diffeomorphism it may not be a Poisson map (see §4.2.8 for an example).

The term canonical is also used in the literature (see for instance LANDAU AND LIFSHITZ (1976)) to denote transformations that preserve the canonical form of the Hamilton equations (4.1.3). This definition is not equivalent to (4.1.12) since there are transformations that respect the form of the equations (4.1.3) but do not satisfy (4.1.12) as shown by the following elementary counterexample due to ARNOLD (1989). Consider $\mathbb{R}^2$ with the canonical symplectic form. The transformation $\varphi : \mathbb{R}^2 \to \mathbb{R}^2$ given by $\varphi(q, p) := (q, 2p)$, $(q, p) \in \mathbb{R}^2$, preserves the form (4.1.3) of Hamilton's equations but it is clearly not canonical in our sense.

4.1.20 Example. Let G be a Lie group and T^*G its cotangent bundle. The map $\lambda : T^*G \to G \times \mathfrak{g}^*$ defined by $\lambda(\alpha) := (g, (T_eL_g)^*\alpha)$, for any $\alpha \in T_g^*G$, defines a vector bundle isomorphism usually referred to as the ***left trivialization*** of T^*G. In the context of mechanics of rigid bodies, the components of the image by λ of an element $\alpha \in T_g^*G$ are called the ***body coordinates*** of α. If we consider T^*G as a symplectic manifold with the canonical form introduced in the example 4.1.3, then the left trivialization map λ is a symplectomorphism if we take on $G \times \mathfrak{g}^*$ the symplectic form ω_B defined by

$$\omega_B(g, \nu)((T_eL_g\xi, \rho), (T_eL_g\eta, \sigma)) := \langle \sigma, \xi \rangle - \langle \rho, \eta \rangle + \langle \nu, [\xi, \eta] \rangle, \tag{4.1.13}$$

with $(g, \nu) \in G \times \mathfrak{g}^*$, $\xi, \eta \in \mathfrak{g}$, and $\rho, \sigma \in \mathfrak{g}^*$. The proof of this and related formulas will be carried out in Theorem 6.2.4.

4.1.21 Poisson and symplectic submanifolds. Let $(S, \{\cdot,\cdot\}^S)$ and $(M, \{\cdot,\cdot\}^M)$ be two Poisson manifolds such that $S \subset M$ and the inclusion $i_S : S \hookrightarrow M$ is an immersion. The Poisson manifold $(S, \{\cdot,\cdot\}^S)$ is called a ***Poisson submanifold*** of $(M, \{\cdot,\cdot\}^M)$ if i_S is a canonical map.

An immersed submanifold Q of M is called a ***quasi-Poisson submanifold*** of $(M, \{\cdot,\cdot\}^M)$ if for any $q \in Q$, any open neighborhood U of q in M, and any $f \in C_M^\infty(U)$, we have

$$X_f(i_Q(q)) \in T_q i_Q(T_q Q),$$

where $i_Q : Q \hookrightarrow M$ is the inclusion and X_f is the Hamiltonian vector field of f on U with respect to the restricted Poisson bracket $\{\cdot,\cdot\}_U^M$.

If $(S, \{\cdot,\cdot\}^S)$ is a Poisson submanifold of $(M, \{\cdot,\cdot\}^M)$, then there is no other bracket $\{\cdot,\cdot\}'$ on S making the inclusion $i : S \hookrightarrow M$ into a canonical map. Indeed, for any $f, g \in C^\infty(S)$ and any $s \in S$, there exist by Proposition 1.1.24 two local extensions $F, G \in C_M^\infty(U)$ of f and g, respectively, such that

$$\{f, g\}^S(s) = \{F, G\}_U^M(s) = \{f, g\}'(s),$$

since both inclusions $(S, \{\cdot,\cdot\}') \hookrightarrow (M, \{\cdot,\cdot\}^M)$ and $(S, \{\cdot,\cdot\}^S) \hookrightarrow (M, \{\cdot,\cdot\}^M)$ are Poisson.

4.1.22 Remark The reader should be warned that even though the concepts of Poisson and quasi-Poisson submanifold are used interchangeably in the literature, these notions are *not* equivalent. Indeed, despite the fact that Proposition 4.1.19 readily implies that any Poisson submanifold is quasi-Poisson, the converse is not true, as the following counterexample shows.

Let $M = \mathbb{R}^2$ be endowed with the Poisson tensor

$$B(x, y) := \begin{pmatrix} 0 & y \\ -y & 0 \end{pmatrix}$$

and $Q = \mathbb{R}^2$ be endowed with the canonical symplectic form $\omega(x, y) := \mathbf{d}x \wedge \mathbf{d}y$. Let $i_Q : (Q, \omega) \hookrightarrow (M, B)$ be the identity map. This diffeomorphism is clearly not Poisson. However (Q, ω) is trivially a quasi-Poisson submanifold of (M, B).

4.1.23 Proposition *Let $(M, \{\cdot, \cdot\})$ be a Poisson manifold and S a quasi-Poisson submanifold of $(M, \{\cdot, \cdot\})$. There exists a unique Poisson structure $\{\cdot, \cdot\}^S$ on S that makes it into a Poisson submanifold of $(M, \{\cdot, \cdot\})$.*

Proof. Let $f, g \in C^\infty(S)$ and $s \in S$ be arbitrary. Define

$$\{f, g\}^S(s) := \{F, G\}_U(s), \tag{4.1.14}$$

where $F, G \in C^\infty_M(U)$ are two local extensions of f and g, respectively, around $s \in S$. In order to prove that $\{\cdot, \cdot\}^S$ is a Poisson bracket on S, it suffices to show that the right-hand side of the expression (4.1.14) does not depend on the local extensions that we used in its definition. Let $F, \overline{F} \in C^\infty_M(U_s)$ be two local extensions of f and $i : S \hookrightarrow M$ the inclusion. Then

$$\begin{aligned}\{\overline{F}, G\}(i(s)) &= \mathbf{d}\overline{F}(i(s)) \cdot X_G(i(s)) = \mathbf{d}\overline{F}(i(s)) \cdot T_s i(v) \\ &= \mathbf{d}F(i(s)) \cdot T_s i(v) = \mathbf{d}F(i(s)) \cdot X_G(i(s)) = \{F, G\}(i(s)),\end{aligned}$$

where $v \in T_s S$ is the unique vector satisfying $X_G(i(s)) = T_s i(v)$ whose existence is a consequence of the quasi-Poisson submanifold character of S. Skew symmetry shows that (4.1.14) does not depend on the local extension G. ■

Given two symplectic manifolds (M, ω) and (S, ω_S) such that $S \subset M$ and the inclusion $i : S \hookrightarrow M$ is an immersion, the manifold (S, ω_S) is a symplectic submanifold of (M, ω) if and only if i is a symplectic map. We emphasize that the symplectic submanifolds of a symplectic manifold (M, ω) are in general neither Poisson nor quasi-Poisson manifolds of M. The only quasi-Poisson submanifolds of a symplectic manifold are its open sets which are, in fact, Poisson submanifolds.

4.1.24 Canonical flows and infinitesimal Poisson automorphisms. It is easy to show that a vector field X on a symplectic manifold is locally Hamiltonian if and only if its flow F_t consists of local symplectomorphisms, that is, if $\mathrm{Dom}(F_t)$ is the domain of definition of F_t for a fixed value of t, then $F_t : \mathrm{Dom}(F_t) \to F_t(\mathrm{Dom}(F_t))$ satisfies

$$F_t^* \omega|_{F_t(\mathrm{Dom}(F_t))} = \omega|_{\mathrm{Dom}(F_t)}.$$

In the Poisson category the flow of any Hamiltonian vector field is a canonical map. However, if a vector field has a canonical flow this does not imply, in general, that the vector field is locally Hamiltonian. The interplay between all these concepts in the Poisson category is explained in detail in the following subsections.

4.1.25 Definition *Let $(M, \{\cdot, \cdot\})$ be a Poisson manifold and $X \in \mathfrak{X}(M)$ a vector field on it. We say that:*

(i) *X is an **infinitesimal Poisson automorphism** provided that its flow F_t consists of local Poisson maps, that is, if $\mathrm{Dom}(F_t)$ is the domain of definition of F_t for a fixed value of t, then for any $f, g \in C^\infty(F_t(\mathrm{Dom}(F_t)))$, the map $F_t : \mathrm{Dom}(F_t) \to F_t(\mathrm{Dom}(F_t))$ satisfies*

$$F_t^*\left(\{f, g\}|_{F_t(\mathrm{Dom}(F_t))}\right) = \{f \circ F_t, g \circ F_t\}|_{\mathrm{Dom}(F_t)}.$$

The set of infinitesimal Poisson automorphisms on $(M, \{\cdot, \cdot\})$ will be denoted by $\mathfrak{X}_{LPA}(M)$.

(ii) *X is a* ***locally Hamiltonian vector field*** *if for each $z \in M$ there exists an open neighborhood U_z and a smooth function $f \in C^\infty(U_z)$ such that $X|_{U_z} = X_f$. The set of locally Hamiltonian vector fields on $(M, \{\cdot,\cdot\})$ will be denoted by $\mathfrak{X}_{LH}(M)$.*

(iii) *X is a* ***Hamiltonian vector field*** *if there exists a smooth function $f \in C^\infty(M)$ such that $X = X_f$. The set of Hamiltonian vector fields on $(M, \{\cdot,\cdot\})$ will be denoted by $\mathfrak{X}_H(M)$.*

The proof of the following proposition can be found in LIBERMANN AND MARLE (1987) (Chapter III, Proposition 10.3).

4.1.26 Proposition *Let X be a smooth vector field on the Poisson manifold $(M, \{\cdot,\cdot\})$. Then X is an infinitesimal Poisson automorphism if and only if one of the following two equivalent conditions are satisfied:*

(i) *X is a derivation of the Poisson algebra $(C^\infty(M), \{\cdot,\cdot\})$, that is, for any $f, g \in C^\infty(M)$, we have that*

$$X[\{f, g\}] = \{X[f], g\} + \{f, X[g]\}. \tag{4.1.15}$$

(ii) *Let $B \in \Lambda^2(T^*M)$ be the Poisson tensor of $(M, \{\cdot,\cdot\})$. Then, we have that $\pounds_X \Lambda = 0$.*

4.1.27 The Symplectic Foliation Theorem. As we already mentioned, any symplectic manifold is a Poisson manifold with the bracket introduced in Definition 4.1.7. The "converse" of this statement is provided by the Symplectic Foliation Theorem. The main idea behind this result is that the characteristic distribution presented in §4.1.14 is an integrable generalized distribution. More importantly, we will prove that its maximal integral manifolds are symplectic with the unique symplectic form that makes the inclusion into the Poisson manifold a Poisson map. The maximal integral manifolds of the characteristic distribution are called ***symplectic leaves*** of $(M, \{\cdot,\cdot\})$.

4.1.28 Theorem (Symplectic Foliation Theorem) *Let $(M, \{\cdot,\cdot\})$ be a Poisson manifold and D the associated characteristic distribution. D is a smooth and integrable generalized distribution and its maximal integral leaves form a generalized foliation decomposing M into initial submanifolds $\mathcal{L}$, each of which is symplectic with the unique symplectic form that makes the inclusion $i : \mathcal{L} \hookrightarrow M$ into a Poisson map, that is, $\mathcal{L}$ is a Poisson submanifold of $(M, \{\cdot,\cdot\})$.*

Proof. We start by studying the properties of the characteristic distribution D. By definition $D := B^\sharp(T^*M)$, where B is the Poisson tensor. Since for any point $m \in M$ we have $T_m^*M = \operatorname{span}\{\mathbf{d}f(m) \mid f \in C_c^\infty(M)\}$, with $C_c^\infty(M)$ the set of compactly supported functions on M, it follows that

$$D = \operatorname{span}\{X_f \mid f \in C_c^\infty(M)\}. \tag{4.1.16}$$

This equality automatically implies that D is a smooth generalized distribution since it is spanned by a family of smooth vector fields.

We now show that D is integrable by checking that it is invariant with respect to the action of the pseudogroup associated to the flows of the vector fields that span it. Let $f, g \in C_c^\infty(M)$. Since f is compactly supported, so is the associated Hamiltonian vector field X_f, which guarantees that its flow F_t is complete. Then, as the Hamiltonian flow F_t is Poisson, we have by Proposition 4.1.19

$$TF_t \circ X_g = TF_t \circ X_{g \circ F_{-t} \circ F_t} = X_{g \circ F_{-t}} \circ F_t.$$

As $g \circ F_{-t} \in C_c^\infty(M)$, the invariance of D is guaranteed and, by the Stefan–Sussmann Theorem, also its integrability.

Let $\mathcal{L} \subset M$ be a maximal integral manifold of D. The theory of generalized distributions reviewed in Section 1.4 guarantees that $\mathcal{L}$ is an initial submanifold of M and that the set of these submanifolds forms a generalized foliation of M. Let $i : \mathcal{L} \hookrightarrow M$ be the inclusion. We will now define a symplectic form for $\mathcal{L}$. First, notice that by (4.1.16), for any $f, g \in C^\infty(M)$ the Hamiltonian vector fields $X_f, X_g \in \mathfrak{X}(M)$ are tangent to $\mathcal{L}$ in the sense of §1.2.4. Consequently, there exist unique vector fields $X'_f, X'_g \in \mathfrak{X}(\mathcal{L})$ such that

$$Ti \circ X'_f = X_f \circ i \quad \text{and} \quad Ti \circ X'_g = X_g \circ i. \tag{4.1.17}$$

Second, given that for any point $z \in \mathcal{L}$, any two vectors $v_z, w_z \in T_z\mathcal{L}$ can be written as $v_z = X'_f(z)$, $w_z = X'_g(z)$ for suitable $f, g \in C^\infty(M)$, we can define

$$\omega_{\mathcal{L}}(z)(v_z, w_z) := \{f, g\}(z), \quad z \in \mathcal{L} \text{ and } v_z, w_z \in T_z\mathcal{L}. \tag{4.1.18}$$

We check that (4.1.18) is a good definition. Suppose, for instance, that we have two different functions $f, h \in C^\infty(M)$ whose associated vector fields $X'_f, X'_h \in \mathfrak{X}(\mathcal{L})$ uniquely determined by the prescription in (4.1.17) are such that $v_z = X'_f(z) = X'_h(z)$, and consequently $X_f(z) = X_h(z)$. Then

$$\begin{aligned}\omega_{\mathcal{L}}(z)(X'_f(z), w_z) &= \{f, g\}(z) = -\mathbf{d}g(z) \cdot X_f(z)\\ &= -\mathbf{d}g(z) \cdot X_h(z) = \{h, g\}(z) = \omega_{\mathcal{L}}(z)(X'_h(z), w_z).\end{aligned}$$

Since the same procedure can be carried out for w_z, we can conclude that (4.1.18) defines an antisymmetric two-form on $\mathcal{L}$. It is easy to see that the Jacobi identity satisfied by the bracket implies that $\omega_{\mathcal{L}}$ is closed. We now check that $\omega_{\mathcal{L}}$ is nondegenerate. Let $z \in \mathcal{L}$ and $v_z = X'_f(z) \in T_z\mathcal{L}$, $f \in C^\infty(M)$, be such that

$$\omega_{\mathcal{L}}(z)(v_z, w_z) = 0, \quad \text{for any } w_z \in T_z\mathcal{L}.$$

The definition of $\omega_{\mathcal{L}}$ implies that $\mathbf{d}g(z) \cdot X_f(z) = 0$ for all $g \in C^\infty(M)$. Consequently, $X_f(z) = 0$ and, as $T_z i$ is injective, $X'_f(z) = v_z = 0$, as required.

We finally show that the inclusion $i : \mathcal{L} \hookrightarrow M$ is a Poisson map. Let $f, g \in C^\infty(M)$ arbitrary. On the one hand we have that

$$(\{f, g\} \circ i)(m) = \{f, g\}(m) = \omega_{\mathcal{L}}(m)(X'_f(m), X'_g(m)). \tag{4.1.19}$$

At the same time, the relation between the vector fields X_g and X'_g given by (4.1.17) implies that their flows F_t and F'_t, respectively, satisfy $i \circ F'_t = F_t \circ i$. We can use this fact to write

$$
\begin{aligned}
(\{f, g\} \circ i)(m) &= \{f, g\}(m) = \mathbf{d}f(m) \cdot X_g(m) = \left.\frac{d}{dt}\right|_{t=0} (f \circ F_t)(m) \\
&= \left.\frac{d}{dt}\right|_{t=0} (f \circ F_t \circ i)(m) = \left.\frac{d}{dt}\right|_{t=0} (f \circ i \circ F'_t)(m) = \mathbf{d}(f \circ i)(m) \cdot X'_g(m) \\
&= \omega_{\mathcal{L}}(m)(X_{f\circ i}(m), X'_g(m)). \qquad (4.1.20)
\end{aligned}
$$

Since the point m and the functions $f, g \in C^\infty(M)$ are arbitrary, we can use relations (4.1.19) and (4.1.20) to conclude that

$$\omega_{\mathcal{L}}(X'_f - X_{f\circ i}, X'_g) = 0 \quad \text{for all } g \in C^\infty(M).$$

The nondegeneracy of $\omega_{\mathcal{L}}$ implies that $X'_f = X_{f\circ i}$. If we apply Ti to both sides of this equality and we look at (4.1.17) we obtain that $X_f \circ i = Ti \circ X_{f\circ i}$, for any $f \in C^\infty(M)$. Proposition 4.1.19 guarantees that i is Poisson, as required. The inclusion i being Poisson determines $\omega_{\mathcal{L}}$ in a unique way. ■

Note that an immersed submanifold S of a Poisson manifold $(M, \{\cdot, \cdot\})$ is a quasi-Poisson submanifold of $(M, \{\cdot, \cdot\})$ if and only if for any $s \in S$

$$T_s\mathcal{L}_s \subset T_s i(T_s S) \qquad (4.1.21)$$

where $i : S \hookrightarrow M$ is the inclusion and $\mathcal{L}_s$ is the symplectic leaf of $(M, \{\cdot, \cdot\})$ that contains $s \in S$.

4.1.29 Example Let $\mathfrak{g}^*$ be the dual a Lie algebra $\mathfrak{g}$ endowed with one of the Lie–Poisson structures introduced in §4.1.13. The symplectic leaves of the Poisson manifolds $(\mathfrak{g}^*, \{\cdot, \cdot\}_\pm)$ coincide with the connected components of the orbits of the elements in $\mathfrak{g}^*$ under the coadjoint action. In this situation, the symplectic form for the leaves defined in (4.1.18) is given by

$$\omega^{\pm}_{\mathcal{O}}(\nu)(\xi_{\mathfrak{g}^*}(\nu), \eta_{\mathfrak{g}^*}(\nu)) = \pm\langle \nu, [\xi, \eta]\rangle,$$

for any coadjoint orbit $\mathcal{O}$, arbitrary $\nu \in \mathcal{O}$, and $\xi, \eta \in \mathfrak{g}$. The symplectic structures $\omega^{\pm}_{\mathcal{O}}$ on $\mathcal{O}$ are usually called the ***±-orbit*** or ***Kostant–Kirillov–Souriau (KKS) symplectic forms***. All of the statements above will be proved in Theorem 4.5.31.

4.1.30 A first taste of symplectic reduction. The proof of the Symplectic Foliation Theorem and the characterization of the leaves of an integrable distribution as pseudogroup orbits (see Theorem 3.2.1) show that each symplectic leaf of a Poisson manifold is the set of points that can be joined to a given point by a finite number of smooth curves, each of which is a piece of an integral curve of a locally defined Hamiltonian vector field. We emphasize that the symplectic leaves $\mathcal{L}$ are connected initial submanifolds in M (relative to the inclusion $i : \mathcal{L} \hookrightarrow M$), carrying a symplectic structure whose Poisson bracket coincides with that of M. The tangent space $T_z\mathcal{L}$ at $z \in \mathcal{L}$ to a symplectic leaf $\mathcal{L}$ consists of all vectors that are equal to the value of some Hamiltonian vector field at z and, therefore, the symplectic leaves are invariant under the flow of any Hamiltonian vector field on M.

The last statement, together with the canonical character of the inclusion $i : \mathcal{L} \hookrightarrow M$, allows us to formulate a first reduction result. Let X_h be a Hamiltonian vector field on the Poisson manifold $(M, \{\cdot, \cdot\})$ for which we want to compute the integral curve $F_t(z)$ starting at the point $z \in M$. Let $\mathcal{L}$ be the unique symplectic leaf of $(M, \{\cdot, \cdot\})$ that contains the point z and let $i : \mathcal{L} \hookrightarrow M$ be the inclusion. Consider now the Hamiltonian dynamical system $(\mathcal{L}, \omega_{\mathcal{L}}, h \circ i)$ and let $F_t^{\mathcal{L}}$ be the flow of $X_{h \circ i}$. As the inclusion i is canonical, we have that $F_t \circ i = i \circ F_t^{\mathcal{L}}$. In other words, if we know the flow $F_t^{\mathcal{L}}$ we have solved our problem since $F_t(z) = (i \circ F_t^{\mathcal{L}})(z)$. Consequently, we have *reduced* the problem of finding $F_t(z)$ to that of determining $F_t^{\mathcal{L}}(z)$. We speak of $(\mathcal{L}, \omega_{\mathcal{L}}, h \circ i)$ as the ***reduced system*** of $(M, \{\cdot, \cdot\}, h)$ at $z \in M$ since, $\mathcal{L}$ being dimensionally smaller than M, acquiring information about $F_t^{\mathcal{L}}$ should be easier than it would be for F_t.

This reduction technique will appear later on in the book as a particular case of a more general procedure having to do with the additional symmetries of a Hamiltonian dynamical system.

For the next proposition we need an easy lemma from linear algebra.

4.1.31 Lemma *Let V be a finite-dimensional vector space and $A, B \subset V$ two subspaces. Then $(A \cap B)^\circ = A^\circ + B^\circ$.*

Proof. Since $A \cap B \subset A, B$, it follows that $A^\circ, B^\circ \subset (A \cap B)^\circ$. Therefore, $A^\circ + B^\circ \subset (A \cap B)^\circ$.

To prove the converse, let $\alpha \in (A \cap B)^\circ$. Denote by W a complement of $A + B$ in V; thus $V = (A + B) \oplus W$. Define $\beta_A, \beta_B \in V^*$ by

$$\beta_A(a + b + w) := \alpha(b) + \frac{1}{2}\alpha(w)$$

$$\beta_B(a + b + w) := \alpha(a) + \frac{1}{2}\alpha(w)$$

for $a \in A$, $b \in B$, $w \in W$. The functions β_A and β_B are well defined. Indeed, if $a + b = a' + b'$, with $a, a' \in A$ and $b, b' \in B$, then $a - a' = b' - b \in A \cap B$ and thus $\alpha(a) = \alpha(a')$, $\alpha(b) = \alpha(b')$. This shows that $\beta_A(a' + b' + w) = \alpha(b') + \frac{1}{2}\alpha(w) = \alpha(b) + \frac{1}{2}\alpha(w) = \beta_A(a + b + w)$ and similarly for β_B. The functions β_A and β_B are linear. Moreover,

$$\begin{aligned}\alpha(a + b + w) &= \alpha(b) + \frac{1}{2}\alpha(w) + \alpha(a) + \frac{1}{2}\alpha(w) \\ &= \beta_A(a + b + w) + \beta(a + b + w)\end{aligned}$$

which shows that $\alpha = \beta_A + \beta_B$. Finally, the definition of β_A implies that $\beta_A(a) = 0$ for all $a \in A$, that is, $\beta_A \in A^\circ$. Similarly, $\beta_B \in B^\circ$. Thus, $\alpha \in A^\circ + B^\circ$. ■

4.1.32 Proposition *Let $(M, \{\cdot, \cdot\})$ be a Poisson manifold and $B \in \Lambda^2(T^*M)$ the associated Poisson tensor. Then for any $m \in M$ and any vector subspace $V \subset T_m M$*

$$B^\sharp(m)(V^\circ) = (V \cap T_m\mathcal{L})^{\omega_{\mathcal{L}}(m)}$$

where $V^\circ := \{\alpha_m \in T_m^*M \mid \langle \alpha_m, v\rangle = 0$*, for all* $v \in V\} \subset T_m^*M$ *is the annihilator of* V *in* T_m^*M*,* $\mathcal{L}$ *is the symplectic leaf of* $(M, \{\cdot,\cdot\})$ *that contains the point* m*, and* $\omega_{\mathcal{L}}$ *is the symplectic form on* $\mathcal{L}$*.*

Proof. Let $\alpha_m \in V^\circ$ and $f \in C^\infty(M)$ be such that $\alpha_m = \mathbf{d}f(m)$. Let $v \in V \cap T_m\mathcal{L}$ be arbitrary. Hence

$$\omega_{\mathcal{L}}(m)(X_f(m), v) = \mathbf{d}f(m)\cdot v = \langle \alpha_m, v\rangle = 0.$$

Conversely, let $X_f(m) \in (V \cap T_m\mathcal{L})^{\omega_{\mathcal{L}}(m)} \subset T_m\mathcal{L}$. This means that

$$\omega_{\mathcal{L}}(m)(X_f(m), v) = \mathbf{d}f(m)\cdot v = 0,$$

for any $v \in V \cap T_m\mathcal{L}$. Hence, by Lemma 4.1.31, $\mathbf{d}f(m) \in (V \cap T_m\mathcal{L})^\circ = V^\circ + (T_m\mathcal{L})^\circ$, and consequently $X_f(m) = B^\sharp(m)(\mathbf{d}f(m)) \in B^\sharp(m)\,(V^\circ + (T_m\mathcal{L})^\circ) = B^\sharp(m)\,(V^\circ)$ since $B^\sharp(m)\,((T_m\mathcal{L})^\circ) = \{0\}$.

4.1.33 Proposition *Let* $(M, \{\cdot,\cdot\})$ *be a Poisson manifold and* $B \in \Lambda^2(T^*M)$ *the associated Poisson tensor. Then for any* $m \in M$ *and any vector subspace* $V \subset T_mM$*,*

$$B^\sharp(m)\left(\left(B^\sharp(m)(V^\circ)\right)^\circ\right) = V \cap T_m\mathcal{L},$$

where $\mathcal{L}$ *is the symplectic leaf containing the point* $m \in M$*.*

Proof. Let $v \in B^\sharp(m)\left(\left(B^\sharp(m)(V^\circ)\right)^\circ\right)$ be arbitrary. The relation

$$v \in B^\sharp(m)\left(\left(B^\sharp(m)(V^\circ)\right)^\circ\right)$$

is equivalent to $v = X_f(m)$, where $f \in C^\infty(M)$ is such that

$$\mathbf{d}f(m)|_{B^\sharp(m)(V^\circ)} = 0. \tag{4.1.22}$$

Since

$$\begin{aligned} B^\sharp(m)(V^\circ) &= \{B^\sharp(m)(\alpha) \mid \alpha \in V^\circ\} \\ &= \{X_g(m) \mid g \in C^\infty(M) \text{ such that } \mathbf{d}g(m)|_V = 0\}, \end{aligned}$$

the expression (4.1.22) is equivalent to

$$0 = \mathbf{d}f(m)\cdot X_g(m) = \{f, g\}(m) = -\mathbf{d}g(m)\cdot X_f(m) = -\mathbf{d}g(m)\cdot v,$$

for any $g \in C^\infty(M)$ such that $\mathbf{d}g(m)|_V = 0$. This in turn is equivalent to $v \in (V^\circ)^\circ = V$, which proves the claim. ■

4.1.34 More about Poisson automorphisms. Let $(M, \{\cdot,\cdot\})$ be a Poisson manifold and $f : M \to M$ a Poisson automorphism. The map f does *not* necessarily leave invariant the symplectic leaves of M. See §4.2.5 for an example.

At the infinitesimal level the situation is similar, which motivates the following definition. The vector field $X \in \mathfrak{X}(M)$ is an ***infinitesimal leaf preserving Poisson automorphism*** if it is an infinitesimal Poisson automorphism such that its flow F_t satisfies

$F_t(\mathcal{L} \cap \mathrm{Dom}(F_t)) \subset \mathcal{L}$ for any symplectic leaf $\mathcal{L}$ of M. The set of infinitesimal leaf preserving Poisson automorphisms on $(M, \{\cdot, \cdot\})$ will be denoted by $\mathfrak{X}_{LPA}(M, \{\cdot, \cdot\})$, or simply by $\mathfrak{X}_{LPA}(M)$. It is easy to see (see §1.2.4) that $X \in \mathfrak{X}_{LPA}(M)$ if and only if $X(z) \in T_z\mathcal{L}$, for any symplectic leaf $\mathcal{L}$ and any $z \in \mathcal{L}$.

The following inclusions are easy to prove (see Definition 4.1.25)

$$\mathfrak{X}_H(M) \subset \mathfrak{X}_{LH}(M) \subset \mathfrak{X}_{LPA}(M) \subset \mathfrak{X}_{PA}(M). \tag{4.1.23}$$

If M is symplectic, then $\mathfrak{X}_{LH}(M) = \mathfrak{X}_{LPA}(M) = \mathfrak{X}_{PA}(M)$. In addition,

$$[\mathfrak{X}_{LH}(M), \mathfrak{X}_{LH}(M)] = \mathfrak{X}_H(M),$$

a result due to CALABI (1970) and LICHNEROWICZ (1973).

All inclusions in (4.1.23) are *strict* as we show below.

(i) If $M = \mathbb{T}^2 = S^1 \times S^1$ is the two-dimensional torus whose symplectic form is given by the area form $\mathbf{d}\psi_1 \wedge \mathbf{d}\psi_2$, where ψ_1, ψ_2, are the angle coordinates on each circle, then the vector field whose flow $(\psi_1, \psi_2) \mapsto (\psi_1 + t, \psi_2)$ is clearly canonical, is a locally but not globally Hamiltonian vector field.

(ii) If $M = \mathbb{R}^2$ endowed with the Poisson bracket $\{x, y\} = x^2 + y^2$, the vector field $X(x, y) = x\partial/\partial x + y\partial/\partial y$ is a leaf preserving infinitesimal Poisson automorphism, but it is not a locally Hamiltonian vector field because in any neighborhood of the origin there is no function $f \in C^\infty(\mathbb{R}^2)$ such that $X = X_f$. Indeed, it is easy to check that for any $f, g \in C^\infty(\mathbb{R}^2)$ we have that

$$\{f, g\} = (x^2 + h^2)\left(\frac{\partial f}{\partial x}\frac{\partial g}{\partial y} - \frac{\partial f}{\partial y}\frac{\partial g}{\partial x}\right)$$

and

$$X_f = (x^2 + y^2)\left[\frac{\partial f}{\partial y}\frac{\partial}{\partial x} - \frac{\partial f}{\partial x}\frac{\partial}{\partial y}\right]. \tag{4.1.24}$$

Thus, if $X = X_f$, then at any point $(x, y) \neq (0, 0)$ we have from (4.1.24) that

$$\frac{\partial f}{\partial x} = -\frac{y}{x^2 + y^2}, \qquad \frac{\partial f}{\partial y} = \frac{x}{x^2 + y^2}$$

and hence if we integrate $\mathbf{d}f$ on the circle $c(t) = r(\cos t, \sin t)$, $t \in [0, 2\pi]$, Stokes' Theorem yields

$$\begin{aligned} 0 &= \int_c \mathbf{d}f = \int_c \left(\frac{\partial f}{\partial x}dx + \frac{\partial f}{\partial y}dy\right) = \int_c \left(-\frac{y}{x^2 + y^2}dx + \frac{x}{x^2 + y^2}dy\right) \\ &= \frac{1}{r^2}\int_0^{2\pi} (-\sin t d(\cos t) + \cos t d(\sin t)) = \frac{2\pi}{r^2}, \end{aligned}$$

which is clearly a contradiction.

(iii) If M is a manifold with trivial Poisson bracket, that is, the bracket of any two functions is zero, then any nonzero vector field on M is an infinitesimal Poisson automorphism but it is not leaf preserving.

4.2 Canonical Lie group and algebra actions

The symmetries of a Hamiltonian system are usually encoded via the action of a Lie group or a Lie algebra consistent with the underlying structure of the given dynamical system.

4.2.1 Canonical Lie group and algebra actions. Let $(M, \{\cdot, \cdot\})$ be a Poisson manifold (respectively, (M, ω) a symplectic manifold), let G be a Lie group, and let $\Phi : G \times M \to M$ be a smooth left action of G on M. We say that the action Φ is ***canonical*** or ***Poisson*** if Φ acts by canonical transformations, that is, for any $f, h \in C^\infty(M)$ and any $g \in G$ we have

$$\Phi_g^*\{f, h\} = \{\Phi_g^* f, \Phi_g^* h\} \qquad (\text{resp. } \Phi_g^*\omega = \omega).$$

We say that the Hamiltonian system $(M, \{\cdot, \cdot\}, h)$ is ***G-symmetric*** when the Lie group G acts canonically on $(M, \{\cdot, \cdot\})$ and the Hamiltonian function h is G-invariant, that is, $h \in C^\infty(M)^G$.

The infinitesimal analog of this concept is the ***canonical*** or ***Poisson action of a Lie algebra*** $\mathfrak{g}$. We say that an action of the Lie algebra $\mathfrak{g}$ on the Poisson (respectively, symplectic) manifold $(M, \{\cdot, \cdot\})$ (respectively, (M, ω)) is canonical if the vector fields $\xi_M \in \mathfrak{X}(M)$, $\xi \in \mathfrak{g}$, are infinitesimal Poisson automorphisms, that is, if B is the Poisson tensor we have that $\pounds_{\xi_M} B = 0$ (respectively, $\pounds_{\xi_M}\omega = 0$). Using Proposition 4.1.26, this condition amounts to

$$\xi_M[\{f, g\}] = \{\xi_M[f], g\} + \{f, \xi_M[g]\}, \tag{4.2.1}$$

for all $\xi \in \mathfrak{g}$ and $f, g \in C^\infty(M)$. We say that the Hamiltonian system $(M, \{\cdot, \cdot\}, h)$ is $\mathfrak{g}$-***symmetric*** if the Lie algebra $\mathfrak{g}$ acts canonically on $(M, \{\cdot, \cdot\})$ and the Hamiltonian function h is $\mathfrak{g}$-invariant, that is, $\xi_M[h] = 0$, for any $\xi \in \mathfrak{g}$.

4.2.2 Proposition *Let (M, ω) be a symplectic manifold and $\mathfrak{g}$ a Lie algebra acting canonically on it. Then for any $\xi, \eta, \zeta \in \mathfrak{g}$ we have*

$$\omega([\xi, \eta]_M, \zeta_M) + \omega([\eta, \zeta]_M, \xi_M) + \omega([\zeta, \xi]_M, \eta_M) = 0. \tag{4.2.2}$$

Proof. Since

$$\begin{aligned}\mathbf{d}\omega(\xi_M, \eta_M, \zeta_M) &= \pounds_{\xi_M}(\omega(\eta_M, \zeta_M)) - \pounds_{\eta_M}(\omega(\xi_M, \zeta_M)) \\ &\quad + \pounds_{\zeta_M}(\omega(\xi_M, \eta_M)) - \omega([\xi_M, \eta_M], \zeta_M) \\ &\quad + \omega([\xi_M, \zeta_M], \eta_M) - \omega([\eta_M, \zeta_M], \xi_M),\end{aligned}$$

and the $\mathfrak{g}$-action is symplectic we have

$$\begin{aligned}\pounds_{\xi_M}(\omega(\eta_M, \zeta_M)) &= (\pounds_{\xi_M}\omega)(\eta_M, \zeta_M)) + \omega(\pounds_{\xi_M}\eta_M, \zeta_M) + \omega(\eta_M, \pounds_{\xi_M}\zeta_M) \\ &= \omega([\xi_M, \eta_M], \zeta_M) + \omega(\eta_M, [\xi_M, \zeta_M]),\end{aligned}$$

and similarly for $\pounds_{\eta_M}(\omega(\xi_M, \zeta_M))$ and $\pounds_{\zeta_M}(\omega(\xi_M, \eta_M))$. Thus we get

$$\mathbf{d}\omega(\xi_M, \eta_M, \zeta_M) = -(\omega([\xi, \eta]_M, \zeta_M) + \omega([\eta, \zeta]_M, \xi_M) + \omega([\zeta, \xi]_M, \eta_M)).$$

As $\mathbf{d}\omega = 0$, the relation (4.2.2) follows.

4.2.3 Proposition *Let $(M, \{\cdot, \cdot\})$ be a Poisson manifold and denote by $B \in \Lambda^2(T^*M)$ the associated Poisson tensor. Let G be a Lie group acting canonically on M. Then, for any $m \in M$ and any vector subspace $V \subset T_m^*M$:*

(i) *$B^\sharp(m) : T_m^*M \to T_mM$ is G_m-equivariant.*

(ii) *If the Poisson bracket $\{\cdot, \cdot\}$ is induced by a symplectic form ω, then*

$$B^\sharp(m)(V^{G_m}) = (B^\sharp(m)(V))^{G_m},$$

where the G_m-superscript denotes the set of G_m-fixed points in the corresponding spaces.

Proof. Part **(i)** is a trivial consequence of the canonical character of the action. Part **(ii)** follows from Part **(i)** and the nondegeneracy of $B^\sharp(m)$ in the symplectic case. ■

4.2.4 Invariant Hamiltonians and equivariant vector fields. Notice that, by Proposition 4.1.19, for a G-invariant system $(M, \{\cdot, \cdot\}, h)$ the associated Hamiltonian vector field X_h is G-equivariant and, consequently, the associated flow F_t is G-equivariant. Indeed, for any $g \in G$ we have that

$$T\Phi_g \circ X_h = T\Phi_g \circ X_{h\circ\Phi_g} = X_h \circ \Phi_g, \tag{4.2.3}$$

where the first equality is a consequence of the G-invariance of h and the second one follows from the canonical character of the action. The equivariance of the flow F_t follows from Proposition 3.3.2.

The converse implication is only true up to Casimir functions. Indeed, let X_h be a G-equivariant Hamiltonian vector field on $(M, \{\cdot, \cdot\})$, that is, $T\Phi_g \circ X_h = X_h \circ \Phi_g$ for any $g \in G$. Hence $X_h \circ \Phi_g = T\Phi_g \circ X_h = T\Phi_g \circ X_{h\circ\Phi_{g^{-1}}\circ\Phi_g} = X_{h\circ\Phi_{g^{-1}}} \circ \Phi_g$ since the G-action is canonical (see Proposition 4.1.19). Thus $X_{h\circ\Phi_{g^{-1}}} = X_h$ or $X_{h\circ\Phi_{g^{-1}}-h} \equiv 0$ for any $g \in G$, which means that there exists a family of Casimir functions $C_g \in \mathcal{C}(M)$ parametrized by $g \in G$, such that

$$h \circ \Phi_g = h + C_g.$$

4.2.5 Canonical actions and symplectic leaves. A very important point that we should emphasize is that a canonical action on a Poisson manifold does not necessarily preserve its symplectic leaves as the following elementary example shows. Let $(\mathbb{R}^3, \{\cdot, \cdot\})$ be the Poisson manifold formed by the Euclidean three-dimensional space $\mathbb{R}^3$ together with the Poisson structure induced by the Poisson tensor B given in Euclidean coordinates by

$$B = \begin{pmatrix} 0 & 1 & 0 \\ -1 & 0 & 1 \\ 0 & -1 & 0 \end{pmatrix}.$$

With this Poisson bracket, the Hamiltonian vector field X_f associated to any smooth function $f \in C^\infty(\mathbb{R}^3)$ is given by

$$X_f(x, y, z) = \frac{\partial f}{\partial y}\frac{\partial}{\partial x} + \left(\frac{\partial f}{\partial z} - \frac{\partial f}{\partial x}\right)\frac{\partial}{\partial y} - \frac{\partial f}{\partial y}\frac{\partial}{\partial z}. \tag{4.2.4}$$

A quick calculation using the previous expression shows that the Casimir functions of $(\mathbb{R}^3, \{\cdot, \cdot\})$ are the functions $f \in C^\infty(\mathbb{R}^3)$ of the form $f(x, y, z) := g(x + z)$, with $g \in C^\infty(\mathbb{R})$. Also, the symplectic leaves of this Poisson manifold are given by the planes of the form $x + z =$ constant.

Consider the action of the additive group $(\mathbb{R}, +)$ on $\mathbb{R}^3$ given by $\lambda \cdot (x, y, z) := (x + \lambda, y, z)$, for any $\lambda \in \mathbb{R}$ and any $(x, y, z) \in \mathbb{R}^3$. It is clear that this action is Poisson but, obviously, the symplectic leaves $x + z =$ constant are not invariant under this group action.

4.2.6 Canonical actions and isotropy type manifolds. Let $(M, \{\cdot, \cdot\}, h)$ be a G-invariant Hamiltonian system. The G-equivariance of X_h we proved in (4.2.3) guarantees that its G-equivariant flow F_t leaves the connected components of the isotropy type manifolds corresponding to the G-action invariant (see Proposition 3.3.4). We will refer to this elementary fact as the ***law of conservation of the isotropy***.

If the manifold M is symplectic and the action is proper it turns out that these type submanifolds are actually symplectic. In the following proposition we use the notation and conventions of § 2.4.10.

4.2.7 Proposition *Let $\Phi : G \times M \to M$ be a canonical proper action of the Lie group G on the symplectic manifold (M, ω). Let $M^m_{G_m}$ be the connected component of the G_m-isotropy type submanifold M_{G_m} that contains the point $m \in M$ and let $i : M^m_{G_m} \hookrightarrow M$ be the inclusion. Then*

(i) *$(M^m_{G_m}, i^*\omega)$ is a symplectic submanifold of (M, ω);*

(ii) *the group $L^m := N(G_m)^m / G_m$ acts, freely, properly, and canonically on $M^m_{G_m}$ via the map*

$$\begin{array}{rcl} \Psi : N(G_m)^m / G_m \times M^m_{G_m} & \longrightarrow & M^m_{G_m} \\ (nG_m, z) & \longmapsto & n \cdot z. \end{array}$$

Proof. **(i)** We reproduce the proof in GUILLEMIN AND STERNBERG (1984c). First, the properness of the G-action and Proposition 2.4.7 ensure that $M^m_{G_m}$ is a (embedded) submanifold of M and that $T_m M^m_{G_m} = (T_m M)^{G_m}$. Second, the form $i^*\omega$ is closed and antisymmetric since ω is. Consequently, we just need to show that $i^*\omega$ is nondegenerate.

We proceed by using the compactness of G_m to choose a G_m-invariant positive definite inner product $\ll \cdot, \cdot \gg$ on $T_z M$, with $z \in M^m_{G_m}$ arbitrary. There exists a unique G_m-equivariant linear map $J : T_z M \to T_z M$ such that $\ll u, v \gg = \omega(z)(u, Jv)$, for any $u, v \in T_z M$. We now prove the nondegeneracy of $i^*\omega$. Suppose that $u \in (T_z M)^{G_m}$ is such that $(i^*\omega)(z)(u, v) = \omega(z)(u, v) = 0$ for all $v \in (T_z M)^{G_m}$. As J is G_m-equivariant, we have that $J((T_z M)^{G_m}) \subset (T_z M)^{G_m}$ and hence the previous identity implies that $\ll u, v \gg = \omega(z)(u, Jv) = 0$ for all $v \in (T_z M)^{G_m}$. As $\ll \cdot, \cdot \gg$ is positive definite on $T_z M$, it is also on $(T_z M)^{G_m}$, and thus $u = 0$. The arbitrary character of $z \in M^m_{G_m}$ implies the nondegeneracy of $i^*\omega$.

(ii) The L^m-action is clearly free. It is also proper because $N(G_m)^m$ is closed in G. We check that the action is canonical. For any $l = nG_m \in L^m$ we have

$$\Psi_l^*(i^*\omega) = (i \circ \Psi_l)^*\omega = (\Phi_n \circ i)^*\omega = i^*(\Phi_n^*\omega) = i^*\omega,$$

as required. ■

4.2.8 An important point that the reader should keep in mind is that even though $(M^m_{G_m}, i^*\omega)$ is a symplectic submanifold of (M, ω), the inclusion $i : M^m_{G_m} \hookrightarrow M$ is *not*, in general, a Poisson map. This imposes restrictions when we want to carry out a reduction procedure similar to the one described in §4.1.30 regarding the symplectic leaves of a Poisson manifold. The following elementary example shows that this is actually the case.

Let $(\mathbb{R}^4, \omega)$ be the symplectic manifold with $\omega = \mathbf{d}q^1 \wedge \mathbf{d}p_1 + \mathbf{d}q^2 \wedge \mathbf{d}p_2$. A straightforward computation shows that the action of the circle on $(\mathbb{R}^4, \omega)$ consisting of rotating the (q^1, p_1)-plane is symplectic. The isotropy type submanifold $\mathbb{R}^4_{S^1}$ corresponding to the entire circle S^1 equals the (q^2, p_2)-plane that, by the previous proposition, is a symplectic submanifold; of course, in this case, this is trivial to verify directly. Notice that the inclusion map that takes the (q^2, p_2)-plane into $(\mathbb{R}^4, \omega)$ is symplectic but not Poisson since (4.1.7) (or, equivalently, the condition in Proposition 4.1.19) is not satisfied.

The following result shows that $i : M^m_{G_m} \hookrightarrow M$ can be used to reduce Hamiltonian vector fields associated to G-invariant functions and that if we restrict the Poisson algebra to only those functions, i is then a Poisson map.

4.2.9 Proposition *Let $\Phi : G \times M \to M$ be a canonical proper action of the Lie group G on the symplectic manifold (M, ω). Let $M^m_{G_m}$ be the connected component of the G_m-isotropy type submanifold M_{G_m} that contains the point $m \in M$ and let $i : M^m_{G_m} \hookrightarrow M$ be the inclusion.*

(i) *Let $h \in C^\infty(M)^G$. The Hamiltonian vector field $X_{h\circ i}$ on $(M^m_{G_m}, i^*\omega)$ associated to the restriction $h \circ i \in C^\infty(M^m_{G_m})$ is i-related to X_h, that is,*

$$Ti \circ X_{h\circ i} = X_h \circ i.$$

(ii) *The map $i : M^m_{G_m} \hookrightarrow M$ induces a Poisson map between the Poisson algebras $(C^\infty(M^m_{G_m})^{L^m}, \{\cdot,\cdot\}_{M^m_{G_m}})$ and $(C^\infty(M)^G, \{\cdot,\cdot\})$.*

Proof. **(i)** Recall that, as we showed in §4.2.4, the G-invariance of h implies that X_h is an equivariant vector field with equivariant flow (Proposition 3.3.2). This automatically implies that X_h is tangent to $M^m_{G_m}$ (Proposition 3.3.4) and therefore, for any $z \in M^m_{G_m}$ and any $v_z \in T_z M^m_{G_m}$, we have

$$\mathbf{d}(h \circ i)(z) \cdot v_z = \mathbf{d}h(z) \cdot (T_z i \cdot v_z) = \omega(z)(X_h(z), T_z i \cdot v_z). \tag{4.2.5}$$

We can also write

$$\mathbf{d}(h \circ i)(z) \cdot v_z = (i^*\omega)(z)(X_{h\circ i}(z), v_z) = \omega(z)(T_z i \cdot X_{h\circ i}(z), T_z i \cdot v_z). \tag{4.2.6}$$

Expressions (4.2.5) and (4.2.6) imply that

$$\omega(z)(X_h(z) - T_z i \cdot X_{h\circ i}(z), w_z) = 0, \quad \text{for any} \quad w_z \in T_z M^m_{G_m}. \tag{4.2.7}$$

We now recall that because $M^m_{G_m}$ is a symplectic submanifold of M, we can write

$$T_zM = T_zM^m_{G_m} \oplus (T_zM^m_{G_m})^\omega.$$

Consequently, as $X_h(z) - T_zi \cdot X_{h\circ i}(z) \in T_zM^m_{G_m}$, we have that

$$\omega(z)(X_h(z) - T_zi \cdot X_{h\circ i}(z), u_z) = 0, \quad \text{for any} \quad u_z \in T_zM,$$

which guarantees that $X_h(z) = T_zi \cdot X_{h\circ i}(z)$, as required.
(ii) Let $f, g \in C^\infty(M)^G$ arbitrary. The functions $f \circ i, g \circ i \in C^\infty(M^m_{G_m})$ are clearly L^m-invariant. Now, since $i : (M^m_{G_m}, i^*\omega) \hookrightarrow (M, \omega)$ is a symplectic map by Proposition 4.2.7 **(i)**, for any $z \in M^m_{G_m}$, we have

$$\begin{aligned}
\{f \circ i, g \circ i\}_{M^m_{G_m}}(z) &= (i^*\omega)(z)(X_{f\circ i}(z), X_{g\circ i}(z)) \\
&= \omega(z)(T_zi \cdot X_{f\circ i}(z), T_zi \cdot X_{g\circ i}(z))) \\
&= \omega(z)(X_f(z), X_g(z)) \quad \text{(by part (i))} \\
&= (\{f, g\} \circ i)(z).
\end{aligned}$$

■

4.2.10 The nature of the isotropy type submanifolds in the purely Poisson case depends very strongly on how the canonical action interacts with the symplectic leaves of the manifold. For instance, a proper canonical action for which the leaves are invariant is such that each symplectic leaf is a proper G-symplectic manifold to which Proposition 4.2.7 can be applied. However, as we saw in §4.2.5, this is a very strong hypothesis.

4.3 Momentum maps

There are many notions in the literature related to the concept of momentum (also referred to as moment) map. Among the manifold of ideas underlying all of them we will single out the quest after the conservation laws associated to the symmetries of a dynamical system, which in our case will always be Hamiltonian.

In the previous section we learned how to encode the symmetries of a Hamiltonian system. The rest of this chapter as well as the next one will use the invariance properties of a symmetric system to construct certain maps, namely the momentum maps, which in the sense that we make precise in the following definition, will yield some of the conservation laws of these systems.

4.3.1 Definition *Let $(M, \{\cdot, \cdot\})$ be a Poisson manifold and G (respectively, $\mathfrak{g}$) a Lie group (respectively, Lie algebra) acting canonically on it. Let S be a set and $\mathbf{J} : M \to S$ a map. We say that $\mathbf{J}$ is a* ***Noether momentum map*** *for the G-action (respectively, $\mathfrak{g}$-action) on $(M, \{\cdot, \cdot\})$ when the flow F_t of any Hamiltonian vector field associated to any G-invariant (respectively, $\mathfrak{g}$-invariant) Hamiltonian function $h \in C^\infty(M)$ preserves the fibers of $\mathbf{J}$, that is,*

$$\mathbf{J} \circ F_t = \mathbf{J}|_{\mathrm{Dom}(F_t)}. \tag{4.3.1}$$

We will refer to condition (4.3.1) *as* ***Noether's condition.***

In most cases, the set S in the previous definition has additional structure. As we will see, S is sometimes a Poisson manifold itself and the Noether momentum map $\mathbf{J}$ is a Poisson map. In other situations, S is a G-space and $\mathbf{J}$ is G-equivariant.

If G is a Lie group acting canonically on $(M, \{\cdot, \cdot\})$ and $\mathbf{J} : M \to S$ is a Noether momentum map for the associated canonical Lie algebra $\mathfrak{g}$-action, then $\mathbf{J}$ is obviously also a Noether momentum map for the G-action.

4.3.2 Example Let $(M, \{\cdot, \cdot\})$ be a Poisson manifold and G a Lie group acting canonically and properly on it. Let $\mathcal{J}_F : M \to M/A_F$ be the projection onto the leaf space of the distribution introduced in Theorem 3.5.1 and spanned by all the G-equivariant vector fields. As was proved in Theorem 3.5.1, the level sets of this map are the connected components of the isotropy type submanifolds that, as we mentioned in §4.2.6, are preserved by any G-equivariant Hamiltonian flow. This statement ensures that $\mathcal{J}_F$ satisfies Noether's condition and thus $\mathcal{J}_F$ is a Noether momentum map.

4.3.3 Remark While Definition 4.3.1 was motivated by the idea that the symmetries of the Hamiltonian generate conservation laws, topological considerations linked to cobordism lead to a totally different generalization, as introduced by GINZBURG, GUILLEMIN, AND KARSHON (1999). An ***abstract moment map*** is an equivariant map $\Psi : M \to \mathfrak{g}^*$ from a G-manifold M to the dual $\mathfrak{g}^*$ of the Lie algebra $\mathfrak{g}$ of G such that $\iota_{\mathfrak{h}}^* \circ \Psi : M \to \mathfrak{h}^*$ is constant on connected components of M^H for every closed subgroup H of G. In this definition $\mathfrak{h}$ denotes the Lie algebra of H, $\iota_{\mathfrak{h}} : \mathfrak{h} \hookrightarrow \mathfrak{g}$ is the inclusion map, $\iota_{\mathfrak{h}}^* : \mathfrak{g}^* \to \mathfrak{h}^*$ is its dual, and M^H is the submanifold of H-fixed points in M. The standard momentum map that will be studied in Section 4.5 on a symplectic manifold satisfies this property; see (4.5.3). The abstract moment map is not a Noether momentum map in the sense of Definition 4.3.1, it will not be studied further in this book, and the reader is referred to GINZBURG, GUILLEMIN, AND KARSHON (2002) for the development and applications of this theory.

4.4 The Chu momentum map

In this section we present a first example of a Noether momentum map whose definition makes an essential use of the symplectic structure. Apart from its intrinsic interest as a Noether momentum map, this construction will be extremely important in the statement and proof of a symplectic version of the Slice Theorem, presented in Chapter 7.

Let (M, ω) be a symplectic manifold and $\mathfrak{g}$ a Lie algebra acting canonically on it. The ***Chu map*** (see CHU (1975)) is defined as the map $\Psi : M \to Z^2(\mathfrak{g})$ (we use the notation introduced in §2.1.18) given by

$$\Psi(m)(\xi, \eta) = \omega(m)(\xi_M(m), \eta_M(m)), \qquad \text{for all} \quad \xi, \eta \in \mathfrak{g}.$$

The fact that Ψ maps into $Z^2(\mathfrak{g})$ is a consequence of the closedness of the symplectic form ω and of the canonical character of the $\mathfrak{g}$-action. Indeed, the evaluation of equality (4.2.2) at any point $m \in M$ is equivalent to $\mathbf{d}(\Psi(m)) = 0$ (see (2.1.17)), which shows that Ψ maps into $Z^2(\mathfrak{g})$.

The main properties of the Chu map are summarized in the following proposition. Before giving its statement, we recall that if G is a Lie group with Lie algebra $\mathfrak{g}$, then

G acts on $\Lambda^2(\mathfrak{g})$ via the dual of the diagonal adjoint representation of G on $\mathfrak{g} \times \mathfrak{g}$, that is, for any $g \in G$, $\xi, \eta, \zeta \in \mathfrak{g}$, and $\alpha \in \Lambda^2(\mathfrak{g})$, we define

$$\begin{aligned} \langle \mathrm{Ad}^*_{g^{-1}} \alpha, (\xi, \eta) \rangle &:= \langle \alpha, (\mathrm{Ad}_{g^{-1}} \xi, \mathrm{Ad}_{g^{-1}} \eta) \rangle, \\ \langle \mathrm{ad}^*_\zeta \alpha, (\xi, \eta) \rangle &:= \langle \alpha, (\mathrm{ad}_\zeta \xi, \eta) \rangle + \langle \alpha, (\xi, \mathrm{ad}_\zeta \eta) \rangle. \end{aligned}$$

4.4.1 Proposition *Let (M, ω) be a symplectic manifold and $\mathfrak{g}$ a Lie algebra acting canonically on it. Then the corresponding Chu map $\Psi : M \to Z^2(\mathfrak{g})$ satisfies the following properties:*

(i) $T_m \Psi(\cdot)(\xi, \eta) = \omega(m)([\xi, \eta]_M(m), \cdot)$, *for any $m \in M$, $\xi, \eta \in \mathfrak{g}$.*

(ii) $\ker T_m \Psi = ([\mathfrak{g}, \mathfrak{g}]_M(m))^\omega$, *for any $m \in M$.*

(iii) *Ψ is a Noether momentum map for the $\mathfrak{g}$-action on M.*

(iv) *For any $m \in M$ and any $\xi, \eta, \zeta \in \mathfrak{g}$ we have*

$$\begin{aligned} T_m \Psi(\zeta_M(m)) \cdot (\xi, \eta) &= \Psi(m)(\mathrm{ad}_\xi \zeta, \eta) + \Psi(m)(\xi, \mathrm{ad}_\eta \zeta) \\ &= -\langle \mathrm{ad}^*_\zeta \Psi(m), (\xi, \eta) \rangle. \end{aligned} \tag{4.4.1}$$

Let G be a Lie group that acts canonically on (M, ω) via the map $\Phi : G \times M \to M$ and whose Lie algebra is $\mathfrak{g}$. Suppose that the canonical Lie algebra action associated to this group action coincides with the $\mathfrak{g}$-action in the statement. Then the Chu map is G-equivariant; that is, for any $m \in M$ and any $g \in G$ we have that

$$\Psi(g \cdot m) = \mathrm{Ad}^*_{g^{-1}} \Psi(m). \tag{4.4.2}$$

Additionally, for any $m \in M$, $h \in G_m$, and $\xi \in \mathfrak{g}$ we have that

$$\mathrm{Ad}^*_{h^{-1}}(\Psi(m)(\xi, \cdot)) = \Psi(m)(\mathrm{Ad}_h \xi, \cdot). \tag{4.4.3}$$

Proof. **(i)** Let $X, Y \in \mathfrak{X}_{LH}(M)$, that is, $\pounds_X \omega = \pounds_Y \omega = 0$. Then,

$$-\mathbf{i}_{[X,Y]} \omega = -\pounds_X \mathbf{i}_Y \omega + \mathbf{i}_Y \pounds_X \omega = -\pounds_X \mathbf{i}_Y \omega,$$

since $\pounds_X \omega = 0$. Now, as $-\pounds_X \mathbf{i}_Y \omega = -\mathbf{i}_X \mathbf{d} \mathbf{i}_Y \omega - \mathbf{d} \mathbf{i}_X \mathbf{i}_Y \omega$, and $\mathbf{d} \mathbf{i}_Y \omega = \mathbf{d} \mathbf{i}_Y \omega + \mathbf{i}_Y \mathbf{d} \omega = \pounds_Y \omega = 0$, we have that

$$-\mathbf{i}_{[X,Y]} \omega = -\mathbf{d}(\mathbf{i}_X \mathbf{i}_Y \omega) = \mathbf{d}(\omega(X, Y)).$$

If we put $X = \xi_M$ and $Y = \eta_M$ in the previous expression, we get

$$\mathbf{d}(\omega(\xi_M, \eta_M)) = -\mathbf{i}_{[\xi_M, \eta_M]} \omega = \mathbf{i}_{[\xi, \eta]_M} \omega = \omega([\xi, \eta]_M, \cdot).$$

Consequently, for any $v_m \in T_m M$, we can write

$$T_m \Psi(v_m)(\xi, \eta) = \mathbf{d}(\omega(\xi_M, \eta_M))(m) \cdot v_m = \omega(m)([\xi, \eta]_M(m), v_m)$$

and hence the expression in **(i)** follows.

(ii) It is a straightforward consequence of **(i)**.

(iii) We check that Noether's condition is satisfied for Ψ. Let $h \in C^\infty(M)^{\mathfrak{g}}$ and let X_h be the corresponding Hamiltonian vector field with flow F_t. Then, for any $m \in M$ and $\xi, \eta \in \mathfrak{g}$, we have

$$\begin{aligned}\{h, \Psi(\cdot)(\xi, \eta)\}(m) &= -\mathbf{d}[\Psi(\cdot)(\xi, \eta)](m) \cdot X_h(m) \\ &= -\omega(m)([\xi, \eta]_M(m), X_h(m)) \\ &= \mathbf{d}h(m) \cdot [\xi, \eta]_M(m) = [\xi, \eta]_M[h](m) = 0\end{aligned}$$

where we have used point **(i)** and the $\mathfrak{g}$-invariance of h. This computation shows that the function $\Psi(\cdot)(\xi, \eta)$ is constant along the flow F_t of X_h. As $\xi, \eta \in \mathfrak{g}$ are arbitrary, we have that $\Psi \circ F_t = \Psi|_{\mathrm{Dom}(F_t)}$, as required.

(iv) Identity (4.4.1) is a consequence of point **(i)** and (4.2.2) . Indeed, for any $m \in M$ and any $\xi, \eta, \zeta \in \mathfrak{g}$ we have that

$$\begin{aligned}T_m\Psi(\zeta_M(m)) \cdot (\xi, \eta) &= \omega(m)([\xi, \eta]_M(m), \zeta_M(m)) \\ &= \omega(m)([\zeta, \eta]_M(m), \xi_M(m)) + \omega(m)([\xi, \zeta]_M(m), \eta_M(m)) \\ &= \Psi(m)(\mathrm{ad}_\xi\zeta, \eta) + \Psi(m)(\xi, \mathrm{ad}_\eta\zeta).\end{aligned}$$

As to (4.4.2), let $g \in G$, $\xi, \eta \in \mathfrak{g}$, and $m \in M$ be arbitrary. Then

$$\begin{aligned}\Psi(g \cdot m)(\xi, \eta) &= \omega(g \cdot m)(\xi_M(g \cdot m), \eta_M(g \cdot m)) \\ &= \omega(g \cdot m)(T_m\Phi_g \cdot (\mathrm{Ad}_{g^{-1}}\xi)_M(m), T_m\Phi_g \cdot (\mathrm{Ad}_{g^{-1}}\eta)_M(m)) \\ &= (\Phi_g^*\omega)(m)((\mathrm{Ad}_{g^{-1}}\xi)_M(m), (\mathrm{Ad}_{g^{-1}}\eta)_M(m)) \\ &= \omega(m)((\mathrm{Ad}_{g^{-1}}\xi)_M(m), (\mathrm{Ad}_{g^{-1}}\eta)_M(m)) \\ &= \Psi(m)(\mathrm{Ad}_{g^{-1}}\xi, \mathrm{Ad}_{g^{-1}}\eta) = (\mathrm{Ad}^*_{g^{-1}}\Psi(m))(\xi, \eta).\end{aligned}$$

Since this equality holds for any $\xi, \eta \in \mathfrak{g}$, we have that $\Psi(g \cdot m) = \mathrm{Ad}^*_{g^{-1}}\Psi(m)$, as required. We now prove (4.4.3). Let $\eta \in \mathfrak{g}$ arbitrary. Then

$$\begin{aligned}\langle \mathrm{Ad}^*_{h^{-1}}(\Psi(m)(\xi, \cdot)), \eta\rangle &= \Psi(m)(\xi, \mathrm{Ad}_{h^{-1}}\eta) \\ &= \omega(m)(\xi_M(m), (\mathrm{Ad}_{h^{-1}}\eta)_M(m)) \\ &= \omega(m)(T_m\Phi_{h^{-1}} \cdot (\mathrm{Ad}_h\xi)_M(m), T_m\Phi_{h^{-1}} \cdot \eta_M(m)) \\ &= \omega(\Phi_{h^{-1}}(m))(T_m\Phi_{h^{-1}} \cdot (\mathrm{Ad}_h\xi)_M(m), T_m\Phi_{h^{-1}} \cdot \eta_M(m)) \\ &= \omega(m)((\mathrm{Ad}_h\xi)_M(m), \eta_M(m)) = \Psi(m)(\mathrm{Ad}_h\xi, \eta) \\ &= \langle \Psi(m)(\mathrm{Ad}_h\xi, \cdot), \eta\rangle. \quad \blacksquare\end{aligned}$$

4.4.2 The Evens–Lu momentum map. This momentum map, introduced in EVENS AND LU (2001), could be seen as a generalization of the Chu map. This construction takes place in the category of Poisson Lie groups and associates to each such object G a momentum map into some Poisson space (the Lagrangian subalgebras of the Drinfeld double of the Lie bialgebra obtained as the Lie algebra of the double group DG) for every Poisson action of the given group.

4.5 The standard momentum map

The Chu map is always well-defined as soon as we have a canonical action on a symplectic manifold. However, it presents the disadvantage of not reproducing well-known conservations laws and of giving trivial results in some relevant situations. For instance, when the Lie algebra $\mathfrak{g}$ is Abelian, part **(ii)** of Proposition 4.4.1 implies that $\ker T_m \Psi = T_m M$ and hence Ψ is just a constant map on the connected components of M.

Some of these drawbacks are overcome with the use of the ***standard momentum map*** to which we will refer just as the ***momentum map*** if there is no danger of confusion. The ideas underlying the momentum map go back to S. Lie (see the second volume of LIE (1890)). The modern definition of this object is due to Kostant, Souriau, and Kirillov (see MARSDEN AND RATIU (1999) for a brief account of the history of this definition as well as for valuable references). It was actually J.-M. Souriau who first linked it to the momentum maps appearing in physics and who coined the name (in French) *"application moment"* (see SOURIAU (1965)). Following the French denomination, several authors (see for instance GUILLEMIN AND STERNBERG (1984c)) use the term ***moment map*** to refer to the momentum map. Our decision to use the word *momentum* is based on the fact that, as we will see below, the momentum map is a generalization of the standard linear and angular momenta in classical mechanics and that the physics terminology uses *momentum* for these conserved quantities as opposed to *moment*.

The definition of the momentum map only requires, as it was already the case for the Chu map, a canonical Lie algebra action. Its existence is guaranteed when the infinitesimal generators of this action are Hamiltonian vector fields. In other words, if the Lie algebra $\mathfrak{g}$ acts canonically on the Poisson manifold $(M, \{\cdot, \cdot\})$, then, for each $\xi \in \mathfrak{g}$, we require the existence of a globally defined function $\mathbf{J}^\xi \in C^\infty(M)$, such that

$$\xi_M = X_{\mathbf{J}^\xi}.$$

4.5.1 Definition *Let $\mathfrak{g}$ be a Lie algebra acting canonically on the Poisson manifold $(M, \{\cdot, \cdot\})$. Suppose that for any $\xi \in \mathfrak{g}$, the vector field ξ_M is globally Hamiltonian, with Hamiltonian function $\mathbf{J}^\xi \in C^\infty(M)$. The map $\mathbf{J} : M \to \mathfrak{g}^*$ defined by the relation*

$$\langle \mathbf{J}(z), \xi \rangle = \mathbf{J}^\xi(z),$$

for all $\xi \in \mathfrak{g}$ and $z \in M$, is called a ***standard momentum map*** *of the $\mathfrak{g}$-action.*

Notice that momentum maps are not uniquely determined; indeed, $\mathbf{J}_1$ and $\mathbf{J}_2$ are momentum maps for the same canonical action if and only if for any $\xi \in \mathfrak{g}$

$$\mathbf{J}_1^\xi - \mathbf{J}_2^\xi \in \mathcal{C}(M).$$

Obviously, if M is symplectic and connected, then $\mathbf{J}$ is determined up to a constant in $\mathfrak{g}^*$.

4.5.2 Example: linear momentum. Take the phase space of the N-particle system, that is, $T^*\mathbb{R}^{3N}$. The additive group $\mathbb{R}^3$ acts on it by spatial translation on each factor:

$v \cdot (q_i, p^i) = (q_i + v, p^i)$, with $i = 1, \ldots, N$. This action is canonical and has an associated momentum map that coincides with the classical ***linear momentum***

$$\begin{aligned} \mathbf{J} : T^*\mathbb{R}^{3N} &\longrightarrow \mathrm{Lie}(\mathbb{R}^3) \simeq \mathbb{R}^3 \\ (q_i, p^i) &\longmapsto \textstyle\sum_{i=1}^N p_i. \end{aligned}$$

4.5.3 Example: angular momentum. Let SO(3) act on $\mathbb{R}^3$ and then, by lift, on $T^*\mathbb{R}^3$, that is, $A \cdot (q, p) = (Aq, Ap)$. This action is canonical and has an associated momentum map

$$\begin{aligned} \mathbf{J} : T^*\mathbb{R}^3 &\longrightarrow \mathfrak{so}(3)^* \simeq \mathbb{R}^3 \\ (q, p) &\longmapsto q \times p, \end{aligned}$$

which is the classical ***angular momentum***.

4.5.4 Example: lifted actions on cotangent bundles. The previous two examples are particular cases of the following situation. Let G be a Lie group acting on the manifold Q and then by lift on its cotangent bundle T^*Q. Any such lifted action is canonical with respect to the canonical symplectic form on T^*Q and has an associated momentum map $\mathbf{J} : T^*Q \to \mathfrak{g}^*$ given by

$$\langle \mathbf{J}(\alpha_q), \xi \rangle = \langle \alpha_q, \xi_Q(q) \rangle, \tag{4.5.1}$$

for any $\alpha_q \in T^*Q$ and any $\xi \in \mathfrak{g}$.

4.5.5 Example: The lifted action on the cotangent bundle of a Lie group. This is a particular case of the previous example that we study in light of the body coordinates introduced in §4.1.20. Consider the action of a Lie group on its cotangent bundle T^*G by the lift of the left translations. In body coordinates this action takes the expression $g \cdot (h, \mu) := (gh, \mu)$, for any $g, h \in G$, $\mu \in \mathfrak{g}^*$. By the previous example, this action is canonical with respect to the symplectic form in (4.1.13) and has an associated momentum map given by the expression (4.5.1). In body coordinates this momentum map takes the form

$$\mathbf{J}(g, \mu) = \mathrm{Ad}^*_{g^{-1}} \mu.$$

4.5.6 Example: lifted actions on tangent bundles. Let G be a Lie group acting on the manifold Q and then by lift on its tangent bundle TQ. Let $L : TQ \to \mathbb{R}$ be a function (a Lagrangian) such that its ***fiber derivative*** or ***Legendre transform*** $\mathbb{F}L : TQ \to T^*Q$, defined by

$$\langle \mathbb{F}L(v_q), w_q \rangle = \left.\frac{d}{dt}\right|_{t=0} L(v_q + t w_q), \qquad v_q, w_q \in T_q Q,$$

makes the form $\omega_L := (\mathbb{F}L)^*\omega$ into a symplectic form for TQ, where ω is the canonical symplectic form on T^*Q (such a Lagrangian is said to be ***regular*** or ***nondegenerate***). Then any such lifted action of G on TQ is canonical with respect to ω_L and has an associated momentum map given by

$$\langle \mathbf{J}(v_q), \xi \rangle = \langle \mathbb{F}L(v_q), \xi_Q(q) \rangle,$$

for any $v_q \in TQ$ and any $\xi \in \mathfrak{g}$.

4.5.7 Example: symplectic linear actions. Let (V, ω) be a symplectic linear space and let G be a subgroup of the linear symplectic group, acting naturally on V. By the choice of G this action is canonical and has a momentum map given by the expression

$$\langle \mathbf{J}(v), \xi \rangle = \frac{1}{2}\omega(\xi_V(v), v),$$

for $\xi \in \mathfrak{g}$ and $v \in V$ arbitrary.

4.5.8 Momentum maps induced by Lie subgroups. Let G be a Lie group acting canonically on the Poisson manifold $(M, \{\cdot, \cdot\})$ and H a Lie subgroup of G. Let $i : \mathfrak{h} \to \mathfrak{g}$ be the injection induced by the tangent space at the identity of the inclusion $H \hookrightarrow G$ and let $i^* : \mathfrak{g}^* \to \mathfrak{h}^*$ be the dual projection. The H-action on $(M, \{\cdot, \cdot\})$ by restriction of the G-action is canonical and, moreover, if the G-action has an associated momentum map $\mathbf{J} : M \to \mathfrak{g}^*$, then the composition $i^* \circ \mathbf{J} : M \to \mathfrak{h}^*$ is a momentum map for the H-action.

4.5.9 Momentum maps and equivariant Poisson diffeomorphisms. Let $(M_1, \{\cdot, \cdot\}_1)$ and $(M_2, \{\cdot, \cdot\}_2)$ be two Poisson manifolds and let $\mathfrak{g}$ be a Lie algebra that acts canonically on both M_1 and M_2. Let $\phi : M_1 \to M_2$ be a canonical $\mathfrak{g}$-equivariant diffeomorphism. If $\mathbf{J}_1 : M_1 \to \mathfrak{g}^*$ is a momentum map for the $\mathfrak{g}$-action on M_1, then the map $\mathbf{J}_2 : M_2 \to \mathfrak{g}^*$ defined by $\mathbf{J}_2 := \mathbf{J}_1 \circ \phi^{-1}$ is a momentum map for the $\mathfrak{g}$-action on M_2. Conversely, if $\phi : M_1 \to M_2$ is a canonical $\mathfrak{g}$-equivariant immersion and $\mathbf{J}_2 : M_2 \to \mathfrak{g}^*$ is a momentum map for the $\mathfrak{g}$-action on M_2, then $\mathbf{J}_1 := \mathbf{J}_2 \circ \phi$ is a momentum map for the $\mathfrak{g}$-action on M_1.

4.5.10 Properties of the standard momentum maps. The first feature of standard momentum maps is that they satisfy Noether's condition. This fact was identified by NOETHER (1918) for Lagrangian systems and by SOURIAU (1953) and SMALE (1970) for Hamiltonian systems. Due to its relevance we frame it as a theorem.

4.5.11 Theorem *Let $\mathfrak{g}$ be a Lie algebra acting canonically on the Poisson manifold $(M, \{\cdot, \cdot\})$. Assume that this action admits a momentum map $\mathbf{J} : M \to \mathfrak{g}^*$. Then the momentum map is a constant of the motion for the Hamiltonian vector field associated to any $\mathfrak{g}$-invariant function $h \in C^\infty(M)^{\mathfrak{g}}$, that is, it satisfies Noether's condition* (4.3.1).

Proof. It suffices to notice that for any $\xi \in \mathfrak{g}$ and any $m \in M$ we have

$$\{f, \mathbf{J}^\xi\}(m) = \mathbf{d}f(m) \cdot X_{\mathbf{J}^\xi}(m) = \mathbf{d}f(m) \cdot \xi_M(m) = \xi_M[f](m) = 0. \quad \blacksquare$$

4.5.12 Proposition *Let $\mathfrak{g}$ be a Lie algebra acting canonically on the Poisson manifold $(M, \{\cdot, \cdot\})$ and $\mathbf{J} : M \to \mathfrak{g}^*$ an associated momentum map. Let $m \in M$ and let $\mathcal{L} \subset M$ be the symplectic leaf containing m. Then*

$$T_m\mathbf{J} \cdot T_m\mathcal{L} = (\mathfrak{g}_m)^\circ, \tag{4.5.2}$$

where $(\mathfrak{g}_m)^\circ$ denotes the annihilator in $\mathfrak{g}^*$ of the isotropy subalgebra $\mathfrak{g}_m$ of m. When M is a symplectic manifold the expression (4.5.2) *can be rewritten as*

$$\text{range}\,(T_m\mathbf{J}) = (\mathfrak{g}_m)^\circ. \tag{4.5.3}$$

Relation (4.5.3) appears for the first time in ARMS et al. (1981) and it is sometimes referred to in the literature as the ***bifurcation lemma*** since it establishes a link between the symmetry of a point and the rank of the momentum map at that point.

Proof. Let $v_m \in T_m\mathcal{L}$, $\xi \in \mathfrak{g}_m$, and let $f \in C^\infty(M)$ be such that $v_m = X_f(m)$. Then,

$$\begin{aligned}\langle T_m\mathbf{J}\cdot v_m, \xi\rangle &= \langle T_m\mathbf{J}\cdot X_f(m), \xi\rangle = \mathbf{dJ}^\xi(m)\cdot X_f(m)\\ &= \{\mathbf{J}^\xi, f\}(m) = -\mathbf{d}f(m)\cdot X_{\mathbf{J}^\xi}(m) = -\mathbf{d}f(m)\cdot \xi_M(m) = 0,\end{aligned}$$

which proves that $T_m\mathbf{J}\cdot T_m\mathcal{L}\subset(\mathfrak{g}_m)^\circ$.

We now show that $(\mathfrak{g}_m)^\circ\subset T_m\mathbf{J}\cdot T_m\mathcal{L}$ or, equivalently, that $(T_m\mathbf{J}\cdot T_m\mathcal{L})^\circ\subset\mathfrak{g}_m$. Let $\xi\in(T_m\mathbf{J}\cdot T_m\mathcal{L})^\circ$, that is, $\xi\in\mathfrak{g}$ is such that for any $f\in C^\infty(M)$ we have that

$$0 = \langle T_m\mathbf{J}\cdot X_f(m), \xi\rangle = \mathbf{dJ}^\xi(m)\cdot X_f(m) = \{\mathbf{J}^\xi, f\}(m) = -\mathbf{d}f(m)\cdot\xi_M(m).$$

Since the function f is arbitrary $\xi_M(m) = 0$ necessarily, and hence $\xi\in\mathfrak{g}_m$.

Relation (4.5.3) is a straightforward corollary of (4.5.2) if we take into account that whenever M is a symplectic manifold $T_m\mathcal{L} = T_mM$. ■

4.5.13 Corollary *Let $\mathbf{J}: M\to\mathfrak{g}^*$ be a momentum map associated to a locally free canonical $\mathfrak{g}$-action on the Poisson manifold $(M, \{\cdot,\cdot\})$. Then $\mathbf{J}$ is a submersion onto some open subset of $\mathfrak{g}^*$.*

Proof. Since the action is locally free we have that $\mathfrak{g}_m = \{0\}$ for any $m\in M$. In this situation Proposition 4.5.12 guarantees that $\mathbf{J}$ is a submersion and thereby an open map. In particular $\mathbf{J}(M)$ is an open subset of $\mathfrak{g}_m^*$. ■

4.5.14 Proposition *Let $\mathfrak{g}$ be a Lie algebra acting canonically on the Poisson manifold $(M, \{\cdot,\cdot\})$, $B\in\Lambda^2(T^*M)$ the corresponding Poisson tensor, and $\mathbf{J}: M\to\mathfrak{g}^*$ an associated momentum map. Let $m\in M$ and let $(\mathcal{L}, \omega_\mathcal{L})$ be the symplectic leaf containing m. Then*

(i) $\langle T_m\mathbf{J}\cdot v_m, \xi\rangle = \omega_\mathcal{L}(\xi_M(m), v_m)$, *for any* $v_m\in T_m\mathcal{L}$,

(ii) $\ker T_m\mathbf{J}\cap T_m\mathcal{L} = (\mathfrak{g}\cdot m)^{\omega_\mathcal{L}}$, *and*

(iii) $B^\sharp(m)((\mathfrak{g}\cdot m)^\circ)\subset\ker T_m\mathbf{J}$.

Proof. **(i)** is straightforward. As to **(ii)** notice that by **(i)** the vector $v_m\in\ker T_m\mathbf{J}\cap T_m\mathcal{L}$ if and only if $\omega_\mathcal{L}(\xi_M(m), v_m) = 0$, for all $\xi\in\mathfrak{g}$, and hence $v_m\in(\mathfrak{g}\cdot m)^{\omega_\mathcal{L}}$. To prove **(iii)** consider $v_m\in B^\sharp(m)((\mathfrak{g}\cdot m)^\circ)$ arbitrary. By definition v_m can be written as $v_m = X_f(m)$ with $f\in C^\infty(M)$ such that $\mathbf{d}f(m)|_{\mathfrak{g}\cdot m} = 0$. Consequently, for any $\xi\in\mathfrak{g}$

$$\langle T_m\mathbf{J}\cdot v_m, \xi\rangle = \mathbf{dJ}^\xi(m)\cdot X_f(m) = -\mathbf{d}f(m)\cdot\xi_M(m) = 0,$$

which implies that $v_m\in\ker T_m\mathbf{J}$. ■

4.5.15 Proposition *Let $\mathfrak{g}$ be a Lie algebra acting canonically on the Poisson manifold $(M, \{\cdot,\cdot\})$ and $\mathbf{J}: M\to\mathfrak{g}^*$ an associated momentum map. Let $\mathcal{L}$ be a symplectic leaf of $(M, \{\cdot,\cdot\})$. Then*

(i) *The* $\mathfrak{g}$*-action on* $(M, \{\cdot, \cdot\})$ *restricts to a canonical* $\mathfrak{g}$*-action on* $(\mathcal{L}, \omega_{\mathcal{L}})$.

(ii) $\mathbf{J}_{\mathcal{L}} := \mathbf{J}|_{\mathcal{L}} : \mathcal{L} \to \mathfrak{g}^*$ *is a momentum map for this action.*

Proof. **(i)** Let $m \in \mathcal{L}$ and $\xi \in \mathfrak{g}$ arbitrary. Then $\xi_M(m) = X_{\mathbf{J}^\xi}(m) \in T_m\mathcal{L}$, hence the $\mathfrak{g}$-action leaves $\mathcal{L}$ invariant and thus $\mathfrak{g}$ acts canonically on $\mathcal{L}$ since the inclusion $\mathcal{L} \hookrightarrow M$ is a Poisson map. **(ii)** follows from the canonical character of the inclusion $\mathcal{L} \hookrightarrow M$ and the discussion in §4.5.9. ■

4.5.16 The existence of standard momentum maps. As we saw in Proposition 4.4.1 the Chu momentum map presents the advantage of always existing in the presence of a canonical Lie algebra action and of being automatically equivariant. These two features are an issue in the framework of standard momentum maps that we will address in the following subsections.

We start with the problem of the existence of standard momentum maps; in order to show that this is not guaranteed we consider the following well-known example. Let $\mathbb{T}^2 = \{(e^{i\theta_1}, e^{i\theta_2})\}$ be the two-torus considered as a symplectic manifold with the area form $\omega := \mathbf{d}\theta_1 \wedge \mathbf{d}\theta_2$. The circle $S^1 = \{e^{i\phi}\}$ acts canonically on $\mathbb{T}^2$ by $e^{i\phi} \cdot (e^{i\theta_1}, e^{i\theta_2}) := (e^{i(\theta_1+\phi)}, e^{i\theta_2})$. However, a straightforward verification shows that the infinitesimal generators associated to this action cannot be integrated to a Hamiltonian vector field.

In the case of canonical actions on Poisson manifolds, the situation is even more complicated as far as the existence of standard momentum maps is concerned since we have to superimpose to the integration problem the condition on the action being leaf preserving. In other words, the condition on an action being canonical does not necessarily imply that it preserves the symplectic leaves, which is a necessary condition for the existence of a momentum map by its very definition. An elementary example that illustrates this point very well is the canonical group action presented in §4.2.5. In that case, the group orbits are transversal to the symplectic leaves and therefore a momentum map cannot possibly exist.

The following result fully characterizes the existence of momentum maps in the symplectic case.

4.5.17 Proposition *Let* (M, ω) *be a symplectic manifold and* $\mathfrak{g}$ *a Lie algebra acting canonically on it. There exists a momentum map associated to this action if and only if the linear map*

$$\begin{array}{rccc} \rho: & \mathfrak{g}/[\mathfrak{g}, \mathfrak{g}] & \longrightarrow & H^1(M, \mathbb{R}) \\ & [\xi] & \longmapsto & [\mathbf{i}_{\xi_M}\omega] \end{array}$$

is identically zero.

Proof. We start by showing that the map ρ is well defined. It suffices to show that if $\eta \in [\mathfrak{g}, \mathfrak{g}]$, then $\mathbf{i}_{\eta_M}\omega$ is exact. The element η can be written as a sum of brackets of elements in $\mathfrak{g}$. For simplicity suppose that $\eta = [\xi, \zeta]$. Now, as the Lie algebra action is canonical we have that $\pounds_{\xi_M}\omega = \pounds_{\zeta_M}\omega = 0$, that is, ξ_M and ζ_M are locally Hamiltonian vector fields and consequently their Lie bracket $[\xi_M, \zeta_M]$ is Hamiltonian with associated Hamiltonian function $\omega(\zeta_M, \xi_M)$. Therefore, $\mathbf{i}_{\eta_M}\omega = \mathbf{i}_{[\xi,\zeta]_M}\omega = -\mathbf{i}_{[\xi_M, \zeta_M]}\omega = -\mathbf{d}(\omega(\zeta_M, \xi_M))$, which is an exact form, as required. The map ρ is clearly linear.

Now, notice that there exists a momentum map $\mathbf{J} : M \to \mathfrak{g}^*$ if and only if for any $\xi \in \mathfrak{g}$ we can write $\mathbf{i}_{\xi_M}\omega = \mathbf{d}\mathbf{J}^\xi$. This is equivalent to $[\mathbf{i}_{\xi_M}\omega] = 0$, which in turn can be rewritten as $\rho[\xi] = 0$, for any $\xi \in \mathfrak{g}$. ■

This result allows us to immediately identify several situations where standard momentum maps automatically exist, namely $H^1(M, \mathbb{R}) = 0$ or $\mathfrak{g}/[\mathfrak{g}, \mathfrak{g}] \simeq H^1(\mathfrak{g}, \mathbb{R}) = 0$ trivially imply that $\rho \equiv 0$. In particular, if $\mathfrak{g}$ is semisimple, the First Whitehead Lemma implies that $H^1(\mathfrak{g}, \mathbb{R}) = 0$ and therefore a standard momentum map always exists.

One can find in the literature other results that, in more specific situations, characterize the existence of standard momentum maps in different terms. A particularly beautiful one, due to FRANKEL (1959), says that a canonical circle action on a Kähler manifold has a momentum map if and only if it has fixed points. MCDUFF (1988) has proved that this phenomenon is purely Kähler, that is, it does not hold for a general symplectic manifold unless the manifold in question has dimension four.

4.5.18 The equivariance properties of standard momentum maps. Let $(M, \{\cdot, \cdot\})$ be a Poisson manifold and $\mathfrak{g}$ a Lie algebra acting canonically on it with a momentum map $\mathbf{J} : M \to \mathfrak{g}^*$. A natural question to ask is when the map $(\mathfrak{g}, [\cdot, \cdot]) \to (C^\infty(M), \{\cdot, \cdot\})$ defined by $\xi \mapsto \mathbf{J}^\xi$, $\xi \in \mathfrak{g}$, is a Lie algebra homomorphism, that is,

$$\mathbf{J}^{[\xi, \eta]} = \{\mathbf{J}^\xi, \mathbf{J}^\eta\}, \quad \xi, \eta \in \mathfrak{g}. \tag{4.5.4}$$

A straightforward computation shows that this is the case if and only if

$$T_z\mathbf{J} \cdot \xi_M(z) = -\operatorname{ad}^*_\xi \mathbf{J}(z), \tag{4.5.5}$$

for any $\xi \in \mathfrak{g}$ and any $z \in M$. A momentum map that satisfies this relation in called ***infinitesimally equivariant***. The reason behind this terminology is in the fact that this is the infinitesimal version of the ***global*** or ***coadjoint equivariance*** that can be formulated when the Lie algebra action is associated to the action of a Lie group G. In this situation we say that $\mathbf{J}$ is G***-equivariant*** when

$$\operatorname{Ad}^*_{g^{-1}} \circ \mathbf{J} = \mathbf{J} \circ \Phi_g, \tag{4.5.6}$$

for all $g \in G$ or, equivalently $\mathbf{J}^{\operatorname{Ad}_g \xi}(g \cdot z) = \mathbf{J}^\xi(z)$, for all $g \in G$, $\xi \in \mathfrak{g}$, and $z \in M$. Notice that (4.5.5) is the derivative of (4.5.6), with respect to g at $g = e$ in the direction ξ.

Lie algebra actions admitting infinitesimally equivariant momentum maps appear frequently in the literature as ***Hamiltonian actions*** and Lie group actions with coadjoint equivariant momentum maps are called ***globally Hamiltonian actions***. It can be proved (see for instance GUILLEMIN AND STERNBERG (1984c); MARSDEN AND RATIU (1999); SOURIAU (1969)) that if the symmetry group G is connected, then global and infinitesimal equivariance of the momentum map are equivalent concepts.

The reader should notice that all momentum maps in the examples 4.5.2 through 4.5.7 are equivariant.

Since momentum maps are not uniquely defined, one may ask whether one can choose them to be equivariant. It turns out that if the momentum map in question is associated to the action of a compact Lie group this can always be done, as we show below.

4.5.19 Proposition (MONTALDI (1997a)) *Let G be a compact Lie group acting canonically on the Poisson manifold $(M, \{\cdot, \cdot\})$ with an associated momentum map $\mathbf{J} : M \to \mathfrak{g}^*$. Then there exists a momentum map that is equivariant.*

Proof. For each $g \in G$, $z \in M$ define

$$\mathbf{J}_g(z) = \mathrm{Ad}^*_{g^{-1}} \mathbf{J}(g^{-1} \cdot z).$$

or, equivalently,

$$\mathbf{J}_g^{\xi} = \mathbf{J}^{\mathrm{Ad}_{g^{-1}}\xi} \circ \Phi_{g^{-1}}.$$

Then $\mathbf{J}_g$ is also a momentum map for the G-action on M. Indeed, if $z \in M$, $\xi \in \mathfrak{g}$, and $f \in C^{\infty}(M)$ we have

$$\begin{aligned}
\{f, \mathbf{J}_g^{\xi}\}(z) &= -\mathbf{d}\mathbf{J}_g^{\xi}(z) \cdot X_f(z) \\
&= -\mathbf{d}\mathbf{J}^{\mathrm{Ad}_{g^{-1}}\xi}(g^{-1} \cdot z) \cdot T_z\Phi_{g^{-1}} \cdot X_f(z) \\
&= -\mathbf{d}\mathbf{J}^{\mathrm{Ad}_{g^{-1}}\xi}(g^{-1} \cdot z) \cdot (\Phi_g^* X_f)(g^{-1} \cdot z) \\
&= -\mathbf{d}\mathbf{J}^{\mathrm{Ad}_{g^{-1}}\xi}(g^{-1} \cdot z) \cdot X_{\Phi_g^* f}(g^{-1} \cdot z) \\
&= \{\Phi_g^* f, \mathbf{J}^{\mathrm{Ad}_{g^{-1}}\xi}\}(g^{-1} \cdot z) \\
&= (\mathrm{Ad}_{g^{-1}}\xi)_M[\Phi_g^* f](g^{-1} \cdot z) \\
&= (\Phi_g^* \xi_M)[\Phi_g^* f](g^{-1} \cdot z) \\
&= \mathbf{d}f(z) \cdot \xi_M(z) \\
&= \{f, \mathbf{J}^{\xi}\}(z).
\end{aligned}$$

Therefore, $\{f, \mathbf{J}_g^{\xi} - \mathbf{J}^{\xi}\} = 0$ for every $f \in C^{\infty}(M)$, that is, $\mathbf{J}_g^{\xi} - \mathbf{J}^{\xi}$ is a Casimir function on M for every $g \in G$ and every $\xi \in \mathfrak{g}$. Now define

$$\langle \mathbf{J} \rangle = \int_G \mathbf{J}_g dg$$

where dg denotes the Haar measure on G normalized such that the total volume of G is one. Equivalently, this definition states that

$$\langle \mathbf{J} \rangle^{\xi} = \int_G \mathbf{J}_g^{\xi} dg$$

for every $\xi \in \mathfrak{g}$. By the linearity of the Poisson bracket in each factor, it follows that

$$\{f, \langle \mathbf{J} \rangle^{\xi}\} = \int_G \{f, \mathbf{J}_g^{\xi}\} dg = \int_G \{f, \mathbf{J}^{\xi}\} dg = \{f, \mathbf{J}^{\xi}\}.$$

Thus $\langle \mathbf{J} \rangle$ is also a momentum map for the G-action on M and $\langle \mathbf{J} \rangle^{\xi} - \mathbf{J}^{\xi}$ is a Casimir function on M for every $\xi \in \mathfrak{g}$, that is, $\langle \mathbf{J} \rangle - \mathbf{J} \in L(\mathfrak{g}, \mathcal{C}(M))$.

The momentum map $\langle \mathbf{J} \rangle$ is equivariant. Noting that $\mathbf{J}_g(h \cdot z) = \mathrm{Ad}^*_{h^{-1}} \mathbf{J}_{h^{-1}g}(z)$ and using invariance of the Haar measure on G under translations, for any $h \in G$, we indeed have, after changing variables $k := h^{-1}g$ in the third equality below,

$$\begin{aligned}\langle \mathbf{J} \rangle (h \cdot z) &= \int_G \mathrm{Ad}^*_{h^{-1}} \mathbf{J}_{h^{-1}g}(z) dg = \mathrm{Ad}^*_{h^{-1}} \int_G \mathbf{J}_{h^{-1}g}(z) dg \\ &= \mathrm{Ad}^*_{h^{-1}} \int_G \mathbf{J}_k(z) dk = \mathrm{Ad}^*_{h^{-1}} \langle \mathbf{J} \rangle (z). \quad \blacksquare\end{aligned}$$

Another result related to the possibility of choosing equivariant momentum maps is the one that states that a *canonical action of a semisimple Lie algebra on a symplectic manifold admits an infinitesimally equivariant momentum map.* The equivariance part of this result uses the Second Whitehead lemma which says that $H^2(\mathfrak{g}, \mathbb{R}) = 0$ if $\mathfrak{g}$ is semisimple (see ABRAHAM AND MARSDEN (1978); GUILLEMIN AND STERNBERG (1984c); MARSDEN AND RATIU (1999); SOURIAU (1969)). We shall identify below a specific element of $H^2(\mathfrak{g}, \mathbb{R})$ which is the obstruction to the equivariance of a momentum map (assuming it exists).

4.5.20 Cocycles and affine actions. Given that, in general, it is not possible to choose a coadjoint equivariant momentum map, we could ask if we can define another action on this space with respect to which we have equivariance. In the following lines we shall see that the answer is affirmative. All along this subsection we will restrict our discussion to canonical actions on connected symplectic manifolds. For the Poisson case see MARSDEN AND RATIU (1999).

4.5.21 Proposition *Let (M, ω) be a connected symplectic manifold and G a Lie group acting on M in a canonical fashion with an associated momentum map $\mathbf{J} : M \to \mathfrak{g}^*$. Define the* ***non-equivariance one-cocycle*** *associated to $\mathbf{J}$ as the map*

$$\begin{array}{rcl}\sigma : G & \longrightarrow & \mathfrak{g}^* \\ g & \longmapsto & \mathbf{J}(\Phi_g(z)) - \mathrm{Ad}^*_{g^{-1}}(\mathbf{J}(z)).\end{array}$$

Then:

(i) *The definition of σ does not depend on the choice of $z \in M$;*

(ii) *The mapping σ is a $\mathfrak{g}^*$-valued one-cocycle on G with respect to the coadjoint representation of G on $\mathfrak{g}^*$ (see §2.1.19).*

(iii) *If $\mathbf{J}'$ is another momentum map for the same canonical action of G on M, then its non-equivariance one-cocycle σ' is in the same Lie group cohomology class as σ; that is, $\sigma - \sigma'$ is a one-coboundary.*

Proof. **(i)** Let $g \in G$ be fixed and $\tau_g : M \to \mathfrak{g}^*$ the mapping defined by $\tau_g(z) = \mathbf{J}(\Phi_g(z)) - \mathrm{Ad}^*_{g^{-1}}(\mathbf{J}(z))$, $z \in M$. Now, for any $\xi \in \mathfrak{g}$ and $v_z \in T_zM$ we have that

$$
\begin{aligned}
\langle T_z\tau_g \cdot v_z, \xi\rangle &= \mathbf{d}\mathbf{J}^{\xi}(g\cdot z)\cdot T_z\Phi_g\cdot v_z - \mathbf{d}\mathbf{J}^{\mathrm{Ad}_{g^{-1}}\xi}(z)\cdot v_z \\
&= \omega(g\cdot z)(\xi_M(g\cdot z), T_z\Phi_g\cdot v_z) - \omega(z)((\mathrm{Ad}_{g^{-1}}\xi)_M(z), v_z) \\
&= \omega(g\cdot z)(T_z\Phi_g\cdot(\mathrm{Ad}_{g^{-1}}\xi)_M(z), T_z\Phi_g\cdot v_z) \\
&\quad - \omega(z)((\mathrm{Ad}_{g^{-1}}\xi)_M(z), v_z) \\
&= (\Phi_g^*\omega)(z)((\mathrm{Ad}_{g^{-1}}\xi)_M(z), v_z) - \omega(z)((\mathrm{Ad}_{g^{-1}}\xi)_M(z), v_z) \\
&= \omega(z)((\mathrm{Ad}_{g^{-1}}\xi)_M(z), v_z) - \omega(z)((\mathrm{Ad}_{g^{-1}}\xi)_M(z), v_z) = 0.
\end{aligned}
$$

Since v_z and ξ are arbitrary, this shows that $T\tau_g = 0$ and hence, as M is connected, the function τ_g is constant. This proves that the definition of σ does not depend on the choice of $z \in M$.
(ii) On the one hand we have that for any $g, h \in G$

$$\sigma(gh) = \mathbf{J}(gh\cdot z) - \mathrm{Ad}^*_{(gh)^{-1}}\mathbf{J}(z). \tag{4.5.7}$$

Using the independence of the definition of σ on the choice of the point in the manifold (proved in point **(i)**) we take the point $h\cdot z$ and we write $\sigma(g) = \mathbf{J}(gh\cdot z) - \mathrm{Ad}^*_{g^{-1}}\mathbf{J}(h\cdot z)$. We now take the point $z \in M$ and write $\sigma(h) = \mathbf{J}(h\cdot z) - \mathrm{Ad}^*_{h^{-1}}\mathbf{J}(z)$. Hence $\sigma(g) + \mathrm{Ad}^*_{g^{-1}}\sigma(h) = \mathbf{J}(gh\cdot z) - \mathrm{Ad}^*_{(gh)^{-1}}\mathbf{J}(z)$ which, by (4.5.7), coincides with $\sigma(gh)$ establishing the cocycle identity.
(iii) The defining property of a momentum map implies that for any $\xi \in \mathfrak{g}$, $\mathbf{d}(\mathbf{J}^{\xi} - \mathbf{J}'^{\xi}) = \mathbf{i}_{\xi_M}\omega - \mathbf{i}_{\xi_M}\omega = 0$. The connectedness of M implies that $\mathbf{J} - \mathbf{J}'$ is a constant function. Now, using this fact, we have that for any $g \in G$:

$$
\begin{aligned}
\sigma(g) - \sigma'(g) &= \mathbf{J}(g\cdot z) - \mathbf{J}'(g\cdot z) - \mathrm{Ad}^*_{g^{-1}}(\mathbf{J}(z) - \mathbf{J}'(z)) \\
&= \mathbf{J}(z) - \mathbf{J}'(z) - \mathrm{Ad}^*_{g^{-1}}(\mathbf{J}(z) - \mathbf{J}'(z)).
\end{aligned}
$$

Hence, if we set $\mu = \mathbf{J}(z) - \mathbf{J}'(z)$ we have that $\sigma(g) - \sigma(g') = \mu - \mathrm{Ad}^*_{g^{-1}}\mu$, which is a coboundary. ■

4.5.22 Remark This proposition identifies the cohomology class $[\sigma]$ in the first group cohomology as the obstruction to the equivariance of a momentum map. If the Lie group G is semisimple, Whitehead's Lemma for groups implies that any non-equivariance one-cocycle is actually a coboundary. Therefore, by part **(iii)** of the proposition, any momentum map can be modified in this case to be G-equivariant.

Using the non-equivariance one-cocycle we can define a new action of G on $\mathfrak{g}^*$, with respect to which a given momentum map $\mathbf{J}$ is equivariant.

4.5.23 Definition *Let G be a Lie group acting canonically on the connected symplectic manifold (M, ω) with associated momentum map $\mathbf{J} : M \to \mathfrak{g}^*$. If $\sigma : G \to \mathfrak{g}^*$ is the non-equivariance one-cocycle of $\mathbf{J}$, we define the* ***affine action*** *of G on $\mathfrak{g}^*$ with cocycle σ by*

$$
\begin{aligned}
\Theta : G \times \mathfrak{g}^* &\longrightarrow \mathfrak{g}^* \\
(g, \mu) &\longmapsto \mathrm{Ad}^*_{g^{-1}}\mu + \sigma(g).
\end{aligned}
$$

4.5.24 Proposition *The map Θ in the previous definition determines a left action of G on $\mathfrak{g}^*$. The momentum map $\mathbf{J} : M \to \mathfrak{g}^*$ is equivariant with respect to the symplectic action Φ on M and the affine action Ψ on $\mathfrak{g}^*$.*

Proof. Θ is trivially an action as a consequence of σ satisfying the cocycle identity. The equivariance of $\mathbf{J}$ with respect to this action is obvious. ■

In the previous proposition we defined a group cocycle that measured the lack of coadjoint equivariance; it was used to define a new action on $\mathfrak{g}^*$ with respect to which the momentum map is equivariant. Something similar can be done at the infinitesimal level. Indeed, we saw before that the infinitesimal coadjoint equivariance of a momentum map is equivalent to this object defining a Lie algebra homomorphism. In the next result we show that the mathematical object that measures the lack of infinitesimal equivariance is a Lie algebra two-cocycle that, in the presence of a canonical group action, can be obtained as the derivative of the non-equivariance group cocycle.

4.5.25 Theorem *Let $\mathfrak{g}$ be a Lie algebra acting canonically on the connected symplectic manifold (M, ω) with momentum map $\mathbf{J} : M \to \mathfrak{g}^*$. Define the* ***infinitesimal non-equivariance two-cocycle*** *associated to $\mathbf{J}$ as the element $\Sigma \in \Lambda^2(\mathfrak{g})$ given by*

$$\Sigma(\xi, \eta) = \mathbf{J}^{[\xi, \eta]}(z) - \{\mathbf{J}^{\xi}, \mathbf{J}^{\eta}\}(z), \quad z \in M, \quad \xi, \eta \in \mathfrak{g}. \tag{4.5.8}$$

Then:

(i) *The definition of Σ does not depend on the choice of $z \in M$.*

(ii) *For any $\xi, \eta \in \mathfrak{g}$ we have $X_{\{\mathbf{J}^{\xi}, \mathbf{J}^{\eta}\}} = X_{\mathbf{J}^{[\xi, \eta]}}$.*

(iii) *$\Sigma \in Z^2(\mathfrak{g}, \mathbb{R})$, that is, Σ is a Lie algebra two-cocycle (see* (2.1.18)*), that is, it satisfies the* ***two-cocycle identity***

$$\Sigma([\xi, \eta], \zeta) + \Sigma([\eta, \zeta], \xi) + \Sigma([\zeta, \xi], \eta) = 0 \tag{4.5.9}$$

for all $\xi, \eta, \zeta \in \mathfrak{g}$.

(iv) *For arbitrary $z \in M$ and $\eta \in \mathfrak{g}$, we have*

$$T_z\mathbf{J} \cdot \eta_M(z) = -\mathrm{ad}_{\eta}^*\mathbf{J}(z) + \Sigma(\eta, \cdot). \tag{4.5.10}$$

(v) *If $\mathbf{J} : M \to \mathfrak{g}^*$ is a momentum map associated to the canonical action of a Lie group G that has $\sigma : G \to \mathfrak{g}^*$ as non-equivariance cocycle, then the infinitesimal non-equivariance cocycle $\Sigma \in Z^2(\mathfrak{g}, \mathbb{R})$ is given by*

$$\begin{aligned} \Sigma : \mathfrak{g} \times \mathfrak{g} &\longrightarrow \mathbb{R} \\ (\xi, \eta) &\longmapsto \Sigma(\xi, \eta) = \mathbf{d}\widehat{\sigma}_{\eta}(e) \cdot \xi, \end{aligned} \tag{4.5.11}$$

where $\widehat{\sigma}_{\eta} : G \to \mathbb{R}$ is defined by $\widehat{\sigma}_{\eta}(g) = \langle \sigma(g), \eta \rangle$.

(vi) *In the hypotheses of* **(v)** *for any $\xi, \eta \in \mathfrak{g}$ and any $g \in G$, the following identity holds:*

$$\Sigma(\mathrm{Ad}_g \xi, \mathrm{Ad}_g \eta) = \Sigma(\xi, \eta) + \langle \mathrm{Ad}_g^* \sigma(g), [\xi, \eta] \rangle. \tag{4.5.12}$$

Proof. **(i)** Define the function $\tau_{\xi,\eta} \in C^\infty(M)$ by $\tau_{\xi,\eta}(z) = \mathbf{J}^{[\xi,\eta]}(z) - \{\mathbf{J}^\xi, \mathbf{J}^\eta\}(z)$. We now compute its derivative:

$$\begin{aligned}\mathbf{d}\tau_{\xi,\eta} &= \mathbf{dJ}^{[\xi,\eta]} - \mathbf{d}(\{\mathbf{J}^\xi, \mathbf{J}^\eta\}) = \mathbf{i}_{[\xi,\eta]_M}\omega - \mathbf{i}_{X_{\{\mathbf{J}^\xi,\mathbf{J}^\eta\}}}\omega \\ &= -\mathbf{i}_{[\xi_M,\eta_M]}\omega + \mathbf{i}_{[X_{\mathbf{J}^\xi},X_{\mathbf{J}^\eta}]}\omega = -\mathbf{i}_{[\xi_M,\eta_M]}\omega + \mathbf{i}_{[\xi_M,\eta_M]}\omega = 0.\end{aligned}$$

The connectedness of M implies that $\tau_{\xi,\eta}$ is constant and that therefore the definition of Σ does not depend on the point $z \in M$ used in its definition.
(ii) By point **(i)** the functions $\{\mathbf{J}^\xi, \mathbf{J}^\eta\}$ and $\mathbf{J}^{[\xi,\eta]}$ differ by a constant, and hence $X_{\{\mathbf{J}^\xi,\mathbf{J}^\eta\}} = X_{\mathbf{J}^{[\xi,\eta]}}$.
(iii) It is a straightforward consequence of the Jacobi identities satisfied by the brackets $[\cdot,\cdot]$ and $\{\cdot,\cdot\}$, as well as of point **(ii)**.
(iv) For any $z \in M$ and $\xi, \eta \in \mathfrak{g}$, we have that

$$\begin{aligned}\langle T_z\mathbf{J} \cdot \eta_M(z), \xi\rangle &= \mathbf{dJ}^\xi(z) \cdot \eta_M(z) = \{\mathbf{J}^\xi, \mathbf{J}^\eta\}(z) = \mathbf{J}^{[\xi,\eta]}(z) - \Sigma(\xi,\eta) \\ &= -\langle \mathbf{J}(z), \mathrm{ad}_\eta \xi\rangle + \Sigma(\eta,\xi) = -\langle \mathrm{ad}^*_\eta \mathbf{J}(z), \xi\rangle + \Sigma(\eta,\xi).\end{aligned}$$

Since the element ξ is arbitrary, the relation follows.
(v) Using the relation in the previous point, as well as the definition of the non-equivariance cocycle, we can write, for any $\xi, \eta \in \mathfrak{g}$

$$\begin{aligned}\Sigma(\xi,\eta) &= \langle T_z\mathbf{J} \cdot \xi_M(z), \eta\rangle + \langle \mathrm{ad}^*_\xi \mathbf{J}(z), \eta\rangle \\ &= \left.\frac{d}{dt}\right|_{t=0} \langle \mathbf{J}(\exp t\xi \cdot z), \eta\rangle + \langle \mathrm{ad}^*_\xi \mathbf{J}(z), \eta\rangle \\ &= \left.\frac{d}{dt}\right|_{t=0} (\langle \sigma(\exp t\xi), \eta\rangle + \langle \mathrm{Ad}^*_{\exp(-t\xi)}\mathbf{J}(z), \eta\rangle) + \langle \mathrm{ad}^*_\xi \mathbf{J}(z), \eta\rangle \\ &= \left.\frac{d}{dt}\right|_{t=0} \langle \sigma(\exp t\xi), \eta\rangle = \mathbf{d}\widehat{\sigma}_\eta(e) \cdot \xi.\end{aligned}$$

(vi) The defining property of a momentum map and the identity (2.2.3) yield

$$X_{\mathbf{J}^{\mathrm{Ad}_g \eta}} = (\mathrm{Ad}_g \eta)_M = \Phi^*_{g^{-1}}\eta_M = T\Phi_g \circ \eta_M \circ \Phi_{g^{-1}},$$

where $\Phi : G \times M \to M$ denotes the given canonical Lie group action. Thus for any $z \in M$ we get

$$\begin{aligned}\{\mathbf{J}^{\mathrm{Ad}_g \xi}, \mathbf{J}^{\mathrm{Ad}_g \xi}\}(z) &= \mathbf{dJ}^{\mathrm{Ad}_g \xi}(z)\left(T_{g\cdot z}\Phi_g(\eta_M(g^{-1}\cdot z))\right) \\ &= \mathbf{d}\left(\mathbf{J}^{\mathrm{Ad}_g \xi} \circ \Phi_g\right)(g^{-1}\cdot z)\left(\eta_M(g^{-1}\cdot z)\right).\end{aligned} \tag{4.5.13}$$

Since $\mathbf{J}$ is equivariant relative to the affine action of G on $\mathfrak{g}^*$, we have

$$\begin{aligned}\mathbf{J}^{\mathrm{Ad}_g \xi}(g \cdot z) &= \langle \mathbf{J}(g \cdot z), \mathrm{Ad}_g \xi\rangle = \langle g \cdot \mathbf{J}(z), \mathrm{Ad}_g \xi\rangle \\ &= \langle \mathrm{Ad}^*_{g^{-1}}(\mathbf{J}(z)) + \sigma(g), \mathrm{Ad}_g \xi\rangle \\ &= \mathbf{J}^\xi(z) + \langle \sigma(g), \mathrm{Ad}_g \xi\rangle,\end{aligned}$$

which shows that $\mathbf{d}\left(\mathbf{J}^{\mathrm{Ad}_g \xi} \circ \Phi_g\right) = \mathbf{d}\mathbf{J}^{\xi}$, since the second summand is a constant function on M. Therefore we continue (4.5.13) to get, again from the defining property of a momentum map,

$$\mathbf{d}\mathbf{J}^{\xi}(g^{-1} \cdot z)\left(X_{\mathbf{J}^{\eta}}(g^{-1} \cdot z)\right) = \{\mathbf{J}^{\xi}, \mathbf{J}^{\eta}\}(g^{-1} \cdot z).$$

We have obtained the following identity

$$\{\mathbf{J}^{\mathrm{Ad}_g \xi}, \mathbf{J}^{\mathrm{Ad}_g \xi}\}(z) = \{\mathbf{J}^{\xi}, \mathbf{J}^{\eta}\}(g^{-1} \cdot z). \tag{4.5.14}$$

Next we compute the other term in the two-cocycle using the definition of the affine action and the equivariance of $\mathbf{J}$ relative to it. We have

$$\begin{aligned}
\mathbf{J}^{[\mathrm{Ad}_g \xi, \mathrm{Ad}_g \eta]}(z) &= \langle \mathbf{J}(z), \mathrm{Ad}_g[\xi, \eta]\rangle = \langle \mathrm{Ad}_g^* \mathbf{J}(z), [\xi, \eta]\rangle \\
&= \langle \mathrm{Ad}_g^* \mathbf{J}(z) + \sigma(g^{-1}) - \sigma(g^{-1}), [\xi, \eta]\rangle \\
&= \langle g^{-1} \cdot \mathbf{J}(z) - \sigma(g^{-1}), [\xi, \eta]\rangle \\
&= \langle \mathbf{J}(g^{-1} \cdot z), [\xi, \eta]\rangle - \langle \sigma(g^{-1}), [\xi, \eta]\rangle \\
&= \mathbf{J}^{[\xi, \eta]}(g^{-1} \cdot z) - \langle \sigma(g^{-1}), [\xi, \eta]\rangle \\
&= \mathbf{J}^{[\xi, \eta]}(g^{-1} \cdot z) + \langle \mathrm{Ad}_g^* \sigma(g), [\xi, \eta]\rangle.
\end{aligned} \tag{4.5.15}$$

The last equality is obtained in the following way. Recall that $\sigma(e) = 0$ and the cocycle identity $\sigma(hg) = \sigma(h) + \mathrm{Ad}_{h^{-1}}^* \sigma(g)$. Put here $h = g^{-1}$ to get $\mathrm{Ad}_g^* \sigma(g) = -\sigma(g^{-1})$.

Thus, from (4.5.14), (4.5.15), and the fact that Σ does not depend at what point of the connected manifold M it is computed (proved in **(i)**), we get for any $z \in M$

$$\begin{aligned}
\Sigma(\mathrm{Ad}_g \xi, \mathrm{Ad}_g \eta) &= \mathbf{J}^{[\mathrm{Ad}_g \xi, \mathrm{Ad}_g \eta]}(z) - \{\mathbf{J}^{\mathrm{Ad}_g \xi}, \mathbf{J}^{\mathrm{Ad}_g \xi}\}(z) \\
&= \mathbf{J}^{[\xi, \eta]}(g^{-1} \cdot z) + \langle \mathrm{Ad}_g^* \sigma(g), [\xi, \eta]\rangle - \{\mathbf{J}^{\xi}, \mathbf{J}^{\eta}\}(g^{-1} \cdot z) \\
&= \Sigma(\xi, \eta) + \langle \mathrm{Ad}_g^* \sigma(g), [\xi, \eta]\rangle
\end{aligned}$$

which proves the desired formula. ■

4.5.26 Remark The previous proposition identifies the cohomology class $[\Sigma] \in H^2(\mathfrak{g}, \mathbb{R})$ as the obstruction to the infinitesimal equivariance of a momentum map. If Σ is a coboundary, then a given momentum map can be modified to be infinitesimally equivariant. This happens if the Lie algebra $\mathfrak{g}$ is semisimple, because then the Second Whitehead Lemma states that $H^2(\mathfrak{g}, \mathbb{R}) = 0$. Thus, the two Whitehead Lemmas ensure that any semisimple canonical Lie algebra action admits an infinitesimally equivariant momentum map.

4.5.27 The affine Lie–Poisson space and the momentum map as a canonical mapping. Let $\Sigma \in Z^2(\mathfrak{g}; \mathbb{R})$ be a two-cocycle on $\mathfrak{g}$. It defines the central extension $\mathfrak{g}_\Sigma := \mathfrak{g} \oplus \mathbb{R}$ by setting

$$[(\xi, s), (\eta, t)] = ([\xi, \eta], -\Sigma(\xi, \eta)), \tag{4.5.16}$$

where $\xi, \eta \in \mathfrak{g}$ and $s, t \in \mathbb{R}$. Formula (4.5.16) defines a Lie bracket if and only if Σ is a two-cocycle, as an easy verification shows. The pairing

$$\langle (\mu, a), (\xi, t) \rangle := \langle \mu, \xi \rangle + at,$$

where $\xi \in \mathfrak{g}$, $\mu \in \mathfrak{g}^*$, and $t, a \in \mathbb{R}$ identifies $\mathfrak{g}_\Sigma^*$ with $\mathfrak{g}^* \oplus \mathbb{R}$. The $\pm$-Lie–Poisson bracket of $\bar{f}, \bar{h} : \mathfrak{g}_\Sigma^* \to \mathbb{R}$ is therefore given by

$$\begin{aligned}
\{\bar{f}, \bar{h}\}_\pm(\mu, a) &= \pm \left\langle (\mu, a), \left[\frac{\delta \bar{f}}{\delta(\mu, a)}, \frac{\delta \bar{h}}{\delta(\mu, a)} \right] \right\rangle \\
&= \pm \left\langle (\mu, a), \left[\left(\frac{\delta \bar{f}}{\delta \mu}, \frac{\partial \bar{f}}{\partial a} \right), \left(\frac{\delta \bar{h}}{\delta \mu}, \frac{\partial \bar{h}}{\partial a} \right) \right] \right\rangle \\
&= \pm \left\langle (\mu, a), \left(\left[\frac{\delta \bar{f}}{\delta \mu}, \frac{\delta \bar{h}}{\delta \mu} \right], -\Sigma \left(\frac{\delta \bar{f}}{\delta \mu}, \frac{\delta \bar{h}}{\delta \mu} \right) \right) \right\rangle \\
&= \pm \left\langle \mu, \left[\frac{\delta \bar{f}}{\delta \mu}, \frac{\delta \bar{h}}{\delta \mu} \right] \right\rangle \mp a \Sigma \left(\frac{\delta \bar{f}}{\delta \mu}, \frac{\delta \bar{h}}{\delta \mu} \right),
\end{aligned} \tag{4.5.17}$$

where $\delta \bar{f}/\delta \mu, \delta \bar{h}/\delta \mu \in \mathfrak{g}$ denote the partial functional derivatives of $\bar{f}$ and $\bar{h}$ relative to $\mu \in \mathfrak{g}^*$ (see §4.1.13) and $\partial \bar{f}/\partial a, \partial \bar{h}/\partial a \in \mathbb{R}$ denote the usual partial derivatives of $\bar{f}$ and $\bar{h}$ relative to the real variable a. Thus the Hamiltonian vector field of $\bar{h} : \mathfrak{g}_\Sigma^* \to \mathbb{R}$ has the expression

$$X_{\bar{h}}(\mu, a) = \left(\mp \operatorname{ad}^*_{\frac{\delta \bar{h}}{\delta \mu}} \mu \pm a \Sigma \left(\frac{\delta \bar{h}}{\delta \mu}, \cdot \right), 0 \right). \tag{4.5.18}$$

Notice that for each $a \in \mathbb{R}$, $\mathfrak{g}^* \oplus \{a\}$ is a Poisson submanifold of $\mathfrak{g}_\Sigma^*$ if on the first factor one defines the Poisson bracket

$$\{f, h\}_\pm^{a\Sigma}(\mu) := \pm \left\langle \mu, \left[\frac{\delta f}{\delta \mu}, \frac{\delta h}{\delta \mu} \right] \right\rangle \mp a \Sigma \left(\frac{\delta f}{\delta \mu}, \frac{\delta h}{\delta \mu} \right) \tag{4.5.19}$$

for any two smooth functions $f, h : \mathfrak{g}^* \to \mathbb{R}$. The case $a = 1$ will be important in what follows and (4.5.19) will be called in this case the ***affine Lie–Poisson bracket***. The Hamiltonian vector field defined by $h : \mathfrak{g}^* \to \mathbb{R}$ is hence given (for $a = 1$) by

$$X_h^\Sigma(\mu) = \mp \operatorname{ad}^*_{\frac{\delta h}{\delta \mu}} \mu \pm \Sigma \left(\frac{\delta h}{\delta \mu}, \cdot \right). \tag{4.5.20}$$

The ***affine Lie–Poisson space*** determined by the two-cocycle $\Sigma \in Z^2(\mathfrak{g}; \mathbb{R})$ is defined as the vector space $\mathfrak{g}^*$ endowed with the Poisson bracket

$$\{f, g\}_\pm^\Sigma(\mu) := \pm \left\langle \mu, \left[\frac{\delta f}{\delta \mu}, \frac{\delta g}{\delta \mu} \right] \right\rangle \mp \Sigma \left(\frac{\delta f}{\delta \mu}, \frac{\delta g}{\delta \mu} \right), \tag{4.5.21}$$

for $f, g \in C^\infty(\mathfrak{g}^*)$ and $\mu \in \mathfrak{g}^*$. The brackets (4.5.21) are also called the $\pm\Sigma$-***Lie–Poisson structures***.

4.5.28 Theorem *Let $\mathfrak{g}$ be a Lie algebra acting canonically on the connected symplectic manifold (M, ω) with momentum map $\mathbf{J} : M \to \mathfrak{g}^*$. Suppose that this momentum map has an associated infinitesimal non-equivariance two-cocycle $\Sigma \in Z^2(\mathfrak{g}, \mathbb{R})$. Denote by $\{\cdot, \cdot\}$ the Poisson bracket defined by the symplectic form ω. Then the momentum map $\mathbf{J} : (M, \{\cdot, \cdot\}) \longrightarrow (\mathfrak{g}^*, \{\cdot, \cdot\}_+^\Sigma)$ is canonical. Conversely, if M is connected, any canonical map $\mathbf{J} : (M, \{\cdot, \cdot\}) \longrightarrow (\mathfrak{g}^*, \{\cdot, \cdot\}_+^\Sigma)$ is a momentum map of a Poisson action of $\mathfrak{g}$ on M with Σ as its infinitesimal non-equivariance two-cocycle.*

Proof. For $f, g \in C^\infty(\mathfrak{g}^*)$ and $m \in M$, let $\mu := \mathbf{J}(m)$ and $\xi := \delta f/\delta\mu$, $\eta := \delta g/\delta\mu$. Then, using (4.5.8) in the last equality below, we get

$$\begin{aligned}
\mathbf{J}^*\{f, h\}_+^\Sigma(m) &= \{f, h\}_+^\Sigma(\mu) = \left\langle \mu, \left[\frac{\delta f}{\delta\mu}, \frac{\delta g}{\delta\mu}\right]\right\rangle - \Sigma\left(\frac{\delta f}{\delta\mu}, \frac{\delta g}{\delta\mu}\right) \\
&= \langle \mu, [\xi, \eta]\rangle - \Sigma(\xi, \eta) = \langle \mathbf{J}(m), [\xi, \eta]\rangle - \Sigma(\xi, \eta) \\
&= \mathbf{J}^{[\xi,\eta]}(m) - \Sigma(\xi, \eta) \\
&= \{\mathbf{J}^\xi, \mathbf{J}^\eta\}(m). \qquad (4.5.22)
\end{aligned}$$

However, for any $m \in M$ and $v_m \in T_m M$,

$$\mathbf{d}(f \circ \mathbf{J})(m)(v_m) = \mathbf{d}f(\mu)\big(T_m\mathbf{J}(v_m)\big) = \left\langle \frac{\delta f}{\delta\mu}, T_m\mathbf{J}(v_m)\right\rangle = \mathbf{d}\mathbf{J}^\xi(m)(v_m),$$

that is, $f \circ \mathbf{J}$ and $\mathbf{J}^\xi$ have identical derivatives at the point m. Since the Poisson bracket depends only on the point values of the first derivatives, this implies that

$$\{f \circ \mathbf{J}, g \circ \mathbf{J}\}(m) = \{\mathbf{J}^\xi, \mathbf{J}^\eta\}(m). \qquad (4.5.23)$$

Equations (4.5.22) and (4.5.23) prove that $\mathbf{J} : (M, \{\cdot, \cdot\}) \longrightarrow (\mathfrak{g}^*, \{\cdot, \cdot\}_+^\Sigma)$ is canonical.

Conversely, assume that $\mathbf{J} : (M, \{\cdot, \cdot\}) \longrightarrow (\mathfrak{g}^*, \{\cdot, \cdot\}_+^\Sigma)$ is canonical, that is, $\mathbf{J}^*\{f, h\}_+^\Sigma = \{f \circ \mathbf{J}, g \circ \mathbf{J}\}$ for any $f, g \in C^\infty(\mathfrak{g}^*)$. Define for any $\xi \in \mathfrak{g}$ the vector field $\xi_M := X_{\mathbf{J}^\xi}$. By (4.1.6), (4.5.8), and Theorem 4.5.25**(ii)**, for any $\xi, \eta \in \mathfrak{g}$ we have $[\xi_M, \eta_M] = [X_{\mathbf{J}^\xi}, X_{\mathbf{J}^\eta}] = -X_{\{\mathbf{J}^\xi, \mathbf{J}^\eta\}} = -X_{\mathbf{J}^{[\xi,\eta]}} = -[\xi, \eta]_M$, thereby proving that the prescription $\xi \in \mathfrak{g} \mapsto \xi_M \in \mathfrak{X}(M)$ is a Lie algebra action of $\mathfrak{g}$ on M. This action is canonical since for any $\xi \in \mathfrak{g}$ and any $F_1, F_2 \in C^\infty(M)$ we have by the Jacobi identity

$$\begin{aligned}
\xi_M[\{F_1, F_2\}] &= X_{\mathbf{J}^\xi}[\{F_1, F_2\}] = \{\{F_1, F_2\}, \mathbf{J}^\xi\} \\
&= -\{\{F_2, \mathbf{J}^\xi\}, F_1\} - \{\{\mathbf{J}^\xi, F_1\}, F_2\} \\
&= \{F_1, X_{\mathbf{J}^\xi}[F_2]\} + \{X_{\mathbf{J}^\xi}[F_1], F_2\}.
\end{aligned}$$

By definition, this canonical action admits $\mathbf{J}$ as a momentum map. Finally, to compute the infinitesimal non-equivariance two-cocycle, let $\xi, \eta \in \mathfrak{g}$ be arbitrary, fix an element $m \in M$, define $\mu = \mathbf{J}(m)$, and choose two functions $f, g \in C^\infty(M)$ such that $\xi := \delta f/\delta\mu$, $\eta := \delta g/\delta\mu$. Then (4.5.22), (4.5.23), and the hypothesis that $\mathbf{J}$ is canonical show that

$$\begin{aligned}
&\mathbf{J}^{[\xi,\eta]}(m) - \{\mathbf{J}^\xi, \mathbf{J}^\eta\}(m) \\
&\qquad = \mathbf{J}^*\{f, h\}_+^\Sigma(m) + \Sigma(\xi, \eta) - \{f \circ \mathbf{J}, g \circ \mathbf{J}\}(m) = \Sigma(\xi, \eta).
\end{aligned}$$

Since Theorem 4.5.25(**i**) guarantees via the connectedness of M that the left-hand side does not depend on the point $m \in M$, this shows that Σ is indeed the infinitesimal non-equivariance two-cocycle associated to $\mathbf{J}$. ■

4.5.29 Symplectic structures of coadjoint and affine orbits. Due to their importance in the geometry of Lie–Poisson spaces we now study the properties of the orbits of the affine action introduced in Definition 4.5.23.

Assume that Σ is a real valued Lie algebra two-cocycle, that is, Σ is skew symmetric and satisfies the cocycle identity $\Sigma(\xi, [\eta, \zeta]) + \Sigma(\eta, [\zeta, \xi]) + \Sigma(\zeta, [\xi, \eta]) = 0$ for all $\xi, \eta, \zeta \in \mathfrak{g}$. As was shown in Theorem 4.5.25, if σ is the non-equivariance one-cocycle and Σ is the infinitesimal non-equivariance two-cocycle associated to a momentum map $\mathbf{J} : M \to \mathfrak{g}^*$ on a connected symplectic manifold M, this condition holds.

4.5.30 Lemma *Let G be a Lie group, $\sigma : G \to \mathfrak{g}^*$ a coadjoint one-cocycle, and $\mu \in \mathfrak{g}^*$. Let $\mathcal{O}_\mu$ the orbit through the point μ of the affine G-action on $\mathfrak{g}^*$ associated to σ. Then the orbit $\mathcal{O}_\mu$ is an initial submanifold of $\mathfrak{g}^*$ and for any $\nu \in \mathcal{O}_\mu$ we have that*

$$T_\nu \mathcal{O}_\mu = \{\xi_{\mathfrak{g}^*}(\nu) := -\mathrm{ad}^*_\xi \nu + \Sigma(\xi, \cdot) \mid \xi \in \mathfrak{g}\}, \tag{4.5.24}$$

where $\xi_{\mathfrak{g}^*}$ denotes the infinitesimal generator associated to $\xi \in \mathfrak{g}$ and Σ defined by (4.5.11) *is assumed to be a real valued Lie algebra two-cocycle.*

Proof. The initial character of $\mathcal{O}_\mu$ follows from Proposition 2.3.12. The expression for the infinitesimal generator $\xi_{\mathfrak{g}^*}$ is a straightforward consequence of the definition in (4.5.11). ■

The symplectic structure described in the following theorem is associated to various names: Lie, Borel, Weil and, more recently, ARNOLD (1966a), KIRILLOV (1976a), KOSTANT (1965), and SOURIAU (1969).

4.5.31 Theorem *Let G be a Lie group, $\sigma : G \to \mathfrak{g}^*$ a coadjoint one-cocycle, and $\mu \in \mathfrak{g}^*$. Let $\mathcal{O}_\mu$ be the orbit through the point μ of the affine G-action on $\mathfrak{g}^*$ associated to σ. Assume that the bilinear map $\Sigma : \mathfrak{g} \times \mathfrak{g} \to \mathbb{R}$ defined by* (4.5.11) *is a real valued Lie algebra two-cocycle. Then the affine orbit $\mathcal{O}_\mu$ is a symplectic manifold with G-invariant symplectic structure $\omega^{\pm}_{\mathcal{O}_\mu}$ given by*

$$\omega^{\pm}_{\mathcal{O}_\mu}(\nu)(\xi_{\mathfrak{g}^*}(\nu), \eta_{\mathfrak{g}^*}(\nu)) = \pm\langle \nu, [\xi, \eta]\rangle \mp \Sigma(\xi, \eta), \tag{4.5.25}$$

for arbitrary $\nu \in \mathcal{O}_\mu$, and $\xi, \eta \in \mathfrak{g}$. The symplectic structures $\omega^{\pm}_{\mathcal{O}_\mu}$ on $\mathcal{O}_\mu$ are usually called the **±-orbit** *or* **Kostant–Kirillov–Souriau (KKS) symplectic forms**. *The connected components of $(\mathcal{O}_\mu, \omega^{\pm}_{\mathcal{O}_\mu})$ are the symplectic leaves of $(\mathfrak{g}^*, \{\cdot, \cdot\}^{\Sigma}_{\pm})$. The symbols $\mathcal{O}^+_\mu$ and $\mathcal{O}^-_\mu$ will denote the pairs $(\mathcal{O}_\mu, \omega^{+}_{\mathcal{O}_\mu})$ and $(\mathcal{O}_\mu, \omega^{-}_{\mathcal{O}_\mu})$, respectively.*

Proof. The theorem will be proved for the +-orbit symplectic structure by following the standard method found, for example, in MARSDEN AND RATIU (1999). The proof consists of several steps.

First, one shows that (4.5.25) yields a well-defined two-form on $\mathcal{O}_\mu$, that is, one proves that the right-hand side is independent of the choices $\xi, \eta \in \mathfrak{g}$ that define the tangent vectors $\xi_{\mathfrak{g}^*}(\nu), \eta_{\mathfrak{g}^*}(\nu) \in T_\nu \mathcal{O}_\mu$. From (4.5.24) it follows that $\xi_{\mathfrak{g}^*}(\nu) = \xi'_{\mathfrak{g}^*}(\nu)$ if and only if $-\langle \nu, [\xi, \zeta]\rangle + \Sigma(\xi, \zeta) = -\langle \nu, [\xi', \zeta]\rangle + \Sigma(\xi', \zeta)$, for any $\zeta \in \mathfrak{g}$, which shows that $\omega^+_{\mathcal{O}_\mu}(\nu)(\xi_{\mathfrak{g}^*}(\nu), \zeta_{\mathfrak{g}^*}(\nu)) = \omega^+_{\mathcal{O}_\mu}(\nu)(\xi'_{\mathfrak{g}^*}(\nu), \zeta_{\mathfrak{g}^*}(\nu))$ for any $\zeta \in \mathfrak{g}$. This proves the independence of (4.5.25) on ξ and, by skew symmetry, also on η.

Second, one shows that $\omega^+_{\mathcal{O}_\mu}(\nu)$ is not a degenerate two-form on $T_\nu \mathcal{O}_\mu$. Indeed, $\omega^+_{\mathcal{O}_\mu}(\nu)(\xi_{\mathfrak{g}^*}(\nu), \eta_{\mathfrak{g}^*}(\nu)) = \langle \nu, [\xi, \eta]\rangle - \Sigma(\xi, \eta) = 0$, for all $\eta \in \mathfrak{g}$, is equivalent by (4.5.24) to $0 = \operatorname{ad}^*_\xi \nu - \Sigma(\xi, \cdot) = -\xi_{\mathfrak{g}^*}(\nu)$, which proves the claim.

Third, one shows that $\omega^+_{\mathcal{O}_\mu}$ is G-invariant. Let $\Theta : (g, \nu) \in G \times \mathfrak{g}^* \mapsto \operatorname{Ad}^*_{g^{-1}} \nu + \sigma(g) \in \mathfrak{g}^*$ be the affine action associated to the non-equivariance group one-cocycle $\sigma : G \to \mathfrak{g}^*$ given in Definition 4.5.23. Then by (2.2.3), $T_\nu \Theta_g(\xi_{\mathfrak{g}^*}(\nu)) = (\operatorname{Ad}_g \xi)_{\mathfrak{g}^*}(g \cdot \nu)$, so using (4.5.12), we get

$$
\begin{aligned}
(\Theta_g^* \omega^+_{\mathcal{O}_\mu})(\nu)(\xi_{\mathfrak{g}^*}(\nu), \eta_{\mathfrak{g}^*}(\nu)) &= \omega^+_{\mathcal{O}_\mu}(g \cdot \nu)\left(T_\nu \Phi_g(\xi_{\mathfrak{g}^*}(\nu)), T_\nu \Phi_g(\eta_{\mathfrak{g}^*}(\nu))\right) \\
&= \omega^+_{\mathcal{O}_\mu}(g \cdot \nu)\left((\operatorname{Ad}_g \xi)_{\mathfrak{g}^*}(g \cdot \nu), (\operatorname{Ad}_g \eta)_{\mathfrak{g}^*}(g \cdot \nu)\right) \\
&= \langle g \cdot \nu, [\operatorname{Ad}_g \xi, \operatorname{Ad}_g \eta]\rangle - \Sigma(\operatorname{Ad}_g \xi, \operatorname{Ad}_g \eta) \\
&= \langle \operatorname{Ad}^*_{g^{-1}} \nu + \sigma(g), \operatorname{Ad}_g[\xi, \eta]\rangle - \Sigma(\xi, \eta) - \langle \operatorname{Ad}^*_g \sigma(g), [\xi, \eta]\rangle \\
&= \langle \nu, [\xi, \eta]\rangle - \Sigma(\xi, \eta) \\
&= \omega^+_{\mathcal{O}_\mu}(\nu)(\xi_{\mathfrak{g}^*}(\nu), \eta_{\mathfrak{g}^*}(\nu)).
\end{aligned}
$$

Fourth, for $\nu \in \mathcal{O}_\mu$ define the smooth surjective map $\Theta^\nu : G \to \mathcal{O}_\mu$ by $\Theta^\nu(g) := \Theta(g, \nu) = \operatorname{Ad}^*_{g^{-1}} \nu + \sigma(g)$ whose derivative has, by (4.5.11), the expression $T_e\Theta^\nu(\xi) = -\operatorname{ad}^*_\xi \nu + \Sigma(\xi, \cdot) = \xi_{\mathfrak{g}^*}(\nu)$. Therefore Θ^ν has surjective derivative at the identity element. In addition, since $\Theta : G \times \mathcal{O}_\mu \to \mathcal{O}_\mu$ is a left action, the map Θ^ν is equivariant, that is $\Theta^\nu \circ L_g = \Theta_g \circ \Theta^\nu$. Taking the derivative of this relation at the identity element and using the chain rule shows that the tangent map $T_g \Theta^\nu$ is also surjective. Thus Θ^ν is a surjective submersion.

Finally, one shows that $\omega^+_{\mathcal{O}_\mu}$ is closed. This is done in the following manner. For each $\nu \in \mathfrak{g}^*$ define the left invariant one-form $\nu_L \in \Omega^1(G)$ by $\nu_L(g) := (T_g^* L_{g^{-1}})(\nu)$, that is, $L_g^* \nu_L = \nu_L$ for all $g \in G$. If $\xi \in \mathfrak{g}$, denote by $\xi_L \in \mathfrak{X}(G)$ the left invariant vector field whose value at the identity element is ξ, that is, $\xi_L(g) := T_e L_g(\xi)$. In particular, $\nu_L(\xi_L) = \langle \nu, \xi\rangle$ is a constant function on G. Define $\lambda := \Theta^{\nu *} \omega^+_{\mathcal{O}_\mu}$. Then, for any $\xi, \eta \in \mathfrak{g}$ we have

$$
\begin{aligned}
\lambda(\xi_L, \eta_L)(e) &= \left(\Theta^{\nu *} \omega^+_{\mathcal{O}_\mu}\right)(e)(\xi, \eta) = \omega^+_{\mathcal{O}_\mu}(\Theta^\nu(e))\left(T_e \Theta^\nu(\xi), T_e \Theta^\nu(\eta)\right) \\
&= \omega^+_{\mathcal{O}_\mu}(\nu)(\xi_{\mathfrak{g}^*}(\nu), \eta_{\mathfrak{g}^*}(\nu)) = \langle \nu, [\xi, \eta]\rangle = \nu_L([\xi_L, \eta_L])(e)
\end{aligned}
$$

since $[\xi_L, \eta_L] = [\xi, \eta]_L$. Using the equivariance of Θ^ν and the G-invariance of $\omega^+_{\mathcal{O}_\mu}$ we can conclude that for any $g \in G$

$$
\begin{aligned}
L_g^* \lambda = L_g^* \Theta^{\nu *} \omega^+_{\mathcal{O}_\mu} &= (\Theta^\nu \circ L_g)^* \omega^+_{\mathcal{O}_\mu} = (\Theta_g \circ \Theta^\nu)^* \omega^+_{\mathcal{O}_\mu} \\
&= \Theta^{\nu *} \Theta_g^* \omega^+_{\mathcal{O}_\mu} = \Theta^{\nu *} \omega^+_{\mathcal{O}_\mu} = \lambda,
\end{aligned}
$$

that is, λ is G-invariant. This immediately implies that

$$\lambda(\xi_L, \eta_L) = \nu_L([\xi_L, \eta_L])$$

since both sides of this equation are G-invariant functions and they coincide at the identity. The proof is finished if one shows that $\lambda = -\mathbf{d}\nu_L$. Indeed

$$\Theta^{\nu *}\mathbf{d}\omega^+_{\mathcal{O}_\mu} = \mathbf{d}\Theta^{\nu *}\omega^+_{\mathcal{O}_\mu} = \mathbf{d}\lambda = -\mathbf{d}^2\nu_L = 0.$$

Since Θ^ν is a surjective submersion, the map $\Theta^{\nu *}$ on forms is injective and one concludes that $\mathbf{d}\omega^+_{\mathcal{O}_\mu} = 0$.

To show that $\lambda = -\mathbf{d}\nu_L$ one recalls the formula of the exterior derivative of a one-form α

$$(\mathbf{d}\alpha)(X, Y) = X[\alpha(Y)] - Y[\alpha(X)] - \alpha([X, Y]),$$

for any vector fields X and Y. We apply this identity at $g \in G$ for $\alpha = \nu_L$ and where we denote in the computation below $\xi := T_g L_{g^{-1}}(X(g))$ and $\eta := T_g L_{g^{-1}}(Y(g))$. Note that $\xi_L[\nu_L(\eta_L)] = 0$ since $\nu_L(\eta_L)$ is constant and similarly $\eta_L[\nu_L(\xi_L)] = 0$. Thus we get, by G-invariance of λ and ν_L,

$$\begin{aligned}
\lambda(X, Y)(g) &= (L^*_{g^{-1}}\lambda)(g)(X(g), Y(g)) = \lambda(e)(\xi, \eta) = \lambda(\xi_L, \eta_L)(e) \\
&= \nu_L([\xi_L, \eta_L])(e) \\
&= -\xi_L[\nu_L(\eta_L)](e) + \eta_L[\nu_L(\xi_L)](e) + \nu_L([\xi_L, \eta_L])(e) \\
&= -(\mathbf{d}\nu_L)(\xi_L, \eta_L)(e) = -(L^*_g\mathbf{d}\nu_L)(\xi_L, \eta_L)(e) \\
&= -\mathbf{d}\nu_L(g)\left(T_e L_g(\xi), T_e L_g(\eta)\right) = -\mathbf{d}\nu_L(g)\left(X(g), Y(g)\right) \\
&= -\mathbf{d}\nu_L\,(X, Y)\,(g),
\end{aligned}$$

which concludes the proof that $\omega^+_{\mathcal{O}_\mu}$ is a symplectic form on $\mathcal{O}^+_\mu$.

It remains to show that the connected components of $\mathcal{O}^+_\mu$ are the symplectic leaves of $(\mathfrak{g}^*, \{\cdot, \cdot\}^\Sigma_+)$. For a smooth function $h : \mathfrak{g}^* \to \mathbb{R}$ we shall compute the Hamiltonian vector field associated to the restriction $h_\mu := h|_{\mathcal{O}_\mu}$ relative to the affine orbit symplectic form given by (4.5.25). Since X_{h_μ} is tangent to $\mathcal{O}_\mu$, for any $\nu = \mathrm{Ad}^*_{g^{-1}}\mu + \sigma(g) \in \mathcal{O}_\mu$, we necessarily have $X_{h_\mu}(\nu) = -\mathrm{ad}^*_{\eta(\nu)}\nu + \Sigma(\eta(\nu), \cdot)$ for some $\eta(\nu) \in \mathfrak{g}$ by (4.5.24). For any $\zeta \in \mathfrak{g}$ we have

$$\begin{aligned}
&\mathbf{d}h_\mu(\nu)\left(-\mathrm{ad}^*_\zeta \nu + \Sigma(\zeta, \cdot)\right) \\
&\quad = \omega^+_{\mathcal{O}_\mu}(\nu)(-\mathrm{ad}^*_{\eta(\nu)}(\nu) + \Sigma(\eta(\nu), \cdot),\ -\mathrm{ad}^*_\zeta(\nu) + \Sigma(\zeta, \cdot)) \\
&\quad = \langle \nu,\ [\eta(\nu), \zeta]\rangle - \Sigma(\eta(\nu), \zeta).
\end{aligned}$$

Therefore,

$$\begin{aligned}
\left\langle -\mathrm{ad}^*_\zeta(\nu) + \Sigma(\zeta, \cdot), \frac{\delta h}{\delta \nu}\right\rangle &= \mathbf{d}h(\nu)\left(-\mathrm{ad}^*_\zeta(\nu) + \Sigma(\zeta, \cdot)\right) \\
&= \langle \nu, [\eta(\nu), \zeta]\rangle - \Sigma(\eta(\nu), \zeta) \\
&= \left\langle -\mathrm{ad}^*_\zeta(\nu) + \Sigma(\zeta, \cdot), \eta(\nu)\right\rangle,
\end{aligned}$$

and hence

$$\eta(\nu) - \frac{\delta h}{\delta \nu} \in \{-\operatorname{ad}^*_\zeta(\nu) + \Sigma(\zeta, \cdot) \mid \zeta \in \mathfrak{g}\}^\circ = (T_\mu \mathcal{O}_\mu)^\circ = \mathfrak{g}_\nu$$

by (4.5.26). Thus

$$\eta(\nu) = \frac{\delta h}{\delta \nu} + \xi, \qquad \text{where } \xi \in \mathfrak{g}_\nu$$

which implies that $X_{h_\mu}(\nu) = -\operatorname{ad}^*_{\delta h/\delta \nu}\nu + \Sigma(\frac{\delta h}{\delta \nu}, \cdot)$, the expression of the Hamiltonian vector field generated by $h : \mathfrak{g}^* \to \mathbb{R}$ relative to the $+\Sigma$-Lie–Poisson bracket (see (4.5.20)). This shows that the affine orbit $(\mathcal{O}_\mu, \omega^+_{\mathcal{O}_\mu})$ is a Poisson initial submanifold of $(\mathfrak{g}^*, \{\cdot, \cdot\}^\Sigma_+)$ and concludes the proof of the theorem. ■

4.5.32 Corollary *Let $\mathfrak{g}$ be a Lie algebra, $\Sigma \in Z^2(\mathfrak{g}; \mathbb{R})$ a two-cocycle on $\mathfrak{g}$, and $\{\cdot, \cdot\}^\Sigma_\pm$ the $\pm\Sigma$-Lie–Poisson structures on $\mathfrak{g}^*$. For any $\mu \in \mathfrak{g}^*$, the symplectic leaf $\mathcal{L}_\mu$ of $(\mathfrak{g}^*, \{\cdot, \cdot\}^\Sigma_\pm)$ containing μ is determined by*

$$\mathcal{L}_\mu = \pi_{\mathfrak{g}^*}(G_\Sigma \cdot (\mu, 1)),$$

where $\pi_{\mathfrak{g}^*} : \mathfrak{g}^* \oplus \mathbb{R} \to \mathfrak{g}^*$ is the projection onto the $\mathfrak{g}^*$-factor, G_Σ is any connected Lie group that has as Lie algebra the central extension $\mathfrak{g}_\Sigma$ defined on (4.5.16) (available by Lie's Third Fundamental Theorem), and $G_\Sigma \cdot (\mu, 1)$ is the G_Σ-coadjoint orbit of $(\mu, 1)$.

Proof. We will carry out the proof for the structure $(\mathfrak{g}^*, \{\cdot, \cdot\}^\Sigma_+)$; the proof for the negative counterpart is totally identical.

Let $\nu \in \mathcal{L}_\mu$. By the definition of symplectic leaf $\nu = \mathcal{F}_T(\mu)$, with $\mathcal{F}_T$ a finite composition of Hamiltonian flows on $(\mathfrak{g}^*, \{\cdot, \cdot\}^\Sigma_+)$ associated to smooth real valued functions on $\mathfrak{g}^*$. Suppose by simplicity that $\mathcal{F}_T = F_t$, with F_t the flow of X_h, $h \in C^\infty(\mathfrak{g}^*)$. By (4.5.18) and (4.5.20) it is easy to see that the map $\overline{F}_t(\nu, a) := (F_t(\nu), a)$ is the Hamiltonian flow on $(\mathfrak{g}^*_\Sigma, \{\cdot, \cdot\}_+)$ associated to the function $\overline{h} : \mathfrak{g}^*_\Sigma \to \mathbb{R}$ defined by $\overline{h}(\nu, a) := h(\nu)$, $(\nu, a) \in \mathfrak{g}^*_\Sigma$. Hence $(\nu, 1) = (F_t(\mu), 1)$ belongs to the symplectic leaf of $(\mathfrak{g}^*_\Sigma, \{\cdot, \cdot\}_+)$ that contains $(\mu, 1)$, which by Theorem 4.5.31 and the connectedness of G_Σ equals $G_\Sigma \cdot (\mu, 1)$. This proves one inclusion.

Conversely, let $(\nu, a) \in G_\Sigma \cdot (\mu, 1)$. Since by Theorem 4.5.31 $G_\Sigma \cdot (\mu, 1)$ is the symplectic leaf of $(\mathfrak{g}^*_\Sigma, \{\cdot, \cdot\}_+)$ containing $(\mu, 1)$ we can write $(\nu, a) = \overline{\mathcal{F}}_T(\mu, 1)$, with $\overline{\mathcal{F}}_T$ a finite composition of Hamiltonian flows on $(\mathfrak{g}^*_\Sigma, \{\cdot, \cdot\}_+)$. By simplicity take $\overline{\mathcal{F}}_T = \overline{F}_t$ with $\overline{F}_t$ the flow of a smooth function $\overline{h} : \mathfrak{g}^*_\Sigma \to \mathbb{R}$. The zero on the right-hand side of (4.5.18) implies that $a = 1$ and hence $\overline{F}_t(\mu, 1) = (F_t(\mu), 1)$ with F_t the Hamiltonian flow of the vector field X^Σ_h determined by (4.5.20) and associated to the function $h : \mathfrak{g}^* \to \mathbb{R}$ given by $h(\nu) := \overline{h}(\nu, 1)$, $\nu \in \mathfrak{g}^*$. This concludes the proof. ■

4.5.33 The Reduction Lemma. We now introduce a result that, in spite of its simplicity, will be of paramount importance later when we study symplectic reduction. We give some notation first. Let $\mathfrak{g}$ be a Lie algebra acting canonically on the manifold

(M, ω) with a momentum map $\mathbf{J} : M \to \mathfrak{g}^*$ that has an infinitesimal non-equivariance two-cocycle Σ. For any $\mu \in \mathfrak{g}^*$ define its ***isotropy subalgebra*** $\mathfrak{g}_\mu$ relative to Σ by

$$\mathfrak{g}_\mu := \{\xi \in \mathfrak{g} \mid -\mathrm{ad}^*_\xi \mu + \Sigma(\xi, \cdot) = 0\}.$$

If Σ is the infinitesimal non-equivariance two-cocycle associated to a non-equivariance group cocycle σ, the subalgebra $\mathfrak{g}_\mu$ coincides with the Lie algebra of the isotropy subgroup G_μ of μ with respect to the affine action defined using σ. Moreover, it is easy to check that if $\mathcal{O}_\mu$ is the affine orbit containing the element $\mu \in \mathfrak{g}^*$, then

$$\mathfrak{g}_\mu^\circ = T_\mu \mathcal{O}_\mu. \tag{4.5.26}$$

Indeed, by (4.5.30) any element in $T_\mu \mathcal{O}_\mu$ can be written as $-\mathrm{ad}^*_\xi \mu + \Sigma(\xi, \cdot)$ with $\xi \in \mathfrak{g}$. Hence, for any $\eta \in \mathfrak{g}_\mu$, we have that

$$\langle -\mathrm{ad}^*_\xi \mu + \Sigma(\xi, \cdot), \eta \rangle = \langle \mathrm{ad}^*_\eta \mu - \Sigma(\eta, \cdot), \xi \rangle = 0,$$

which proves that $T_\mu \mathcal{O}_\mu \subset \mathfrak{g}_\mu^\circ$. Now, since $\dim T_\mu \mathcal{O}_\mu = \dim \mathcal{O}_\mu = \dim G - \dim G_\mu = \dim \mathfrak{g} - \dim \mathfrak{g}_\mu = \dim \mathfrak{g}_\mu^\circ$, the identity (4.5.26) holds.

4.5.34 Lemma (Reduction Lemma) *Let $\mathfrak{g}$ be a Lie algebra acting canonically on the manifold (M, ω) with a momentum map $\mathbf{J} : M \to \mathfrak{g}^*$ that has an infinitesimal non-equivariance two-cocycle Σ. Then for any $m \in M$ such that $\mathbf{J}(m) = \mu$ we have*

$$\mathfrak{g}_\mu \cdot m = \mathfrak{g} \cdot m \cap \ker T_m \mathbf{J} = \mathfrak{g} \cdot m \cap (\mathfrak{g} \cdot m)^\omega. \tag{4.5.27}$$

Proof. The second equality is a consequence of Proposition 4.5.14. As to the first one (the actual Reduction Lemma) it is a straightforward consequence of (4.5.10). Indeed, the vector $\xi_M(m) \in \mathfrak{g} \cdot m \cap \ker T_m \mathbf{J}$ if and only if $0 = T_m \mathbf{J} \cdot \xi_M(m) = -\mathrm{ad}^*_\eta \mathbf{J}(m) + \Sigma(\xi, \cdot) = -\mathrm{ad}^*_\eta \mu + \Sigma(\xi, \cdot)$ which is, by definition, equivalent to $\xi \in \mathfrak{g}_\mu$. ■

4.5.35 The first equality in (4.5.27) is still true in the Poisson context. More specifically, let $(M, \{\cdot, \cdot\})$ be a Poisson manifold acted canonically upon by a Lie algebra $\mathfrak{g}$ admitting an infinitesimally equivariant standard momentum map $\mathbf{J} : M \to \mathfrak{g}^*$ (see (4.5.5)). Then for any $m \in M$ such that $\mathbf{J}(m) = \mu$ we have

$$\mathfrak{g}_\mu \cdot m = \mathfrak{g} \cdot m \cap \ker T_m \mathbf{J}. \tag{4.5.28}$$

4.6 Momentum maps and isotropy type submanifolds

Let (M, ω) be a connected symplectic manifold and G a Lie group acting on M in a canonical and proper fashion with an associated momentum map $\mathbf{J} : M \to \mathfrak{g}^*$ that has $\sigma : G \to \mathfrak{g}^*$ as a non-equivariance one-cocycle. In Proposition 4.2.7 we saw that for any point $m \in M$, the connected component $M_{G_m}^m$ of the isotropy type submanifold M_{G_m} that contains it is a symplectic submanifold of M acted properly and canonically upon by $L^m := N(G_m)^m / G_m$. The statements that we present in the following theorem basically say that the canonical L^m-action on $M_{G_m}^m$ inherits a momentum map for the G-action on M. The ideas in this section will be of great importance in what follows, specially in the context of singular reduction and the so-called Sjamaar Principle (see Chapter 8).

4.6.1 Theorem *Let (M, ω) be a connected symplectic manifold and G a Lie group acting on M in a canonical and proper fashion. Suppose that this action has a momentum map $\mathbf{J} : M \to \mathfrak{g}^*$ with non-equivariance one-cocycle $\sigma : G \to \mathfrak{g}^*$. For any $m \in M$ such that $\mathbf{J}(m) = \mu$, the free, proper, and canonical action of $L^m := N(G_m)^m / G_m$ on $M^m_{G_m}$ has an associated momentum map $\mathbf{J}_{L^m} : M^m_{G_m} \to (\mathrm{Lie}(L^m))^*$ given by*

$$\mathbf{J}_{L^m}(z) := \Lambda(\mathbf{J}|_{M^m_{G_m}}(z) - \mu), \quad z \in M^m_{G_m}. \tag{4.6.1}$$

In this expression $\Lambda : (\mathfrak{g}^\circ_m)^{G_m} \to (\mathrm{Lie}(L^m))^*$ denotes the natural L^m-equivariant isomorphism given by

$$\left\langle \Lambda(\beta), \left.\frac{d}{dt}\right|_{t=0} \exp t\xi G_m \right\rangle = \langle \beta, \xi \rangle, \tag{4.6.2}$$

for any $\beta \in (\mathfrak{g}^\circ_m)^{G_m}$, $\xi \in \mathrm{Lie}(N(G_m)^m) = \mathrm{Lie}(N(G_m))$. The non-equivariance one-cocycle $\tau : M^m_{G_m} \to (\mathrm{Lie}(L^m))^*$ of the momentum map $\mathbf{J}_{L^m}$ is given by the map

$$\tau(l) = \Lambda(\sigma(n) + n \cdot \mu - \mu), \quad \textit{for any} \quad l = nG_m \in L^m, n \in N(G_m)^m. \tag{4.6.3}$$

Let Σ_σ and Σ_τ be the infinitesimal non-equivariance cocycles associated to σ and τ respectively. If $\xi, \eta \in \mathrm{Lie}(L^m)$ and $\xi_1, \eta_1 \in \mathrm{Lie}(N(G_m))$ are such that $\xi = \left.\frac{d}{dt}\right|_{t=0} \exp t\xi_1 G_m$ and $\eta = \left.\frac{d}{dt}\right|_{t=0} \exp t\eta_1 G_m$, then

$$\Sigma_\tau(\xi, \eta) = \Sigma_\sigma(\xi_1, \eta_1) - \langle \mu, [\xi_1, \eta_1] \rangle. \tag{4.6.4}$$

Proof. We adapt the proof given in ORTEGA (1998). We denote $L := N(G_m)/G_m$ and $\mathfrak{l} = \mathrm{Lie}(L)$. Recall that by (2.4.9) we have that $\mathfrak{l} = \mathrm{Lie}(L) = \mathrm{Lie}(L^m)$ and hence Λ can be considered as a map between $(\mathfrak{g}^\circ_m)^{G_m}$ and $\mathfrak{l}^*$. We will proceed by showing first that $\Lambda : (\mathfrak{g}^\circ_m)^{G_m} \to \mathfrak{l}^*$ is L-equivariant and later on we will show that it is a well-defined isomorphism.

Λ is L-equivariant. The L-actions that we consider are the L-coadjoint action on $\mathfrak{l}^*$ and the L-action on $(\mathfrak{g}^\circ_m)^{G_m}$ defined by

$$\begin{aligned} \varphi : \; L \times (\mathfrak{g}^\circ_m)^{G_m} &\longrightarrow (\mathfrak{g}^\circ_m)^{G_m} \\ (l = nG_m, \beta) &\longmapsto \mathrm{Ad}^*_{n^{-1}} \beta, \end{aligned}$$

where $n \in N(G_m)$. In order to check that this action is well defined, consider $n, n' \in N(G_m)$ such that $nG_m = n'G_m$. There exists an element h in G_m such that $n' = nh$ and therefore

$$\mathrm{Ad}^*_{(n')^{-1}} \beta = \mathrm{Ad}^*_{(nh)^{-1}} \beta = \mathrm{Ad}^*_{h^{-1}n^{-1}} \beta = \mathrm{Ad}^*_{n^{-1}} \mathrm{Ad}^*_{h^{-1}} \beta = \mathrm{Ad}^*_{n^{-1}} \beta.$$

We now verify that φ really maps into $(\mathfrak{g}^\circ_m)^{G_m}$. If $n \in N(G_m)$ and $\beta \in (\mathfrak{g}^\circ_m)^{G_m}$, the element $\mathrm{Ad}^*_{n^{-1}} \beta$ clearly belongs to $\mathfrak{g}^\circ_m$, because for any $\nu \in \mathfrak{g}_m$, if we denote $\langle \cdot, \cdot \rangle$ the natural pairing between $\mathfrak{g}^*$ and $\mathfrak{g}$, we have

$$\left\langle \mathrm{Ad}^*_{n^{-1}} \beta, \nu \right\rangle = \left\langle \beta, \mathrm{Ad}_{n^{-1}} \nu \right\rangle = 0,$$

since $\beta \in \mathfrak{g}_m^\circ$, and $\mathrm{Ad}_{n^{-1}}\nu \in \mathfrak{g}_m$. We now show that $\mathrm{Ad}^*_{n^{-1}}\beta$ is also G_m-fixed. For $h \in G_m$ arbitrary, we have $hn = nh'$ for some $h' \in G_m$ since G_m is normal in $N(G_m)$ and thus

$$\begin{aligned} h \cdot \mathrm{Ad}^*_{n^{-1}}\beta &= \mathrm{Ad}^*_{h^{-1}}\mathrm{Ad}^*_{n^{-1}}\beta = \mathrm{Ad}^*_{(hn)^{-1}}\beta = \mathrm{Ad}^*_{(nh')^{-1}}\beta \\ &= \mathrm{Ad}^*_{n^{-1}}\mathrm{Ad}^*_{(h')^{-1}}\beta = \mathrm{Ad}^*_{n^{-1}}\beta, \end{aligned}$$

as required.

We now check that Λ is L-equivariant with respect to these actions. Let $l = nG_m \in L$, $\beta \in (\mathfrak{g}_m^\circ)^{G_m}$, and $\xi \in \mathrm{Lie}(N(G_m))$ be arbitrary. Then, by the defining relation (4.6.2) of Λ, we have

$$\begin{aligned} \left\langle \Lambda(l\cdot\beta), \left.\frac{d}{dt}\right|_{t=0} \exp t\xi G_m \right\rangle &= \langle l\cdot\beta, \xi\rangle = \langle \mathrm{Ad}^*_{n^{-1}}\beta, \xi\rangle = \langle \beta, \mathrm{Ad}_{n^{-1}}\xi\rangle \\ &= \left\langle \Lambda(\beta), \left.\frac{d}{dt}\right|_{t=0} \exp t(\mathrm{Ad}_{n^{-1}}\xi)G_m \right\rangle \\ &= \left\langle \Lambda(\beta), \left.\frac{d}{dt}\right|_{t=0} n^{-1}(\exp t\xi)nG_m \right\rangle \\ &= \left\langle \Lambda(\beta), \left.\frac{d}{dt}\right|_{t=0} (n^{-1}G_m)(\exp t\xi G_m)(nG_m) \right\rangle \\ &= \left\langle \Lambda(\beta), \mathrm{Ad}_{l^{-1}} \left.\frac{d}{dt}\right|_{t=0} (\exp t\xi G_m) \right\rangle \\ &= \left\langle \mathrm{Ad}^*_{l^{-1}}\Lambda(\beta), \left.\frac{d}{dt}\right|_{t=0} (\exp t\xi G_m) \right\rangle, \end{aligned}$$

which proves that $\Lambda(l\cdot\beta) = \mathrm{Ad}^*_{l^{-1}}\Lambda(\beta)$, as required.

Λ is a well-defined isomorphism. We now show that the map $\Lambda : (\mathfrak{g}_m^\circ)^{G_m} \to \mathfrak{l}^*$ defined in (4.6.2) is an isomorphism by proving that it is given by the composition of three isomorphisms to be described separately in the following three points.

(i) $(\mathfrak{g}_m^\circ)^{G_m}$ is isomorphic to $((\mathfrak{g}/\mathfrak{g}_m)^*)^{G_m}$. This is a consequence of the well-known linear algebraic statement stating that if W is a vector subspace of the vector space V, the map $g : W^\circ \to (V/W)^*$ defined by $\langle g(\beta), v + W\rangle := \langle \beta, v\rangle$, for any $v \in V$, $\beta \in W^\circ$, is an isomorphism.

In our particular situation this fact allows us to conclude that $(\mathfrak{g}_m)^\circ$ is isomorphic to $(\mathfrak{g}/\mathfrak{g}_m)^*$. We now notice that $\mathfrak{g}/\mathfrak{g}_m$ is a G_m-space when we consider the G_m-action defined by

$$h\cdot(\xi + \mathfrak{g}_m) := \mathrm{Ad}_h\xi + \mathfrak{g}_m, \quad h \in G_m,\ \xi \in \mathfrak{g}. \tag{4.6.5}$$

Hence, the dual space $(\mathfrak{g}/\mathfrak{g}_m)^*$ is automatically a G_m-space when we consider the corresponding contragredient action. Moreover, a straightforward verification shows that the isomorphism $(\mathfrak{g}_m)^\circ \to (\mathfrak{g}/\mathfrak{g}_m)^*$ is G_m-equivariant with respect to this action and therefore it restricts to an isomorphism

$$A : ((\mathfrak{g}_m)^\circ)^{G_m} \longrightarrow ((\mathfrak{g}/\mathfrak{g}_m)^*)^{G_m}$$

given by $\langle A(\alpha), \xi + \mathfrak{g}_m \rangle := \langle \alpha, \xi \rangle, \alpha \in ((\mathfrak{g}_m)^\circ)^{G_m}, \xi + \mathfrak{g}_m \in \mathfrak{g}/\mathfrak{g}_m$.

(ii) $((\mathfrak{g}/\mathfrak{g}_m)^*)^{G_m}$ **is isomorphic to** $((\mathfrak{g}/\mathfrak{g}_m)^{G_m})^*$. Lemma 2.5.11 provides the needed isomorphism $B : ((\mathfrak{g}/\mathfrak{g}_m)^*)^{G_m} \to ((\mathfrak{g}/\mathfrak{g}_m)^{G_m})^*$.

(iii) $((\mathfrak{g}/\mathfrak{g}_m)^{G_m})^*$ **is isomorphic to** $\mathfrak{l}^*$. A quick inspection shows that

$$L = N(G_m)/G_m = (G/G_m)^{G_m} = (G/G_m)_{G_m}.$$

Consequently, by Proposition 2.4.7, we have that

$$\mathfrak{l} = T_{G_m}(N(G_m)/G_m) = (T_{G_m}(G/G_m))^{G_m}. \tag{4.6.6}$$

At the same time, if we denote by $\pi : G \to G/G_m$ the projection, its derivative $T_e\pi : \mathfrak{g} \to T_{G_m}(G/G_m)$ defines a linear isomorphism $\mathfrak{g}/\mathfrak{g}_m \to T_{G_m}(G/G_m)$ given by $\xi + \mathfrak{g}_m \longmapsto \frac{d}{dt}\big|_{t=0} \exp t\xi \cdot G_m$, for any $\xi \in \mathfrak{g}$. The group G_m acts on $\mathfrak{g}/\mathfrak{g}_m$ and on $T_{G_m}(G/G_m)$ by (4.6.5) and by $h \cdot \frac{d}{dt}\big|_{t=0} \exp t\xi \cdot G_m := \frac{d}{dt}\big|_{t=0} \exp(t\mathrm{Ad}_h\xi) \cdot G_m$, $h \in G_m$, $\xi \in \mathfrak{g}$, respectively. Given that the isomorphism $\mathfrak{g}/\mathfrak{g}_m \to T_{G_m}(G/G_m)$ is G_m-equivariant it restricts to another isomorphism $(\mathfrak{g}/\mathfrak{g}_m)^{G_m} \to (T_{G_m}(G/G_m))^{G_m}$. Since $(T_{G_m}(G/G_m))^{G_m} = \mathfrak{l}$ by (4.6.6), the inverse of this map is another isomorphism $\mathfrak{l} \to (\mathfrak{g}/\mathfrak{g}_m)^{G_m}$ whose dual map $C : ((\mathfrak{g}/\mathfrak{g}_m)^{G_m})^* \to \mathfrak{l}^*$ is the isomorphism that we need.

Finally $\Lambda : (\mathfrak{g}_m^\circ)^{G_m} \to \mathfrak{l}^*$ is an isomorphism since it can be written as the composition of the isomorphisms A, B, and C, that is,

$$\Lambda = C \circ B \circ A.$$

Proof of Expression (4.6.1). We next show that the mapping

$$\mathbf{J}_{L^m}(z) := \Lambda(\mathbf{J}|_{M^m_{G_m}}(z) - \mu), \quad z \in M^m_{G_m},$$

is a well-defined momentum map for the L^m-action on $M^m_{G_m}$. We start by noticing that $T_z\mathbf{J}|_{M^m_{G_m}}(T_zM^m_{G_m}) \subset \mathfrak{g}_m^\circ$ for each $z \in M^m_{G_m}$, because if $v_z \in T_zM^m_{G_m}$ and $\xi \in \mathfrak{g}_m$ are arbitrary, the defining property of a momentum map implies that

$$\langle T_z\mathbf{J}|_{M^m_{G_m}} \cdot v_z, \xi \rangle = \omega(z)(\xi_M(z), v_z) = 0,$$

since $\xi_M(z) = 0$ for any $z \in M^m_{G_m}$. Let now $c(t)$ be a curve in $M^m_{G_m}$ joining m to an arbitrary point $z \in M^m_{G_m}$ (for instance $c(0) = m$ and $c(1) = z$) and denote

$$v(t) = \frac{d}{dt}\mathbf{J}|_{M^m_{G_m}}(c(t)) \in \mathfrak{g}_m^\circ.$$

Then

$$\int_0^1 v(t)dt = \int_0^1 \frac{d}{dt}\mathbf{J}|_{M^m_{G_m}}(c(t))dt = \mathbf{J}(z) - \mathbf{J}(m) = \mathbf{J}(z) - \mu$$

is an element of $\mathfrak{g}_m^\circ$. Therefore, $\mathbf{J}(z) \in \mu + \mathfrak{g}_m^\circ$ for any $z \in M^m_{G_m}$. Moreover, the G-equivariance of the momentum map $\mathbf{J}$ with respect to the affine action defined with the non-equivariance cocycle σ of $\mathbf{J}$ implies that

$$\mathbf{J}(z) \in \mu + (\mathfrak{g}_m^\circ)^{G_m}. \tag{4.6.7}$$

Indeed, suppose that $\mathbf{J}(z) = \mu + \beta$, with $\beta \in \mathfrak{g}_m^\circ$. Since $z \in M_{G_m}^m$, we have for any $h \in G_m = G_z$

$$\mu + \beta = \mathbf{J}(z) = \mathbf{J}(h \cdot z) = h \cdot \mathbf{J}(z) = \mathrm{Ad}_{h^{-1}}^* \mu + \mathrm{Ad}_{h^{-1}}^* \beta + \sigma(h). \tag{4.6.8}$$

At the same time

$$\mu = \mathbf{J}(m) = \mathbf{J}(h \cdot m) = h \cdot \mathbf{J}(m) = \mathrm{Ad}_{h^{-1}}^* \mu + \sigma(h),$$

that is , $\mathrm{Ad}_{h^{-1}}^* \mu = \mu - \sigma(h)$, which substituted into the right-hand side of (4.6.8) proves that $\mathrm{Ad}_{h^{-1}}^* \beta = \beta$ and hence $\beta \in (\mathfrak{g}_m^\circ)^{G_m}$, as required.

Expression (4.6.7) justifies why the combination $\mathbf{J}|_{M_{G_m}^m}(z) - \mu$ falls in the domain of Λ which makes of (4.6.1) a good definition.

We now show that $\mathbf{J}_{L^m}$ is the sought after momentum map. If true, for any $\lambda = \frac{d}{dt}\big|_{t=0} \exp t\xi G_m \in \mathfrak{l}$, $\xi \in \mathrm{Lie}N(G_m)$ arbitrary, the mapping $\mathbf{J}_{L^m}$ needs to satisfy

$$\mathbf{dJ}_{L^m}^\lambda(z) \cdot v_z = (i^*\omega)(z)(\lambda_{M_{G_m}^m}(z), v_z), \tag{4.6.9}$$

for each $z \in M_{G_m}^m$ and $v_z = \frac{d}{dt}\big|_{t=0} c(t) \in T_z M_{G_m}^m$. The map $i : M_{G_m}^m \hookrightarrow M$ is the inclusion which is symplectic and equivariant with respect to the L^m-action on $M_{G_m}^m$ and the G-action on M. Using expression (4.6.1), the left-hand side of (4.6.9) can be written as

$$\begin{aligned}
\mathbf{dJ}_{L^m}^\lambda(z) \cdot v_z &= \frac{d}{dt}\bigg|_{t=0} \langle \Lambda(\mathbf{J}|_{M_{G_m}^m}(c(t)) - \mu), \lambda\rangle \\
&= \frac{d}{dt}\bigg|_{t=0} \langle \mathbf{J}|_{M_{G_m}^m}(c(t)) - \mu, \xi\rangle = \langle T_z\mathbf{J} \cdot (T_z i \cdot v_z), \xi\rangle \\
&= \mathbf{dJ}^\xi(z) \cdot (T_z i \cdot v_z) = \omega(z)(\xi_M(z), T_z i \cdot v_z) \\
&= \omega(z)(T_z i \cdot \lambda_{M_{G_m}^m}(z), T_z i \cdot v_z) \\
&= (i^*\omega)(z)(\lambda_{M_{G_m}^m}(z), v_z),
\end{aligned}$$

which proves that $\mathbf{J}_{L^m}(z)$ is a well-defined momentum map for the L^m-action on $M_{G_m}^m$.

The non-equivariance one-cocycle of $\mathbf{J}_{L^m}$. The non-equivariance one-cocycle $\tau : L^m \to \mathfrak{l}^*$ of $\mathbf{J}_{L^m}$ is very easy to compute. Let $l = nH \in L^m$. By definition, we have

$$\begin{aligned}
\tau(l) &= \mathbf{J}_{L^m}(l \cdot m) - l \cdot \mathbf{J}_{L^m}(m) \\
&= \Lambda(\mathbf{J}|_{M_{G_m}^m}(l \cdot m) - \mu) - l \cdot \Lambda(\mathbf{J}|_{M_{G_m}^m}(m) - \mu) \\
&= \Lambda(\mathbf{J}|_{M_{G_m}^m}(n \cdot m) - \mu) - \Lambda(n \cdot \mathbf{J}|_{M_{G_m}^m}(m) - n \cdot \mu) \\
&= \Lambda(\sigma(n) + \mathrm{Ad}_{n^{-1}}^* \mathbf{J}(m) - \mu) \\
&= \Lambda(\sigma(n) + n \cdot \mu - \mu).
\end{aligned}$$

The infinitesimal non-equivariance cocycle of $\mathbf{J}_{L^m}$. Let $\xi, \eta \in \mathrm{Lie}(L^m)$ and $\xi_1, \eta_1 \in \mathrm{Lie}(N(G_m))$ are such that $\xi = \frac{d}{dt}\big|_{t=0} \exp t\xi_1 G_m$ and $\eta = \frac{d}{dt}\big|_{t=0} \exp t\eta_1 G_m$ then by (4.5.11), (4.6.2), and (4.6.3) we have that

$$\begin{aligned}
\Sigma_\tau(\xi, \eta) &= \left.\frac{d}{dt}\right|_{t=0} \langle \tau(\exp t\xi), \eta\rangle \\
&= \left.\frac{d}{dt}\right|_{t=0} \langle \Lambda(\sigma(\exp t\xi_1) + \exp t\xi_1 \cdot \mu - \mu), \eta\rangle \\
&= \left.\frac{d}{dt}\right|_{t=0} \langle \sigma(\exp t\xi_1) + \exp t\xi_1 \cdot \mu - \mu, \eta_1\rangle \\
&= \Sigma_\sigma(\xi_1, \eta_1) - \langle \mathrm{ad}^*_{\xi_1}\mu, \eta_1\rangle = \Sigma_\sigma(\xi_1, \eta_1) - \langle \mu, [\xi_1, \eta_1]\rangle. \quad \blacksquare
\end{aligned}$$

4.6.2 Corollary *Let (M, ω) be a connected symplectic manifold and G a Lie group acting on M in a canonical and proper fashion. Suppose that this action has an associated momentum map $\mathbf{J} : M \to \mathfrak{g}^*$. Then for any $m \in M$, denoting $\mathbf{J}(m) = \mu$, the intersection*

$$\mathbf{J}^{-1}(\mu) \cap M^m_{G_m}$$

is an embedded submanifold of M.

Proof. First, notice that by Corollary 4.5.13 the momentum map $\mathbf{J}_{L^m} : M^m_{G_m} \to \mathfrak{l}^*$ corresponding to the L^m-action on $M^m_{G_m}$ is a submersion because this action is free. Using the expression (4.6.1) we obtain

$$\mathbf{J}_{L^m}(m) = \Lambda(\mathbf{J}|_{M^m_{G_m}}(m) - \mu) = \Lambda(0) = 0. \tag{4.6.10}$$

Therefore, since $\mathbf{J}_{L^m}$ is a submersion the entire level set $\mathbf{J}^{-1}_{L^m}(0) \subset M^m_{G_m}$ is an embedded submanifold of $M^m_{G_m}$ and thereby of M. We conclude the proof by showing that $\mathbf{J}^{-1}_{L^m}(0) = \mathbf{J}^{-1}(\mu) \cap M^m_{G_m}$. By definition, $\mathbf{J}^{-1}_{L^m}(0) = \{z \in M^m_{G_m} \mid \mathbf{J}_{L^m}(z) = 0\} = \{z \in M^m_{G_m} \mid \Lambda(\mathbf{J}|_{M^m_{G_m}}(z) - \mu) = 0\} = \{z \in M^m_{G_m} \mid \mathbf{J}|_{M^m_{G_m}}(z) = \mu\} = \mathbf{J}^{-1}(\mu) \cap M^m_{G_m}$. ■

Theorem 4.6.1 can be easily used to provide an expression for a momentum map associated to the natural action on the *local* isotropy type manifolds introduced in §2.4.9.

4.6.3 Corollary *Let (M, ω) be a symplectic manifold and G a Lie group acting on M in a canonical and proper fashion. Suppose that this action has a momentum map $\mathbf{J} : M \to \mathfrak{g}^*$. Let H be an isotropy subgroup of G and M^l_H a local isotropy manifold associated to it. Let I be a set of labels for the different connected components $(M^l_H)_i$ of M^l_H and $\{z_i\}_{i\in I}$ a subset of M^l_H such that for each $i \in I$ the element z_i belongs to $(M^l_H)_i$. Then the free, proper, and canonical action of $L := N(H)/H$ on M^l_H has an associated momentum map $\mathbf{J}_L : M^l_H \to \mathfrak{l}^*$ given by*

$$\mathbf{J}_L(z) := \Lambda(\mathbf{J}|_{(M^l_H)_i}(z) - \mu_i), \quad z \in (M^l_H)_i \text{ and } \mu_i := \mathbf{J}(z_i). \tag{4.6.11}$$

In this expression $\Lambda : (\mathfrak{h}^\circ)^H \to \mathfrak{l}^*$ denotes the natural L-equivariant isomorphism introduced in (4.6.2).

4.7 The convexity properties of momentum maps

In this chapter we have reviewed some of the most elementary properties of the momentum map that by no means reflect all that is known about this object in less general situations. Momentum maps have been used over the years to study mathematical problems that go way beyond its applications to mechanics and that range from representation theory to spectral problems in matrices. This makes it impossible for us to review them all.

Nevertheless, from the geometrical point of view there is a particularly beautiful result that describes the image of a momentum map associated to a compact globally Hamiltonian action on a compact symplectic manifold. This result says that the intersection of the image of a coadjoint equivariant momentum map with a Weyl chamber of the Lie algebra of a maximal torus of the symmetry group is a *compact and convex polytope*. This polytope is referred to as the ***momentum polytope***.

This result is due to ATIYAH (1982) and GUILLEMIN AND STERNBERG (1982, 1984a), for Abelian actions and to KIRWAN (1984b) for the general case. The reader is also encouraged to check with CONDEVAUX et al. (1988); HILGERT et al. (1994); PRATO (1994); SJAMAAR (1998), and references therein for additional information.

There is a remarkable situation in which the momentum polytope is particularly important. DELZANT (1988) has proved that ***toric manifolds*** are completely characterized by their associated momentum polytopes. We recall that a symplectic toric manifold is a compact connected symplectic $2n$-dimensional manifold (M, ω) equipped with an effective canonical action of an n-torus $\mathbb{T}^n$ that has an associated invariant momentum map $\mathbf{J} : M \to \mathbb{R}^n$. Delzant's theorem proves that the symplectic toric manifolds are classified by their momentum polytopes. In order to formalize this statement we need to introduce the notion of ***Delzant polytope***. A Delzant polytope in $\mathbb{R}^n$ is a convex polytope that is also:

(i) Simple: there are n edges meeting at each vertex.

(ii) Rational: the edges meeting at a vertex p are of the form $p + tu_i$, $0 \leq t < \infty$, $u_i \in \mathbb{Z}^n$, $i \in \{1, \ldots, n\}$.

(iii) Smooth: the vectors $\{u_1, \ldots, u_n\}$ can be chosen to be an integral basis of $\mathbb{Z}^n$.

Delzant's Theorem can be stated by saying that the map

$$\begin{array}{ccc} \{\text{symplectic toric manifolds}\} & \longrightarrow & \{\text{Delzant polytopes}\} \\ (M, \omega, \mathbb{T}^n, \mathbf{J} : M \to \mathbb{R}^n) & \longmapsto & \mathbf{J}(M) \end{array}$$

is a bijection. The constructions employed in the proof of this result use very strongly the symplectic reduction techniques that we will introduce in Chapters 6 and 8. The reader interested in this topic is encouraged to check with AUDIN (1991), and references therein.

A feature of the standard momentum maps closely related to the convexity of its image is the connectedness of its fibers. Indeed, KIRWAN (1984a,b) (see also HILGERT et al. (1994)) has proved that if a momentum map arising from an action of a compact

Lie group on a connected symplectic manifold is a proper map (which happens for instance when the manifold in question is compact), then its fibers are connected. LERMAN (1995) has used his ***symplectic cuts*** technique to prove the same result for linear compact group actions on symplectic vector spaces whose associated momentum map is not necessarily proper.

Chapter 5

Generalizations of the Momentum Map

5.1 A short interlude on connections and holonomy

Before proceeding with more Noether momentum maps we briefly recall standard background material on connections on principal fiber bundles and holonomy groups that we will need in the following section. The reader interested in the proofs can find them in KOBAYASHI AND NOMIZU (1963); GREUB et al. (1970b), or in HUSEMOLLER (1994).

5.1.1 Connections on principal fiber bundles. Let (P, M, π, G) be a principal fiber bundle. Denote by $R : P \times G \rightarrow P$ the right G-action whose orbit space is $P/G = M$. A ***connection*** on (P, M, π, G) is a $\mathfrak{g}$-valued one-form $A \in \Omega^1(P, \mathfrak{g})$ such that for any $\xi \in \mathfrak{g}$, $g \in G$, $z \in P$, and $v_z \in T_zP$, we have that

(i) $A(z) : T_zP \rightarrow \mathfrak{g}$ is linear,

(ii) $A(z) \cdot \xi_M(z) = \xi$,

(iii) $A(R_g z) \cdot (T_z R_g \cdot v_z) = \mathrm{Ad}_{g^{-1}}(A(z) \cdot v_z)$.

The connection A provides a splitting of the tangent bundle $TP = V \oplus H$, where V is the bundle of ***vertical vectors*** defined by $V =: \ker T\pi$ and H that of the ***horizontal vectors*** given by $H := \ker A$, that is, $H(z) := \{v_z \in T_zP \mid A(z) \cdot v_z = 0\}$, for all $z \in M$. A curve $c : I \subset \mathbb{R} \rightarrow P$ is ***horizontal*** if $c'(t) \in H(c(t))$ for any $t \in I$. Given a curve $d : [0, 1] \rightarrow M$ on M and a point $z \in P$ there exists a unique horizontal curve $c : [0, 1] \rightarrow P$ such that $c(0) = z$ and $(\pi \circ c)(t) = d(t)$, for all $t \in [0, 1]$. This unique curve c is called the ***horizontal lift*** of the curve d through z.

Each point $z \in P$ and each loop $d : [0, 1] \rightarrow M$ at the point $\pi(z)$ determine an element in G. Indeed, let $c : [0, 1] \rightarrow M$ be the horizontal lift of d through z. Since $d(0) = d(1) = \pi(z)$, we have that $z = c(0), c(1) \in \pi^{-1}(\pi(z))$ and hence there exists a unique element $g \in G$ such that $c(1) = R_g(c(0))$. The elements in G determined by all the loops at $\pi(z)$ form a (not necessarily) closed subgroup $\mathcal{H}(z)$ of G called the ***holonomy group*** of A with reference point $z \in P$. If two points $z_1, z_2 \in P$ can be joined by a horizontal curve, then $\mathcal{H}(z_1) = \mathcal{H}(z_2)$. If two points $z_1, z_2 \in P$ are in the same fiber of π, then there exists $g \in G$ such that $z_2 = R_g(z_1)$ and hence $\mathcal{H}(z_2) = g^{-1}\mathcal{H}(z_1)g$.

5.1.2 The Reduction Theorem. Let (P, M, π, G) be a principal fiber bundle. Let $\iota : Q \hookrightarrow P$ be an injectively immersed submanifold of P and H a Lie subgroup of G (not necessarily embedded) that leaves Q invariant. If (Q, M, π', H) is a principal fiber bundle, where $\pi' : Q \to Q/H = M$ is the projection, we say that (Q, M, π', H) is a ***reduction*** of (P, M, π, G). Given a reduction (Q, M, π', H) of the principal bundle (P, M, π, G), a connection $A \in \Omega(P; \mathfrak{g})$ is said to be ***reducible*** to the connection $A' \in \omega(Q; \mathfrak{h})$, where $\mathfrak{h}$ is the Lie algebra of H, if $A'(q)(u_q, v_q) = A(\iota(q))(T_q\iota(u_q), T_q\iota(v_q))$ for all $q \in Q$ and all $u_q, v_q \in T_qQ$.

Let now $A \in \Omega(P; \mathfrak{g})$ be a connection on (P, M, π, G) where M is connected and paracompact. For $z \in P$ denote by $P(z)$ the set of points in P which can be joined to z by a horizontal curve. $P(z)$ is called the ***holonomy bundle*** through z. The ***Reduction Theorem*** states that $(P(z), M, \pi', \mathcal{H}(z))$ is a reduction of (P, M, π, G) and that the connection A is reducible to a connection in $(P(z), M, \pi', \mathcal{H}(z))$.

5.1.3 Properties of the holonomy bundle. The tangent space to the holonomy bundle $P(z)$ at any point $y \in P(z)$ can be written as the direct sum

$$T_yP(z) = H(y) \oplus \mathrm{Lie}(\mathcal{H}(z)) \cdot y,$$

where $H(y)$ is the horizontal space at $y \in P(z) \subset P$ of the connection $A \in \Omega(P; \mathfrak{g})$. The collection of the tangent spaces to the holonomy bundles forms a smooth and involutive distribution on P whose maximal integral manifolds are the holonomy bundles themselves. This implies that the holonomy bundles are not only injectively immersed submanifolds but initial submanifolds of P.

All the holonomy bundles are isomorphic as principal bundles via the group action, that is, given any two points $z_1, z_2 \in P$ there exists an element $g \in G$ such that $R_g : P(z_1) \to P(z_2)$ is a principal bundle isomorphism whose associated structure group isomorphism is conjugation by g^{-1}. Indeed, three possibilities can occur:

(i) $z_1, z_2 \in P$ are two points that can be joined by a horizontal curve; then $P(z_1) = P(z_2)$ by definition.

(ii) $z_1, z_2 \in P$ are two points in the same fiber, that is, there exists $g \in G$ such that $z_2 = R_g(z_1)$. Since the group action maps horizontal curves to horizontal curves one has $R_g(P(z_1)) = P(z_2)$. In addition, $R_g : P(z_1) \to P(z_2)$ is a principal bundle isomorphism relative to the group isomorphism $\mathcal{H}(z_1) \to \mathcal{H}(z_2)$ implemented by conjugation using the element $g^{-1} \in G$.

(iii) If none of the above possibilities hold, any two points $z_1, z_2 \in P$ are such that $\pi(z_1)$ and $\pi(z_2)$ can be joined by a smooth curve (connectedness of the base). Horizontally lift this curve through z_1. Its endpoint z_3 lies in the fiber of z_2. Therefore, $P(z_1) = P(z_3)$ by point **(i)** and there exists a $g \in G$ such that $z_2 = R_g(z_3)$. Therefore, by point **(ii)** $R_g : P(z_3) \to P(z_2)$ is a principal bundle isomorphism between $P(z_1) = P(z_3)$ and $P(z_2)$.

5.1.4 The Holonomy Theorem. Let (P, M, π, G) be a principal fiber bundle where M is a connected and paracompact manifold. Let $A \in \Omega(P, \mathfrak{g})$ be a connection on P. Given any vector $v_z \in T_z P$ we will denote by $v_z^H \in H(z)$ its horizontal part. Recall that the ***curvature form*** $\Omega \in \Omega^2(P; \mathfrak{g})$ of the connection A is defined as

$$\Omega(z)(v_z, w_z) = \mathbf{d}A(z)(v_z^H, w_z^H). \tag{5.1.1}$$

A connection A is said to be ***flat*** when its curvature form is identically zero. The de Rham cohomology class $[\Omega] \in H^2(P; \mathfrak{g})$ corresponding to the curvature form Ω is an algebraic invariant of the bundle (P, M, π, G) that does not depend on the connection A that we used to define it. This result is due to A. Weil and $[\Omega]$ is referred to as the ***Chern class*** of the principal fiber bundle (P, M, π, G).

The ***Holonomy Theorem*** of AMBROSE AND SINGER (1953) states that given any point $z \in P$, the Lie algebra of the holonomy group $\mathcal{H}(z)$ of A with reference point z equals the subspace of $\mathfrak{g}$ spanned by all the elements of the form $\Omega(p)(v_p, w_p)$, $p \in P(z)$, $v_p, w_p \in H(p)$.

This theorem implies that a connection is flat if and only if its holonomy groups are discrete. In turn this is equivalent to the horizontal subbundle being an involutive distribution that has the holonomy bundles as maximal integral manifolds.

5.1.5 Proposition *Let A be a flat connection on the principal bundle (P, M, π, G) with connected and paracompact base M and let $(P(z), M, \pi', \mathcal{H}(z))$ be the holonomy reduced bundle at a point $z \in P$. Then $\pi' : P(z) \to M$ is a covering map.*

Proof. Since the connection is flat, the Lie algebra $\operatorname{Lie}(\mathcal{H}(z))$ of the holonomy group is trivial by the Holonomy Theorem and hence $\mathcal{H}(z)$ is a discrete group. As $(P(z), M, \pi', \mathcal{H}(z))$ is a locally trivial bundle, any point $m \in M$ has an open neighborhood U such that $(\pi')^{-1}(U)$ is diffeomorphic to $U \times \mathcal{H}(z)$. Since $\mathcal{H}(z)$ is discrete, each subset $U \times \{g\}$, $g \in \mathcal{H}(z)$, is an open subset diffeomorphic to U. Hence, π' is a covering map. ■

5.1.6 Holonomy Computations. Let (P, M, π, G) be a right principal bundle, $A \in \Omega(P; \mathfrak{g})$ a connection on P, and $\Omega \in \Omega^2(P; \mathfrak{g})$ its curvature. Let $d : [0, 1] \to M$ be a closed smooth path contained in the open subset $U \subset M$ and let $\sigma : U \to P$ be a smooth local section. Then $\sigma^* A$ is a $\mathfrak{g}$-valued one-form on U and hence applied to $\dot{d}(t) \in T_{d(t)} M$ gives a curve in $\mathfrak{g}$. Let $g(t) \in G$ be the solution of the differential equation

$$\frac{dg(t)}{dt} = -T_e R_{g(t)} \left((\sigma^* A) \left(\dot{d}(t) \right) \right)$$

with initial condition $g(t) = e$. Then *the holonomy of the path $c(t)$ measured from $\sigma(c(0))$ is $g(1)$.* This statement follows easily from the local expression of a connection one-form and its transformation laws under the group action.

If G is an Abelian, then it is a cylinder $\mathbb{R}^k \times \mathbb{T}^{n-k}$, where $n = \dim G$. In particular, $g(t) \in G$ is of the form $\exp \eta(t)$ for some smooth function $\eta(t) \in \mathfrak{g}$. Thus $T_{g(t)} R_{g(t)^{-1}} \dot{g}(t) = \dot{\eta}(t)$ and the previous statement gives the holonomy of the closed path $d : [0, 1] \to M$ as

$$g(1) = \exp \eta(1) = \exp \left(- \int_0^1 (\sigma^* A)(\dot{d}(s)) ds \right).$$

If the loop bounds a two dimensional submanifold S of M then, using the Stokes Theorem and the structure equations of A, the holonomy has the alternate expression

$$g(1) = \exp \eta(1) = \exp\left(-\iint_S \sigma^* \mathbf{d}A\right) = \exp\left(-\iint_S \Omega\right). \tag{5.1.2}$$

5.2 Cylinder valued momentum maps

In this section we study a construction due to CONDEVAUX, DAZORD, AND MOLINO (1988) that, in the context of connected symplectic manifolds acted canonically upon by a Lie algebra, generalizes the standard momentum map and has the important property that, unlike the standard momentum map, it is always defined.

5.2.1 The setup. From now on and until the end of the section we will be concerned by a connected and paracompact symplectic manifold (M, ω) acted canonically upon by a Lie algebra $\mathfrak{g}$.

Let $\pi : M \times \mathfrak{g}^* \to M$ be the projection onto M. Consider π as the bundle map of the trivial principal fiber bundle $(M \times \mathfrak{g}^*, M, \pi, \mathfrak{g}^*)$ that has $(\mathfrak{g}^*, +)$ as an Abelian structure group. The group $(\mathfrak{g}^*, +)$ acts on $M \times \mathfrak{g}^*$ by $\nu \cdot (m, \mu) := (m, \mu - \nu)$, with $m \in M$ and $\mu, \nu \in \mathfrak{g}^*$. Let $\alpha \in \Omega^1(M \times \mathfrak{g}^*; \mathfrak{g}^*)$ be the connection one-form defined by

$$\langle \alpha(m, \mu) \cdot (v_m, \nu), \xi \rangle := (\mathbf{i}_{\xi_M}\omega)(m) \cdot v_m - \langle \nu, \xi \rangle, \tag{5.2.1}$$

where $(m, \mu) \in M \times \mathfrak{g}^*$, $(v_m, \nu) \in T_m M \times \mathfrak{g}^*$, $\xi \in \mathfrak{g}$, and $\langle \cdot, \cdot \rangle$ denotes the natural pairing between $\mathfrak{g}^*$ and $\mathfrak{g}$.

We briefly check that the defining properties **(i)** through **(iii)** introduced in §5.1.1 are satisfied by α. First notice that $\alpha(m, \mu) : T_m M \times \mathfrak{g}^* \to \mathfrak{g}^*$ is a linear map for all $(m, \mu) \in M \times \mathfrak{g}^*$. Second, the infinitesimal generator $\nu_{M \times \mathfrak{g}^*}$ associated to an element $\nu \in \mathfrak{g}^*$ is given by

$$\nu_{M \times \mathfrak{g}^*}(m, \mu) = \left.\frac{d}{dt}\right|_{t=0} (m, \mu - t\nu) = (0, -\nu).$$

Consequently, for any $\xi \in \mathfrak{g}$, $\langle \alpha(m, \mu) \cdot \nu_{M \times \mathfrak{g}^*}(m, \mu), \xi \rangle = \langle \alpha(m, \mu) \cdot (0, -\nu), \xi \rangle = \langle \nu, \xi \rangle$, that is, $\alpha(m, \mu) \cdot \nu_{M \times \mathfrak{g}^*}(m, \mu) = \nu$. Third, it is obvious that for any $\rho \in \mathfrak{g}^*$ we have $\langle \alpha(m, \mu - \rho) \cdot (v_m, \nu), \xi \rangle = (\mathbf{i}_{\xi_M}\omega)(m) \cdot v_m - \langle \nu, \xi \rangle = \langle \alpha(m, \mu) \cdot (v_m, \nu), \xi \rangle$, which proves that α is a well-defined connection one-form on $M \times \mathfrak{g}^*$.

5.2.2 The horizontal bundle of α. The vertical subbundle $V \subset T(M \times \mathfrak{g}^*)$ of $\pi : M \times \mathfrak{g}^* \to M$ is given for any $(m, \mu) \in M \times \mathfrak{g}^*$ by

$$V(m, \mu) := \{(0, \rho) \in T_{(m,\mu)}(M \times \mathfrak{g}^*) \mid \rho \in \mathfrak{g}^*\}. \tag{5.2.2}$$

By definition, the horizontal subspace $H(m, \mu)$ determined by α at the point $(m, \mu) \in M \times \mathfrak{g}^*$ is given by

$$H(m, \mu) = \{(v_m, \nu) \in T_{(m,\mu)}(M \times \mathfrak{g}^*) \mid (\mathbf{i}_{\xi_M}\omega)(m) \cdot v_m - \langle \nu, \xi \rangle = 0, \text{ for all } \xi \in \mathfrak{g}\}. \tag{5.2.3}$$

Consequently, given any vector $(v_m, \nu) \in T_{(m,\mu)}(M \times \mathfrak{g}^*)$, its horizontal $(v_m, \nu)^H$ and vertical $(v_m, \nu)^V$ parts are such that

$$(v_m, \nu)^H = (v_m, \rho) \quad \text{and} \quad (v_m, \nu)^V = (0, \rho'),$$

where $\rho, \rho' \in \mathfrak{g}^*$ are uniquely determined by the relations

$$\langle \rho, \xi \rangle = (\mathbf{i}_{\xi_M}\omega)(m) \cdot v_m \quad \text{and} \quad \rho' = \nu - \rho, \quad \text{for any} \quad \xi \in \mathfrak{g}.$$

5.2.3 α **is a flat connection.** We compute the curvature form Ω associated to α using the prescription given in (5.1.1). Let $(m, \mu) \in M \times \mathfrak{g}^*$, $v_m, u_m \in T_mM$, $\xi \in \mathfrak{g}$, and $\nu, \rho \in \mathfrak{g}^*$ be arbitrary. By definition,

$$\langle \Omega(m, \mu)((v_m, \nu), (u_m, \rho)), \xi \rangle = \langle \mathbf{d}\alpha(m, \mu)((v_m, \nu)^H, (u_m, \rho)^H), \xi \rangle. \tag{5.2.4}$$

Let now (X_1, Y_1) and (X_2, Y_2) be vector fields on $M \times \mathfrak{g}^*$ such that $(X_1(m), Y_1(\mu)) = (v_m, \nu)$ and $(X_2(m), Y_2(\mu)) = (u_m, \rho)$. Using these vector fields, the right-hand side of (5.2.4) can be rewritten as

$$\begin{aligned} \langle (X_1, Y_1)[\alpha(X_2, Y_2)](m, \mu), \xi \rangle - \langle (X_2, Y_2)[\alpha(X_1, Y_1)](m, \mu), \xi \rangle \\ - \langle \alpha([X_1, X_2], 0)(m, \mu), \xi \rangle. \end{aligned} \tag{5.2.5}$$

Let (m_t^1, μ_t^1) and (m_t^2, μ_t^2) be the flows of (X_1, Y_1) and (X_2, Y_2), respectively. Choose Y_1 and Y_2 with flows $\mu_t^1(\mu) = \mu + t\nu$ and $\mu_t^2(\mu) = \mu + t\rho$. We can use these flows to compute

$$\begin{aligned} &\langle (X_1, Y_1)[\alpha(X_2, Y_2)](m, \mu), \xi \rangle \\ &\quad = \frac{d}{dt}\Big|_{t=0} \left\langle \alpha(m_t^1(m), \mu_t^1(\mu)) \cdot (X_2(m_t^1(m)), Y_2(\mu_t^1(\mu))), \xi \right\rangle \\ &\quad = \frac{d}{dt}\Big|_{t=0} \left((\mathbf{i}_{\xi_M}\omega)(m_t^1(m)) \cdot X_2(m_t^1(m)) - \left\langle Y_2(\mu_t^1(\mu)), \xi \right\rangle \right) \\ &\quad = X_1 \left[\mathbf{i}_{\xi_M}\omega(X_2) \right](m) - \frac{d}{dt}\Big|_{t=0} \frac{d}{ds}\Big|_{s=0} \langle \mu + t\nu + s\rho, \xi \rangle \\ &\quad = X_1 \left[\mathbf{i}_{\xi_M}\omega(X_2) \right](m). \end{aligned}$$

Analogously, we have

$$\langle (X_2, Y_2)[\alpha(X_1, Y_1)](m, \mu), \xi \rangle = X_2 \left[\mathbf{i}_{\xi_M}\omega(X_1) \right](m).$$

Consequently, the expression (5.2.5) equals

$$\begin{aligned} X_1 \left[\mathbf{i}_{\xi_M}\omega(X_2) \right](m) - X_2 \left[\mathbf{i}_{\xi_M}\omega(X_1) \right](m) - \mathbf{i}_{\xi_M}\omega(m)([X_1, X_2](m)) \\ = \mathbf{d}(\mathbf{i}_{\xi_M}\omega)(m)(X_1(m), X_2(m)) \\ = (\pounds_{\xi_M}\omega)(m)(X_1(m), X_2(m)) - (\mathbf{i}_{\xi_M}\mathbf{d}\omega)(m)(X_1(m), X_2(m)) = 0, \end{aligned}$$

which guarantees the flatness of α.

5.2.4 Cylinder valued momentum maps. For $(z, \mu) \in M \times \mathfrak{g}^*$, let $(M \times \mathfrak{g}^*)(z, \mu)$ be the holonomy bundle through (z, μ) and let $\mathcal{H}(z, \mu)$ be the holonomy group of α with reference point (z, μ). The Reduction Theorem guarantees that $((M \times \mathfrak{g}^*)(z, \mu), M, \pi|_{(M\times\mathfrak{g}^*)(z,\mu)}, \mathcal{H}(z, \mu))$ is a reduction of $(M \times \mathfrak{g}^*, M, \pi, \mathfrak{g}^*)$. To simplify notation, we will write $(\widetilde{M}, M, \widetilde{p}, \mathcal{H})$ instead of $((M \times \mathfrak{g}^*)(z, \mu), M, \pi|_{(M\times\mathfrak{g}^*)(z,\mu)}, \mathcal{H}(z, \mu))$. Let $\widetilde{\mathbf{K}} : \widetilde{M} \subset M \times \mathfrak{g}^* \to \mathfrak{g}^*$ be the projection into the $\mathfrak{g}^*$-factor.

Consider now the closure $\overline{\mathcal{H}}$ of $\mathcal{H}$ in $\mathfrak{g}^*$. Since $\overline{\mathcal{H}}$ is a closed subgroup of $(\mathfrak{g}^*, +)$, the quotient $C := \mathfrak{g}^*/\overline{\mathcal{H}}$ is a cylinder (that is, it is isomorphic to the Abelian Lie group $\mathbb{R}^a \times \mathbb{T}^b$ for some $a, b \in \mathbb{N}$). Let $\pi_C : \mathfrak{g}^* \to \mathfrak{g}^*/\overline{\mathcal{H}} = C$ be the projection. Define $\mathbf{K} : M \to C$ to be the map that makes the following diagram commutative:

$$\begin{array}{ccc} \widetilde{M} & \xrightarrow{\widetilde{\mathbf{K}}} & \mathfrak{g}^* \\ {\scriptstyle \widetilde{p}}\downarrow & & \downarrow{\scriptstyle \pi_C} \\ M & \xrightarrow{\mathbf{K}} & \mathfrak{g}^*/\overline{\mathcal{H}}. \end{array} \tag{5.2.6}$$

In other words, $\mathbf{K}$ is defined by $\mathbf{K}(m) = \pi_C(\nu)$, where $\nu \in \mathfrak{g}^*$ is any element such that $(m, \nu) \in \widetilde{M}$. This is a good definition because if we have two points $(m, \nu), (m, \nu') \in \widetilde{M}$, this implies that $(m, \nu), (m, \nu') \in \widetilde{p}^{-1}(m)$ and, as $\mathcal{H}$ is the structure group of the principal fiber bundle $\widetilde{p} : \widetilde{M} \to M$, there exists an element $\rho \in \mathcal{H}$ such that $\nu' = \nu + \rho$. Consequently, $\pi_C(\nu) = \pi_C(\nu')$.

We will refer to $\mathbf{K} : M \to \mathfrak{g}^*/\overline{\mathcal{H}} =: C$ as a ***cylinder valued momentum map*** associated to the canonical $\mathfrak{g}$-action on (M, ω). CONDEVAUX, DAZORD, AND MOLINO (1988) call $\mathbf{K}$ the ***reduced momentum map***.

A natural question is how the definition of $\mathbf{K}$ depends on the choice of the holonomy bundle $\widetilde{M}$. In order to answer it, let $\widetilde{M_1}$ and $\widetilde{M_2}$ be two holonomy bundles of $(M \times \mathfrak{g}^*, M, \pi, \mathfrak{g}^*)$. By §5.1.3 there exists $\tau \in \mathfrak{g}^*$ such that $\widetilde{M_2} = R_\tau(\widetilde{M_1})$, where $R_\tau(m, \mu) := (m, \mu + \tau)$, for any $(m, \mu) \in M \times \mathfrak{g}^*$. Notice first that since $(\mathfrak{g}^*, +)$ is Abelian, all the holonomy groups based at any point are the same and hence the projection $\pi_C : \mathfrak{g}^* \to \mathfrak{g}^*/\overline{\mathcal{H}}$ in (5.2.6) does not depend on the choice of $\widetilde{M}$. Additionally, π_C is a group homomorphism. Let now $\widetilde{p}_{\widetilde{M_i}} : \widetilde{M_i} \to M$, $\widetilde{\mathbf{K}}_{\widetilde{M_i}} : \widetilde{M_i} \to \mathfrak{g}^*$, and $\mathbf{K}_{\widetilde{M_i}} : M \to \mathfrak{g}^*$ be the maps in the diagram (5.2.6) constructed using the holonomy bundles $\widetilde{M_i}$, $i \in \{1, 2\}$. Let $m \in M$. By definition $\mathbf{K}_{\widetilde{M_2}}(m) = \mathbf{K}_{\widetilde{M_2}}(\widetilde{p}_{\widetilde{M_2}}(m, \nu))$, where $(m, \nu) \in \widetilde{M_2}$. Since $\widetilde{M_2} = R_\tau(\widetilde{M_1})$ there exists an element $\nu' \in \mathfrak{g}^*$ such that $(m, \nu') \in \widetilde{M_1}$ and $(m, \nu) = (m, \nu' + \tau)$. Hence,

$$\begin{aligned} \mathbf{K}_{\widetilde{M_2}}(m) &= \mathbf{K}_{\widetilde{M_2}}(\widetilde{p}_{\widetilde{M_2}}(m, \nu)) = \mathbf{K}_{\widetilde{M_2}}(\widetilde{p}_{\widetilde{M_2}}(m, \nu' + \tau)) \\ &= \pi_C(\widetilde{\mathbf{K}}_{\widetilde{M_2}}(m, \nu' + \tau)) = \pi_C(\nu' + \tau) = \pi_C(\nu') + \pi_C(\tau) \\ &= \mathbf{K}_{\widetilde{M_1}}(m) + \pi_C(\tau). \end{aligned}$$

Given that in the previous chain of equalities the point $m \in M$ is arbitrary and $\tau \in \mathfrak{g}^*$ depends only on $\widetilde{M_1}$ and $\widetilde{M_2}$ we have that

$$\mathbf{K}_{\widetilde{M_2}} = \mathbf{K}_{\widetilde{M_1}} + \pi_C(\tau).$$

5.2.5 Example. One of the advantages of cylinder valued momentum maps when compared with the standard ones is that the former are always available. Let us compute a cylinder valued momentum map for the example introduced in §4.5.16 of a canonical circle action that does not have a standard momentum map. Recall that the setup in that case consisted of a torus $\mathbb{T}^2 = \{(e^{i\theta_1}, e^{i\theta_2})\}$ considered as a symplectic manifold with the area form $\omega := \mathbf{d}\theta_1 \wedge \mathbf{d}\theta_2$ and a circle $S^1 = \{e^{i\phi}\}$ acting canonically on it by $e^{i\phi} \cdot (e^{i\theta_1}, e^{i\theta_2}) := (e^{i(\theta_1+\phi)}, e^{i\theta_2})$.

In order to construct a cylinder valued momentum map associated to this action consider the trivial principal bundle $\mathbb{T}^2 \times \mathbb{R} \to \mathbb{T}^2$ with $(\mathbb{R}, +)$ as structure group. A straightforward computation shows that the horizontal vectors in $T(\mathbb{T}^2 \times \mathbb{R})$ with respect to the connection α defined in (5.2.1) are of the form $((a, b), b)$, with $a, b \in \mathbb{R}$. Since any surface $\widetilde{\mathbb{T}^2}_\tau := \{((e^{i\theta_1}, e^{i\theta_2}), \tau + \theta_2) \in \mathbb{T}^2 \times \mathbb{R} \mid \theta_1, \theta_2 \in \mathbb{R}\} \subset \mathbb{T}^2 \times \mathbb{R}$ integrates the horizontal distribution spanned by these vectors, it is immediately clear that the holonomy groups constructed at any point equal $(\mathbb{Z}, +)$. Therefore, we can define a cylinder valued momentum map $\mathbf{K} : \mathbb{T}^2 \to \mathbb{R}/\mathbb{Z} \simeq S^1$ by using the diagram

$$\begin{array}{ccc} \widetilde{\mathbb{T}^2}_0 & \xrightarrow{\widetilde{\mathbf{K}}} & \mathbb{R} \\ {\scriptstyle \widetilde{p}}\downarrow & & \downarrow{\scriptstyle \pi_C} \\ \mathbb{T}^2 & \xrightarrow{\mathbf{K}} & \mathbb{R}/\mathbb{Z} \simeq S^1. \end{array}$$

More specifically, we have that $\mathbf{K}(e^{i\theta_1}, e^{i\theta_2}) = e^{i\theta_2}$, for any $(e^{i\theta_1}, e^{i\theta_2}) \in \mathbb{T}^2$.

5.2.6 Example. The reader may be wondering why in the definition of the cylinder valued momentum maps we used the closure $\overline{\mathcal{H}}$ of the holonomy subgroup instead of just the holonomy subgroup $\mathcal{H}$. Our next example shows that $\mathcal{H}$ is, in general, not closed in $\mathfrak{g}^*$, and thus taking $\overline{\mathcal{H}}$ is the only way to guarantee that the cylinder valued momentum map has a smooth manifold as target and can be chosen to be itself a smooth map.

Let $M := \mathbb{T}^2 \times \mathbb{T}^2$ be the product of two tori whose elements will be denoted by the four-tuples $(e^{i\theta_1}, e^{i\theta_2}, e^{i\psi_1}, e^{i\psi_2})$. Endow M with the symplectic structure ω defined by

$$\omega := \mathbf{d}\theta_1 \wedge \mathbf{d}\theta_2 + \sqrt{2}\,\mathbf{d}\psi_1 \wedge \mathbf{d}\psi_2.$$

Consider the canonical circle action given by $e^{i\phi} \cdot (e^{i\theta_1}, e^{i\theta_2}, e^{i\psi_1}, e^{i\psi_2}) := (e^{i(\theta_1+\phi)}, e^{i\theta_2}, e^{i(\psi_1+\phi)}, e^{i\psi_2})$ and the trivial principal bundle $(M \times \mathbb{R}) \to M$ with $(\mathbb{R}, +)$ as structure group. It is easy to see that the horizontal vectors in $T(M \times \mathbb{R})$ with respect to the connection α defined in (5.2.1) are of the form $((a_1, a_2, b_1, b_2), -a_2 - \sqrt{2}b_2)$, with $a_1, a_2, b_1, b_2 \in \mathbb{R}$. The surfaces $\widetilde{M}_\tau \subset M \times \mathbb{R}$ of the form $\widetilde{M}_\tau := \{((e^{i\theta_1}, e^{i\theta_2}, e^{i\psi_1}, e^{i\psi_2}), \tau - \theta_2 - \sqrt{2}\psi_2) \in M \times \mathbb{R} \mid \theta_1, \theta_2, \psi_1, \psi_2 \in \mathbb{R}\}$ integrate the horizontal distribution spanned by these vectors.

Take now $\widetilde{M} := \widetilde{M}_0$ and consider the projection $\widetilde{p} : \widetilde{M} \to M$. It is clear that $\widetilde{p}^{-1}(e^{i\theta_1}, e^{i\theta_2}, e^{i\psi_1}, e^{i\psi_2}) = \{(e^{i\theta_1}, e^{i\theta_2}, e^{i\psi_1}, e^{i\psi_2}, -(\theta_2 + 2n\pi) - \sqrt{2}(\psi_2 + 2m\pi)) \mid m, n \in \mathbb{Z}\}$. Since the holonomy group $\mathcal{H}$ of α at any point in $\widetilde{M}$ coincides with the structure group of the fibration $\widetilde{p} : \widetilde{M} \to M$, it follows that $\mathcal{H} = \mathbb{Z} + \sqrt{2}\mathbb{Z} \subset \mathbb{R}$. This is an example of a nonclosed holonomy subgroup since $\mathcal{H}$ is dense in $\mathbb{R}$, that is,

$\overline{\mathcal{H}} = \mathbb{R}$. Therefore, in this case, the cylinder valued momentum map is trivial since its range is just a point.

5.2.7 Properties of cylinder valued momentum maps. We summarize them in the following theorem.

5.2.8 Theorem *Let (M, ω) be a connected and paracompact symplectic manifold and $\mathfrak{g}$ a Lie algebra acting canonically on it. Then any cylinder valued momentum map $\mathbf{K} : M \to C$ associated to this action has the following properties:*

(i) *$\mathbf{K}$ is a smooth Noether momentum map.*

(ii) *For any $v_m \in T_m M$, $m \in M$, we have that*

$$T_m\mathbf{K}(v_m) = T_\mu \pi_C(\nu),$$

where $\mu \in \mathfrak{g}^*$ is any element such that $\mathbf{K}(m) = \pi_C(\mu)$ and $\nu \in \mathfrak{g}^*$ is uniquely determined by

$$\langle \nu, \xi \rangle = (\mathbf{i}_{\xi_M}\omega)(m) \cdot v_m, \quad \textit{for any} \quad \xi \in \mathfrak{g}. \tag{5.2.7}$$

(iii) $\ker(T_m\mathbf{K}) = \left(\left(\mathrm{Lie}(\overline{\mathcal{H}})\right)^\circ \cdot m\right)^\omega.$

(iv) Bifurcation Lemma:

$$\mathrm{range}\,(T_m\mathbf{K}) = T_\mu \pi_C\left((\mathfrak{g}_m)^\circ\right),$$

where $\mu \in \mathfrak{g}^*$ is any element such that $\mathbf{K}(m) = \pi_C(\mu)$.

Proof. As $\mathfrak{g}^*/\overline{\mathcal{H}}$ is a homogeneous manifold, we have that $\pi_C : \mathfrak{g}^* \to \mathfrak{g}^*/\overline{\mathcal{H}}$ is a surjective submersion. Moreover, since by (5.2.6), $\mathbf{K} \circ \widetilde{p} = \pi_C \circ \widetilde{\mathbf{K}}$ is a smooth map and $\widetilde{p}$ is a surjective submersion, the map $\mathbf{K}$ is necessarily smooth.

We start by proving **(ii)**. Let $m \in M$ and $(m, \mu) \in \widetilde{p}^{-1}(m)$. If $v_m = T_{(m,\mu)}\widetilde{p}(v_m, \nu)$, then (5.2.6) gives

$$T_m\mathbf{K}(v_m) = T_m\mathbf{K}\left(T_{(m,\mu)}\widetilde{p}(v_m, \nu)\right) = T_\mu \pi_C\left(T_{(m,\mu)}\widetilde{\mathbf{K}}(v_m, \nu)\right) = T_\mu \pi_C(\nu).$$

(i) We now check that $\mathbf{K}$ satisfies Noether's condition. Let $h \in C^\infty(M)^{\mathfrak{g}}$ and let F_t be the flow of the associated Hamiltonian vector field X_h. Using the expression for the derivative $T_m\mathbf{K}$ in **(ii)** it follows that $T_m\mathbf{K}(X_h(m)) = T_\mu\pi_C(\nu)$, where $\mu \in \mathfrak{g}^*$ is any element such that $\mathbf{K}(m) = \pi_C(\mu)$ and $\nu \in \mathfrak{g}^*$ is uniquely determined by

$$\langle \nu, \xi \rangle = (\mathbf{i}_{\xi_M}\omega)(m)(X_h(m)) = -\mathbf{d}h(m)(\xi_M(m)) = \xi_M[h](m) = 0,$$

for all $\xi \in \mathfrak{g}$, which proves that $\nu = 0$ and, consequently, $T_m\mathbf{K}(X_h(m)) = 0$, for all $m \in M$. Finally, as

$$\frac{d}{dt}(\mathbf{K} \circ F_t)(m) = T_{F_t(m)}\mathbf{K}\left(X_h(F_t(m))\right) = 0,$$

we have that $\mathbf{K} \circ F_t = \mathbf{K}|_{\mathrm{Dom}(F_t)}$, as required.

(iii) Due to the expression in **(ii)**, a vector $v_m \in \ker T_m\mathbf{K}$ if and only if the unique element $\nu \in \mathfrak{g}^*$ determined by (5.2.7) satisfies $T_\nu \pi_C(\nu) = 0$, that is, $\nu \in \mathrm{Lie}(\overline{\mathcal{H}})$. Equivalently, we have that $\langle \nu, \xi \rangle = 0$, for any $\xi \in (\mathrm{Lie}(\overline{\mathcal{H}}))^\circ \subset (\mathfrak{g}^*)^* = \mathfrak{g}$ which in terms of v_m yields that $(\mathbf{i}_{\xi_M}\omega)(m) \cdot v_m = 0$ for any $\xi \in (\mathrm{Lie}(\overline{\mathcal{H}}))^\circ$. This can obviously be rewritten by saying that $v_m \in \left(\left(\mathrm{Lie}(\overline{\mathcal{H}})\right)^\circ \cdot m \right)^\omega$.

(iv) We start by checking that $\mathrm{range}\,(T_m\mathbf{K}) \subset T_\mu \pi_C\left((\mathfrak{g}_m)^\circ\right)$. Let $T_m\mathbf{K}(v_m) \in \mathrm{range}\,(T_m\mathbf{K})$. Let $\nu \in \mathfrak{g}^*$ be the element determined by (5.2.7) which hence satisfies $T_m\mathbf{K}(v_m) = T_\mu\pi_C(\nu)$. Now, notice that for any $\xi \in \mathfrak{g}_m$ we have that

$$\langle \nu, \xi \rangle = \omega(m)(\xi_M(m), v_m) = 0,$$

which implies that $\nu \in (\mathfrak{g}_m)^\circ$. This proves the inclusion $\mathrm{range}\,(T_m\mathbf{K}) \subset T_\mu\pi_C\left((\mathfrak{g}_m)^\circ\right)$. Hence, the equality will be proved if we show that

$$\mathrm{rank}\,(T_m\mathbf{K}) = \dim\left(T_\mu\pi_C\left((\mathfrak{g}_m)^\circ\right)\right). \tag{5.2.8}$$

On the one hand we can use the equality in **(iii)** to obtain

$$\begin{aligned} \mathrm{rank}\,(T_m\mathbf{K}) &= \dim M - \dim(\ker T_m\mathbf{K}) \\ &= \dim M - \dim M + \dim\left(\left(\mathrm{Lie}(\overline{\mathcal{H}})\right)^\circ \cdot m\right) \\ &= \dim\left(\left(\mathrm{Lie}(\overline{\mathcal{H}})\right)^\circ\right) - \dim\left(\mathfrak{g}_m \cap \left(\mathrm{Lie}(\overline{\mathcal{H}})\right)^\circ\right). \end{aligned} \tag{5.2.9}$$

On the other hand,

$$\begin{aligned} \dim\left(T_\mu\pi_C\left((\mathfrak{g}_m)^\circ\right)\right) &= \dim(\mathfrak{g}_m)^\circ - \dim(\ker T_\mu\pi_C|_{(\mathfrak{g}_m)^\circ}) \\ &= \dim\mathfrak{g} - \dim\mathfrak{g}_m - \dim(\ker T_\mu\pi_C \cap (\mathfrak{g}_m)^\circ) \\ &= \dim\mathfrak{g} - \dim\mathfrak{g}_m - \dim(\mathrm{Lie}(\overline{\mathcal{H}}) \cap (\mathfrak{g}_m)^\circ) \\ &= \dim\mathfrak{g} - \dim\mathfrak{g}_m - \dim(\mathrm{Lie}(\overline{\mathcal{H}})) - \dim(\mathfrak{g}_m)^\circ \\ &\qquad + \dim(\mathrm{Lie}(\overline{\mathcal{H}}) + (\mathfrak{g}_m)^\circ) \\ &= -\dim(\mathrm{Lie}(\overline{\mathcal{H}})) + \dim\mathfrak{g} - \dim([\mathrm{Lie}(\overline{\mathcal{H}}) + (\mathfrak{g}_m)^\circ]^\circ) \\ &= \dim\left(\mathrm{Lie}(\overline{\mathcal{H}}))^\circ\right) - \dim\left(\mathfrak{g}_m \cap \left(\mathrm{Lie}(\overline{\mathcal{H}})\right)^\circ\right), \end{aligned}$$

which coincides with (5.2.9), thereby establishing (5.2.8). ■

5.2.9 The relation between standard and cylinder valued momentum maps. We now address in what sense cylinder valued momentum maps generalize standard ones. As we shall see in the following result, it can be proved that given a canonical Lie algebra action on a connected symplectic manifold, there exists a standard momentum map if and only if the holonomy $\mathcal{H}$ is trivial. Moreover, in such a situation the cylinder valued momentum maps *are* standard momentum maps. This shows that cylinder valued momentum maps are truly a generalization of the standard ones. Incidentally, this result provides us with an alternative characterization of the existence of standard momentum maps in different terms than those in Proposition 4.5.17.

5.2.10 Proposition *Let (M, ω) be a connected paracompact symplectic manifold and $\mathfrak{g}$ a Lie algebra acting canonically on it. Let $\mathbf{K} : M \to \mathfrak{g}^*/\overline{\mathcal{H}}$ be a cylinder valued momentum map (always available). Then there exists a standard momentum map if and only if $\mathcal{H} = \{0\}$. In this case, $\mathbf{K}$ is a standard momentum map.*

Proof. Suppose that the cylinder valued momentum map $\mathbf{K} : M \to \mathfrak{g}^*/\overline{\mathcal{H}}$ has been constructed using the reduced bundle $((M \times \mathfrak{g}^*)(m, \nu) =: \widetilde{M}, M, \pi|_{(M\times\mathfrak{g}^*)(m,\nu)} =: \widetilde{p}, \mathcal{H}(m, \nu) := \mathcal{H})$, $(m, \nu) \in M \times \mathfrak{g}^*$. We now show that if there exists a standard momentum map $\mathbf{J} : M \to \mathfrak{g}^*$ associated to this action, then $\mathcal{H} = \{0\}$. Indeed, if $\mu \in \mathcal{H}$, the very definition of $\mathcal{H}$ implies that there exists a loop $c : [0, 1] \to M$ at m, that is, $c(0) = c(1) = m$, such that one of its horizontal lifts $\widetilde{c} : [0, 1] \to \widetilde{M}$ given by the function $t \mapsto (c(t), \rho(t))$ is such that $\rho(0) = \rho \in \mathfrak{g}^*$ and $\rho(1) = \rho + \mu$, $\rho \in \mathfrak{g}^*$. Now, since $\widetilde{c}$ is horizontal we have that $\langle \alpha(c(t), \rho(t))(c'(t), \rho'(t)), \xi\rangle = 0$, for any $\xi \in \mathfrak{g}$. Equivalently,

$$(\mathbf{i}_{\xi_M}\omega)(c(t))\left(c'(t)\right) = \langle \rho'(t), \xi\rangle.$$

We can use the momentum map $\mathbf{J} : M \to \mathfrak{g}^*$ to express this identity as

$$\mathbf{dJ}^\xi(c(t))\left(c'(t)\right) = \langle \rho'(t), \xi\rangle,$$

which is the same as

$$\frac{d}{dt}\mathbf{J}^\xi(c(t)) = \frac{d}{dt}\langle \rho(t), \xi\rangle.$$

Integrating we obtain

$$\mathbf{J}^\xi(c(t)) - \mathbf{J}^\xi(m) = \int_0^t \frac{d}{ds}\mathbf{J}^\xi(c(s))\,ds = \int_0^t \frac{d}{ds}\langle \rho(s), \xi\rangle\,ds = \langle \rho(t), \xi\rangle - \langle \rho, \xi\rangle.$$

If we take $t = 1$ in this expression we obtain that $\langle \mu, \xi\rangle = \langle \rho(1) - \rho(0), \xi\rangle = \mathbf{J}^\xi(c(1)) - \mathbf{J}^\xi(c(0)) = \mathbf{J}^\xi(m) - \mathbf{J}^\xi(m) = 0$. Since $\xi \in \mathfrak{g}$ is arbitrary, it follows that $\mu = 0$ and, consequently, $\mathcal{H} = \{0\}$.

Conversely, suppose that $\mathcal{H} = \{0\}$. Let $c : [0, 1] \to M$ be a loop at an arbitrary point $z \in M$, that is, $c(0) = c(1) = z$. Let $\nu \in \widetilde{p}^{-1}(z)$ and let $\widetilde{c}(t) = (c(t), \nu(t))$ be the horizontal lift of c starting at the point $(z, \nu) \in \widetilde{M}$. Since (z, ν) belongs to the same holonomy bundle as (m, ν) we have that the holonomy group with reference at that point is zero. This implies that

$$0 = \langle \nu(1) - \nu(0), \xi\rangle = \int_0^1 \frac{d}{ds}\langle \nu(s), \xi\rangle\,ds = \int_0^1 \mathbf{i}_{\xi_M}\omega(c(s)) \cdot c'(s)\,ds = \int_c \mathbf{i}_{\xi_M}\omega.$$

Since the equality $\int_c \mathbf{i}_{\xi_M}\omega = 0$ holds for any loop c at any point m, the de Rham Theorem (see for instance CHOQUET-BRUHAT AND DEWITT-MORETTE (1982, page 226)) implies that the cohomology class $[\mathbf{i}_{\xi_M}\omega]$ of the form $\mathbf{i}_{\xi_M}\omega$ is trivial, that is, there exists a function, call it $\mathbf{J}^\xi \in C^\infty(M)$, such that $\mathbf{i}_{\xi_M}\omega = \mathbf{dJ}^\xi$. Since this is true for any $\xi \in \mathfrak{g}$, the existence of a standard momentum map is guaranteed by choosing $\mathbf{J} : M \to \mathfrak{g}^*$ such that $\langle \mathbf{J}(m), \xi\rangle = \mathbf{J}^\xi(m)$, for any $\xi \in \mathfrak{g}$ and $m \in M$.

We finally show that, in this case, the momentum map $\mathbf{J}$ can be chosen to coincide with $\mathbf{K}$, which makes $\mathbf{K}$ a standard momentum map. This is easy to see once we realize that the graph Graph $(\mathbf{J}) := \{(m, \mathbf{J}(m)) \in M \times \mathfrak{g}^* \mid m \in M\}$ integrates the horizontal distribution associated to $\tilde{\alpha}$. Indeed, choose $\mathbf{J}$ such that $\mathbf{J}(m) = \nu$. Then, by (5.2.3), we have for any $(z, \rho) \in \widetilde{M}$

$$\begin{aligned} H(z,\rho) &= \{(v_z, \tau) \in T_{(z,\rho)}(M \times \mathfrak{g}^*) \mid \langle \tau, \xi \rangle = (\mathbf{i}_{\xi_M}\omega)(z)(v_z), \text{ for all } \xi \in \mathfrak{g}\} \\ &= \{(v_z, \tau) \in T_{(z,\rho)}(M \times \mathfrak{g}^*) \mid \langle \tau, \xi \rangle = \mathbf{d}\mathbf{J}^{\xi}(z)(v_z), \text{ for all } \xi \in \mathfrak{g}\} \\ &= \{(v_z, \tau) \in T_{(z,\rho)}(M \times \mathfrak{g}^*) \mid \langle \tau, \xi \rangle = \langle T_z\mathbf{J}(v_z), \xi \rangle, \text{ for all } \xi \in \mathfrak{g}\} \\ &= \{(v_z, T_z\mathbf{J}(v_z)) \mid v_z \in T_zM\} = T_{(z,\mathbf{J}(z))}\text{Graph }(\mathbf{J}). \end{aligned}$$

Consequently, since $\mathbf{J}$ is defined up to a constant in $\mathfrak{g}^*$, it can be chosen so that Graph $(\mathbf{J}) = \widetilde{M}$ and hence $\mathbf{J} = \mathbf{K}$. ■

5.3 The universal covering and covered spaces of a symplectic Lie algebra action

Some of the tools that we used to introduce cylinder valued momentum maps can also be utilized to construct an object in the category whose objects are the ***Hamiltonian coverings*** of a symplectic manifold acted canonically upon by a Lie algebra and that satisfies interesting universality properties. The results of this section are due to ORTEGA AND RATIU (2003c). A Hamiltonian covering of a symplectic manifold acted canonically upon by a Lie algebra is a standard covering of that manifold by another symplectic manifold acted upon by the same Lie algebra but, this time, the action is required to have a standard momentum map associated. More explicitly, we have the following definition.

5.3.1 Definition *Let (M, ω) be a connected symplectic manifold and $\mathfrak{g}$ a Lie algebra acting symplectically on it. We say that the map $p_N : N \to M$ is a **Hamiltonian covering map** of (M, ω) if it satisfies the following conditions:*

(i) *p_N is a smooth covering map*

(ii) *(N, ω_N) is a connected symplectic manifold*

(iii) *p_N is a symplectic map*

(iv) *$\mathfrak{g}$ acts symplectically on (N, ω_N) and admits a standard momentum map $\mathbf{K}_N : N \to \mathfrak{g}^*$*

(v) *p_N is $\mathfrak{g}$-equivariant, that is, $\xi_M(p_N(n)) = T_n p_N \cdot \xi_N(n)$, for any $n \in N$ and any $\xi \in \mathfrak{g}$.*

The connectedness hypothesis on N assumed in the previous definition implies that the momentum maps of the $\mathfrak{g}$-action on N differ from $\mathbf{K}_N$ by a constant element in $\mathfrak{g}^*$. We will denote by $[\mathbf{K}_N]$ the equivalence class consisting of all the maps $N \to \mathfrak{g}^*$ that differ from $\mathbf{K}_N$ by a constant map.

5.3.2 Definition *Let (M, ω) be a connected symplectic manifold and $\mathfrak{g}$ a Lie algebra acting symplectically on it. Let $\mathfrak{H}$ be the category whose objects* $\mathrm{Ob}(\mathfrak{H})$ *are the four-tuples $(p_N : N \to M, \omega_N, \mathfrak{g}, [\mathbf{K}_N])$ with $p_N : N \to M$ a Hamiltonian covering map of (M, ω) and whose morphisms* $\mathrm{Mor}(\mathfrak{H})$ *are the smooth maps $q : (N_1, \omega_1) \to (N_2, \omega_2)$ that satisfy the following properties:*

(i) *q is a symplectic covering map*

(ii) *q is $\mathfrak{g}$-equivariant*

(iii) *the diagram*

$$\begin{array}{ccccc} & & \mathfrak{g}^* & & \\ & \mathbf{K}_{N_1} \nearrow & & \nwarrow \mathbf{K}_{N_2} & \\ N_1 & & \xrightarrow{\quad q \quad} & & N_2 \\ & p_{N_1} \searrow & & \swarrow p_{N_2} & \\ & & M & & \end{array}$$

commutes for some $\mathbf{K}_{N_1} \in [\mathbf{K}_{N_1}]$ and $\mathbf{K}_{N_2} \in [\mathbf{K}_{N_2}]$.

We will refer to $\mathfrak{H}$ as the category of ***Hamiltonian covering maps****.*

The main goal of this section is to show that the category $\mathfrak{H}$ admits universal covering and covered spaces. More explicitly, we will show that there exist two objects $(\widehat{p} : \widehat{M} \to M, \omega_{\widehat{M}}, \mathfrak{g}, [\widehat{\mathbf{K}}])$ (the ***universal Hamiltonian covering space***) and $(\widetilde{p} : \widetilde{M} \to M, \omega_{\widetilde{M}}, \mathfrak{g}, [\widetilde{\mathbf{K}}])$ (the ***universal Hamiltonian covered space***) in $\mathfrak{H}$ such that for any other object $(p_N : N \to M, \omega_N, \mathfrak{g}, [\mathbf{K}_N])$ in $\mathfrak{H}$, there exist morphisms (not necessarily unique) $\widehat{q} : \widehat{M} \to N$ and $\widetilde{q} : N \to \widetilde{M}$ in $\mathrm{Mor}(\mathfrak{H})$. Even though the objects that satisfy these properties are not necessarily unique, they are all isomorphic to $(\widehat{p} : \widehat{M} \to M, \omega_{\widehat{M}}, \mathfrak{g}, [\widehat{\mathbf{K}}])$ and $(\widetilde{p} : \widetilde{M} \to M, \omega_{\widetilde{M}}, \mathfrak{g}, [\widetilde{\mathbf{K}}])$ in $\mathfrak{H}$, respectively, which justifies the adjective "universal" when we will refer to the Hamiltonian covering and covered spaces.

The universal Hamiltonian covering space will be easily obtained in §5.3.3 from a standard simply connected universal covering manifold. The universal Hamiltonian covered space is constructed in Proposition 5.3.7 using the holonomy bundles of the connection that we introduced to construct the cylinder valued momentum maps. We will prove its universality properties in Theorem 5.3.9.

5.3.3 The standard universal covering as a Hamiltonian covering. Let (M, ω) be a connected symplectic manifold and $\mathfrak{g}$ a Lie algebra acting symplectically on it. Let $\widehat{p} : \widehat{M} \to M$ be the simply connected universal covering of M. This can be made into a Hamiltonian covering map in a straightforward manner, as will be shown below.

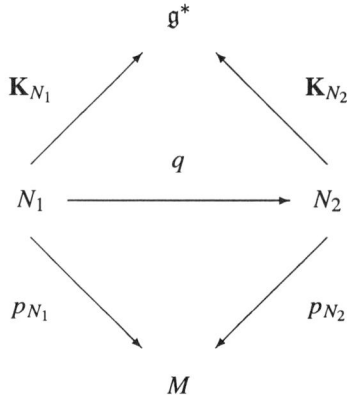

Since $\widehat{p}$ is a local diffeomorphism, the two-form $\omega_{\widehat{M}} := \widehat{p}^*\omega$ is a symplectic form on $\widehat{M}$. Thus, properties **(i)**, **(ii)**, and **(iii)** of Definition 5.3.1 hold. We can use the $\mathfrak{g}$-action on M to define a symplectic $\mathfrak{g}$-action on $(\widehat{M}, \omega_{\widehat{M}})$ by

$$\xi_{\widehat{M}}(z) := (T_{(z)}\widehat{p})^{-1}\xi_M(\widehat{p}(z)), \quad \text{for any} \quad \xi \in \mathfrak{g} \quad \text{and} \quad z \in \widehat{M}. \tag{5.3.1}$$

This is a good definition since $\widehat{p}$ is a covering map and hence a local diffeomorphism. Moreover, the map $(z, \xi) \in \widehat{M} \times \mathfrak{g} \mapsto \xi_{\widehat{M}}(z) \in T\widehat{M}$ is clearly smooth. Note that by definition, the vector fields $\xi_{\widehat{M}}$ and ξ_M are $\widehat{p}$-related for all $\xi \in \mathfrak{g}$. This immediately shows that $[\xi, \eta]_{\widehat{M}} = -[\xi_{\widehat{M}}, \eta_{\widehat{M}}]$ for any $\xi, \eta \in \mathfrak{g}$ and that $\pounds_{\xi_{\widehat{M}}}\omega_{\widehat{M}} = \pounds_{\xi_{\widehat{M}}}\widehat{p}^*\omega = \widehat{p}^*\pounds_{\xi_M}\omega = 0$, for any $\xi \in \mathfrak{g}$. Thus, expression (5.3.1) defines a symplectic action of $\mathfrak{g}$ on $(\widehat{M}, \omega_{\widehat{M}})$ relative to which $\widehat{p}$ is equivariant by construction. Finally, the $\mathfrak{g}$-action on $\widehat{M}$ admits a momentum map $\widehat{\mathbf{K}} : \widehat{M} \to \mathfrak{g}^*$ because $\widehat{M}$ is simply connected and therefore $H^1(\widehat{M}, \mathbb{R}) = 0$ (see Proposition 4.5.17). Thus, conditions **(iv)** and **(v)** are also satisfied which makes $(\widehat{p} : \widehat{M} \to M, \omega_{\widehat{M}}, \mathfrak{g}, [\widehat{\mathbf{K}}])$ an object of $\mathfrak{H}$.

5.3.4 Proposition *Let (M, ω) be a connected symplectic manifold and $\mathfrak{g}$ a Lie algebra acting symplectically on it. Let $(\widehat{p} : \widehat{M} \to M, \omega_{\widehat{M}}, \mathfrak{g}, [\widehat{\mathbf{K}}])$ be the object in $\mathfrak{H}$ constructed above using the universal covering of M. Then for any other object $(p_N : N \to M, \omega_N, \mathfrak{g}, [\mathbf{K}_N])$ of $\mathfrak{H}$, there exists a morphism $q : \widehat{M} \to N$ in* Mor($\mathfrak{H}$). *Any other object in $\mathfrak{H}$ that satisfies the same universality property is isomorphic to $(\widehat{p} : \widehat{M} \to M, \omega_{\widehat{M}}, \mathfrak{g}, [\widehat{\mathbf{K}}])$.*

Proof. Since $\widehat{M}$ is the universal covering space of M, there exists a smooth covering map $q : \widehat{M} \to M$ (in general not unique) such that $p_N \circ q = \widehat{p}$. We shall prove that this is a morphism in $\mathfrak{H}$. Indeed, since p_N and $\widehat{p}$ are symplectic maps we have

$$\omega_{\widehat{M}} = \widehat{p}^*\omega = (p_N \circ q)^*\omega = q^* p_N^*\omega = q^*\omega_N,$$

so condition **(i)** in Definition 5.3.2 is satisfied. Additionally, since $\widehat{p}$ and p_N are $\mathfrak{g}$-equivariant we have, for any $z \in \widehat{M}$ and $\xi \in \mathfrak{g}$,

$$\begin{aligned} T_{q(z)}p_N\left(T_z q(\xi_{\widehat{M}}(z))\right) &= T_z\widehat{p}\left(\xi_{\widehat{M}}(z)\right) = \xi_M(\widehat{p}(z)) \\ &= \xi_M(p_N(q(z))) = T_{q(z)}p_N\left(\xi_N(q(z))\right). \end{aligned}$$

Since $T_{q(z)}p_N$ is an isomorphism, it follows that $T_z q(\xi_{\widehat{M}}(z)) = \xi_N(q(z))$ and so **(ii)** is satisfied. To verify **(iii)** it suffices to note that $\mathbf{K}_N \circ q$ is a momentum map for the $\mathfrak{g}$-action on $\widehat{M}$.

In order to prove the last sentence in the statement let $(\widehat{p}' : \widehat{M}' \to M, \omega_{\widehat{M}'}, \mathfrak{g}, [\widehat{\mathbf{K}}'])$ be another object in $\mathfrak{H}$ satisfying the same universality property as $(\widehat{p} : \widehat{M} \to M, \omega_{\widehat{M}}, \mathfrak{g}, [\widehat{\mathbf{K}}])$. Let $q : \widehat{M} \to \widehat{M}'$ and $q' : \widehat{M}' \to \widehat{M}$ be the corresponding morphisms. Since both q and q' are symplectic covering maps their composition $q' \circ q : \widehat{M} \to \widehat{M}$ is also a symplectic covering map (see Theorems 3, 5, 6 in Section 2.2 and Theorem 10 in Section 2.4 of SPANIER (1966)). Thus $q' \circ q$ is a local diffeomorphism. Since $\widehat{M}$ is simply connected, this map is also injective (SPANIER, 1966, Theorem 9, page 73). Consequently, $\varphi := q' \circ q$ is a bijective local diffeomorphism, hence a diffeomorphism. Finally, this proves that both q and q' are isomorphisms in $\mathfrak{H}$ with inverses $\varphi^{-1} \circ q'$ and $q \circ \varphi^{-1}$, respectively. ■

5.3.5 Remark It should be noticed that the universality property for $(\widehat{p} : \widehat{M} \to M, \omega_{\widehat{M}}, \mathfrak{g}, [\widehat{\mathbf{K}}])$ stated in the previous proposition does not imply that it is an initial object in $\mathfrak{H}$ due to the nonuniqueness of the morphism q. This agrees with the situation encountered for general manifolds.

5.3.6 The holonomy bundles of α are Hamiltonian coverings of $(M, \omega, \mathfrak{g})$. In this section we show that the holonomy bundles of the connection α defined in (5.2.1) constitute an object in the category $\mathfrak{H}$. From now on and until the end of the section we will assume that the manifold M is paracompact. We now prove the following proposition.

5.3.7 Proposition *Let (M, ω) be a connected paracompact symplectic manifold and let $\mathfrak{g}$ be a Lie algebra acting symplectically on it. Let α be the connection on the trivial bundle $(\pi : M \times \mathfrak{g}^* \to M, \mathfrak{g}^*)$ introduced in* (5.2.1) *and $(\widetilde{p} : \widetilde{M} \to M, \mathcal{H})$ one of its holonomy bundles. The pair $(\widetilde{M}, \omega_{\widetilde{M}} := \widetilde{p}^*\omega)$ is a symplectic manifold on which $\mathfrak{g}$ acts symplectically by*

$$\xi_{\widetilde{M}}(m, \mu) := (\xi_M(m), -\Psi(m)(\xi, \cdot)), \text{ for any } \quad \xi \in \mathfrak{g}, \ (m, \mu) \in \widetilde{M}, \tag{5.3.2}$$

where $\Psi : M \to Z^2(\mathfrak{g})$ is the Chu map. Finally, the projection $\widetilde{\mathbf{K}} : \widetilde{M} \to \mathfrak{g}^*$ of $\widetilde{M}$ into $\mathfrak{g}^*$ is a momentum map for this action. Moreover, the four tuple $(\widetilde{p} : \widetilde{M} \to M, \omega_{\widetilde{M}}, \mathfrak{g}, [\widetilde{\mathbf{K}}])$ is an object in the category $\mathfrak{H}$ introduced in Definition 5.3.2.

Proof. We start by noticing that the projection $\widetilde{p} : \widetilde{M} \to M$ is a smooth covering projection as a consequence of the flatness of α and Proposition 5.1.5. Now, as $\widetilde{p}$ is a local diffeomorphism, the equality $\omega_{\widetilde{M}} := \widetilde{p}^*\omega$ defines a symplectic form on $\widetilde{M}$ with respect to which $\widetilde{p}$ is a symplectic map. We have hence shown that $\widetilde{p} : \widetilde{M} \to M$ satisfies properties **(i)**, **(ii)**, and **(iii)** in Definition 5.3.1.

We now define a $\mathfrak{g}$-action on $\widetilde{M}$ by

$$\xi_{\widetilde{M}}(m, \mu) := (T_{(m,\mu)}\widetilde{p})^{-1}\xi_M(m), \quad \text{for any} \quad \xi \in \mathfrak{g}, \ (m, \mu) \in \widetilde{M}. \tag{5.3.3}$$

This is a good definition since $\widetilde{p}$ is a covering map and hence a local diffeomorphism. Moreover, the map $((m, \mu), \xi) \in \widetilde{M} \times \mathfrak{g} \mapsto \xi_{\widetilde{M}}(m, \mu) \in T\widetilde{M}$ is clearly smooth. Note that, by definition, the vector fields $\xi_{\widetilde{M}}$ and ξ_M are $\widetilde{p}$-related for all $\xi \in \mathfrak{g}$. This immediately shows that $[\xi, \eta]_{\widetilde{M}} = -[\xi_{\widetilde{M}}, \eta_{\widetilde{M}}]$ for any $\xi, \eta \in \mathfrak{g}$ and that $\pounds_{\xi_{\widetilde{M}}}\omega_{\widetilde{M}} = \pounds_{\xi_{\widetilde{M}}}\widetilde{p}^*\omega = \widetilde{p}^*\pounds_{\xi_M}\omega = 0$, for any $\xi \in \mathfrak{g}$. Thus, expression (5.3.3) defines a symplectic action of $\mathfrak{g}$ on $(\widetilde{M}, \omega_{\widetilde{M}})$. We now show that (5.3.3) can be rewritten as (5.3.2), that is,

$$\xi_{\widetilde{M}}(m, \mu) := (\xi_M(m), -\Psi(m)(\xi, \cdot)), \text{ for any } \quad \xi \in \mathfrak{g}, \ (m, \mu) \in \widetilde{M}. \tag{5.3.4}$$

We start by checking that the right-hand side of this expression is a horizontal vector with respect to α and thereby tangent to $\widetilde{M}$, which means that $\langle \alpha(m, \mu)(\xi_{\widetilde{M}}(m, \mu)), \eta \rangle = 0$, for any $\eta \in \mathfrak{g}$. The definition of α implies

$$\begin{aligned}\langle \alpha(m, \mu)(\xi_{\widetilde{M}}(m, \mu)), \eta \rangle &= (\mathbf{i}_{\eta_M}\omega)(m)(\xi_M(m)) + \langle \Psi(m)(\xi, \cdot), \eta \rangle \\ &= \omega(m)(\eta_M(m), \xi_M(m)) + \omega(m)(\xi_M(m), \eta_M(m)) = 0.\end{aligned}$$

Consequently $(\xi_M(m), -\Psi(m)(\xi, \cdot))$ is horizontal and therefore it suffices to note that $\widetilde{p}$ is the projection onto M to prove the equivalence between (5.3.3) and (5.3.4). The same remark makes evident the $\mathfrak{g}$-equivariance of $\widetilde{p} : \widetilde{M} \to M$.

We conclude by showing that the projection $\widetilde{\mathbf{K}} : \widetilde{M} \to \mathfrak{g}^*$ is a standard momentum map for the $\mathfrak{g}$-action on $\widetilde{M}$ defined in (5.3.2). Let $\xi \in \mathfrak{g}$ be arbitrary and $\widetilde{\mathbf{K}}^\xi := \langle \widetilde{\mathbf{K}}, \xi \rangle$. On the one hand, we have that $\mathbf{d}\widetilde{\mathbf{K}}^\xi(m, \mu)(v_m, \nu) = \langle \nu, \xi \rangle$, for any $(m, \mu) \in \widetilde{M}$ and any $(v_m, \nu) \in T_{(m,\mu)}\widetilde{M} = H(m, \mu)$. On the other hand,

$$\begin{aligned}
\mathbf{i}_{\xi_{\widetilde{M}}} \omega_{\widetilde{M}}(m, \mu)(v_m, \nu) &= \mathbf{i}_{\xi_{\widetilde{M}}} (\widetilde{p}^* \omega)(m, \mu)(v_m, \nu) \\
&= (\widetilde{p}^* \omega)(m, \mu)(\xi_{\widetilde{M}}(m, \mu), (v_m, \nu)) \\
&= (\widetilde{p}^* \omega)(m, \mu)((\xi_M(m), -\Psi(m)(\xi, \cdot)), (v_m, \nu)) \\
&= \omega(m)(\xi_M(m), v_m) = \langle \nu, \xi \rangle,
\end{aligned}$$

which proves the claim. ■

5.3.8 Remark We emphasize that the momentum map $\widetilde{\mathbf{K}}$ in the previous proposition is not equivariant in general. Indeed its infinitesimal non-equivariance cocycle is given by

$$\begin{aligned}
\Sigma(\xi, \eta) &:= \widetilde{\mathbf{K}}^{[\xi,\eta]}(m, \mu) - \{\widetilde{\mathbf{K}}^\xi, \widetilde{\mathbf{K}}^\eta\}(m, \mu) \\
&= \langle \mu, [\xi, \eta] \rangle - (\widetilde{p}^* \omega)(m, \mu)(\xi_{\widetilde{M}}(m, \mu), \eta_{\widetilde{M}}(m, \mu)) \\
&= \langle \mu, [\xi, \eta] \rangle - \omega(m)(\xi_M(m), \eta_M(m)) \\
&= \langle \mu, [\xi, \eta] \rangle - \Psi(m)(\xi, \eta), \qquad (5.3.5)
\end{aligned}$$

for any $\xi, \eta \in \mathfrak{g}$. The value of Σ does not depend on the point $(m, \mu) \in \widetilde{M}$ used to define it because for any $(v_m, \nu) \in T_{(m,\mu)}\widetilde{M}$ the function $f(m, \mu) := \langle \mu, [\xi, \eta] \rangle - \Psi(m)(\xi, \eta)$ has vanishing differential. Indeed,

$$\begin{aligned}
\mathbf{d}f(m, \mu)(v_m, \mu) &= \langle \nu, [\xi, \eta] \rangle - T_m\Psi(v_m)(\xi, \eta) \\
&= \langle \nu, [\xi, \eta] \rangle - \omega(m)([\xi, \eta]_M(m), v_m) = 0,
\end{aligned}$$

where we used the horizontality of (v_m, ν) in the last equality. The connectedness of $\widetilde{M}$ concludes the argument.

5.3.9 Theorem *Let (M, ω) be a connected paracompact symplectic manifold and let $\mathfrak{g}$ be a Lie algebra acting symplectically on it. Then the Hamiltonian covering $(\widetilde{p} : \widetilde{M} \to M, \omega_{\widetilde{M}}, \mathfrak{g}, [\widetilde{\mathbf{K}}])$ constructed in Proposition 5.3.7 is a universal Hamiltonian covered space in the category $\mathfrak{H}$ of Hamiltonian covering maps, that is, given any other object $(p_N : N \to M, \omega_N, \mathfrak{g}, [\mathbf{K}_N])$ in $\mathfrak{H}$, there exists a (not necessarily unique) morphism $q : N \to \widetilde{M}$ in* Mor$(\mathfrak{H})$*. Any other object of $\mathfrak{H}$ that satisfies this universality property is isomorphic to $(\widetilde{p} : \widetilde{M} \to M, \omega_{\widetilde{M}}, \mathfrak{g}, [\widetilde{\mathbf{K}}])$.*

Proof. Let $(p_N : N \to M, \omega_N, \mathfrak{g}, [\mathbf{K}_N]) \in \mathfrak{H}$ and $n_0 \in N$. Define $\widetilde{m}_0 := (p_N(n_0), \mathbf{K}_N(n_0)) \in M \times \mathfrak{g}^*$. Given that $M \times \mathfrak{g}^*$ is foliated by the holonomy bundles of the connection α in (5.2.1), the point $\widetilde{m}_0$ lies in one of them, say $\widetilde{M}'$. Let $\tau \in \mathfrak{g}^*$ be such that $\widetilde{M}' = R_\tau(\widetilde{M})$ and define $\overline{\mathbf{K}}_N := \mathbf{K}_N - \tau$. The map $\overline{\mathbf{K}}_N : N \to \mathfrak{g}^*$ is also a momentum map for the $\mathfrak{g}$-action on N, $[\overline{\mathbf{K}}_N] = [\mathbf{K}_N]$, and, moreover, $(p_N(n_0), \overline{\mathbf{K}}_N(n_0)) \in \widetilde{M}$. Hence we can assume, without loss of generality, that $(p_N : N \to M, \omega_N, \mathfrak{g}, [\mathbf{K}_N])$ is such that $(p_N(n_0), \mathbf{K}_N(n_0)) \in \widetilde{M}$. Using this choice we define the map $g : N \to M \times \mathfrak{g}^*$ by $n \longmapsto (p_N(n), \mathbf{K}_N(n))$, $n \in N$.

We will now show that $g(N) \subset \widetilde{M}$. We start by proving that $T_n g(v_n) \in H(p_N(n), \mathbf{K}_N(n))$ for all $n \in N$ and $v_n \in T_n N$. Indeed, since $T_n g(v_n) = (T_n p_N(v_n), T_n \mathbf{K}_N(v_n))$ we have for any $\xi \in \mathfrak{g}$

$$\begin{aligned}
\langle \alpha(g(n))(T_n g(v_n)), \xi \rangle &= \omega(p_N(n))(\xi_M(p_N(n)), T_n p_N(v_n)) \\
&\quad - \langle T_n \mathbf{K}_N(v_n), \xi \rangle \\
&= \omega(p_N(n))(T_n p_N(\xi_N(n)), T_n p_N(v_n)) \\
&\quad - \mathbf{d}\mathbf{K}_N^{\xi}(n)(v_n) \\
&= \omega_N(n)(\xi_N(n), v_n) - \mathbf{d}\mathbf{K}_N^{\xi}(n)(v_n) = 0,
\end{aligned}$$

where we used the $\mathfrak{g}$-equivariance and the symplectic character of p_N. Let now $n \in N$ be arbitrary. As N is connected, there exists a smooth curve $c : [0, 1] \to N$ such that $c(0) = n_0$ and $c(1) = n$. Since the derivative Tg of g maps into the horizontal bundle of α, the chain rule implies that $g(c(t))$ is a horizontal curve starting at $g(c(0)) = g(n_0) \in \widetilde{M}$. Hence, by the definition of the holonomy bundle, $g(c(1)) = g(n) \in \widetilde{M}$. This argument and the arbitrary character of $n \in N$ show that $g(N) \subset \widetilde{M}$.

Let $q : N \to \widetilde{M}$ be the map obtained from g by restriction of the range. We will show that q is the unique morphism needed to prove the statement of the theorem. First, the map q is smooth since g is smooth and $\widetilde{M}$ is an initial submanifold of $M \times \mathfrak{g}^*$. Second, we verify that q satisfies the three conditions in Definition 5.3.2 that characterize an element in $\mathrm{Mor}(\mathfrak{H})$.

(i) q is a symplectic covering projection: Since $p_N : N \to M$ and $\widetilde{p} : \widetilde{M} \to M$ are covering projections and $\widetilde{p} \circ q = p_N$ it follows that $q : N \to \widetilde{M}$ is a covering projection (SPANIER, 1966, Lemma 1, page 79). Since p_N and $\widetilde{p}$ are symplectic, so is q.

(ii) q is $\mathfrak{g}$-equivariant: Let $\xi \in \mathfrak{g}$, $n \in N$ be arbitrary. On the one hand

$$\xi_{\widetilde{M}}(q(n)) = \xi_{\widetilde{M}}(p_N(n), \mathbf{K}_N(n)) = (\xi_M(p_N(n)), -\Psi(p_N(n))(\xi, \cdot)).$$

On the other hand

$$\begin{aligned}
T_n q\,(\xi_N(n)) &= (T_n p_N\,(\xi_N(n))\,, T_n \mathbf{K}_N\,(\xi_N(n))) \\
&= (\xi_M(p_N(n)), T_n \mathbf{K}_N\,(\xi_N(n))).
\end{aligned}$$

Consequently, the map q is $\mathfrak{g}$-equivariant if and only if

$$T_n \mathbf{K}\,(\xi_N(n)) = -\Psi(p_N(n))(\xi, \cdot).$$

This identity holds because for any $\eta \in \mathfrak{g}$ we have

$$\begin{aligned}\langle T_n \mathbf{K}_N(\xi_N(n)), \eta\rangle &= \mathbf{dK}_N^\eta(n)(\xi_N(n)) = \omega_N(n)(\eta_N(n), \xi_N(n)) \\ &= (p_N^*\omega)(n)(\eta_N(n), \xi_N(n)) \\ &= \omega(p_N(n))(T_n p_N(\eta_N(n)), T_n p_N(\xi_N(n))) \\ &= \omega(p_N(n))(\eta_M(p_N(n)), \xi_M(p_N(n))) \\ &= -\Psi(p_N(n))(\xi, \eta).\end{aligned}$$

(iii) The diagram in Definition 5.3.2 commutes since $\widetilde{p} \circ q = p_N$ and $\widetilde{\mathbf{K}} \circ q = \mathbf{K}_N$ by the definition of q.

We conclude by showing that any other object $(\widetilde{p}' : \widetilde{M}' \to M, \omega_{\widetilde{M}'}, \mathfrak{g}, [\widetilde{\mathbf{K}}'])$ that satisfies the universality property that we just proved for $(\widetilde{p} : \widetilde{M} \to M, \omega_{\widetilde{M}}, \mathfrak{g}, [\widetilde{\mathbf{K}}])$ is necessarily isomorphic to it. Indeed, the universality property satisfied by $(\widetilde{p} : \widetilde{M} \to M, \omega_{\widetilde{M}}, \mathfrak{g}, [\widetilde{\mathbf{K}}])$ and $(\widetilde{p}' : \widetilde{M}' \to M, \omega_{\widetilde{M}'}, \mathfrak{g}, [\widetilde{\mathbf{K}}'])$ implies the existence of two morphisms $q : \widetilde{M}' \to \widetilde{M}$ and $q' : \widetilde{M} \to \widetilde{M}'$ in $\operatorname{Mor}(\mathfrak{H})$ and of an element $\tau \in \mathfrak{g}^*$ such that the following diagram commutes

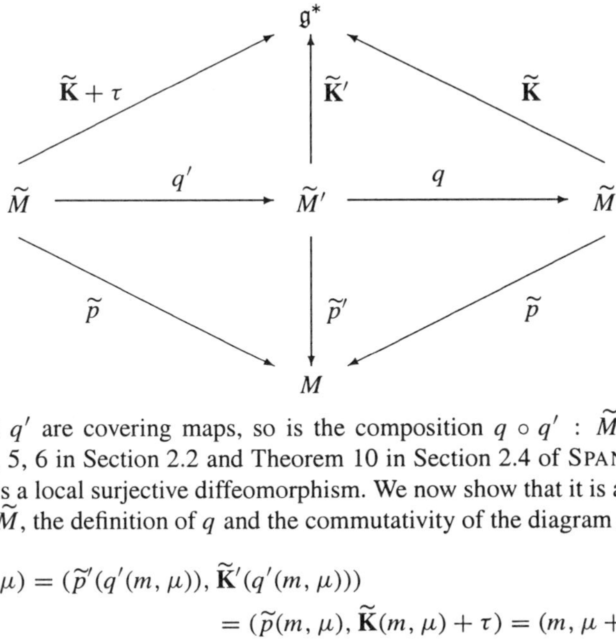

Since q and q' are covering maps, so is the composition $q \circ q' : \widetilde{M} \to \widetilde{M}$ (see Theorems 3, 5, 6 in Section 2.2 and Theorem 10 in Section 2.4 of SPANIER (1966)). Thus $q \circ q'$ is a local surjective diffeomorphism. We now show that it is also injective. If $(m, \mu) \in \widetilde{M}$, the definition of q and the commutativity of the diagram above yield

$$\begin{aligned} q \circ q'(m, \mu) &= (\widetilde{p}'(q'(m, \mu)), \widetilde{\mathbf{K}}'(q'(m, \mu))) \\ &= (\widetilde{p}(m, \mu), \widetilde{\mathbf{K}}(m, \mu) + \tau) = (m, \mu + \tau). \end{aligned} \quad (5.3.6)$$

Hence if $(m, \mu), (m', \mu') \in \widetilde{M}$ satisfy $q \circ q'(m, \mu) = q \circ q'(m', \mu')$, then (5.3.6) implies that $(m, \mu) = (m', \mu')$. Consequently, $\varphi := q \circ q'$ is a bijective local diffeomorphism and hence a diffeomorphism. This proves that both q and q' are isomorphisms is $\mathfrak{H}$. ■

5.3.10 Remark It should be noticed that the universality property for $(\widetilde{p} : \widetilde{M} \to M, \omega_{\widetilde{M}}, \mathfrak{g}, [\widetilde{\mathbf{K}}])$ stated in the theorem does not imply that it is a final object in $\mathfrak{H}$ due to the nonuniqueness of the morphism q.

5.3.11 Remark One could consider larger categories than $\mathfrak{H}$ in which case the universality result in Theorem 5.3.9 would be weaker. For example, if we drop the condition that $p_N : N \to M$ is a covering map in the definition of the objects of $\mathfrak{H}$, then the morphism $q : N \to \widetilde{M}$ is not necessarily a covering map.

5.3.12 Example We shall illustrate the difference between the universal Hamiltonian covering and covered spaces by considering the following elementary example.

Let $\mathbb{T}^2 = \{(e^{i\theta_1}, e^{i\theta_2})\}$ be the two torus considered as a symplectic manifold with its area form $\omega := \mathbf{d}\theta_1 \wedge \mathbf{d}\theta_2$ and the circle $S^1 = \{e^{i\phi}\}$ acting canonically on it by $e^{i\phi} \cdot (e^{i\theta_1}, e^{i\theta_2}) := (e^{i(\theta_1+\phi)}, e^{i\theta_2})$. Proposition 5.3.4 guarantees that the universal covering space $\mathbb{R}^2$ of $\mathbb{T}^2$ can be endowed with the necessary structure to make it the universal Hamiltonian covering space of $(\mathbb{T}^2, \omega)$.

A straightforward computation shows that, in this case, the horizontal vectors in $T(\mathbb{T}^2 \times \mathbb{R})$ with respect to the connection α defined in (5.2.1) are of the form $((a, b), b)$, with $a, b \in \mathbb{R}$. Given that any surface $\widetilde{\mathbb{T}^2}_\tau := \{((e^{i\theta_1}, e^{i\theta_2}), \tau + \theta_2) \in \mathbb{T}^2 \times \mathbb{R} \mid \theta_1, \theta_2 \in \mathbb{R}\}$ integrates the horizontal distribution, it is immediately clear that the universal Hamiltonian covered space is given in this example by any of the cylinders $\widetilde{\mathbb{T}^2}_\tau$.

5.4 Lie group valued momentum maps

The cylinder valued momentum maps presented in the previous sections are closely related to the ***Lie group valued momentum maps*** introduced by McDuff (1988); Ginzburg (1992);Huebschmann and Jeffrey (1994);Huebschmann (1995a); and Alekseev, Malkin, and Meinrenken (1998). The motivation behind the construction shown below rests more in the understanding of certain moduli spaces rather than in the study of Hamiltonian conservation laws. We give the definition of these objects only for Abelian symmetry groups because in the non-Abelian case these momentum maps are defined on spaces that are neither symplectic nor Poisson (they are referred to as ***quasi-Hamiltonian spaces***) thereby leaving the categories on which we focus in this book.

5.4.1 Definition *Let G be an Abelian Lie group whose Lie algebra $\mathfrak{g}$ acts canonically on the symplectic manifold (M, ω). Let $(\cdot, \cdot)$ be some bilinear symmetric nondegenerate form on the Lie algebra $\mathfrak{g}$. The map $\mathbf{J} : M \to G$ is called a G-**valued momentum map** for the $\mathfrak{g}$-action on M whenever*

$$\mathbf{i}_{\xi_M}\omega(m) \cdot v_m = \left(T_m(L_{\mathbf{J}(m)^{-1}} \circ \mathbf{J})(v_m), \xi\right), \tag{5.4.1}$$

for any $\xi \in \mathfrak{g}$, $m \in M$, and $v_m \in T_m M$.

5.4.2 Proposition *Let G be an Abelian Lie group whose Lie algebra $\mathfrak{g}$ acts canonically on the symplectic manifold (M, ω). Let $\mathbf{J} : M \to G$ be a G-valued momentum map for this action. Then*

(i) $\mathbf{J} : M \to G$ *is a Noether momentum map in the sense of Definition* 4.3.1.

(ii) $\ker T_m\mathbf{J} = (\mathfrak{g} \cdot m)^\omega$ *for any $m \in M$.*

Proof. **(i)** Let F_t be the flow of the Hamiltonian vector field X_h associated to a $\mathfrak{g}$-invariant function $h \in C^\infty(M)^{\mathfrak{g}}$. By the defining relation (5.4.1) of the Lie group valued momentum maps we have for any $m \in M$ and any $\xi \in \mathfrak{g}$

$$
\begin{aligned}
&((T_{\mathbf{J}(F_t(m))}L_{\mathbf{J}(F_t(m))^{-1}} \circ T_{F_t(m)}\mathbf{J})(X_h(F_t(m))), \xi) \\
&\qquad = (T_{F_t(m)}(L_{\mathbf{J}(F_t(m))^{-1}} \circ \mathbf{J}))(X_h(F_t(m))), \xi) \\
&\qquad = \mathbf{i}_{\xi_M}\omega(F_t(m)))(X_h(F_t(m))) \\
&\qquad = -\mathbf{d}h(F_t(m)))(\xi_M(F_t(m))) = 0.
\end{aligned}
$$

Consequently,

$$(T_{\mathbf{J}(F_t(m))}L_{\mathbf{J}(F_t(m))^{-1}} \circ T_{F_t(m)}\mathbf{J})(X_h(F_t(m))) = 0$$

and hence $T_{F_t(m)}\mathbf{J}(X_h(F_t(m))) = 0$, which can be rewritten as

$$\frac{d}{dt}(\mathbf{J} \circ F_t)(m) = 0.$$

The arbitrary character of t and m implies that $\mathbf{J} \circ F_t = \mathbf{J}|_{\mathrm{Dom}(F_t)}$, as required.

(ii) A vector $v_m \in \ker T_m\mathbf{J}$ if and only if $T_m\mathbf{J}(v_m) = 0$. This identity is equivalent to $((T_{\mathbf{J}(m)}L_{\mathbf{J}(m)^{-1}} \circ T_m\mathbf{J})(v_m), \xi) = 0$, for any $\xi \in \mathfrak{g}$ and, by (5.4.1), to $\mathbf{i}_{\xi_M}\omega(m)(v_m) = 0$, for all $\xi \in \mathfrak{g}$, which in turn amounts to $v_m \in (\mathfrak{g} \cdot m)^\omega$. ■

5.4.3 Lie group and cylinder valued momentum maps. A hint of the close relation between cylinder and group valued momentum maps is given by the simple example in §5.2.5. It is easy to see that the cylinder valued momentum map for the circle action on the torus is also a circle valued momentum map in the sense of Definition 5.4.1. In the rest of the section we shall see how this example is a particular case of a more general phenomenon. We start with a very suggestive proposition that states that any cylinder valued momentum map associated to an Abelian Lie algebra action whose corresponding holonomy group is closed can be understood as a Lie group valued momentum map.

5.4.4 Proposition *Let (M, ω) be a connected paracompact symplectic manifold and $\mathfrak{g}$ an Abelian Lie algebra acting canonically on it. Let $\mathcal{H} \subset \mathfrak{g}^*$ be the holonomy group associated to the connection α in (5.2.1) and $(\cdot, \cdot) : \mathfrak{g} \times \mathfrak{g} \to \mathbb{R}$ some bilinear symmetric nondegenerate form on $\mathfrak{g}$. Let $f : \mathfrak{g} \to \mathfrak{g}^*$ be the isomorphism given by $\xi \longmapsto (\xi, \cdot)$, $\xi \in \mathfrak{g}$ and $\mathcal{T} := f^{-1}(\mathcal{H})$. The map f induces an Abelian group isomorphism $\bar{f} : \mathfrak{g}/\mathcal{T} \to \mathfrak{g}^*/\mathcal{H}$ by $\bar{f}(\xi + \mathcal{T}) := (\xi, \cdot) + \mathcal{H}$. Suppose that $\mathcal{H}$ is closed in $\mathfrak{g}^*$ and define $\mathbf{J} := \bar{f}^{-1} \circ \mathbf{K} : M \to \mathfrak{g}/\mathcal{T}$, where $\mathbf{K} : M \to \mathfrak{g}^*/\mathcal{H}$ is a cylinder valued momentum map for the $\mathfrak{g}$-action on (M, ω). Then*

$$\mathbf{i}_{\xi_M}\omega(m) \cdot v_m = \left(T_m(L_{\mathbf{J}(m)^{-1}} \circ \mathbf{J})(v_m), \xi\right), \tag{5.4.2}$$

for any $\xi \in \mathfrak{g}$ and $v_m \in T_mM$. Consequently, the map $\mathbf{J} : M \to \mathfrak{g}/\mathcal{T}$ constitutes a $\mathfrak{g}/\mathcal{T}$-valued momentum map for the canonical action of the Lie algebra $\mathfrak{g}$ of $(\mathfrak{g}/\mathcal{T}, +)$ on (M, ω).

Proof. We start by noticing that the right-hand side of (5.4.2) makes sense due to the closedness hypothesis on $\mathcal{H}$. Indeed, this condition and the fact that $\mathcal{H}$ is discrete due to the flatness of α imply that $\mathfrak{g}^*/\mathcal{H}$, and therefore $\mathfrak{g}/\mathcal{T}$, are Abelian Lie groups whose Lie algebras can be naturally identified with $\mathfrak{g}^*$ and $\mathfrak{g}$, respectively. This identification is used in (5.4.2), where we think of $T_m(L_{\mathbf{J}(m)^{-1}} \circ \mathbf{J})(v_m) \in \operatorname{Lie}(\mathfrak{g}/\mathcal{T})$ as an element of $\mathfrak{g}$.

In what follows we will use the following notation: given $\mu \in \mathfrak{g}^*$ arbitrary, we denote by $\xi_\mu \in \mathfrak{g}$ the unique element such that $\mu = (\xi_\mu, \cdot)$.

We now compute $T_m\mathbf{J}(v_m)$. Let $\mu + \mathcal{H} := \mathbf{K}(m)$ and hence $\mathbf{J}(m) = \xi_\mu + \mathcal{T}$. Now, by Theorem 5.2.8 **(ii)** we have

$$T_m\mathbf{J}(v_m) = T_m(\bar{f}^{-1} \circ \mathbf{K})(v_m) = T_{\mu+\mathcal{H}}\bar{f}^{-1}(T_m\mathbf{K}(v_m)) = T_{\mu+\mathcal{H}}\bar{f}^{-1}(T_\mu\pi_C(\nu)),$$

where the element $\nu \in \mathfrak{g}^*$ is given by

$$\langle \nu, \eta \rangle = \mathbf{i}_{\eta_M}\omega(m)(v_m), \quad \text{for all} \quad \eta \in \mathfrak{g}. \tag{5.4.3}$$

Since $(\bar{f}^{-1} \circ \pi_C)(\rho) = \xi_\rho + \mathcal{T}$ for any $\rho \in \mathfrak{g}^*$, we can write

$$\begin{aligned} T_{\mu+\mathcal{H}}\bar{f}^{-1}(T_\mu\pi_C(\nu)) &= T_\mu\left(\bar{f}^{-1} \circ \pi_C\right)(\nu) \\ &= \left.\frac{d}{dt}\right|_{t=0} \left(\bar{f}^{-1} \circ \pi_C\right)(\mu + t\nu) = \left.\frac{d}{dt}\right|_{t=0} (\xi_\mu + t\xi_\nu + \mathcal{T}). \end{aligned}$$

Hence,

$$T_m\mathbf{J}(v_m) = \left.\frac{d}{dt}\right|_{t=0} (\xi_\mu + t\xi_\nu + \mathcal{T}) \in T_{\xi_\mu+\mathcal{T}}(\mathfrak{g}/\mathcal{T}).$$

Now,

$$\begin{aligned} \left(T_m(L_{\mathbf{J}(m)^{-1}} \circ \mathbf{J})(v_m), \xi\right) &= \left(T_{\mathbf{J}(m)}L_{\mathbf{J}(m)^{-1}}(T_m\mathbf{J}(v_m)), \xi\right) \\ &= \left(\left.\frac{d}{dt}\right|_{t=0} (-\xi_\mu + \mathcal{T}) + (\xi_\mu + t\xi_\nu + \mathcal{T}), \xi\right) \\ &= (\xi_\nu, \xi) = \langle \nu, \xi \rangle = \mathbf{i}_{\xi_M}\omega(m)(v_m), \end{aligned}$$

where the last equality is a consequence of (5.4.3). ■

5.4.5 Lie group valued momentum maps produce closed holonomy subgroups. So far in this section we have investigated how cylinder valued momentum maps can be viewed as a Lie group valued momentum maps. Now we shall focus on the converse relation, that is, we shall isolate hypotheses that guarantee that a Lie group valued momentum map naturally induces a cylinder valued momentum map.

5.4.6 Theorem *Let (M, ω) be a connected paracompact symplectic manifold and $\mathfrak{g}$ an Abelian Lie algebra acting canonically on it. Let $\mathcal{H} \subset \mathfrak{g}^*$ be the holonomy group associated to the connection α in (5.2.1) associated to the $\mathfrak{g}$-action and let $(\cdot, \cdot)$: $\mathfrak{g} \times \mathfrak{g} \to \mathbb{R}$ be a bilinear symmetric nondegenerate form on $\mathfrak{g}$. Let $f : \mathfrak{g} \to \mathfrak{g}^*$,*

$\bar{f} : \mathfrak{g}/\mathcal{T} \to \mathfrak{g}^*/\mathcal{H}$, and let $\mathcal{T} := f^{-1}(\mathcal{H})$ be as in the statement of Proposition 5.4.4. Let G be a connected Abelian Lie group whose Lie algebra is $\mathfrak{g}$ and suppose that there exists a G-valued momentum map $\mathbf{A} : M \to G$ associated to the $\mathfrak{g}$-action whose definition uses the form $(\cdot, \cdot)$.

(i) *If* $\exp : \mathfrak{g} \to G$ *is the exponential map, then*

$$\mathcal{H} \subset f(\ker \exp). \tag{5.4.4}$$

(ii) *$\mathcal{H}$ is closed in $\mathfrak{g}^*$.*

Let $\mathbf{J} := \bar{f}^{-1} \circ \mathbf{K} : M \to \mathfrak{g}/\mathcal{T}$, where $\mathbf{K} : M \to \mathfrak{g}^*/\mathcal{H}$ is a cylinder valued momentum map for the $\mathfrak{g}$-action on (M, ω). If $f(\ker \exp) \subset \mathcal{H}$, then $\mathbf{J} : M \to \mathfrak{g}/\mathcal{T} = \mathfrak{g}/\ker\exp \simeq G$ is a G-valued momentum map that differs from $\mathbf{A}$ by a constant in G.

Conversely, if $\mathcal{H} = f(\ker \exp)$, then $\mathbf{J} : M \to \mathfrak{g}/\ker\exp \simeq G$ is a G-valued momentum map.

5.4.7 Remark The presence of a Lie group valued momentum map associated to a canonical Lie algebra action does not imply the reverse inclusion in (5.4.4). A simple example that illustrates this statement is the canonical action of a two torus $\mathbb{T}^2$ on itself via

$$(e^{i\phi_1}, e^{i\phi_2}) \cdot (e^{i\theta_1}, e^{i\theta_2}) := (e^{i(\theta_1+\phi_1)}, e^{i\theta_2}),$$

where we consider the torus as a symplectic manifold with the area form. A straightforward computation shows that the surface

$$\widetilde{\mathbb{T}^2} := \left\{((e^{i\theta_1}, e^{i\theta_2}), (\theta_2, 0)) \in \mathbb{T}^2 \times \mathbb{R}^2 \mid \theta_1, \theta_2 \in \mathbb{R}\right\},$$

is the holonomy bundle containing the point $((e, e), (0, 0)) \in \mathbb{T}^2 \times \mathbb{R}^2$ associated to the connection that defines the corresponding cylinder valued momentum map. This immediately shows that $\mathcal{H} = \mathbb{Z} \times \{0\}$ while $f(\ker \exp) = \mathbb{Z} \times \mathbb{Z}$ which is clearly not contained in $\mathcal{H}$.

Proof of the theorem. We start by assuming that the $\mathfrak{g}$-action on (M, ω) has an associated G-valued momentum map $\mathbf{A} : M \to G$ and we will show that $\mathcal{H} \subset f(\ker \exp)$.

Let $\mu \in \mathcal{H}$. The definition of the holonomy group $\mathcal{H}$ implies the existence of a piecewise smooth loop $m : [0, 1] \to M$ at the point m, that is, $m(0) = m(1) = m \in M$, whose horizontal lift $\widetilde{m}(t) = (m(t), \mu(t))$ starting at the point $(m, 0)$ satisfies $\mu = \mu(1)$. The horizontality of $\widetilde{m}(t)$ implies that

$$\langle \dot{\mu}(t), \xi \rangle = \mathbf{i}_{\xi_M} \omega(m(t))(\dot{m}(t)) = \left(T_{m(t)} \left(L_{\mathbf{A}(m(t))^{-1}} \circ \mathbf{A}\right)(\dot{m}(t)), \xi\right),$$

for any $\xi \in \mathfrak{g}$ or, equivalently,

$$\dot{\mu}(t) = f\left(\left.\frac{d}{ds}\right|_{s=0} \mathbf{A}(m(t))^{-1}\mathbf{A}(m(s))\right). \tag{5.4.5}$$

Fix $t_0 \in [0, 1]$. Since the exponential map $\exp : \mathfrak{g} \to G$ is a local diffeomorphism, there exists a smooth curve $\xi : I_{t_0} := (t_0 - \epsilon, t_0 + \epsilon) \to \mathfrak{g}$, for $\epsilon > 0$ sufficiently small, such that for any $s \in (-\epsilon, \epsilon)$

$$\mathbf{A}(m(t_0 + s)) = \exp \xi(t_0 + s) \mathbf{A}(m(t_0)). \tag{5.4.6}$$

We now reformulate locally the expression (5.4.5) using the function $\xi : I_{t_0} \to \mathfrak{g}$. Let $\tau, \sigma \in (-\epsilon, \epsilon)$ be such that $t = t_0 + \tau$ and $s = t_0 + \sigma$. Expression (5.4.5) can be rewritten in I_{t_0} as

$$\begin{aligned}
\frac{d}{d\tau}\mu(t_0+\tau) &= f\left(\left.\frac{d}{d\sigma}\right|_{\sigma=\tau} \mathbf{A}(m(t_0 + \tau))^{-1}\mathbf{A}(m(t_0 + \sigma))\right) \\
&= f\left(\left.\frac{d}{d\sigma}\right|_{\sigma=\tau} \exp(-\xi(t_0 + \tau)) \exp \xi(t_0 + \sigma)\mathbf{A}(m(t_0))^{-1}\mathbf{A}(m(t_0))\right) \\
&= f\left(\left.\frac{d}{d\sigma}\right|_{\sigma=\tau} \exp\left(\xi(t_0 + \sigma) - \xi(t_0 + \tau)\right)\right) \\
&= f\left(T_0 \exp\left(\left.\frac{d}{d\sigma}\right|_{\sigma=\tau} (\xi(t_0 + \sigma) - \xi(t_0 + \tau))\right)\right) \\
&= f\left(\left.\frac{d}{d\sigma}\right|_{\sigma=\tau} (\xi(t_0 + \sigma) - \xi(t_0 + \tau))\right) \\
&= f\left(\left.\frac{d}{d\sigma}\right|_{\sigma=\tau} \xi(t_0 + \sigma)\right) = f\left(\frac{d}{d\tau}\xi(t_0 + \tau)\right),
\end{aligned}$$

which shows that for any $t \in I_{t_0}$

$$\dot{\mu}(t) = f(\dot{\xi}(t)). \tag{5.4.7}$$

We now cover the interval $[0, 1]$ with a finite number of intervals $I_1, \ldots, I_n$ such that in each of them we define a function $\xi_i : I_i \to \mathfrak{g}$ that satisfies (5.4.6) and (5.4.7). We now write $I_i = [a_i, a_{i+1}]$, with $i \in \{1, \ldots, n\}$, $a_1 = 0$, and $a_{n+1} = 1$. Using these intervals, since $\mu(0) = 0$, we can write

$$\begin{aligned}
\mu = \mu(1) &= \int_0^1 \dot{\mu}(t)dt \\
&= \int_{I_1} \dot{\mu}(t)dt + \cdots + \int_{I_n} \dot{\mu}(t)dt \\
&= f\left(\int_{I_1} \dot{\xi}_1(t)dt + \cdots + \int_{I_n} \dot{\xi}_n(t)dt\right) \\
&= f\left(\xi_1(a_2) - \xi_1(a_1) + \cdots + \xi_n(a_{n+1}) - \xi_n(a_n)\right).
\end{aligned} \tag{5.4.8}$$

The construction of the intervals $I_i, i \in \{1, \ldots, n\}$ implies that $\mathbf{A}(m(a_i)) = \exp \xi_i(a_i) \times \mathbf{A}(m(a_i))$. Hence

$$\exp \xi_i(a_i) = e \tag{5.4.9}$$

and hence $\xi_i(a_i) \in \ker \exp$ for all $i \in \{1, \ldots, n\}$. We also have that

$$\begin{aligned}\mathbf{A}(m(1)) &= \mathbf{A}(m(a_{n+1})) = \exp \xi_n(a_{n+1})\mathbf{A}(m(a_n))\\ &= \exp \xi_n(a_{n+1}) \exp \xi_{n-1}(a_n)\mathbf{A}(m(a_{n-1}))\\ &= \exp \xi_n(a_{n+1}) \exp \xi_{n-1}(a_n) \cdots \exp \xi_1(a_2)\mathbf{A}(m(a_1))\\ &= \exp(\xi_1(a_2) + \cdots + \xi_n(a_{n+1}))\mathbf{A}(m(0)).\end{aligned}$$

Since $m(0) = m(1) = m$ we have $\mathbf{A}(m(0)) = \mathbf{A}(m(1))$ and therefore $\exp(\xi_1(a_2) + \cdots + \xi_n(a_{n+1})) = e$, which implies that $\xi_1(a_2) + \cdots + \xi_n(a_{n+1}) \in \ker \exp$. If we substitute this relation and (5.4.9) in (5.4.8) we obtain that $\mu \in f(\ker \exp)$, which proves the inclusion $\mathcal{H} \subset f(\ker \exp)$.

We now show that $\mathcal{H}$ is closed in $\mathfrak{g}^*$. The inclusion $\mathcal{H} \subset f(\ker \exp)$, the closedness of $\ker \exp$ in $\mathfrak{g}$, and the fact that f is an isomorphism imply that

$$\overline{\mathcal{H}} \subset \overline{f(\ker \exp)} = f(\ker \exp).$$

Because G is Abelian, $\ker \exp$ is a discrete subgroup of $(\mathfrak{g}, +)$ (see §2.1.4) and hence $\overline{\mathcal{H}}$ is a discrete subgroup of $\mathfrak{g}^*$. This implies that $\overline{\mathcal{H}} \subset \mathcal{H}$. Indeed, for any element $\mu \in \overline{\mathcal{H}}$ there exists an open neighborhood $U_\mu \subset \mathfrak{g}^*$ such that $U_\mu \cap \overline{\mathcal{H}} = \{\mu\}$ ($\overline{\mathcal{H}}$ is discrete). As $\mu \in \overline{\mathcal{H}}$ we have that $\emptyset \neq U_\mu \cap \mathcal{H} \subset U_\mu \cap \overline{\mathcal{H}} = \{\mu\}$, which implies that $\mu \in \mathcal{H}$. This shows that $\mathcal{H} = \overline{\mathcal{H}}$ and therefore that $\mathcal{H}$ is closed in $\mathfrak{g}^*$.

Assume now that $f(\ker \exp) \subset \mathcal{H}$. The hypothesis on the existence of a Lie group valued momentum map implies, via the inclusion (5.4.4) that we just proved, that $f(\ker \exp) = \mathcal{H}$ and that $\mathcal{H}$ is closed in $\mathfrak{g}^*$. Proposition 5.4.4 implies that $\mathbf{J} : M \to \mathfrak{g}/\ker \exp \simeq G$ is a G-valued momentum map for the $\mathfrak{g}$-action on (M, ω). We now show that $\mathbf{A}$ and $\mathbf{J}$ differ by a constant in G. The expression (5.4.1) for $\mathbf{A}$ and (5.4.2) for $\mathbf{J}$ imply that for any $\xi \in \mathfrak{g}$ and any $v_m \in T_m M$ we have

$$\left(T_m(L_{\mathbf{A}(m)^{-1}} \circ \mathbf{A})(v_m), \xi\right) = \mathbf{i}_{\xi_M}\omega(m)(v_m) = \left(T_m(L_{\mathbf{J}(m)^{-1}} \circ \mathbf{J})(v_m), \xi\right),$$

which implies that $T\mathbf{J} = T\mathbf{A}$. Since the manifold M is connected we have that $\mathbf{A}$ and $\mathbf{J}$ coincide up to a constant element in $\mathbf{G}$.

The last claim in the theorem is a straightforward corollary of Proposition 5.4.4. ■

5.5 The optimal momentum map

The optimal momentum map has been introduced in Ortega and Ratiu (2002a) as an approach, based on generalized distributions, to the problem of finding and describing the conservation laws associated to a canonical symmetry.

This object is related to global rather than to infinitesimal symmetries. Actually, one of the main goals behind its study consists of capturing the conservation laws that cannot be detected by the previously described momentum maps, all of which become trivial when the Lie algebra of the symmetry group is zero, thereby leaving aside the treatment of discrete symmetries, which are of great importance in applications.

Another particularly convenient feature of the optimal momentum map is its generality. Most of the constructions presented so far are very symplectic in nature. We shall see in this section that the optimal momentum map always exists for any canonical group action on a Poisson manifold.

The justification for the use of the term "optimal" lies in the fact that the object to be introduced later is capable of capturing in its level sets the smallest submanifolds of the phase space preserved by the Hamiltonian vector fields associated to the functions that are invariant with respect to a given symmetry group. To be more specific, recall from §4.2.4 that the Hamiltonian vector field associated to an invariant Hamiltonian is automatically equivariant and therefore satisfies the law of conservation of the isotropy introduced in §4.2.6, that is, the local isotropy type manifolds are invariant under its flow. This conservation law, which in principle cannot be detected by any of the momentum maps that we introduced so far, will appear in our study of the optimal momentum map.

5.5.1 The polar distribution and its integrability. The optimal momentum map has much to do with the notion of ***polarity*** introduced in ORTEGA (2003b).

5.5.2 Definition *Let $(M, \{\cdot, \cdot\})$ be a Poisson manifold and $A \subset \mathcal{P}_L(M)$ a pseudogroup of local Poisson diffeomorphisms of M. Let F_A be the set of Hamiltonian vector fields associated to all the elements of $C^\infty(U)^A$, for all the open A-invariant subsets U of M, that is,*

$$F_A = \left\{X_f \mid f \in C^\infty(U)^A, \text{ with } U \subset M \text{ open and } A\text{-invariant}\right\}. \tag{5.5.1}$$

*The distribution $\mathcal{D}_{F_A}$ associated to the family F_A will be called the **polar distribution** defined by A. Any generating family of vector fields for $\mathcal{D}_{F_A}$ will be called a **polar family** of A. The family F_A will be called the **standard polar family** of A. A pseudogroup of local Poisson diffeomorphisms associated to the flows of any polar family of A will be referred to as a **polar pseudogroup** induced by A. The polar pseudogroup $A_{F_A} \subset \mathcal{P}_L(M)$ induced by the standard polar family F_A will be called the **standard polar pseudogroup** and will be denoted by A'.*

5.5.3 Proposition *Let $(M, \{\cdot, \cdot\})$ be a Poisson manifold and $A \subset \mathcal{P}_L(M)$ a pseudogroup of local Poisson diffeomorphisms of M. If A has the extension property (see* §3.1.2*), then the family*

$$F := \left\{X_f \mid f \in C^\infty(M)^A\right\}$$

is a polar family.

Proof. We will show that for any $m \in M$ we have $\mathcal{D}_{F_A}(m) = \mathcal{D}_F(m)$. Indeed, by the finite dimensionality of M, there exists a $r \in \mathbb{N}$ such that

$$\mathcal{D}_{F_A}(m) = \operatorname{span}\{X_{f_1}(m), \ldots, X_{f_r}(m)\},$$

where $f_i \in C^\infty(U_i)^A$ for any $i \in \{1, \ldots, r\}$. The extension property ensures the existence, for each $i \in \{1, \ldots, r\}$, of an open A-invariant subset $V_i \subset U_i$ containing the point m and an A-invariant function $F_i \in C^\infty(M)^A$ such that $f_i|_{V_i} = F_i|_{V_i}$. Consequently,

$$\mathcal{D}_{F_A}(m) = \operatorname{span}\{X_{f_1}(m), \ldots, X_{f_r}(m)\} = \operatorname{span}\{X_{F_1}(m), \ldots, X_{F_r}(m)\},$$

which shows that $\mathcal{D}_{F_A}(m) \subset \mathcal{D}_F(m)$. As the other inclusion is trivially true, equality holds. ■

A very important point regarding the polar distribution of a group of Poisson diffeomorphisms is that it is automatically Poisson and integrable, as we see in the next proposition.

5.5.4 Proposition *Let $(M, \{\cdot, \cdot\})$ be a Poisson manifold and $A \subset \mathcal{P}(M)$ a group of Poisson diffeomorphisms of M. Then the following hold:*

(i) *The polar pseudogroup A' acts canonically and is integrable.*

(ii) *The group A commutes with its polar A', that is, for any $(\mathcal{F}_T, \mathrm{Dom}(\mathcal{F}_T)) \in A'$ the domain $\mathrm{Dom}(\mathcal{F}_T)$ is an open A-invariant set, and for any $\phi \in A$, we have that $(\mathcal{F}_T \circ \phi, \mathrm{Dom}(\mathcal{F}_T)) = (\phi \circ \mathcal{F}_T, \mathrm{Dom}(\mathcal{F}_T))$.*

(iii) *Any $(\mathcal{F}_T, \mathrm{Dom}(\mathcal{F}_T)) \in A'$ induces $(\bar{\mathcal{F}}_T, \pi_A(\mathrm{Dom}(\mathcal{F}_T)))$ of $(M/A, C^\infty_{M/A})$, which is a local diffeomorphism uniquely determined by the relation $\bar{\mathcal{F}}_T \circ \pi_A = \pi_A \circ \mathcal{F}_T$, where $\pi_A : M \to M/A$ is the projection. In other words, the standard polar pseudogroup A' acts on $(M/A, C^\infty_{M/A})$.*

(iv) *The group A acts naturally on the orbit space M/A'. More specifically, for any $\phi \in A$, there is a diffeomorphism $\bar{\phi}$ of the quotient space $(M/A', C^\infty_{M/A'})$ uniquely determined by the relation $\bar{\phi} \circ \pi_{A'} = \pi_{A'} \circ \phi$, where $\pi_{A'} : M \to M/A'$ is the projection.*

Proof. **(i)** A' acts canonically since its elements are finite compositions of Hamiltonian flows and therefore they are local Poisson diffeomorphisms. As to its integrability, according to Theorem 3.2.1 we have to show that for any $(\mathcal{F}_T, \mathrm{Dom}(\mathcal{F}_T)) \in A'$ and any $m \in \mathrm{Dom}(\mathcal{F}_T)$ we have $T_m\mathcal{F}_T(\mathcal{D}_{F_A}(m)) = \mathcal{D}_{F_A}(\mathcal{F}_T(m))$. In order to establish this equality we take $h \in C^\infty(U)^A$ with U an open A-invariant subset of M. Let $V := U \cap \mathrm{Dom}(\mathcal{F}_T)$ be such that $m \in V$ and define $\mathcal{F}_T^V := \mathcal{F}_T|_V : V \to \mathcal{F}_T(V)$ and $h^V := h|_V$. Since V is an A-invariant open subset of M and $\mathcal{F}_T^V$ is a Poisson map we can write

$$\begin{aligned} T_m\mathcal{F}_T\,(X_h(m)) &= T_m\mathcal{F}_T^V\left(X_{h^V}(m)\right) \\ &= T_m\mathcal{F}_T^V\left(X_{h^V\circ(\mathcal{F}_T^V)^{-1}\circ\mathcal{F}_T^V}(m)\right) = X_{h^V\circ(\mathcal{F}_T^V)^{-1}}(\mathcal{F}_T^V(m)) \end{aligned}$$

which belongs to $A'(\mathcal{F}_T(m))$ since by §4.2.4 and Proposition 3.3.2, $h^V \circ (\mathcal{F}_T^V)^{-1}$ belongs to $C^\infty(\mathcal{F}_T(V))^A$. Consequently, $T_m\mathcal{F}_T(\mathcal{D}_{F_A}(m)) \subset \mathcal{D}_{F_A}(\mathcal{F}_T(m))$.

Conversely, let $f \in C^\infty(W)^A$ be such that $\mathcal{F}_T(m) \in W$. When we define $S := \mathcal{F}_T(\mathrm{Dom}(\mathcal{F}_T)) \cap W$, $f^S := f|_S$, $\mathcal{F}_T^S := \mathcal{F}_T|_{\mathcal{F}^{-1}(S)}$, we then have that $X_f(\mathcal{F}_T(m)) = X_{f^S}(\mathcal{F}_T^S(m)) = T_m\mathcal{F}_T^S(X_{f^S\circ\mathcal{F}_T^S}(m))$, which belongs to $T_m\mathcal{F}_T(\mathcal{D}_{F_A}(m))$, as required.

(ii) The elements $\mathcal{F}_T \in A'$ are finite compositions of Hamiltonian flows associated to A-invariant Hamiltonians. The statement follows from §4.2.4 and part **(i)** in Proposition 3.3.2.

(iii) Given $(\mathcal{F}_T, \mathrm{Dom}(\mathcal{F}_T)) \in A'$, the well-defined map $\left(\bar{\mathcal{F}}_T, \pi_A(\mathrm{Dom}(\mathcal{F}_T))\right) = (\bar{\mathcal{F}}_T, \mathrm{Dom}(\mathcal{F}_T)/A)$ that satisfies $\bar{\mathcal{F}}_T \circ \pi_A = \pi_A \circ \mathcal{F}_T$ is guaranteed to exist by **(ii)**. Since $\mathcal{F}_T$ is a local diffeomorphism of M and the projection π_A is open (see §3.1.2) and continuous, $\bar{\mathcal{F}}_T$ is necessarily continuous. We also have $\bar{\mathcal{F}}_T^* C^\infty(\mathcal{F}_T(\mathrm{Dom}(\mathcal{F}_T))/A) \subset C^\infty(\mathrm{Dom}(\mathcal{F}_T)/A)$ since for any $f \in C^\infty(\mathcal{F}_T(\mathrm{Dom}(\mathcal{F}_T))/A)$ the map $f \circ \bar{\mathcal{F}}_T \circ \pi_A|_{\mathrm{Dom}(\mathcal{F}_T)} = f \circ \pi_A \circ \mathcal{F}_T|_{\mathrm{Dom}(\mathcal{F}_T)} \in C^\infty(\mathrm{Dom}(\mathcal{F}_T))^A$, and hence $f \circ \bar{\mathcal{F}}_T \in C^\infty(\mathrm{Dom}(\mathcal{F}_T)/A)$. Since we could do the same with $\bar{\mathcal{F}}_T^{-1}$, we conclude that the map $\bar{\mathcal{F}}_T$ is a local diffeomorphism.

(iv) This follows directly from **(ii)**. ■

5.5.5 The optimal momentum map. Let G be a Lie group acting canonically on the Poisson manifold $(M, \{\cdot, \cdot\})$ via the map $\Phi : G \times M \to M$. Let $A_G \subset \mathcal{P}(M)$ be the associated group of Poisson diffeomorphisms, that is,

$$A_G := \{\Phi_g \mid g \in G\}.$$

The ***optimal momentum map*** $\mathcal{J} : M \to M/A'_G$ is defined as the projection of M onto the orbit space M/A'_G of the pseudogroup A'_G, polar to A_G that, by Proposition 5.5.4, is integrable. We will refer to the quotient M/A'_G as the ***momentum space*** of $\mathcal{J}$.

Notice that the integrability of A'_G allows us to provide a definition of momentum map equivalent to the one introduced in Definition 4.3.1 by saying that a map $\mathbf{J} : M \to S$ is a Noether momentum map for the G-action on $(M, \{\cdot, \cdot\})$ when $\mathbf{J}$ is a first integral of the generalized foliation associated to integrable polar pseudogroup A'_G.

5.5.6 Notation. In order to lighten up the notation, we shall use in what follows interchangeably the symbol A'_G to denote both the standard polar pseudogroup to A_G and the polar distribution. The context will make the distinction of these two concepts clear. Moreover, we will denote by $A'_G \cdot m$ the polar pseudogroup acting on the element $m \in M$ and by $A'_G(m)$ the polar distribution evaluated at m.

5.5.7 The optimal momentum map for proper actions and compact group representations. If the G-action on M is proper, the subgroup A_G has the extension property by Proposition 2.5.6. Consequently, by Proposition 5.5.3, the optimal momentum map can be defined in this case as the projection $\mathcal{J} : M \to M/\mathcal{D}_F$ onto the leaf space of the integrable distribution spanned by the family of vector fields

$$F := \left\{X_f \mid f \in C^\infty(M)^G\right\}. \tag{5.5.2}$$

Also, since the polar distribution A'_G is spanned by G-equivariant vector fields, Corollary 3.3.6 guarantees that when the G-action is proper A'_G is necessarily complete.

A particular case of the situation presented above is that of the canonical linear actions of compact groups on Poisson vector spaces. Under these hypotheses one can use the Schwarz–Mather Theorem presented in §2.5.3 to further simplify the polar family in (5.5.2). Let G be a compact Lie group that acts linearly and canonically on the Poisson vector space $(V, \{\cdot, \cdot\})$ and let $\mathcal{B} := \{\sigma_1, \ldots, \sigma_n\}$ be a Hilbert basis for this action. By the Schwarz–Mather Theorem, any G-invariant function can be written as $f(\sigma_1, \ldots, \sigma_n)$, with $f \in C^\infty(\mathbb{R}^n)$ arbitrary, so the chain rule guarantees that the distribution spanned by the family F in (5.5.2) is the same as the one spanned by the *finite* family

$$\{X_{\sigma_1}, \ldots, X_{\sigma_n}\}.$$

5.5.8 Example. Consider the canonical action of $(\mathbb{R}, +)$ on the Poisson manifold $(\mathbb{R}^3, \{\cdot, \cdot\})$ introduced in §4.2.5. This action does not admit any of the previously introduced momentum maps since it takes place in the Poisson context and, additionally, it does not preserve the symplectic leaves. However, as we shall see below, the optimal momentum map produces a conservation law associated to this symmetry.

As the $(\mathbb{R}, +)$-action is proper, we can use the distribution in (5.5.2) to define the corresponding optimal momentum map. Notice first that the invariant functions $f \in C^\infty(M)^{\mathbb{R}}$ in this case are all of the form $f(x, y, z) \equiv \bar{f}(y, z)$, with $\bar{f} \in C^\infty(\mathbb{R}^2)$ arbitrary. The expression (4.2.4) of the Hamiltonian vector fields defined by this bracket shows that the $A'_{\mathbb{R}}$-orbits on $\mathbb{R}^3$ coincide with those of the $\mathbb{R}^2$-action on $\mathbb{R}^3$ given by $(\mu, \nu) \cdot (x, y, z) := (x + \mu, y + \nu, z - \mu)$, for any $(\mu, \nu) \in \mathbb{R}^2$ and any $(x, y, z) \in \mathbb{R}^3$. Therefore, M/A'_G can be identified with $\mathbb{R}$ and the associated optimal momentum map takes the form

$$\begin{array}{rccc} \mathcal{J}: & \mathbb{R}^3 & \longrightarrow & \mathbb{R} \\ & (x, y, z) & \longmapsto & x + z. \end{array}$$

A straightforward verification shows that the Hamiltonian flow associated to any invariant function $f(x, y, z) \equiv \bar{f}(y, z)$ preserves the level sets of $\mathcal{J}$; moreover, $\mathcal{J}$ is a Casimir function of the Poisson manifold $(\mathbb{R}^3, \{\cdot, \cdot\})$.

5.5.9 Example. Consider the example presented in §4.5.16 regarding a canonical circular symmetry on a torus to which it is impossible to associate a globally defined standard momentum map. We compute the optimal momentum map in this case.

The properness of the action allows us to use again the leaf space of the distribution (5.5.2) It is easy to see that in this case, every S^1-invariant function $f \in C^\infty(\mathbb{T}^2)^{S^1}$ can be written as

$$f(e^{i\theta_1}, e^{i\theta_2}) = g(e^{i\theta_2}),$$

with $g \in C^\infty(S^1)$ arbitrary. The Hamiltonian vector field associated to any of these invariant functions is given by $X_f = \frac{\partial g}{\partial \theta_2} \frac{\partial}{\partial \theta_1}$. Since g is an arbitrary function on the circle, we can identify the quotient M/A'_G with the second circle S^1 in the torus $\mathbb{T}^2$. The optimal momentum map is therefore given by the expression

$$\begin{array}{rccc} \mathcal{J}: & \mathbb{T}^2 & \longrightarrow & S^1 \\ & (e^{i\theta_1}, e^{i\theta_2}) & \longmapsto & e^{i\theta_2}. \end{array}$$

In this case, the optimal momentum map is S^1-valued and, moreover, it coincides with the Lie group valued momentum map introduced in §5.4.

5.5.10 The momentum space. The examples considered above are somewhat misleading in the sense that the resulting optimal momentum maps are smooth maps between manifolds. This feature, common to all the momentum maps studied in the preceding sections, is not usually an attribute of the optimal momentum map. Indeed, its target is, by construction, the leaf space of a distribution spanned by equivariant vector fields whose topology could be extremely convoluted. All one can say, in general, about the optimal momentum map $\mathcal{J} : M \to M/A'_G$, which is a projection on a quotient space, is that it is a *continuous* and *open* map (see §3.1.2). The following example shows that even when the G-action is very simple and the corresponding orbit space $M/G = M/A_G$ is a quotient regular manifold, the associated momentum space M/A'_G does not have to share those properties.

5.5.11 Example. Let $M := \mathbb{T}^2 \times \mathbb{T}^2$ be the product of two tori whose elements we will denote by the four-tuples $(e^{i\theta_1}, e^{i\theta_2}, e^{i\psi_1}, e^{i\psi_2})$. We endow M with the symplectic structure ω defined by $\omega := \mathbf{d}\theta_1 \wedge \mathbf{d}\theta_2 + \sqrt{2}\,\mathbf{d}\psi_1 \wedge \mathbf{d}\psi_2$. We now consider the canonical circle action given by $e^{i\phi} \cdot (e^{i\theta_1}, e^{i\theta_2}, e^{i\psi_1}, e^{i\psi_2}) := (e^{i(\theta_1+\phi)}, e^{i\theta_2}, e^{i(\psi_1+\phi)}, e^{i\psi_2})$. First, notice that since the circle is compact and acts freely on M, the corresponding orbit space M/A_{S^1} is a smooth manifold such that the projection $\pi_{A_{S^1}} : M \to M/A_{S^1}$ is a surjective submersion. The polar distribution A'_{S^1} does not have that property. Indeed, $C^\infty(M)^{S^1}$ comprises all the functions f of the form $f \equiv f(e^{i\theta_2}, e^{i\psi_2}, e^{i(\theta_1-\psi_1)})$. An inspection of the Hamiltonian flows associated to such functions readily shows that the leaves of A'_{S^1} fill densely the manifold M and that the leaf space M/A'_{S^1} can be identified with the leaf space $\mathbb{T}^2/\mathbb{R}$ of a Kronecker (irrational) foliation of a two-torus $\mathbb{T}^2$. Under these circumstances M/A'_{S^1} cannot possibly be a regular quotient manifold.

5.5.12 Example. We now consider $\mathbb{C}^3$ with the symplectic form ω given by

$$\omega((z_1, z_2, z_3), (z'_1, z'_2, z'_3)) = -\operatorname{Im}\langle (z_1, z_2, z_3), (z'_1, z'_2, z'_3)\rangle.$$

Consider now the natural action of the Lie group SU(3) on $\mathbb{C}^3$ via matrix multiplication. This action is canonical and linear and therefore has a standard associated momentum map. Since we are in the situation described in §5.5.7, the polar distribution $A'_{\mathrm{SU}(3)}$ is spanned by the Hamiltonian vector fields associated to the elements of a Hilbert basis of invariant polynomials. In this case, the polynomial

$$f(z_1, z_2, z_3) = \frac{1}{2}\left(|z_1|^2 + |z_2|^2 + |z_3|^2\right)$$

constitutes such a basis. The Hamiltonian flow of X_f is given by

$$F_t(z_1, z_2, z_3) = (z_1 e^{-it}, z_2 e^{-it}, z_3 e^{-it}).$$

Therefore, the momentum space $\mathbb{C}^3/A'_{\mathrm{SU}(3)}$ coincides with $\mathbb{C}^3/S^1$, where S^1 acts on $\mathbb{C}^3$ by

$$e^{i\phi} \cdot (z_1, z_2, z_3) = (e^{i\phi} z_1, e^{i\phi} z_2, e^{i\phi} z_3). \tag{5.5.3}$$

This quotient space can be identified with $(\mathbb{CP}(2) \times \mathbb{R}^+) \cup \{*\}$, where $\{*\}$ denotes a singleton or, said differently, with the cone $C(\mathbb{CP}(2))$ based on $\mathbb{CP}(2)$. Indeed, if $\pi : \mathbb{C}^3 \to \mathbb{C}^3/S^1$ is the canonical projection and $\mathbf{z} = (z_1, z_2, z_3)$, then the mapping that assigns $\pi(z_1, z_2, z_3)$ to $([\mathbf{z}/\|\mathbf{z}\|], \|\mathbf{z}\|)$ if $\mathbf{z} \neq 0$, and to $*$ if $\mathbf{z} = \mathbf{0}$, provides the needed identification (the symbol $[\mathbf{z}/\|\mathbf{z}\|]$ denotes the element $\pi(\mathbf{z}/\|\mathbf{z}\|) \in \mathbb{CP}(2)$). We have the following expression for the optimal momentum map:

$$\begin{aligned} \mathcal{J} : \mathbb{C}^3 &\longrightarrow (\mathbb{CP}(2) \times \mathbb{R}^+) \cup \{*\} \\ \mathbf{z} &\longmapsto \begin{cases} \left(\left[\frac{\mathbf{z}}{\|\mathbf{z}\|} \right], \|\mathbf{z}\| \right) & \text{if } \mathbf{z} \neq 0 \\ * & \text{if } \mathbf{z} = 0. \end{cases} \end{aligned}$$

5.5.13 The G-action on the momentum space. An ingredient that will be important later on is the unique G-action on the momentum space that makes the optimal momentum map G-equivariant. The existence of this smooth action has already been stated in part **(iv)** of Proposition 5.5.4. This action is defined by

$$g \cdot \mathcal{J}(m) := \mathcal{J}(g \cdot m), \quad \text{for any} \quad g \in G, m \in M,$$

and coincides with the smooth G-action that is always available on the leaf space of any distribution spanned by G-equivariant vector fields, as we saw in §3.4.2. In the previous sentence, the term smooth refers to the group action as a map between topological spaces with a presheaf of functions.

5.5.14 The optimal momentum map is universal. The optimal momentum map $\mathcal{J} : M \to M/A'_G$ associated to a canonical G-action on a Poisson manifold $(M, \{\cdot, \cdot\})$ is by its very definition a Noether momentum map, that is, it satisfies Noether's condition (4.3.1). Indeed, due to the integrability of the polar distribution A'_G proved in Proposition 5.5.4, the Stefan–Sussmann Theorem implies that the level sets of the optimal momentum map, that is, the leaves of the polar distribution, coincide with the orbits of the polar pseudogroup. More specifically, if $m \in M$ is such that $\mathbf{J}(m) = \rho \in M/A'_G$, then $\mathcal{J}^{-1}(\rho) = A'_G \cdot m$. As the polar pseudogroup consists of finite compositions of flows of Hamiltonian vector fields associated to all the possible invariant Hamiltonians, Noether's condition for $\mathcal{J}$ follows trivially.

More importantly, $\mathcal{J}$ is universal with respect to this condition. In other words, any other Noether momentum map associated to the G-action factors through $\mathcal{J}$, as we prove in the following theorem.

5.5.15 Theorem (Universality of the optimal momentum map.) *The optimal momentum map is a universal object in the category of Hamiltonian symmetric systems with a Noether momentum map. More specifically, if*

$$(M, \{\cdot, \cdot\}, G, \mathbf{K} : M \to P)$$

is any Hamiltonian G-space with a Noether momentum map $\mathbf{K} : M \to P$ and $\mathcal{J} : M \to M/A'_G$ is the optimal momentum map defined using the canonical G-action on M, then there exists a unique map $\varphi : M/A'_G \to P$ such that the following diagram commutes:

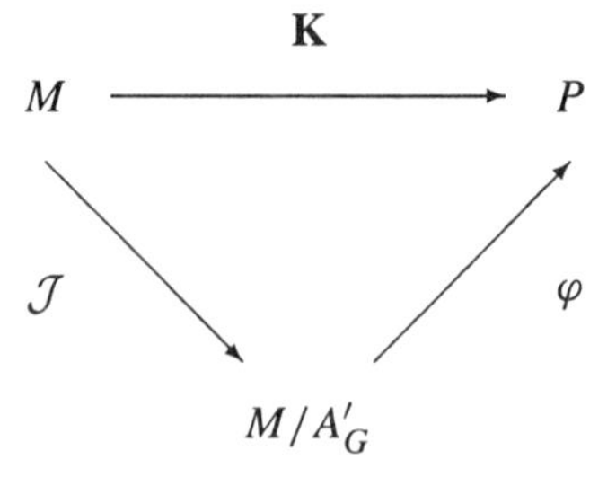

If $\mathbf{K}$ *is smooth and* G*-equivariant with respect to some presheaf of functions on* P *and some* G*-action on this space, then* φ *is smooth and* G*-equivariant.*

Proof. The function φ is given, for any $\rho = \mathcal{J}(m) \in M/A'_G$, by the expression

$$\varphi(\rho) := \mathbf{K}(m).$$

The map φ is well-defined since if $m' \in \mathcal{J}^{-1}(\rho)$, then there exists a finite composition $\mathcal{F}_T$ of flows associated to G-invariant Hamiltonians such that $m' = \mathcal{F}_T(m)$. Given that $\mathbf{K}$ is a Noether momentum map we have that

$$\mathbf{K}(m') = \mathbf{K}(\mathcal{F}_T(m)) = \mathbf{K}(m) = \varphi(\rho).$$

The eventual smoothness of φ, as well as its G-equivariance, are a simple diagram chasing exercise. The uniqueness is guaranteed by the fact that the diagram that defines φ commutes and by the surjectivity of $\mathcal{J}$. ■

5.5.16 The optimal momentum map in the symplectic category. Let G be a Lie group acting properly and canonically on the symplectic manifold (M, ω). In this situation the polar distribution A'_G admits a very compact expression.

5.5.17 Theorem (Ortega and Ratiu (2002a)) *Let* G *be a Lie group acting properly and canonically on the symplectic manifold* (M, ω)*. Then for any* $m \in M$*, we have that*

$$A'_G(m) = (\mathfrak{g} \cdot m)^{\omega} \cap T_m M^m_{G_m}, \tag{5.5.4}$$

where $M^m_{G_m}$ *is the connected component of the isotropy type submanifold* M_{G_m} *that contains the point* m*.*

Proof. Expression (5.5.4) is a consequence of the following chain of equalities:

$$\begin{aligned}
A'_G(m) &= \operatorname{span}\{X_f(m) \mid f \in C^\infty(M)^G\} \\
&= B^\sharp(m)\left(\operatorname{span}\{\mathbf{d}f(m) \mid f \in C^\infty(M)^G\}\right) \\
&= B^\sharp(m)\left(\left((T_m(G \cdot m))^\circ\right)^{G_m}\right) && \text{(by Theorem 2.5.10)} \\
&= \left(B^\sharp(m)\left((T_m(G \cdot m))^\circ\right)\right)^{G_m} && \text{(by Proposition 4.2.3)} \\
&= \left((T_m(G \cdot m))^\omega\right)^{G_m} && \text{(by Proposition 4.1.32)} \\
&= \left((\mathfrak{g} \cdot m)^\omega\right)^{G_m} = (\mathfrak{g} \cdot m)^\omega \cap T_m M^m_{G_m},
\end{aligned}$$

as required. The symbol $B^\sharp : T^*M \to TM$ denotes the vector bundle isomorphism associated to the symplectic form ω. ■

Our next result provides an interesting link between the optimal momentum map and the standard momentum map, provided that the latter exists.

5.5.18 Corollary *Let G be a Lie group acting properly and canonically on the connected symplectic manifold (M, ω) and admitting a standard momentum map $\mathbf{J} : M \longrightarrow \mathfrak{g}^*$. Let $\mathcal{J} : M \to M/A'_G$ be the optimal momentum map. Then, for any $m \in M$ such that $\mathbf{J}(m) = \mu$ and $\mathcal{J}(m) = \rho$, we have that*

$$\mathcal{J}^{-1}(\rho) = (\mathbf{J}^{-1}(\mu) \cap M^m_{G_m})^m, \tag{5.5.5}$$

where $(\mathbf{J}^{-1}(\mu) \cap M^m_{G_m})^m$ denotes the connected component of $\mathbf{J}^{-1}(\mu) \cap M^m_{G_m}$ that contains the point m.

The isotropy subgroup G_ρ of the point $\rho \in M/A'_G$ with respect to the action defined in §5.5.13 is such that

$$G_\rho = N_{G_\mu}(G_m)^{c(m)},$$

where $N_{G_\mu}(G_m)^{c(m)}$ is the closed subgroup of $N_{G_\mu}(G_m)$ that leaves the connected component $(\mathbf{J}^{-1}(\mu) \cap M^m_{G_m})^m$ of $\mathbf{J}^{-1}(\mu) \cap M^m_{G_m}$ invariant.

Proof. Let $\mathbf{J}_{L^m} : M^m_{G_m} \to \mathfrak{l}^*$ be the momentum map corresponding to the L^m-action on the symplectic manifold $(M^m_{G_m}, \omega_{M^m_{G_m}})$ and let $\lambda = \mathbf{J}_{L^m}(m)$. In Corollary 4.6.2 we saw that the freeness of the L^m-action on $M^m_{G_m}$ implies that the set $\mathbf{J}^{-1}_{L^m}(\lambda) = \mathbf{J}^{-1}(\mu) \cap M^m_{G_m}$ is an embedded submanifold of M that contains the point m. Therefore, for any $z \in \mathbf{J}^{-1}(\mu) \cap M^m_{G_m}$, we can write

$$T_z(\mathbf{J}^{-1}(\mu) \cap M^m_{G_m}) = T_z(\mathbf{J}^{-1}_{L^m}(\lambda)) = \ker T_z\mathbf{J}_{L^m} = \ker T_z\mathbf{J}|_{M^m_{G_m}}, \tag{5.5.6}$$

where the last equality is a consequence of Expression (4.6.1) whose derivative implies that $T\mathbf{J}_{L^m} = \Lambda \circ T\mathbf{J}|_{M^m_{G_m}}$. Since Λ is an isomorphism, we have that $\ker T\mathbf{J}_{L^m} = \ker T\mathbf{J}|_{M^m_{G_m}}$. Consequently, by Theorem 5.5.17 we have that

$$\begin{aligned} T_z(\mathbf{J}^{-1}(\mu) \cap M^m_{G_m}) = \ker T_z\mathbf{J}|_{M^m_{G_m}} &= \ker T_z\mathbf{J} \cap T_z M^m_{G_m} \\ &= (\mathfrak{g} \cdot z)^\omega \cap T_z M^m_{G_m} = A'_G(z), \end{aligned} \tag{5.5.7}$$

which guarantees that $(\mathbf{J}^{-1}(\mu) \cap M^m_{G_m})^m$ is the integral manifold of the polar distribution A'_G containing the point m. At the same time, the characterization of the level sets of $\mathcal{J}$ as A'_G-orbits implies, via Noether's Theorem for standard momentum maps and the law of conservation of the isotropy, that $\mathcal{J}^{-1}(\rho) = A'_G \cdot m \subset (\mathbf{J}^{-1}(\mu) \cap M^m_{G_m})^m$. Since $\mathcal{J}^{-1}(\rho)$ is also a maximal integral leaf of A'_G that contains m, it follows that $\mathcal{J}^{-1}(\rho) = (\mathbf{J}^{-1}(\mu) \cap M^m_{G_m})^m$.

We now show that $G_\rho = N_{G_\mu}(G_m)^{c(m)}$. If $g \in G_\rho$, we have that $g \cdot \mathcal{J}^{-1}(\rho) = \mathcal{J}^{-1}(\rho)$ and consequently by (5.5.5) we have that $g \cdot m \in (\mathbf{J}^{-1}(\mu) \cap M^m_{G_m})^m$. This immediately implies that $g \in N_{G_\mu}(G_m)$ and hence $G_\rho \subset N_{G_\mu}(G_m)^{c(m)}$. The reverse inclusion is obvious. It remains to be shown that $N_{G_\mu}(G_m)^{c(m)}$ is closed in $N_{G_\mu}(G_m)$. Indeed, let $g \in \overline{N_{G_\mu}(G_m)^{c(m)}}$; take a sequence $\{g_n\}_{n\in\mathbb{N}} \subset N_{G_\mu}(G_m)^{c(m)}$ such that $g_n \to g$. By the continuity of the action, for any $z \in (\mathbf{J}^{-1}(\mu) \cap M^m_{G_m})^m$ we have that $g_n \cdot z \to g \cdot z$, which shows that $g \cdot z \in \overline{(\mathbf{J}^{-1}(\mu) \cap M^m_{G_m})^m} = (\mathbf{J}^{-1}(\mu) \cap M^m_{G_m})^m$. The last equality is a consequence of the fact that the connected components of a topological space are closed. Given that $z \in (\mathbf{J}^{-1}(\mu) \cap M^m_{G_m})^m$ is arbitrary we have that $g \in N_{G_\mu}(G_m)^{c(m)}$ necessarily. ■

5.5.19 Remark Even though in general the level sets of the optimal momentum map are just initial submanifolds of M, the previous result together with Corollary 4.6.2 imply that in the symplectic case and in the presence of a standard momentum map these level set are actually embedded submanifolds.

5.6 Momentum maps and groupoid moment maps

In this section we show that some of the momentum maps introduced in the last two chapters can be interpreted as the moment maps associated to some groupoid action naturally defined on the manifold in question. Recall that even though the terms moment and momentum are similar, the notion of moment map introduced in §3.6.8 is intrinsic to the concept of a groupoid action and hence it is not a priori related to the various momentum maps introduced in the last two chapters. Nevertheless, the similarities presented below, among others, led MIKAMI AND WEINSTEIN (1988) to choose the terminology introduced in§3.6.8.

The results presented in this section are admittedly incomplete. They suggest various possible improvements that are the subject of ongoing research.

5.6.1 Action groupoids and momentum maps. We start with some elementary remarks suggested by example **(ii)** in §3.6.9. Let (M, ω) be a connected symplectic manifold acted canonically upon by a Lie group G. Suppose that this action admits a standard momentum map $\mathbf{J} : M \to \mathfrak{g}^*$ with non-equivariance one-cocycle $\sigma : G \to \mathfrak{g}^*$. Let $\Theta : G \times \mathfrak{g}^* \to \mathfrak{g}^*$ be the affine action on $\mathfrak{g}^*$ constructed with this cocycle and let $G \times \mathfrak{g}^* \rightrightarrows \mathfrak{g}^*$ be the associated action groupoid. Notice that this is a generalization of the symplectic groupoid introduced in §3.6.6. Now, as the momentum map $\mathbf{J}$ is equivariant with respect to the G-action on M and the affine G-action on $\mathfrak{g}^*$, it naturally induces an action of the groupoid $T^*G \simeq G \times \mathfrak{g}^*$ on M whose associated moment map is $\mathbf{J}$ itself (see **(ii)** in §3.6.9).

The same remark can be made regarding the optimal momentum map $\mathcal{J} : M \to M/A'_G$ associated to a G-canonical action on the Poisson manifold $(M, \{\cdot, \cdot\})$. In this case the groupoid in question is the action groupoid $G \times M/A'_G \rightrightarrows M/A'_G$ associated to the G-action on M/A'_G introduced in §5.5.13. This groupoid acts naturally on M with associated moment map $\mathcal{J}$.

5.6.2 A groupoid model for the optimal momentum map. The ideas leading to the material in this subsection are due to WEINSTEIN (2002a). Expression (5.5.5) suggests that in the presence of a standard momentum map, the level sets of the optimal momentum map can be "parametrized", up to connected components, by the isotropy subgroups of the group action and the momentum values. This idea, made more precise in what follows, allows one, if certain connectedness hypotheses are satisfied, to model the momentum space M/A'_G as a subset of a much more manageable space. This is implemented by noting that the optimal momentum map appears as the moment map associated to a natural groupoid action that arises in the problem.

Let (M, ω) be a connected symplectic manifold acted canonically upon by a Lie group G and suppose that this action admits a standard momentum map $\mathbf{J} : M \to \mathfrak{g}^*$ with non-equivariance one-cocycle $\sigma : G \to \mathfrak{g}^*$. Let $T^*G \simeq G \times \mathfrak{g}^* \rightrightarrows \mathfrak{g}^*$ be the action groupoid associated to the affine action of G on $\mathfrak{g}^*$ and $\mathfrak{B}(G) \rightrightarrows \mathfrak{S}(G)$ the Baer groupoid of G (see §3.6.7). Let $T^*G \times \mathfrak{B}(G) \rightrightarrows \mathfrak{g}^* \times \mathfrak{S}(G)$ be the product groupoid and $\Gamma \rightrightarrows \mathfrak{g}^* \times \mathfrak{S}(G)$ the wide subgroupoid defined by

$$\Gamma := \{((g, \mu), gH) \mid g \in G,\ \mu \in \mathfrak{g}^*,\ H \in \mathfrak{S}(G)\}.$$

We will now prove that $\Gamma \rightrightarrows \mathfrak{g}^* \times \mathfrak{S}(G)$ acts naturally on M with moment map

$$\begin{aligned} \mathfrak{J} : M &\longrightarrow \mathfrak{g}^* \times \mathfrak{S}(G) \\ m &\longmapsto (\mathbf{J}(m), G_m). \end{aligned} \tag{5.6.1}$$

First, notice that

$$\Gamma \times_{\mathfrak{J}} M := \{(((g, \mathbf{J}(m)), gG_m), m) \in \Gamma \times M \mid g \in G,\ m \in M\}.$$

Second, define the action of Γ on M by

$$\begin{aligned} \Psi : \qquad \Gamma \times_{\mathfrak{J}} M &\longrightarrow M \\ (((g, \mathbf{J}(m)), gG_m), m) &\longmapsto g \cdot m \end{aligned}$$

and check that Ψ verifies the axioms of a groupoid action:
(i) Let $(((g, \mathbf{J}(m)), gG_m), m) \in \Gamma \times_{\mathfrak{J}} M$ be arbitrary. Then, $\mathfrak{J}(((g, \mathbf{J}(m)), gG_m) \cdot m) = \mathfrak{J}(g \cdot m) = (\mathbf{J}(g \cdot m), G_{g \cdot m}) = (g \cdot \mathbf{J}(m), gG_m g^{-1}) = \alpha(((g, \mathbf{J}(m)), gG_m))$.
(ii) Let $((g, l \cdot \mu), gH), ((l, \mu), l(l^{-1}Hl)) \in \Gamma$ be two composable elements in Γ. Notice that their product $m(((g, l \cdot \mu), gH), ((l, \mu), l(l^{-1}Hl))) = ((gl, \mu), gl(l^{-1}Hl))$ is an element in $\Gamma \times_{\mathfrak{J}} M$ if and only if $\mu = \mathbf{J}(m)$ and $l^{-1}Hl = G_m$, for some $m \in M$. Hence, on the one hand, we have

$$(((g, l \cdot \mathbf{J}(m)), g(lG_m l^{-1})) \cdot ((l, \mathbf{J}(m)), lG_m)) \cdot m = gl \cdot m.$$

On the other hand, since $((l, \mathbf{J}(m)), lG_m) \cdot m = l \cdot m$, we have

$$((g, l \cdot \mathbf{J}(m)), g(lG_m l^{-1})) \cdot (l \cdot m) = ((g, \mathbf{J}(l \cdot m)), gG_{l \cdot m}) \cdot (l \cdot m) = gl \cdot m,$$

as required.
(iii) For any $m \in M$ we have that

$$\epsilon(\mathfrak{J}(m)) \cdot m = \epsilon((\mathbf{J}(m), G_m)) \cdot m = ((e, \mathbf{J}(m)), G_m) \cdot m = e \cdot m = m.$$

This proves that $\Gamma \rightrightarrows \mathfrak{g}^* \times \mathfrak{S}(G)$ acts on M with moment map $\mathfrak{J} : M \to \mathfrak{g}^* \times \mathfrak{S}(G)$.

The importance of the moment map $\mathfrak{J}$ is that it is a Noether momentum map in the sense of Definition 4.3.1 and encodes through its two components the conservation of the standard momentum and the law of conservation of the isotropy which was one of the guiding principles behind the introduction of the optimal momentum map. Indeed, both objects are closely related since the universality property of the optimal momentum map (Theorem 5.5.15) implies that there exists a unique map $\varphi : M/A'_G \to \mathfrak{g}^* \times \mathfrak{S}(G)$ such that the diagram

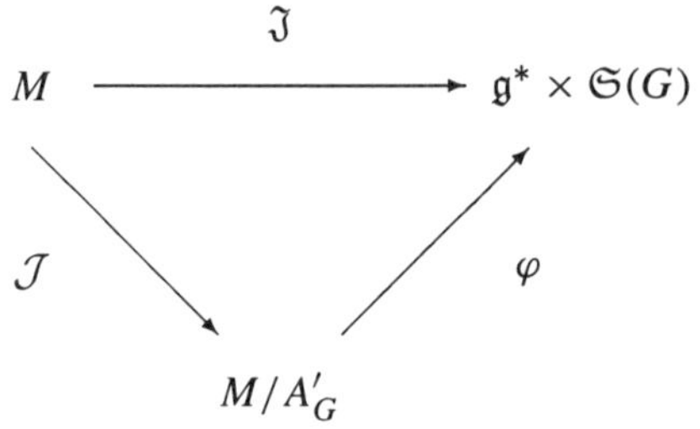

commutes. Recall that the map φ is defined by the equality $\varphi(\mathcal{J}(m)) = \mathfrak{J}(m) = (\mathbf{J}(m), G_m)$, $m \in M$. This expression, together with (5.5.5), allows us to injectively embed the quotient space M/A'_G into $\mathfrak{g}^* \times \mathfrak{S}(G)$ provided that both the isotropy orbit type manifolds M_{G_m} as well as the intersections $\mathbf{J}^{-1}(\mu) \cap M_{G_m}$ are connected. Indeed, let $m, m' \in M$ be such that $\varphi(\mathcal{J}(m)) = \varphi(\mathcal{J}(m'))$ or, equivalently, $(\mathbf{J}(m), G_m) = (\mathbf{J}(m'), G_{m'})$. Expression (5.5.5) in Corollary 5.5.18, together with the connectedness hypotheses, implies that there is a unique element $\rho \in M/A'_G$ such that $m, m' \in \mathcal{J}^{-1}(\rho)$, which establishes that φ is injective.

Chapter 6

Regular Symplectic Reduction Theory

In this chapter we present the most elementary version of symplectic reduction using standard momentum maps. The symplectic reduction method represents a generalization and synthesis of various techniques of elimination of variables from classical mechanics that are based on the existence of conserved quantities. Early specific examples are the reduction to the center of mass frame in the n-body problem using translational invariance and Jacobi's elimination of the node that allows the elimination of four variables using rotational invariance.

Precursors of modern reduction theory can be found in the works of CARTAN (1922), SMALE (1970), SOURIAU (1969), ROBBIN (1973), and MEYER (1973). The definitive formulation is due to MARSDEN AND WEINSTEIN (1974). The reader interested in the history behind this subject is encouraged to check with MARSDEN, RATIU, AND SCHEURLE (2000); MARSDEN AND WEINSTEIN (2001). In all the works that we just quoted symplectic reduction is carried out in the presence of strong regularity hypotheses. This chapter discusses only this situation, while the singular case is presented in Chapter 8. We will not follow the historical development of the subject but will present the theory in its most condensed and general form, creating a template that will turn out to be of much value in the singular case.

We will restrict our exposition to the general symplectic case, leaving aside the very important situations in which the manifold that represents the phase space of the system in question has additional structure:

- The reduction of Kähler manifolds has been explored in KIRWAN (1984a).
- The case in which the symplectic manifold in question is the cotangent bundle of a configuration space manifold Q, and the canonical action is the lift to T^*Q of an action on Q (see examples in §4.5.4 and 4.5.6) is reviewed without proofs. The additional structures present in this case make reduction a very rich and complex phenomenon, still under study, even in the regular case. We refer the reader to MARSDEN AND RATIU (2003) for the proofs and applications. The sources for these results can be found in ABRAHAM AND MARSDEN (1978);

MARSDEN et al. (1998, 2002); MARSDEN AND PERLMUTTER (2000); PERLMUTTER AND RATIU (2004), and references therein.

- The reduction of nonholonomic systems with symmetry is being subjected to intense study. We will not treat this topic because the theory is still in development even in the regular case. Regarding the progress in this area we refer to BLOCH (2003); BLOCH et al. (1996); BATES AND SNIATYCKI (1993); BATES et al. (1996); SNIATYCKI (1998); CANTRIJN et al. (1998); BATES (1998), and references therein.

The reduction strategies described in this chapter are mainly used in applications that involve the Hamiltonian formalism. The reader interested in the relation of these ideas to quantization is encouraged to check with LANDSMAN (1998), LANDSMAN et al. (2001), and references therein.

6.1 Point reduction

The symplectic reduction procedure presented in this section consists of constructing a symplectic manifold out of a given symmetric one on which the conservation laws and degeneracies associated to the symmetry have been eliminated. This strategy allows for the reduction of a symmetric Hamiltonian dynamical system to a dimensionally smaller one. We emphasize that this reduction procedure preserves the symplectic category, that is, if we start with a Hamiltonian system on a symplectic manifold, the reduced space is also a Hamiltonian system on a symplectic manifold, which is usually referred to as the ***symplectic*** or ***Marsden–Weinstein reduced space***.

Throughout this chapter (M, ω) is a connected symplectic manifold acted properly, freely, and canonically upon by a Lie group G. We will also assume that this action has an associated momentum map $\mathbf{J} : M \rightarrow \mathfrak{g}^*$ with non-equivariance one-cocycle $\sigma : G \rightarrow \mathfrak{g}^*$. The regularity assumption that we referred to in the introduction to this chapter is hidden in the freeness of the G-action; in this situation, the Bifurcation Lemma (see Proposition 4.5.12) guarantees that the momentum maps associated to this action are surjective submersions onto some open subset of $\mathfrak{g}^*$.

The following classical result is due to MARSDEN AND WEINSTEIN (1974).

6.1.1 Theorem (Symplectic point reduction) *Let $\Phi : G \times M \rightarrow M$ be a free proper canonical action of the Lie group G on the connected symplectic manifold (M, ω). Suppose that this action has an associated momentum map $\mathbf{J} : M \rightarrow \mathfrak{g}^*$, with non-equivariance one-cocycle $\sigma : G \rightarrow \mathfrak{g}^*$. Let $\mu \in \mathfrak{g}^*$ be a value of $\mathbf{J}$ and denote by G_μ the isotropy of μ under the affine action of G on $\mathfrak{g}^*$. Then:*

(i) *The space $M_\mu := \mathbf{J}^{-1}(\mu)/G_\mu$ is a regular quotient manifold and, moreover, it is a symplectic manifold with symplectic form ω_μ uniquely characterized by the relation*

$$\pi_\mu^* \omega_\mu = i_\mu^* \omega. \tag{6.1.1}$$

*The maps $i_\mu : \mathbf{J}^{-1}(\mu) \hookrightarrow M$ and $\pi_\mu : \mathbf{J}^{-1}(\mu) \rightarrow \mathbf{J}^{-1}(\mu)/G_\mu$ denote the inclusion and the projection, respectively. The pair (M_μ, ω_μ) is called the **symplectic point reduced space**.*

(ii) *Let $h \in C^\infty(M)^G$ be a G-invariant Hamiltonian. The flow F_t of the Hamiltonian vector field X_h leaves the connected components of $\mathbf{J}^{-1}(\mu)$ invariant and commutes with the G-action, so it induces a flow F_t^μ on M_μ defined by*

$$\pi_\mu \circ F_t \circ i_\mu = F_t^\mu \circ \pi_\mu. \tag{6.1.2}$$

(iii) *The vector field generated by the flow F_t^μ on (M_μ, ω_μ) is Hamiltonian with associated* ***reduced Hamiltonian function*** *$h_\mu \in C^\infty(M_\mu)$ defined by*

$$h_\mu \circ \pi_\mu = h \circ i_\mu.$$

The vector fields X_h and X_{h_μ} are π_μ-related. The triple $(M_\mu, \omega_\mu, h_\mu)$ is called the ***reduced Hamiltonian system****.*

(iv) *Let $k \in C^\infty(M)^G$ be another G-invariant function. Then $\{h, k\}$ is also G-invariant and $\{h, k\}_\mu = \{h_\mu, k_\mu\}_{M_\mu}$, where $\{\cdot, \cdot\}_{M_\mu}$ denotes the Poisson bracket associated to the symplectic form ω_μ on M_μ.*

Proof. **(i)** The freeness of the action and Proposition 4.5.12 imply that $\mathbf{J}$ is a submersion onto some open subset of $\mathfrak{g}^*$ and, consequently, the level set $\mathbf{J}^{-1}(\mu)$ is a G_μ-invariant (by the equivariance of $\mathbf{J}$), closed, and embedded submanifold of M. The free and proper G-action on M restricts to a free and proper G_μ-action on $\mathbf{J}^{-1}(\mu)$ (see §2.3.7) and consequently the quotient $M_\mu := \mathbf{J}^{-1}(\mu)/G_\mu$ is a regular quotient manifold by Proposition 2.3.8.

We continue by showing that for any $m \in \mathbf{J}^{-1}(\mu)$, every vector $v_m \in T_m\mathbf{J}^{-1}(\mu)$ can be written as $T_m i_\mu(v_m) = X_f(m)$, where $f \in C^\infty(M)^G$, and $i_\mu : \mathbf{J}^{-1}(\mu) \hookrightarrow M$ is the canonical injection. This is indeed a consequence of the freeness of the action, Theorem 5.5.17, and Proposition 4.5.14 which, in the hypotheses of our theorem show that

$$\begin{aligned} T_m\mathbf{J}^{-1}(\mu) &= \ker T_m\mathbf{J} = (\mathfrak{g} \cdot m)^\omega = A'_G(m) \\ &= \{X_f(m) \mid f \in C^\infty(M)^G\}. \end{aligned} \tag{6.1.3}$$

We use this fact to define a two-form ω_μ on M_μ by

$$\omega_\mu([m]_\mu)(T_m\pi_\mu(v), T_m\pi_\mu(w)) = \{f, g\}(m), \tag{6.1.4}$$

where $[m]_\mu := \pi_\mu(m)$, $\{\cdot, \cdot\}$ is the Poisson bracket associated to the symplectic form ω on M, and $f, g \in C^\infty(M)^G$ are two G-invariant functions such that $T_m i_\mu(v) = X_f(m)$ and $T_m i_\mu(w) = X_g(m)$.

We now check that expression (6.1.4) is a good definition for the form ω_μ on the quotient M_μ. Let $m, m' \in \mathbf{J}^{-1}(\mu)$ be such that $\pi_\mu(m) = \pi_\mu(m')$ and let $v', w' \in T_{m'}\mathbf{J}^{-1}(\mu)$ be such that $T_m\pi_\mu(v) = T_{m'}\pi_\mu(v')$, $T_m\pi_\mu(w) = T_{m'}\pi_\mu(w)'$. Let $f', g' \in C^\infty(M)^G$ be such that $v' = X_{f'}(m')$ and $w' = X_{g'}(m')$. The condition $\pi_\mu(m) = \pi_\mu(m')$ implies the existence of an element $k \in G_\mu$ such that $m' = \Phi_k(m)$. We also have that $T_m\pi_\mu = T_{m'}\pi_\mu \circ T_m\Phi_k$. Analogously, because of the equalities $T_m\pi_\mu(v) = T_{m'}\pi_\mu(v')$, $T_m\pi_\mu(w) = T_{m'}\pi_\mu(w')$, there exist two elements $\xi^1, \xi^2 \in \mathfrak{g}_\mu$ such that $X_{f'}(m') - T_m\Phi_k \cdot X_f(m) = \xi^1_{\mathbf{J}^{-1}(\mu)}(m') = X_{\mathbf{J}^{\xi^1}}(m')$, and $X_{g'}(m') - T_m\Phi_k \cdot X_g(m) =$

$\xi^2_{\mathbf{J}^{-1}(\mu)}(m') = X_{\mathbf{J}^{\xi^2}}(m')$, or, analogously $X_{f'}(m') = X_{\mathbf{J}^{\xi^1}+f\circ\Phi_{k-1}}(m') = X_{\mathbf{J}^{\xi^1}+f}(m')$, and $X_{g'}(m') = X_{\mathbf{J}^{\xi^2}+g\circ\Phi_{k-1}}(m') = X_{\mathbf{J}^{\xi^2}+g}(m')$. Hence

$$\begin{aligned}\omega_\mu(\pi_\mu(m'))(v', w') &= \{f', g'\}(m') = \{\mathbf{J}^{\xi^1} + f, \mathbf{J}^{\xi^2} + g\}(m') \\ &= \{\mathbf{J}^{\xi^1} + f, \mathbf{J}^{\xi^2} + g\}(\Phi_k(m)) \\ &= \{f \circ \Phi_k, g \circ \Phi_k\}(m) + \{f \circ \Phi_k, \mathbf{J}^{\xi^2} \circ \Phi_k\}(m) \\ &\quad + \{\mathbf{J}^{\xi^1} \circ \Phi_k, g \circ \Phi_k\}(m) + \{\mathbf{J}^{\xi^1} \circ \Phi_k, \mathbf{J}^{\xi^2} \circ \Phi_k\}(m). \qquad (6.1.5)\end{aligned}$$

Given that for any $\xi \in \mathfrak{g}$ and any $k \in G$

$$\mathbf{J}^\xi \circ \Phi_k = \mathbf{J}^{\mathrm{Ad}_{k^{-1}}\xi} + \langle \sigma(k), \xi \rangle,$$

we can conclude that the functions $\mathbf{J}^\xi \circ \Phi_k$ and $\mathbf{J}^{\mathrm{Ad}_{k^{-1}}\xi}$ differ by a constant and, as the Poisson bracket depends only on the first derivative and the functions f and g are G-invariant, (6.1.5) can be rewritten as

$$\{f, g\}(m) + \{f, \mathbf{J}^{\mathrm{Ad}_{k^{-1}}\xi^2}\}(m) + \{\mathbf{J}^{\mathrm{Ad}_{k^{-1}}\xi^1}, g\}(m) + \{\mathbf{J}^{\mathrm{Ad}_{k^{-1}}\xi^1}, \mathbf{J}^{\mathrm{Ad}_{k^{-1}}\xi^2}\}(m).$$

The G-invariance of the functions f and g implies, by Noether's Theorem 4.5.11, that $\{f, \mathbf{J}^{\mathrm{Ad}_{k^{-1}}\xi^2}\} = \{\mathbf{J}^{\mathrm{Ad}_{k^{-1}}\xi^1}, g\} = 0$. Moreover, as the Lie algebra $\mathfrak{g}_\mu$ is invariant by the adjoint action of the group G_μ, we have that $\mathrm{Ad}_{k^{-1}}\xi^2 \in \mathfrak{g}_\mu$ and hence $(\mathrm{Ad}_{k^{-1}}\xi^2)_M(m) \in T_m\mathbf{J}^{-1}(\mu) = \ker T_m\mathbf{J}$. Consequently,

$$\begin{aligned}\{\mathbf{J}^{\mathrm{Ad}_{k^{-1}}\xi^1}, \mathbf{J}^{\mathrm{Ad}_{k^{-1}}\xi^2}\}(m) &= \mathbf{d}\mathbf{J}^{\mathrm{Ad}_{k^{-1}}\xi^1}(m)\left(X_{\mathbf{J}^{\mathrm{Ad}_{k^{-1}}\xi^2}}(m)\right) \\ &= \mathbf{d}\mathbf{J}^{\mathrm{Ad}_{k^{-1}}\xi^1}(m)\left(\left(\mathrm{Ad}_{k^{-1}}\xi^2\right)_{\mathbf{J}^{-1}(\mu)}(m)\right) \\ &= \left\langle T_m\mathbf{J}\left(\left(\mathrm{Ad}_{k^{-1}}\xi^2\right)_{\mathbf{J}^{-1}(\mu)}(m)\right), \mathrm{Ad}_{k^{-1}}\xi^1\right\rangle = 0.\end{aligned}$$

All these facts inserted in (6.1.5) imply that $\omega_\mu(\pi_\mu(m'))(v', w') = \{f, g\}(m) = \omega_\mu(\pi_\mu(m))(v, w)$, and hence (6.1.4) defines indeed a two-form on $T_{[m]_\mu}M_\mu$.

Next we verify (6.1.1). Let m be an arbitrary point in $\mathbf{J}^{-1}(\mu)$ and let v, $w \in T_m\mathbf{J}^{-1}(\mu)$. We write $T_m i_\mu(v) = X_f(m)$ and $T_m i_\mu(w) = X_g(m)$, for some f, $g \in C^\infty(M)^G$. Then, by (6.1.4),

$$(\pi_\mu^*\omega_\mu)(m)(v, w) = \{f, g\}(m) = \omega(m)(X_f(m), X_g(m)) = (i_\mu^*\omega)(m)(v, w).$$

Conversely, if (6.1.1) is valid, the above chain of equalities shows that (6.1.4) holds. Since π_μ is a surjective submersion, this shows that (6.1.1), which is equivalent to (6.1.4), uniquely determines ω_μ and that ω_μ is a smooth two-form on the quotient M_μ. The Jacobi identity for the bracket $\{\cdot, \cdot\}$ on M and its antisymmetry imply that ω_μ is closed and antisymmetric.

Finally, we show that ω_μ is nondegenerate. Indeed, if

$$\omega_\mu([m]_\mu)(T_m\pi_\mu(v), T_m\pi_\mu(w)) = 0$$

for any $w \in T_m\mathbf{J}^{-1}(\mu)$, then $\omega(m)(v, w) = 0$ for any $w \in T_m\mathbf{J}^{-1}(\mu)$ which implies, by Proposition 4.5.14 and the Reduction Lemma 4.5.34 that $w \in \ker T_m\mathbf{J} \cap (\ker T_m\mathbf{J})^{\omega} = (\mathfrak{g} \cdot m)^{\omega} \cap ((\mathfrak{g} \cdot m)^{\omega})^{\omega} = (\mathfrak{g} \cdot m)^{\omega} \cap \mathfrak{g} \cdot m = \mathfrak{g}_{\mu} \cdot m$. This shows that $T_m\pi_{\mu} \cdot v = 0$, as required. The pair (M_{μ}, ω_{μ}) is therefore a symplectic manifold.

(ii) By Noether's Theorem 4.5.11, the flow F_t leaves invariant the connected components of $\mathbf{J}^{-1}(\mu)$. The G-invariance of h and the canonical character of the action imply that F_t is G-equivariant (see §4.2.4). Let F_t^{μ} be the flow on M_{μ} that makes the following diagram commutative:

$$\begin{array}{ccc} \mathbf{J}^{-1}(\mu) & \xrightarrow{F_t \circ i_{\mu}} & \mathbf{J}^{-1}(\mu) \\ \pi_{\mu}\downarrow & & \downarrow\pi_{\mu} \\ M_{\mu} & \xrightarrow{F_t^{\mu}} & M_{\mu}. \end{array}$$

(iii) Due to the G-invariance of h, the function $h_{\mu} \in C^{\infty}(M_{\mu})$ is uniquely determined by the identity $h_{\mu} \circ \pi_{\mu} = h \circ i_{\mu}$. Let $Y \in \mathfrak{X}(M_{\mu})$ be the vector field on M_{μ} whose flow is F_t^{μ}. By construction, Y is π_{μ}-related to X_h. Indeed, differentiating relation (6.1.2) with respect to the time t, we obtain

$$T\pi_{\mu} \circ X_h \circ i_{\mu} = Y \circ \pi_{\mu}.$$

We now verify that Y is a Hamiltonian vector field with Hamiltonian function h_{μ}, that is, $Y = X_{h_{\mu}}$. Let $m \in \mathbf{J}^{-1}(\mu)$ and $v \in T_m\mathbf{J}^{-1}(\mu)$ be arbitrary. Then

$$\begin{aligned} \omega_{\mu}(\pi_{\mu}(m))\left(Y(\pi_{\mu}(m)), T_m\pi_{\mu}(v)\right) &= \omega_{\mu}(\pi_{\mu}(m))\left(T_m\pi_{\mu}(X_h(m)), T_m\pi_{\mu}(v)\right) \\ &= \omega(m)(X_h(m), v) = \mathbf{d}h(m) \cdot v \\ &= \mathbf{d}(h_{\mu} \circ \pi_{\mu})(m) \cdot v \\ &= \mathbf{d}h_{\mu}(\pi_{\mu}(m))\left(T_m\pi_{\mu}(v)\right) \\ &= \omega_{\mu}(\pi_{\mu}(m))\left(X_{h_{\mu}}(\pi_{\mu}(m)), T_m\pi_{\mu}(v)\right), \end{aligned}$$

which shows that $Y = X_{h_{\mu}}$.

(iv) The G-invariance of $\{h, k\}$ is a straightforward corollary of the canonical character of the action. Indeed, for any $g \in G$ we have

$$\{h, k\} \circ \Phi_g = \Phi_g^*\{h, k\} = \{\Phi_g^*h, \Phi_g^*k\} = \{h, k\}.$$

The function $\{h, k\}_{\mu} \in C^{\infty}(M_{\mu})$ is uniquely characterized by the identity $\{h, k\}_{\mu} \circ \pi_{\mu} = \{h, k\} \circ i_{\mu}$. By the definition of the Poisson bracket on (M_{μ}, ω_{μ}), π_{μ}-relatedness of the relevant Hamiltonian vector fields, and (6.1.4), we have for any $m \in \mathbf{J}^{-1}(\mu)$

$$\begin{aligned} \{h_{\mu}, k_{\mu}\}_{M_{\mu}}([m]_{\mu}) &= \omega_{\mu}([m]_{\mu})\left(X_{h_{\mu}}([m]_{\mu}), X_{k_{\mu}}([m]_{\mu})\right) \\ &= \omega_{\mu}([m]_{\mu})\left(T_m i_{\mu}(X_h(m)), T_m i_{\mu}(X_k(m))\right) = \{h, k\}(m), \end{aligned}$$

that is, the function $\{h_{\mu}, k_{\mu}\}_{M_{\mu}}$ also satisfies the relation $\{h_{\mu}, k_{\mu}\}_{M_{\mu}} \circ \pi_{\mu} = \{h, k\} \circ i_{\mu}$, which proves the desired equality $\{h_{\mu}, k_{\mu}\}_{M_{\mu}} = \{h, k\}_{\mu}$. ■

6.1.2 The connectedness assumption on the symplectic manifold (M, ω) imposed in Theorem 6.1.1 is motivated by the need to have a well-defined non-equivariance cocycle $\sigma : M \to \mathfrak{g}^*$ (see Proposition 4.5.21) that later is used to define an affine action on $\mathfrak{g}^*$ with respect to which the momentum map $\mathbf{J} : M \to \mathfrak{g}^*$ is equivariant and whose isotropies satisfy the Reduction Lemma 4.5.34. Consequently, if the momentum map $\mathbf{J} : M \to \mathfrak{g}^*$ in our setup is coadjoint equivariant, the connectedness hypothesis in the statement of Theorem 6.1.1 can be dropped.

6.1.3 Let us apply Theorem 6.1.1 to $M = T^*G$, where G is a Lie group with Lie algebra $\mathfrak{g}$, the G-action being the cotangent lift of *left* translation, and the associated momentum map $\mathbf{J}_L : \alpha_g \in T^*G \mapsto T_e^*R_g(\alpha_g) \in \mathfrak{g}^*$ which is right invariant. All hypotheses in Theorem 6.1.1 are satisfied and for each $\mu \in \mathfrak{g}^*$ we can form the symplectic point reduced space $((T^*G)_\mu, \omega_\mu)$. Recall also that the momentum map for the lift of *right* translations is left invariant and is given by $\mathbf{J}_R : \alpha_g \in T^*G \mapsto T_e^*L_g(\alpha_g) \in \mathfrak{g}^*$. Finally, recall from Theorem 4.5.31 that the orbit symplectic form on $\mathcal{O}_\mu = \{\mathrm{Ad}_g\, \mu \mid g \in G\}$ has the expression

$$\omega^{\pm}_{\mathcal{O}_\mu}(\nu)(\mathrm{ad}^*_\xi(\nu),\, \mathrm{ad}^*_\eta(\nu)) = \pm\langle \nu,\, [\xi,\, \eta]\rangle,$$

for arbitrary $\nu \in \mathcal{O}_\mu$ and $\xi, \eta \in \mathfrak{g}$. The following result is standard (see, e.g., ABRAHAM AND MARSDEN (1978); GUILLEMIN AND STERNBERG (1984c); LIBERMANN AND MARLE (1987); MARSDEN AND RATIU (2003); MCDUFF AND SALAMON (1998)).

6.1.4 Theorem (Coadjoint orbits as point reduced spaces) *The momentum map* $\mathbf{J}_R :$ $T^*G \to \mathfrak{g}^*$ *induces for each* $\mu \in \mathfrak{g}^*$ *a symplectic diffeomorphism* $\overline{\mathbf{J}}_R : ((T^*G)_\mu, \omega_\mu)$ $\to (\mathcal{O}_\mu, \omega^-_{\mathcal{O}_\mu})$.

Proof. By the characterization (6.1.1) of ω_μ it suffices to show that $\mathbf{J}_R^*\omega^-_{\mathcal{O}_\mu} = -\mathbf{d}\Theta$, where $\Theta \in \Omega^1(T^*G)$ is the canonical one-form of the cotangent bundle (see Example 4.1.3). Since $\mathbf{J}^{-1}(\mu) = \{T_g^*R_{g^{-1}}\mu \mid g \in G\}$ is the graph of the right invariant one-form on G whose value at e is μ, any tangent vector at $T_g^*R_{g^{-1}}\mu$ to the submanifold $\mathbf{J}^{-1}(\mu)$ is tangent to a curve of the form $\alpha_\xi(t) := T^*_{g\exp(t\xi)}R_{\exp(-t\xi)g^{-1}}\mu = T^*_{g\exp(t\xi)}L_{g^{-1}}(T^*_{\exp(t\xi)}R_{\exp(-t\xi)}\nu) = \overline{L}_g\overline{R}_{\exp(t\xi)}\nu$ for some $\xi \in \mathfrak{g}$, where $\nu := \mathrm{Ad}^*_g\,\mu$ and where $\overline{L}_g := T^*L_{g^{-1}}$, $\overline{R}_g := T^*R_{g^{-1}} : T^*G \to T^*G$ are the cotangent lifts of the left and right translations, respectively. Therefore, $\alpha'_\xi(0) = T_\nu\overline{L}_g(\xi_{T^*G}(\nu))$, where ξ_{T^*G} denotes the infinitesimal generator defined by $\xi \in \mathfrak{g}$ relative to the *right* action $\overline{R} : G \times T^*G \to T^*G$. Since $T_g^*R_{g^{-1}}\mu = \overline{R}_g\mu = \overline{L}_g\nu$ and the action $\overline{L}$ preserves Θ we have, for any $\xi, \eta \in \mathfrak{g}$,

$$\begin{aligned}
-\mathbf{d}\Theta(\overline{L}_g\nu)\left(\alpha'_\xi(0), \alpha'_\eta(0)\right) &= -\mathbf{d}\Theta(\overline{L}_g\nu)\left(T_\nu\overline{L}_g(\xi_{T^*G}(\nu)), T_\nu\overline{L}_g(\eta_{T^*G}(\nu))\right)\\
&= -\mathbf{d}\Theta(\nu)\left(\xi_{T^*G}(\nu), \eta_{T^*G}(\nu)\right)\\
&= -\xi_{T^*G}[\Theta(\eta_{T^*G})](\nu) + \eta_{T^*G}[\Theta(\xi_{T^*G})](\nu) + \Theta([\xi_{T^*G}, \eta_{T^*G}])(\nu).
\end{aligned}$$

However, denoting by $\pi : T^*G \to G$ the cotangent bundle projection, by the definition of the canonical one-form Θ on T^*G we have for any $\alpha_g \in T_g^*G$

$$\Theta(\eta_{T^*G})(\alpha_g) = \left\langle \alpha_g, T_{\alpha_g}\pi(\eta_{T^*G})\right\rangle = \left\langle \alpha_g, \eta_G(g)\right\rangle = \mathbf{J}_R^\eta(\alpha_g)$$

since η_{T^*G} and η_G are π-related. Thus, using the identity $[\xi_{T^*G}, \eta_{T^*G}] = [\xi, \eta]_{T^*G}$ valid for *right* actions, we can continue the computation above and write

$$\begin{aligned}
&- \xi_{T^*G}[\Theta(\eta_{T^*G})](\nu) + \eta_{T^*G}[\Theta(\xi_{T^*G})](\nu) + \Theta([\xi, \eta]_{T^*G})(\nu) \\
&= -X_{\mathbf{J}_R^\xi}[\mathbf{J}_R^\eta](\nu) + X_{\mathbf{J}_R^\eta}[\mathbf{J}_R^\xi](\nu) + \mathbf{J}_R^{[\xi,\eta]}(\nu) \\
&= -\{\mathbf{J}_R^\eta, \mathbf{J}_R^\xi\}(\nu) + \{\mathbf{J}_R^\xi, \mathbf{J}_R^\eta\}(\nu) - \{\mathbf{J}_R^\xi, \mathbf{J}_R^\eta\}(\nu) = \{\mathbf{J}_R^\xi, \mathbf{J}_R^\eta\}(\nu)
\end{aligned}$$

since for *right* actions infinitesimal equivariance is equivalent to $\mathbf{J}_R^{[\xi,\eta]} = -\{\mathbf{J}_R^\xi, \mathbf{J}_R^\eta\}$. Thus, using this relation again, we finally obtain

$$\begin{aligned}
-\mathbf{d}\Theta(\overline{L}_g\nu)\left(\alpha'_\xi(0), \alpha'_\eta(0)\right) &= \{\mathbf{J}_R^\xi, \mathbf{J}_R^\eta\}(\nu) = -\mathbf{J}_R^{[\xi,\eta]}(\nu) = -\langle \nu, [\xi, \eta]\rangle \\
&= \omega^-_{\mathcal{O}_\mu}(\nu)\left(\mathrm{ad}^*_\xi(\nu), \mathrm{ad}^*_\eta(\nu)\right) \\
&= (\mathbf{J}_R^*\omega^-_{\mathcal{O}_\mu})(\overline{L}_g\nu)\left(\alpha'_\xi(0), \alpha'_\eta(0)\right),
\end{aligned}$$

since $\mathbf{J}_R(\overline{L}_g\nu) = \nu$ and $T_{\overline{L}_g\nu}\mathbf{J}_R(\alpha'_\xi(0)) = \mathrm{ad}^*_\xi(\nu)$. This last equality is proved in the following way. First notice that

$$\begin{aligned}
\mathbf{J}_R(\alpha_\xi(t)) &= T_e^* L_{g\exp(t\xi)}\alpha_\xi(t) = \overline{L}_{(g\exp(t\xi))^{-1}}\alpha_\xi(t) \\
&= \overline{L}_{\exp(-t\xi)}\overline{L}_{g^{-1}}\overline{L}_g\overline{R}_{\exp(t\xi)}\nu = \overline{L}_{\exp(-t\xi)}\overline{R}_{\exp(t\xi)}\nu = \mathrm{Ad}^*_{\exp(t\xi)}\nu.
\end{aligned}$$

Second, using this identity we get

$$T_{\overline{L}_g\nu}\mathbf{J}_R(\alpha'_\xi(0)) = \left.\frac{d}{dt}\right|_{t=0}\mathbf{J}_R(\alpha_\xi(t)) = \left.\frac{d}{dt}\right|_{t=0}\mathrm{Ad}^*_{\exp(t\xi)}\nu = \mathrm{ad}^*_\xi(\nu)$$

which proves the claim. ■

6.1.5 Foliation reduction. The notion of symplectic reduction due to MARSDEN AND WEINSTEIN (1974) presented above can be seen as a particular case of a more general notion that goes back to CARTAN (1922). Consider a manifold N endowed with a closed two-form ω; the pair (N, ω) is called a ***presymplectic manifold***. The ***characteristic*** or ***null distribution*** D_ω of N relative to ω is defined at each point $n \in N$ by

$$D_{\omega,n} := \{u \in T_nN \mid \omega(n)(u, v) = 0 \text{ for all } v \in T_nN\}. \qquad (6.1.6)$$

Assume that the characteristic distribution is differentiable (see §1.4.4 for all the notions used below). Let us show that it is involutive, that is, if $X \in \mathfrak{X}(M)$, with (local) flow F_t, has values in D_ω, then $T_nF_t(D_{\omega,n}) \subset D_{\omega,F_t(n)}$ for all t for which the flow is defined. To see this, note that X satisfies, by definition, $\mathbf{i}_X\omega = 0$ and thus, since ω is closed, this implies that $\pounds_X\omega = 0$ which is equivalent to $F_t^*\omega = \omega$. Now let $u \in T_nN$. Then for any $v \in T_{F_t(n)}N$ we have

$$\begin{aligned}
\omega(F_t(n))\,(T_nF_t(u), v) &= \omega(F_t(n))\left(T_nF_t(u), T_nF_t(T_{F_t(n)}(F_{-t}(v))\right) \\
&= (F_t^*\omega)(n)\left(u, T_{F_t(n)}F_{-t}(v)\right) = \omega(n)\left(u, T_{F_t(n)}F_{-t}(v)\right) = 0,
\end{aligned}$$

which shows that $T_nF_t(u) \in D_{\omega, F_t(n)}$. Therefore, by the Stefan–Sussman Theorem 3.2.1, the distribution D_ω is integrable and thus it defines a generalized foliation $\mathfrak{F}_\omega$ on N. Assume that the leaf space $N/\mathfrak{F}_\omega$ is a smooth manifold such that the projection $\pi_{\mathfrak{F}_\omega} : N \to N/\mathfrak{F}_\omega$ is a surjective submersion, that is, one assumes that the generalized foliation is ***regular***.

Define a two-form $\Omega_{\mathfrak{F}_\omega} \in \Omega^2(N/\mathfrak{F}_\omega)$ by

$$\Omega_{\mathfrak{F}_\omega}(\pi_{\mathfrak{F}_\omega}(n))\left(T_n\pi_{\mathfrak{F}_\omega}(u), T_n\pi_{\mathfrak{F}_\omega}(v)\right) := \omega(n)(u, v) \tag{6.1.7}$$

where $u, v \in T_nN$. Let us prove that $\Omega_{\mathfrak{F}_\omega}(\pi_{\mathfrak{F}_\omega}(n))$ is well defined on the vector space $T_{\pi_{\mathfrak{F}_\omega}(n)}(N/\mathfrak{F}_\omega)$. If $\pi_{\mathfrak{F}_\omega}(n) = \pi_{\mathfrak{F}_\omega}(n')$, then n and n' lie on the same leaf of the generalized foliation $\mathfrak{F}_\omega$, that is, there is a diffeomorphism $\varphi : N \to N$, obtained as a finite composition of flows of smooth sections of $D_{\mathfrak{F}_\omega}$ such that $n' = \varphi(n)$ (see 1.4.4). If $X \in \mathfrak{X}(N)$ is a smooth section of D_ω with flow F_t, then, as we have already shown, $F_t^*\omega = \omega$ and therefore $\varphi^*\omega = \omega$.

Let $u'' := T_n\varphi(u)$, $v'' := T_n\varphi(v) \in T_{n'}N$ and let $u', v' \in T_{n'}N$ be arbitrary vectors such that $T_{n'}\pi_{\mathfrak{F}_\omega}(u') = T_n\pi_{\mathfrak{F}_\omega}(u)$, $T_{n'}\pi_{\mathfrak{F}_\omega}(v') = T_n\pi_{\mathfrak{F}_\omega}(v)$. Then, since $\pi_{\mathfrak{F}_\omega} \circ \varphi = \pi_{\mathfrak{F}_\omega}$ we have $T_{n'}\pi_{\mathfrak{F}_\omega}(u'') = \left(T_{n'}\pi_{\mathfrak{F}_\omega} \circ T_n\varphi\right)(u) = T_n(\pi_{\mathfrak{F}_\omega} \circ \varphi)(u) = T_n\pi_{\mathfrak{F}_\omega}(u) = T_{n'}\pi_{\mathfrak{F}_\omega}(u')$, that is, $u' - u'' \in \ker T_{n'}\pi_{\mathfrak{F}_\omega} = D_{\omega,n'}$ and similarly, $v' - v'' \in D_{\omega,n'}$. Therefore,

$$\begin{aligned}\omega(n')(u', v') &= \omega(n')((u' - u'') + u'', (v' - v'') + v'') = \omega(n')(u'', v'')\\ &= \omega(\varphi(n))(T_n\varphi(u), T_n\varphi(v)) = (\varphi^*\omega)(n)(u, v) = \omega(n)(u, v),\end{aligned}$$

which shows that $\Omega_{\mathfrak{F}_\omega}(\pi_{\mathfrak{F}_\omega}(n))$ given in (6.1.7) is a well-defined two-form on the vector space $T_{\pi_{\mathfrak{F}_\omega}(n)}(N/\mathfrak{F}_\omega)$ and that $\pi_{\mathfrak{F}_\omega}^*\Omega_{\mathfrak{F}_\omega} = \omega$.

Since $\pi_{\mathfrak{F}_\omega}$ is a surjective submersion, it follows that $\Omega_{\mathfrak{F}_\omega}$ is smooth and that (6.1.7) uniquely determines it. Moreover, $\pi_{\mathfrak{F}_\omega}^*\mathbf{d}\Omega_{\mathfrak{F}_\omega} = \mathbf{d}\pi_{\mathfrak{F}_\omega}^*\Omega_{\mathfrak{F}_\omega} = \mathbf{d}\omega = 0$ which implies, again because $\pi_{\mathfrak{F}_\omega}$ is a surjective submersion, that $\mathbf{d}\Omega_{\mathfrak{F}_\omega} = 0$.

Finally, note that $\Omega_{\mathfrak{F}_\omega}$ is nondegenerate. Indeed if for any $n \in N$ and any $v \in T_nN$ we have

$$0 = \Omega_{\mathfrak{F}_\omega}(\pi_{\mathfrak{F}_\omega}(n))\left(T_n\pi_{\mathfrak{F}_\omega}(u), T_n\pi_{\mathfrak{F}_\omega}(v)\right) = \omega(n)(u, v),$$

then $u \in D_{\omega,n}$, that is, $T_n\pi_{\mathfrak{F}_\omega}(u) = 0$. This proves the following theorem.

6.1.6 Theorem (Presymplectic foliation reduction) *Let (N, ω) be a presymplectic manifold whose characteristic distribution is differentiable and whose associated generalized foliation $\mathfrak{F}_\omega$ is regular. Then there is a unique symplectic form $\Omega_{\mathfrak{F}_\omega} \in \Omega^2(N/\mathfrak{F}_\omega)$ on the quotient manifold $N/\mathfrak{F}_\omega$ characterized by the relation $\pi_{\mathfrak{F}_\omega}^*\Omega_{\mathfrak{F}_\omega} = \omega$. The symbol $\pi_{\mathfrak{F}_\omega} : N \to N/\mathfrak{F}_\omega$ denotes the projection onto the leaf space of $\mathfrak{F}$.*

Note that if the presymplectic form has constant rank, then the distribution D_ω defines a vector subbundle of TN and is, in particular, differentiable.

The idea presented in the theorem above has been taken as definition of reduction by some authors such as SNIATYCKI AND TULCZYJEW (1972). For them, symplectic reduction means the construction of new symplectic manifolds out of a given one (M, ω) using the combination of two operations: restriction to an injectively immersed manifold N of M on which ω induces a two-form whose associated characteristic distribution D_ω, defined at each $n \in N$ by $D_{\omega,n} := T_nN \cap (T_nN)^{i^*\omega}$ is differentiable. If the associated foliation is regular, then the quotient manifold is symplectic by the previous theorem and we have obtained the following statement.

6.1.7 Corollary (Symplectic foliation reduction) *Let (M, ω) be a symplectic manifold and $i : N \hookrightarrow M$ an injectively immersed manifold. Assume that the characteristic distribution of $i^*\omega$ on N is differentiable and that the associated foliation $\mathfrak{F}_{i^*\omega}$ is regular. Then there is a unique symplectic form $\Omega_{\mathfrak{F}_{i^*\omega}} \in \Omega^2(N/\mathfrak{F}_{i^*\omega})$ on the quotient manifold $\pi_{\mathfrak{F}_{i^*\omega}} : N \to N/\mathfrak{F}_{i^*\omega}$ characterized by the relation $\pi^*_{\mathfrak{F}_{i^*\omega}} \Omega_{\mathfrak{F}_{i^*\omega}} = i^*\omega$. If $TN \cap (TN)^{i^*\omega}$ is a subbundle of TN, then the characteristic distribution is differentiable and $TN/[TN \cap (TN)^{i^*\omega}] \to N$ is a symplectic vector bundle, that is, each fiber has an induced symplectic form varying smoothly over N.*

An even more general notion of symplectic reduction has been formulated by BENENTI AND TULCZYJEW (1982) and BENENTI (1983) for whom symplectic reduction of a symplectic manifold (M, ω) is a surjective submersion $\pi : N \to P$ of a submanifold (which may be just immersed) N of M onto another symplectic manifold (P, Ω) which satisfies $\pi^*\Omega = i^*\omega$, where $i : N \hookrightarrow M$ is the inclusion. A review of the results around these notions can be found in LIBERMANN AND MARLE (1987).

6.1.8 Point reduction as foliation reduction. One can obtain the Symplectic Point Reduction Theorem 6.1.1**(i)** from the Symplectic Foliation Reduction Theorem 6.1.7 under some (severe) connectedness hypotheses. To see this, let (M, ω) be a connected symplectic manifold acted upon canonically and freely by a Lie group G and suppose that the action admits a (not necessarily equivariant) momentum map $\mathbf{J} : M \to \mathfrak{g}^*$. Take in Corollary 6.1.7 $N = \mathbf{J}^{-1}(\mu)$ for $\mu \in \mathfrak{g}^*$ and compute the characteristic distribution of $i_\mu^*\omega$ on $\mathbf{J}^{-1}(\mu)$. For any $z \in \mathbf{J}^{-1}(\mu)$, the identities (6.1.3), Proposition 4.5.14**(ii)**, and the Reduction Lemma 4.5.34 yield

$$\begin{aligned} T_z\mathbf{J}^{-1}(\mu) \cap (T_z\mathbf{J}^{-1}(\mu))^\omega &= \ker T_z\mathbf{J} \cap (\ker T_z\mathbf{J})^\omega = (\mathfrak{g} \cdot z)^\omega \cap [(\mathfrak{g} \cdot z)^\omega]^\omega \\ &= (\mathfrak{g} \cdot z)^\omega \cap \mathfrak{g} \cdot z = \mathfrak{g}_\mu \cdot z = T_z(G_\mu \cdot z), \end{aligned}$$

which says that if *the group G_μ is connected*, then the characteristic foliation has its leaves equal to the orbits of of G_μ in $\mathbf{J}^{-1}(\mu)$. Thus the leaf space of the foliation induced by the characteristic distribution on $\mathbf{J}^{-1}(\mu)$ coincides with the quotient manifold $\mathbf{J}^{-1}(\mu)/G_\mu = M_\mu$. Therefore, the Symplectic Point Reduction Theorem 6.1.1**(i)** follows directly from the Symplectic Foliation Reduction Theorem 6.1.7.

6.1.9 Other symplectic reduction results. In Theorem 6.1.1 we carried out symplectic reduction using standard momentum maps. There exist similar results for some of the other momentum maps introduced in the preceding chapters. For instance, Chapter 9 will be dedicated to carry out reduction using as main tool the optimal momentum map.

There exists a symplectic reduction result in the context of Lie group valued momentum maps ALEKSEEV et al. (1998) that shows that under certain regularity hypotheses the analog of the symplectic reduced space in the quasi-Hamiltonian context is a quasi-Hamiltonian space on which equivariant quasi-Hamiltonian dynamics can be naturally reduced.

A similar situation is encountered when dealing with the canonical actions of a symplectic groupoid on a symplectic manifold. The reader interested in this topic is encouraged to check with MIKAMI AND WEINSTEIN (1988).

6.1.10 Reconstruction of dynamics. Let us return to the general setup of point reduction. Denote, as usual, by $\Phi : G \times M \rightarrow M$ a free proper canonical action of the Lie group G on the connected symplectic manifold (M, ω) and assume that this action has an associated momentum map $\mathbf{J} : M \rightarrow \mathfrak{g}^*$, with non-equivariance one-cocycle $\sigma : G \rightarrow \mathfrak{g}^*$. Let $\mu \in \mathfrak{g}^*$ be a value of $\mathbf{J}$ and denote by G_μ the isotropy of μ under the affine action of G on $\mathfrak{g}^*$. Theorem 6.1.1**(ii), (iii)** shows how the dynamics generated by a G-invariant Hamiltonian $h \in C^\infty(M)^G$ behaves under the reduction process. Namely, if F_t is the flow of the Hamiltonian vector field X_h, then it leaves the connected components of $\mathbf{J}^{-1}(\mu)$ invariant and it commutes with the G-action, so it induces a flow F_t^μ on the point reduced space M_μ defined by $\pi_\mu \circ F_t \circ i_\mu = F_t^\mu \circ \pi_\mu$, where $\pi_\mu : \mathbf{J}^{-1}(\mu) \rightarrow M_\mu$ is the canonical projection and $i_\mu : \mathbf{J}^{-1}(\mu) \hookrightarrow M$ is the inclusion. In addition, the vector field generated by the flow F_t^μ on (M_μ, ω_μ) is Hamiltonian relative to the reduced Hamiltonian $h_\mu \in C^\infty(M_\mu)$ defined by $h_\mu \circ \pi_\mu = h \circ i_\mu$. Thus the vector fields X_h and X_{h_μ} are π_μ-related.

We pose now the converse question. Assume that an integral curve $c_\mu(t)$ of the reduced Hamiltonian system X_{h_μ} on (M_μ, ω_μ) is known. Let $m_0 \in \mathbf{J}^{-1}(\mu)$ be given. Can one determine from this data the integral curve of the Hamiltonian system X_h with initial condition m_0? In other words, can one ***reconstruct*** the solution of the given system knowing the corresponding reduced solution? We shall show below that the answer to this question is affirmative in the sense that there is a method that will determine, in principle, the solution of the original system.

The general method of reconstruction is the following (ABRAHAM AND MARSDEN (1978) §4.3, MARSDEN et al. (1990), MARSDEN AND RATIU (2003)). Pick a smooth curve $d(t)$ in $\mathbf{J}^{-1}(\mu)$ such that $d(0) = m_0$ and $\pi_\mu(d(t)) = c_\mu(t)$. We shall give later concrete choices for such curves in terms of connections. Then, if $c(t)$ denotes the integral curve of X_h with $c(0) = m_0$, we can write $c(t) = g(t) \cdot d(t)$ for some smooth curve $g(t)$ in G_μ. We shall determine now $g(t)$ and therefore $c(t)$. We have

$$
\begin{aligned}
X_h(c(t)) = \dot{c}(t) &= T_{d(t)}\Phi_{g(t)}\dot{d}(t) + T_{d(t)}\Phi_{g(t)}\left(T_{g(t)}L_{g(t)^{-1}}\dot{g}(t)\right)_M(d(t)) \\
&= T_{d(t)}\Phi_{g(t)}\left(\dot{d}(t) + (T_{g(t)}L_{g(t)^{-1}}\dot{g}(t))_M(d(t))\right)
\end{aligned}
$$

which implies

$$
\begin{aligned}
\dot{d}(t) + (T_{g(t)}L_{g(t)^{-1}}\dot{g}(t))_M(d(t)) &= T_{g(t)\cdot d(t)}\Phi_{g(t)^{-1}}X_h(c(t)) \\
&= T_{g(t)\cdot d(t)}\Phi_{g(t)^{-1}}X_h(g(t)\cdot d(t)) \\
&= \left(\Phi_{g(t)}^* X_h\right)(d(t)) = X_h(d(t))
\end{aligned}
$$

since, by hypothesis, $h = \Phi_g^* h$ and thus $X_h = X_{\Phi_g^* h} = \Phi_g^* X_h$ for any $g \in G$. This equation is solved in two steps as follows:

- **Step 1:** *Find a smooth curve* $\xi(t)$ *in* $\mathfrak{g}_\mu$ *such that*

$$\xi(t)_M(d(t)) = X_h(d(t)) - \dot{d}(t); \tag{6.1.8}$$

- **Step 2:** *with* $\xi(t) \in \mathfrak{g}_\mu$ *determined above, solve the nonautonomous differential equation on* G_μ

$$\dot{g}(t) = T_e L_{g(t)}\xi(t), \qquad \textit{with} \qquad g(0) = e. \tag{6.1.9}$$

The first step is typically of algebraic nature. For example, if G is a matrix Lie group (6.1.8) is just a matrix equation. We shall show below how one can solve this algebraic equation in terms of a connection on the principal G_μ-bundle $\pi_\mu : \mathbf{J}^{-1}(\mu) \to M_\mu$. The second step is the main difficulty in finding a complete answer to the reconstruction problem; equation (6.1.9) cannot be solved explicitly, in general. However, if G is Abelian, this equation can be solved *by quadratures* in the following way. Since the connected component of the p-dimensional Lie group G is a cylinder $\mathbb{R}^k \times \mathbb{T}^{p-k}$, the exponential map $\exp(\xi_1, \ldots, \xi_p) = (\xi_1, \ldots, \xi_k, \xi_{k+1} (\mathrm{mod}\, 2\pi), \ldots, \xi_p (\mathrm{mod}\, 2\pi))$ is onto, so we can write $g(t) = \exp \eta(t)$ for some smooth curve $\eta(t) \in \mathfrak{g}$ satisfying $\eta(0) = 0$. Equation (6.1.9) gives then $\xi(t) = T_{g(t)} L_{g(t)^{-1}} \dot{g}(t) = \dot{\eta}(t)$ since $\dot{g}(t) = T_{\eta(t)} \exp \dot{\eta}(t) = T_e L_{\exp \eta(t)} \dot{\eta}(t)$ by (2.1.2) and the fact that $\mathrm{ad}_{\eta(t)} = 0$ since $\mathfrak{g}$ is Abelian. Therefore, in this case, the solution of (6.1.9) is given by

$$g(t) = \exp\left(\int_0^t \xi(s) ds\right).$$

Let us now show how one can solve (6.1.8) in a natural geometric way using a connection on the left principal G_μ-bundle $\pi_\mu : \mathbf{J}^{-1}(\mu) \to M_\mu$. Recall that such a connection is given by a $\mathfrak{g}_\mu$-valued one-form $A \in \Omega^1(\mathbf{J}^{-1}(\mu); \mathfrak{g}_\mu)$ satisfying for all $m \in \mathbf{J}^{-1}(\mu)$ the relations

$$A(m)\left(\xi_{\mathbf{J}^{-1}(\mu)}(m)\right) = \xi, \quad \text{for all} \quad \xi \in \mathfrak{g}_\mu$$

and

$$A(g \cdot m)(T_m \Phi_g(v_m)) = \mathrm{Ad}_g(A(m)(v_m)), \quad \text{for all} \quad g \in G_\mu, v_m \in T_m\mathbf{J}^{-1}(\mu).$$

Let G_μ act on M by restriction so that $\xi_M = \xi_{\mathbf{J}^{-1}(\mu)}$ for any $\xi \in \mathfrak{g}_\mu$. Choose in Step 1 of the reconstruction method the curve $d(t)$ to be the horizontal lift of $c_\mu(t)$ through m_0, that is, $d(t)$ is uniquely characterized by the conditions $A(d(t))(\dot{d}(t)) = 0$, $\pi_\mu(d(t)) = c_\mu(t)$, for all t, and $d(0) = m_0$. Then the solution of (6.1.8) is given by

$$\xi(t) = A(d(t))\big(X_h(d(t))\big).$$

6.1.11 Remark In §7.7 we shall consider equations that also give the dynamics of the original system in terms of the reduced dynamics. Those equations, called *reconstruction equations*, are only valid in a G-invariant neighborhood of a given point, whereas the equations obtained above are valid globally. While giving more information in a G-invariant neighborhood of the initial condition, the reconstruction equations of §7.7 are crucial in the proof of various persistence and bifurcation theorems for relative equilibria. For similar problems for relative periodic orbits, the reconstruction method given above is more appropriate.

Also, when studying phases in mechanical problems, the reconstruction method just discussed is of paramount importance. We shall not address this subject here and refer to MARSDEN et al. (1990), MARSDEN, RATIU, AND SCHEURLE (2000), BLAOM (2000), and references therein for further information on this subject. In §6.6.27 we shall recall without proof several formulas for reconstruction phases that appear naturally in the study of mechanical systems on cotangent bundles.

6.1.12 Holonomy in Quantum Mechanics. The same method used in the reconstruction of dynamics allows one to easily prove a formula of AHARONOV AND ANANDAN (1987) as noted by MARSDEN et al. (1990).

Any complex Hilbert space $(\mathcal{H}, \langle \cdot, \cdot \rangle)$ is an infinite-dimensional strong symplectic manifold relative to the quantum mechanical symplectic form

$$\omega(\psi_1, \psi_2) := -2\hbar \operatorname{Im}\langle \psi_1, \psi_2 \rangle,$$

where $\psi_1, \psi_2 \in \mathcal{H}$ and $\hbar \in \mathbb{R}$ is a constant (usually taken to be Planck's constant). The circle S^1 acts freely and canonically on the symplectic manifold $(\mathcal{H}, \omega)$ by usual multiplication. For $\xi \in \mathbb{R}$, the Lie algebra of S^1, the infinitesimal generator is given by $\xi_{\mathcal{H}}(\psi) = i\xi\psi$. An associated equivariant momentum map $\mathbf{J} : \mathcal{H} \to \mathbb{R}$ has the expression $\mathbf{J}(\psi)\xi = \frac{1}{2}\omega(i\xi\psi, \psi) = -\hbar \operatorname{Im}\langle i\xi\psi, \psi\rangle = -\xi\hbar\|\psi\|^2$, that is, $\mathbf{J}(\psi) = -\hbar\|\psi\|^2$. The value $-\hbar$ is regular for $\mathbf{J}$, the level set $\mathbf{J}^{-1}(-\hbar)$ is the unit sphere in $\mathcal{H}$ and thus the reduced space $\mathcal{H}_{-\hbar} = \mathbb{P}\mathcal{H}$, the projectivized Hilbert space, has the reduced symplectic form

$$\omega_{-\hbar}([\psi])(T_\psi \pi_{-\hbar}(\varphi_1), T_\psi \pi_{-\hbar}(\varphi_2)) = -2\hbar \operatorname{Im}\langle \varphi_1, \varphi_2 \rangle$$

for any $[\psi] \in \mathcal{H}_{-\hbar}$ and any $\varphi_1, \varphi_2 \in \mathcal{H}$ orthogonal to ψ.

There is a natural connection on the principal S^1-bundle $\mathbf{J}^{-1}(-\hbar) \to \mathbb{P}\mathcal{H}$ given by $A(\psi)(\varphi) := \hbar \operatorname{Re}\langle -i\psi, \varphi \rangle = -\hbar \operatorname{Im}\langle \psi, \varphi \rangle$. The curvature of this connection is just the usual differential of A since the structure group is Abelian, that is, $\Omega(\psi)(\varphi_1, \varphi_2) = 2\hbar \operatorname{Im}\langle \varphi_1, \varphi_2 \rangle = -\omega(\varphi_1, \varphi_2)$.

If $[\psi] : [0, 1] \to \mathbb{P}\mathcal{H}$ is a smooth loop, for example a closed curve of a Hamiltonian vector field on $\mathbb{P}\mathcal{H}$, then (5.1.2) immediately gives its holonomy. Since $-\iint \Omega = \iint \omega$ we conclude that *the holonomy of a loop in projective complex Hilbert space is the exponential of the symplectic area of any two-dimensional submanifold in the projective space bounding the loop.* This is the result of AHARONOV AND ANANDAN (1987).

6.2 Coadjoint orbits as point reduced spaces

This section carries out the point reduction program for the cotangent bundle of a Lie group G. It will be shown that the point reduced spaces of T^*G are symplectically diffeomorphic to coadjoint orbits in $\mathfrak{g}^*$. The left and right trivializations of T^*G induce symplectic structures on $G \times \mathfrak{g}^*$ that will be determined explicitly (thereby proving (4.1.13) in §4.1.20) as well as the corresponding Poisson brackets. These formulas will be useful in the reconstruction of the reduced dynamics that turns out to be much simpler than in the general case due to the rich additional structure present in this setup. The formula for the symplectic structure on $G \times \mathfrak{g}^*$ will also be crucial in the proof of the Symplectic Slice Theorem in §7.2.

6.2.1 Let us apply Theorem 6.1.1 to $M = T^*G$, where G is a Lie group with Lie algebra $\mathfrak{g}$, the G-action being the cotangent lift of *left* translation, and the associated momentum map $\mathbf{J}_L : \alpha_g \in T^*G \mapsto T_e^*R_g(\alpha_g) \in \mathfrak{g}^*$ which is right invariant. All hypotheses in Theorem 6.1.1 are satisfied and for each $\mu \in \mathfrak{g}^*$ we can form the symplectic point reduced space $((T^*G)_\mu, \omega_\mu)$. Recall also that the momentum map for the lift of *right* translations is left invariant and is given by $\mathbf{J}_R : \alpha_g \in T^*G \mapsto T_e^*L_g(\alpha_g) \in \mathfrak{g}^*$. Finally, recall from Theorem 4.5.31 that the orbit symplectic form on $\mathcal{O}_\mu = \{\mathrm{Ad}_g\,\mu \mid g \in G\}$ has the expression

$$\omega^{\pm}_{\mathcal{O}_\mu}(\nu)(\mathrm{ad}^*_\xi(\nu),\ \mathrm{ad}^*_\eta(\nu)) = \pm\langle \nu,\ [\xi,\ \eta]\rangle,$$

for arbitrary $\nu \in \mathcal{O}_\mu$ and $\xi, \eta \in \mathfrak{g}$. The following result is standard (see, e.g., Abraham and Marsden (1978); Guillemin and Sternberg (1984c); Libermann and Marle (1987), Marsden and Ratiu (2003), McDuff and Salamon (1998)).

6.2.2 Theorem (Coadjoint orbits as point reduced spaces) *The momentum map $\mathbf{J}_R$: $T^*G \to \mathfrak{g}^*$ induces for each $\mu \in \mathfrak{g}^*$ a symplectic diffeomorphism $\overline{\mathbf{J}}_R : ((T^*G)_\mu, \omega_\mu) \to (\mathcal{O}_\mu, \omega^-_{\mathcal{O}_\mu})$ given by $\overline{\mathbf{J}}_R([T_g^*R_{g^{-1}}\mu]) = \mathrm{Ad}_g^*\,\mu$.*

Proof. Since $\mathbf{J}_L^{-1}(\mu) = \{T_g^*R_{g^{-1}}\mu \mid g \in G\}$ is the graph of the right invariant one-form on G whose value at e is μ, it follows that $\mathbf{J}_R(T_g^*R_{g^{-1}}\mu) = T_e^*L_g(T_g^*R_{g^{-1}}\mu) = \mathrm{Ad}_g^*\,\mu$, which implies that $\overline{\mathbf{J}}_R([T_g^*R_{g^{-1}}\mu]) = \mathrm{Ad}_g^*\,\mu$. Also, since $g \in G \mapsto T_g^*R_{g^{-1}}\mu \in T^*G$ is a diffeomorphism onto its image (the graph of the above mentioned right invariant one-form on G) and the manifold structure of the coadjoint orbit $\mathcal{O}_\mu$ is by definition obtained by declaring the bijection $G_\mu g \in G/G_\mu \mapsto \mathrm{Ad}_g^*\,\mu \in \mathcal{O}_\mu$ to be a diffeomorphism, it follows that $\overline{\mathbf{J}}_R$ is a diffeomorphism.

By the characterization (6.1.1) of ω_μ it suffices to show that $\mathbf{J}_R^*\omega^-_{\mathcal{O}_\mu} = -\mathbf{d}\Theta$, where $\Theta \in \Omega^1(T^*G)$ is the canonical one-form of the cotangent bundle (see Example 4.1.3). Since $\mathbf{J}_L^{-1}(\mu) = \{T_g^*R_{g^{-1}}\mu \mid g \in G\}$ is the graph of the right invariant one-form on G whose value at e is μ, any tangent vector at $T_g^*R_{g^{-1}}\mu$ to the submanifold $\mathbf{J}_L^{-1}(\mu)$ is tangent to a curve of the form $\alpha_\xi(t) := T^*_{g\exp(t\xi)}R_{\exp(-t\xi)g^{-1}}\mu = T^*_{g\exp(t\xi)}L_{g^{-1}}(T^*_{\exp(t\xi)}R_{\exp(-t\xi)}\nu) = \overline{L}_g\overline{R}_{\exp(t\xi)}\nu$ for some $\xi \in \mathfrak{g}$, where $\nu := \mathrm{Ad}_g^*\,\mu$ and where $\overline{L}_g := T^*L_{g^{-1}}$, $\overline{R}_g := T^*R_{g^{-1}} : T^*G \to T^*G$ are the cotangent lifts of the left and right translations respectively. Therefore, $\alpha'_\xi(0) = T_\nu\overline{L}_g(\xi_{T^*G}(\nu))$, where ξ_{T^*G} denotes the infinitesimal generator defined by $\xi \in \mathfrak{g}$ relative to the *right* action $\overline{R} : G \times T^*G \to T^*G$. Since $T_g^*R_{g^{-1}}\mu = \overline{R}_g\mu = \overline{L}_g\nu$ and the action $\overline{L}$ preserves Θ

we have, for any $\xi, \eta \in \mathfrak{g}$,

$$
\begin{aligned}
-\mathbf{d}\Theta(\overline{L}_g\nu)\left(\alpha'_\xi(0), \alpha'_\eta(0)\right) &= -\mathbf{d}\Theta(\overline{L}_g\nu)\left(T_\nu\overline{L}_g(\xi_{T^*G}(\nu)), T_\nu\overline{L}_g(\eta_{T^*G}(\nu))\right) \\
&= -\mathbf{d}\Theta(\nu)\left(\xi_{T^*G}(\nu), \eta_{T^*G}(\nu)\right) \\
&= -\xi_{T^*G}[\Theta(\eta_{T^*G})](\nu) + \eta_{T^*G}[\Theta(\xi_{T^*G})](\nu) + \Theta([\xi_{T^*G}, \eta_{T^*G}])(\nu).
\end{aligned}
$$

However, denoting by $\pi : T^*G \to G$ the cotangent bundle projection, by the definition of the canonical one-form Θ on T^*G we have for any $\alpha_g \in T^*_gG$

$$
\Theta(\eta_{T^*G})(\alpha_g) = \left\langle \alpha_g, T_{\alpha_g}\pi(\eta_{T^*G})\right\rangle = \left\langle \alpha_g, \eta_G(g))\right\rangle = \mathbf{J}_R^\eta(\alpha_g)
$$

since η_{T^*G} and η_G are π-related. Thus, using the identity $[\xi_{T^*G}, \eta_{T^*G}] = [\xi, \eta]_{T^*G}$ valid for *right* actions, we can continue the computation above and write

$$
\begin{aligned}
&- \xi_{T^*G}[\Theta(\eta_{T^*G})](\nu) + \eta_{T^*G}[\Theta(\xi_{T^*G})](\nu) + \Theta([\xi, \eta]_{T^*G})(\nu) \\
&\quad = -X_{\mathbf{J}_R^\xi}[\mathbf{J}_R^\eta](\nu) + X_{\mathbf{J}_R^\eta}[\mathbf{J}_R^\xi](\nu) + \mathbf{J}_R^{[\xi,\eta]}(\nu) \\
&\quad = -\{\mathbf{J}_R^\eta, \mathbf{J}_R^\xi\}(\nu) + \{\mathbf{J}_R^\xi, \mathbf{J}_R^\eta\}(\nu) - \{\mathbf{J}_R^\xi, \mathbf{J}_R^\eta\}(\nu) = \{\mathbf{J}_R^\xi, \mathbf{J}_R^\eta\}(\nu)
\end{aligned}
$$

since for *right* actions infinitesimal equivariance is equivalent to $\mathbf{J}_R^{[\xi,\eta]} = -\{\mathbf{J}_R^\xi, \mathbf{J}_R^\eta\}$. Thus, using this relation again, we finally obtain

$$
\begin{aligned}
-\mathbf{d}\Theta(\overline{L}_g\nu)\left(\alpha'_\xi(0), \alpha'_\eta(0)\right) &= \{\mathbf{J}_R^\xi, \mathbf{J}_R^\eta\}(\nu) = -\mathbf{J}_R^{[\xi,\eta]}(\nu) = -\langle \nu, [\xi, \eta]\rangle \\
&= \omega^-_{\mathcal{O}_\mu}(\nu)\left(\mathrm{ad}^*_\xi(\nu), \mathrm{ad}^*_\eta(\nu)\right) \\
&= (\mathbf{J}_R^*\omega^-_{\mathcal{O}_\mu})(\overline{L}_g\nu)\left(\alpha'_\xi(0), \alpha'_\eta(0)\right),
\end{aligned}
$$

since $\mathbf{J}_R(\overline{L}_g\nu) = \nu$ and $T_{\overline{L}_g\nu}\mathbf{J}_R(\alpha'_\xi(0)) = \mathrm{ad}^*_\xi(\nu)$. This last equality is proved in the following way. First notice that

$$
\begin{aligned}
\mathbf{J}_R(\alpha_\xi(t)) &= T_e^*L_{g\exp(t\xi)}\alpha_\xi(t) = \overline{L}_{(g\exp(t\xi))^{-1}}\alpha_\xi(t) \\
&= \overline{L}_{\exp(-t\xi)}\overline{L}_{g^{-1}}\overline{L}_g\overline{R}_{\exp(t\xi)}\nu = \overline{L}_{\exp(-t\xi)}\overline{R}_{\exp(t\xi)}\nu = \mathrm{Ad}^*_{\exp(t\xi)}\,\nu.
\end{aligned}
$$

Second, using this identity we get

$$
T_{\overline{L}_g\nu}\mathbf{J}_R(\alpha'_\xi(0)) = \left.\frac{d}{dt}\right|_{t=0}\mathbf{J}_R(\alpha_\xi(t)) = \left.\frac{d}{dt}\right|_{t=0}\mathrm{Ad}^*_{\exp(t\xi)}\,\nu = \mathrm{ad}^*_\xi(\nu)
$$

which proves the claim. ■

6.2.3 Left and right trivializations of T^*G. Left and right translation induce two trivializations of T^*G in the following manner. Define $\lambda, \rho : T^*G \to G \times \mathfrak{g}^*$ by

$$
\lambda(\alpha_g) := \left(g, T_e^*L_g(\alpha_g)\right) \qquad \text{and} \qquad \rho(\alpha_g) := \left(g, T_e^*R_g(\alpha_g)\right), \tag{6.2.1}
$$

where $\alpha_g \in T_g^*G$. These smooth maps are diffeomorphisms since their inverses are given respectively by

$$
\lambda^{-1}(g, \mu) := T_g^*L_{g^{-1}}\mu \qquad \text{and} \qquad \rho^{-1}(g, \mu) := T_g^*R_{g^{-1}}\mu, \tag{6.2.2}
$$

where $g \in G$ and $\mu \in \mathfrak{g}^*$. Following standard terminology from continuum mechanics, objects on T^*G will be said to be in the ***material*** or ***Lagrangian representation***, their push-forward by λ in ***body*** or ***convective representation***, and their push-forward by ρ in ***space*** or ***Eulerian representation***. In what follows we shall explicitly compute the canonical symplectic form, the Poisson bracket, and the Hamiltonian vector field in the body and space representations. We begin with the computation (Abraham and Marsden (1978), §4.4, Cushman and Bates (1997), Marsden and Ratiu (2003)) of the symplectic form on $G \times \mathfrak{g}^*$ in body and space representations.

6.2.4 Theorem (Trivialized canonical forms) *Denote by* $\theta_B := \lambda_*\theta$, $\theta_S := \rho_*\theta$, $\omega_B := \lambda_*\omega$, *and* $\omega_S := \rho_*\omega$ *the body and space representations of the canonical one-form* θ *and the symplectic form* ω *on* T^*G. *Then*

$$\theta_B(g,\mu)(v_g,\beta) = \langle \mu, T_gL_{g^{-1}}v_g\rangle \tag{6.2.3}$$

$$\theta_S(g,\mu)(v_g,\beta) = \langle \mu, T_gR_{g^{-1}}v_g\rangle \tag{6.2.4}$$

$$\begin{aligned}&\omega_B(g,\mu)\big((u_g,\alpha),(v_g,\beta)\big) = \\ &\qquad \langle \beta, T_gL_{g^{-1}}u_g\rangle - \langle \alpha, T_gL_{g^{-1}}v_g\rangle + \langle \mu, [T_gL_{g^{-1}}u_g, T_gL_{g^{-1}}v_g]\rangle\end{aligned} \tag{6.2.5}$$

$$\begin{aligned}&\omega_S(g,\mu)\big((u_g,\alpha),(v_g,\beta)\big) = \\ &\qquad \langle \beta, T_gR_{g^{-1}}u_g\rangle - \langle \alpha, T_gR_{g^{-1}}v_g\rangle - \langle \mu, [T_gR_{g^{-1}}u_g, T_gR_{g^{-1}}v_g]\rangle,\end{aligned} \tag{6.2.6}$$

where $g \in G$, $\mu, \alpha, \beta \in \mathfrak{g}^*$, $u_g, v_g \in T_gG$.

Proof. Denote by $\pi : T^*G \to G$ the cotangent bundle projection. Then the definition of the canonical one-form $\theta(\alpha_g)(w_{\alpha_g}) = \alpha_g(T_{\alpha_g}\pi(w_{\alpha_g}))$, for $\alpha_g \in T^*_gG$, $w_{\alpha_g} \in T_{\alpha_g}(T^*G)$, yields

$$\begin{aligned}\theta_B(g,\mu)(v_g,\beta) &= (\lambda_*\theta)(g,\mu)(v_g,\beta) = \theta(\lambda^{-1}(g,\mu))\big(T_{(g,\mu)}\lambda^{-1}(v_g,\beta)\big)\\ &= \lambda^{-1}(g,\mu)\big((T_{\lambda^{-1}(g,\mu)}\pi \circ T_{(g,\mu)}\lambda^{-1})(v_g,\beta)\big)\\ &= (T_g^*L_{g^{-1}}\mu)\big((T_{(g,\mu)}(\pi \circ \lambda^{-1})(v_g,\beta)\big)\\ &= \langle \mu, T_gL_{g^{-1}}v_g\rangle\end{aligned}$$

since $(\pi \circ \lambda^{-1})(g,\mu) = g$. This proves (6.2.3). Formula (6.2.4) has an identical proof.

Pushing the formula $\omega = -\mathbf{d}\theta$ forward by the diffeomorphism λ yields $\omega_B = -\mathbf{d}\theta_B$. Fix in what follows $g \in G$, $\mu, \alpha, \beta \in \mathfrak{g}^*$, and $u_g, v_g \in T_gG$. Define $X = (X^1, X^2)$, $Y = (Y^1, Y^2) \in \mathfrak{X}(G) \times \mathfrak{X}(\mathfrak{g}^*) \subset \mathfrak{X}(G \times \mathfrak{g}^*)$ by $X^2(\nu) := \alpha$, $Y^2(\nu) := \beta$, $X^1(h) := T_eL_h\xi$, $Y^1(h) := T_eL_h\eta$, where $\xi := T_gL_{g^{-1}}u_g$, $\eta := T_gL_{g^{-1}}v_g$, for all $h \in G$ and $\nu \in \mathfrak{g}^*$. The flow of X is hence given by $(t,h,\nu) \mapsto (\varphi_t^1(h), \varphi_t^2(\nu)) := (h\exp t\xi, \nu + t\alpha)$. We shall compute $\omega_B(g,\mu)\big((u_g,\alpha),(v_g,\beta)\big)$ by calculating the formula

$$\omega_B(X,Y) = -\mathbf{d}\theta_B(X,Y) = -X[\theta_B(Y)] + Y[\theta_B(X)] + \theta_B([X,Y])$$

at the point (g,μ) and taking into account that by the definition of X and Y, we have $X(g,\mu) = (u_g,\alpha)$, $Y(g,\mu) = (v_g,\beta)$.

To compute the first term, use (6.2.3) and the definition of Y^1 to get

$$\begin{aligned} X[\theta_B(Y)](g,\mu) &= \big(\pounds_X(\theta_B(Y))\big)(g,\mu) \\ &= \frac{d}{dt}\bigg|_{t=0} \theta_B(\varphi_t^1(g),\varphi_t^2(\mu))\big(Y^1(\varphi_t^1(g),\varphi_t^2(\mu)), Y^2(\varphi_t^1(g),\varphi_t^2(\mu))\big) \\ &= \frac{d}{dt}\bigg|_{t=0} \Big\langle \varphi_t^2(\mu), T_{\varphi_t^1(g)}L_{\varphi_t^1(g)^{-1}}Y^1(\varphi_t^1(g),\varphi_t^2(\mu))\Big\rangle \\ &= \frac{d}{dt}\bigg|_{t=0} \Big\langle \varphi_t^2(\mu), \eta\Big\rangle = \langle \alpha, \eta\rangle = \big\langle \alpha, T_gL_{g^{-1}}v_g\big\rangle . \end{aligned}$$

Similarly,

$$Y[\theta_B(X)](g,\mu) = \big\langle \beta, T_gL_{g^{-1}}u_g\big\rangle .$$

This already determines the first two terms in (6.2.5).

To calculate the third, notice first that since $X^1, Y^1 \in \mathfrak{X}(G)$ are left invariant vector fields, we have $[X^1,Y^1](g) = T_eL_g([X^1,Y^1](e)) = [\xi,\eta]$, by the definition of the Lie bracket in $\mathfrak{g}$. Therefore, by (6.2.3) we get

$$\begin{aligned} \theta_B([X,Y])(g,\mu) &= \theta_B(g,\mu)\big([X^1,Y^1](g),[X^2,Y^2](\mu)\big) \\ &= \langle \mu, T_g, L_{g^{-1}}[X^1,Y^1](g)\rangle = \langle \mu, [\xi,\eta]\rangle \\ &= \langle \mu, [T_g, L_{g^{-1}}u_g, T_g, L_{g^{-1}}v_g]\rangle, \end{aligned}$$

which proves (6.2.5).

Formula (6.2.6) is proved in the same way with the sole exception that $X^1, Y^1 \in \mathfrak{X}(G)$ are taken to be right invariant. This is the cause of the minus sign in the third term: $[X^1,Y^1](g) = T_eR_g([X^1,Y^1](e)) = -[\xi,\eta]$, by the definition of the Lie bracket in $\mathfrak{g}$. ■

Next, we compute the body and space representations of the Hamiltonian vector field for a left, respectively right, invariant Hamiltonian.

6.2.5 Theorem (Trivialized Hamiltonian vector field) *Let F_t be the flow of the Hamiltonian vector field X_h, where $h : T^*G \to \mathbb{R}$ is a left invariant function. For $\mu \in \mathfrak{g}^*$ define $g_\mu(t) := (\pi \circ F_t)(\mu) \in G$, where $\pi : T^*G \to G$ is the cotangent bundle projection. Then the Hamiltonian vector field $\lambda_* X_h \in \mathfrak{X}(G \times \mathfrak{g}^*)$ in body representation is given by*

$$(\lambda_* X_h)(g,\mu) = \left(T_eL_g \frac{\delta h|_{\mathfrak{g}^*}}{\delta \mu}, \mu, \mathrm{ad}^*_{\frac{\delta h|_{\mathfrak{g}^*}}{\delta\mu}} \mu \right) \in T_gG \times T_\mu\mathfrak{g}^*. \tag{6.2.7}$$

This vector field on $G \times \mathfrak{g}^*$ is Hamiltonian relative to the symplectic form ω_B given by (6.2.5) and the Hamiltonian function $h_B := h \circ \lambda^{-1}$. The flow of the μ-dependent vector field on G given by the first factor in (6.2.7) is given by $(t,g) \mapsto gg_\mu(t)$. The flow of the vector field on $\mathfrak{g}^*$ given by the last factor in (6.2.7) is given by $(t,\mu) \mapsto T_e^*L_{g_\mu(t)}(F_t(\mu))$.

A similar statement holds for a right invariant Hamiltonian $h : T^*G \to \mathbb{R}$. The expression $\rho_* X_h \in \mathfrak{X}(G \times \mathfrak{g}^*)$ in space representation is

$$(\lambda_* X_h)(g, \mu) = \left(T_e R_g \frac{\delta h|_{\mathfrak{g}^*}}{\delta \mu}, \mu, -\operatorname{ad}^*_{\frac{\delta h|_{\mathfrak{g}^*}}{\delta \mu}} \mu \right) \in T_g G \times T_\mu \mathfrak{g}^*. \tag{6.2.8}$$

This vector field on $G \times \mathfrak{g}^*$ is Hamiltonian relative to the symplectic form ω_S given by (6.2.6) *and the Hamiltonian function $h_S := h \circ \rho^{-1}$. Let F_t be the flow of the right invariant Hamiltonian vector field X_h. Then the flow of the μ-dependent vector field on G given by the first factor in* (6.2.8) *is given by $(t, g) \mapsto g_\mu(t) g$, where, as before, $g_\mu(t) := (\pi \circ F_t)(\mu)$. The flow of the vector field on $\mathfrak{g}^*$ given by the last factor in* (6.2.8) *is given by $(t, \mu) \mapsto T_e^* R_{g_\mu(t)}(F_t(\mu))$.*

Proof. Note that the lift of left translation in body representation has the expression $g \cdot (k, \mu) = (gk, \mu)$, for $g, k \in G$ and $\mu \in \mathfrak{g}^*$. Therefore, the lift of this action to $T(G \times \mathfrak{g}^*)$ is given by

$$g \cdot (v_k, \mu, \nu) = (T_k L_g(v_k), \mu, \nu), \tag{6.2.9}$$

where $v_k \in T_k G$ and $(\mu, \nu) \in T_\mu \mathfrak{g}^* = \{\mu\} \times \mathfrak{g}^*$. Write

$$\lambda_* X_h(g, \mu) = (X(g, \mu), \mu, Y(g, \mu)) \in T_g G \times \{\mu\} \times \mathfrak{g}^*$$

and notice that left invariance of the Hamiltonian function h implies left invariance of X_h and therefore also of $\lambda_* X_h$, which means that $g \cdot \lambda_* X_h(g^{-1}k, \mu) = X_h(k, \mu)$ for any $g \in G$. In view of (6.2.9), this says that $(X(k, \mu), \mu, Y(k, \mu)) = g \cdot (X(g^{-1}k, \mu), \mu, Y(g^{-1}k, \mu)) = (T_{g^{-1}k} L_g X(g^{-1}k, \mu), \mu, Y(g^{-1}k, \mu))$ for any $g, k \in G$ and $\mu \in \mathfrak{g}^*$. In particular, this implies that $Y(g^{-1}k, \mu) = Y(k, \mu)$ for any $g, k \in G$; taking here $g = k$ we get $Y(k, \mu) = Y(e, \mu)$ for any $k \in G$, that is, Y is independent of the variable in G and we can write

$$\lambda_* X_h(g, \mu) = (X^\mu(g), \mu, Y(\mu)), \tag{6.2.10}$$

where $X^\mu \in \mathfrak{X}(G)$ is a left invariant vector field on G depending on the parameter μ and $Y : \mathfrak{g}^* \to \mathfrak{g}^*$.

Since F_t is the flow of X_h, $\lambda \circ F_t \circ \lambda^{-1}$ is the flow of the push-forward $\lambda_* X_h$. Left invariance of h and hence of X_h, implies that F_t commutes with the lift of left translation to T^*G, and thus we get from (6.2.2), the definition of $g_\mu(t) = (\pi \circ F_t)(\mu)$, and (6.2.1)

$$\begin{aligned}(\lambda \circ F_t \circ \lambda^{-1})(g, \mu) &= (\lambda \circ F_t)\left(T_g^* L_{g^{-1}} \mu\right) = \lambda\left(T_g^* L_{g^{-1}} F_t(\mu)\right) \\ &= \left(g g_\mu(t), \left(T_e^* L_{g g_\mu(t)} \circ T_g^* L_{g^{-1}}\right)(F_t(\mu))\right) \\ &= \left(g g_\mu(t), T_e^* L_{g_\mu(t)}(F_t(\mu))\right).\end{aligned}$$

This shows that the flow of Y is given by $(t, \mu) \mapsto T_e^* L_{g_\mu(t)}(F_t(\mu))$ and that the flow of the μ-dependent vector field $X^\mu \in \mathfrak{X}(G)$ is $(t, g) \mapsto g g_\mu(t)$. This proves the last part of the theorem for left invariant Hamiltonians.

To prove formula (6.2.7) we shall use the expression of the flows just found. Since $t \mapsto gg_\mu(t)$ is the integral curve of $X^\mu \in \mathfrak{X}(G)$ through $g \in G$, its derivative at $t = 0$ equals $X^\mu(g)$, so that taking into account that $g_\mu(t) = (\pi \circ F_t)(\mu)$, we get

$$\begin{aligned} X^\mu(g) &= \left.\frac{d}{dt}\right|_{t=0} gg_\mu(t) = T_e L_g \left.\frac{d}{dt}\right|_{t=0} g_\mu(t) \\ &= T_e L_g \left.\frac{d}{dt}\right|_{t=0} (\pi \circ F_t)(\mu) \\ &= T_e L_g \big(T_\mu \pi (X_h(\mu))\big). \end{aligned} \tag{6.2.11}$$

However, an easy direct verification in canonical local charts on T^*G shows that

$$T_\mu \pi (X_h(\mu)) = \frac{\delta h|_{\mathfrak{g}^*}}{\delta \mu} \in \mathfrak{g} \tag{6.2.12}$$

which, together with (6.2.10) and (6.2.11) gives the expression of the first component in (6.2.7).

To compute $Y : \mathfrak{g}^* \to \mathfrak{g}^*$ we use the conservation of the momentum map $\mathbf{J}_L : T^*G \to \mathfrak{g}^*$ of the lifted left action of G on T^*G given by $\mathbf{J}_L(\alpha_g) = T_e^* R_g(\alpha_g)$ for $\alpha_g \in T_g^*G$, the just proved fact that the flow of Y is given by $(t, \mu) \mapsto F_t(\mu) \circ T_e L_{g_\mu(t)}$, and the identity $(d/dt)|_{t=0} g_\mu(t) = \delta h|_{\mathfrak{g}^*}/\delta\mu$ shown in (6.2.11), to conclude that for any $\xi \in \mathfrak{g}$ and any $\mu \in \mathfrak{g}^*$ we have

$$\begin{aligned} 0 &= \left(\pounds_{X_h} \mathbf{J}_L^\xi\right)(\mu) = \left.\frac{d}{dt}\right|_{t=0} \left(\mathbf{J}_L^\xi \circ F_t\right)(\mu) = \left.\frac{d}{dt}\right|_{t=0} \langle \mathbf{J}_L(F_t(\mu)), \xi\rangle \\ &= \left.\frac{d}{dt}\right|_{t=0} \left\langle F_t(\mu),\ T_e R_{(\pi\circ F_t)(\mu)} \xi \right\rangle = \left.\frac{d}{dt}\right|_{t=0} \left\langle F_t(\mu) \circ T_e L_{g_\mu(t)},\ \mathrm{Ad}_{g_\mu(t)^{-1}} \xi \right\rangle \\ &= \langle Y(\mu), \xi\rangle + \left\langle \mu,\ \left.\frac{d}{dt}\right|_{t=0} \mathrm{Ad}_{g_\mu(t)^{-1}} \xi \right\rangle \\ &= \langle Y(\mu), \xi\rangle - \left\langle \mu,\ \left[\left.\frac{d}{dt}\right|_{t=0} g_\mu(t),\ \xi \right] \right\rangle \\ &= \langle Y(\mu), \xi\rangle - \left\langle \mu,\ \left[\frac{\delta h|_{\mathfrak{g}^*}}{\delta\mu},\ \xi \right] \right\rangle \\ &= \left\langle Y(\mu) - \mathrm{ad}^*_{\frac{\delta h|_{\mathfrak{g}^*}}{\delta\mu}} \mu,\ \xi \right\rangle. \end{aligned}$$

This is equivalent to

$$Y(\mu) = \mathrm{ad}^*_{\frac{\delta h|_{\mathfrak{g}^*}}{\delta\mu}} \mu$$

which, together with (6.2.10), gives the expression of the last component in (6.2.7).

Since $\lambda : (T^*G, \omega) \to (G \times \mathfrak{g}^*, \omega_B)$ is a symplectic diffeomorphism by construction, the Hamiltonian vector field of $h_B = h \circ \lambda^{-1}$ relative to ω_B equals $\lambda_* X_h$, which proves the remaining statement for left invariant functions on T^*G.

The proof for the right invariant case is carried out in an identical manner. ■

6.2.6 Reconstruction of dynamics on T^*G. The previous theorem gives an easy method to reconstruct the dynamics on T^*G from that on a coadjoint orbit. Let $h \in C^\infty(T^*G)$ be a left invariant Hamiltonian. The integral curve $\alpha(t)$ of X_h with initial condition $\alpha(0) \in T^*_{g_0}G$ is found in the following way:

- **Step 1:** *find the integral curve* $\mu(t) \in \mathfrak{g}^*$ *of*

$$\dot{\mu}(t) = \operatorname{ad}^*_{\frac{\delta h|_{\mathfrak{g}^*}}{\delta\mu(t)}} \mu(t), \qquad \textit{with} \qquad \mu(0) = T_e^*L_{g_0}\alpha(0);$$

- **Step 2:** *with* $\mu(t) \in \mathfrak{g}^*$ *determined above, solve the nonautonomous differential equation on* G

$$\dot{g}(t) = T_eL_{g(t)}\frac{\delta h|_{\mathfrak{g}^*}}{\delta\mu(t)}, \qquad \textit{with} \qquad g(0) = g_0;$$

- **Step 3:** *set* $\alpha(t) := T^*_{g(t)}L_{g(t)^{-1}}\mu(t)$.

Indeed, $\alpha(t)$ defined above is the integral curve of X_h passing through $\alpha(0)$ if and only if $\lambda(\alpha(t)) = (\pi(\alpha(t)), T_e^*L_{\pi(\alpha(t))}\alpha(t))$ is the integral curve of λ_*X_h passing through $(\pi(\alpha(0)), T_e^*L_{\pi(\alpha(0))}\alpha(0)) = (g_0, T_e^*L_{g_0}\alpha(0))$. The curve $\alpha(t)$ defined in Step 3 has the property that $\lambda(\alpha(0)) = (g_0, T_e^*L_{g_0}\alpha(0))$. Since $\pi(\alpha(t)) = g(t)$ and $T_e^*L_{\pi(\alpha(t))}\alpha(t)) = \mu(t)$ satisfy the differential equations in the first two steps by construction, it follows from (6.2.7) that $\lambda(\alpha(t)) = (g(t), \mu(t))$ is an integral curve of λ_*X_h.

If $h \in C^\infty(T^*G)$ is right invariant, the integral curve $\alpha(t)$ of X_h with initial condition $\alpha(0) \in T^*_{g_0}G$ is found in the following way:

- **Step 1:** *find the integral curve* $\mu(t) \in \mathfrak{g}^*$ *of*

$$\dot{\mu}(t) = -\operatorname{ad}^*_{\frac{\delta h|_{\mathfrak{g}^*}}{\delta\mu(t)}} \mu(t), \qquad \textit{with} \qquad \mu(0) = T_e^*R_{g_0}\alpha(0);$$

- **Step 2:** *with* $\mu(t) \in \mathfrak{g}^*$ *determined above, solve the nonautonomous differential equation on* G

$$\dot{g}(t) = T_eR_{g(t)}\frac{\delta h|_{\mathfrak{g}^*}}{\delta\mu(t)}, \qquad \textit{with} \qquad g(0) = g_0;$$

- **Step 3:** *set* $\alpha(t) := T^*_{g(t)}R_{g(t)^{-1}}\mu(t)$.

Theorems 6.2.4 and 6.2.5 immediately yield the expression of the canonical Poisson bracket in body and space representations.

6.2.7 Corollary (Trivialized Poisson bracket) *Let* $\varphi, \psi \in C^\infty(G \times \mathfrak{g}^*)$. *Then the Poisson brackets determined by* ω_B *and* ω_S *respectively are given by*

$$\{\varphi, \psi\}_B(g, \mu) = \mathbf{d}\varphi^\mu(g)\left(T_eL_g\frac{\delta\psi^g}{\delta\mu}\right) - \mathbf{d}\psi^\mu(g)\left(T_eL_g\frac{\delta\varphi^g}{\delta\mu}\right) - \left\langle\mu, \left[\frac{\delta\varphi^g}{\delta\mu}, \frac{\delta\psi^g}{\delta\mu}\right]\right\rangle \tag{6.2.13}$$

$$
\begin{aligned}
\{\varphi, \psi\}_S(g, \mu) = \mathbf{d}\varphi^{\mu}(g)\left(T_e R_g \frac{\delta \psi^g}{\delta \mu}\right) - \mathbf{d}\psi^{\mu}(g)\left(T_e R_g \frac{\delta \varphi^g}{\delta \mu}\right) \\
+\left\langle \mu, \left[\frac{\delta \varphi^g}{\delta \mu}, \frac{\delta \psi^g}{\delta \mu}\right]\right\rangle,
\end{aligned} \tag{6.2.14}
$$

where $\varphi^g, \psi^g : \mathfrak{g}^* \to \mathbb{R}$ *and* $\varphi^{\mu}, \psi^{\mu} : G \to \mathbb{R}$ *are defined by* $\varphi^g(\mu) := \varphi^{\mu}(g) := \varphi(g, \mu)$, $\psi^g(\mu) := \psi^{\mu}(g) := \psi(g, \mu)$, *for all* $g \in G$ *and all* $\mu \in \mathfrak{g}^*$. *The Hamiltonian vector fields* X_{φ}^B *and* X_{φ}^S *of* $\varphi \in C^{\infty}(G \times \mathfrak{g}^*)$ *relative to* ω_B *and* ω_S *are given respectively by*

$$
X_{\varphi}^B(g, \mu) = \left(T_e L_g \frac{\delta \varphi^g}{\delta \mu}, -T_e^* L_g \left(\mathbf{d}\varphi^{\mu}(g)\right) + \mathrm{ad}^*_{\frac{\delta \varphi^g}{\delta \mu}} \mu\right) \tag{6.2.15}
$$

$$
X_{\varphi}^S(g, \mu) = \left(T_e R_g \frac{\delta \varphi^g}{\delta \mu}, -T_e^* R_g \left(\mathbf{d}\varphi^{\mu}(g)\right) - \mathrm{ad}^*_{\frac{\delta \varphi^g}{\delta \mu}} \mu\right) \tag{6.2.16}
$$

for $(g, \mu) \in G \times \mathfrak{g}^*$.

6.3 Orbit reduction

There is an alternative approach to symplectic reduction that has been developed by Marle MARLE (1976), KAZHDAN, KOSTANT, AND STERNBERG (1978), MARSDEN (1981), and BLAOM (2001) which consists of choosing as numerator of the symplectic reduced space the group invariant saturation of the level sets of a standard momentum map of the action. This option produces as a result a space that is symplectomorphic to the Marsden–Weinstein quotient but presents the advantage of being more appropriate in the context of quantization problems and of making easier the comparison of the symplectic reduced spaces corresponding to different values of the momentum map. This will become very important when comparing symplectic and Poisson reduction.

6.3.1 Theorem (Symplectic orbit reduction) *Let* $\Phi : G \times M \to M$ *be a free proper canonical action of the Lie group* G *on the connected symplectic manifold* (M, ω). *Suppose that this action has an associated momentum map* $\mathbf{J} : M \to \mathfrak{g}^*$, *with non-equivariance one-cocycle* $\sigma : G \to \mathfrak{g}^*$. *Let* $\mathcal{O}_{\mu} := G \cdot \mu \subset \mathfrak{g}^*$ *be the* G*-orbit of the point* $\mu \in \mathfrak{g}^*$ *with respect to the affine action of* G *on* $\mathfrak{g}^*$ *associated to* σ. *Then:*

(i) *The set* $M_{\mathcal{O}_{\mu}} := \mathbf{J}^{-1}(\mathcal{O}_{\mu})/G$ *is a regular quotient symplectic manifold with the symplectic form* $\omega_{\mathcal{O}_{\mu}}$ *uniquely characterized by the relation*

$$
i_{\mathcal{O}_{\mu}}^* \omega = \pi_{\mathcal{O}_{\mu}}^* \omega_{\mathcal{O}_{\mu}} + \mathbf{J}_{\mathcal{O}_{\mu}}^* \omega_{\mathcal{O}_{\mu}}^+, \tag{6.3.1}
$$

where $\mathbf{J}_{\mathcal{O}_{\mu}}$ *is the restriction of* $\mathbf{J}$ *to* $\mathbf{J}^{-1}(\mathcal{O}_{\mu})$ *and* $\omega_{\mathcal{O}_{\mu}}^+$ *is the* $+$*-symplectic structure on the affine orbit* $\mathcal{O}_{\mu}$ *introduced in Theorem 4.5.31. The maps* $i_{\mathcal{O}_{\mu}} : \mathbf{J}^{-1}(\mathcal{O}_{\mu}) \hookrightarrow M$ *and* $\pi_{\mathcal{O}_{\mu}} : \mathbf{J}^{-1}(\mathcal{O}_{\mu}) \to M_{\mathcal{O}_{\mu}}$ *are natural injection and the projection, respectively. The pair* $(M_{\mathcal{O}_{\mu}}, \omega_{\mathcal{O}_{\mu}})$ *is called the* ***symplectic orbit reduced space****.*

(ii) *Let $h \in C^\infty(M)^G$ be a G-invariant Hamiltonian. The Hamiltonian flow F_t of h leaves the connected components of $\mathbf{J}^{-1}(\mathcal{O}_\mu)$ invariant and commutes with the G-action, so it induces a flow $F_t^{\mathcal{O}_\mu}$ on $M_{\mathcal{O}_\mu}$, uniquely determined by*

$$\pi_{\mathcal{O}_\mu} \circ F_t \circ i_{\mathcal{O}_\mu} = F_t^{\mathcal{O}_\mu} \circ \pi_{\mathcal{O}_\mu}. \tag{6.3.2}$$

(iii) *The vector field generated by the flow $F_t^{\mathcal{O}_\mu}$ on $(M_{\mathcal{O}_\mu}, \omega_{\mathcal{O}_\mu})$ is Hamiltonian with associated* **reduced Hamiltonian function** *$h_{\mathcal{O}_\mu} \in C^\infty(M_{\mathcal{O}_\mu})$ defined by*

$$h_{\mathcal{O}_\mu} \circ \pi_{\mathcal{O}_\mu} = h \circ i_{\mathcal{O}_\mu}.$$

The vector fields X_h and $X_{h_{\mathcal{O}_\mu}}$ are $\pi_{\mathcal{O}_\mu}$-related.

(iv) *Let $k \in C^\infty(M)^G$ be another G-invariant function. Then $\{h, k\}$ is also G-invariant and $\{h, k\}_{\mathcal{O}_\mu} = \{h_{\mathcal{O}_\mu}, k_{\mathcal{O}_\mu}\}_{M_{\mathcal{O}_\mu}}$, where $\{\cdot,\cdot\}_{M_{\mathcal{O}_\mu}}$ denotes the Poisson bracket associated to the symplectic form $\omega_{\mathcal{O}_\mu}$ on $M_{\mathcal{O}_\mu}$.*

Proof. (i) We start by spelling out the smooth structure of the quotient $\mathbf{J}^{-1}(\mathcal{O}_\mu)/G$. We first study the set $\mathbf{J}^{-1}(\mathcal{O}_\mu)$. Recall that by Proposition 2.3.12 the affine orbit $\mathcal{O}_\mu$ is an initial submanifold of $\mathfrak{g}^*$. At the same time, the freeness of the action and Proposition 4.5.12 imply that $\mathbf{J}$ is a submersion onto some open subset of $\mathfrak{g}^*$, which guarantees that the map $\mathbf{J}$ is transversal to $\mathcal{O}_\mu$. More explicitly, for any $z \in \mathbf{J}^{-1}(\mathcal{O}_\mu)$, we have that

$$(T_z\mathbf{J})(T_zM) + T_{\mathbf{J}(z)}\mathcal{O}_\mu = \mathfrak{g}^*.$$

Theorem 1.1.15 implies that $\mathbf{J}^{-1}(\mathcal{O}_\mu)$ is an initial submanifold of M of dimension

$$\dim(\mathbf{J}^{-1}(\mathcal{O}_\mu)) = \dim M - \dim G_\mu \tag{6.3.3}$$

whose tangent space at $z \in \mathbf{J}^{-1}(\mathcal{O}_\mu)$ equals

$$T_z(\mathbf{J}^{-1}(\mathcal{O}_\mu)) = (T_z\mathbf{J})^{-1}(T_{\mathbf{J}(z)}\mathcal{O}_\mu) = \mathfrak{g} \cdot z + \ker(T_z\mathbf{J}). \tag{6.3.4}$$

The last equality can be easily proved by noticing that $\mathfrak{g}\cdot z + \ker(T_z\mathbf{J}) \subset T_z(\mathbf{J}^{-1}(\mathcal{O}_\mu))$ and that $\dim(\mathfrak{g}\cdot z + \ker(T_z\mathbf{J})) = \dim \mathfrak{g}\cdot z + \dim\ker(T_z\mathbf{J}) - \dim(\mathfrak{g}\cdot z \cap \ker(T_z\mathbf{J})) = \dim \mathfrak{g}\cdot z + \dim(\mathfrak{g}\cdot z)^\omega - \dim \mathfrak{g}_{\mathbf{J}(z)}\cdot z$. Since $\mathbf{J}(z) = g \cdot \mu$ for some $g \in G$ we have $\dim(\mathfrak{g}\cdot z + \ker(T_z\mathbf{J})) = \dim M - \dim \mathfrak{g}_{g\cdot\mu} = \dim M - \dim G_\mu = \dim T_z(\mathbf{J}^{-1}(\mathcal{O}_\mu))$, by (6.3.3) which proves (6.3.4). If we now combine (6.3.4) with (6.1.3) we have that

$$T_z(\mathbf{J}^{-1}(\mathcal{O}_\mu)) = \mathfrak{g}\cdot z + A'_G(z). \tag{6.3.5}$$

The free and proper G-action on M restricts to a free and proper G-action on the G-invariant initial submanifold $\mathbf{J}^{-1}(\mathcal{O}_\mu)$ and consequently the quotient $M_{\mathcal{O}_\mu} := \mathbf{J}^{-1}(\mu)/G$ is a regular quotient manifold by Proposition 2.3.8 with $\pi_{\mathcal{O}_\mu} : \mathbf{J}^{-1}(\mathcal{O}_\mu) \to M_{\mathcal{O}_\mu}$ a surjective submersion. The smoothness of the G-action on $\mathbf{J}^{-1}(\mathcal{O}_\mu)$ is guaranteed by the fact that the map $\Phi : G \times M \to M$ restricts to a smooth map $\Phi^{\mathcal{O}_\mu}$:

$G \times \mathbf{J}^{-1}(\mathcal{O}_\mu) \to M$ such that $\Phi^{\mathcal{O}_\mu}(G \times \mathbf{J}^{-1}(\mathcal{O}_\mu)) \subset \mathbf{J}^{-1}(\mathcal{O}_\mu)$. As $\mathbf{J}^{-1}(\mathcal{O}_\mu)$ is an initial submanifold of M the G-action on $\mathbf{J}^{-1}(\mathcal{O}_\mu)$ obtained from $\Phi^{\mathcal{O}_\mu}$ by restriction of its range to $\mathbf{J}^{-1}(\mathcal{O}_\mu)$ is a smooth map.

We now prove that $M_{\mathcal{O}_\mu}$ is a symplectic manifold. We begin by showing that each of its tangent spaces is a symplectic vector space. If $m \in \mathbf{J}^{-1}(\mathcal{O}_\mu)$, denote $[m]_{\mathcal{O}_\mu} := \pi_{\mathcal{O}_\mu}(m)$. We assume, without loss of generality, that $m \in \mathbf{J}^{-1}(\mu)$. Let $v, w \in T_m(\mathbf{J}^{-1}(\mathcal{O}_\mu))$ be arbitrary tangent vectors. By (6.3.5) we can write v and w as

$$v = \xi_M(m) + X_f(m), \quad \text{with} \quad \xi \in \mathfrak{g} \text{ and } f \in C^\infty(M)^G \tag{6.3.6}$$

$$w = \eta_M(m) + X_g(m), \quad \text{with} \quad \eta \in \mathfrak{g} \text{ and } g \in C^\infty(M)^G. \tag{6.3.7}$$

We define a two-form $\omega_{\mathcal{O}_\mu}([m]_{\mathcal{O}_\mu})$ on $T_{[m]_{\mathcal{O}_\mu}} M_{\mathcal{O}_\mu}$ by

$$\omega_{\mathcal{O}_\mu}([m]_{\mathcal{O}_\mu})(T_m\pi_{\mathcal{O}_\mu}(v),\, T_m\pi_{\mathcal{O}_\mu}(w)) := \{f, g\}(m). \tag{6.3.8}$$

Our first job is to show that (6.3.8) is a good definition. Firstly, suppose that $[m]_{\mathcal{O}_\mu} = [m']_{\mathcal{O}_\mu}$ and that $v', w' \in T_{m'}M$ are such that $T_m\pi_{\mathcal{O}_\mu}(v) = T_{m'}\pi_{\mathcal{O}_\mu}(v')$ and $T_m\pi_{\mathcal{O}_\mu}(w) = T_{m'}\pi_{\mathcal{O}_\mu}(w')$. Let $h \in G$ be such that $m' = h \cdot m$. Because $T_m\Phi_h : T_mM \to T_{m'}M$ is an isomorphism, there exist $v'', w'' \in T_mM$ such that $v' = T_m\Phi_h(v'')$ and $w' = T_m\Phi_h(w'')$. Since by hypothesis $T_m\pi_{\mathcal{O}_\mu}(v) = T_{m'}\pi_{\mathcal{O}_\mu}(v')$, $T_m\pi_{\mathcal{O}_\mu}(w) = T_{m'}\pi_{\mathcal{O}_\mu}(w')$, and the projection $\pi_{\mathcal{O}_\mu}$ is G-invariant, we have

$$T_m\pi_{\mathcal{O}_\mu}(v - v'') = T_m\pi_{\mathcal{O}_\mu}(w - w'') = 0.$$

Thus there exist Lie algebra elements $\zeta^1, \zeta^2 \in \mathfrak{g}$ such that $v'' = v + \zeta^1_M(m) = \xi_M(m)+X_f(m)+\zeta^1_M(m) = (\xi+\zeta^1)_M(m)+X_f(m)$ by (6.3.6) and $w'' = w+\zeta^2_M(m) = \eta_M(m) + X_g(m) + \zeta^2_M(m) = (\eta + \zeta^2)_M(m) + X_g(m)$ by (6.3.7). Therefore,

$$\begin{aligned}
&\omega_{\mathcal{O}_\mu}([m']_{\mathcal{O}_\mu})(T_{m'}\pi_{\mathcal{O}_\mu}(v'),\, T_{m'}\pi_{\mathcal{O}_\mu}(w')) \\
&\quad = \omega_{\mathcal{O}_\mu}([m']_{\mathcal{O}_\mu})(T_m(\pi_{\mathcal{O}_\mu} \circ \Phi_h)(v''),\, T_m(\pi_{\mathcal{O}_\mu} \circ \Phi_h)(w'')) \\
&\quad = \omega_{\mathcal{O}_\mu}([m]_{\mathcal{O}_\mu})(T_m\pi_{\mathcal{O}_\mu}((\xi + \zeta^1)_M(m) + X_f(m)), \\
&\qquad\qquad\qquad\qquad T_m\pi_{\mathcal{O}_\mu}((\eta + \zeta^2)_M(m) + X_g(m))) \\
&= \{f, g\}(m) = \omega_{\mathcal{O}_\mu}([m]_{\mathcal{O}_\mu})(T_m\pi_{\mathcal{O}_\mu}(v),\, T_m\pi_{\mathcal{O}_\mu}(w)).
\end{aligned}$$

Secondly we have to verify that (6.3.8) does not depend on the functions and on the Lie algebra elements utilized in (6.3.6) and (6.3.7) to define the tangent vectors $v, w \in T_m(\mathbf{J}^{-1}(\mathcal{O}_\mu))$. Suppose that

$$v = \xi_M(m) + X_f(m) = \xi'_M(m) + X_{f'}(m), \text{ with } \xi, \xi' \in \mathfrak{g},\ f, f' \in C^\infty(M)^G$$
$$w = \eta_M(m) + X_g(m) = \eta'_M(m) + X_{g'}(m), \text{ with } \eta, \eta' \in \mathfrak{g},\ g, g' \in C^\infty(M)^G.$$

These two equalities and the defining property of the momentum map imply that $X_f(m) - X_{f'}(m) = X_{\mathbf{J}^{\xi'-\xi}}(m)$ and $X_g(m) - X_{g'}(m) = X_{\mathbf{J}^{\eta'-\eta}}(m)$. Thus, by Noether's

Theorem

$$\begin{aligned}\{f',\, g'\}(m) &= X_{g'}[f'](m) = \langle \mathbf{d}f'(m),\, X_{g'}(m)\rangle \\ &= \langle \mathbf{d}f'(m),\, X_g(m) - X_{\mathbf{J}^{\eta'-\eta}}(m)\rangle = \langle \mathbf{d}f'(m),\, X_g(m)\rangle \\ &= -\langle \mathbf{d}g(m),\, X_{f'}(m)\rangle = -\langle \mathbf{d}g(m),\, X_f(m) - X_{\mathbf{J}^{\xi'-\xi}}(m)\rangle \\ &= -\langle \mathbf{d}g(m),\, X_f(m)\rangle = \{f,\, g\}(m).\end{aligned}$$

The considerations above show that the two-form $\omega_{\mathcal{O}_\mu}([m]_{\mathcal{O}_\mu})$ on $T_{[m]_{\mathcal{O}_\mu}} M_{\mathcal{O}_\mu}$ given by (6.3.8) is well defined.

We now show that $\omega_{\mathcal{O}_\mu}([m]_{\mathcal{O}_\mu})$ is nondegenerate. Indeed, if we write the arbitrary vectors $v,\ w \in T_m\mathbf{J}^{-1}(\mathcal{O}_\mu)$ as in (6.3.6), (6.3.7) and suppose that

$$\omega_{\mathcal{O}_\mu}([m]_{\mathcal{O}_\mu})(T_m\pi_{\mathcal{O}_\mu}(v),\, T_m\pi_{\mathcal{O}_\mu}(w)) = 0$$

for any $w \in T_m\mathbf{J}^{-1}(\mathcal{O}_\mu)$, then we have

$$\{f,\, g\}(m) = \langle \mathbf{d}f(m),\, X_g(m)\rangle = \omega(m)(X_f(m),\, X_g(m)) = 0,$$

for any $g \in C^\infty(M)^G$ which implies that $X_f(m) \in \ker T_m\mathbf{J} \cap (\ker T_m\mathbf{J})^\omega = \mathfrak{g}\cdot m \cap (\mathfrak{g}\cdot m)^\omega = \mathfrak{g}_\mu \cdot m$. Consequently, there exists an element $\eta \in \mathfrak{g}_\mu$ such that $v = \xi_M(m) + \eta_M(m)$ and hence $T_m\pi_{\mathcal{O}_\mu}(v) = 0$, as required. The pair $(T_{[m]_{\mathcal{O}_\mu}} M_{\mathcal{O}_\mu},\, \omega_{\mathcal{O}_\mu}([m]_{\mathcal{O}_\mu}))$ is therefore a symplectic vector space.

We next show that $\omega_{\mathcal{O}_\mu}$ is determined by expression (6.3.1). Let m be an arbitrary point that again we assume, without loss of generality, to be in $\mathbf{J}^{-1}(\mu)$ and let $v,\ w \in T_m\mathbf{J}^{-1}(\mathcal{O}_\mu)$. We write $v = X_f(m) + \xi_M(m)$ and $w = X_g(m) + \eta_M(m)$, with $f,\ g \in C^\infty(M)^G$ and $\xi,\ \eta \in \mathfrak{g}$. On the one hand, by the definition of $\omega_{\mathcal{O}_\mu}$ in (6.3.8) we have

$$\begin{aligned}(\pi^*_{\mathcal{O}_\mu}\omega_{\mathcal{O}_\mu})(m)(v,\, w) &= \omega_{\mathcal{O}_\mu}([m]_{\mathcal{O}_\mu})(T_m\pi_{\mathcal{O}_\mu}(v),\, T_m\pi_{\mathcal{O}_\mu}\cdot w) \\ &= \{f,\, g\}(m) = \omega(m)(X_f(m),\, X_g(m)) \\ &= \omega(m)(v - \xi_M(m),\, w - \eta_M(m)).\end{aligned}$$

Since both $v - \xi_M(m) = X_f(m)$ and $w - \eta_M(m) = X_g(m)$ belong to $\ker T_m\mathbf{J} = (\mathfrak{g}\cdot m)^\omega$, we can continue writing

$$\begin{aligned}\omega(m)(v - \xi_M(m),\, w - \eta_M(m)) &= \omega(m)(v - \xi_M(m),\, w) \\ &= \omega(m)(v,\, w) - \omega(m)(\xi_M(m),\, w) \\ &= \omega(m)(v,\, w) - \omega(m)(\xi_M(m),\, X_g(m) + \eta_M(m)) \\ &= \omega(m)(v,\, w) - \omega(m)(\xi_M(m),\, \eta_M(m)) \\ &= (i^*_{\mathcal{O}_\mu}\omega)(m)(v,\, w) - \omega(m)(\xi_M(m),\, \eta_M(m)).\end{aligned}$$

In order to conclude the required equality we just need to show that

$$\omega(m)(\xi_M(m),\, \eta_M(m)) = \mathbf{J}^*_{\mathcal{O}_\mu}\omega^+_{\mathcal{O}_\mu}(m)(v,\, w).$$

Indeed, by Noether's Theorem $T_m\mathbf{J}(X_f(m)) = T_m\mathbf{J}(X_g(m)) = 0$ and hence

$$\begin{aligned}(\mathbf{J}^*_{\mathcal{O}_\mu}\omega^+_{\mathcal{O}_\mu})(m)(v, w) &= \omega^+_{\mathcal{O}_\mu}(\mu)(T_m\mathbf{J}(v), T_m\mathbf{J}(w)) \\ &= \omega^+_{\mathcal{O}_\mu}(\mu)(T_m\mathbf{J}(\xi_M(m)), T_m\mathbf{J}(\eta_M(m))) \\ &= \omega^+_{\mathcal{O}_\mu}(\mu)(-\mathrm{ad}^*_\xi\mu + \Sigma(\xi, \cdot), -\mathrm{ad}^*_\eta\mu + \Sigma(\eta, \cdot)) \\ &= \omega^+_{\mathcal{O}_\mu}(\mu)(\xi_{\mathfrak{g}^*}(\mu), \eta_{\mathfrak{g}^*}(\mu)) = \langle\mu, [\xi, \eta]\rangle - \Sigma(\xi, \eta),\end{aligned}$$

where we used the expression for $T_m\mathbf{J}(\xi_M(m))$, given in Theorem 4.5.25. On the other hand, we compute $\omega(m)(\xi_M(m), \eta_M(m))$ directly to get

$$\begin{aligned}\omega(m)(\xi_M(m), \eta_M(m)) &= \mathbf{dJ}^\xi(m)(\eta_M(m)) = \langle T_m\mathbf{J}(\eta_M(m)), \xi\rangle \\ &= \langle\mu, [\xi, \eta]\rangle - \Sigma(\xi, \eta),\end{aligned}$$

which proves the desired equality (6.3.1).

Since $\pi_{\mathcal{O}_\mu} : \mathbf{J}^{-1}(\mathcal{O}_\mu) \to \mathbf{J}^{-1}(\mathcal{O}_\mu)/G = M_{\mathcal{O}_\mu}$ is a surjective submersion, formula (6.3.1) shows that $\omega_{\mathcal{O}_\mu}$ is a smooth two-form on the quotient manifold $M_{\mathcal{O}_\mu}$. The Jacobi identity for the bracket $\{\cdot, \cdot\}$ and its antisymmetry guarantee that $\omega_{\mathcal{O}_\mu}$ is closed and antisymmetric, which proves that the pair $(M_{\mathcal{O}_\mu}, \omega_{\mathcal{O}_\mu})$ is a symplectic manifold whose symplectic form is characterized by (6.3.1).

The remaining points in the theorem are proved by mimicking the steps in the equivalent statements for point reduction in Theorem 6.1.1. ■

The following proposition provides a useful characterization of the smooth structure for $\mathbf{J}^{-1}(\mathcal{O}_\mu)$ introduced in the proof of part **(i)** of Theorem 6.3.1, which makes $\mathbf{J}^{-1}(\mathcal{O}_\mu)$ into an initial submanifold of M.

6.3.2 Proposition *In the hypotheses of Theorem* 6.3.1, *consider the set* $\mathbf{J}^{-1}(\mathcal{O}_\mu)$ *with the smooth structure that makes it into an initial submanifold of* M. *Then the map*

$$\begin{array}{rccc} f : & G \times_{G_\mu} \mathbf{J}^{-1}(\mu) & \longrightarrow & \mathbf{J}^{-1}(\mathcal{O}_\mu) \\ & [g, z] & \longmapsto & g \cdot z \end{array}$$

is a diffeomorphism.

Proof. The map $G \times \mathbf{J}^{-1}(\mu) \to M$ given by $(g, z) \to g \cdot z$, with $(g, z) \in G \times \mathbf{J}^{-1}(\mu)$, is smooth and invariant with respect to the twisted action of G_μ on the product $G \times \mathbf{J}^{-1}(\mu)$. Hence it drops to a smooth map $\varphi : G \times_{G_\mu} \mathbf{J}^{-1}(\mu) \to M$ that satisfies that $\varphi(G \times_{G_\mu} \mathbf{J}^{-1}(\mu)) \subset \mathbf{J}^{-1}(\mathcal{O}_\mu)$ and that $\varphi([g, z]) = f([g, z])$. The initial character of $\mathbf{J}^{-1}(\mathcal{O}_\mu)$ implies that f is smooth.

We now show that the inverse $f^{-1} : \mathbf{J}^{-1}(\mathcal{O}_\mu) \to G \times_{G_\mu} \mathbf{J}^{-1}(\mu)$ of f given by $g \cdot z \mapsto [g, z]$ is smooth. Let $j : \mathcal{O}_\mu \to G/G_\mu$ be the standard diffeomorphism and $\sigma : U_{gG_\mu} \subset G/G_\mu \to G$ a local smooth section of the submersion $G \to G/G_\mu$ in a neighborhood U_{gG_μ} of $gG_\mu \in G/G_\mu$. Let $V := \mathbf{J}^{-1}_{\mathcal{O}_\mu}(j^{-1}(U_{gG_\mu}))$. V is an open neighborhood of z in $\mathbf{J}^{-1}(\mathcal{O}_\mu)$ because $j \circ \mathbf{J}_{\mathcal{O}_\mu}(z) = j(g \cdot \mu) = gG_\mu \in U_{gG_\mu}$. We now note that for any $m \in V$ we can write that

$$f^{-1}(m) = [\sigma \circ j \circ \mathbf{J}_{\mathcal{O}_\mu}(m), (\sigma \circ j \circ \mathbf{J}_{\mathcal{O}_\mu}(m))^{-1} \cdot m],$$

and consequently the smoothness of f^{-1} is guaranteed. ■

6.3.3 Analogous to the situation encountered in Section 6.1 we shall apply now the Symplectic Foliation Reduction Theorem (given in Corollary 6.1.7) to the initial submanifold $\mathbf{J}^{-1}(\mathcal{O}_\mu)$ of M under the hypotheses, notations, and conventions of the Symplectic Orbit Reduction Theorem 6.3.1. Recall from (6.3.4) that $T_z(\mathbf{J}^{-1}(\mathcal{O}_\mu)) = \mathfrak{g} \cdot z + \ker(T_z\mathbf{J})$. Thus, by Proposition 4.5.14**(ii)** and the Reduction Lemma 4.5.34 we get

$$\begin{aligned}\left[T_z(\mathbf{J}^{-1}(\mathcal{O}_\mu))\right]^\omega &= [\mathfrak{g} \cdot z + \ker(T_z\mathbf{J})]^\omega = [\mathfrak{g} \cdot z]^\omega \cap [\ker(T_z\mathbf{J})]^\omega \\ &= [\mathfrak{g} \cdot z]^\omega \cap \mathfrak{g} \cdot z = \mathfrak{g}_{\mathbf{J}(z)} \cdot z,\end{aligned}$$

Therefore the characteristic distribution of $i^*_{\mathcal{O}_\mu}\omega$ on $\mathbf{J}^{-1}(\mathcal{O}_\mu)$ is given by

$$\begin{aligned}T_z(\mathbf{J}^{-1}(\mathcal{O}_\mu)) \cap \left[T_z(\mathbf{J}^{-1}(\mathcal{O}_\mu))\right]^\omega &= [\mathfrak{g} \cdot z + \ker(T_z\mathbf{J})] \cap \mathfrak{g}_{\mathbf{J}(z)} \cdot z \\ &= \mathfrak{g}_{\mathbf{J}(z)} \cdot z = T_z(G_{\mathbf{J}(z)} \cdot z). \qquad (6.3.9)\end{aligned}$$

Since $\nu \in \mathcal{O}_\mu$, the isotropy groups G_ν and G_μ are conjugate. Thus the null distribution of $i^*_{\mathcal{O}_\mu}\omega$ on $\mathbf{J}^{-1}(\mathcal{O}_\mu)$ has constant rank. This implies that it is a subbundle of $T(\mathbf{J}^{-1}(\mathcal{O}_\mu))$ which, by Corollary 6.1.7, determines a smooth regular foliation $\mathfrak{N}$ whose leaf space $\mathbf{J}^{-1}(\mathcal{O}_\mu)/\mathfrak{N}$ is a smooth symplectic manifold whose associated projection $\pi_\mathfrak{N} : \mathbf{J}^{-1}(\mathcal{O}_\mu) \to \mathbf{J}^{-1}(\mathcal{O}_\mu)/\mathfrak{N}$ is a surjective submersion. Denote by $\Omega_\mathfrak{N}$ the reduced symplectic form on $\mathbf{J}^{-1}(\mathcal{O}_\mu)/\mathfrak{N}$. Recall from Corollary 6.1.7 that it is characterized by the equality $\pi^*_\mathfrak{N}\Omega_\mathfrak{N} = i^*_{\mathcal{O}_\mu}\omega$.

However, (6.3.9) shows that the integral manifold of the foliation $\mathfrak{N}$ through $z \in \mathbf{J}^{-1}(\mathcal{O}_\mu)$ equals the orbit $G_{\mathbf{J}(z)} \cdot z$, if G_μ (and hence $G_{\mathbf{J}(z)}$) *is connected*. Thus, unlike the situation encountered in §6.1.8, the reduced symplectic manifold $\left(\mathbf{J}^{-1}(\mathcal{O}_\mu)/\mathfrak{N}, \Omega_\mathfrak{N}\right)$ *is different* from the orbit reduced space $\left(M_{\mathcal{O}_\mu} := \mathbf{J}^{-1}(\mathcal{O}_\mu)/G, \omega_{\mathcal{O}_\mu}\right)$: the equivalence class of $z \in \mathbf{J}^{-1}(\mathcal{O}_\mu)$ for the first space is the $G_{\mathbf{J}(z)}$-orbit through z, whereas for the second space it is the G-orbit through z. The next theorem will relate $\left(\mathbf{J}^{-1}(\mathcal{O}_\mu)/\mathfrak{N}, \Omega_\mathfrak{N}\right)$ to $\left(M_{\mathcal{O}_\mu} := \mathbf{J}^{-1}(\mathcal{O}_\mu)/G, \omega_{\mathcal{O}_\mu}\right)$.

The following notation will be used below. If (M_i, ω_i), $i = 1, 2$, are two symplectic manifolds, the product $M_1 \times M_2$ is also a symplectic manifold in two different ways, by considering on it the symplectic forms $\omega_1 \oplus \omega_2 := \pi_1^*\omega_1 + \pi_2^*\omega_2$ and $\omega_1 \ominus \omega_2 := \pi_1^*\omega_1 - \pi_2^*\omega_2$, where $\pi_i : M_1 \times M_2 \to M_i$ are the projections. Define the ***symplectic sum*** by $M_1 \oplus M_2 := (M_1 \times M_2, \omega_1 \oplus \omega_2)$ and the ***symplectic difference*** by $M_1 \ominus M_2 := (M_1 \times M_2, \omega_1 \ominus \omega_2)$. Denote $\mathcal{O}^\pm_\mu := (\mathcal{O}_\mu, \omega^\pm_{\mathcal{O}_\mu})$, where $\omega^\pm_{\mathcal{O}_\mu}$ is the $\pm$-symplectic structure on the affine orbit $\mathcal{O}_\mu$ introduced in Theorem 4.5.31.

6.3.4 Theorem (Symplectic foliation orbit reduction) *Let $\Phi : G \times M \to M$ be a free proper canonical action of the Lie group G on the connected symplectic manifold (M, ω). Suppose that this action has an associated momentum map $\mathbf{J} : M \to \mathfrak{g}^*$, with non-equivariance one-cocycle $\sigma : G \to \mathfrak{g}^*$. Let $\mathcal{O}_\mu := G \cdot \mu \subset \mathfrak{g}^*$ be the G-orbit of the point $\mu \in \mathfrak{g}^*$ with respect to the affine action of G on $\mathfrak{g}^*$ associated to σ, define $\mathbf{J}_{\mathcal{O}_\mu} := \mathbf{J}|_{\mathbf{J}^{-1}(\mathcal{O}_\mu)} : \mathbf{J}^{-1}(\mathcal{O}_\mu) \to \mathcal{O}_\mu$, and assume that the isotropy group G_μ is connected. Then the map* $\operatorname{id} \times \mathbf{J}_{\mathcal{O}_\mu} : z \in \mathbf{J}^{-1}(\mathcal{O}_\mu) \mapsto (z, \mathbf{J}(z)) \in \mathbf{J}^{-1}(\mathcal{O}_\mu) \times \mathcal{O}_\mu$ *induces a symplectic diffeomorphism $\chi : \left(\mathbf{J}^{-1}(\mathcal{O}_\mu)/\mathfrak{N}, \Omega_\mathfrak{N}\right) \to M_{\mathcal{O}_\mu} \oplus \mathcal{O}^+_\mu$.*

$$
\begin{array}{ccc}
\mathbf{J}^{-1}(\mathcal{O}_\mu) & \xrightarrow{\ \mathrm{id} \times \mathbf{J}_{\mathcal{O}_\mu}\ } & \mathbf{J}^{-1}(\mathcal{O}_\mu) \times \mathcal{O}_\mu \\
\Big\downarrow \pi_{\mathfrak{N}} & & \Big\downarrow \pi_{\mathcal{O}_\mu} \times \mathrm{id}_{\mathcal{O}_\mu} \\
(\mathbf{J}^{-1}(\mathcal{O}_\mu)/\mathfrak{N}, \Omega_{\mathfrak{N}}) & \xrightarrow{\ \chi\ } & M_{\mathcal{O}_\mu} \oplus \mathcal{O}_\mu^+
\end{array}
$$

Proof. The map id $\times \mathbf{J}_{\mathcal{O}_\mu}$ is compatible with the foliation $\mathfrak{N}$ and the G-action on the first factor of $\mathbf{J}^{-1}(\mathcal{O}_\mu) \times \mathcal{O}_\mu$, that is, if $z \in \mathbf{J}^{-1}(\mathcal{O}_\mu)$ and $h \in G_{\mathbf{J}(z)}$, then $(\mathrm{id} \times \mathbf{J}_{\mathcal{O}_\mu})(h \cdot z) = (h \cdot z, \mathbf{J}(h \cdot z)) = (h \cdot z, h \cdot \mathbf{J}(z)) = (h \cdot z, \mathbf{J}(z)) = h \cdot (z, \mathbf{J}(z))$. Thus it induces a smooth map $\chi : \mathbf{J}^{-1}(\mathcal{O}_\mu)/\mathfrak{N} \to M_{\mathcal{O}_\mu} \times \mathcal{O}_\mu$ such that the diagram above is commutative, that is,

$$
\chi\left(\pi_{\mathfrak{N}}(z)\right) = \left(\pi_{\mathcal{O}_\mu}(z), \mathbf{J}(z)\right).
$$

If $v \in T_z\left(\mathbf{J}^{-1}(\mathcal{O}_\mu)\right)$, then $T_{\pi_{\mathfrak{N}}(z)}\chi\left(T_z\pi_{\mathfrak{N}}(v)\right) = \left(T_z\pi_{\mathcal{O}_\mu}(v), T_z\mathbf{J}(v)\right) = (0, 0)$ if and only if $v \in \ker T_z\pi_{\mathcal{O}_\mu} \cap \ker T_z\mathbf{J} = \mathfrak{g} \cdot z \cap (\mathfrak{g} \cdot z)^\omega = \mathfrak{g}_{\mathbf{J}(z)} \cdot z$ by Proposition 4.5.14**(ii)** and the Reduction Lemma 4.5.34. Thus $T_z\pi_{\mathfrak{N}}(v) = 0$ which shows that χ is an immersion. We have $\dim(M_{\mathcal{O}_\mu} \times \mathcal{O}_\mu) = \dim(\mathbf{J}^{-1}(\mathcal{O}_\mu)) - \dim G + \dim \mathcal{O}_\mu = \dim(\mathbf{J}^{-1}(\mathcal{O}_\mu)) - \dim G_\mu = \dim(\mathbf{J}^{-1}(\mathcal{O}_\mu)/\mathfrak{N})$ since the dimension of the leaves of $\mathfrak{N}$ equals $\dim G_\mu$. Thus χ is an immersion between equal-dimensional manifolds and is hence a local diffeomorphism.

We shall prove that χ is invertible. If $(\pi_{\mathcal{O}_\mu}(z), \nu) \in \mathbf{J}^{-1}(\mathcal{O}_\mu)/G \times \mathcal{O}_\mu$, then there is an element $g \in G$ such that $\nu = g \cdot \mathbf{J}(z)$. Note that if $h \in G$ also satisfies $\nu = h \cdot \mathbf{J}(z)$, then $h^{-1}g \in G_{\mathbf{J}(z)}$ and thus $gG_{\mathbf{J}(z)}g^{-1} = hG_{\mathbf{J}(z)}h^{-1}$. Therefore the leaf through $g \cdot z$, which equals the $G_{\mathbf{J}(g \cdot z)} = G_{g \cdot \mathbf{J}(z)} = gG_{\mathbf{J}(z)}g^{-1}$-orbit, is identical to the leaf through $h \cdot z$, since it is equal to the $G_{\mathbf{J}(h \cdot z)} = G_{h \cdot \mathbf{J}(z)} = hG_{\mathbf{J}(z)}h^{-1}$-orbit. This shows that the map $(\pi_{\mathcal{O}_\mu}(z), \nu) \in \mathbf{J}^{-1}(\mathcal{O}_\mu)/G \times \mathcal{O}_\mu \mapsto \pi_{\mathfrak{N}}(g \cdot z) \in \mathbf{J}^{-1}(\mathcal{O}_\mu)/\mathfrak{N}$, where $g \in G$ is such that $\nu = g \cdot \mathbf{J}(z)$, is well defined. It is straightforward to check that this is the inverse of χ.

Therefore $\chi : \mathbf{J}^{-1}(\mathcal{O}_\mu)/\mathfrak{N} \to M_{\mathcal{O}_\mu} \times \mathcal{O}_\mu$ is an invertible local diffeomorphism which implies that it is a diffeomorphism.

We shall show now that χ is a symplectic diffeomorphism. Recall that $\Omega_{\mathfrak{N}}$ and $\omega_{\mathcal{O}_\mu}$ are uniquely characterized by the relations

$$
\pi_{\mathfrak{N}}^*\Omega_{\mathfrak{N}} = i_{\mathcal{O}_\mu}^*\omega \quad \text{and} \quad i_{\mathcal{O}_\mu}^*\omega = \pi_{\mathcal{O}_\mu}^*\omega_{\mathcal{O}_\mu} + \mathbf{J}_{\mathcal{O}_\mu}^*\omega_{\mathcal{O}_\mu}^+.
$$

Thus, from the commutativity of the diagram and the obvious identity $\left(\pi_{\mathcal{O}_\mu} \times \mathrm{id}_{\mathcal{O}_\mu}\right) \circ \left(\mathrm{id} \times \mathbf{J}_{\mathcal{O}_\mu}\right) = \pi_{\mathcal{O}_\mu} \times \mathbf{J}_{\mathcal{O}_\mu}$ we get from the definition of the symplectic sum $M_{\mathcal{O}_\mu} \oplus \mathcal{O}_\mu^+$

$$
\begin{aligned}
\pi_{\mathfrak{N}}^*\chi^*(\omega_{\mathcal{O}_\mu} \oplus \omega_{\mathcal{O}_\mu}^+) &= \left(\mathrm{id} \times \mathbf{J}_{\mathcal{O}_\mu}\right)^* \left(\pi_{\mathcal{O}_\mu} \times \mathrm{id}_{\mathcal{O}_\mu}\right)^* (\omega_{\mathcal{O}_\mu} \oplus \omega_{\mathcal{O}_\mu}^+) \\
&= \left(\pi_{\mathcal{O}_\mu} \times \mathbf{J}_{\mathcal{O}_\mu}\right)^* (\omega_{\mathcal{O}_\mu} \oplus \omega_{\mathcal{O}_\mu}^+) = \pi_{\mathcal{O}_\mu}^*\omega_{\mathcal{O}_\mu} + \mathbf{J}_{\mathcal{O}_\mu}^*\omega_{\mathcal{O}_\mu}^+ \\
&= i_{\mathcal{O}_\mu}^*\omega = \pi_{\mathfrak{N}}^*\Omega_{\mathfrak{N}}.
\end{aligned}
$$

Since $\pi_{\mathfrak{N}}$ is a surjective submersion, this implies $\chi^*(\omega_{\mathcal{O}_\mu} \oplus \omega^+_{\mathcal{O}_\mu}) = \Omega_{\mathfrak{N}}$, which proves the theorem. ■

The proof of Theorem 6.3.4 shows that $\chi : \mathbf{J}^{-1}(\mathcal{O}_\mu)/\mathfrak{N} \to M_{\mathcal{O}_\mu} \times \mathcal{O}_\mu$ is a diffeomorphism. It turns out that it can be used to give an alternative characterization of the orbit reduced symplectic form $\omega_{\mathcal{O}_\mu}$ on $M_{\mathcal{O}_\mu}$ if G_μ is connected.

6.3.5 Theorem *Assume the hypotheses of Theorem* 6.3.4. *Then the orbit reduced symplectic form $\omega_{\mathcal{O}_\mu}$ is the unique symplectic form on on $M_{\mathcal{O}_\mu}$ such that the diffeomorphism $\chi : \left(\mathbf{J}^{-1}(\mathcal{O}_\mu)/\mathfrak{N}, \Omega_{\mathfrak{N}}\right) \to M_{\mathcal{O}_\mu} \oplus \mathcal{O}^+_\mu$ is symplectic.*

Proof. Theorem 6.3.1 endows the quotient manifold $M_{\mathcal{O}_\mu} := \mathbf{J}^{-1}(\mathcal{O}_\mu)/G$ with a symplectic form $\omega_{\mathcal{O}_\mu}$ uniquely characterized by the relation (6.3.1). Theorem 6.3.4 shows that χ is always a diffeomorphism and that, if $M_{\mathcal{O}_\mu}$ carries the symplectic form $\omega_{\mathcal{O}_\mu}$, it is symplectic.

To prove uniqueness, let $\overline{\omega}_{\mathcal{O}_\mu}$ be another symplectic form on $M_{\mathcal{O}_\mu}$ such that $\chi^*(\overline{\omega}_{\mathcal{O}_\mu} \oplus \omega^+_{\mathcal{O}_\mu}) = \Omega_{\mathfrak{N}}$. But then

$$\begin{aligned} i^*_{\mathcal{O}_\mu}\omega &= \pi^*_{\mathfrak{N}}\Omega_{\mathfrak{N}} = \pi^*_{\mathfrak{N}}\chi^*(\overline{\omega}_{\mathcal{O}_\mu} \oplus \omega^+_{\mathcal{O}_\mu}) \\ &= \left(\mathrm{id} \times \mathbf{J}_{\mathcal{O}_\mu}\right)^* \left(\pi_{\mathcal{O}_\mu} \times \mathrm{id}_{\mathcal{O}_\mu}\right)^* (\overline{\omega}_{\mathcal{O}_\mu} \oplus \omega^+_{\mathcal{O}_\mu}) \\ &= \left(\pi_{\mathcal{O}_\mu} \times \mathbf{J}_{\mathcal{O}_\mu}\right)^* (\overline{\omega}_{\mathcal{O}_\mu} \oplus \omega^+_{\mathcal{O}_\mu}) = \pi^*_{\mathcal{O}_\mu}\overline{\omega}_{\mathcal{O}_\mu} + \mathbf{J}^*_{\mathcal{O}_\mu}\omega^+_{\mathcal{O}_\mu} \end{aligned}$$

which is precisely (6.3.1) for $\overline{\omega}_{\mathcal{O}_\mu}$ instead of $\omega_{\mathcal{O}_\mu}$. Thus $\pi^*_{\mathcal{O}_\mu}(\overline{\omega}_{\mathcal{O}_\mu} - \omega_{\mathcal{O}_\mu}) = 0$ which implies that $\overline{\omega}_{\mathcal{O}_\mu} = \omega_{\mathcal{O}_\mu}$ since $\pi_{\mathcal{O}_\mu}$ is a surjective submersion. ■

6.4 The regular reduction diagram

As already announced, the two approaches to reduction, namely, point and orbit reduction, are equivalent, as the following theorem shows.

6.4.1 Theorem (The Reduction Diagram) *Let (M, ω) be a connected symplectic manifold and let G be a Lie group acting canonically, freely, and properly on M. Suppose that this action has an associated momentum map $\mathbf{J} : M \to \mathfrak{g}^*_m$ with nonequivariance cocycle σ. Let $\mu \in \mathfrak{g}^*$ be a value of $\mathbf{J}$. Let $l_\mu : \mathbf{J}^{-1}(\mu) \hookrightarrow \mathbf{J}^{-1}(\mathcal{O}_\mu)$ be the canonical injection. Then the induced map $L_\mu : M_\mu \to M_{\mathcal{O}_\mu}$ defined by the commutative diagram*

$$\begin{array}{ccc} \mathbf{J}^{-1}(\mu) & \xrightarrow{\quad l_\mu \quad} & \mathbf{J}^{-1}(\mathcal{O}_\mu) \\ \pi_\mu \Big\downarrow & & \Big\downarrow \pi_{\mathcal{O}_\mu} \\ M_\mu & \xrightarrow{\quad L_\mu \quad} & M_{\mathcal{O}_\mu} \end{array}$$

is a symplectic diffeomorphism between (M_μ, ω_μ) and $(M_{\mathcal{O}_\mu}, \omega_{\mathcal{O}_\mu})$.

Proof Notice first that the smoothness of l_μ implies that of L_μ. We now show that the map L_μ is bijective. First, it is one-to-one because if $[z]_\mu := \pi_\mu(z)$, $[z']_\mu := \pi_\mu(z') \in M_\mu$ are such that $[z]_{\mathcal{O}_\mu} = L_\mu([z]_\mu) = L_\mu([z']_\mu) = [z']_{\mathcal{O}_\mu}$, then there exists an element $g \in G$ such that $z = g \cdot z'$. If we apply $\mathbf{J}$ to both sides of this equality, we obtain that $\mu = g \cdot \mu$, hence $g \in G_\mu$ and hence $[z]_\mu = [z']_\mu$. The surjectivity of L_μ is guaranteed by the fact that $\mathbf{J}^{-1}(\mathcal{O}_\mu) = G \cdot \mathbf{J}^{-1}(\mu)$.

We now check that L_μ is a symplectic map, that is, it satisfies $L_\mu^* \omega_{\mathcal{O}_\mu} = \omega_\mu$. Since π_μ is a surjective submersion it suffices to show that $\pi_\mu^* L_\mu^* \omega_{\mathcal{O}_\mu} = \pi_\mu^* \omega_\mu$ or, equivalently, that $(\pi_{\mathcal{O}_\mu} \circ l_\mu)^* \omega_{\mathcal{O}_\mu} = \pi_\mu^* \omega_\mu$. Since $i_{\mathcal{O}_\mu} \circ l_\mu = i_\mu : \mathbf{J}^{-1}(\mu) \hookrightarrow M$ and $\mathbf{J}_{\mathcal{O}_\mu} \circ l_\mu = \mu$, formula (6.3.1) yields for any $m \in \mathbf{J}^{-1}(\mu)$ and $v, w \in T_m\mathbf{J}^{-1}(\mu)$

$$
\begin{aligned}
\left((\pi_{\mathcal{O}_\mu} \circ l_\mu)^* \omega_{\mathcal{O}_\mu}\right)(m)(v, w) &= l_\mu^*(i_{\mathcal{O}_\mu}^* \omega - \mathbf{J}_{\mathcal{O}_\mu}^* \omega_{\mathcal{O}_\mu}^+)(m)(v, w) \\
&= \omega(m)(v, w) = \left(\pi_\mu^* \omega_\mu\right)(m)(v, w)
\end{aligned}
$$

which proves the claim.

Since L_μ is a symplectic map, it is necessarily an immersion. However, from (6.3.3) it follows that the spaces M_μ and $M_{\mathcal{O}_\mu}$ have both dimension equal to $\dim M - \dim G - \dim G_\mu$. Thus, the symplectic bijection L_μ is necessarily a local diffeomorphism and therefore is a diffeomorphism. ■

The smooth inverse $F_\mu : M_{\mathcal{O}_\mu} \to M_\mu$ of L_μ is defined in the following manner. Let $z \in M_{\mathcal{O}_\mu}$ be arbitrary. Since $\mathbf{J}^{-1}(\mathcal{O}_\mu) = G \cdot \mathbf{J}^{-1}(\mu)$ there is a point $m \in \mathbf{J}^{-1}(\mu)$ such that $z = \pi_{\mathcal{O}_\mu}(m)$. Define $F_\mu(z) := \pi_\mu(m)$. An immediate inspection shows that F_μ is well defined and that it is the inverse of L_μ. Since L_μ is a diffeomorphism, so is F_μ.

6.5 Reduction by shift

Reduction at a general point can be replaced by reduction at zero at the expense of enlarging the manifold by the affine orbit. This result is often quoted when saying that it suffices to perform reduction at zero and that one need not worry about reduction at an arbitrary point. This is only partially true since the enlarging of the manifold by the affine orbit changes the original phase space of the problem under consideration and it also introduces topological complications inherent to the affine orbit. This section presents this result in the regular case.

6.5.1 A momentum map on the affine orbit. Let $\Phi : G \times M \to M$ be a canonical action of the Lie group G on the connected symplectic manifold (M, ω). Suppose that this action has an associated momentum map $\mathbf{J} : M \to \mathfrak{g}^*$, with non-equivariance group one-cocycle $\sigma : G \to \mathfrak{g}^*$ and corresponding non-equivariance Lie algebra two-cocycle $\Sigma : \mathfrak{g} \times \mathfrak{g} \to \mathbb{R}$ (see Proposition 4.5.21 and 4.5.25). Recall from Definition 4.5.23 that the affine action of G on $\mathfrak{g}^*$ is given by $g \cdot \mu := \mathrm{Ad}_{g^{-1}}^* \mu + \sigma(g)$. By Theorem 4.5.31, the orbit $\mathcal{O}_\mu$ of the affine action admits the G-invariant symplectic forms

$$
\omega_{\mathcal{O}_\mu}^\pm(\nu)(\xi_{\mathfrak{g}^*}(\nu), \eta_{\mathfrak{g}^*}(\nu)) = \pm\langle \nu, [\xi, \eta]\rangle \mp \Sigma(\xi, \eta),
$$

where, by (4.5.24), $\xi_{\mathfrak{g}^*}(\nu) = -\operatorname{ad}^*_\xi \nu + \Sigma(\xi, \cdot)$ and $\eta_{\mathfrak{g}^*}(\nu) = -\operatorname{ad}^*_\eta \nu + \Sigma(\eta, \cdot)$ are the infinitesimal generators of the affine action defined by $\xi, \eta \in \mathfrak{g}$, respectively.

A momentum map of the affine G-action on $\mathcal{O}^\pm_\mu$ is $\pm$ the inclusion. Indeed, if $i : \mathcal{O}_\mu \hookrightarrow \mathfrak{g}^*$, then for any $\xi, \eta \in \mathfrak{g}$ we have $i^\xi(\nu) = \langle \nu, \xi \rangle$ so that

$$\begin{aligned}\left(\mathbf{i}_{\xi_{\mathfrak{g}^*}} \omega^\pm_{\mathcal{O}_\mu}\right)(\nu)(\eta_{\mathfrak{g}^*}(\nu)) &= \pm\langle \nu, [\xi, \eta]\rangle \mp \Sigma(\xi, \eta) = \mp\langle \nu, [\eta, \xi]\rangle \pm \Sigma(\eta, \xi)\\ &= \pm\langle -\operatorname{ad}^*_\eta \nu + \Sigma(\eta, \cdot), \xi\rangle = \pm\langle \eta_{\mathfrak{g}^*}(\nu), \xi\rangle\\ &= \pm\mathbf{d}i^\xi(\nu)\left(\eta_{\mathfrak{g}^*}(\nu)\right),\end{aligned}$$

which proves that $\xi_{\mathfrak{g}^*}(\nu) = X_{\pm i^\xi}$.

In what follows, we shall work only with $\mathcal{O}^+_\mu$ for simplicity. In this case, we will take the inclusion $i : \mathcal{O}_\mu \hookrightarrow \mathfrak{g}^*$ as the momentum map of the affine action on the affine orbit $\mathcal{O}^+_\mu$.

Consider the canonical diagonal action of G on the symplectic difference $M \ominus \mathcal{O}^+_\mu$. A momentum map is hence given by $\mathbf{J} \circ \pi_1 - \pi_2 : M \ominus \mathcal{O}^+_\mu \to \mathfrak{g}^*$, where $\pi_1 : M \times \mathcal{O}_\mu \to M$ and $\pi_2 : M \times \mathcal{O}_\mu \to \mathcal{O}_\mu$ are the projections. Let $(M \ominus \mathcal{O}^+_\mu)_0 := ((\mathbf{J} \circ \pi_1 - \pi_2)^{-1}(0)/G, (\omega \ominus \omega^+_{\mathcal{O}_\mu})_0)$ be the symplectic point reduced space at zero.

6.5.2 Theorem (Shifting theorem) *Under the hypotheses of the Symplectic Orbit Reduction Theorem* 6.3.1, *the symplectic orbit reduced space $M_{\mathcal{O}_\mu}$, the point reduced spaces M_μ, and $(M \ominus \mathcal{O}^+_\mu)_0$ are symplectically diffeomorphic.*

Proof Note that if $z \in \mathbf{J}^{-1}(\mathcal{O}_\mu)$, then $(z, \mathbf{J}(z)) \in (\mathbf{J} \circ \pi_1 - \pi_2)^{-1}(0)$. Conversely, if $(z, \nu) \in (\mathbf{J} \circ \pi_1 - \pi_2)^{-1}(0)$, then $\mathbf{J}(z) = \nu \in \mathcal{O}_\mu$, that is, $z \in \mathbf{J}^{-1}(\mathcal{O}_\mu)$. Thus there are smooth G-equivariant maps $\varphi : z \in \mathbf{J}^{-1}(\mathcal{O}_\mu) \mapsto (z, \mathbf{J}(z)) \in (\mathbf{J} \circ \pi_1 - \pi_2)^{-1}(0)$ and $\pi_1|_{(\mathbf{J}\circ\pi_1 - \pi_2)^{-1}(0)} : (\mathbf{J} \circ \pi_1 - \pi_2)^{-1}(0) \to \mathbf{J}^{-1}(\mathcal{O}_\mu)$ which are inverses of each other. They induce therefore smooth maps on the quotients $\Phi : M_{\mathcal{O}_\mu} \to (M \ominus \mathcal{O}^+_\mu)_0$ and $\Psi : (M \ominus \mathcal{O}^+_\mu)_0 \to M_{\mathcal{O}_\mu}$ which are also inverses of each other. The mappings needed in the subsequent computation are summarized in the diagram below; they respect the conventions of the Symplectic Point and Orbit Reduction Theorems 6.1.1 and 6.3.1.

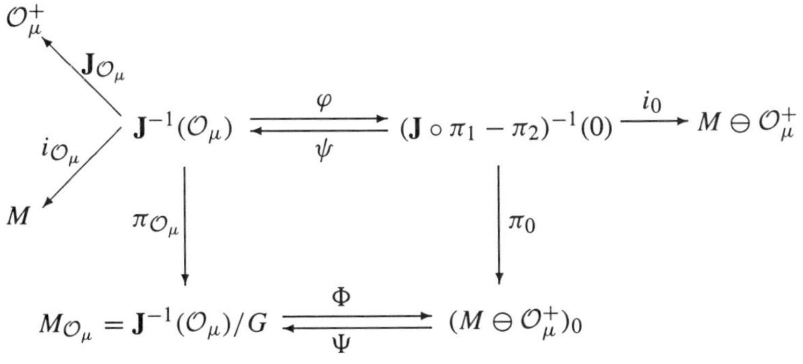

We shall prove now that $\Phi^*(\omega \ominus \omega^+_{\mathcal{O}_\mu})_0 = \omega_{\mathcal{O}_\mu}$ which, by the Symplectic Orbit Reduction Theorem 6.3.1, is equivalent to $\pi^*_{\mathcal{O}_\mu}\Phi^*(\omega \ominus \omega^+_{\mathcal{O}_\mu})_0 = \pi^*_{\mathcal{O}_\mu}\omega_{\mathcal{O}_\mu} = i^*_{\mathcal{O}_\mu}\omega -$

$\mathbf{J}^*_{\mathcal{O}_\mu}\omega^+_{\mathcal{O}_\mu}$, since $\pi_{\mathcal{O}_\mu}$ is a surjective submersion. Since $\Phi \circ \pi_{\mathcal{O}_\mu} = \pi_0 \circ \varphi$, this in turn is equivalent to $i^*_{\mathcal{O}_\mu}\omega - \mathbf{J}^*_{\mathcal{O}_\mu}\omega^+_{\mathcal{O}_\mu} = \varphi^*\pi_0^*(\omega \ominus \omega^+_{\mathcal{O}_\mu})_0 = \varphi^* i_0^*(\omega \ominus \omega^+_{\mathcal{O}_\mu})$, by the Symplectic Point Reduction Theorem 6.1.1. This last formula is proved in the following way. If $z \in \mathbf{J}^{-1}(\mathcal{O}_\mu)$ and $u, v \in T_z(\mathbf{J}^{-1}(\mathcal{O}_\mu))$, then

$$
\begin{aligned}
\left(i^*_{\mathcal{O}_\mu}\omega - \mathbf{J}^*_{\mathcal{O}_\mu}\omega^+_{\mathcal{O}_\mu}\right)(z)(u, v) &= \omega(z)(u, v) - \omega^+_{\mathcal{O}_\mu}(\mathbf{J}(z))\left(T_z\mathbf{J}(u), T_z\mathbf{J}(v)\right) \\
&= \left(\omega \ominus \omega^+_{\mathcal{O}_\mu}\right)(z, \mathbf{J}(z))\left((u, T_z\mathbf{J}(u)), (v, T_z\mathbf{J}(v))\right) \\
&= \left(\varphi^* i_0^*(\omega \ominus \omega^+_{\mathcal{O}_\mu})\right)(z)(u, v).
\end{aligned}
$$

This shows that $\Phi : M_{\mathcal{O}_\mu} \to (M \ominus \mathcal{O}^+_\mu)_0$ is a symplectic diffeomorphism. Theorem 6.4.1 implies the rest of the proof. ■

6.6 Cotangent bundle reduction

In this section we record, for completeness and without any proofs, the theory of point reduction for the special case of cotangent bundles and lifted actions. The general reduction theory presented above becomes much richer in this case and is extremely important in applications. We shall not pursue the singular case in this book because the general theory of singular cotangent bundle reduction has not yet been developed. From several examples scattered in the literature, it is already clear that the theorems in this section should have analogues in the singular case. The full proofs of the statements below can be found in MARSDEN AND RATIU (2003).

Let $\Phi : G \times Q \to Q$ be a left smooth action of the Lie group G on the manifold and Q. The lifted left action $G \times T^*Q \to T^*Q$, given by $g \cdot \alpha_q = T^*_{g\cdot q}\Phi_{g^{-1}}(\alpha_q)$ for $g \in G$ and $\alpha_q \in T^*_q Q$, admits the equivariant momentum map $\mathbf{J} : T^*Q \to \mathfrak{g}^*$ whose expression is $\langle \mathbf{J}(\alpha_q), \xi\rangle = \alpha_q((\xi_Q(q))$, where $\xi \in \mathfrak{g}$, the Lie algebra of G. Adopting the conventions introduced in Example 4.1.3, $\Omega_Q = -\mathbf{d}\Theta_Q$ will denote the canonical symplectic form on T^*Q.

To build up to the fully general reduction theorems we shall present first the special but very important case of cotangent bundle reduction at the zero value of the momentum map.

6.6.1 Theorem (Cotangent bundle reduction at zero) *Assume that the action of G on Q is free and proper, so that the quotient Q/G is a smooth manifold and the projection $\rho : Q \to Q/G$ is a submersion. Then 0 is a regular value of $\mathbf{J}$ and the map*

$$
\varphi_0 : \left((T^*Q)_0, (\Omega_Q)_0\right) \to \left(T^*(Q/G), \Omega_{Q/G}\right)
$$

given by $\varphi_0([\alpha_q])(T_q\rho(v_q)) := \alpha_q(v_q)$, where $\alpha_q \in \mathbf{J}^{-1}(0)$ and $v_q \in T_qQ$, is a well-defined symplectic diffeomorphism.

There are two versions of the cotangent bundle reduction theorem, one involving an embedding, the other a fibration. Both are highly nontrivial generalizations of the previous theorem. We begin with the embedding theorem.

6.6.2 Embedding version of cotangent bundle reduction. Assume that $\mu \in \mathfrak{g}^*$ is a regular value of $\mathbf{J}$, that $Q_\mu := Q/G_\mu$ is a smooth manifold, and that the canonical projection $\rho_\mu : Q \to Q_\mu$ is a surjective submersion. This is assured if the coadjoint isotropy group G_μ acts freely and properly on Q, a hypotheses that we shall assume in this subsection. Then G_μ acts also freely and properly on $\mathbf{J}^{-1}(\mu)$ so that the reduced phase space $((T^*Q)_\mu, \omega_\mu)$ is a symplectic manifold.

For $\mu \in \mathfrak{g}^*$, let G_μ be the coadjoint isotropy subgroup, let $\mathfrak{g}_\mu := \{\xi \in \mathfrak{g} \mid \mathrm{ad}_\xi^* \mu = 0\}$ be its Lie algebra, $\mu' := \mu|\mathfrak{g}_\mu \in \mathfrak{g}_\mu^*$ the restriction of μ to $\mathfrak{g}_\mu$, and consider the G_μ-action on Q and its lift to T^*Q. An equivariant momentum map of this action is the map $\mathbf{J}^\mu : T^*Q \to \mathfrak{g}_\mu$ obtained by restricting $\mathbf{J}$, that is, $\mathbf{J}^\mu(\alpha_q) = \mathbf{J}(\alpha_q)|\mathfrak{g}_\mu$. Assume there is a G_μ-invariant one-form α_μ on Q with values in $(\mathbf{J}^\mu)^{-1}(\mu')$.

For $\xi \in \mathfrak{g}_\mu$ and $q \in Q$, the identity $(\mathbf{i}_{\xi_Q}\alpha_\mu)(q) = \alpha_\mu(q)(\xi_Q(q)) = \langle \mathbf{J}(\alpha_\mu(q)), \xi\rangle = \langle \mu', \xi\rangle$ shows that $\mathbf{i}_{\xi_Q}\alpha_\mu$ is a constant function on Q. Therefore, for $\xi \in \mathfrak{g}_\mu$, this implies $\mathbf{i}_{\xi_Q}\mathbf{d}\alpha_\mu = \pounds_{\xi_Q}\alpha_\mu - \mathbf{d}\mathbf{i}_{\xi_Q}\alpha_\mu = 0$, since $\pounds_{\xi_Q}\alpha_\mu = 0$ by G_μ-invariance of α_μ. It follows that there is a unique two-form β_μ on Q_μ such that $\rho_\mu^*\beta_\mu = \mathbf{d}\alpha_\mu$. Since ρ_μ is a submersion, β_μ is closed, but need not be exact. Let $B_\mu = \pi_{Q_\mu}^*\beta_\mu$ where $\pi_{Q_\mu} : T^*Q_\mu \to Q_\mu$ is the cotangent bundle projection. The following theorem is due to Marsden (see ABRAHAM AND MARSDEN (1978), page 300).

6.6.3 Theorem (Embedding cotangent bundle reduction) *Under the above hypotheses, the map*

$$\varphi_\mu : ((T^*Q)_\mu, (\Omega_Q)_\mu) \to (T^*Q_\mu, \Omega_{Q_\mu} - B_\mu),$$

given by $\varphi_\mu([\alpha_q])(T_q\rho_\mu(v_q)) := (\alpha_q - \alpha_\mu(q))(v_q)$, *for* $\alpha_q \in \mathbf{J}^{-1}(\mu)$, $v_q \in T_qQ$, *is a symplectic embedding onto a vector subbundle of* T^*Q_μ. *The map* φ_μ *is onto* T^*Q_μ *if and only if* $\mathfrak{g} = \mathfrak{g}_\mu$. *The additional summand* B_μ *in the symplectic structure of* T^*Q_μ *is called a* ***magnetic term***.

If $\mathfrak{g}$ is Abelian or $\mu = 0$, the embedding φ_μ is onto and thus the reduced space is again, topologically, a cotangent bundle. The Abelian case is due to SATZER (1977).

It should be noted that there is a choice in this theorem, namely the one-form α_μ. Whereas the reduced symplectic space $((T^*Q)_\mu, (\Omega_Q)_\mu)$ is intrinsic, the symplectic structure on the space T^*Q_μ depends on α_μ. The theorem states that no matter how α_μ is chosen, there is a symplectic diffeomorphism, which also depends on α_μ, of the reduced space onto a vector subbundle of T^*Q_μ. We shall comment on possible choices of such a form later.

6.6.4 The Kaluza–Klein construction. To understand why B_μ is called a *magnetic term*, consider the problem of a particle of mass m and charge e moving in $\mathbb{R}^3$ under the influence of a given magnetic field $\mathbf{B} = B_x\mathbf{i} + B_y\mathbf{j} + B_z\mathbf{k}$, $\operatorname{div}\mathbf{B} = 0$. The Lorentz Force Law gives the equations of motion

$$m\frac{\mathbf{v}}{dt} = \frac{e}{c}\mathbf{v} \times \mathbf{B}, \tag{6.6.1}$$

where c is the speed of light, and $\mathbf{v} = (\dot{x}, \dot{y}, \dot{z}) = \dot{\mathbf{q}}$ is the velocity of the particle. What is the Hamiltonian description of these equations?

There are two possible answers to this question. To formulate them, associate to the divergence free vector field $\mathbf{B}$ the closed two-form $B = B_x dy \wedge dz - B_y dx \wedge dz + B_z dx \wedge dy$. Also, write $\mathbf{B} = \operatorname{curl} \mathbf{A}$ for some other vector field $\mathbf{A} = (A_x, A_y, A_z)$, called the ***magnetic potential***

Answer 1. Take on $T^*\mathbb{R}^3$ the symplectic form $\Omega_B = dx \wedge dp_x + dy \wedge dp_y + dz \wedge dp_z - \frac{e}{c} B$, where $(p_x, p_y, p_z) = \mathbf{p} := m\mathbf{v}$ is the momentum of the particle, and the Hamiltonian $h = m(\dot{x}^2 + \dot{y}^2 + \dot{z}^2)/2$. A direct verification shows that $\mathbf{d}h = \Omega_B(X_h, \cdot)$, where

$$\begin{aligned} X_h = \dot{x}\frac{\partial}{\partial x} + \dot{y}\frac{\partial}{\partial y} + \dot{z}\frac{\partial}{\partial x} + \frac{e}{c}(B_z\dot{y} - B_y\dot{z})\frac{\partial}{\partial p_x} \\ + \frac{e}{c}(B_x\dot{z} - B_z\dot{x})\frac{\partial}{\partial p_y} + \frac{e}{c}(B_y\dot{x} - B_z\dot{x})\frac{\partial}{\partial p_z}, \end{aligned} \tag{6.6.2}$$

which gives the equations of motion (6.6.1).

Answer 2. Take on $T^*\mathbb{R}^3$ the canonical symplectic form $\Omega = dx \wedge dp_x + dy \wedge dp_y + dz \wedge dp_z$ and the Hamiltonian $h_A = \|\mathbf{p} - \frac{e}{c}\mathbf{A}\|^2/2m$. A direct verification shows that $\mathbf{d}h_A = \Omega(X_{h_A}, \cdot)$, where X_{h_A} has the *same expression* (6.6.2). In MARSDEN AND RATIU (1999) §6.6, 6.7, it is shown why this is a general phenomenon using the momentum shifting lemma.

Next we show how the magnetic term in the symplectic form Ω_B is obtained by reduction from the ***Kaluza–Klein*** system. Let $Q = \mathbb{R}^3 \times S^1$ with the circle $G = S^1$ acting on Q, only on the second factor. Since the infinitesimal generator of this action defined by $\xi \in \mathfrak{g} = \mathbb{R} = Lie(S^1)$ has the expression $\xi_Q(\mathbf{q}, \theta) = (\mathbf{q}, \theta; \mathbf{0}, \xi)$, if TS^1 is trivialized as $S^1 \times \mathbb{R}$, a momentum map $\mathbf{J} : T^*Q = \mathbb{R}^3 \times S^1 \times \mathbb{R}^3 \times \mathbb{R} \to \mathfrak{g}^* = \mathbb{R}$ is given by $\mathbf{J}(\mathbf{q}, \theta; \mathbf{p}, p)\xi = (\mathbf{p}, p) \cdot (\mathbf{0}, \xi) = p\xi$, that is, $\mathbf{J}(\mathbf{q}, \theta; \mathbf{p}, p) = p$. In this case the coadjoint action is trivial, so for any $\mu \in \mathfrak{g}^* = \mathbb{R}$, we have $G_\mu = S^1$, $\mathfrak{g}_\mu = \mathbb{R}$, and $\mu' = \mu$. The one-form $\alpha_\mu = \mu(A_x dx + A_y dy + A_z dz + d\theta) \in \Omega^1(Q)$, where $d\theta$ denotes the length one-form on S^1, is clearly $G_\mu = S^1$-invariant, has values in $\mathbf{J}^{-1}(\mu) = \{(\mathbf{q}, \theta; \mathbf{p}, \mu) \mid \mathbf{q}, \mathbf{p} \in \mathbb{R}^3, \theta \in S^1\}$, and its exterior differential equals $\mathbf{d}\alpha_\mu = \mu B$. Thus the closed two-form β_μ on the base $Q_\mu = Q/G_\mu = Q/S^1 = \mathbf{R}^3$ equals μB and hence the magnetic term, that is, the closed two-form $B_\mu = \pi^*_{Q_\mu}\beta_\mu$ in Theorem 6.6.3, is also μB since $\pi_{Q_\mu} : Q = \mathbb{R}^3 \times S^1 \to Q/G_\mu = \mathbb{R}^3$ is the projection. Therefore, the reduced space $(T^*Q)_\mu$ is symplectically diffeomorphic to $(T^*\mathbb{R}^3, dx \wedge dp_x + dy \wedge dp_y + dz \wedge dp_z - \mu B)$, which coincides with the phase space in *Answer 1* if we put $\mu = e/c$. *The magnetic term in the symplectic form is, up to a factor, the magnetic field.*

The kinetic energy Hamiltonian $h(\mathbf{q}, \theta; \mathbf{p}, p) := \frac{1}{2m}\|\mathbf{p}\|^2 + \frac{1}{2}p^2$ of the Kaluza–Klein metric, that is, the Riemannian metric obtained by keeping the standard metrics on each factor and declaring $\mathbb{R}^3$ and S^1 orthogonal, induces the reduced Hamiltonian $h_\mu(\mathbf{q}) = \frac{1}{2m}\|\mathbf{p}\|^2 + \frac{1}{2}\mu^2$ which, up to the constant $\mu^2/2$, equals the kinetic energy Hamiltonian in *Answer 1*. Note that this reduced system is *not* the geodesic flow of the Euclidean metric because of the presence of the magnetic term in the symplectic form. However, *the equations of motion of a charged particle in a magnetic field are obtained by reducing the geodesic flow of the Kaluza–Klein metric.*

A similar construction is carried out in Yang–Mills theory where A is a connection on a principal bundle and B is its curvature. Magnetic terms appear also in classical mechanics. For example, in rotating systems the Coriolis force plays this role.

6.6.5 Connections and magnetic terms. Prior to Theorem 6.6.3 it has been shown that $\mathbf{i}_{\xi_Q}\mathbf{d}\alpha_\mu = 0$ for any $\xi \in \mathfrak{g}_\mu$. This means that $\mathbf{d}\alpha_\mu$ is a *horizontal* two-form on the principal G_μ-bundle $\rho_\mu : Q \to Q_\mu$. Thus, relative to *any* connection on this bundle, the two-form $\mathbf{d}\alpha_\mu$ coincides with the covariant derivative of α_μ, that is, it equals the μ-component of the curvature of the connection (see §5.1.1 for a review of connections). It turns out that this observation gives a wide class of choices for the form α_μ in Theorem 6.6.3. The following result is due to KUMMER (1981).

6.6.6 Proposition *Let $\mathcal{A} \in \Omega^1(Q; \mathfrak{g}_\mu)$ be a connection one-form on the principal G_μ-bundle $\rho_\mu : Q \to Q_\mu$. Then the one-form $\alpha_\mu \in \Omega^1(Q)$ appearing in the statement of Theorem* 6.6.3 *can be chosen to equal $\alpha_\mu(q) := \mathcal{A}(q)^*\mu'$, where $\mu' := \mu|_{\mathfrak{g}_\mu}$. Thus the magnetic term B_μ is the pull-back to T^*Q_μ of the μ'-component of the curvature of $\mathcal{A}$, thought of as a two-form on the base Q_μ.*

Proof We need to show that the one-form $\alpha_\mu(q) := \mathcal{A}(q)^*\mu'$ is G_μ-invariant and has values in $(\mathbf{J}^\mu)^{-1}(\mu')$, where $\mathbf{J}^\mu : T^*Q \to \mathfrak{g}_\mu$ is given by $\mathbf{J}^\mu(\alpha_q) = \mathbf{J}(\alpha_q)|_{\mathfrak{g}_\mu}$.

Denote also by $\Phi : G_\mu \times Q \to Q$ the induced G_μ-action on Q. The equivariance property of a left connection one-form is

$$\mathcal{A}(g \cdot q)(T_q\Phi_g(v_q)) = \mathrm{Ad}_g(\mathcal{A}(q)(v_q)),$$

where $g \in G_\mu$ and $v_q \in T_qQ$; see §5.1.1 for the definition of a *right* connection. Then, for any $g \in G_\mu$ and $v_q \in T_qQ$ we have

$$\begin{aligned}(\Phi_g^*\alpha_\mu)(v_q) &= \alpha_\mu(g \cdot q)(T_q\Phi_g(v_q)) = \mathcal{A}(g \cdot q)^*(\mu')(T_q\Phi_g(v_q))\\ &= \left\langle \mu', \mathcal{A}(g \cdot q)(T_q\Phi_g(v_q))\right\rangle = \left\langle \mu', \mathrm{Ad}_g(\mathcal{A}(q)(v_q))\right\rangle\\ &= \left\langle \mathrm{Ad}_g^*\,\mu', \mathcal{A}(q)(v_q)\right\rangle = \left\langle \mu', \mathcal{A}(q)(v_q)\right\rangle = \mathcal{A}(q)^*\mu'(v_q)\\ &= \alpha_\mu(q)(v_q),\end{aligned}$$

which shows that α_μ is G_μ-invariant.

Next, the first property of a connection yields for any $\xi \in \mathfrak{g}_\mu$,

$$\begin{aligned}\left\langle \mathbf{J}^\mu(\alpha_\mu(q)), \xi\right\rangle &= \left\langle \alpha_\mu(q), \xi_Q(q)\right\rangle = \left\langle \mathcal{A}(q)^*(\mu'), \xi_Q(q)\right\rangle\\ &= \left\langle \mu', \mathcal{A}(q)(\xi_Q(q))\right\rangle = \left\langle \mu', \xi\right\rangle,\end{aligned}$$

which shows that α_μ takes values in $(\mathbf{J}^\mu)^{-1}(\mu')$.

The μ'-component of the connection $\mathcal{A}$ is by definition the one-form $\langle \mu', \mathcal{A}\rangle(q)(v_q) := \langle \mu', \mathcal{A}(q)(v_q)\rangle = \mathcal{A}(q)^*\mu'(v_q) = \alpha_\mu(q)(v_q)$, which shows that it equals α_μ. The curvature $\mathcal{B} \in \Omega^2(Q; \mathfrak{g}_\mu)$ of $\mathcal{A}$ is defined by formula (5.1.1), that is, $\mathcal{B}(q)(u_q, v_q) = \mathbf{d}\mathcal{A}(q)(u_q^H, v_q^H)$, where $u_q, v_q \in T_qQ$ and u_q^H, v_q^H denote the horizontal components of the vectors u_q and v_q respectively. Thus, since $\langle \mu', \mathcal{A}\rangle = \alpha_\mu$, the μ'-component of the curvature equals

$$\begin{aligned}\langle \mu', \mathcal{B}\rangle(q)(u_q, v_q) :&= \langle \mu, \mathcal{B}(q)(u_q, v_q)\rangle = \langle \mu', \mathbf{d}\mathcal{A}(q)(u_q^H, v_q^H)\rangle\\ &= \langle \mu', \mathbf{d}\mathcal{A}\rangle(q)(u_q^H, v_q^H) = \mathbf{d}(\langle \mu', \mathcal{A}\rangle)(q)(u_q^H, v_q^H)\\ &= \mathbf{d}\alpha_\mu(q)(u_q^H, v_q^H) = \mathbf{d}\alpha_\mu(q)(u_q, v_q)\end{aligned}$$

since $\mathbf{d}\alpha_\mu$ is a horizontal two-form. This shows that $\langle \mu', \mathcal{B} \rangle = \mathbf{d}\alpha_\mu$, which proves the proposition. ■

6.6.7 The amended potential. If the manifold Q is Riemannian, then a ***simple mechanical system*** is defined by requiring that the Hamiltonian h is equal to the kinetic energy k of the given metric plus a potential $v : Q \to \mathbb{R}$, that is, $h = k + v \circ \pi_Q$, where $\pi_Q : T^*Q \to Q$ is the cotangent bundle projection. Denoting the Riemannian metric on Q by $\langle\!\langle \cdot, \cdot \rangle\!\rangle$, define the vector bundle isomorphism $\flat : TQ \to T^*Q$ by $u_q^\flat := \langle\!\langle u_q, \cdot \rangle\!\rangle \in T_q^*Q$ for any $u_q \in T_qQ$. Let $\sharp : T^*Q \to TQ$ be the inverse of $\flat$. Then the kinetic energy $k : T^*Q \to \mathbb{R}$ is given by $k(\alpha_q) = \frac{1}{2}\langle\!\langle \alpha_q^\sharp, \alpha_q^\sharp \rangle\!\rangle$, for any $\alpha_q \in T_q^*Q$.

A ***simple mechanical system with symmetry*** is a simple mechanical system for which there is a Lie group G that acts smoothly on Q by isometries and the potential v is G-invariant. The ***amended*** or ***effective potential*** is defined by $v_\mu := h \circ \alpha_\mu$, where α_μ is the one-form associated to the ***mechanical connection***, which is the left G_μ-connection on the principal bundle $Q \to Q/G_\mu$ whose horizontal subbundle is the metric orthogonal to the vertical subbundle. The mechanical connection one-form $\mathcal{A}_{\text{mech}} \in \Omega^1(Q; \mathfrak{g}_\mu)$ is given by

$$\mathcal{A}_{\text{mech}}(q)(u_q) = \mathbb{I}_\mu(q)^{-1}\mathbf{J}(u_q^\flat), \tag{6.6.3}$$

where $q \in Q$, $u_q \in T_qQ$, $u_q^\flat := \langle\!\langle u_q, \cdot \rangle\!\rangle \in T_q^*Q$ is the covector associated to u_q by the Riemannian metric on Q, and the ***μ-locked inertia tensor*** $\mathbb{I}_\mu(q) : \mathfrak{g}_\mu \to \mathfrak{g}_\mu^*$ is defined for each $q \in Q$ by $\mathbb{I}_\mu(q)(\zeta)(\eta) := \langle\!\langle \zeta_Q(q), \eta_Q(q) \rangle\!\rangle$, for $\zeta, \eta \in \mathfrak{g}_\mu$. SMALE (1970) has uniquely characterized the one-form α_μ by the minimization problem

$$k(\alpha_\mu(q)) = \inf\{k(\alpha_q) \mid \alpha_q \in \mathbf{J}^{-1}(\mu)\}.$$

The amended potential is given in terms of the locked inertia tensor by

$$v_\mu(q) = \frac{1}{2}\langle \mu', \mathbb{I}_\mu(q)^{-1}\mu' \rangle + v(q);$$

it is G_μ-invariant so it induces a smooth function $\widehat{v}_\mu : Q/G_\mu \to \mathbb{R}$. The fundamental result about simple mechanical systems with symmetry is the following: *the push-forward by the embedding φ_μ given in Theorem* 6.6.3 *of the reduced Hamiltonian h_μ of a simple mechanical system $h = k + v \circ \pi_Q$ on T^*Q is the restriction to the vector subbundle $\varphi_\mu((T^*Q)_\mu) \subset T^*(Q/G_\mu)$, which is also a symplectic submanifold of $(T^*(Q/G_\mu), \Omega_{Q/G_\mu} - B_\mu)$, of the simple mechanical system on $T^*(Q/G_\mu)$ whose kinetic energy is given by the quotient Riemannian metric on Q/G_μ and whose potential is $\widehat{v}_\mu$. However, Hamilton's equations on $T^*(Q/G_\mu)$ for this simple mechanical system are computed relative to the magnetic symplectic form $\Omega_{Q/G_\mu} - B_\mu$.*

There is a wealth of applications starting from this classical theorem to mechanical systems, spanning such diverse areas as topological characterization of the level sets of the energy-momentum map to methods of proving nonlinear stability of relative equilibria; see SMALE (1970); ABRAHAM AND MARSDEN (1978); SIMO, LEWIS, AND MARSDEN (1991) and references therein.

6.6.8 Theorem (Fibration cotangent bundle reduction) *Assume that G acts freely and properly on Q. Then the reduced symplectic manifold $(T^*Q)_\mu$ is a fiber bundle over $T^*(Q/G)$ with fiber the coadjoint orbit $\mathcal{O}_\mu$.*

The diagram below illustrates and compares the two theorems. The left side of the diagram is the embedding cotangent bundle reduction theorem, whereas the right side is the fibration cotangent bundle theorem. It also gives a sketch of the proof of the fibration cotangent bundle theorem: the symplectic orbit reduction method in Theorem 6.3.1 and the reduction diagram in Theorem 6.4.1 are needed. Note that the arrows labeled respectively by "injection" and "projection" are *not* symplectic maps. For the embedding, the magnetic term will make the injection a symplectic map. For the projection, the situation is more involved. We shall briefly review the theory in the following two sections.

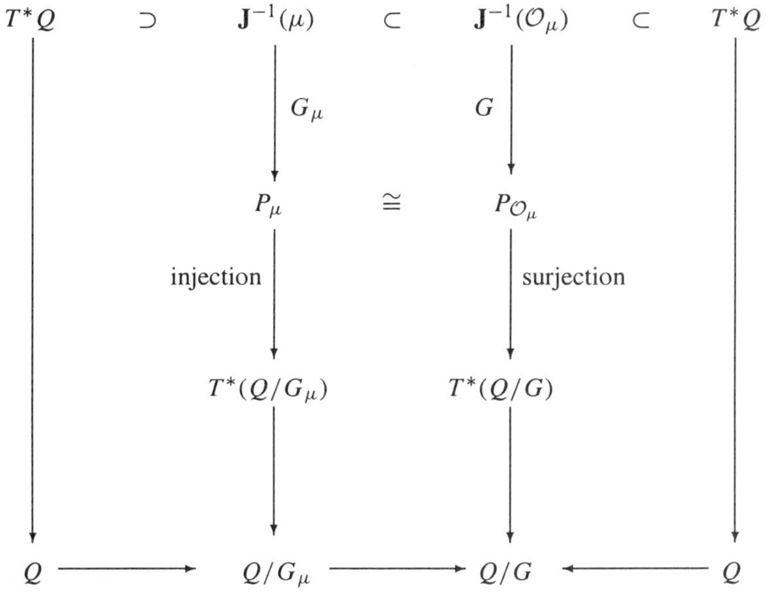

6.6.9 Gauged Lie–Poisson brackets. In the following subsections, the following question will be answered. Given is a Lie group G acting freely and properly on a smooth manifold Q. The lift of the G-action to T^*Q is also a free and proper action and thus both Q/G and $(T^*Q)/G$ are smooth manifolds. In addition, the manifold $(T^*Q)/G$ carries a natural Poisson bracket uniquely determined by the requirement that the canonical projection $T^*Q \to (T^*Q)/G$ be a Poisson map. Is there a way to give an *explicit* formula for this Poisson bracket, globally on this manifold or one diffeomorphic to it? It turns out that this question has two possible answers, once a connection on the principal bundle $\pi : Q \to Q/G$ is introduced. The results presented below are due to PERLMUTTER AND RATIU (2004); some formulas of the same type can already be found in MONTGOMERY et al. (1984) and ZAALANI (1999).

6.6.10 Pull-back commutes with association. In order to present these answers, a general remark about two-bundle constructions is in order. Let $\rho : P \to M$ be a left principal bundle with structure group G, where G acts freely and properly on P. The group operation will be denoted by concatenation: for $g \in G$ and $p \in P$ the symbol gp denotes the action of g on the point p. Let $\tau : N \to M$ be a surjective submersion and let $\tilde{P}$ denote the pull-back bundle over N given by

$$\tilde{P} = \{(n, p) \in N \times P \mid \rho(p) = \tau(n)\},$$

providing the following commutative diagram:

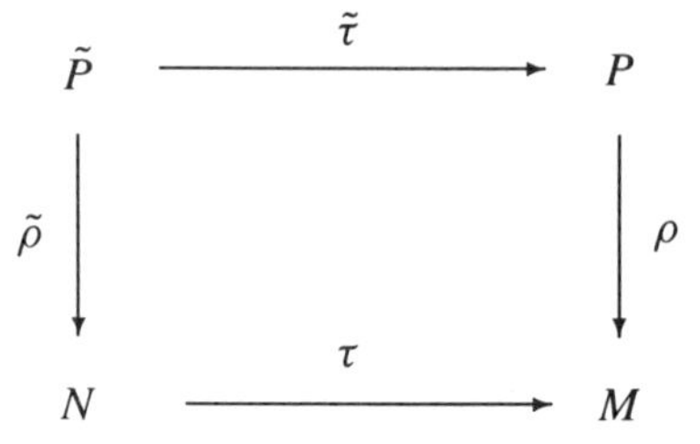

where $\tilde{\rho} : \tilde{P} \to N$ and $\tilde{\tau} : \tilde{P} \to P$ are the projections on the first and second factors. It is easily shown that $\tilde{P}$ is a smooth manifold of dimension $\dim P + \dim N - \dim M$ and that the free and proper G-action on P induces a free and proper G-action on $\tilde{P}$ given by

$$g \cdot (n, p) = (n, gp)$$

with respect to which $\tilde{\rho}$ is the quotient projection. Consequently, $\tilde{P}$ is a G-principal bundle with base N, $\tilde{\rho} : \tilde{P} \to N$ is the bundle projection, and the map $\tilde{\tau}$ is a submersion with fiber over the point $p \in P$ equal to $\tilde{\tau}^{-1}(p) = \{(n, p) \in N \times P \mid \rho(p) = \tau(n)\} = \tau^{-1}(\rho(p)) \times \{p\} \subset \tilde{P}$, and hence diffeomorphic to $\tau^{-1}(\rho(p))$.

Now suppose that there is a left action of G on a manifold V, also denoted by concatenation. There are two associated bundles that one can construct: $P \times_G V$ and $\tilde{P} \times_G V$. They are fiber bundles over M and N respectively, both with fibers diffeomorphic to V. The following conventions will be used for associated bundles: the Lie group G acts on the product $P \times V$ freely and properly by the diagonal action $g \cdot (p, v) = (gp, gv)$ for $g \in G$, $p \in P$, and $v \in V$. The associated bundle $P \times_G V$ is defined as the quotient manifold $(P \times V)/G$.

Summarizing, the associated bundle $\tilde{P} \times_G V \to N$ is obtained from the principal bundle $\rho : P \to M$, the surjective submersion $\tau : N \to M$, and the G-manifold V by pull-back and association, in this order.

These operations can be reversed. First, one-forms the associated bundle $\rho_E : [p, v] \in E := P \times_G V \mapsto \rho(p) \in M$ and then one pulls it back by the surjective submersion $\tau : N \to M$ to obtain the pull-back bundle $\tilde{\rho}_E : \tilde{E} \to N$, whose fibers are all diffeomorphic to V, defined by the following commutative diagram:

$$
\begin{array}{ccc}
\tilde{E} & \xrightarrow{\ \tilde{\tau}_E\ } & E = P \times_G V \\
{\scriptstyle \tilde{\rho}_E}\downarrow & & \downarrow{\scriptstyle \rho_E} \\
N & \xrightarrow{\ \tau\ } & M
\end{array}
$$

where $\tilde{E} = \{(n, [p, v]) \mid \tau(n) = \rho_E([p, v]) = \rho(p)\}$, $\tilde{\rho}_E(n, [p, v]) := n$, and $\tilde{\tau}_E(n, [p, v]) := [p, v]$. The fibers of $\tilde{\tau}_E$ equal $\tilde{\tau}_E^{-1}([p, v]) = \{(n, [p, v]) \mid \tau(n) = \rho_E([p, v]) = \rho(p)\} = \tau^{-1}(\rho(p)) \times \{[p, v]\} \simeq \tau^{-1}(\rho(p))$.

These two constructions are bundle isomorphic. More precisely, the map $\Phi : \tilde{P} \times_G V \to \tilde{E}$ defined by

$$\Phi([(n, p), v]) := (n, [p, v]) \tag{6.6.4}$$

is a fiber bundle isomorphism over N. Its inverse is given by $(n, [p, v]) \in \tilde{E} \mapsto ([n, p], v) \in \tilde{P} \times_G V$.

6.6.11 Sternberg space. Let Q be a manifold and G a Lie group with Lie algebra $\mathfrak{g}$ acting freely and properly on it. Let $\mathcal{A} \in \Omega^1(Q; \mathfrak{g})$ be a connection one-form on the left G-principal bundle $\pi : Q \to Q/G$.

The construction of the Sternberg space proceeds in two steps. First one pulls back the configuration space bundle $\pi : Q \to Q/G$ by the cotangent bundle projection $\tau : T^*(Q/G) \to Q/G$ to obtain the G-principal bundle

$$\tilde{Q} = \{(\alpha_{[q]}, q) \mid [q] = \pi(q), q \in Q\}$$

with base $T^*(Q/G)$ whose fiber over $\alpha_{[q]}$ is diffeomorphic to $\pi^{-1}([q])$. The diagram defining this pull-back bundle and the associated maps is given below:

$$
\begin{array}{ccc}
\tilde{Q} & \xrightarrow{\ \tilde{\tau}\ } & Q \\
{\scriptstyle \tilde{\pi}}\downarrow & & \downarrow{\scriptstyle \pi} \\
T^*(Q/G) & \xrightarrow{\ \tau\ } & Q/G
\end{array}
$$

where $\tilde{\pi}$ and $\tilde{\tau}$ are the projections on the first and second factors, respectively. It is easy to see that $\tilde{Q}$ is also a vector bundle over Q which is isomorphic to the annihilator $V(Q)^0 \subset T^*Q$ of the vertical bundle $V(Q) := \ker T\pi \subset TQ$. The fibers of these vector subbundles are given by $V(Q)_q := \ker T_q\pi = \{\xi_Q(q) \mid \xi \in \mathfrak{g}\} \subset T_qQ$ and $V(Q)^0_q := \{\alpha_q \in T_q^*Q \mid \langle \alpha_q, \xi_Q(q) \rangle = 0\} \subset T_q^*Q$ for each $q \in Q$.

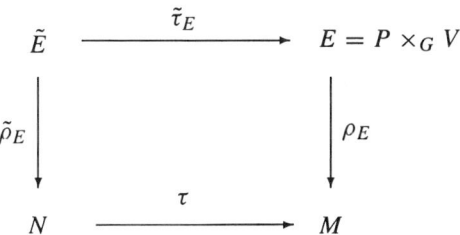

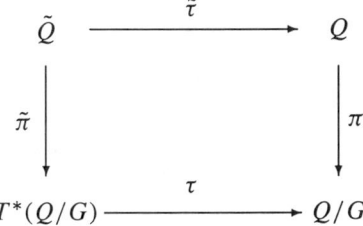

The second step is to form the coadjoint bundle of $\tilde{Q}$, that is, the associated vector bundle to the G-principal bundle $\tilde{Q} \to T^*(Q/G)$ given by the coadjoint representation of G on $\mathfrak{g}^*$. The ***Sternberg space***, denoted by S, is thus defined by

$$S = \tilde{Q} \times_G \mathfrak{g}^*.$$

The connection-dependent map $\Phi_{\mathcal{A}} : S \to (T^*Q)/G$ defined by

$$\Phi_{\mathcal{A}}([(\alpha_{[q]}, q), \mu]) := [T_q^*\pi(\alpha_{[q]}) + \mathcal{A}(q)^*\mu], \tag{6.6.5}$$

where $q \in Q$, $\alpha_q \in T_q^*Q$, and $\mu \in \mathfrak{g}^*$, is a vector bundle isomorphism over Q/G whose inverse is given by $[T_q^*\pi(\alpha_{[q]}) + \mathcal{A}(q)^*\mu] \in (T^*Q)/G \mapsto [(\alpha_{[q]}, q), \mu] \in S$. The smooth manifold $(T^*Q)/G$ carries a unique natural Poisson bracket making the projection $\rho : T^*Q \to (T^*Q)/G$ a canonical map. The ***Sternberg space Poisson bracket*** $\{\cdot, \cdot\}_S$ is defined as the pull-back by $\Phi_{\mathcal{A}}$ of the Poisson bracket of $(T^*Q)/G$.

6.6.12 Weinstein space. Construct first the coadjoint bundle $\widetilde{\mathfrak{g}}^* := Q \times_G \mathfrak{g}^*$ associated to the principal bundle $\pi : Q \to Q/G$ and then pull it back over the cotangent bundle projection $\tau : T^*(Q/G) \to Q/G$ to $T^*(Q/G)$. This gives the ***Weinstein space***

$$W := \{(\alpha_{[q]}, [q, \mu]) \mid \tau(\alpha_{[q]}) = \pi_{\widetilde{\mathfrak{g}}^*}([q, \mu]) := [q]\} \tag{6.6.6}$$

which satisfies the commutative diagram

$$\begin{array}{ccc} W & \xrightarrow{\;\tilde{\tau}\;} & \widetilde{\mathfrak{g}}^* \\ {\scriptstyle \tilde{\pi}_{\widetilde{\mathfrak{g}}^*}}\Big\downarrow & & \Big\downarrow{\scriptstyle \pi_{\widetilde{\mathfrak{g}}^*}} \\ T^*(Q/G) & \xrightarrow{\;\tau\;} & Q/G \end{array}$$

where $\tilde{\pi}_{\widetilde{\mathfrak{g}}^*}$ and $\tilde{\tau}$ are the projections on the first and second factors, respectively.

Notice that W is a vector bundle over $T^*(Q/G)$ with fiber over $\alpha_{[q]}$ given by $\tilde{\pi}_{\widetilde{\mathfrak{g}}^*}^{-1}(\alpha_{[q]}) = \{\alpha_{[q]}\} \times \{[q, \mu] \mid \mu \in \mathfrak{g}^*\}$. Furthermore, W is a vector bundle over Q/G; the fiber over $[q]$ equals $W_{[q]} = T_{[q]}^*(Q/G) \oplus \widetilde{\mathfrak{g}}_{[q]}^*$. That is, we have the immediate identification,

$$W = T^*(Q/G) \oplus \widetilde{\mathfrak{g}}^* \tag{6.6.7}$$

as vector bundles with base Q/G. We shall use this identification later in §6.6.18 to determine the symplectic leaves in W.

Let $H(Q)$ be the horizontal subbundle defined by the connection $\mathcal{A}$; thus $TQ = H(Q) \oplus V(Q)$, where $H(Q)_q := \ker \mathcal{A}(q)$. For each $q \in Q$ the linear map $T_q\pi|_{H(Q)_q} : H(Q)_q \to T_{[q]}(Q/G)$ is an isomorphism. Let $\operatorname{hor}_q := (T_q\pi|_{H(Q)_q})^{-1} : T_{[q]}(Q/G) \to H(Q)_q \subset T_qQ$ be the horizontal lift operator induced by the connection $\mathcal{A}$. Thus

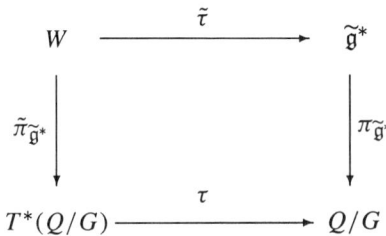

$\text{hor}_q^* : T_q^*Q \to T_{[q]}^*(Q/G)$ is a linear surjective map whose kernel is the annihilator $H(Q)_q^0$ of the horizontal space. Using these notations, it can be shown that the connection-dependent map $\Psi_{\mathcal{A}} : (T^*Q)/G \to W$ defined by

$$\Psi_{\mathcal{A}}([\alpha_q]) := (\text{hor}_q^*(\alpha_q), [q, \mathbf{J}(\alpha_q)]) \tag{6.6.8}$$

where $q \in Q$, $\alpha_q \in T_q^*Q$, and $\mathbf{J} : T^*Q \to \mathfrak{g}^*$ is the momentum map of the lifted action, $\langle \mathbf{J}(\alpha_q), \xi \rangle = \alpha_q((\xi_Q(q))$ for $\xi \in \mathfrak{g}$, is a vector bundle isomorphism over Q/G whose inverse is given by $(\alpha_{[q]}, [q, \mu]) \in W \mapsto [T_q^*\pi(\alpha_{[q]}) + \mathcal{A}(q)^*\mu] \in (T^*Q)/G$. The ***Weinstein space Poisson bracket*** $\{\cdot, \cdot\}_W$ is defined as the push-forward by $\Psi_{\mathcal{A}}$ of the Poisson bracket of $(T^*Q)/G$.

The vector bundle isomorphisms Φ, $\Phi_{\mathcal{A}}$, and $\Psi_{\mathcal{A}}$ defined respectively by (6.6.4), (6.6.5), and (6.6.8) make the following diagram

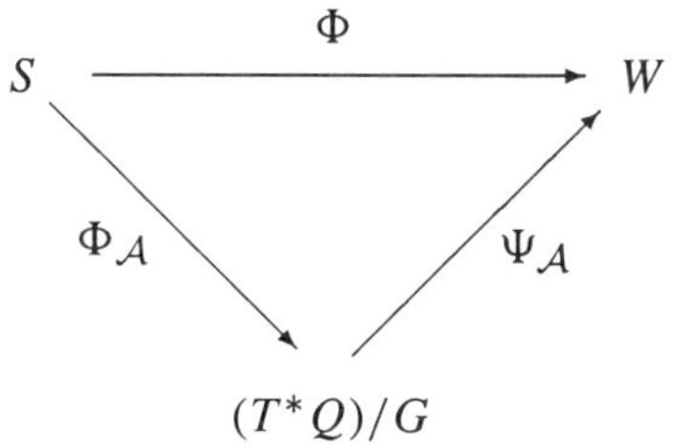

commutative. In addition, $\Phi : (S, \{\cdot, \cdot\}_S) \to (W, \{\cdot, \cdot\}_W)$ is an isomorphism of Poisson manifolds and $\Phi^* : W^* \to S^*$ is an isomorphism of Lie algebra bundles, that is, the restriction of Φ^* to every fiber is a Lie algebra isomorphim depending smoothly on the base point. In what follows we shall use the bundle structure on the Sternberg and Weinstein spaces to explicitly write out the Poisson brackets on S and W.

6.6.13 Affine connections on vector bundles. In what follows we shall need some elementaty facts about affine connections on vector bundles. If $\rho : \mathcal{V} \to M$ is a vector bundle, denote by $\Lambda^k(M; \mathcal{V})$ the smooth vector bundle whose fiber at $m \in M$ consists of all skew symmetric k-linear maps $T_mM \times \cdots \times T_mM \to \mathcal{V}_m$. Let $\Omega^k(M; \mathcal{V})$ denote the space of its sections, that is, $\Omega^k(M; \mathcal{V})$ consists of all $\mathcal{V}$-valued k-forms on M. Elements of $\Omega^k(M; \mathcal{V})$ are called ***vector bundle valued forms*** on M. Let $\Gamma(\mathcal{V})$ denote the space of ***sections*** of the vector bundle $\rho : \mathcal{V} \to M$, that is, $\sigma \in \Gamma(M)$ if and only if $\sigma : M \to \mathcal{V}$ is a smooth map satisfying $\rho \circ \sigma = \textit{identity}$ on M. For $U \subset M$ open, denote by $\Gamma_U(\mathcal{V})$ and $\Omega_U^1(M; \mathcal{V})$ the space of sections of $\mathcal{V}$ and the space of $\mathcal{V}$-valued one-forms on U, respectively.

A collection of $\mathbb{R}$-linear maps $\nabla : \Gamma_U(\mathcal{V}) \to \Omega_U^1(M; \mathcal{V})$, defined for each open set $U \subset M$ satisfying $\nabla(f\sigma) = (\mathbf{d}f)\sigma + f\nabla\sigma$ for any $f \in C^\infty(U)$ and $\sigma \in \Gamma_U(\mathcal{V})$ is called an ***affine*** or ***linear connection*** and ∇ is its ***covariant derivative***. For each $X \in \mathfrak{X}(U)$ one obtains by contraction an operator ∇_X on $\Gamma(U)$ which therefore satisfies the identities $\nabla_{fX}\sigma = f\nabla_X\sigma$ and $\nabla_X(f\sigma) = X[f]\sigma + f\nabla_X\sigma$, for any $\sigma \in \Gamma_U(\mathcal{V})$, $X \in \mathfrak{X}(U)$, and $f \in C^\infty(U)$. From these properties it follows that ∇_X depends only on the point values of X and hence each $v_m \in T_mM$ defines an operator $\nabla_{v_m} : \Gamma(\mathcal{V}) \to \mathcal{V}_m$ given by $\nabla_{v_m}\sigma := (\nabla\sigma)(m)(v_m)$. In particular, $(\nabla_X\sigma)(m) = \nabla_{X(m)}\sigma$.

The covariant derivative defines the ***connector*** $\kappa : T\mathcal{V} \to \mathcal{V}$ by

$$\nabla_X \sigma = \kappa \circ T\sigma \circ X \tag{6.6.9}$$

for any locally defined section $\sigma \in \Gamma_U(\mathcal{V})$ and vector field $X \in \mathfrak{X}(U)$. The connector satisfies the following properties:

- $\kappa : T\mathcal{V} \to \mathcal{V}$ is a vector bundle homomorphism relative to both vector bundle structures of $T\mathcal{V}$, with bases $\mathcal{V}$ and TM, respectively; that is, the following diagrams

$$\begin{array}{ccc} T\mathcal{V} & \xrightarrow{\kappa} & \mathcal{V} \\ {\scriptstyle \tau_{\mathcal{V}}}\downarrow & & \downarrow{\scriptstyle \rho} \\ \mathcal{V} & \xrightarrow{\rho} & M \end{array}$$

$$\begin{array}{ccc} T\mathcal{V} & \xrightarrow{\kappa} & \mathcal{V} \\ {\scriptstyle T\rho}\downarrow & & \downarrow{\scriptstyle \rho} \\ TM & \xrightarrow{\tau_M} & M \end{array}$$

  commute and κ is linear when restricted to every fiber; the maps $\tau_M : TM \to M$ and $\tau_{\mathcal{V}} : T\mathcal{V} \to \mathcal{V}$ denote the tangent bundle projections;

- $(K \circ \mathrm{ver})(u_m, v_m) = v_m$ for any $u_m, v_m \in T_m M$, where ver is the ***vertical lift operator*** defined by

$$\mathrm{ver}(u_m, v_m) := \left.\frac{d}{dt}\right|_{t=0} (u_m + v_m).$$

Conversely, given a smooth map $\kappa : T\mathcal{V} \to \mathcal{V}$ satisfying the three propoerties above, formula (6.6.9) determines the covariant derivative of a linear connection. The two definitions of a linear connection, via a covariant derivative or a connector, are thus equivalent.

The ***horizontal subbundle*** of the linear connection determined by a connector $\kappa : T\mathcal{V} \to \mathcal{V}$ is defined by $H(\mathcal{V}) := \ker \kappa$. The ***vertical subbundle*** is defined as $V(\mathcal{V}) := \ker T\rho$. Then $T\mathcal{V} = H(\mathcal{V}) \oplus V(\mathcal{V})$ and the vector bundle map $T\rho : H(\mathcal{V}) \to TM$ covering ρ is an isomorphism on every fiber. Given $v \in \mathcal{V}$, the linear map

$$\mathrm{hor}_v := (T_v\rho|_{H(\mathcal{V})})^{-1} : T_{\rho(v)}M \to H_v(\mathcal{V}) \tag{6.6.10}$$

is called the ***horizontal lift operator*** of the linear connection determined by the connector κ.

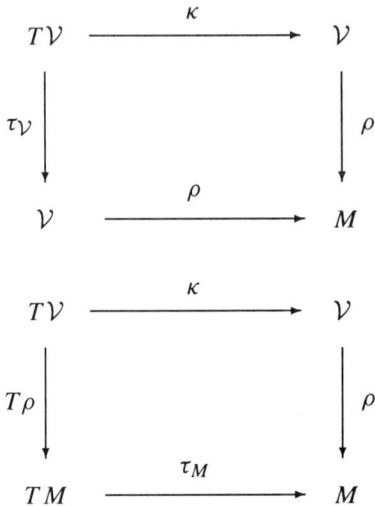

6.6.14 Covariant derivative on S. Returning to the general notation of §6.6.10, let $\rho : P \to M$ be a left principal G-bundle, V a left representation space of the Lie group G, and $\mathrm{hor}_p : T_{\pi(p)}M \to T_pP$ the horizontal lift operator of the given connection $\mathcal{A} \in \Omega^1(P; \mathfrak{g})$ at $p \in P$. The horizontal lift operator of the affine connection on the associated vector bundle $\rho_E : E = P \times_G V \to M$ induced by $\mathcal{A}$ is defined by

$$\mathrm{hor}_{[p,v]}(v_m) := T_{(p,v)}\pi_{P\times V}\left(\mathrm{hor}_p(v_m), 0\right), \tag{6.6.11}$$

where $p \in P$, $v \in V$, $m = \rho(p) = [p]$, $v_m \in T_mM$, $\pi_{P\times V} : P \times V \to E$ is the quotient map on the space of orbits, and $[p, v] := \pi_{P\times V}(p, v) \in E$ (see, for instance CENDRA et al. (2001a) or GREUB et al. (1970b)). The covariant derivative $\mathbf{d}_{\mathcal{A}} f([p, v]) \in T_m^*M$ of a real valued function $f \in C^\infty(P \times_G V)$ relative to the affine connection given by the horizontal lift operator (6.6.11) at a point $[p, v] \in P \times_G V$ is defined by

$$\mathbf{d}_{\mathcal{A}} f([p, v])(v_m) := \mathbf{d}f([p, v])\left(\mathrm{hor}_{[p,v]}(v_m)\right). \tag{6.6.12}$$

We shall use this general formula to induce a covariant derivative of functions on S. For this, note that $\tilde{\mathcal{A}} := \tilde{\tau}^*\mathcal{A} \in \Omega^1(\tilde{Q}; \mathfrak{g})$, where $\tilde{\tau} : \tilde{Q} \to Q$ is the projection on the second factor, is a connection one-form on the principal G-bundle $\tilde{\pi} : \tilde{Q} \to T^*(Q/G)$. Its horizontal spaces are thus given by

$$H_{(\alpha_{[q]},q)}(\tilde{Q}) = T_{\alpha_{[q]}}(T^*(Q/G)) \times H_q(Q) \tag{6.6.13}$$

and the horizontal lift operator $\mathrm{hor}_{(\alpha_{[q]},q)} : T_{\alpha_{[q]}}(T^*(Q/G)) \to T_{(\alpha_{[q]},q)}\tilde{Q}$ has the expression

$$\mathrm{hor}_{(\alpha_{[q]},q)}\left(v_{\alpha_{[q]}}\right) = \left(v_{\alpha_{[q]}}, \mathrm{hor}_q\left(T_{\alpha_{[q]}}\tau(v_{\alpha_{[q]}})\right)\right). \tag{6.6.14}$$

Thus, for the case of the associated bundle $\tilde{\pi}_{\tilde{Q}} : S \to T^*(Q/G)$, $\tilde{\pi}_{\tilde{Q}}([(\alpha_{[q]}, q), \mu]) = \alpha_{[q]}$, formula (6.6.11) becomes, in view of (6.6.14),

$$\begin{aligned} \mathrm{hor}_s(v_{\alpha_{[q]}}) &= T_{((\alpha_{[q]},q),\mu)}\pi_{\tilde{Q}\times\mathfrak{g}^*}\left(\mathrm{hor}_{(\alpha_{[q]},q)}\, v_{\alpha_{[q]}}, 0\right) \\ &= T_{((\alpha_{[q]},q),\mu)}\pi_{\tilde{Q}\times\mathfrak{g}^*}\left(\left(v_{\alpha_{[q]}}, \mathrm{hor}_q(T_{\alpha_{[q]}}\tau(v_{\alpha_{[q]}}))\right), 0\right), \end{aligned} \tag{6.6.15}$$

where $s = [(\alpha_{[q]}, q), \mu] \in S$ and $\pi_{\tilde{Q}\times\mathfrak{g}^*} : \tilde{Q} \times \mathfrak{g}^* \to S$ is the quotient projection. Combining now the general formula (6.6.12) with (6.6.15) yields the expression of the covariant derivative $\mathbf{d}^S_{\tilde{\mathcal{A}}} f(s) \in T^*_{\tilde{\pi}_{\tilde{Q}}(s)}T^*(Q/G)$ of the function $f \in C^\infty(S)$ at the point $s \in S$, namely

$$\begin{aligned} \mathbf{d}^S_{\tilde{\mathcal{A}}} f(s)\left(v_{\alpha_{[q]}}\right) &:= \mathbf{d}f(s)\left(\mathrm{hor}_s\left(v_{\alpha_{[q]}}\right)\right) \\ &= \mathbf{d}f(s)\left(T_{((\alpha_{[q]},q),\mu)}\pi_{\tilde{Q}\times\mathfrak{g}^*}\left(\left(v_{\alpha_{[q]}}, \mathrm{hor}_q(T_{\alpha_{[q]}}\tau(v_{\alpha_{[q]}}))\right), 0\right)\right), \end{aligned} \tag{6.6.16}$$

where $s = [(\alpha_{[q]}, q), \mu] \in S$, $\tilde{\pi}_{\tilde{Q}}(s) = \alpha_{[q]}$, and $v_{\alpha_{[q]}} \in T_{\alpha_{[q]}}T^*(Q/G)$.

6.6.15 Covariant derivative on W. We begin by recalling how one induces a covariant derivative on sections of the pull-back of a vector bundle.

Let $\gamma : \mathcal{V} \to M$ be a vector bundle with an affine connection whose covariant derivative is denoted by ∇. Denote for any $v \in \mathcal{V}$ by $\operatorname{hor}_v : T_{\gamma(v)}M \to T_v\mathcal{V}$ the horizontal lift operator of the connection ∇. Let N be another manifold and $\tau : N \to M$ a surjective submersion. Denote by $\tilde{\mathcal{V}} := \{(n, v) \mid \tau(n) = \gamma(v)\}$ the pull-back bundle over N; this is a vector bundle $\tilde{\gamma} : \tilde{\mathcal{V}} \to N$ whose fibers are isomorphic to those of $\mathcal{V}$, where $\tilde{\gamma}$ is the projection on the first factor N. Denote by $\tilde{\tau} : \tilde{\mathcal{V}} \to \mathcal{V}$ the projection on the second factor $\mathcal{V}$ and recall that $\gamma \circ \tilde{\tau} = \tau \circ \tilde{\gamma}$. The horizontal lift operator $\operatorname{hor}_{(n,v)} : T_nN \to T_{(n,v)}\tilde{\mathcal{V}}$ is defined by

$$\operatorname{hor}_{(n,v)}(u_n) := (v_n, \operatorname{hor}_v T_n\tau(u_n)), \tag{6.6.17}$$

for $(n, v) \in \tilde{\mathcal{V}}$ and $u_n \in T_nN$ (see, e.g., CENDRA et al. (2001a) or GREUB et al. (1970b)). If $f \in C^\infty(\tilde{\mathcal{V}})$, its covariant exterior derivative $\tilde{\nabla} f(n, v) \in T_n^*N$ is defined by

$$\tilde{\nabla} f(n, v)(u_n) := \mathbf{d}f(n, v)\big(\operatorname{hor}_{(n,v)}(u_n)\big), \tag{6.6.18}$$

where $(n, v) \in \tilde{\mathcal{V}}$ and $u_n \in T_nN$.

We shall apply these formulas to the bundle $\tilde{\pi}_{\tilde{\mathfrak{g}}^*} : W \to T^*(Q/G)$, which is the pull-back to $T^*(Q/G)$ of the coadjoint bundle $\pi_{\tilde{\mathfrak{g}}^*} : \tilde{\mathfrak{g}}^* = Q \times_G \mathfrak{g}^* \to Q/G$ over the cotangent bundle projection $\tau : T^*(Q/G) \to Q/G$. Formula (6.6.17) becomes for $w = (\alpha_{[q]}, [q, \mu]) \in W$ and $v_{\alpha_{[q]}} \in T_{\alpha_{[q]}}T^*(Q/G)$

$$\begin{aligned} \operatorname{hor}_w(v_{\alpha_{[q]}}) &= \big(v_{\alpha_{[q]}}, \operatorname{hor}_{[q,\mu]}(T_{\alpha_{[q]}}\tau(v_{\alpha_{[q]}}))\big) \\ &= \big(v_{\alpha_{[q]}}, T_{(q,\mu)}\pi_{Q\times\mathfrak{g}^*}\left(\operatorname{hor}_q(T_{\alpha_{[q]}}\tau(v_{\alpha_{[q]}})), 0\right)\big), \end{aligned} \tag{6.6.19}$$

in view of (6.6.11), where $\operatorname{hor}_q : T_{\pi(q)}(Q/G) \to T_qQ$ is the horizontal lift defined by the principal connection $\mathcal{A} \in \Omega^1(Q; \mathfrak{g})$. The exterior covariant derivative $\mathbf{d}^W_{\tilde{\mathcal{A}}} f(w) \in T^*_{\alpha_{[q]}}T^*(Q/G)$ of a function $f \in C^\infty(W)$ at a point $w \in W$ is then defined, according to (6.6.18), on the vector $v_{\alpha_{[q]}} \in T_{\alpha_{[q]}}T^*(Q/G)$, by

$$\begin{aligned} \mathbf{d}^W_{\tilde{\mathcal{A}}} f(w)\left(v_{\alpha_{[q]}}\right) &:= \mathbf{d}f(w)\left(\operatorname{hor}_w\left(v_{\alpha_{[q]}}\right)\right) \\ &= \mathbf{d}f(w)\left(v_{\alpha_{[q]}}, T_{(q,\mu)}\pi_{Q\times\mathfrak{g}^*}\left(\operatorname{hor}_q(T_{\alpha_{[q]}}\tau(v_{\alpha_{[q]}})), 0\right)\right). \end{aligned} \tag{6.6.20}$$

The covariant derivatives on S and W are linked by the following formula: if $f \in C^\infty(W)$, then

$$\mathbf{d}^S_{\tilde{\mathcal{A}}}(f \circ \Phi) = (\mathbf{d}^W_{\tilde{\mathcal{A}}} f) \circ \Phi, \tag{6.6.21}$$

where $\Phi : ([(\alpha_{[q]}, q), \mu]) \in S \mapsto (\alpha_{[q]}, [q, \mu]) \in W$ is the canonical (connection independent) vector bundle isomorphism over $T^*(Q/G)$.

6.6.16 Gauged Poisson bracket on S**.** Recall from (5.1.1) that the curvature two-form $\operatorname{Curv}_{\mathcal{A}} \in \Omega^2(Q; \mathfrak{g})$ of the connection $\mathcal{A}$ on the principal bundle $\pi : Q \to Q/G$ is defined by $\operatorname{Curv}_{\mathcal{A}}(q)(u_q, v_q) = \mathbf{d}\mathcal{A}(q)(u_q^H, v_q^H)$, where u_q^H, v_q^H denote the horizontal components of $u_q, v_q \in T_qQ$, respectively. Unlike the connection one-form

$\mathcal{A}$, the curvature two-form $\mathrm{Curv}_{\mathcal{A}}$ induces a $\tilde{\mathfrak{g}}$-valued two-form $B \in \Omega^2(Q/G; \tilde{\mathfrak{g}})$ on the base Q/G by the formula

$$B([q])(u_{[q]}, v_{[q]}) = [q, \mathrm{Curv}_{\mathcal{A}}(u_q, v_q)] \tag{6.6.22}$$

for any $u_q, v_q \in T_qQ$ that satisfy $T_q\pi(u_q) = u_{[q]}$ and $T_q\pi(v_q) = v_{[q]}$.

With these notations and definitions we can present the formula for the Poisson bracket on the Sternberg space S which, recall, is a connection-dependent realization of the reduced Poisson manifold $(T^*Q)/G$. The Poisson bracket of $f, g \in C^\infty(S)$ at the point $s = [(\alpha_{[q]}, q), \mu] \in S$ is given by

$$\begin{aligned}\{f, g\}_S(s) &= \omega_{Q/G}(\alpha_{[q]})\left(\mathbf{d}^S_{\tilde{\mathcal{A}}} f(s)^\sharp, \mathbf{d}^S_{\tilde{\mathcal{A}}} g(s)^\sharp\right) \\ &\quad + \left\langle v, \tilde{B}(\alpha_{[q]})\left(\mathbf{d}^S_{\tilde{\mathcal{A}}} f(s)^\sharp, \mathbf{d}^S_{\tilde{\mathcal{A}}} g(s)^\sharp\right)\right\rangle - \left\langle s, \left[\frac{\delta f}{\delta s}, \frac{\delta g}{\delta s}\right]\right\rangle,\end{aligned} \tag{6.6.23}$$

where $\omega_{Q/G}$ is the canonical symplectic form on $T^*(Q/G)$, $\tilde{B} \in \Omega^2(T^*(Q/G); \tilde{\mathfrak{g}})$ is the curvature two-form of the connection $\tilde{\mathcal{A}} \in \Omega^1(\tilde{Q}; \mathfrak{g})$ and is thus the $\tilde{\mathfrak{g}}$-valued two-form on $T^*(Q/G)$ given by

$$\tilde{B} = \tau^* B, \tag{6.6.24}$$

with B defined in (6.6.22), $\sharp : T^*(T^*(Q/G)) \to T(T^*(Q/G))$ is the vector bundle isomorphism induced by $\omega_{Q/G}$, $v := [q, \mu] \in \tilde{\mathfrak{g}}^*$, and $\delta f/\delta s \in S^*_{\alpha_{[q]}}$ is the usual fiber derivative of f at the point $s \in S$, that is,

$$\left\langle s', \frac{\delta f}{\delta s}\right\rangle := \left.\frac{d}{dt}\right|_{t=0} f\left([(\alpha_{[q]}, q), \mu + t\nu]\right) \tag{6.6.25}$$

for any $s' := [(\alpha_{[q]}, q), \nu)] \in S$.

Note that the third term in (6.6.23) can be written more conveniently in the following manner. Denoting by $\delta f/\delta v \in \tilde{\mathfrak{g}}$ the unique element in the fiber at $[q]$ of the adjoint bundle $\tilde{\mathfrak{g}}$ defined by the equality

$$\left\langle [q, \nu], \frac{\delta f}{\delta v}\right\rangle = \left.\frac{d}{dt}\right|_{t=0} f\left([(\alpha_{[q]}, q), \mu + t\nu]\right) \tag{6.6.26}$$

for any $\nu \in \mathfrak{g}^*$; then

$$\left\langle s, \left[\frac{\delta f}{\delta s}, \frac{\delta g}{\delta s}\right]\right\rangle = \left\langle v, \left[\frac{\delta f}{\delta v}, \frac{\delta g}{\delta v}\right]\right\rangle.$$

Thus $\delta f/\delta v$ is an element in $\tilde{\mathfrak{g}}$ over the point $[q] \in Q/G$ and can therefore be paired with $v \in \tilde{\mathfrak{g}}^*$. It should be remarked here that we abuse the symbol $\delta f/\delta v$ which should denote the usual fiber derivative of a function on the vector bundle $\tilde{\mathfrak{g}}^*$; however, this makes no a priori sense in this case, since $f \in C^\infty(S)$ is not a function on $\tilde{\mathfrak{g}}^*$.

6.6.17 Gauged Poisson bracket on W. With the same notations and definitions as in 6.6.16, the Poisson bracket of $f, g \in C^\infty(W)$ at a point $w \in W$ is given by

$$\begin{aligned}\{f, g\}_W(w) &= \omega_{Q/G}(\alpha_{[q]})\left(\mathbf{d}^W_{\tilde{\mathcal{A}}} f(w)^\sharp, \mathbf{d}^W_{\tilde{\mathcal{A}}} g(w)^\sharp\right) \\ &+ \left\langle v, \tilde{B}(\alpha_{[q]})\left(\mathbf{d}^W_{\tilde{\mathcal{A}}} f(w)^\sharp, \mathbf{d}^W_{\tilde{\mathcal{A}}} g(w)^\sharp\right)\right\rangle - \left\langle w, \left[\frac{\delta f}{\delta w}, \frac{\delta g}{\delta w}\right]\right\rangle, \end{aligned} \tag{6.6.27}$$

where $w = (\alpha_{[q]}, [q, \mu]) \in W$, $v = [q, \mu] \in \tilde{\mathfrak{g}}^*$, and $\delta f/\delta v \in W^*_{\alpha_{[q]}}$ denotes the usual fiber derivative of f in W, that is,

$$\left\langle w', \frac{\delta f}{\delta w}\right\rangle := \left.\frac{d}{dt}\right|_{t=0} f(w + tw')$$

for any $w' \in W_{\alpha_{[q]}}$. Recall that W is a coadjoint bundle and hence is a vector bundle over $T^*(Q/G)$ with fiber isomorphic to $\mathfrak{g}^*$, so that W^* is an adjoint bundle and thus the fibers, which are all isomorphic to $\mathfrak{g}$, have a Lie algebra structure depending smoothly on the base point; it is this bracket that appears in the last summand of (6.6.27).

6.6.18 The symplectic leaves of $(T^*Q)/G$. This subsection presents the results of MARSDEN AND PERLMUTTER (2000). As before, it is assumed that the Lie group G acts freely and properly on the manifold Q and that $\mathcal{A} \in \Omega^1(Q; \mathfrak{g})$ is a connection one-form on the principal bundle $\pi : Q \to Q/G$. Denote by $\tilde{\mathfrak{g}} := (Q \times \mathfrak{g})/G$ the adjoint bundle of Q, that is, the associated vector bundle defined by the adjoint representation of G on $\mathfrak{g}$ whose base is Q/G and fiber is isomorphic to $\mathfrak{g}$. Then there is a a vector bundle isomorphism

$$\alpha_{\mathcal{A}} : (TQ)/G \to T(Q/G) \oplus \tilde{\mathfrak{g}}$$

given by $\alpha_{\mathcal{A}}([v_q]) := T_q\pi \oplus [q, \mathcal{A}(q)(v_q)]$, where corner brackets denote equivalence classes in the respective quotient manifolds. Thus $(\alpha_{\mathcal{A}}^{-1})^* : (T^*Q)/G \to T^*(Q/G) \oplus \tilde{\mathfrak{g}}^* = W$ (see (6.6.7)) is a diffeomorphism, where $\tilde{\mathfrak{g}}^*$ denotes the coadjoint bundle of Q, that is, the dual of the adjoint bundle.

Let $\mathcal{O} \subset \mathfrak{g}^*$ be a coadjoint orbit of G. Then using the Symplectic Orbit Reduction Theorem 6.3.1 it is straightforward to verify that

$$(\alpha_{\mathcal{A}}^{-1})^*\left(\mathbf{J}^{-1}(\mathcal{O})/G\right) = T^*(Q/G) \times_{Q/G} \widetilde{\mathcal{O}} \subset T^*(Q/G) \oplus \tilde{\mathfrak{g}},$$

where $\widetilde{\mathcal{O}} := (Q \times \mathcal{O})/G \to Q/G$ is the associated fiber bundle using the coadjoint action of G on $\mathcal{O}$ and

$$T^*(Q/G) \times_{Q/G} \widetilde{\mathcal{O}} := \{(\alpha_{[q]}, [q, \nu]) \mid \alpha_{[q]} \in T^*_{[q]}(Q/G), q \in Q, \nu \in \mathcal{O}\}$$

is the fibered product of $T^*(Q/G)$ with $\widetilde{\mathcal{O}}$ which is a fiber bundle over Q/G whose fiber over $[q] \in Q/G$ equals $T^*_{[q]}(Q/G) \times \widetilde{\mathcal{O}}_{[q]}$. Therefore, the reduced orbit symplectic form $\Omega_{\mathcal{O}}$ of the orbit reduced space $(T^*Q)_{\mathcal{O}} = \mathbf{J}^{-1}(\mathcal{O})/G$ pushes forward by $(\alpha_{\mathcal{A}}^{-1})^*$ to a symplectic form $\Omega_{\mathcal{A}}$ on $T^*(Q/G) \times_{Q/G} \widetilde{\mathcal{O}}$ that has an *explicit* formula, namely

$$\Omega_{\mathcal{A}} = \Omega_{Q/G} - \beta,$$

where $\Omega_{Q/G}$ is the canonical symplectic form on the cotangent bundle $T^*(Q/G)$ and β is the unique two-form on $\widetilde{\mathcal{O}}$ determined by

$$\pi_G^*\beta = \mathbf{d}\alpha + \pi_2^*\omega_{\mathcal{O}}^+,$$

where $\pi_G : Q \times \mathcal{O} \to (Q \times \mathcal{O})/G = \widetilde{\mathcal{O}}$ is the projection on the quotient, $\pi_2 : Q \times \mathcal{O} \to \mathcal{O}$ is the projection on the second factor, $\omega_{\mathcal{O}}^+$ is the $+$-orbit symplectic two-form on $\mathcal{O}$, and $\alpha \in \Omega^1(Q \times \mathcal{O})$ is given by

$$\alpha(q, \nu)(v_q, -\operatorname{ad}^*_{\xi'} \nu) := \langle \nu, \mathcal{A}(q)(v_q)\rangle$$

for $q \in Q$, $\nu \in \mathcal{O}$, $v_q \in T_qQ$, $-\operatorname{ad}^*_{\xi'}\nu \in T_\nu\mathcal{O}$, and $\xi' \in \mathfrak{g}$. In addition, $\mathbf{d}\alpha$ also has the *explicit* expression

$$\begin{aligned}\mathbf{d}\alpha(q, \nu)&\Big((u_q, -\operatorname{ad}^*_{\xi'}\nu), (v_q, -\operatorname{ad}^*_{\eta'}\nu)\Big)\\ &= \langle \nu, [\eta', \xi] + [\eta, \xi'] + [\xi, \eta] + \operatorname{Curv}_{\mathcal{A}}(q)(u_q, v_q)\rangle,\end{aligned}$$

where $\operatorname{Curv}_{\mathcal{A}} \in \Omega^2(Q; \mathfrak{g})$ is the curvature two-form of $\mathcal{A}$, $q \in Q$, $\nu \in \mathcal{O}$, $\xi, \xi', \eta, \eta' \in \mathfrak{g}$, $u_q, v_q \in T_qQ$, and

$$u_q = \xi_Q(q) + u_q^H, \qquad v_q = \eta(q) + v_q^H$$

is the decomposition of u_q, v_q in the vertical plus horizontal components.

The theorem of MARSDEN AND PERLMUTTER (2000) states that *the leaves of $T^*(Q/G) \oplus \tilde{\mathfrak{g}}^* = W$ relative to the gauged Lie–Poisson bracket given in* §6.6.17 *are the symplectic manifolds* $(T^*(Q/G) \times_{Q/G} \widetilde{\mathcal{O}}, \Omega_{\mathcal{A}})$, where $\Omega_{\mathcal{A}}$ is computed by the procedure given above. Thus the symplectic form $\Omega_{\mathcal{O}}$ on the leaves $(T^*Q)_{\mathcal{O}}$ of $(T^*Q)/G$ is the pull-back of $\Omega_{\mathcal{A}}$ by $(\alpha_{\mathcal{A}}^{-1})^*$.

6.6.19 Reconstruction of dynamics for cotangent bundles. In the case of cotangent bundles, the reconstruction of the dynamics on $\mathbf{J}^{-1}(\mu) \subset T^*Q$ from the reduced one on $\mathbf{J}^{-1}(\mu)/G_\mu = (T^*Q)_\mu$ presents certain particularities that shall be reviewed in this subsection following MARSDEN et al. (1990). In certain cases explicit formulas can be obtained.

A general reconstruction method of the dynamics from the reduced dynamics was given in §6.1.10. Let $h : T^*Q \to \mathbb{R}$ be a G-invariant Hamiltonian, $\mu \in \mathfrak{g}^*$, $\alpha_q \in \mathbf{J}^{-1}(\mu)$, and $c_\mu(t)$ the integral curve of the reduced system with initial condition $[\alpha_q] \in (T^*Q)_\mu$ given by the Hamiltonian function $h_\mu : (T^*Q)_\mu \to \mathbb{R}$. In terms of a connection $A \in \Omega^1(\mathbf{J}^{-1}(\mu); \mathfrak{g}_\mu)$ on the left G_μ-principal bundle $\mathbf{J}^{-1}(\mu) \to (T^*Q)_\mu$ this procedure involved four steps:

- **Step 1:** *Horizontally lift the curve $c_\mu(t) \in (T^*Q)_\mu$ to a curve $d(t) \in \mathbf{J}^{-1}(\mu)$ with $d(0) = \alpha_q$;*
- **Step 2:** *set $\xi(t) = A(d(t))\,(X_h(d(t))) \in \mathfrak{g}_\mu$;*
- **Step 3:** *with $\xi(t) \in \mathfrak{g}_\mu$ determined in Step 2, solve the nonautonomous differential equation $\dot{g}(t) = T_{g(t)}\xi(t)$ with initial condition $g(0) = e$;*

- **Step 4:** *the curve* $c(t) = g(t) \cdot d(t)$, *with* $d(t)$ *found in Step 1 and* $g(t)$ *found in Step 3 is the integral curve of* X_h *with initial condition* $c(0) = \alpha_q$.

This method depends on the choice of the connection $A \in \Omega^1(\mathbf{J}^{-1}(\mu); \mathfrak{g}_\mu)$. Due to the fact that the phase space of the Hamiltonian system is a cotangent bundle there are some instances when such a choice is natural.

6.6.20 One-dimensional coadjoint isotropy group. If $G_\mu = S^1$ or $G_\mu = \mathbb{R}$, then there is a natural choice of a connection $A \in \Omega^1(\mathbf{J}^{-1}(\mu); \mathfrak{g}_\mu)$. In this case, the Lie algebra $\mathfrak{g}_\mu$ is one dimensional. Pick a generator $\zeta \in \mathfrak{g}_\mu$, $\zeta \neq 0$, and identify $\mathfrak{g}_\mu$ with $\mathbb{R}$ by the isomorphism $a \in \mathbb{R} \mapsto a\zeta \in \mathfrak{g}_\mu$. Then a connection one-form on the S^1 (or $\mathbb{R}$) principal bundle $\mathbf{J}^{-1}(\mu) \to (T^*Q)_\mu$ is just a one-form A on $\mathbf{J}^{-1}(\mu)$ that can be chosen to equal

$$A = \frac{1}{\langle \mu, \zeta \rangle} \theta_\mu,$$

where θ_μ is the pull-back of the canonical one-form $\theta \in \Omega^1(T^*Q)$ to the submanifold $\mathbf{J}^{-1}(\mu)$. The curvature of this connection is the two-form on $(T^*Q)_\mu$ given by

$$\operatorname{curv}(A) = -\frac{1}{\langle \mu, \zeta \rangle} \omega_\mu,$$

where ω_μ is the reduced symplectic form on $(T^*Q)_\mu$. In this case, the curve $\xi(t) \in \mathfrak{g}_\mu$ in Step 2 is given by

$$\xi(t) = \Lambda[h](d(t)),$$

where $\Lambda \in \mathfrak{X}(T^*Q)$ is the ***Liouville vector field*** characterized by the property of being the only vector field on T^*Q that satisfies the relation $\mathbf{i}_\Lambda \mathbf{d}\theta = \theta$. In canonical local coordinates (q^i, p_i) on T^*Q, $\Lambda = p_i \frac{\partial}{\partial p_i}$.

6.6.21 Induced connection. Any connection $\mathcal{A} \in \Omega^1(Q; \mathfrak{g}_\mu)$ on the left principal bundle $Q \to Q/G_\mu$ induces a connection $A \in \Omega^1(\mathbf{J}^{-1}(\mu); \mathfrak{g}_\mu)$ by

$$A(\alpha_q)\left(V_{\alpha_q}\right) := \mathcal{A}(q)\left(T_{\alpha_q}\pi_Q\left(V_{\alpha_q}\right)\right),$$

where $q \in Q$, $\alpha_q \in T_q^*Q$, $V_{\alpha_q} \in T_{\alpha_q}(T^*Q)$, and $\pi_Q : T^*Q \to Q$ is the cotangent bundle projection. In this case, the curve $\xi(t) \in \mathfrak{g}_\mu$ in Step 2 is given by

$$\xi(t) = \mathcal{A}(q(t))\big(\mathbb{F}h(d(t))\big),$$

where $q(t) := \pi_Q(d(t))$ is the base integral curve and the vector bundle morphism $\mathbb{F}h : T^*Q \to TQ$ is the fiber derivative of h given by

$$\mathbb{F}h(\alpha_q)(\beta_q) := \left.\frac{d}{dt}\right|_{t=0} h(\alpha_q + t\beta_q)$$

for any $\alpha_q, \beta_q \in T_a^*Q$.

Two particular instances of this situation are noteworthy.

A. Assume that the Hamiltonian h is that of a simple mechanical system with symmetry. In this case, choosing $\mathcal{A} = \mathcal{A}_{\rm mech}$ given in (6.6.3), the curve $\xi(t) \in \mathfrak{g}_\mu$ in Step 2 is given by

$$\xi(t) = \mathcal{A}_{\rm mech}(q(t)) \left(d(t)^\sharp\right), \tag{6.6.28}$$

where $d(t)^\sharp \in TQ$ is the curve of tangent vectors associated to the curve of one-forms $d(t) \in T^*Q$ by the kinetic energy metric $\langle\!\langle \cdot, \cdot \rangle\!\rangle$ on Q.

B. If $Q = G$ is a Lie group, $\dim G_\mu = 1$, and ζ is a generator of $\mathfrak{g}_\mu$, then the connection $\mathcal{A} \in \Omega^1(G)$ can be chosen to equal

$$\mathcal{A}(g) := \frac{1}{\langle \mu, \zeta \rangle} T_g^* R_{g^{-1}}(\mu).$$

6.6.22 Reconstruction of dynamics for simple mechanical systems with symmetry. The case of simple mechanical systems with symmetry deserves special attention since several steps in the reconstruction method presented in §6.6.19 can be simplified. For example, taking advantage of the explicit expression of the mechanical connection, (6.6.28) can be made more explicit. Also, for simple mechanical systems the knowledge of the base integral curve $q(t)$ suffices to determine the entire integral curve on T^*Q. Indeed, if $h = k + v \circ \pi_q$ is the Hamiltonian, the Legendre transformation $\mathbb{F}h : T^*Q \to TQ$ determines the Lagrangian system on TQ given by $\ell(u_q) = \frac{1}{2}\|u_q\|^2 - v(u_q)$, for $u_q \in T_qQ$. Lagrange's equations are second order and thus the evolution of the velocities is given by the time derivative $q'(t)$ of the base integral curve. However, for simple mechanical systems, the Legendre transformation is a vector bundle isomorphism, $\mathbb{F}h = (\mathbb{F}\ell)^{-1}$, and hence the solution of the Hamiltonian system is given by $\mathbb{F}\ell(q'(t))$. Taking into account all of these facts and applying the reconstruction method in §6.6.19 yields the following method.

To find the integral curve $c(t)$ of the simple mechanical system with G-symmetry $h = k + v \circ \pi_Q$ on T^*Q with initial condition $c(0) = \alpha_q \in T_q^*Q$, knowing the integral curve $c_\mu(t)$ of the reduced Hamiltonian system on $(T^*Q)_\mu$ given by the reduced Hamiltonian function $h_\mu : (T^*Q)_\mu \to \mathbb{R}$ with initial condition $c_\mu(0) = [\alpha_q]$ one proceeds in the following manner:

- **Step 1.** *Compute the curve* $\varphi_\mu(c_\mu(t)) \in T^*(Q/G_\mu)$, *where* $\varphi_\mu : \left((T^*Q)_\mu, (\Omega_Q)_\mu\right) \to (T^*(Q/G_\mu), \Omega_{Q/G_\mu} - B_\mu)$ *is the symplectic embedding given in Theorem* 6.6.3 *and the magnetic term* B_μ *is induced by the mechanical connection* (6.6.3). *Then* $\varphi_\mu(c_\mu(t))$ *is an integral curve of the Hamiltonian system on* $\left(T^*(Q/G_\mu), \Omega_{Q/G_\mu} - B_\mu\right)$ *given by the function that is the sum of the kinetic energy of the quotient Riemannian metric and the quotient amended potential* $\widehat{v}_\mu$. *Let* $q_\mu(t) := \pi_{Q/G_\mu}(c_\mu(t))$ *be the base integral curve of this system, where* $\pi_{Q/G_\mu} : T^*(Q/G_\mu) \to Q/G_\mu$ *is the cotangent bundle projection.*

- **Step 2.** *Relative to the mechanical connection* $\mathcal{A}_{\rm mech} \in \Omega^1(Q; \mathfrak{g}_\mu)$, *horizontally lift* $q_\mu(t) \in Q/G_\mu$ *to a curve* $q_h(t) \in Q$ *passing through* $q_h(0) = q$.

- **Step 3.** *Determine* $\xi(t) \in \mathfrak{g}_\mu$ *from the algebraic system* $\langle\!\langle \xi(t)_Q(q_h(t)), \eta_Q(q_h(t)) \rangle\!\rangle = \langle \mu, \eta \rangle$ *for all* $\eta \in \mathfrak{g}_\mu$, *where* $\langle\!\langle \cdot, \cdot \rangle\!\rangle$ *is the G-invariant kinetic energy Riemannian metric on Q. This implies that* $q_h'(0)$ *and* $\xi(0)_Q(q)$ *are the horizontal and* *vertical components of the vector $\alpha_q^\sharp \in T_qQ$ which is asociated by the metric $\langle\!\langle \cdot, \cdot \rangle\!\rangle$ to the initial condition α_q.*

- **Step 4.** *Solve the nonautonomous equation $\dot{g}(t) = T_eL_{g(t)}\xi(t)$ in G_μ with initial condition $g(0) = e$.*

- **Step 5.** *The curve $q(t) := g(t) \cdot q_h(t)$, with $q_h(t)$ and $g(t)$ determined in Steps 2 and 4 respectively, is the base integral curve of the simple mechanical system with symmetry defined by the function h satisfying $q(0) = 0$. The curve $(\mathbb{F}h)^{-1}(q'(t)) \in T^*Q$ is the integral curve of this system with initial condition $c(0) = \alpha_q$. In addition, $q'(t) = g(t) \cdot \left(q_h'(t) + \xi(t)_Q(q_h(t))\right)$ is the horizontal plus vertical decomposition relative to the connection induced on $\mathbf{J}^{-1}(\mu) \to (T^*Q)_\mu$ by the mechanical connection $\mathcal{A}_{\mathrm{mech}} \in \Omega^1(Q; \mathfrak{g}_\mu)$.*

There are several important situations when Step 4, the main obstruction to an explicit solution of the reconstruction problem, can be carried out. We shall review some of them below.

6.6.23 The case $G_\mu = S^1$. If G_μ is Abelian, the equation in Step 4 has the solution $g(t) = \exp \int_0^t \xi(s)ds$. If, in addition, $G_\mu = S^1$, then $\xi(s)$ can be explicitly determined by Step 3. Indeed, if $\zeta \in \mathfrak{g}_\mu$ is a generator of $\mathfrak{g}_\mu$, writing $\xi(s) = a(s)\zeta$ for some smooth real valued function a defined on some open interval around the origin, the algebraic equation in Step 3 implies that $\langle\!\langle a(s)\xi(t)_Q(q_h(t)), \zeta_Q(q_h(t)) \rangle\!\rangle = \langle \mu, \zeta \rangle$ which gives

$$a(s) = \frac{\langle \mu, \zeta \rangle}{\|\zeta_Q(q_h(s))\|^2}.$$

Therefore, the base integral curve of the solution of the simple mechanical system with symmetry on T^*Q passing through q is

$$q(t) = \exp\left(\langle \mu, \zeta \rangle \int_0^t \frac{ds}{\|\zeta_Q(q_h(s))\|^2} \zeta \right) \cdot q_h(t)$$

and

$$q'(t) = \exp\left(\langle \mu, \zeta \rangle \int_0^t \frac{ds}{\|\zeta_Q(q_h(s))\|^2} \zeta \right) \cdot \left(q_h'(t) + \frac{\langle \mu, \zeta \rangle}{\|\zeta_Q(q_h(s))\|^2} \zeta_Q(q_h(t)) \right).$$

6.6.24 The case of compact Lie groups. An obvious situation when the differential equation in Step 4 can be solved is if $\xi(t) = \xi$ for all t, where ξ is a given element of $\mathfrak{g}_\mu$. Then the solution is $g(t) = \exp(t\xi)$. However, Step 3 puts certain restrictions under this hypothesis, because it requires that $\langle\!\langle \xi(t)_Q(q_h(t)), \eta_Q(q_h(t)) \rangle\!\rangle = \langle \mu, \eta \rangle$ for any $\eta \in \mathfrak{g}_\mu$. This is satisfied if there is a bilinear nondegenerate form $(\cdot, \cdot)$ on $\mathfrak{g}$ satisfying

$$(\zeta, \eta) = \langle\!\langle \zeta_Q(q), \eta_Q(q) \rangle\!\rangle \quad \text{for all} \quad q \in Q \quad \text{and} \quad \zeta, \eta \in \mathfrak{g}. \tag{6.6.29}$$

This is a very strong condition. It implies that $(\cdot, \cdot)$ is positive definite and invariant under the adjoint action of G on $\mathfrak{g}$. This excludes semisimple Lie algebras of noncompact type. Worse, if G is compact, which insures the existence of a positive adjoint invariant inner product on $\mathfrak{g}$, and $Q = G$, this condition implies that the kinetic energy metric is invariant under the adjoint action. Nevertheless, there are examples in which such conditions are natural, such as in Kaluza–Klein theories.

Concluding, if G is a compact Lie group and $(\cdot, \cdot)$ is a positive definite metric invariant under the adjoint action of G on $\mathfrak{g}$ satisfying (6.6.29), then the element $\xi(t)$ in Step 3 can be chosen to be constant and is determined by the identity $(\xi, \cdot) = \mu|_{\mathfrak{g}_\mu}$ on $\mathfrak{g}_\mu$. The solution of the equation on Step 4 is $g(t) = \exp(t\xi)$.

6.6.25 The case when $\dot{\xi}(t)$ is proportional to $\xi(t)$. Try to find a real valued function $f(t)$ such that $g(t) = \exp(f(t)\xi(t))$ is a solution of the equation $\dot{g}(t) = T_e L_{g(t)}\xi(t)$. Since $g(0) = e$, we require that $f(0) = 0$. Taking the time derivative of $g(t)$ yields $\dot{g}(t) = \left(T_{f(t)\xi(t)} \exp\right)(\dot{f}(t)\xi(t) + f(t)\dot{\xi}(t))$. On the other hand, by (2.1.2), we have $T_\zeta \exp(\zeta) = T_e T_{\exp\zeta}(\zeta)$, for any $\zeta \in \mathfrak{g}$, so we get

$$\begin{aligned} T_e L_{g(t)}\xi(t) &= T_e L_{\exp(f(t)\xi(t))}\xi(t) \\ &= \frac{1}{f(t)} T_e L_{\exp(f(t)\xi(t))}(f(t)\xi(t)) \\ &= \frac{1}{f(t)} T_{f(t)\xi(t)} \exp(f(t)\xi(t)) = T_{f(t)\xi(t)} \exp(\xi(t)). \end{aligned}$$

Thus, for t small enough, using the fact that the exponential map is a diffeomorphism around the origin, the equation $\dot{g}(t) = T_e L_{g(t)}\xi(t)$ holds for $g(t)$ of the form $g(t) = \exp(f(t)\xi(t))$ if and only if $\dot{f}(t)\xi(t) + f(t)\dot{\xi}(t) = \xi(t)$. This requires that $\xi(t)$ and $\dot{\xi}(t)$ be proportional. Thus if $\dot{\xi}(t) = \alpha(t)\xi(t)$ for some known smooth function $\alpha(t)$, then this equation can be solved in the following manner. Replacing this proportionality expression in the prior differential equation gives $\dot{f}(t) + \alpha(t)f(t) = 1$ with $f(0) = 0$, whose solution is

$$f(t) = \int_0^t \exp\left(\int_t^s \alpha(r)dr\right) ds.$$

6.6.26 The case of G_μ solvable. Write

$$g(t) = \exp(f_1(t)\xi_1) \exp(f_2(t)\xi_2) \ldots \exp(f_n(t)\xi_n),$$

for some basis $\{\xi_1, \xi_2, \ldots, \xi_n\}$ of $\mathfrak{g}_\mu$ and some smooth real valued functions f_i, $i = 1, 2, \ldots, n$, defined around zero. WEI AND NORMAN (1964) have shown that if G_μ is solvable, the equation in Step 4 can be solved by qudaratures for the f_i.

6.6.27 Reconstruction phases for simple mechanical systems with S^1 symmetry. Returning to the setting of 6.6.7, let $(Q, \langle\!\langle \cdot, \cdot \rangle\!\rangle)$ be a Riemannian manifold on which a Lie group G acts freely and properly by isometries. Consider the Hamiltonian h of a simple symmetric mechanical system, that is, $h = k + v \circ \pi_Q$, where $\pi_Q : T^*Q \to Q$ is the cotangent bundle projection, $v \in C^\infty(Q)$ is a G-invariant function, and k is the kinetic energy of the metric on T^*Q. If $\mu \in \mathfrak{g}^*$ is fixed, the μ-locked inertia tensor $\mathbb{I}_\mu(q) : \mathfrak{g}_\mu \to \mathfrak{g}_\mu^*$ is given for each $q \in Q$ by $\langle \mathbb{I}_\mu(q)(\zeta), \eta \rangle := \langle\!\langle \zeta_Q(q), \eta_Q(q) \rangle\!\rangle$, where $\zeta, \eta \in \mathfrak{g}_\mu$ are arbitrary, the mechanical connection $\mathcal{A}_{\text{mech}} \in \Omega^1(Q; \mathfrak{g}_\mu)$ is given by $\mathcal{A}_{\text{mech}}(q)(u_q) = \mathbb{I}_\mu(q)^{-1}\mathbf{J}(u_q^\flat)$, where the vector bundle isomorphism $\flat$:

$TQ \to T^*Q$ is induced by the metric via $u_q^\flat := \langle\langle u_q, \cdot \rangle\rangle \in T_q^*Q$ for any $u_q \in T_qQ$, and $\mathbf{J} : T^*Q \to \mathfrak{g}^*$ is the momentum map of the lifted G-action on T^*Q given by $\langle \mathbf{J}(\alpha_q), \xi \rangle = \alpha_q(\xi_Q(q))$ for any $\alpha_q \in T_q^*Q$ and $\xi \in \mathfrak{g}$, and the amended potential is given by $v_\mu := h \circ \alpha_\mu$, where $\alpha_\mu := \langle \mu, \mathcal{A}_{\mathrm{mech}} \rangle \in \Omega^1(Q)$. Denote by $\overline{v}_\mu : Q/G_\mu \to \mathbb{R}$ the induced function on the base.

Let $c : [0, T] \to T^*Q$ be an integral curve of the system with Hamiltonian h and suppose that its projection $c_\mu : [0, T] \to (T^*Q)_\mu$ to the reduced space is a closed integral curve of the reduced system with Hamiltonian h_μ. The ***reconstruction phase*** associated to the loop $c_\mu(t)$ is the group element $g \in G_\mu$, satisfying the identity $c(T) = g \cdot c(0)$.

In this subsection we shall present two explicit formulas of the reconstruction phase for the case when $G_\mu = S^1$. We let $\zeta \in \mathfrak{g}_\mu = \mathbb{R}$ be a generator of the coadjoint isotropy algebra and write $c(T) = \exp(\varphi\zeta) \cdot c(0)$; in this case φ is identified with the reconstruction phase and, as we shall see in concrete mehcanical examples, it truly represents an angle.

If $G_\mu = S^1$, the G_μ-principal bundle $\pi_\mu : \mathbf{J}^{-1}(\mu) \to (T^*Q)_\mu := \mathbf{J}^{-1}(\mu)/G_\mu$ admits two natural connections:

$$A = \frac{1}{\mu\zeta}\theta_\mu \in \Omega^1(\mathbf{J}^{-1}(\mu)),$$

where θ_μ is the pull-back of the canonical one-form on the cotangent bundle to the momentum level submanifold $\mathbf{J}^{-1}(\mu)$ discussed in §6.6.20 and

$$A_{\mathrm{mech}} = \pi_Q^* \mathcal{A}_{\mathrm{mech}} \in \Omega^1(\mathbf{J}^{-1}(\mu))$$

presented in §6.6.21. There is no reason to choose one connection over the other and thus there are two natural formulas for the reconstruction phase in this case.

Let $c_\mu(t)$ be a periodic orbit of period T of the reduced system and denote also by h_μ the value of the Hamiltonian function on it. Assume that D is a two-dimensional surface in $(T^*Q)_\mu$ whose boundary is the loop $c_\mu(t)$. Since, as a manifold, $(T^*Q)_\mu = T^*(Q/S^1)$ by Theorem 6.6.3, it makes sense to consider the base integral curve $q_\mu(t)$ obtained by projecting $c_\mu(t)$ to the base Q/S^1, which is a closed curve of period T. Denote by

$$\langle \overline{v}_\mu \rangle := \frac{1}{T}\int_0^T \overline{v}_\mu(q_\mu(t))dt$$

the average of $\overline{v}_\mu$ over the loop $q_\mu(t)$. Let $q_h(t) \in Q$ be the $\mathcal{A}_{\mathrm{mech}}$-horizontal lift of $q_\mu(t)$ to Q and let χ be the $\mathcal{A}_{\mathrm{mech}}$-holonomy of the loop $q_\mu(t)$ measured from $q(0)$, the base point of $c(0)$, computed, for example, by formula (5.1.2). Finally, denote, as usual, by ω_μ the reduced symplectic form on $(T^*Q)_\mu$. With these notations MARSDEN et al. (1990) found the following formulas for the phase φ:

$$\varphi = \frac{1}{\mu\zeta}\iint_D \omega_\mu + \frac{2(h_\mu - \langle \overline{v}_\mu \rangle)T}{\mu\zeta} = \chi + \mu\zeta \int_0^T \frac{ds}{\|\zeta_Q(q_h(s))\|^2}. \tag{6.6.30}$$

The first terms in both formulas are the so-called ***geometric phases*** because they carry only geometric information given by the connection, whereas the second terms are called the ***dynamic phases*** since they encapsulate information directly linked to the Hamiltonian. The expression of the total phase as a sum of a geometric and a dynamic phase is not intrinsic and is connection dependent. It can even happen that one of these summands vanishes.

6.6.28 Reconstruction phases for the free rigid body. The motion of the free rigid body is a geodesic with repsect to a left invariant Riemannian metric on SO(3) given by the moment of inertia of the body. For a detailed presentation of the facts below see MARSDEN AND RATIU (1999), Chapter 15. The phase space of the free rigid body motion is $T^* \mathrm{SO}(3)$ and a momentum map $\mathbf{J} : T^* \mathrm{SO}(3) \to \mathbb{R}^3$ of the lift of left translation to the cotangent bundle is given by right translation to the identity element. We have identified here $\mathfrak{so}(3)$ with $\mathbb{R}^3$ by the Lie algebra isomorphism $\mathbf{x} \in (\mathbb{R}^3, \times) \mapsto \hat{\mathbf{x}} \in (\mathfrak{so}(3), [\cdot, \cdot])$, where $\hat{\mathbf{x}}(\mathbf{y}) = \mathbf{x} \times \mathbf{y}$, and $\mathfrak{so}(3)^*$ with $\mathbb{R}^3$ by the inner product on $\mathbb{R}^3$. The reduced manifold $\mathbf{J}^{-1}(\mu)/G_\mu$ is identified with the sphere $S^2_{\|\mu\|}$ in $\mathbb{R}^3$ of radius $\|\mu\|$ with the symplectic form $\omega_\mu = -dS/\|\mu\|$, where dS is the standard area form on $S^2_{\|\mu\|}$ and where $G_\mu \cong S^1$ is the group of rotations around the axis μ. These concentric spheres are the coadjoint orbits of the Lie–Poisson space $\mathfrak{so}(3)^*$ and represent the level sets of the Casimir functions that are all smooth functions of $\|\Pi\|^2$, where $\Pi \in \mathbb{R}^3$ denotes the body angular momentum.

The Hamiltonian of the rigid body on the Lie–Poisson space $T^* \mathrm{SO}(3)/\mathrm{SO}(3) \cong \mathbb{R}^3$ is given by

$$h(\Pi) := \frac{1}{2}\left(\frac{\Pi_1^2}{I_1} + \frac{\Pi_2^2}{I_2} + \frac{\Pi_3^2}{I_3}\right)$$

where $I_1, I_2, I_3 > 0$ are the principal moments of inertia of the body. Let $\mathbb{I} := \operatorname{diag}(I_1, I_2, I_3)$ denote the moment of inertia tensor diagonalized in a principal axis body frame. The Lie–Poisson bracket on $\mathbb{R}^3$ is given by

$$\{f, g\}(\Pi) = -\Pi \cdot (\nabla f(\Pi) \times \nabla g(\Pi))$$

and the equation of motions are

$$\dot{\Pi} = \Pi \times \Omega,$$

where $\Omega \in \mathbb{R}^3$ is the body angular velocity given in terms of Π by $\Omega_i := \Pi/I_i$, for $i = 1, 2, 3$, that is, $\Omega = \mathbb{I}^{-1}\Pi$. The trajectories of the these equations are found by intersecting a family of homothetic energy ellipsoids with the angular momentum concentric spheres. If $I_1 > I_2 > I_3$, one immediately sees that all orbits are periodic with the exception of four centers (the two possible rotations about the long and the short moment of inertia axis of the body), two saddles (the two rotations about the middle moment of inertia axis of the body), and four heteroclinic orbits connecting the two saddles.

Suppose that $\Pi(t)$ is a periodic orbit on the sphere $S^2_{\|\mu\|}$ with period T. After time T by how much has the rigid body rotated in space? The answer to this question follows directly from (6.6.30). Taking $\zeta = \mu/\|\mu\|$ and the potiential $v \equiv 0$ we get

$$\begin{aligned}\varphi &= -\Lambda + \frac{2h_\mu T}{\|\mu\|}\\ &= \iint_D \frac{2\|\mathbb{I}\Pi(s)\|^2 - (\Pi(s)\cdot\mathbb{I}\Pi(s))(\operatorname{tr}\mathbb{I})}{(\Pi(s)\cdot\mathbb{I}\Pi(s))^2} ds + \|\mu\|^3 \int_0^T \frac{ds}{(\Pi(s)\cdot\mathbb{I}\Pi(s))},\end{aligned}$$

where D is one of the two spherical caps on $S^2_{\|\mu\|}$ whose boundary is the periodic orbit $\Pi(t)$, h_μ is the value of the total energy on the solution $\Pi(t)$, and Λ is the oriented solid angle, that is,

$$\Lambda := -\frac{1}{\|\mu\|}\iint_D \omega_\mu, \qquad |\Lambda| = \frac{\text{area } D}{\|\mu\|^2}.$$

An elementary proof of this formula is due to MONTGOMERY (1991a). The second formula is obtained by a direct computation of all the terms in (6.6.30) involving the mechanical connection for this case.

6.6.29 Reconstruction phases for the heavy top. We keep the same notations and conventions as in the previous subsection and refer to MARSDEN et al. (1984a), MARSDEN AND RATIU (1999), Chapter 14, and MARSDEN AND RATIU (2003) for proofs of the following general facts. The heavy top is a Hamiltonian system on $T^*\mathrm{SO}(3)$ whose Hamiltonian function is given by

$$h(\alpha_h) := \frac{1}{2}\|\alpha_h^\sharp\|^2 + Mg\ell\mathbf{k}\cdot h\chi,$$

where $h \in \mathrm{SO}(3)$, $\alpha_h \in T_h^*\mathrm{SO}(3)$, $\mathbf{k}$ is the unit vector of the spatial Oz axis (pointing in the opposite direction of the gravity force), $M \in \mathbb{R}$ is the total mass of the body, $g \in \mathbb{R}$ is the value of the gravitational acceleration, the fixed point about which the body moves is the origin, and χ is the unit vector of the straight line segment of length ℓ connecting the origin to the center of mass of the body. This Hamiltonian is left invariant under rotations about the spatial Oz axis. A momentum map induced by this S^1-action is given by $\mathbf{J} : T^*\mathrm{SO}(3) \to \mathbb{R}$, $\mathbf{J}(\alpha_h) = hT_e^*L_h(\alpha_h)\cdot\mathbf{k}$; recall that $T_e^*L_h(\alpha_h) =: \Pi \in \mathbb{R}^3$ is the body angular momentum. The reduced space $\mathbf{J}^{-1}(\mu)/S^1$ is generically the cotangent bundle of the unit sphere endowed with the symplectic structure given by the sum of the canonical form plus a magnetic term; equivalently, this is the coadjoint orbit in the dual of the Euclidean Lie algebra $\mathfrak{se}(3)^* = \mathbb{R}^3 \times \mathbb{R}^3$ given by $\mathcal{O}_\mu = \{(\Pi, \Gamma) \mid \Pi\cdot\Gamma = \mu, \|\Gamma\|^2 = 1\}$. The projection map $\mathbf{J}^{-1}(\mu) \to \mathcal{O}_\mu$ implementing the symplectic diffeomorphism between the reduced space and the coadjoint orbit in $\mathfrak{se}(3)^*$ is given by $\alpha_h \mapsto (\Pi, \Gamma) := (T_e^*L_h(\alpha_h), h^{-1}\mathbf{k})$. The orbit symplectic form ω_μ on $\mathcal{O}_\mu$ has the expression

$$\begin{aligned}&\omega_\mu(\Pi,\Gamma)((\Pi\times\mathbf{x}+\Gamma\times\mathbf{y}, \Gamma\times\mathbf{x}), (\Pi\times\mathbf{x}'+\Gamma\times\mathbf{y}', \Gamma\times\mathbf{x}'))\\ &\quad = -\Pi\cdot(\mathbf{x}\times\mathbf{x}') - \Gamma\cdot(\mathbf{x}\times\mathbf{y}' - \mathbf{x}'\times\mathbf{y})\end{aligned}$$

for any $\mathbf{x}, \mathbf{x}', \mathbf{y}, \mathbf{y}' \in \mathbb{R}^3$. The heavy top equations

$$\dot{\Pi} = \Pi\times\Omega + Mg\ell\Gamma\times\chi \qquad \dot{\Gamma} = \Gamma\times\Omega$$

are Lie–Poisson equations on $\mathfrak{se}(3)^*$ for the Hamiltonian

$$h(\Pi, \Gamma) = \frac{1}{2}\Pi \cdot \Omega + Mg\ell\Gamma \cdot \chi$$

and the Lie–Poisson bracket

$$\{f, g\}(\Pi, \Gamma) = -\Pi \cdot (\nabla_\Pi f \times \nabla_\Pi g) - \Gamma \cdot (\nabla_\Pi f \times \nabla_\Gamma g - \nabla_\Pi g \times \nabla_\Gamma f),$$

where ∇_Π and ∇_Γ denote the partial gradients.

Let $(\Pi(t), \Gamma(t))$ be a periodic orbit of period T of the heavy top equations. After time T by how much has the heavy top rotated in space? The answer is provided by (6.6.30). A direct computation carried out in MARSDEN et al. (1990) shows that in this case the two expressions in this formula become

$$\begin{aligned}\varphi &= \frac{1}{\mu}\iint_{\mathcal{D}} \omega_\mu + \frac{1}{\mu}\left(2h_\mu T - 2Mg\ell \int_0^T \Gamma(s) \cdot \chi ds\right) \\ &= \iint_D \frac{2\|\mathbb{I}\Gamma(s)\|^2 - (\Gamma(s) \cdot \mathbb{I}\Gamma(s))(\operatorname{tr}\mathbb{I})}{(\Gamma(s) \cdot \mathbb{I}\Gamma(s))^2} ds + \int_0^T \frac{ds}{\Gamma(s) \cdot \mathbb{I}\Gamma(s)},\end{aligned}$$

where D is the spherical cap on the unit sphere whose boundary is the closed curve $\Gamma(t)$ and $\mathcal{D}$ is a two-dimensional submanifold of the orbit $\mathcal{O}_\mu$ bounded by the closed integral curve $(\Pi(t), \Gamma(t))$. As usual, the first terms in each summand represent the geometric phase, and the second terms the dynamic phase.

6.7 Reduction by stages

Let (M, ω) be a symplectic manifold and G a Lie group acting canonically, freely, and properly on M via the map $\Phi : G \times M \to M$. Suppose that this action has a coadjoint equivariant momentum map $\mathbf{J}_G : M \to \mathfrak{g}^*$. Assume that G has a closed normal subgroup N. A natural question is whether the point symplectic reduction procedure described in Theorem 6.1.1 can be carried out in two steps, namely, first a reduction of M by N followed by a reduction by a group related to the quotient group G/N. This problem has been extensively studied in MARSDEN et al. (1998, 2002) and profusely illustrated with a number of interesting examples. In this section we will restrict ourselves to the description of the main theorem in MARSDEN et al. (2002). A generalization of this result to the singular case was given by ORTEGA (2003c) and will be presented in Chapter 9.

6.7.1 The reduction by stages strategy. The goal in this section is to provide a condition, called the ***stages hypothesis***, which allows the breaking up of the construction of the ***full reduced space***

$$M_\mu = \mathbf{J}_G^{-1}(\mu)/G_\mu, \quad \mu \in \mathfrak{g}^*,$$

into two steps. First, a point reduction for the N-action is carried out. Then a group related to the quotient group G/N is found that leaves this first reduced space invariant and it is shown that it admits a (not necessarily equivariant) momentum map, naturally induced by $\mathbf{J}_G$. Second, a point reduced space relative to this new action at an appropriately chosen point is carried out. Finally, it is shown that this second point reduced space is symplectomorphic to the full reduced space.

6.7.2 The action of N and the first symplectic reduced space. We start by studying the action of the closed normal subgroup $N \subset G$. Denote its Lie algebra by $\mathfrak{n}$. Let $i : \mathfrak{n} \hookrightarrow \mathfrak{g}$ be the inclusion and let $i^* : \mathfrak{g}^* \to \mathfrak{n}^*$ be its dual, which is the natural projection given by restriction of linear functionals. Since N is a normal subgroup, G acts on N by conjugation and therefore also on $\mathfrak{n}$ by the derivative of the conjugation at the identity element. This representation will be denoted by $(g, \xi) \in G \times \mathfrak{n} \mapsto g \cdot \xi := \mathrm{Ad}_g\, i(\xi) \in \mathfrak{n}$. Dualizing it, one obtains a representation of G on $\mathfrak{n}^*$ which satisfies

$$g \cdot \nu := g \cdot i^*(\mu) = i^*(\mathrm{Ad}^*_{g^{-1}}\, \mu),$$

where $\mu \in \mathfrak{g}^*$ satisfies $i^*(\mu) = \nu$.

A momentum map $\mathbf{J}_N : M \to \mathfrak{n}^*$ for the action of the group N on M is given by

$$\mathbf{J}_N(z) = i^*(\mathbf{J}_G(z)),$$

for any $z \in M$. $\mathbf{J}_N$ is not only N-equivariant, as one would expect, but also G-equivariant with respect to the actions of G on M and on $\mathfrak{n}^*$. Indeed,

$$\mathbf{J}_N(g \cdot z) = i^*(\mathbf{J}_G(g \cdot z)) = i^*(\mathrm{Ad}^*_{g^{-1}}(\mathbf{J}_G(z))) = g \cdot i^*(\mathbf{J}_G(z)) = g \cdot \mathbf{J}_N(z),$$

for any $z \in M$ and $g \in G$.

Let $\nu \in \mathfrak{n}^*$ be a value of $\mathbf{J}_N$ (which by the freeness of the action will be a regular value) and let N_ν be the isotropy subgroup of ν for the coadjoint action of N on the dual of its Lie algebra. We form the ***first symplectic reduced space***

$$M_\nu = \mathbf{J}_N^{-1}(\nu)/N_\nu.$$

In what follows $\pi_\nu : \mathbf{J}_N^{-1}(\nu) \to M_\nu$ denotes the natural projection, $i_\nu : \mathbf{J}_N^{-1}(\nu) \hookrightarrow M$ the inclusion, and ω_ν the reduced symplectic form on M_ν.

6.7.3 The action of the quotient group and its momentum map. We now proceed to symplectically reduce the space M_ν. In order to prepare for this second reduced space, we start with some general remarks. Since N is a normal subgroup, the adjoint action of G on its Lie algebra $\mathfrak{g}$ leaves the subalgebra $\mathfrak{n}$ invariant, and so it induces a dual action of G on $\mathfrak{n}^*$. By construction, the inclusion map $i : \mathfrak{n} \hookrightarrow \mathfrak{g}$ is equivariant with respect to the action of G on the domain and range. Thus, the dual $i^* : \mathfrak{g}^* \to \mathfrak{n}^*$ is equivariant with respect to the dual action of G. Because N is a subgroup of G, *the adjoint action of N on $\mathfrak{n}$ coincides with the restriction of the G-action on $\mathfrak{n}$ to the subgroup N*. Dualizing this, one sees that *the restriction of the G-action on $\mathfrak{n}^*$ to the subgroup N coincides with the coadjoint action of N on $\mathfrak{n}^*$*.

Let G_ν denote the isotropy subgroup of $\nu \in \mathfrak{n}^*$ for the G-action on $\mathfrak{n}^*$. It follows from the preceding remarks and normality of N in G that

$$N_\nu = G_\nu \cap N,$$

where N_ν is the isotropy subgroup at ν of the coadjoint action of N on $\mathfrak{n}^*$. Indeed, to see that $G_\nu \cap N \subset N_\nu$, let $n \in G_\nu \cap N$ so that, regarded as an element of G, it fixes ν. But since the action of N on $\mathfrak{n}^*$ induced by the action of G on $\mathfrak{n}^*$ coincides with the coadjoint action by the above remarks, this means that n fixes ν using the coadjoint action. The other inclusion is obvious.

It is an elementary fact that the intersection of a normal subgroup N with another subgroup is normal in that subgroup, and so the subgroup $N_\nu \subset G$ is normal in G_ν. Thus we can form the quotient group $H_\nu := G_\nu / N_\nu$. This quotient group will play an important role in what follows.

6.7.4 Lemma *There is a well-defined induced symplectic action of $H_\nu = G_\nu/N_\nu$ on the reduced space M_ν. This action will be denoted $\Psi_\nu : H_\nu \times M_\nu \to M_\nu$.*

Proof First, using the equivariance of $\mathbf{J}_G$ and i^*, we note that the action of G_ν on M leaves the set $\mathbf{J}_N^{-1}(\nu)$ invariant. The action of a group element $g \in G_\nu$ on this space will be denoted $\Phi_g^\nu : \mathbf{J}_N^{-1}(\nu) \to \mathbf{J}_N^{-1}(\nu)$.

Second, it is a general fact that when a group K acts on a manifold Q and $L \subset K$ is a normal subgroup, then the quotient group K/L acts on the quotient space Q/L. If $\pi_L : Q \to Q/L$ is the projection and $\Psi_k : Q \to Q$ denotes the given action of a group element $k \in K$, then the quotient action $\Psi_{[k]}^L : Q/L \to Q/L$ of the element $[k] \in K/L$ is defined through the identity

$$\Psi_{[k]}^L \circ \pi_L = \pi_L \circ \Psi_k,$$

that is, the projection π_L onto the quotient is equivariant with respect to the K-action on Q and the K/L-action on Q/L via the group projection $k \in K \mapsto [k] \in K/L$.

These general considerations show that the group G_ν/N_ν has a well-defined action on the manifold M_ν. The action of a group element $[g] \in G_\nu/N_\nu$ will be denoted by $(\Psi_\nu)_{[g]} : M_\nu \to M_\nu$. We shall now show that this action is symplectic.

Let $\pi_\nu : \mathbf{J}_N^{-1}(\nu) \to M_\nu$ denote the natural projection and let $i_\nu : \mathbf{J}_N^{-1}(\nu) \hookrightarrow M$ be the inclusion. By the equivariance of the projection, we have

$$(\Psi_\nu)_{[g]} \circ \pi_\nu = \pi_\nu \circ \Phi_g^\nu,$$

for all $g \in G_\nu$. Since the action Φ^ν is the restriction of the action Φ of G, we get

$$\Phi_g \circ i_\nu = i_\nu \circ \Phi_g^\nu$$

for each $g \in G_\nu$.

Recall from the reduction theorem that $i_\nu^*\omega = \pi_\nu^*\omega_\nu$. Therefore,

$$\pi_\nu^*(\Psi_\nu)_{[g]}^*\omega_\nu = (\Phi_g^\nu)^*\pi_\nu^*\omega_\nu = (\Phi_g^\nu)^* i_\nu^*\omega = i_\nu^*\Phi_g^*\omega = i_\nu^*\omega = \pi_\nu^*\omega_\nu.$$

Since π_ν is a surjective submersion, this implies that

$$(\Psi_\nu)_{[g]}^*\omega_\nu = \omega_\nu.$$

Thus, we have a symplectic action of G_ν/N_ν on M_ν. ■

In the next few results we prove that there is a momentum map $\mathbf{J}_{H_\nu} : M_\nu \to (\mathfrak{g}_\nu/\mathfrak{n}_\nu)^*$ for the action of $H_\nu = G_\nu/N_\nu$ on M_ν and we study its properties.

6.7.5 Lemma *Suppose N_ν is connected. Then a map $\mathbf{J}_{H_\nu} : M_\nu \to (\mathfrak{g}_\nu/\mathfrak{n}_\nu)^*$ is well-defined by the relation*

$$(r_\nu')^* \circ \mathbf{J}_{H_\nu} \circ \pi_\nu = k_\nu^* \circ \mathbf{J}_G \circ i_\nu - \bar{\nu} \tag{6.7.1}$$

where

$$r_\nu : G_\nu \to G_\nu/N_\nu$$

is the canonical projection,

$$r_\nu' : \mathfrak{g}_\nu \to \mathfrak{g}_\nu/\mathfrak{n}_\nu$$

is the induced Lie algebra homomorphism,

$$k_\nu : \mathfrak{g}_\nu \hookrightarrow \mathfrak{g}$$

is the inclusion,

$$\pi_\nu : \mathbf{J}_N^{-1}(\nu) \to M_\nu$$

is the projection,

$$i_\nu : \mathbf{J}_N^{-1}(\nu) \hookrightarrow M$$

is the inclusion, and $\bar{\nu}$ is some chosen extension of $\nu|_{\mathfrak{n}_\nu}$ to $\mathfrak{g}_\nu$. Equivalently, $\mathbf{J}_{H_\nu}$ is defined by

$$\left\langle \mathbf{J}_{H_\nu}([z]), [\xi] \right\rangle = \langle \mathbf{J}_G(z), \xi \rangle - \langle \bar{\nu}, \xi \rangle , \tag{6.7.2}$$

where $z \in \mathbf{J}_N^{-1}(\nu)$, $\xi \in \mathfrak{g}_\nu$, $\nu \in \mathfrak{n}^*$, $[z] = \pi_\nu(z)$ denotes the equivalence class of z in $M_\nu = \mathbf{J}_N^{-1}(\nu)/N_\nu$, and $[\xi] = r_\nu'(\xi)$ denotes the equivalence class of ξ in $\mathfrak{g}_\nu/\mathfrak{n}_\nu$.

Proof. First, we show that the definition is independent of the representative of $[\xi]$. To do this, it suffices to show that the right-hand side of (6.7.2) vanishes when $\xi \in \mathfrak{n}_\nu$. However, for $\xi \in \mathfrak{n}$, we have $\langle \mathbf{J}_G(z), \xi \rangle = \langle \mathbf{J}_N(z), \xi \rangle = \langle \nu, \xi \rangle$, since $\mathbf{J}_N(z) = \nu$. Therefore, if $\xi \in \mathfrak{n}_\nu \subset \mathfrak{n}$, this implies that $\langle \mathbf{J}_G(z), \xi \rangle - \langle \bar{\nu}, \xi \rangle = \langle \nu, \xi \rangle - \langle \nu, \xi \rangle = 0$.

Second, we show that the right-hand side is independent of the representative of the class $[z]$, which is the same as proving that $\langle \mathbf{J}_G(n \cdot z), \xi \rangle - \langle \bar{\nu}, \xi \rangle = \langle \mathbf{J}_G(z), \xi \rangle - \langle \bar{\nu}, \xi \rangle$ for any $n \in N_\nu$. This is clearly equivalent to showing that $\langle \mathbf{J}_G(n \cdot z), \xi \rangle = \langle \mathbf{J}_G(z), \xi \rangle$ for all $n \in N_\nu$. By equivariance of $\mathbf{J}_G$, this in turn is equivalent to the identity $\left\langle \mathbf{J}_G(z), \mathrm{Ad}_n^{-1} \xi \right\rangle = \langle \mathbf{J}_G(z), \xi \rangle$ or to $\left\langle \mathbf{J}_G(z), \mathrm{Ad}_n^{-1} \xi - \xi \right\rangle = 0$, for all $n \in N_\nu$. However, since $\xi \in \mathfrak{g}_\nu$ and N_ν is normal in G_ν, Lemma 2.1.13 yields $\mathrm{Ad}_n^{-1} \xi - \xi \in \mathfrak{n}_\nu$, so that the above identity becomes $\left\langle \nu, \mathrm{Ad}_n^{-1} \xi - \xi \right\rangle = 0$ because $z \in \mathbf{J}_N(\nu)$.

Thus one needs to show that for $z \in \mathbf{J}_N(\nu)$ and $\xi \in \mathfrak{g}_\nu$ fixed, the function $f : N_\nu \to \mathbb{R}$ given by

$$f(n) := \left\langle \mathbf{J}_G(z), \mathrm{Ad}_n^{-1} \xi - \xi \right\rangle = \left\langle \nu, \mathrm{Ad}_n^{-1} \xi - \xi \right\rangle$$

is identically zero. To show that f vanishes, note first that $f(e) = 0$. Second, the differential of f at the identity in the direction $\eta \in \mathfrak{n}_\nu$ is given by

$$\mathbf{d} f(e)(\eta) = \left\langle \nu, -\mathrm{ad}_\eta \xi \right\rangle = -\left\langle \mathrm{ad}_\eta^* \nu, \xi \right\rangle = 0$$

since $\mathrm{ad}^*_\eta \nu = 0$ because $\eta \in \mathfrak{n}_\nu$.

Third, we show that $f(n_1 n_2) = f(n_1) + f(n_2)$. To do this, we write

$$f(n_1 n_2) = \left\langle \nu, \mathrm{Ad}^{-1}_{n_1 n_2} \xi - \xi \right\rangle = \left\langle \nu, \mathrm{Ad}^{-1}_{n_2} \mathrm{Ad}^{-1}_{n_1} \xi - \mathrm{Ad}^{-1}_{n_2} \xi + \mathrm{Ad}^{-1}_{n_2} \xi - \xi \right\rangle .$$

However, since $n_2 \in N_\nu$, we have

$$\left\langle \nu, \mathrm{Ad}^{-1}_{n_2} \mathrm{Ad}^{-1}_{n_1} \xi - \mathrm{Ad}^{-1}_{n_2} \xi \right\rangle = \left\langle \mathrm{Ad}^*_{n_2^{-1}} \nu, \mathrm{Ad}^{-1}_{n_1} \xi - \xi \right\rangle = \left\langle \nu, \mathrm{Ad}^{-1}_{n_1} \xi - \xi \right\rangle .$$

This calculation shows that $f(n_1 n_2) = f(n_1) + f(n_2)$, as we desired.

Differentiating this relation with respect to n_1 at the identity gives

$$\mathbf{d} f(n_2) \circ T_e R_{n_2} = \mathbf{d} f(e) = 0$$

and hence $\mathbf{d} f = 0$ on N_ν. Since N_ν is connected by hypothesis, we conclude that $f = 0$ on N_ν, which is what we desired to show. ■

6.7.6 Proposition *The map $\mathbf{J}_{H_\nu} : M_\nu \to (\mathfrak{g}_\nu/\mathfrak{n}_\nu)^*$ is a momentum map for the action of G_ν/N_ν on M_ν. This momentum map is in general not coadjoint equivariant, and if M_ν is connected, it has a well-defined non-equivariance one-cocycle $\sigma : M_\nu \to (\mathfrak{g}_\nu/\mathfrak{n}_\nu)^*$ determined by the relation*

$$(r'_\nu)^*(\sigma([g])) = \mathrm{Ad}^*_{g^{-1}} \bar{\nu} - \bar{\nu}, \tag{6.7.3}$$

where $(r'_\nu)^* : (\mathfrak{g}_\nu/\mathfrak{n}_\nu)^* \to \mathfrak{g}^*_\nu$ is the dual of r'_ν. That is, we have the identity

$$\mathbf{J}_{H_\nu}([g] \cdot [z]) - \mathrm{Ad}^*_{[g]^{-1}} \mathbf{J}_{H_\nu}([z]) = \sigma([g]).$$

Proof. We first observe that $\mathbf{J}_{H_\nu}$ *depends on the extension $\bar{\nu}$ to $\mathfrak{g}_\nu$ of $\nu|\mathfrak{n}_\nu$ but only up to a constant element of $(\mathfrak{g}_\nu/\mathfrak{n}_\nu)^*$.* If $\bar{\nu}_1$ and $\bar{\nu}_2$ are two such extensions, then $(\bar{\nu}_1 - \bar{\nu}_2)|_{\mathfrak{n}_\nu} = 0$, and so it equals $(r'_\nu)^*(\rho)$ for some $\rho \in (\mathfrak{g}_\nu/\mathfrak{n}_\nu)^*$. Formula (6.7.2) shows that $\mathbf{J}_{H_{\nu_1}}([z]) - \mathbf{J}_{H_{\nu_2}}([z]) = -\rho$, for any $[z] \in M_\nu$. (This is precisely the ambiguity in the definition of momentum maps; recall that momentum maps are defined only up to the addition of constant elements in the dual of the Lie algebra.)

Secondly, we compute the infinitesimal generator given by $[\xi] = r'_\nu(\xi) \in \mathfrak{g}_\nu/\mathfrak{n}_\nu$. Since

$$\exp_\nu t r'_\nu(\xi) = r_\nu(\exp t\xi),$$

where $\exp_\nu : \mathfrak{g}_\nu/\mathfrak{n}_\nu \to G_\nu/N_\nu$ is the exponential map of the Lie group G_ν/N_ν and exp is that of G_ν, we get for $z \in \mathbf{J}_N^{-1}(\nu)$, using the definition of the G_ν/N_ν-action on M_ν,

$$\begin{aligned}
[\xi]_{M_\nu}([z]) &= \left.\frac{d}{dt}\right|_{t=0} \exp_\nu t r'_\nu(\xi) \cdot \pi_\nu(z) \\
&= \left.\frac{d}{dt}\right|_{t=0} r_\nu(\exp t\xi) \cdot \pi_\nu(z) \\
&= \left.\frac{d}{dt}\right|_{t=0} \pi_\nu(\exp t\xi \cdot z) \\
&= T_z \pi_\nu(\xi_M(z)),
\end{aligned}$$

that is,

$$[\xi]_{M_\nu}([z]) = T_z\pi_\nu(\xi_M(z)). \tag{6.7.4}$$

Thirdly, denote by $\mathbf{J}_G^\xi : M \to \mathbb{R}$, the map $\mathbf{J}_G^\xi(z) = \langle \mathbf{J}_G(z), \xi \rangle$ and similarly for $\mathbf{J}_{H_\nu}^{[\xi]} : M \to \mathbb{R}$, and note that (6.7.2) can be written as

$$\mathbf{J}_{H_\nu}^{[\xi]}(\pi_\nu(z)) = \mathbf{J}_G^\xi(z) - \langle \bar{\nu}, \xi \rangle.$$

Taking the z-derivative of this relation in the direction $v \in T_z\mathbf{J}_N^{-1}(\nu)$, we get

$$\mathbf{dJ}_{H_\nu}^{[\xi]}(\pi_\nu(z))\,(T_z\pi_\nu(v)) = \mathbf{dJ}_G^\xi(z)(v). \tag{6.7.5}$$

Letting ω_ν denote the symplectic form on M_ν, for $z \in \mathbf{J}_N^{-1}(\nu)$, $\xi \in \mathfrak{g}_\nu$, $v \in T_z\mathbf{J}_N^{-1}(\nu)$, we get from (6.7.4) and (6.7.5)

$$\begin{aligned}
\omega_\nu([z])\left([\xi]_{M_\nu}([z]), T_z\pi_\nu(v)\right) &= \omega_\nu(\pi_\nu(z))\,(T_z\pi_\nu(\xi_M(z)), T_z\pi_\nu(v)) \\
&= \left(\pi_\nu^*\omega_\nu\right)(z)\,(\xi_M(z), v) \\
&= \left(i_\nu^*\omega\right)(z)(\xi_M(z), v) = \mathbf{dJ}_G^\xi(z)(v) \\
&= d\mathbf{J}_{H_\nu}^{[\xi]}(\pi_\nu(z))\,(T_z\pi_\nu(v)),
\end{aligned} \tag{6.7.6}$$

which proves that $\mathbf{J}_{H_\nu}$ given by (6.7.1) is a momentum map for the G_ν/N_ν-action on M_ν.

We now compute the non-equivariance cocycle of $\mathbf{J}_{H_\nu}$. Since

$$\mathrm{Ad}_{r_\nu(g)}\, r_\nu'(\xi) = r_\nu'(\mathrm{Ad}_g\, \xi) \tag{6.7.7}$$

for any $g \in G_\nu$, $\xi \in \mathfrak{g}_\nu$, and $z \in \mathbf{J}_N^{-1}(\nu)$, we have

$$\begin{aligned}
&\langle \mathbf{J}_{H_\nu}([g]\cdot[z]) - \mathrm{Ad}^*_{[g]^{-1}}\, \mathbf{J}_{H_\nu}([z]), [\xi]\rangle \\
&\quad = \langle \mathbf{J}_{H_\nu}([g\cdot z]), [\xi]\rangle - \langle \mathbf{J}_{H_\nu}([z]), \mathrm{Ad}_{[g]^{-1}}[\xi]\rangle \\
&\quad = \langle \mathbf{J}_G(g\cdot z), \xi\rangle - \langle \bar{\nu}, \xi\rangle - \langle \mathbf{J}_{H_\nu}([z]), [\mathrm{Ad}_{g^{-1}}\,\xi]\rangle \\
&\quad = \langle \mathrm{Ad}^*_{g^{-1}}\, \mathbf{J}_G(z), \xi\rangle - \langle \bar{\nu}, \xi\rangle - \langle \mathbf{J}_G(z), \mathrm{Ad}_{g^{-1}}\,\xi\rangle + \langle \bar{\nu}, \mathrm{Ad}_{g^{-1}}\,\xi\rangle \\
&\quad = \langle \bar{\nu}, \mathrm{Ad}_{g^{-1}}\,\xi - \xi\rangle \\
&\quad = \langle \mathrm{Ad}^*_{g^{-1}}\,\bar{\nu} - \bar{\nu}, \xi\rangle.
\end{aligned} \tag{6.7.8}$$

Note that if $\xi \in \mathfrak{n}_\nu$, then $\mathrm{Ad}_{g^{-1}}\,\xi \in \mathfrak{n}_\nu$, since N_ν is a normal subgroup of G_ν. Therefore, denoting by $g\cdot\nu$ the action of $g \in G_\nu \subset G$ on $\nu \in \mathfrak{n}^*$, we have

$$\begin{aligned}
\langle \mathrm{Ad}^*_{g^{-1}}\,\bar{\nu}, \xi\rangle &= \langle \bar{\nu}, \mathrm{Ad}_{g^{-1}}\,\xi\rangle \\
&= \langle \nu, \mathrm{Ad}_{g^{-1}}\,\xi\rangle \\
&= \langle g\cdot\nu, \xi\rangle = \langle \nu, \xi\rangle,
\end{aligned}$$

since $g \in G_\nu$. This shows that

$$\mathrm{Ad}^*_{g^{-1}} \bar{\nu} - \bar{\nu} \in \mathfrak{n}^0_\nu$$

where $\mathfrak{n}^0_\nu = \{\lambda \in \mathfrak{g}^*_\nu \mid \lambda|_{\mathfrak{n}_\nu} = 0\}$ is the annihilator of $\mathfrak{n}_\nu$ in $\mathfrak{g}^*_\nu$.

However, since $r'_\nu : \mathfrak{g}_\nu \to \mathfrak{g}_\nu/\mathfrak{n}_\nu$ is surjective, its dual $(r'_\nu)^* : (\mathfrak{g}_\nu/\mathfrak{n}_\nu)^* \to \mathfrak{g}^*_\nu$ is injective and it is easy to verify that

$$(r'_\nu)^* \left((\mathfrak{g}_\nu/\mathfrak{n}_\nu)^*\right) \subset \mathfrak{n}^0_\nu.$$

Since

$$\dim \left((r'_\nu)^* \left((\mathfrak{g}_\nu/\mathfrak{n}_\nu)^*\right)\right) = \dim\, (\mathfrak{g}_\nu/\mathfrak{n}_\nu) = \dim \mathfrak{g}_\nu - \dim \mathfrak{n}_\nu = \dim \mathfrak{n}^0_\nu$$

it follows that

$$(r'_\nu)^* \left((\mathfrak{g}_\nu/\mathfrak{n}_\nu)^*\right) = \mathfrak{n}^0_\nu. \tag{6.7.9}$$

Because of (6.7.9) it follows that there is a unique $\sigma(g) \in (\mathfrak{g}_\nu/\mathfrak{n}_\nu)^*$ such that

$$\mathrm{Ad}^*_{g^{-1}} \bar{\nu} - \bar{\nu} = (r'_\nu)^*(\sigma(g)),$$

which by (6.7.8) yields

$$\mathbf{J}_{H_\nu}([g] \cdot [z]) - \mathrm{Ad}^*_{[g]^{-1}} \mathbf{J}_{H_\nu}([z]) = \sigma(g) \tag{6.7.10}$$

for all $z \in \mathbf{J}_N^{-1}(\nu)$.

Let $g_1, g_2 \in G_\nu$. Dualizing relation (6.7.7) we get

$$\begin{aligned}
(r'_\nu)^*(\sigma(g_1 g_2)) &= \mathrm{Ad}^*_{(g_1 g_2)^{-1}} \bar{\nu} - \bar{\nu} \\
&= \mathrm{Ad}^*_{g_1^{-1}} \mathrm{Ad}^*_{g_2^{-1}} \bar{\nu} - \mathrm{Ad}^*_{g_1^{-1}} \bar{\nu} + \mathrm{Ad}^*_{g_1^{-1}} \bar{\nu} - \bar{\nu} \\
&= \mathrm{Ad}^*_{g_1^{-1}} \left(\mathrm{Ad}^*_{g_2^{-1}} \bar{\nu} - \bar{\nu} \right) + \mathrm{Ad}^*_{g_1^{-1}} \bar{\nu} - \bar{\nu} \\
&= \mathrm{Ad}^*_{g_1^{-1}} (r'_\nu)^*(\sigma(g_2)) + (r'_\nu)^*(\sigma(g_1)) \\
&= (r'_\nu)^*(\sigma(g_1)) + (r'_\nu)^* \left(\mathrm{Ad}^*_{[g_1]^{-1}} \sigma(g_2) \right) \\
&= (r'_\nu)^* \left(\sigma(g_1) + \mathrm{Ad}^*_{[g_1]^{-1}} \sigma(g_2) \right).
\end{aligned}$$

Injectivity of $(r'_\nu)^*$ implies that

$$\sigma(g_1 g_2) = \sigma(g_1) + \mathrm{Ad}^*_{[g_1]^{-1}} \sigma(g_2). \tag{6.7.11}$$

In particular, if $g \in G_\nu$ and $n \in N_\nu$, this relation yields

$$\sigma(ng) = \sigma(n) + \mathrm{Ad}^*_{[n]^{-1}} \sigma(g) = \sigma(n) + \sigma(g),$$

since $[n] = e$. Now we show that $\sigma(n) = 0$. Indeed, if $\xi \in \mathfrak{g}_\nu$,

$$\begin{aligned}\langle (r'_\nu)^*(\sigma(n)), \xi\rangle &= \langle \mathrm{Ad}^*_{n^{-1}} \bar{\nu} - \bar{\nu}, \xi\rangle \\ &= \langle \bar{\nu}, \mathrm{Ad}_{n^{-1}} \xi - \xi\rangle \\ &= \langle \nu, \mathrm{Ad}_{n^{-1}} \xi - \xi\rangle, \end{aligned} \tag{6.7.12}$$

since by Lemma 2.1.13, $\mathrm{Ad}_{n^{-1}} \xi - \xi \in \mathfrak{n}_\nu$. However, we already showed in the proof of Lemma 6.7.5 that for N_ν connected

$$\langle \nu, \mathrm{Ad}_{n^{-1}} \xi - \xi\rangle = 0.$$

Thus, for any $n \in N_\nu$, $g \in G_\nu$, we have $\sigma(ng) = \sigma(g)$, which proves that $\sigma(n)$ does depend on $[g]$ and not on g. Denoting this map by the same letter $\sigma : G_\nu/N_\nu \to (\mathfrak{g}_\nu/\mathfrak{n}_\nu)^*$ formula (6.7.11) shows that it is a one-cocycle on G_ν/N_ν and formula (6.7.10) that it is the non-equivariance cocycle of the momentum map $\mathbf{J}_{H_\nu}$. ■

Although the momentum map $\mathbf{J}_{H_\nu}$ depends on the extension of $\nu|\mathfrak{n}_\nu$ to $\mathfrak{g}_\nu^*$, different choices of extension lead to momentum maps that are equivalent with respect to the cohomology class of their cocycles. That is, if $\bar{\nu}$ and $\bar{\nu}'$ are two choices extending $\nu|_{\mathfrak{n}_\nu}$, then it can be easily shown that for all $[g] \in G_\nu/N_\nu$,

$$\sigma^{\bar{\nu}}([g]) - \sigma^{\bar{\nu}'}([g]) = \lambda - \mathrm{Ad}^*_{[g]^{-1}} \lambda,$$

for the unique $\lambda \in (\mathfrak{g}_\nu/\mathfrak{n}_\nu)^*$ that satisfies

$$(r'_\nu)^*\lambda = \bar{\nu}' - \bar{\nu}.$$

This demonstrates that the two group one-cocycles on G_ν/N_ν differ by a coboundary.

In order to carry out point reduction with the momentum map $\mathbf{J}_{H_\nu}$ we need to compute the isotropy subgroups of the affine action of H_ν on $(\mathfrak{g}_\nu/\mathfrak{n}_\nu)^*$ induced by its non-equivariance cocycle σ. In this particular case, for $\lambda \in (\mathfrak{g}_\nu/\mathfrak{n}_\nu)^*$, the affine action is given by the expression $[g] \cdot \lambda := \mathrm{Ad}^*_{[g]^{-1}} \lambda + \sigma([g])$, for any $[g] \in H_\nu$.

In terms of a given μ (the value at which we will do the full reduction, as above), we can define $\nu = \mu|_\mathfrak{n} \in \mathfrak{n}^*$ and $\rho \in (\mathfrak{g}_\nu/\mathfrak{n}_\nu)^*$ by

$$(r'_\nu)^*(\rho) = \mu|_{\mathfrak{g}_\nu} - \bar{\nu}, \tag{6.7.13}$$

where $\bar{\nu}$ is an arbitrary extension of $\nu|_{\mathfrak{n}_\nu}$ to $\mathfrak{g}_\nu$. Observe that ρ depends on the choice of the extension $\bar{\nu}$ of ν. Equation (6.7.13) makes sense since, for $\eta \in \mathfrak{n}_\nu$, the right-hand side satisfies

$$\langle \mu|_{\mathfrak{g}_\nu} - \bar{\nu}, \eta\rangle = \langle \mu, \eta\rangle - \langle \bar{\nu}, \eta\rangle = \langle \nu, \eta\rangle - \langle \nu, \eta\rangle = 0,$$

that is, $\mu|_\nu - \bar{\nu} \in \mathfrak{n}_\nu^0$.

6.7.7 The second symplectic reduced space. In the first stage of reduction, we have reduced M by the action of N at the point ν to obtain M_ν. Now M_ν can be further reduced by the action of G_ν/N_ν at the value $\rho \in (\mathfrak{g}_\nu/\mathfrak{n}_\nu)^*$ defined in (6.7.13). Let this ***second symplectic reduced space*** be denoted by

$$(M_\nu)_\rho = \mathbf{J}_{H_\nu}^{-1}(\rho)/(G_\nu/N_\nu)_\rho$$

where, as usual, $(G_\nu/N_\nu)_\rho$ is the isotropy subgroup for the affine action of the group G_ν/N_ν on the dual of its Lie algebra.

The different Lie algebraic elements in this expression are linked to each other by the compatibility relation (6.7.13) equivalently written as

$$(r'_\nu)^*(\rho) = k_\nu^*\mu - \bar{\nu}. \tag{6.7.14}$$

6.7.8 Lemma *We have* $N_\nu \subset (G_\nu)_{\mu|_{\mathfrak{g}_\nu}}$ *and*

$$(G_\nu/N_\nu)_\rho = r_\nu\left((G_\nu)_{\mu|_{\mathfrak{g}_\nu}}\right) = (G_\nu)_{\mu|_{\mathfrak{g}_\nu}}/N_\nu.$$

Proof. It is obvious that $N_\nu \subset G_\nu$. We shall prove the lemma by showing that $(G_\nu/N_\nu)_\rho = (G_\nu)_{\mu|_{\mathfrak{g}_\nu}}/N_\nu$. This implies that $N_\nu \subset (G_\nu)_{\mu|_{\mathfrak{g}_\nu}}$ and also the second equality.

We have that $[g] \in (G_\nu/N_\nu)_\rho$ if and only if

$$[g] \cdot \rho = \rho. \tag{6.7.15}$$

Since $r'_\nu : \mathfrak{g}_\nu \to \mathfrak{g}_\nu/\mathfrak{n}_\nu$ is a projection, its dual $(r'_\nu)^* : (\mathfrak{g}_\nu/\mathfrak{n}_\nu)^* \to \mathfrak{g}_\nu^*$ is injective and hence (6.7.15) is equivalent to $(r'_\nu)^*([g]\cdot\rho) = (r'_\nu)^*(\rho)$. However, by (6.7.14), $(r'_\nu)^*(\rho) = k_\nu^*\mu - \bar{\nu} = \mu|_{\mathfrak{g}_\nu} - \overline{\nu}$. On the other hand, by the definition of the coadjoint action of G_ν/N_ν, (6.7.14), and (6.7.3) we have

$$\begin{aligned}
(r'_\nu)^*([g]\cdot\rho) &= (r'_\nu)^*\left(\mathrm{Ad}^*_{[g]^{-1}}\,\rho + \sigma([g])\right)\\
&= \mathrm{Ad}^*_{g^{-1}}(r'_\nu)^*\rho + (r'_\nu)^*(\sigma([g]))\\
&= \mathrm{Ad}^*_{g^{-1}}(\mu|_{\mathfrak{g}_\nu} - \bar{\nu}) + \mathrm{Ad}^*_{g^{-1}}\,\bar{\nu} - \bar{\nu}\\
&= \mathrm{Ad}^*_{g^{-1}}(\mu|_{\mathfrak{g}_\nu}) - \bar{\nu}\\
&= \left(\mathrm{Ad}^*_{g^{-1}}\,\mu\right)|_{\mathfrak{g}_\nu} - \bar{\nu},
\end{aligned}$$

since $g \in G_\nu$. Therefore (6.7.15) is equivalent to

$$\left(\mathrm{Ad}^*_{g^{-1}}\,\mu\right)|_{\mathfrak{g}_\nu} = \mu|_{\mathfrak{g}_\nu},$$

that is, $g \in (G_\nu)_{\mu|_{\mathfrak{g}_\nu}}$. Therefore we showed that $[g] \in (G_\nu/N_\nu)_\rho$ if and only if $g \in (G_\nu)_{\mu|_{\mathfrak{g}_\nu}}$ for all representatives g of $[g]$. This proves that

$$(G_\nu/N_\nu)_\rho = (G_\nu)_{\mu|_{\mathfrak{g}_\nu}}/N_\nu. \quad \blacksquare$$

6.7.9 The Stages Hypothesis. Now we shall require a special hypothesis to state the main reduction by stages theorem. This hypothesis is satisfied by a large number of examples (see MARSDEN et al. (2002)).

Stages Hypothesis: *For all* $\mu_1, \mu_2 \in \mathfrak{g}^*$ *such that*

$$\mu_1|_{\mathfrak{n}} = \mu_2|_{\mathfrak{n}} := \nu \quad \textit{and} \quad \mu_1|_{\mathfrak{g}_\nu} = \mu_2|_{\mathfrak{g}_\nu} := \tau,$$

there exists $n \in (G_\nu)_\tau$ *such that* $\mu_2 = \mathrm{Ad}^*_{n^{-1}}\,\mu_1$.

Notice that the stages hypothesis involves only the structure of the group G and does not involve the action on M.

We are now ready to state the Reduction by Stages Theorem under the Stages Hypothesis assumption.

6.7.10 Theorem (Reduction by Stages) *Let $(M\omega)$ be a symplectic manifold acted freely, properly, and canonically upon by a Lie group G. Suppose that this action has a coadjoint equivariant momentum map $\mathbf{J}_G : M \to \mathfrak{g}^*$. Let N be a normal subgroup of G and $\mathbf{J}_N = i^*\mathbf{J}_G$ an associated momentum map (we use the notations introduced in the previous pages). Let $\mu \in \mathfrak{g}^*$ be a value of $\mathbf{J}_G$ and $\nu = i^*(\mu) \in \mathfrak{n}^*$. Let $M_\nu = \mathbf{J}_N^{-1}(\nu)/N_\nu$ be the symplectic reduced space by the N-action. Assume that the isotropy subgroup N_ν and the reduced space M_ν are connected. Then the quotient group $H_\nu := G_\nu/N_\nu$ acts canonically on M_ν and has an associated momentum map $\mathbf{J}_{H_\nu}$ given by* (6.7.1) *with a well-defined non-equivariance cocycle provided by* (6.7.3).

If the Stages Hypothesis holds, then there is a symplectic diffeomorphism between $M_\mu := \mathbf{J}_G^{-1}(\mu)/G_\mu$ and $(M_\nu)_\rho := \mathbf{J}_{H_\nu}^{-1}(\rho)/(H_\nu)_\rho$, where μ, ν and ρ satisfy (6.7.14).

This theorem states, in particular, that the spaces $(M_\nu)_\rho$ are mutually symplectically diffeomorphic for different choices of extensions $\bar{\nu}$ and so, correspondingly, different choices of ρ.

Proof. Since μ and ν are related by $\nu = \mu|_\mathfrak{n} \in \mathfrak{n}^*$, there is a natural inclusion map

$$j_\mu : \mathbf{J}_G^{-1}(\mu) \to \mathbf{J}_N^{-1}(\nu).$$

Composing this map with π_ν, we get the smooth map

$$\pi_\nu \circ j_\mu : \mathbf{J}_G^{-1}(\mu) \to M_\nu.$$

This map takes values in $\mathbf{J}_{H_\nu}^{-1}(\rho)$ because of the relations

$$(r'_\nu)^* \circ \mathbf{J}_{H_\nu} \circ \pi_\nu = k_\nu^* \circ \mathbf{J}_G \circ i_\nu - \bar{\nu}$$

and $(r'_\nu)^*(\rho) = \mu|_{\mathfrak{g}_\nu} - \bar{\nu}$. Thus, we can regard it as a map

$$\pi_\nu \circ j_\mu : \mathbf{J}_G^{-1}(\mu) \to \mathbf{J}_{H_\nu}^{-1}(\rho).$$

We claim that this map drops on the quotient to a map $[\pi_\nu \circ j_\mu] : M_\mu \to (M_\nu)_\rho$. To see that this is true, define for any element $z \in \mathbf{J}_G^{-1}(\mu)$

$$[\pi_\nu \circ j_\mu][z] := [(\pi_\nu \circ j_\mu)(z)],$$

where $[z] \in \mathbf{J}_\mu^{-1}(\mu)/G_\mu = M_\mu$ and $[(\pi_\nu \circ j_\mu)(z)] \in \mathbf{J}_{H_\nu}^{-1}(\rho)/(G_\nu/N_\nu)_\rho = (M_\nu)_\rho$. We need to show that this is well defined; that is, if $z' = h \cdot z$ for $h \in G_\mu$, then

$$[(\pi_\nu \circ j_\mu)(z')] = [(\pi_\nu \circ j_\mu)(z)].$$

We pause for a moment to show that there is a natural smooth Lie group homomorphism $\psi : G_\mu \to (G_\nu/N_\nu)_\rho$ defined by $g \mapsto [g]$. To explain this, note that $G_\mu \subset G_\nu$

since $\nu = \mu|_{\mathfrak{n}}$ and since G acts on $\mathfrak{n}^*$ by dualizing the restriction of the Ad action of G on $\mathfrak{g}$ to $\mathfrak{n}$. Thus, for $g \in G_\mu$, we can form its class $[g] \in G_\nu/N_\nu$. We claim that this class actually lies in $(G_\nu/N_\nu)_\rho$. To see this, recall from Proposition 6.7.8 that $(G_\nu/N_\nu)_\rho = r_\nu((G_\nu)_{\mu|_{\mathfrak{g}_\nu}})$. Since $[g] = r_\nu(g)$, it suffices to show that $g \in (G_\nu)_{\mu|_{\mathfrak{g}_\nu}}$. But this is clear since g leaves μ invariant.

Next, we show that the map $\pi_\nu \circ j_\mu : \mathbf{J}_G^{-1}(\mu) \to \mathbf{J}_{H_\nu}^{-1}(\rho)$ is ψ-equivariant with respect to the action of G_μ on its domain and $(G_\nu/N_\nu)_\rho$ on the range. By the definition of the action of $(G_\nu/N_\nu)_\rho$ on $\mathbf{J}_{H_\nu}^{-1}(\rho)$ we have, for $h \in G_\mu$,

$$(\pi_\nu \circ j_\mu)(h \cdot z) = \pi_\nu(h \cdot z) = [h \cdot z] = [h] \cdot [z] = \psi(h) \cdot (\pi_\nu \circ j_\mu)(z).$$

Therefore,

$$\begin{aligned}[(\pi_\nu \circ j_\mu)(z')] &= [(\pi_\nu \circ j_\mu)(h \cdot z)] \\ &= [\psi(h)(\pi_\nu \circ j_\mu)(z)] \\ &= [(\pi_\nu \circ j_\mu)(z)]\end{aligned}$$

since $\psi(h) \in (G_\nu/N_\nu)_\rho$. This shows that we have constructed a smooth map

$$F := [\pi_\nu \circ j_\mu] : M_\mu \to (M_\nu)_\rho.$$

We now show that F is actually the symplectomorphism that we need.

***F* is Injective.** Let $z, z' \in \mathbf{J}_G^{-1}(\mu)$ be such that $[(\pi_\nu \circ j_\mu)(z)] = F(\pi_\mu(z)) = F(\pi_\mu(z'))$ $= [(\pi_\nu \circ j_\mu)(z')]$. Thus there is an element $[g] \in (G_\nu/N_\nu)_\rho$ such that $(\pi_\nu \circ j_\mu)(z') = [g] \cdot (\pi_\nu \circ j_\mu)(z) = (\pi_\nu \circ j_\mu)(g \cdot z)$ and so there is an element $n \in N_\nu$ such that $z' = ng \cdot z$. Since both z and z' lie in $\mathbf{J}_G^{-1}(\mu)$ we have that $\mu = \mathbf{J}_G(z') = \mathbf{J}_G(ng \cdot z) = ng \cdot \mathbf{J}_G(z) = ng \cdot \mu$, that is, $ng \in G_\mu$ and therefore $\pi_\mu(z) = \pi_\mu(z')$, as required.

***F* is Surjective.** Let $\pi_\rho([z]) \in (M_\nu)_\rho$ arbitrary, with $[z] = \pi_\nu(z) \in \mathbf{J}_{H_\nu}^{-1}(\rho)$ and $z \in \mathbf{J}_N^{-1}(\nu)$; $\pi_\rho : \mathbf{J}_{H_\nu}^{-1}(\rho) \to (M_\nu)_\rho$ denotes the quotient projection. Let $\mu' := \mathbf{J}_G(z)$. Notice that $\mu'|_{\mathfrak{n}} = i^*\mathbf{J}_G(z) = \mathbf{J}_N(z) = \nu = \mu|_{\mathfrak{n}}$. Also, for any $\xi \in \mathfrak{g}_\nu$ we have that

$$\langle \mu', \xi \rangle = \langle \mathbf{J}_G(z), \xi \rangle = \langle \mathbf{J}_{H_\nu}([z]), r'_\nu(\xi) \rangle + \langle \bar{\nu}, \xi \rangle = \langle \rho, r'_\nu(\xi) \rangle + \langle \bar{\nu}, \xi \rangle = \langle \mu, \xi \rangle.$$

Consequently, $\mu'|_{\mathfrak{g}_\nu} = \mu|_{\mathfrak{g}_\nu} =: \tau$ and therefore, by the Stages Hypothesis, there exists an element $n \in (G_\nu)_\tau$ such that $\mu' = \mathrm{Ad}^*_{n^{-1}} \mu$. As $n \in (G_\nu)_\tau$, we get $r_\nu(n) \in r_\nu((G_\nu)_{\mu|_{\mathfrak{g}_\nu}}) = (G_\nu/N_\nu)_\rho$, by Lemma 6.7.8. Consider now the element $z' := n^{-1} \cdot z \in \mathbf{J}_G^{-1}(\mu)$. By construction, we have

$$F(\pi_\mu(z')) = \pi_\rho(\pi_\nu(j_\mu(z'))) = \pi_\rho([n^{-1}] \cdot \pi_\nu(z)) = \pi_\rho(\pi_\nu(z)) = \pi_\rho([z]),$$

which proves the surjectivity of F.

***F* is Symplectic.** It is a simple diagram chasing exercise. As a consequence of this, it also follows that F is an *immersion*.

Our next task is the construction of an inverse to the map F and prove that this inverse is smooth. We start by considering the G_ν principal bundle, $\mathbf{J}_N^{-1}(\nu) \to \mathbf{J}_N^{-1}(\nu)/G_\nu$ which we can construct from the G equivariance of $\mathbf{J}_N$ and the freeness and properness of the G action. Because N_ν is a normal subgroup of G_ν, we can factor this quotient map into a sequence, $\mathbf{J}_N^{-1}(\nu) \to \mathbf{J}_N^{-1}(\nu)/N_\nu \to \mathbf{J}_N^{-1}(\nu)/G_\nu$ where the first bundle is an N_ν-principal bundle and the second is an G_ν/N_ν-principal bundle. We refine this sequence of bundles as follows. From the submanifold $\mathbf{J}_{H_\nu}^{-1}(\rho)$ in $\mathbf{J}_N^{-1}(\nu)/N_\nu$ and the fact that π_ν is a surjective submersion, we form the submanifold $\pi_\nu^{-1}(\mathbf{J}_{H_\nu}^{-1}(\rho))$ of $\mathbf{J}_N^{-1}(\nu)$. By construction, this submanifold is an N_ν-principal bundle over $\mathbf{J}_{H_\nu}^{-1}(\rho)$. Recall that $(G_\nu)_{\mu|_{\mathfrak{g}_\nu}}/N_\nu = (G_\nu/N_\nu)_\rho$ and, consequently, $r_\nu^{-1}((G_\nu/N_\nu)_\rho) = (G_\nu)_{\mu|_{\mathfrak{g}_\nu}}$. Since $\mathbf{J}_{H_\nu}^{-1}(\rho)$ is the quotient by the N_ν-action of $\pi_\nu^{-1}(\mathbf{J}_{H_\nu}^{-1}(\rho))$, it follows that the subgroup of G_ν that acts on $\pi_\nu^{-1}(\mathbf{J}_{H_\nu}^{-1}(\rho))$ is $(G_\nu)_{\mu|_{\mathfrak{g}_\nu}}$. Since N_ν is a normal subgroup of $(G_\nu)_{\mu|_{\mathfrak{g}_\nu}}$, we can compute the quotient space of this action in stages. We then obtain the sequence of bundles

$$\pi_\nu^{-1}(\mathbf{J}_{H_\nu}^{-1}(\rho)) \to \mathbf{J}_{H_\nu}^{-1}(\rho) \to \mathbf{J}_{H_\nu}^{-1}(\rho)/(G_\nu/N_\nu)_\rho.$$

The first bundle has group N_ν, the second has group $(G_\nu/N_\nu)_\rho$. Clearly this sequence of bundles embeds in the previous sequence with structure groups as subgroups.

Definition of the Inverse. Given $[z] \in \mathbf{J}_{H_\nu}^{-1}(\rho)$, let $z \in \pi_\nu^{-1}(\mathbf{J}_{H_\nu}^{-1}(\rho))$ be in the fiber over $[z]$. Let $\mu' := \mathbf{J}_G(z)$. Since for any representative z of $[z]$ we have that $\mathbf{J}_N(z) = \nu = \mu'|_\mathfrak{n}$, we can write $\mu|_\mathfrak{n} = \mu'|_\mathfrak{n}$. Then, as in the proof that F is surjective, for any $\xi \in \mathfrak{g}_\nu$, we have

$$\langle \mu', \xi \rangle = \langle \mathbf{J}_{H_\nu}([z]), [\xi] \rangle + \langle \bar{\nu}, \xi \rangle$$
$$= \langle (r_\nu')^* \rho, \xi \rangle + \langle \bar{\nu}, \xi \rangle = \langle \mu, \xi \rangle.$$

Therefore, by the Stages Hypothesis, there is an $g \in (G_\nu)_{\mu|_{\mathfrak{g}_\nu}}$ such that $g \cdot \mu' = \mu$, and so $g \cdot z \in \mathbf{J}_G^{-1}(\mu)$. We then define $\phi([z]) := \pi_\mu(g \cdot z)$. This map $\phi : \mathbf{J}_{H_\nu}^{-1}(\rho) \to M_\mu$ will induce a map on the quotient which is clearly, by construction, an inverse to F.

Well-Definedness. To see that ϕ is well defined, take $z' \neq z$ in the fiber over $[z]$, and let $g_{z'} \in (G_\nu)_{\mu|_{\mathfrak{g}_\nu}}$ be such that $g_{z'} \cdot z' \in \mathbf{J}_G^{-1}(\mu)$. We must show that $g \cdot z$ and $g_{z'} \cdot z'$ are on the same G_μ-orbit in $\mathbf{J}_G^{-1}(\mu)$. Since, by construction, both points are in $\mathbf{J}_G^{-1}(\mu)$, it suffices to show they are on the same G-orbit. Since $z' = n \cdot z$ for some $n \in N_\nu \subset (G_\nu)_{\mu|_{\mathfrak{g}_\nu}}$, we have $g_{z'} \cdot z' = g_{z'} \cdot n \cdot z = (g_{z'} n g^{-1}) \cdot g \cdot z$.

Smoothness. We show that ϕ constructed above is smooth. Let Γ be a smooth local section of the N_ν-bundle $\pi_\nu^{-1}(\mathbf{J}_{H_\nu}^{-1}(\rho)) \to \mathbf{J}_{H_\nu}^{-1}(\rho)$ defined on a neighborhood U of $\mathbf{J}_{H_\nu}^{-1}(\rho)$. For each $[z] \in U$, the above construction produces $g_{\Gamma([z])} \in (G_\nu)_{\mu|_{\mathfrak{g}_\nu}}$ such that $g_{\Gamma([z])} \cdot \Gamma([z]) \in \mathbf{J}_G^{-1}(\mu)$. It follows, since $\pi_\nu^{-1}(\mathbf{J}_{H_\nu}^{-1}(\rho))$ is a $(G_\nu)_{\mu|_{\mathfrak{g}_\nu}}$-space, that we have the map

$$\mathbf{J}_G \circ \Gamma : U \to \mathcal{O}_\mu^{(G_\nu)_{\mu|_{\mathfrak{g}_\nu}}} \simeq (G_\nu)_{\mu|_{\mathfrak{g}_\nu}}/G_\mu,$$

where $\mathcal{O}_\mu^{(G_\nu)_{\mu|_{\mathfrak{g}_\nu}}}$ denotes the coadjoint orbit through μ for the coadjoint action restricted to the group $(G_\nu)_{\mu|_{\mathfrak{g}_\nu}}$. Finally, take a smooth lift of this map to $\widetilde{\mathbf{J}_G \circ \Gamma} : U \to (G_\nu)_{\mu|_{\mathfrak{g}_\nu}}$. Then the local representative for ϕ is given by

$$\phi = \pi_\mu \circ ((\widetilde{\mathbf{J}_G \circ \Gamma})^{-1} \cdot \Gamma) : U \to M_\mu,$$

which is clearly a smooth map.

Invariance. Let $[z] \in \mathbf{J}_{H_\nu}^{-1}(\rho)$, $[g] \in (G_\nu/N_\nu)_\rho$, and let $g \in (G_\nu)_{\mu|_{\mathfrak{g}_\nu}}$ be such that $r_\nu(g) = [g]$. Since $\pi_\nu^{-1}(\mathbf{J}_{H_\nu}^{-1}(\rho))$ is a $(G_\nu)_{\mu|_{\mathfrak{g}_\nu}}$-space, we have

$$\phi([g] \cdot [z]) = \phi(\pi_\nu(g \cdot z)) = \pi_\mu(\bar{g}_{gz} \cdot g \cdot z),$$

where $\bar{g}_{gz} \in (G_\nu)_{\mu|_{\mathfrak{g}_\nu}}$ is such that $\bar{g}_{gz} \cdot g \cdot z \in \mathbf{J}_G^{-1}(\mu)$. Let $\bar{g}_z \in (G_\nu)_{\mu|_{\mathfrak{g}_\nu}}$ be such that $\bar{g}_z \cdot z \in \mathbf{J}_G^{-1}(\mu)$. We must show that $\bar{g}_z \cdot z$ and $\bar{g}_{gz}g \cdot z$ are on the same G_μ-orbit. This follows since $\bar{g}_{gz}g \cdot z = \bar{g}_{gz}g\bar{g}_z^{-1} \cdot (\bar{g}_z \cdot z)$.

Finally, since ϕ is smooth and invariant, it induces a smooth map on the quotient $[\phi] : \mathbf{J}_{H_\nu}^{-1}(\rho)/(G_\nu/N_\nu)_\rho \to M_\mu$ which is the inverse of F. ∎

6.7.11 In MARSDEN et al. (2002) the reader can find two more reduction by stages theorems in which the authors prove that the Stages Hypothesis admits a slight weakening if we accept the symplectomorphism in Theorem 6.7.10 to take place between the connected components of the reduced spaces and that this hypothesis can actually be eliminated if the orbit $G^0 \cdot \nu \subset \mathfrak{n}^*$ is closed as a subset of $\mathfrak{n}^*$.

Chapter 7

The Symplectic Slice Theorem

Most of the good technical behavior of proper Lie group actions is a direct consequence of the existence of slices and tubes; they provide a privileged system of semiglobal coordinates in which the group action takes on a particularly simple form. Proper symplectic Lie group actions turn out to behave similarly: the tubular chart can be constructed in such a way that the expression of the symplectic form is very natural and, moreover, if there is a momentum map associated to this canonical action, this construction provides a *normal form* for it. The statement and proof of this ***Symplectic Slice Theorem*** is the main goal of this chapter.

The results presented here have their origins in the works of MARLE (1984, 1985) and GUILLEMIN AND STERNBERG (1984b) who constructed a symplectic slice theorem for canonical Lie group actions that admit associated standard momentum maps. ORTEGA AND RATIU (2002b) noticed that the Chu momentum map suffices to extend these theorems to arbitrary canonical Lie group actions. SCHEERER AND WULFF (2001) have carried out a similar generalization using locally defined standard momentum maps.

7.1 The Witt–Artin decomposition

The decomposition described in this section is the first step towards the Symplectic Slice Theorem. It consists in a four way splitting of the tangent space $T_m M$ at a point $m \in M$ of the symplectic manifold (M, ω) acted upon properly and canonically by a Lie group G. This splitting was first proved by WITT (1937) for symmetric bilinear forms. The reader is encouraged to check with CUSHMAN AND BATES (1997) for this and other very interesting historical remarks.

The first step in the construction of the Witt–Artin decomposition is the splitting of the Lie algebra $\mathfrak{g}$ of G into three parts. The first summand is defined by

$$\mathfrak{k} := \{\xi \in \mathfrak{g} \mid \xi_M(m) \in (\mathfrak{g} \cdot m)^{\omega(m)}\}; \tag{7.1.1}$$

$\mathfrak{k}$ is clearly a vector subspace of $\mathfrak{g}$ that contains the Lie algebra $\mathfrak{g}_m$ of the isotropy subgroup G_m of the point $m \in M$. Hence we can fix an Ad_{G_m}-invariant inner product

$\langle \cdot, \cdot \rangle$ on $\mathfrak{g}$ (always available by the compactness of G_m) and write

$$\mathfrak{k} = \mathfrak{g}_m \oplus \mathfrak{m} \quad \text{and} \quad \mathfrak{g} = \mathfrak{g}_m \oplus \mathfrak{m} \oplus \mathfrak{q}, \tag{7.1.2}$$

where $\mathfrak{m}$ is the $\langle \cdot, \cdot \rangle$-orthogonal complement of $\mathfrak{g}_m$ in $\mathfrak{k}$ and $\mathfrak{q}$ is the $\langle \cdot, \cdot \rangle$-orthogonal complement of $\mathfrak{k}$ in $\mathfrak{g}$. The splittings in (7.1.2) induce similar ones on the duals

$$\mathfrak{k}^* = \mathfrak{g}_m^* \oplus \mathfrak{m}^* \quad \text{and} \quad \mathfrak{g}^* = \mathfrak{g}_m^* \oplus \mathfrak{m}^* \oplus \mathfrak{q}^*. \tag{7.1.3}$$

Each of the spaces in this decomposition should be understood as the set of covectors in $\mathfrak{g}^*$ that can be written as $\langle \xi, \cdot \rangle$, with ξ in the corresponding subspace. For example, $\mathfrak{q}^* = \{\langle \xi, \cdot \rangle \mid \xi \in \mathfrak{q}\}$.

7.1.1 Theorem (Witt–Artin decomposition) *Let (M, ω) be a symplectic manifold and let G be a Lie group acting properly and canonically on it. Then for any $m \in M$*

$$T_m M = \mathfrak{k} \cdot m \oplus \mathfrak{q} \cdot m \oplus V \oplus W. \tag{7.1.4}$$

The definitions and properties of the spaces in this splitting are the following:

(i) $\mathfrak{k} := \{\xi \in \mathfrak{g} \mid \xi_M(m) \in (\mathfrak{g} \cdot m)^{\omega(m)}\}$ *is a Lie subalgebra of* $\mathfrak{g}$.

(ii) $\mathfrak{q} \cdot m := \{\eta_M(m) \mid \eta \in \mathfrak{q}\}$ *is a symplectic vector subspace of* $(T_m M, \omega(m))$.

(iii) *Let $\ll \cdot, \cdot \gg$ be a G_m-invariant inner product in $T_m M$ (available by Lemma 2.3.30). Define V as the orthogonal complement to $\mathfrak{g} \cdot m \cap (\mathfrak{g} \cdot m)^{\omega(m)} = \mathfrak{k} \cdot m$ in $(\mathfrak{g} \cdot m)^{\omega(m)}$ with respect to $\ll \cdot, \cdot \gg$, that is,*

$$(\mathfrak{g} \cdot m)^{\omega(m)} = \left(\mathfrak{g} \cdot m \cap (\mathfrak{g} \cdot m)^{\omega(m)}\right) \oplus V = \mathfrak{k} \cdot m \oplus V.$$

The subspace V is a symplectic G_m-invariant subspace of $(T_m M, \omega(m))$ such that $V \cap \mathfrak{q} \cdot m = \{0\}$.

(iv) $\mathfrak{k} \cdot m := \{\xi_M(m) \mid \xi \in \mathfrak{g}\}$ *is a Lagrangian subspace of* $(V \oplus \mathfrak{q} \cdot m)^{\omega(m)}$.

(v) *W is a G_m-invariant Lagrangian complement to $\mathfrak{k} \cdot m$ in $(V \oplus \mathfrak{q} \cdot m)^{\omega(m)}$.*

(vi) *The map $f : W \to \mathfrak{m}^*$ defined by*

$$\langle f(w), \eta \rangle := \omega(m)(\eta_M(m), w), \quad \textit{for all} \quad \eta \in \mathfrak{m}$$

is a G_m-equivariant isomorphism.

Proof. (i) $\mathfrak{k}$ is clearly a vector subspace of $\mathfrak{g}$. To show it is a Lie subalgebra, we shall prove first that for any $\xi, \eta, \zeta \in \mathfrak{g}$ we have

$$\begin{aligned} &\mathbf{d}\omega(\xi_M, \eta_M, \zeta_M) \\ &\qquad = \omega([\xi, \eta]_M, \zeta_M) + \omega([\eta, \zeta]_M, \xi_M) + \omega([\zeta, \xi]_M, \eta_M). \end{aligned} \tag{7.1.5}$$

To verify this identity one starts by computing $\mathbf{d}\omega$ (see (1.3.13)) for the infinitesimal generators ξ_M, η_M, ζ_M:

$$\begin{aligned}\mathbf{d}\omega&(\xi_M, \eta_M, \zeta_M)\\ &= \xi_M[\omega(\eta_M, \zeta_M)] + \eta_M[\omega(\zeta_M, \xi_M)] + \zeta_M[\omega(\xi_M, \eta_M)]\\ &\quad - \omega([\xi_M, \eta_M], \zeta_M) - \omega([\eta_M, \zeta_M], \xi_M) - \omega([\zeta_M, \xi_M], \eta_M). \end{aligned} \tag{7.1.6}$$

However (see (1.3.1)),

$$\begin{aligned}\xi_M[\omega(\eta_M, \zeta_M)] &= (\pounds_{\xi_M}\omega)(\eta_M, \zeta_M) + \omega([\xi_M, \eta_M], \zeta_M) + \omega(\eta_M, [\xi_M, \zeta_M])\\ &= \omega([\xi_M, \eta_M], \zeta_M) + \omega([\zeta_M, \xi_M], \eta_M)\end{aligned}$$

since $\pounds_{\xi_M}\omega = 0$, the G-action being canonical. Replacing this and the analogous two identities into (7.1.6) immediately yields (7.1.5).

In particular, since $\mathbf{d}\omega = 0$, if $\xi, \eta \in \mathfrak{k}$ and any $\zeta \in \mathfrak{g}$, (7.1.5) gives

$$\omega([\xi, \eta]_M, \zeta_M) = -\omega([\eta, \zeta]_M, \xi_M) - \omega([\zeta, \xi]_M, \eta_M) = 0,$$

which shows that $[\xi, \eta] \in \mathfrak{k}$ and hence that $\mathfrak{k}$ is a Lie subalgebra of $\mathfrak{g}$.

(ii) We show that $\omega(m)|_{\mathfrak{q}\cdot m}$ is nondegenerate. Let $\xi \in \mathfrak{q}$ be such that for any $\eta \in \mathfrak{q}$

$$\omega(m)(\xi_M(m),\ \eta_M(m)) = 0. \tag{7.1.7}$$

Since any element $\zeta \in \mathfrak{g}$ can be written as $\zeta = \eta + \sigma$, with $\eta \in \mathfrak{q}$, $\sigma \in \mathfrak{k}$ we have that

$$\omega(m)(\xi_M(m),\ \zeta_M(m)) = \omega(m)(\xi_M(m),\ \eta_M(m)) + \omega(m)(\xi_M(m),\ \sigma_M(m)) = 0$$

due to (7.1.7) and the definition of $\mathfrak{k}$. Consequently, $\xi_M(m) \in (\mathfrak{g}\cdot m)^{\omega}$ and hence $\xi \in \mathfrak{k} \cap \mathfrak{q} = \{0\}$.

(iii) We start by noting that both $\mathfrak{g}\cdot m$ and $(\mathfrak{g}\cdot m)^{\omega(m)}$ are G_m-invariant. Therefore, the G_m-invariance of $\langle\!\langle \cdot, \cdot \rangle\!\rangle$ guarantees that V is G_m-invariant. We now prove that V is symplectic.

Let $v \in V$ be such that $\omega(m)(v,\ v') = 0$, for any $v' \in V$, that is, $v \in V^{\omega(m)}$. Since $V \subset (\mathfrak{g}\cdot m)^{\omega(m)} \subset (\mathfrak{k}\cdot m)^{\omega(m)}$, we have

$$v \in V^{\omega(m)} \cap (\mathfrak{k}\cdot m)^{\omega(m)} = (V \oplus \mathfrak{k}\cdot m)^{\omega(m)} = \left((\mathfrak{g}\cdot m)^{\omega(m)}\right)^{\omega(m)} = \mathfrak{g}\cdot m.$$

Hence, $v \in V \cap \mathfrak{g}\cdot m \subset (\mathfrak{g}\cdot m)^{\omega(m)} \cap \mathfrak{g}\cdot m = \mathfrak{k}\cdot m$ and, therefore, $v \in \mathfrak{k}\cdot m \cap V = \{0\}$.

We now show that $V \cap \mathfrak{q}\cdot m = \{0\}$. First notice that

$$\mathfrak{k}\cdot m \cap \mathfrak{q}\cdot m = \{0\}. \tag{7.1.8}$$

Indeed, if $\xi \in \mathfrak{k}$ and $\eta \in \mathfrak{q}$ are such that $\xi_M(m) = \eta_M(m)$, then $\xi - \eta \in \mathfrak{g}_m \subset \mathfrak{k}$ and, therefore, $\eta \in \mathfrak{k} \cap \mathfrak{q} = \{0\}$.

As a consequence of (7.1.8) and the direct sum decomposition $\mathfrak{g} = \mathfrak{k} \oplus \mathfrak{q}$, we have

$$\mathfrak{g}\cdot m = \mathfrak{k}\cdot m \oplus \mathfrak{q}\cdot m, \tag{7.1.9}$$

$$(\mathfrak{g}\cdot m)^{\omega(m)} = (\mathfrak{k}\cdot m)^{\omega(m)} \cap (\mathfrak{q}\cdot m)^{\omega(m)}. \tag{7.1.10}$$

Now,

$$
\begin{aligned}
V \cap \mathfrak{q} \cdot m &= V \cap \mathfrak{q} \cdot m \oplus (\mathfrak{k} \cdot m \cap \mathfrak{q} \cdot m) \qquad \text{(by (7.1.8))}\\
&\subset (V \oplus \mathfrak{k} \cdot m) \cap \mathfrak{q} \cdot m\\
&= (\mathfrak{g} \cdot m)^{\omega(m)} \cap \mathfrak{q} \cdot m\\
&= (\mathfrak{k} \cdot m)^{\omega(m)} \cap (\mathfrak{q} \cdot m)^{\omega(m)} \cap (\mathfrak{q} \cdot m) \qquad \text{(by (7.1.10))}\\
&= (\mathfrak{k} \cdot m)^{\omega(m)} \cap \{0\} = \{0\}, \qquad \text{(by part } \textbf{(ii)}\text{)}
\end{aligned}
$$

as required.

(iv) Part **(iii)** guarantees that the sum $V + \mathfrak{q} \cdot m$ is direct. Moreover, since V and $\mathfrak{q} \cdot m$ are symplectic subspaces of $T_m M$, so is $V \oplus \mathfrak{q} \cdot m$. This provides the first step in the Witt–Artin decomposition of $T_m M$ into symplectic subspaces:

$$
T_m M = V \oplus \mathfrak{q} \cdot m \oplus (V \oplus \mathfrak{q} \cdot m)^{\omega(m)}. \tag{7.1.11}
$$

We now show that $\mathfrak{k} \cdot m \subset (V \oplus \mathfrak{q} \cdot m)^{\omega(m)}$, which is equivalent to $V \subset (\mathfrak{k} \cdot m)^{\omega(m)}$ and $\mathfrak{q} \cdot m \subset (\mathfrak{k} \cdot m)^{\omega(m)}$. The first inclusion holds because $V \subset (\mathfrak{g} \cdot m)^{\omega(m)} \subset (\mathfrak{k} \cdot m)^{\omega(m)}$. As to the second one, note that $\mathfrak{q} \cdot m \subset \mathfrak{g} \cdot m \subset (\mathfrak{k} \cdot m)^{\omega(m)}$ since $\mathfrak{g} \cdot m \cap (\mathfrak{g} \cdot m)^{\omega(m)} = \mathfrak{k} \cdot m$.

Finally, we show that $\mathfrak{k} \cdot m$ is a Lagrangian subspace of $(V \oplus \mathfrak{q} \cdot m)^{\omega(m)}$, that is,

$$
(\mathfrak{k} \cdot m)^{\omega(m)|_{(V \oplus \mathfrak{q} \cdot m)^{\omega(m)}}} = \mathfrak{k} \cdot m,
$$

or, equivalently,

$$
(\mathfrak{k} \cdot m)^{\omega(m)} \cap (V \oplus \mathfrak{q} \cdot m)^{\omega(m)} = \mathfrak{k} \cdot m. \tag{7.1.12}
$$

To prove this equality note that the definition of the subspace V implies that

$$
\mathfrak{g} \cdot m = (\mathfrak{k} \cdot m)^{\omega(m)} \cap V^{\omega(m)}. \tag{7.1.13}
$$

Additionally, recall that if A, B, and C are vector subspaces of a vector space E, such that $A \subset B$, and $A \cap C = \{0\}$, then

$$
(B \cap C) \oplus A = B \cap (C \oplus A). \tag{7.1.14}
$$

We now prove equality (7.1.12). Indeed,

$$
\begin{aligned}
(\mathfrak{k} \cdot m)^{\omega(m)} &\cap (V \oplus \mathfrak{q} \cdot m)^{\omega(m)}\\
&= (\mathfrak{k} \cdot m + (V \oplus \mathfrak{q} \cdot m))^{\omega(m)}\\
&= (\mathfrak{g} \cdot m + V)^{\omega(m)} \qquad \text{(by (7.1.9))}\\
&= (\mathfrak{g} \cdot m)^{\omega(m)} \cap V^{\omega(m)}\\
&= \left((\mathfrak{k} \cdot m)^{\omega(m)} \cap V^{\omega(m)}\right)^{\omega(m)} \cap V^{\omega(m)} \qquad \text{(by (7.1.13))}\\
&= \left[\left((\mathfrak{k} \cdot m)^{\omega(m)} \cap V^{\omega(m)}\right) \oplus V\right]^{\omega(m)}\\
&= \left[(\mathfrak{k} \cdot m)^{\omega(m)} \cap (V^{\omega(m)} \oplus V)\right]^{\omega(m)} \qquad \text{(by (7.1.14))}\\
&= ((\mathfrak{k} \cdot m)^{\omega(m)} \cap T_m M)^{\omega(m)} \qquad \text{(by part } \textbf{(iii)}\text{)}\\
&= \mathfrak{k} \cdot m,
\end{aligned}
$$

as required.

(v) The existence of W is a consequence of the following Lie algebraic lemma.

7.1.2 Lemma *Let (E, Ω) be a symplectic representation space of the compact Lie group H. Then any H-invariant Lagrangian subspace of (E, Ω) admits an H-invariant Lagrangian complement.*

Proof. Let $\langle \cdot, \cdot \rangle$ be an H-invariant inner product on E always available by the compactness of H. Let $J : E \to E$ be the H-equivariant map uniquely determined by the relation

$$\langle u, v \rangle = \Omega(u, J(v)) \text{ for any } u, v \in V.$$

By construction $J^2 = -I_E$ and $\Omega(u, J(v)) = -\Omega(J(u), v)$ for all $u, v \in E$. Let F be an H-invariant Lagrangian subspace of E. We will now show that $J(F)$ is an H-invariant Lagrangian complement of F in E.

First, the H-invariance of $J(F)$ is obvious by the H-equivariance of J and the H-invariance of F. Second, we have that

$$\dim J(F)^{\Omega} = \dim E - \dim J(F) = \dim E - \dim F = \dim F, \qquad (7.1.15)$$

where in the last equality we used the Lagrangian character of F. Third, $J(F) \subset J(F)^{\Omega}$. Indeed, for any $u, v \in F$ we have that

$$\Omega(J(u), J(v)) = -\Omega(u, J^2(v)) = \Omega(u, v) = 0.$$

This inclusion, together with (7.1.15), implies that $J(F) = J(F)^{\Omega}$ which shows that $J(F)$ is Lagrangian. Finally, since

$$\begin{aligned} J(F)^{\Omega} &= \{v \in E \mid \Omega(v, J(u)) = 0 \text{ for all } u \in F\} \\ &= \{v \in E \mid \langle v, u \rangle = 0 \text{ for all } u \in F\} = F^{\perp} \end{aligned}$$

it follows that $J(F) = F^{\perp}$ is a complement to F. ▼

Hence, we have that

$$(V \oplus \mathfrak{q} \cdot m)^{\omega(m)} = \mathfrak{k} \cdot m \oplus W,$$

which together with (7.1.11) establishes (7.1.4).

(vi) The map f is clearly linear and H-equivariant. In order to show that f is injective let $w \in W$ be such that $f(w) = 0$. This means that $\omega(m)(\eta_M(m), w) = 0$, for all $\eta \in \mathfrak{m}$ or, equivalently, $w \in (\mathfrak{m} \cdot m)^{\omega(m)} = ((\mathfrak{g}_m \oplus \mathfrak{m}) \cdot m)^{\omega(m)} = (\mathfrak{k} \cdot m)^{\omega(m)}$. Now, the definition of W in **(v)** gives $(V \oplus \mathfrak{q} \cdot m)^{\omega(m)} = \mathfrak{k} \cdot m \oplus W$ and hence

$$w \in (\mathfrak{k} \cdot m)^{\omega(m)} \cap W \subset (\mathfrak{k} \cdot m)^{\omega(m)} \cap (V \oplus \mathfrak{q} \cdot m)^{\omega(m)} = \mathfrak{k} \cdot m$$

by (7.1.12). Therefore $w \in \mathfrak{k} \cdot m \cap W = \{0\}$, again by **(v)**, which concludes the proof of the injectivity of f.

Since W and $\mathfrak{k} \cdot m$ are Lagrangian complements in $(V \oplus \mathfrak{q} \cdot m)^{\omega(m)}$, it follows from (7.1.2) that

$$\dim W = \dim \mathfrak{k} \cdot m = \dim \mathfrak{k} - \dim \mathfrak{g}_m = \dim \mathfrak{m} = \dim \mathfrak{m}^*,$$

which shows that f is an isomorphism. ■

7.2 The symplectic tube

In this section we construct a symplectic tube that, as we shall see in the following sections, will model a G-invariant neighborhood of a symplectic G-manifold. In our presentation we follow the approach in ORTEGA AND RATIU (2002b) which in turn is a generalization of that in SJAMAAR AND LERMAN (1991).

7.2.1 Definition *Let (M, ω) be a symplectic manifold and G a Lie group acting properly and canonically on it. Let $m \in M$ and let V be a symplectic G_m-space constructed in part* **(iii)** *of Theorem* 7.1.1. *Any such space will be called a* ***symplectic normal space*** *at m. Since the G_m-action on $(V, \omega(m)|_V)$ is linear and canonical it has (by Example* 4.5.7*) a standard associated momentum map to be denoted by $\mathbf{J}_V : V \to \mathfrak{g}_m^*$.*

The next statement uses the notation introduced in Theorem 7.1.1 and in the previous definition.

7.2.2 Proposition *Let (M, ω) be a symplectic manifold and G a Lie group acting properly and canonically on it. Let $m \in M$, V be a symplectic normal space at m, and $\mathfrak{m} \subset \mathfrak{g}$ the subspace introduced in the splitting* (7.1.2). *Then there exist G_m-invariant neighborhoods $\mathfrak{m}_r^*$ and V_r of the origin in $\mathfrak{m}^*$ and V, respectively, such that the twisted product*

$$Y_r := G \times_{G_m} \left(\mathfrak{m}_r^* \times V_r\right) \tag{7.2.1}$$

is a symplectic manifold with the two-form ω_{Y_r} defined by

$$\begin{aligned} &\omega_{Y_r}([g, \rho, v])(T_{(g,\rho,v)}\pi(T_eL_g(\xi_1), \alpha_1, u_1), T_{(g,\rho,v)}\pi(T_eL_g(\xi_2), \alpha_2, u_2)) \\ &\quad := \langle \alpha_2 + T_v\mathbf{J}_V(u_2), \xi_1\rangle - \langle \alpha_1 + T_v\mathbf{J}_V(u_1), \xi_2\rangle + \langle \rho + \mathbf{J}_V(v), [\xi_1, \xi_2]\rangle \\ &\quad + \Psi(m)(\xi_1, \xi_2) + \omega(m)(u_1, u_2), \end{aligned} \tag{7.2.2}$$

where $\Psi : M \to Z^2(\mathfrak{g})$ is the Chu map associated to the G-action on (M, ω), $\pi : G \times \left(\mathfrak{m}_r^* \times V_r\right) \to G \times_{G_m} \left(\mathfrak{m}_r^* \times V_r\right)$ is the projection, $[g, \rho, v] \in Y_r$, $\xi_1, \xi_2 \in \mathfrak{g}$, $\alpha_1, \alpha_2 \in \mathfrak{m}^*$, and $u_1, u_2 \in V$.

The Lie group G acts canonically on (Y_r, ω_{Y_r}) by $g \cdot [h, \eta, v] := [gh, \eta, v]$, for any $g \in G$ and any $[h, \eta, v] \in Y_r$.

In what follows we will refer to the symplectic manifold (Y_r, ω_{Y_r}) as a ***symplectic tube*** *of (M, ω) at the point m.*

Proof. We start the proof by constructing a G-invariant symplectic form on a neighborhood of the zero section of the trivial bundle $G \times \mathfrak{k}^*$. First, the splitting (7.1.3) provides an injection $i : G \times \mathfrak{k}^* \hookrightarrow G \times \mathfrak{g}^*$ that will be used to pull-back the canonical symplectic form of $G \times \mathfrak{g}^* \cong T^*G$ (see §4.1.20) to $G \times \mathfrak{k}^*$ in order to obtain a closed two-form ω_1 on $G \times \mathfrak{k}^*$. Second, define on $G \times \mathfrak{k}^*$ the skew-symmetric two-form ω_2 by

$$\omega_2(g, \nu)\left((T_eL_g(\xi), \rho), (T_eL_g(\eta), \sigma)\right) = \omega(m)\left(\xi_M(m), \eta_M(m)\right) = \Psi(m)(\xi, \eta),$$

for any $(g, \nu) \in G \times \mathfrak{k}^*$, $\xi, \eta \in \mathfrak{g}$, and $\rho, \sigma \in \mathfrak{k}^*$. We now prove that there exists a G_m-invariant neighborhood $\mathfrak{k}_r^*$ of the origin in $\mathfrak{k}^*$ such that the restriction of the form

$$\Omega := \omega_1 + \omega_2 \tag{7.2.3}$$

to $T_r := G \times \mathfrak{k}_r^*$ is a symplectic form.

Ω **is closed**: The two-form ω_1 is clearly closed. In order to show that ω_2 is closed we define for any $\xi \in \mathfrak{g}$ and $\rho \in \mathfrak{k}^*$ the vector field $(\xi, \rho) \in \mathfrak{X}(G \times \mathfrak{k}^*)$ given by $(\xi, \rho)(g, \nu) := (T_e L_g(\xi), \nu)$, whose flow is $F_t(g, \nu) = (g \exp t\xi, \rho + t\nu)$. It is easy to see that the Lie bracket of two such vector fields is given by $[(\xi, \rho), (\eta, \sigma)] = ([\xi, \eta], 0)$. Formula (1.3.13) gives

$$\begin{aligned} &\mathbf{d}\omega_2((\xi,\rho),(\eta,\sigma),(\lambda,\tau)) \\ &\quad = (\xi,\rho)[\omega_2((\eta,\sigma),(\lambda,\tau))] - (\eta,\sigma)[\omega_2((\xi,\rho),(\lambda,\tau))] \\ &\quad + (\lambda,\tau)[\omega_2((\xi,\rho),(\eta,\sigma))] - \omega_2([(\xi,\rho),(\eta,\sigma)],(\lambda,\tau)) \\ &\quad + \omega_2([(\xi,\rho),(\lambda,\tau)],(\eta,\sigma)) - \omega_2([(\eta,\sigma),(\lambda,\tau)],(\xi,\rho)) \end{aligned}$$

for any $\xi, \eta, \lambda \in \mathfrak{g}$ and any $\rho, \sigma, \tau \in \mathfrak{k}^*$. Now note that for any $(g, \nu) \in G \times \mathfrak{k}^*$ we have, for instance, that

$$\begin{aligned} &((\xi,\rho)\,[\omega_2((\eta,\sigma),(\lambda,\tau))])(g,\nu) \\ &= \left.\frac{d}{dt}\right|_{t=0} \omega_2(g\exp t\xi, \nu + t\rho)\,((\eta,\sigma)(g\exp t\xi, \nu + t\rho),(\lambda,\tau)(g\exp t\xi, \nu + t\rho)) \\ &= \left.\frac{d}{dt}\right|_{t=0} \omega(m)(\eta_M(m),\lambda_M(m)) = 0, \end{aligned}$$

and also,

$$\omega_2([(\xi,\rho),(\eta,\sigma)],(\lambda,\tau))(g,\nu) = \omega(m)\,([\xi,\eta]_M(m),\lambda_M(m))\,.$$

The last three equalities imply by (7.1.5) that

$$\begin{aligned} &\mathbf{d}\omega_2(g,\nu)\big((T_eL_g(\xi),\rho),(T_eL_g(\eta),\sigma),(T_eL_g(\lambda),\tau)\big) \\ &\quad = -\omega(m)\,([\xi,\eta]_M(m),\lambda_M(m)) + \omega(m)\,([\xi,\lambda]_M(m),\eta_M(m)) \\ &\qquad - \omega(m)\,([\eta,\lambda]_M(m),\xi_M(m)) \\ &\quad = \mathbf{d}\omega(m)(\xi_M(m),\eta_M(m),\lambda_M(m)) = 0, \end{aligned}$$

which guarantees the closedness of ω_2.

Ω **is nondegenerate on** $G \times \{0\}$: Let $\xi = \xi_1 + \xi_2$ and $\eta = \eta_1 + \eta_2$ be arbitrary elements in $\mathfrak{g}$, with $\xi_1, \eta_1 \in \mathfrak{k}$ and $\xi_2, \eta_2 \in \mathfrak{q}$. Let also $(g, \nu) \in G \times \mathfrak{k}^*$, $\rho, \sigma \in \mathfrak{k}^*$, and suppose that for all $\eta \in \mathfrak{g}$ and $\sigma \in \mathfrak{k}^*$ we have that

$$\Omega(g,\nu)\big((T_eL_g(\xi),\rho),(T_eL_g(\eta),\sigma)\big) = 0,$$

or equivalently (recall (4.1.13)):

$$\langle\sigma,\xi_1\rangle - \langle\rho,\eta_1\rangle + \langle\nu,[\xi,\eta]\rangle + \omega(m)\,((\xi_2)_M(m),(\eta_2)_M(m)) = 0. \tag{7.2.4}$$

We show that when $\nu = 0$ this implies that $\xi = 0$ and $\rho = 0$. Indeed, suppose that $\nu = 0$. Setting $\eta = 0$ in (7.2.4) and letting σ vary, we obtain $\xi_1 = 0$. Also, setting $\eta_2 = 0$ and letting η_1 vary we have $\rho = 0$. Finally, since $\omega(m)\,((\xi_2)_M(m), \eta_M(m)) = 0$ for all $\eta \in \mathfrak{g}$, this implies that $(\xi_2)_M(m) \in \mathfrak{g} \cdot m \cap (\mathfrak{g} \cdot m)^\omega$ and hence $\xi_2 \in \mathfrak{k} \cap \mathfrak{q} = \{0\}$ and, consequently, $\xi = 0$ and $\rho = 0$.

Since nondegeneracy is an open condition, we can choose an $\mathrm{Ad}(G_m)$-invariant neighborhood $\mathfrak{k}_r^*$ of zero in $\mathfrak{k}^*$ where the expression in (7.2.4) is nondegenerate. Also, as the expression in (7.2.4) does not depend on G (the form Ω is by construction G-invariant), it follows that Ω is nondegenerate on $G \times \mathfrak{k}_r^*$ and hence (T_r, Ω) is a symplectic manifold.

The symplectic form for ω_{Y_r} on Y_r is obtained via a symplectic reduction of the symplectic forms of T_r and V. First, consider the left action $\mathcal{R}$ of G_m on T_r defined by

$$\mathcal{R}_h(g, \nu) = (gh^{-1}, \mathrm{Ad}^*_{h^{-1}}\nu), \quad h \in G_m, \ (g, \nu) \in T_r.$$

Using the definition of Ω it is straightforward to verify that this action is globally Hamiltonian on T_r with equivariant momentum map $\mathbf{J}_{\mathcal{R}} : T_r \to \mathfrak{g}_m^*$, given by

$$\mathbf{J}_{\mathcal{R}}((g, (\eta, \rho))) = -\eta, \qquad \text{for any} \qquad (\eta, \rho) \in (\mathfrak{g}_m^*)_r \oplus \mathfrak{m}_r^* = \mathfrak{k}_r^*.$$

As we already pointed out, the G_m-action on V is globally Hamiltonian with momentum map $\mathbf{J}_V : V \to \mathfrak{g}_m^*$. Putting together these two actions, we construct a product action of G_m on the symplectic manifold $T_r \times V$, which is Hamiltonian, with G_m-equivariant momentum map $\boldsymbol{K} : T_r \times V \cong G \times \mathfrak{m}_r^* \times (\mathfrak{g}_m^*)_r \times V \to \mathfrak{g}_m^*$, given by the sum $\mathbf{J}_{\mathcal{R}} + \mathbf{J}_V$, that is,

$$\begin{array}{rcl} \boldsymbol{K} : G \times \mathfrak{m}_r^* \times (\mathfrak{g}_m^*)_r \times V & \longrightarrow & \mathfrak{g}_m^* \\ (g, \rho, \eta, v) & \longmapsto & \mathbf{J}_V(v) - \eta. \end{array}$$

The G_m-action on $T_r \times V$ is free and proper and $0 \in \mathfrak{g}_m^*$ is clearly a regular value of $\boldsymbol{K}$. Therefore $\boldsymbol{K}^{-1}(0)/G_m$ is a well–defined symplectic point reduced space in the sense of Theorem 6.1.1 that can be identified with $Y_r = G \times_{G_m} (\mathfrak{m}_r^* \times V_r)$ by means of the quotient diffeomorphism L induced by the G_m-equivariant diffeomorphism l

$$\begin{array}{rcl} l : G \times \mathfrak{m}_r^* \times V_r & \longrightarrow & \boldsymbol{K}^{-1}(0) \subset G \times \mathfrak{m}_r^* \times (\mathfrak{g}_m^*)_r \times V_r \\ (g, \rho, v) & \longmapsto & (g, \rho, \mathbf{J}_{V_m}(v), v), \end{array}$$

where the G_m-invariant neighborhood of the origin V_r has been chosen so that $\mathbf{J}_{V_m}(V_r) \subset (\mathfrak{g}_m^*)_r$. We define the symplectic form ω_{Y_r} on Y_r as the pull-back by L of the reduced symplectic form Ω_0 on $\boldsymbol{K}^{-1}(0)/G_m$. Thus we have the following commutative diagram with the lower arrow a symplectic diffeomorphism:

$$\begin{array}{ccc} G \times \mathfrak{m}_r^* \times V_r & \xrightarrow{\ l\ } & \boldsymbol{K}^{-1}(0) \subset G \times \mathfrak{m}_r^* \times (\mathfrak{g}_m^*)_r \times V_r \\ \pi \Big\downarrow & & \Big\downarrow \pi_0 \\ (G \times_{G_m} (\mathfrak{m}_r^* \times V_r), \omega_{Y_r}) & \xrightarrow{\ L\ } & (\boldsymbol{K}^{-1}(0)/G_m, \Omega_0). \end{array} \tag{7.2.5}$$

We now show that the symplectic form ω_{Y_r} that we just defined coincides with the one given in the statement, namely, with expression (7.2.2). First notice that by definition $\omega_{Y_r} = L^*\Omega_0$. As the projection π is a submersion, this is equivalent to $\pi^*\omega_{Y_r} = \pi^*(L^*\Omega_0)$. Using the maps in the diagram (7.2.5) we can express this equality as

$\pi^*\omega_{Y_r} = l^*\pi_0^*\Omega_0 = l^*i_0^*(\Omega \oplus \omega_V)$, where $\omega_V := \omega(m)|_V$ is the symplectic form on the symplectic normal space V. We now check that this coincides with (7.2.2). Indeed, for any $[g, \rho, v] \in Y_r$, $\xi_1, \xi_2 \in \mathfrak{g}$, $\alpha_1, \alpha_2 \in \mathfrak{m}^*$, and $u_1, u_2 \in V$, we have

$$\begin{aligned}
&\omega_{Y_r}([g, \rho, v])(T_{(g,\rho,v)}\pi(T_eL_g(\xi_1), \alpha_1, u_1), T_{(g,\rho,v)}\pi(T_eL_g(\xi_2), \alpha_2, u_2)) \\
&\quad = (\pi^*\omega_{Y_r})(g, \rho, v)((T_eL_g(\xi_1), \alpha_1, u_1), (T_eL_g(\xi_2), \alpha_2, u_2)) \\
&\quad = ((i_0 \circ l)^*(\Omega \oplus \omega_V))(g, \rho, v)((T_eL_g(\xi_1), \alpha_1, u_1), (T_eL_g(\xi_2), \alpha_2, u_2)) \\
&\quad = (\Omega \oplus \omega_V)(g, \rho, \mathbf{J}_V(v), v)((T_eL_g(\xi_1), \alpha_1, T_v\mathbf{J}_V(u_1), u_1), \\
&\qquad\qquad (T_eL_g(\xi_2), \alpha_2, T_v\mathbf{J}_V(u_2), u_2)) \\
&\quad = \Omega(g, \rho, \mathbf{J}_V(v))((T_eL_g(\xi_1), \alpha_1, T_v\mathbf{J}_V(u_1)), (T_eL_g(\xi_2), \alpha_2, T_v\mathbf{J}_V(u_2))) \\
&\qquad + \omega_V(v)(u_1, u_2) \\
&\quad = \langle \alpha_2 + T_v\mathbf{J}_V(u_2), \xi_1 \rangle - \langle \alpha_1 + T_v\mathbf{J}_V(u_1), \xi_2 \rangle + \langle \rho + \mathbf{J}_V(v), [\xi_1, \xi_2] \rangle \\
&\qquad + \Psi(m)(\xi_1, \xi_2) + \omega(m)(u_1, u_2),
\end{aligned}$$

as required.

Finally, the fact that the G-action on Y_r in the statement is symplectic is a straightforward verification. ■

7.3 The G-relative Darboux Theorem

The importance of the Hamiltonian tube (Y_r, ω_{Y_r}) introduced in the previous section is in the fact that it models the symplectic manifold (M, ω) as a Hamiltonian G-space in a neighborhood of the orbit $G \cdot m$. In order to prove this claim, which we will do in the following section, we need a generalization of the Darboux Theorem that accommodates the presence of a canonical Lie group action.

7.3.1 Theorem (G-relative Darboux Theorem) *Let M be a manifold and ω_0 and ω_1 two symplectic forms on it. Let G be a Lie group acting properly on M and symplectically with respect to both ω_0 and ω_1. Let $m \in M$ and assume that*

$$\omega_0(g \cdot m)(v_{g \cdot m}, w_{g \cdot m}) = \omega_1(g \cdot m)(v_{g \cdot m}, w_{g \cdot m}) \tag{7.3.1}$$

for all $g \in G$ and $v_{g \cdot m}, w_{g \cdot m} \in T_{g \cdot m}M$. Then there exist two open G-invariant neighborhoods U_0 and U_1 of $G \cdot m$ and a G-equivariant diffeomorphism $\Psi : U_0 \to U_1$ such that $\Psi|_{G \cdot m} = Id$ and $\Psi^*\omega_1 = \omega_0$.

Proof. We reproduce here the approach taken in Bates and Lerman (1997). We begin by constructing a smooth map $\phi : [0, 1] \times U \to U$ where U is a G-invariant neighborhood of $G \cdot m$ satisfying the following properties:

(i) $\phi_t := \phi(t, \cdot) : U \to U$ is G-equivariant,

(ii) $\phi_t|_{G \cdot m}$ is the identity map on $G \cdot m$,

(iii) ϕ_0 is the identity map on U,

(iv) $\phi_1(U) = G \cdot m$,

(v) ϕ_t is a diffeomorphism for $t \neq 1$.

We recall that in the proof of the Tube Theorem 2.3.28 we established the existence of a G_m-invariant Riemannian metric on some G_m-invariant neighborhood of m with associated exponential map Exp_m and such that the mapping

$$\begin{array}{rccc} \tau : & G \times_{G_m} V_m & \longrightarrow & M \\ & [g, v] & \longmapsto & g \cdot \mathrm{Exp}_m v, \end{array}$$

is a G-equivariant diffeomorphism onto some open G-invariant neighborhood U of $G \cdot m$. We recall that V_m is some G_m-invariant neighborhood of the origin in the orthogonal complement to $\mathfrak{g} \cdot m$. Define for any $u = g \cdot \mathrm{Exp}_m v \in U$ the map $\phi_t(u) := g \cdot \mathrm{Exp}_m (1-t)v$. This map clearly satisfies properties **(i)** through **(v)**.

We now use the diffeomorphisms ϕ_t, $t \neq 1$, to construct a one-form α on a G-invariant neighborhood W of $G \cdot m$, $W \subset U$ satisfying:

(a) $\omega_0 - \omega_1 = \mathbf{d}\alpha$.

(b) $\alpha(g \cdot m) = 0$, for any $g \in G$.

(c) α is G-invariant.

Let Y_t be the time-dependent vector field whose flow is ϕ_t for $t \neq 1$ (see (1.2.3)). Now, by the property **(iv)** of ϕ_t and the hypothesis (7.3.1) we have that

$$\begin{aligned} \omega_0 - \omega_1 &= \phi_1^*(\omega_1 - \omega_0) - (\omega_1 - \omega_0) \\ &= \int_0^1 \frac{d}{dt} \phi_t^*(\omega_1 - \omega_0) dt \qquad \text{(by (1.3.6))} \\ &= \int_0^1 \phi_t^*(\pounds_{Y_t}(\omega_1 - \omega_0)) dt \\ &= \int_0^1 \phi_t^*(\mathbf{d}\mathbf{i}_{Y_t}(\omega_1 - \omega_0)) dt \\ &= \mathbf{d} \int_0^1 \phi_t^*(\mathbf{i}_{Y_t}(\omega_1 - \omega_0)) dt. \end{aligned}$$

Define $\alpha := \int_0^1 \phi_t^*(\mathbf{i}_{Y_t}(\omega_1 - \omega_0))dt$. This form satisfies **(a)** by construction. Hypothesis (7.3.1) and Property **(ii)** of ϕ_t guarantee that **(b)** is also satisfied. Property **(c)** is trivially verified.

We now define $\omega_t := \omega_0 + t(\omega_1 - \omega_0)$. By (7.3.1) we have that $\omega_t(g \cdot m) = \omega_0(g \cdot m)$ for any $g \in G$ and any $t \in [0, 1]$, and consequently $\omega_t(g \cdot m)$ is nondegenerate. The G-invariance of ω_t implies the existence of a G-invariant neighborhood $U_t \subset U$ of $G \cdot m$ and a real number $\epsilon_t > 0$ such that $\omega_s(z)$ is nondegenerate for every $s \in I_t := (t - \epsilon_t, t + \epsilon_t)$ and $z \in U_t$. Cover the interval $[0, 1]$ with a finite number of such intervals $\{I_{t_1}, \ldots, I_{t_n}\}$ and let $\{U_{t_1}, \ldots, U_{t_n}\}$ be the corresponding G-invariant neighborhoods of $G \cdot m$. Then, the form ω_t is nondegenerate on $W := \bigcap_{i=1}^n U_{t_i}$ for every $t \in [0, 1]$. The nondegeneracy of ω_t on W, for any $t \in [0, 1]$, guarantees the existence of a G-equivariant time-dependent vector field X_t on W satisfying

$$\mathbf{i}_{X_t} \omega_t = \alpha. \qquad (7.3.2)$$

Let Ψ_t be the flow of X_t. By (1.3.6) we have that

$$\begin{aligned}\frac{d}{dt}\Psi_t^*\omega_t &= \Psi_t^*(\pounds_{X_t}\omega_t + \frac{d}{dt}\omega_t)\\ &= \Psi_t^*(\mathbf{i}_{X_t}\mathbf{d}\omega_t + \mathbf{d}\mathbf{i}_{X_t}\omega_t + \omega_1 - \omega_0)\\ &= \Psi_t^*(\mathbf{d}\mathbf{i}_{X_t}\omega_t + \omega_1 - \omega_0)\\ &= \Psi_t^*(\mathbf{d}\alpha + \omega_1 - \omega_0) = 0,\end{aligned}$$

where in the last two equalities we used (7.3.2) and the property **(a)** of the one-form α. Since $\Psi_0 = Id$ we get $\Psi_1^*\omega_1 = \omega_0$. If we take $U_0 = W$, $U_1 = \Psi_1(W)$, and $\Psi = \Psi_1 : U_0 \to U_1$, the theorem is proved. ■

7.3.2 Remark The result just proved can be generalized in an obvious way to the case in which the orbit $G \cdot m$ is replaced by any G-invariant submanifold of M. The proof in that case requires the existence of a G-invariant tubular neighborhood of the submanifold which can be guaranteed, for instance, by requiring M to be paracompact. Note that the proof just presented does not require any such hypothesis.

7.4 The Symplectic Slice Theorem

7.4.1 Theorem (Symplectic Slice Theorem) *Let (M, ω) be a symplectic manifold and let G be a Lie group acting properly and canonically on M. Let $m \in M$ and let (Y_r, ω_{Y_r}) be the G-symplectic tube at that point constructed in Proposition* 7.2.2. *Then there is a G-invariant neighborhood U of m in M and a G-equivariant symplectomorphism $\phi : U \to Y_r$ satisfying $\phi(m) = [e, 0, 0]$.*

Proof. The notation used in the proof below is consistent with that in 7.1 and 7.2. Consider the tube $Y_r = G \times_{G_m} \left(\mathfrak{m}_r^* \times V_r\right)$ and $f : W \to \mathfrak{m}^*$ the G_m-equivariant linear isomorphism $f : W \to \mathfrak{m}^*$ introduced in part **(vi)** of Theorem 7.1.1. Let $f_r^{-1} : \mathfrak{m}_r^* \to W_r := f^{-1}(\mathfrak{m}_r^*) \subset W$ be the restriction of the inverse f^{-1} of f to $\mathfrak{m}_r^*$ and range $f^{-1}(\mathfrak{m}_r^*)$.

We now recall that in the proof of the Tube Theorem 2.3.28 we established the existence of a G_m-invariant Riemannian metric on some G_m-invariant neighborhood of m with associated exponential map Exp_m such that, for r small enough, the mapping (see 2.3.4)

$$\begin{aligned}\tau : G \times_{G_m} (W_r \times V_r) &\longrightarrow M\\ [g, w, v] &\longmapsto g \cdot \mathrm{Exp}_m(w + v),\end{aligned}$$

is a G-equivariant diffeomorphism onto some open G-invariant neighborhood U of the orbit $G \cdot m$. Therefore, the composed map

$$\begin{aligned}\Psi : Y_r = G \times_{G_m} (\mathfrak{m}_r^* \times V_r) &\longrightarrow U\\ [g, \rho, v] &\longmapsto g \cdot \mathrm{Exp}_m(f_r^{-1}(\rho) + v)\end{aligned}$$

has the same properties. Consequently, the open G-invariant neighborhood U can be endowed with two symplectic forms $\omega|_U$ and $\Psi_*\omega_{Y_r}$.

We now prove that these two symplectic forms coincide on $G \cdot m$. Since both forms are G-invariant it suffices to show that $\omega(m) = \Psi_* \omega_{Y_r}(m)$. Let $u_1, u_2 \in T_m M$ arbitrary. The Witt–Artin decomposition (7.1.4) guarantees the existence of $\xi_1, \xi_2 \in \mathfrak{g}$, $w_1, w_2 \in W$, and $v_1, v_2 \in V$ such that $u_i = (\xi_i)_M(m) + w_i + v_i$, $i \in \{1, 2\}$. Hence (7.2.2) and the definition of f_r (see Theorem 7.1.1 **(vi)**) imply

$$\begin{aligned}
&\Psi_* \omega_{Y_r}(m)(u_1, u_2) \\
&\quad = \omega_{Y_r}([e, 0, 0])(T_m \Psi^{-1}((\xi_1)_M(m) + w_1 + v_1), \\
&\qquad T_m \Psi^{-1}((\xi_2)_M(m) + w_2 + v_2)) \\
&\quad = \omega_{Y_r}([e, 0, 0])(T_{(e,0,0)}\pi(\xi_1, f_r(w_1), v_1), T_{(e,0,0)}\pi(\xi_2, f_r(w_2), v_2)) \\
&\quad = \langle f_r(w_2), \xi_1 \rangle - \langle f_r(w_1), \xi_2 \rangle + \omega(m)((\xi_1)_M(m), (\xi_2)_M(m)) \\
&\qquad + \omega(m)(v_1, v_2) \\
&\quad = \omega(m)((\xi_1)_M(m), w_2) - \omega(m)((\xi_2)_M(m), w_1) \\
&\qquad + \omega(m)((\xi_1)_M(m), (\xi_2)_M(m)) + \omega(m)(v_1, v_2) = \omega(m)(u_1, u_2).
\end{aligned}$$

The last equality follows from the fact that $\omega(m)((\xi_1)_M(m), v_2) = \omega(m)((\xi_2)_M(m), v_1) = 0$ since $V \subset (\mathfrak{g} \cdot m)^{\omega}$. Moreover, $\omega(m)(w_1, w_2) = 0$ since W is Lagrangian in $(V \oplus \mathfrak{q} \cdot m)^{\omega(m)}$, and $\omega(m)(w_1, v_2) = \omega(m)(w_2, v_1) = 0$ because $W \subset V^{\omega(m)}$.

In these circumstances, the G-relative Darboux Theorem guarantees the existence of two open G-invariant neighborhoods U_0 and U_1 of $G \cdot m$ in U and a G-equivariant symplectomorphism $\Delta : (U_0, \Psi_* \omega_{Y_r}|_{U_0}) \rightarrow (U_1, \omega|_{U_1})$ which is the identity on $G \cdot m$. Take, without loss of generality, $U_1 = U$. Then the composed map $\Delta \circ \Psi : (Y_r, \omega_{Y_r}) \longrightarrow (U, \omega|_U)$ gives us, for $r > 0$ small enough, the inverse of the map needed in the statement of the theorem. ■

7.5 The Symplectic Slice Theorem and standard momentum maps

One of the features that makes the Symplectic Slice Theorem to be of great importance in geometry and in applications to mechanics is the extremely simple expression of a standard momentum map, when it happens to exist, in these semiglobal coordinates. As already mentioned, the particular case of the Symplectic Slice Theorem in the situation in which there is a momentum map has been treated in MARLE (1984, 1985) and GUILLEMIN AND STERNBERG (1984b); this justifies why the expression of the momentum map in the slice coordinate is usually referred to as the **Marle–Guillemin–Sternberg normal form**.

In the presentation below, this expression will appear as a corollary to a slightly more general approach introduced in the following definition.

7.5.1 Definition *Let (M, ω) be a symplectic manifold acted canonically upon by a Lie group G. For any point $m \in M$, we say that the G-action on M is **tubewise Hamiltonian** at m if there exists a G-invariant open neighborhood of the orbit $G \cdot m$ such that the restriction of the action to the symplectic manifold $(U, \omega|_U)$ has an associated standard momentum map.*

In this section we will use the Symplectic Slice Theorem to provide sufficient conditions that ensure that a given symplectic proper action is tubewise Hamiltonian. This is of much use in the study of singular dual pairs (see Chapter 11).

By Theorem 7.4.1 any orbit of a proper symplectic G-space (M, ω) has an invariant neighborhood around it that can be modeled by a symplectic tube similar to the one presented in Proposition 7.2.2. Thus the canonical proper G-action on (M, ω) is tubewise Hamiltonian if the G-action on each G-invariant model tube has an associated standard momentum map (see §4.5.9). The following result provides a sufficient condition for this to happen.

7.5.2 Proposition *Let (M, ω) be a symplectic manifold and let G be a Lie group with Lie algebra $\mathfrak{g}$ acting properly and canonically on M. For $m \in M$ let $Y_r := G \times_{G_m} (\mathfrak{m}_r^* \times V_r)$ be the slice model around the orbit $G \cdot m$ introduced in Proposition 7.2.2. If the G-equivariant $\mathfrak{g}^*$-valued one-form $\gamma \in \Omega^1(G; \mathfrak{g}^*)$ defined by*

$$\langle \gamma(g)\left(T_e L_g(\eta)\right), \xi\rangle := -\omega(m)\left(\left(\mathrm{Ad}_{g^{-1}}\xi\right)_M(m), \eta_M(m)\right) \tag{7.5.1}$$

for any $g \in G$ and $\xi, \eta \in \mathfrak{g}$ is exact, then the G-action on Y_r given by $g \cdot [h, \eta, v] := [gh, \eta, v]$, for any $g \in G$ and any $[h, \eta, v] \in Y_r$, has an associated standard momentum map and thus the G-action on (M, ω) is tubewise Hamiltonian at m.

Proof. The construction of the symplectic form ω_{Y_r} on Y_r in Proposition 7.2.2 reveals that the existence of a standard momentum map for the G-action on Y_r is guaranteed by the existence of a momentum map for the G-action on the symplectic manifold $(G \times \mathfrak{k}_r^*, \Omega)$ introduced in (7.2.3). This action is given by $g \cdot (h, \eta) := (gh, \eta)$, for any $g, h \in G$, $\eta \in \mathfrak{k}^*$. The existence of this momentum map is in turn equivalent to the vanishing of the map (see Proposition 4.5.17)

$$[\xi] \in \mathfrak{g}/[\mathfrak{g}, \mathfrak{g}] \longmapsto \left[\mathbf{i}_{\xi_{G\times\mathfrak{k}^*}}\Omega\right] \in H^1(G \times \mathfrak{k}^*), \quad \text{for any} \quad \xi \in \mathfrak{g}. \tag{7.5.2}$$

By the definition of Ω we have that for any $\xi, \eta \in \mathfrak{g}$, $g \in G$, and $\nu, \sigma \in \mathfrak{g}^*$

$$\begin{aligned}\mathbf{i}_{\xi_{G\times\mathfrak{k}^*}}\Omega(g, \nu)\left(T_e L_g(\eta), \sigma\right) &= \langle \sigma, \mathrm{Ad}_{g^{-1}}\xi\rangle \\ &\quad + \langle \nu, \left[\mathrm{Ad}_{g^{-1}}\xi, \eta\right]\rangle + \omega(m)\left(\left(\mathrm{Ad}_{g^{-1}}\xi\right)_M(m), \eta_M(m)\right).\end{aligned}$$

The first two terms on the right-hand side of the previous expression are the differential of the real function $f \in C^\infty(G \times \mathfrak{k}^*)$ given by

$$f(g, \nu) := \langle \nu, \mathrm{Ad}_{g^{-1}}\xi\rangle;$$

hence the vanishing of (7.5.2) is equivalent to the exactness of the $\mathfrak{g}^*$-valued one-form γ in the statement. ■

The following proposition provides a characterization of the exactness of (7.5.1) and therefore gives another sufficient condition for the tubewise Hamiltonian character of the action.

7.5.3 Proposition *Assume the hypotheses of Proposition 7.5.2. For $m \in M$, let $Y_r = G \times_{G_m} (\mathfrak{m}_r^* \times V_r)$ be the slice model around the orbit $G \cdot m$. Let $\Sigma_C : \mathfrak{g} \times \mathfrak{g} \longrightarrow \mathbb{R}$ be the two-cocycle induced by the Chu map, that is,*

$$\Sigma_C(\xi, \eta) := \omega(m)\left(\xi_M(m), \eta_M(m)\right), \quad \xi, \eta \in \mathfrak{g},$$

and let $\Sigma_C^\flat : \mathfrak{g} \to \mathfrak{g}^*$ *be defined by* $\Sigma_C^\flat(\xi) = \Sigma_C(\xi, \cdot)$, $\xi \in \mathfrak{g}$. *Then the form* (7.5.1) *is exact if and only if there exists a* $\mathfrak{g}^*$*-valued group one-cocycle* $\theta : G \to \mathfrak{g}^*$ *such that*

$$T_e\theta = \Sigma_C^\flat. \tag{7.5.3}$$

In such a case the action is tubewise Hamiltonian at the point m. *Also, in the presence of this cocycle, the map* $\mathbf{J}_\theta : G \times \mathfrak{k}^* \to \mathfrak{g}^*$ *given by*

$$\mathbf{J}_\theta(g, \nu) := \mathrm{Ad}^*_{g^{-1}} \nu - \theta(g) \tag{7.5.4}$$

is a momentum map for the G*-action on the presymplectic manifold* $G \times \mathfrak{k}^*$ *with non-equivariance cocycle equal to* $-\theta$.

Proof. Suppose first that the form γ in (7.5.1) is exact. In such a case, there exists a function $\theta : G \to \mathfrak{g}^*$ such that

$$\gamma(g) = \mathbf{d}\theta(g), \tag{7.5.5}$$

that is, for any $\xi, \eta \in \mathfrak{g}$ and $g \in G$ we have

$$\begin{aligned}\left\langle T_g\theta\left(T_eL_g(\eta)\right), \xi\right\rangle &= \langle \gamma(g)\left(T_eL_g(\eta)\right), \xi\rangle \\ &= -\omega(m)\left(\left(\mathrm{Ad}_{g^{-1}}\xi\right)_M(m), \eta_M(m)\right).\end{aligned} \tag{7.5.6}$$

This expression determines uniquely the derivative of θ and hence choosing $\theta(e) = 0$ fixes the map $\theta : G \to \mathfrak{g}^*$. We now show that θ is a cocycle by checking that it satisfies the cocycle identity. Indeed, for any $g, h \in G$ and any $\xi, \eta \in \mathfrak{g}$, we have

$$\begin{aligned}\left\langle T_g\left(\theta \circ L_h\right)\left(T_eL_g(\eta)\right), \xi\right\rangle &= \left\langle (T_{hg}\theta \circ T_eL_{hg})(\eta), \xi\right\rangle = \left\langle \gamma(hg)\left(T_eL_{hg}(\eta)\right), \xi\right\rangle \\ &= \left\langle \mathrm{Ad}^*_{h^{-1}}\left(\gamma(g)(T_eL_g(\eta))\right), \xi\right\rangle \\ &= \left\langle T_g\left(\mathrm{Ad}^*_{h^{-1}} \circ\theta\right)(T_eL_g(\eta)), \xi\right\rangle.\end{aligned}$$

Therefore, for any $g, h \in G$ we have $T_g\left(\theta \circ L_h\right) = T_g\left(\mathrm{Ad}^*_{h^{-1}} \circ\theta\right)$ and consequently

$$\theta \circ L_h = \mathrm{Ad}^*_{h^{-1}} \circ\theta + c(h, n),$$

where $c(h, n) \in \mathfrak{g}^*$, for any $h \in G$, and any $n \in [1, \mathrm{Card}\,(G/G^\circ)]$, $n \in \mathbb{N}$. Equivalently, for any $g, h \in G$ we can write

$$\theta(hg) = \mathrm{Ad}^*_{h^{-1}}\theta(g) + c(h, n). \tag{7.5.7}$$

Set in this equality $g = e$ and use $\theta(e) = 0$ to get $\theta(h) = c(h, n)$, for all $h \in G$ and $n \in [1, \mathrm{Card}\,(G/G^\circ)]$, $n \in \mathbb{N}$. Hence (7.5.7) becomes

$$\theta(hg) = \mathrm{Ad}^*_{h^{-1}}\theta(g) + \theta(h),$$

which is the group one-cocycle identity.

Finally, note that from (7.5.6) it is easy to see that $T_e\theta = \gamma(e) = \Sigma_C^\flat$ and therefore θ is the one-cocycle in the statement of the proposition. The converse is straightforward.

The fact that the expression (7.5.4) produces a momentum map for the G-action on $G \times \mathfrak{k}^*$ follows as a straightforward verification of the equality

$$\mathbf{i}_{\xi_{G\times\mathfrak{k}^*}}\Omega = \mathbf{d}\langle \mathbf{J}_\theta, \xi\rangle, \qquad \text{for any} \quad \xi \in \mathfrak{g}. \quad \blacksquare$$

The following corollary presents two situations in which the hypotheses of Proposition 7.5.2 are trivially satisfied.

7.5.4 Corollary *Let (M, ω) be a symplectic manifold and let G be a Lie group with Lie algebra $\mathfrak{g}$ acting properly and canonically on M. If either*

(i) $H^1(G) = 0$, *or*

(ii) *the orbit $G \cdot m$ is isotropic,*

then the G-action on (M, ω) is tubewise Hamiltonian at m.

7.5.5 Theorem (The Marle–Guillemin–Sternberg normal form) *Let (M, ω) be a connected symplectic manifold acting canonically and properly upon by a Lie group G. Suppose that this action has an associated standard momentum map $\mathbf{J} : M \to \mathfrak{g}^*$ with non-equivariance one-cocycle $\sigma : G \to \mathfrak{g}^*$. Let $m \in M$ and (Y_r, ω_{Y_r}) be the symplectic tube at m constructed in Proposition 7.2.2 that models a G-invariant open neighborhood U of the orbit $G \cdot m$ via the G-equivariant symplectomorphism $\phi : (U, \omega|_U) \to (Y_r, \omega_{Y_r})$. Then the canonical left G-action on (Y_r, ω_{Y_r}) admits a momentum map $\mathbf{J}_{Y_r} : Y_r \to \mathfrak{g}^*$ given by the expression*

$$\begin{aligned} \mathbf{J}_{Y_r} : Y_r = G \times_{G_m} (\mathfrak{m}_r^* \times V_r) &\longrightarrow \mathfrak{g}^* \\ [g, \rho, v] &\longmapsto \mathrm{Ad}^*_{g^{-1}}(\mathbf{J}(m) + \rho + \mathbf{J}_V(v)) + \sigma(g). \end{aligned} \tag{7.5.8}$$

The map $\mathbf{J}_{Y_r} \circ \phi$ is a momentum map for the canonical G-action on $(U, \omega|_U)$. Moreover, if the group G is connected, this momentum map satisfies $\mathbf{J}|_U = \mathbf{J}_{Y_r} \circ \phi$.

Proof. We start by using the characterization in Proposition 7.5.3 to ensure that the canonical left G-action on (Y_r, ω_{Y_r}) does indeed have an associated standard momentum map. This is so because the map $\theta : G \to \mathfrak{g}^*$ defined by $\theta(g) := -\mathrm{Ad}^*_{g^{-1}}\mathbf{J}(m) - \sigma(g)$, $g \in G$, satisfies the necessary and sufficient condition (7.5.3). Indeed, for any $\xi, \eta \in \mathfrak{g}$ we have that

$$\begin{aligned} \langle T_e\theta(\xi), \eta\rangle &= -\left.\frac{d}{dt}\right|_{t=0} \langle \mathrm{Ad}^*_{\exp(-t\xi)}\mathbf{J}(m), \eta\rangle - \langle T_e\sigma(\xi), \eta\rangle \\ &= \langle \mathrm{ad}^*_\xi \mathbf{J}(m), \eta\rangle - \Sigma(\xi, \eta) \qquad \text{(by (4.5.11))} \\ &= -\langle T_m\mathbf{J} \cdot \xi_M(m), \eta\rangle \qquad \text{(by (4.5.10))} \\ &= -\omega(m)(\eta_M(m), \xi_M(m)) = \langle \Sigma_C^\flat(\xi), \eta\rangle, \end{aligned}$$

as required.

We now check that $\mathbf{J}_{Y_r}$ is well defined. For any $[g, \rho, v] \in Y_r$ and $h \in G_m$ we have

$$\begin{aligned}\mathbf{J}_{Y_r}([gh, h^{-1} \cdot \rho, h^{-1} \cdot v]) &= \mathrm{Ad}^*_{(gh)^{-1}}(\mathbf{J}(m) + \mathrm{Ad}^*_h \rho + \mathbf{J}_V(h^{-1} \cdot v)) + \sigma(gh) \\ &= \mathrm{Ad}^*_{g^{-1}}(\mathrm{Ad}^*_{h^{-1}}(\mathbf{J}(m) + \mathrm{Ad}^*_h \rho + \mathrm{Ad}^*_h \mathbf{J}_V(v))) + \sigma(g) + \mathrm{Ad}^*_{g^{-1}}\sigma(h) \\ &= \mathrm{Ad}^*_{g^{-1}}(\mathbf{J}(m) + \rho + \mathbf{J}_V(v)) + \sigma(g) = \mathbf{J}_{Y_r}([g, \rho, v]),\end{aligned}$$

where we used the fact that as $h \in G_m$, we have that $\sigma(h) = \mathbf{J}(h \cdot m) - \mathrm{Ad}^*_{h^{-1}}\mathbf{J}(m) = \mathbf{J}(m) - \mathrm{Ad}^*_{h^{-1}}\mathbf{J}(m)$.

It only remains to show that $\mathbf{J}_{Y_r}$ is a momentum map for the left canonical G-action on (Y_r, ω_{Y_r}). Let $[g, \rho, v] \in Y_r$, $\xi, \zeta \in \mathfrak{g}$, $\alpha \in \mathfrak{m}^*$, and $u \in V$ arbitrary. Then, on the one hand we have that

$$\begin{aligned}&\left\langle \mathbf{dJ}_{Y_r}[g, \rho, v](T_e L_g(\xi), \alpha, u), \zeta \right\rangle \\ &= \left.\frac{d}{dt}\right|_{t=0} \left[\langle \mathrm{Ad}^*_{(g \exp t\xi)^{-1}}(\mathbf{J}(m) + \rho + t\alpha + \mathbf{J}_V(v + tu)), \zeta \rangle + \langle \sigma(g \exp t\xi), \zeta \rangle \right] \\ &= \left.\frac{d}{dt}\right|_{t=0} \left[\langle \mathrm{Ad}^*_{g^{-1}} \mathrm{Ad}^*_{(\exp t\xi)^{-1}}(\mathbf{J}(m) + \rho + t\alpha + \mathbf{J}_V(v + tu)), \zeta \rangle \right. \\ &\qquad\qquad \left. + \langle \sigma(g), \zeta \rangle + \langle \mathrm{Ad}^*_{g^{-1}}\sigma(\exp t\xi), \zeta \rangle \right] \\ &= \langle \mathbf{J}(m) + \rho + \mathbf{J}_V(v), [\mathrm{Ad}_{g^{-1}}\zeta, \xi] \rangle + \langle \alpha + T_v \mathbf{J}_V(u), \mathrm{Ad}_{g^{-1}}\zeta \rangle \\ &\quad + \Sigma(\xi, \mathrm{Ad}_{g^{-1}}\zeta).\end{aligned} \tag{7.5.9}$$

On the other hand, notice that the infinitesimal generators associated to the G-action on Y_r take the form

$$\zeta_{Y_r}[g, \rho, v] = \left.\frac{d}{dt}\right|_{t=0} [\exp t\zeta, \rho, v] = T_{(g,\rho,v)}\pi(T_e L_g(\mathrm{Ad}_{g^{-1}}\zeta), 0, 0),$$

where $\pi : G \times (\mathfrak{m}^*_r \times V_r) \to G \times_{G_m} (\mathfrak{m}^*_r \times V_r)$ is the projection. By the expression (7.2.2) for the symplectic form ω_{Y_r} on Y_r, we have

$$\begin{aligned}&\omega_{Y_r}([g, \rho, v])(\zeta_{Y_r}[g, \rho, v], (T_e L_g(\xi), \alpha, u)) \\ &\quad = \langle \alpha + T_v \mathbf{J}_V(u), \mathrm{Ad}_{g^{-1}}\zeta \rangle + \langle \rho + \mathbf{J}_V(v), [\mathrm{Ad}_{g^{-1}}\zeta, \xi] \rangle \\ &\quad\quad + \omega(m)((\mathrm{Ad}_{g^{-1}}\zeta)_M(m), \xi_M(m)).\end{aligned} \tag{7.5.10}$$

By (4.5.10) we can write

$$\begin{aligned}&\omega(m)((\mathrm{Ad}_{g^{-1}}\zeta)_M(m), \xi_M(m)) = \mathbf{dJ}^{\mathrm{Ad}_{g^{-1}}\zeta}(m)(\xi_M(m)) \\ &\quad = \langle -\mathrm{ad}^*_\xi \mathbf{J}(m) + \Sigma(\xi, \cdot), \mathrm{Ad}_{g^{-1}}\zeta \rangle = \langle \mathbf{J}(m), [\mathrm{Ad}_{g^{-1}}\zeta, \xi] \rangle + \Sigma(\xi, \mathrm{Ad}_{g^{-1}}\zeta),\end{aligned}$$

which substituted in (7.5.10) gives an expression identical to (7.5.9) and hence proves that

$$\mathbf{i}_{\zeta_{Y_r}} \omega_{Y_r} = \mathbf{dJ}^\zeta_{Y_r},$$

for any $\zeta \in \mathfrak{g}$, that is, $\mathbf{J}_{Y_r}$ is a momentum map for the G-action on (Y_r, ω_{Y_r}).

Finally, by the remark made in §4.5.9, the map $\mathbf{J}_{Y_r} \circ \phi$ is a momentum map for the canonical G-action on $(U, \omega|_U)$. The map $\mathbf{J}|_U$ also has this property and coincides with $\mathbf{J}_{Y_r} \circ \phi$ at the point m. We now recall that two different momentum maps for the same canonical action on a connected manifold differ by a constant. Consequently, if the group G is connected, so are Y_r and U, which implies that $\mathbf{J}_{Y_r} \circ \phi = \mathbf{J}|_U$. ■

7.6 A normal form for the cylinder valued momentum maps

The goal of this section is to formulate and prove the analogue of Theorem 7.5.5 for the cylinder valued momentum maps, that is, to provide an expression in the symplectic tube at any point $m \in M$ of a connected and paracompact symplectic manifold (M, ω) for the cylinder valued momentum maps associated to the canonical proper action of a connected Lie group G on M. The first tool will be the connection introduced in the following proposition.

7.6.1 Proposition *Let G be a connected Lie group acting canonically on the connected paracompact symplectic manifold (M, ω). For $m \in M$ let $G_m \subset G$ be its isotropy subgroup. Let $G/G_m \times \mathfrak{g}^* \to G/G_m$ be the projection considered as a trivial principal $(\mathfrak{g}^*, +)$-bundle where the action of the vector group $(\mathfrak{g}^*, +)$ on $G/G_m \times \mathfrak{g}^*$ is given by $R_\tau(gG_m, \mu) := (gG_m, \mu - \tau)$, $gG_m \in G/G_m$, $\mu, \tau \in \mathfrak{g}^*$. For any $gG_m \in G/G_m$ and $\mu \in \mathfrak{g}^*$ define*

$$H_{G/G_m}(gG_m, \mu) := \{(T_{gG_m}\pi_{G_m}(T_eL_g(\xi)), -\mathrm{Ad}^*_{g^{-1}}(\Psi(m)(\xi, \cdot))) \mid \xi \in \mathfrak{g}\}, \quad (7.6.1)$$

where $\pi_{G_m} : G \to G/G_m$ denotes the projection and $\Psi : M \to Z^2(\mathfrak{g})$ is the Chu map associated to the $\mathfrak{g}$-action on (M, ω). The collection of spaces $H_{G/G_m}(gG_m, \mu)$, for any $gG_m \in G/G_m$ and $\mu \in \mathfrak{g}^*$, constitutes the horizontal bundle of a principal connection α_{G/G_m} on $G/G_m \times \mathfrak{g}^* \to G/G_m$.

Proof. We start by showing that H_{G/G_m} is the graph of a smooth function. Let $F : T(G/G_m) \times \mathfrak{g}^* \to \mathfrak{g}^*$ be given by

$$F(T_{gG_m}\pi_{G_m}(T_eL_g(\xi)), \mu) := -\mathrm{Ad}^*_{g^{-1}}(\Psi(m)(\xi, \cdot)). \quad (7.6.2)$$

We check that this is a good definition. For any $h \in G_m$ we have

$$T_{gG_m}\pi_{G_m}(T_eL_g(\xi)) = T_{ghG_m}\pi_{G_m}(T_eL_{gh}(\mathrm{Ad}_{h^{-1}}\xi)),$$

we can use (4.4.3) to write

$$\begin{aligned} F(T_{ghG_m}\pi_{G_m}(T_eL_{gh}(\mathrm{Ad}_{h^{-1}}\xi)), \mu) &= -\mathrm{Ad}^*_{(gh)^{-1}}(\Psi(m)(\mathrm{Ad}_{h^{-1}}\xi, \cdot)) \\ &= -\mathrm{Ad}^*_{g^{-1}}(\Psi(m)(\xi, \cdot)) = F(T_{gG_m}\pi_{G_m}(T_eL_g(\xi)), \mu), \end{aligned}$$

which proves that F is well defined. The map F is smooth because for any $gG_m \in G/G_m$ there is a local smooth section σ of π_{G_m} defined on a local neighborhood $U \subset G/G_m$ around gG_m such that for any $v_z \in T_z(G/G_m)$, $z \in U$ we can write

$$F(v_z, \mu) = -\mathrm{Ad}^*_{\sigma(z)^{-1}}(\Psi(m)(T_{\sigma(z)}L_{\sigma(z)^{-1}}(T_z\sigma(v_z)), \cdot)),$$

which being smooth guarantees the smoothness of F. Since H_{G/G_m} is clearly the graph of F, this immediately implies that H_{G/G_m} is a smooth submanifold of $T(G/G_m \times \mathfrak{g}^*)$. Additionally, since for any $(gG_m, \mu) \in G/G_m \times \mathfrak{g}^*$ the value $H_{G/G_m}(gG_m, \mu)$ is a vector subspace of $T_{(gG_m,\mu)}(G/G_m \times \mathfrak{g}^*)$ we have proved that H_{G/G_m} is a subbundle of $T(G/G_m \times \mathfrak{g}^*)$.

We now show that H_{G/G_m} is the horizontal bundle of a principal connection α_{G/G_m} on $G/G_m \times \mathfrak{g}^* \to G/G_m$. Given that the vertical bundle V_{G/G_m} associated to the projection $G/G_m \times \mathfrak{g}^* \to G/G_m$ is given by $V_{G/G_m}(gG_m, \mu) = \{(0, \nu) \mid \nu \in \mathfrak{g}^*\}$, it suffices to check the following two points:

(i) $T_{(gG_m,\mu)}(G/G_m \times \mathfrak{g}^*) = H_{G/G_m}(gG_m, \mu) \oplus V_{G/G_m}(gG_m, \mu)$, for any $(gG_m, \mu) \in G/G_m \times \mathfrak{g}^*$. Obviously $T_{(gG_m,\mu)}(G/G_m \times \mathfrak{g}^*) = H_{G/G_m}(gG_m, \mu) + V_{G/G_m}(gG_m, \mu)$. Now let $v_{(gG_m,\mu)} = (T_{gG_m}\pi_{G_m}(T_eL_g(\xi)), -\mathrm{Ad}^*_{g^{-1}}(\Psi(m)(\xi, \cdot))) \in H_{G/G_m}(gG_m, \mu)$ $\cap\, V_{G/G_m}(gG_m, \mu)$. Since $T_{gG_m}\pi_{G_m}(T_eL_g(\xi)) = 0$, it follows that $\xi \in \mathfrak{g}_m$, which implies that $\Psi(m)(\xi, \cdot) = 0$ and hence $-\mathrm{Ad}^*_{g^{-1}}(\Psi(m)(\xi, \cdot)) = 0$. Consequently, $v_{(gG_m,\mu)} = 0$, as required.

(ii) $T_{(gG_m,\mu)}R_\tau[H_{G/G_m}(gG_m, \mu)] = H_{G/G_m}(R_\tau(gG_m, \mu))$, for any $\tau \in \mathfrak{g}^*$. This equality is a trivial consequence of the identity $T_{(gG_m,\mu)}R_\tau(T_{gG_m}\pi_{G_m}(T_eL_g(\xi)), (\mu, \nu))$ $= (T_{gG_m}\pi_{G_m}(T_eL_g(\xi)), (\mu - \tau, \nu))$, for any $g \in G$, $\xi \in \mathfrak{g}$, and $\mu, \nu, \tau \in \mathfrak{g}^*$. ■

7.6.2 Theorem *Let (M, ω) be a connected paracompact symplectic manifold acted properly and canonically upon by the connected Lie group G. Let $m \in M$ and (Y_r, ω_{Y_r}) be a symplectic tube at m that models a G-invariant neighborhood U of the orbit $G \cdot m$ via the G-equivariant symplectomorphism $\phi : (Y_r, \omega_{Y_r}) \to (U, \omega|_U)$. Let $\mathbf{K} : M \to \mathfrak{g}^*/\overline{\mathcal{H}}$ be a cylinder valued momentum map associated to the G-action on M such that $\mathbf{K}(m) = \pi_C(0)$ and let $\widetilde{G/G_m}$ be the holonomy bundle associated to the connection α_{G/G_m} introduced in Proposition* 7.6.1 *that contains the point* $(G_m, 0)$. *Then for any $[g, \rho, v] \in Y_r$ we have*

$$\mathbf{K}(\phi[g, \rho, v]) = \pi_C(\mathrm{Ad}^*_{g^{-1}}(\rho + \mathbf{J}_V(v)) + \nu), \tag{7.6.3}$$

where $\pi_C : \mathfrak{g}^* \to \mathfrak{g}^*/\overline{\mathcal{H}}$ is the projection and $\nu \in \mathfrak{g}^*$ is any element such that $(gG_m, \nu) \in \widetilde{G/G_m}$.

7.6.3 Remark A straightforward verification shows that the expression (7.6.3) reduces to the Marle–Guillemin–Sternberg normal form (7.5.8) in the presence of a standard momentum map.

Proof of the theorem. In the rest of this section we will denote by α and α_{Y_r} the connections that define cylinder valued momentum maps associated to the G-actions on M and Y_r, respectively. Additionally, $\mathcal{H}$ and $\mathcal{H}_{Y_r}$ will denote the corresponding holonomy groups.

7.6.4 Lemma *In the hypotheses of Theorem 7.6.2, we have that $\mathcal{H}_{Y_r} \subset \mathcal{H}$.*

Proof. Let $\mu \in \mathcal{H}_{Y_r}$. By definition there exists a piecewise smooth loop $y : [0, 1] \to Y_r$ such that $y(0) = y(1) = [e, 0, 0]$ and whose horizontal lift $\widetilde{y}(t) = (y(t), \mu(t))$ is such that $\mu(1) - \mu(0) = \mu$. Consider now the loop $\gamma := \phi \circ y : [0, 1] \to U$ such that

$\gamma(0) = \gamma(1) = m$. We now check that $\widetilde{\gamma}(t) := (\gamma(t), \mu(t))$ is the horizontal lift of γ by verifying that

$$\alpha(\gamma(t), \mu(t))(\dot{\gamma}(t), \dot{\mu}(t)) = 0. \tag{7.6.4}$$

By the definition of the connection α, this relation is equivalent to

$$(\mathbf{i}_{\xi_M}\omega)(\gamma(t)) \cdot \dot{\gamma}(t) = \langle \dot{\mu}(t), \xi \rangle, \quad \text{for all} \quad \xi \in \mathfrak{g},$$

which, by the definition of the loop γ, can be rewritten as

$$\omega(\phi(y(t)))(\xi_M(\phi(y(t))), T_{y(t)}\phi \cdot \dot{y}(t)) = \langle \dot{\mu}(t), \xi \rangle.$$

The G-equivariance of ϕ implies that this expression amounts to

$$\omega(\phi(y(t)))(T_{y(t)}\phi \cdot \xi_{Y_r}(y(t)), T_{y(t)}\phi \cdot \dot{y}(t)) = \langle \dot{\mu}(t), \xi \rangle.$$

Since ϕ is a symplectomorphism, this equality is equivalent to

$$\omega_{Y_r}(y(t))(\xi_{Y_r}(y(t)), \dot{y}(t)) = \langle \dot{\mu}(t), \xi \rangle,$$

which is true since the curve $\widetilde{y}(t) = (y(t), \mu(t))$ is horizontal with respect to α_{Y_r}. This chain of equivalences proves that (7.6.4) is verified and therefore that $\mu = \mu(1) - \mu(0) \in \mathcal{H}$, as required. ▼

We now consider the symplectic manifold (Y_r, ω_{Y_r}) acted canonically upon by the Lie group G and we compute the horizontal distribution H_{Y_r} associated to the connection α_{Y_r}. Let $z = [g, \rho, v] \in Y_r$ and $v_z = T_{(g,\rho,v)}\pi(T_eL_g(\eta), \alpha, u) \in T_zY_r$, with $\pi : G \times \mathfrak{m}_r^* \times V_r \to G \times_{G_m} (\mathfrak{m}_r^* \times V_r)$ the projection. Notice first that for any $\xi \in \mathfrak{g}$

$$\xi_{Y_r}([g, \rho, v]) = \left.\frac{d}{dt}\right|_{t=0} [\exp t\xi g, \rho, v] = T_{(g,\rho,v)}\pi(T_eL_g(\mathrm{Ad}_{g^{-1}}\xi), 0, 0).$$

Hence, for any $\nu \in \mathfrak{g}^*$, $(v_z, \nu) \in H_{Y_r}(z, \mu)$ if and only if

$$\omega_{Y_r}(z)(T_{(g,\rho,v)}\pi(T_eL_g(\mathrm{Ad}_{g^{-1}}\xi), 0, 0), v_z) = \langle \nu, \xi \rangle \quad \text{for any} \quad \xi \in \mathfrak{g}.$$

Using (7.2.2) we can rewrite this expression as

$$\begin{aligned}\langle \alpha + T_v\mathbf{J}_V(u), \mathrm{Ad}_{g^{-1}}\xi \rangle + \langle \rho + \mathbf{J}_V(v), [\mathrm{Ad}_{g^{-1}}\xi, \eta] \rangle& \\ + \Psi(m)(\mathrm{Ad}_{g^{-1}}\xi, \eta) &= \langle \nu, \xi \rangle.\end{aligned}$$

Since in this expression $\xi \in \mathfrak{g}$ is arbitrary we have that

$$\nu = \mathrm{Ad}^*_{g^{-1}}\left(\alpha + T_v\mathbf{J}_V(u) - \mathrm{ad}^*_\eta(\rho + \mathbf{J}_V(v)) - \Psi(m)(\eta, \cdot)\right).$$

Consequently, we have that

$$\begin{aligned} H_{Y_r}(z, \mu) = \Big\{ &(T_{(g,\rho,v)}\pi(T_eL_g(\eta), \alpha, u), \\ &\mathrm{Ad}^*_{g^{-1}}\left(\alpha + T_v\mathbf{J}_V(u) - \mathrm{ad}^*_\eta(\rho + \mathbf{J}_V(v)) - \Psi(m)(\eta, \cdot)\right)) \\ &\mid \eta \in \mathfrak{g}, u \in v, \alpha \in \mathfrak{m}^* \Big\}. \end{aligned} \tag{7.6.5}$$

We emphasize that this expression is well defined because if $(T_eL_g(\zeta), \mathrm{ad}^*_\zeta\rho, -\zeta\cdot v) \in \ker T_{(g,\rho,v)}\pi$, with $\zeta \in \mathfrak{g}_m$, then the second component in (7.6.5) is zero. Indeed,

$$\begin{aligned}\mathrm{Ad}^*_{g^{-1}}\left(\mathrm{ad}^*_\zeta\rho - T_v\mathbf{J}_V(\zeta\cdot v) - \mathrm{ad}^*_\zeta(\rho+\mathbf{J}_V(v)) - \Psi(m)(\zeta,\cdot)\right)&\\ = \mathrm{Ad}^*_{g^{-1}}\left(-T_v\mathbf{J}_V(\zeta\cdot v) - \mathrm{ad}^*_\zeta(\mathbf{J}_V(v))\right)&\end{aligned}$$

since $\Psi(m)(\zeta,\cdot)=0$ because $\zeta\in\mathfrak{g}_m$. Now, since for any $\eta\in\mathfrak{g}_m$ we have $\langle T_v\mathbf{J}_V(\zeta\cdot v),\eta\rangle = -\omega_V(\eta\cdot v,\zeta\cdot v)$ and $-\langle \mathrm{ad}^*_\zeta(\mathbf{J}_V(v)),\eta\rangle = -\langle \mathbf{J}_V(v),[\zeta,\eta]\rangle$ we get

$$\begin{aligned}\langle -T_v\mathbf{J}_V(\zeta\cdot v) - \mathrm{ad}^*_\zeta(\mathbf{J}_V(v)),\eta\rangle &= \omega_V(\zeta\cdot v,\eta\cdot v) - \frac{1}{2}\omega_V([\zeta,\eta]\cdot v, v)\\ &= \omega_V(\zeta\cdot v,\eta\cdot v) - \frac{1}{2}\omega_V(\zeta\cdot(\eta\cdot v)-\eta\cdot(\zeta\cdot v), v)\\ &= \omega_V(\zeta\cdot v,\eta\cdot v) + \frac{1}{2}\omega_V(\eta\cdot v,\zeta\cdot v) - \frac{1}{2}\omega_V(\zeta\cdot v,\eta\cdot v) = 0.\end{aligned}$$

7.6.5 Lemma *Let $\widetilde{G/G_m}$ be the holonomy bundle associated to the connection α_{G/G_m} that contains the point $(G_m,0)\in G/G_m\times\mathfrak{g}^*$. Then the holonomy bundle $\widetilde{Y}_r$ associated to the connection α_{Y_r} and containing $([e,0,0],0)$ is given by*

$$\widetilde{Y}_r = \left\{([g,\rho,v], \mathrm{Ad}^*_{g^{-1}}(\rho+\mathbf{J}_V(v))+\nu) \mid [g,\rho,v]\in Y_r,\ (gG_m,\nu)\in\widetilde{G/G_m}\right\}. \tag{7.6.6}$$

Proof. We start by checking that any point of the form $\widetilde{y} = ([g,\rho,v], \mathrm{Ad}^*_{g^{-1}}(\rho + \mathbf{J}_V(v)) + \nu) \in Y_r\times\mathfrak{g}^*$, satisfying $(gG_m,\nu)\in\widetilde{G/G_m}$, can be joined to $([e,0,0],0)$ via a horizontal curve for the connection α_{Y_r}. This will prove that the right-hand side of (7.6.6) is included in its left-hand side. Let $\gamma(t) = ([g](t),\nu(t))$ be a piecewise smooth horizontal curve for the connection α_{G/G_m} joining $(G_m,0)$ to (gG_m,ν). Using local sections for the projection $G\to G/G_m$, $\gamma(t)$ can be written as a finite concatenation of smooth curves of the form $(g_i(t)G_m,\nu(t))$, $i\in\{1,\ldots,n\}$ such that each $g_i:[0,1]\to G$ is smooth. For the sake of simplicity we shall work with only one such curve. The claim follows by noticing that $y(t) := ([g(t),t\rho,tv], \mathrm{Ad}^*_{g(t)^{-1}}(t\rho + \mathbf{J}_V(tv)) + \nu(t))$ is a horizontal curve for α_{Y_r}.

In order to prove the converse inclusion we take an arbitrary element $([g,\rho,v],\mu)\in\widetilde{Y}_r$. By definition of the holonomy bundle we can construct piecewise smooth curves $g:[0,1]\to G$, $\rho:[0,1]\to\mathfrak{m}^*_r$, $v:[0,1]\to V_r$, and $\mu:[0,1]\to\mathfrak{g}^*$ such that $([g(t),\rho(t),v(t)],\mu(t))$ is a horizontal curve joining $([g,\rho,v],\mu)$ with $([e,0,0],0)$. Horizontality implies that

$$\begin{aligned}\dot\mu(t) &= \mathrm{Ad}^*_{g(t)^{-1}}(\dot\rho(t) + T_{v(t)}\mathbf{J}_V(\dot v(t)) - \mathrm{ad}^*_{g(t)^{-1}\dot g(t)}(\rho(t)+\mathbf{J}_V(v(t)))\\ &\quad - \Psi(m)(g(t)^{-1}\dot g(t),\cdot)) \qquad (7.6.7)\\ &= \frac{d}{dt}\left(\mathrm{Ad}^*_{g(t)^{-1}}(\rho(t)+\mathbf{J}_V(v(t)))\right) - \mathrm{Ad}^*_{g(t)^{-1}}\left(\Psi(m)(g(t)^{-1}\dot g(t),\cdot)\right).\end{aligned}$$

Let now $\widetilde{g}(t) = (g(t)G_m, \nu(t)) \in G/G_m \times \mathfrak{g}^*$ be the horizontal lift through $(G_m, 0)$ of the curve $g(t)G_m \in G/G_m$ with respect to the connection α_{G/G_m}. This means that

$$\dot{\nu}(t) = -\mathrm{Ad}^*_{g(t)^{-1}}(\Psi(m)(g(t)^{-1}\dot{g}(t), \cdot)),$$

which allows us to rewrite (7.6.7) as

$$\dot{\mu}(t) = \frac{d}{dt}\left(\mathrm{Ad}^*_{g(t)^{-1}}(\rho(t) + \mathbf{J}_V(v(t))) + \nu(t)\right).$$

Consequently,

$$\mu(t) = \mathrm{Ad}^*_{g(t)^{-1}}(\rho(t) + \mathbf{J}_V(v(t))) + \nu(t)$$

and hence

$$\begin{aligned}([g, \rho, v], \mu) &= ([g(1), \rho(1), v(1)], \mu(1)) \\ &= ([g(1), \rho(1), v(1)], \mathrm{Ad}^*_{g(1)^{-1}}(\rho(1) + \mathbf{J}_V(v(1)) + \nu(1))).\end{aligned}$$

Since $(g(1)G_m, \nu(1)) \in \widetilde{G/G_m}$ we have that $([g, \rho, v], \mu)$ belongs to the right-hand side of (7.6.6), as required. ▼

7.6.6 Lemma *The holonomy groups $\mathcal{H}_{Y_r}$ and $\mathcal{H}_{G/G_m}$ corresponding to the connections α_{Y_r} and α_{G/G_m}, respectively, coincide, that is,*

$$\mathcal{H}_{Y_r} = \mathcal{H}_{G/G_m}.$$

Proof. Take first $\nu \in \mathcal{H}_{G/G_m}$. By definition there exists a piecewise smooth curve $g : [0, 1] \to G$ such that $g(0)G_m = g(1)G_m = G_m$ and whose horizontal lift $\widetilde{g(t)G_m} = (g(t)G_m, \nu(t))$ with respect to the connection α_{G/G_m} satisfies that $\nu(1) - \nu(0) = \nu$. The horizontality of $\widetilde{g(t)G_m}$ implies that

$$\dot{\nu}(t) = -\mathrm{Ad}^*_{g(t)^{-1}}\left(\Psi(m)(g(t)^{-1}\dot{g}(t), \cdot)\right).$$

Let now $[g(t), 0, 0]$ be a loop on Y_r. The curve $([g(t), 0, 0], \nu(t))$ is its horizontal lift because its derivative lies on the right-hand side of (7.6.5). In addition, this proves that $\nu = \nu(1) - \nu(0) \in \mathcal{H}_{Y_r}$.

In order to prove the converse inclusion take $\mu \in \mathcal{H}_{Y_r}$. Then there exists a piecewise smooth loop $y(t) = [g(t), \rho(t), v(t)]$ in Y_r such that $[g(0), \rho(0), v(0)] = [g(1), \rho(1), v(1)] = [e, 0, 0]$ and whose horizontal lift $\widetilde{y}(t) = ([g(t), \rho(t), v(t)], \mu(t))$ satisfies $\mu = \mu(1) - \mu(0)$. The functions $g : [0, 1] \to G$, $\rho : [0, 1] \to \mathfrak{m}_r^*$, and $v : [0, 1] \to V_r$ can be chosen without loss of generality so that $g(0) = g(1) = e$, $\rho(0) = \rho(1) = 0$, and $v(0) = v(1) = 0$. Let now $(g(t)G_m, \nu(t))$ be the horizontal lift of the loop $g(t)G_m$ with respect to the connection α_{G/G_m}, namely, the function $\nu(t)$

satisfies $\dot{\nu}(t) = -\mathrm{Ad}^*_{g(t)^{-1}}\left(\Psi(m)(g(t)^{-1}\dot{g}(t), \cdot)\right)$. Consequently,

$$\begin{aligned}
\mu = \mu(1) - \mu(0) &= \int_0^1 \dot{\mu}(t)dt \\
&= \int_0^1 \left[\frac{d}{dt}\left(\mathrm{Ad}^*_{g(t)^{-1}}(\rho(t) + \mathbf{J}_V(v(t)))\right) - \mathrm{Ad}^*_{g(t)^{-1}}\left(\Psi(m)(g(t)^{-1}\dot{g}(t), \cdot)\right)\right] dt \\
&= \mathrm{Ad}^*_{g(1)^{-1}}(\rho(1) + \mathbf{J}_V(v(1))) - \mathrm{Ad}^*_{g(0)^{-1}}(\rho(0) + \mathbf{J}_V(v(0))) + (\nu(1) - \nu(0)) \\
&= \nu(1) - \nu(0) \in \mathcal{H}_{G/G_m}. \quad \blacktriangledown
\end{aligned}$$

We are now ready to prove (7.6.3). We start by showing that this expression is well defined. Let $[g, \rho, v], [g', \rho', v'] \in Y_r$ and $\nu, \nu' \in \mathfrak{g}^*$ be such that

$$[g, \rho, v] = [g', \rho', v'] \tag{7.6.8}$$

and

$$(gG_m, \nu), (g'G_m, \nu') \in \widetilde{G/G_m}. \tag{7.6.9}$$

The equality (7.6.8) implies that there exists an element $h \in G_m$ such that $g' = gh$, $\rho' = h^{-1} \cdot \rho$, and $v' = h^{-1} \cdot v$, while (7.6.9) guarantees that $(g'G_m, \nu') = (ghG_m, \nu') = (gG_m, \nu') \in \widetilde{G/G_m}$. As $(gG_m, \nu) \in \widetilde{G/G_m}$ also, there exists an element $\tau \in \mathcal{H}_{G/G_m}$ such that $\nu' = \nu + \tau$. By Lemmas 7.6.4 and 7.6.6 we have $\mathcal{H}_{G/G_m} = \mathcal{H}_{Y_r} \subset \mathcal{H}$, which implies that $\pi_C(\tau) = 0$. Taking this into consideration as well as the G_m-equivariance of the momentum map $\mathbf{J}_V$, we conclude that

$$\begin{aligned}
\mathbf{K}(\phi([g', \rho', v'])) &= \pi_C\left(\mathrm{Ad}^*_{(gh)^{-1}}(h^{-1} \cdot \rho + \mathbf{J}_V(h^{-1} \cdot v)) + \nu'\right) \\
&= \pi_C\left(\mathrm{Ad}^*_{g^{-1}}(\rho + \mathbf{J}_V(v)) + \nu + \tau\right) \\
&= \pi_C\left(\mathrm{Ad}^*_{g^{-1}}(\rho + \mathbf{J}_V(v)) + \nu\right) \\
&= \mathbf{K}(\phi([g, \rho, v])),
\end{aligned}$$

as required.

Let now $\widetilde{Y_r} \subset Y_r \times \mathfrak{g}^*$ and $\widetilde{M} \subset M \times \mathfrak{g}^*$ be the holonomy bundles associated to the connections α_{Y_r} and α and containing the points $([e, 0, 0], 0)$ and $(m, 0)$, respectively. We now show that

$$\text{if} \quad ([g, \rho, v], \mu) \in \widetilde{Y_r} \quad \text{then} \quad (\phi([g, \rho, v]), \mu) \in \widetilde{M}. \tag{7.6.10}$$

Indeed, if $([g, \rho, v], \mu) \in \widetilde{Y_r}$, then there exists a piecewise smooth horizontal curve $\gamma_{Y_r}(t) = ([g(t), \rho(t), v(t)], \mu(t))$ in $Y_r \times \mathfrak{g}^*$ such that $\gamma_{Y_r}(0) = ([e, 0, 0], 0)$ and $\gamma_{Y_r}(1) = ([g, \rho, v], \mu)$. Given that the map ϕ is a G-equivariant symplectomorphism, the curve $\gamma_M(t) := (\phi([g(t), \rho(t), v(t)]), \mu(t))$ is horizontal with respect to α. Indeed, $\gamma_M(t)$ is horizontal if and only if for any $\xi \in \mathfrak{g}$ we have that

$$\omega(\phi(\lambda(t)))(\xi_M(\phi(\lambda(t))), T_{\lambda(t)}\phi \cdot \dot{\lambda}(t)) = \langle \dot{\mu}(t), \xi \rangle,$$

where $\lambda(t) = [g(t), \rho(t), v(t)]$. Since ϕ is G-equivariant this equality is equivalent to

$$\omega(\phi(\lambda(t)))(T_{\lambda(t)}\phi \cdot \xi_{Y_r}(\lambda(t)), T_{\lambda(t)}\phi \cdot \dot{\lambda}(t)) = \langle \dot{\mu}(t), \xi \rangle.$$

Given that ϕ is a symplectic map, this relation amounts to

$$\omega_{Y_r}(\lambda(t))(\xi_{Y_r}(\lambda(t)), \dot{\lambda}(t)) = \langle \dot{\mu}(t), \xi \rangle$$

which is a true relation by the horizontality of γ_{Y_r}. The statement in (7.6.10) implies that we can write

$$\mathbf{K}(\phi([g, \rho, v])) = \pi_C(\mu) = \pi_C(\widetilde{\mathbf{K}_{Y_r}}([g, \rho, v], \mu)),$$

where $\widetilde{\mathbf{K}_{Y_r}} : \widetilde{Y_r} \subset Y_r \times \underline{\mathfrak{g}}^* \rightarrow \mathfrak{g}^*$ is the projection into the $\mathfrak{g}^*$ factor. Expression (7.6.6) guarantees that $\mu = \widetilde{\mathbf{K}_{Y_r}}([g, \rho, v], \mu) = \mathrm{Ad}^*_{g^{-1}}(\rho + \mathbf{J}_V(v)) + \nu)$ with $(gG_m, \nu) \in \widetilde{G/G_m}$ and hence

$$\mathbf{K}(\phi[g, \rho, v]) = \pi_C(\mathrm{Ad}^*_{g^{-1}}(\rho + \mathbf{J}_V(v)) + \nu),$$

as required. ■

7.7 The Reconstruction Equations

The goal of this section is to give the explicit expression of Hamilton's equations for an arbitrary G-invariant Hamiltonian in the semiglobal coordinates provided by the symplectic tube (Y_r, ω_{Y_r}) constructed in Proposition 7.2.2. These equations are known in the literature under the name of ***reconstruction*** (ORTEGA (1998), ORTEGA AND RATIU (2002b)) or ***bundle equations*** (ROBERTS et al. (2002); SCHEERER AND WULFF (2001)) and have been of great use in the study of relative equilibria and relative periodic motions in symmetric Hamiltonian mechanical systems (see, for instance ROBERTS AND DE SOUSA DIAS (1997); ROBERTS et al. (2002); ORTEGA AND RATIU (1999a); WULFF (2003); ORTEGA (2003a), and references therein).

7.7.1 Statement of the problem. In this section we will work on the symplectic G-invariant tube (Y_r, ω_{Y_r}) constructed at a point m of a symplectic manifold (M, ω) acted properly and canonically upon by a Lie group G.

Let $h \in C^\infty(Y_r)^G$ be a G-invariant Hamiltonian on Y_r. Our aim is to compute the differential equations that determine the G-equivariant Hamiltonian vector field $X_h \in \mathfrak{X}(Y_r)^G$ associated to h and characterized by

$$\mathbf{i}_{X_h}\omega_{Y_r} = \mathbf{d}h. \tag{7.7.1}$$

Since the projection $\pi : G \times \mathfrak{m}_r^* \times V_r \rightarrow G \times_H (\mathfrak{m}_r^* \times V_r)$ is a surjective submersion, there are always local sections available that we can use to locally express

$$X_h = T\pi(X_G, X_{\mathfrak{m}_r^*}, X_{V_r}),$$

with X_G, $X_{\mathfrak{m}_r^*}$ and X_{V_r} locally defined smooth maps on Y and having values in TG, $T\mathfrak{m}_r^*$ and TV_r, respectively. Thus, for any $[g, \rho, v] \in Y$, one has $X_G([g, \rho, v]) \in$

T_gG, $X_{\mathfrak{m}_r^*}([g, \rho, v]) \in T_\rho\mathfrak{m}_r^* = \mathfrak{m}^*$, and $X_{V_r}([g, \rho, v]) \in T_vV_r = V$. Moreover, using the Ad_{G_m}-invariant decomposition of the Lie algebra $\mathfrak{g}$

$$\mathfrak{g} = \mathfrak{g}_m \oplus \mathfrak{m} \oplus \mathfrak{q},$$

introduced in the previous section, the mapping X_G can be written, for any $[g, \rho, v] \in Y$, as

$$X_G([g, \rho, v]) = T_eL_g\big(X_{\mathfrak{g}_m}([g, \rho, v]) + X_{\mathfrak{m}}([g, \rho, v]) + X_{\mathfrak{q}}([g, \rho, v])\big),$$

with $X_{\mathfrak{g}_m}$, $X_{\mathfrak{m}}$, and $X_{\mathfrak{q}}$, locally defined smooth maps on Y with values in $\mathfrak{g}_m$, $\mathfrak{m}$, and $\mathfrak{q}$, respectively. There is no hope to completely determine the maps $X_{\mathfrak{g}_m}$, $X_{\mathfrak{m}}$, $X_{\mathfrak{q}}$, $X_{\mathfrak{m}_r^*}$, and X_{V_r} solely from (7.7.1) since this is an equation on the quotient. However, it is possible to gather enough information on $X_{\mathfrak{g}_m}$, $X_{\mathfrak{m}}$, $X_{\mathfrak{q}}$, $X_{\mathfrak{m}_r^*}$, and X_{V_r} such that X_h is uniquely determined by the relation

$$\begin{aligned} X_h([g, \rho, v]) = T_{(g,\rho,v)}\pi\big(&T_eL_g\big(X_{\mathfrak{g}_m}([g, \rho, v]) + X_{\mathfrak{m}}([g, \rho, v]) \\ &+ X_{\mathfrak{q}}([g, \rho, v])\big), X_{\mathfrak{m}_r^*}([g, \rho, v]), X_{V_r}([g, \rho, v])\big). \end{aligned}$$

7.7.2 The use of the symplectic form on the tube. We will now write down explicitly the expression (7.7.1), using the characterization of ω_{Y_r} provided in (7.2.2). Let $[g, \rho, v] \in Y_r$ and $w = T_{(g,\rho,v)}\pi((T_eL_g(v_{\mathfrak{g}_m} + v_{\mathfrak{m}} + v_{\mathfrak{q}}), v_{\mathfrak{m}_r^*}, v_{V_r}) \in T_{[g,\rho,v]}Y$ be arbitrary, with $v_{\mathfrak{g}_m} \in \mathfrak{g}_m$, $v_{\mathfrak{m}} \in \mathfrak{m}$, $v_{\mathfrak{q}} \in \mathfrak{q}$, $v_{\mathfrak{m}_r^*} \in \mathfrak{m}^*$, and $v_{V_r} \in V$. The value of X_h at $[g, \rho, v]$ is uniquely determined by the equality

$$\omega_{Y_r}([g, \rho, v])(X_h([g, \rho, v]), w) = \mathbf{d}h([g, \rho, v])(w),$$

with w arbitrary or, equivalently,

$$\begin{aligned} \omega_{Y_r}([g, \rho, v])(&T_{(g,\rho,v)}\pi(T_eL_g(X_{\mathfrak{g}_m} + X_{\mathfrak{m}} + X_{\mathfrak{q}}), X_{\mathfrak{m}_r^*}, X_{V_r}), \\ &T_{(g,\rho,v)}\pi((T_eL_g(v_{\mathfrak{g}_m} + v_{\mathfrak{m}} + v_{\mathfrak{q}}), v_{\mathfrak{m}_r^*}, v_{V_r})) \\ &= \mathbf{d}h([g, \rho, v])(w), \end{aligned} \tag{7.7.2}$$

where, for economy of notation, we suppressed the dependence of $X_{\mathfrak{g}_m}$, $X_{\mathfrak{m}}$, $X_{\mathfrak{q}}$, $X_{\mathfrak{m}_r^*}$, and X_{V_r} on $[g, \rho, v]$. Note now that since $h \in C^\infty(G \times_{G_m} (\mathfrak{m}_r^* \times V_r))^G$ is G-invariant, the mapping $h \circ \pi \in C^\infty(G \times \mathfrak{m}_r^* \times V_r)^{G_m}$ can be understood as a G_m-invariant function that depends only on the $\mathfrak{m}_r^*$ and V_r variables, that is,

$$h \circ \pi \in C^\infty(\mathfrak{m}_r^* \times V_r)^{G_m}.$$

This implies that $\mathbf{d}h([g, \rho, v])(w)$ can be written as

$$\begin{aligned} \mathbf{d}h([g, \rho, v])(w) &= \mathbf{d}(h \circ \pi)(g, \rho, v)((T_eL_g(v_{\mathfrak{g}_m} + v_{\mathfrak{m}} + v_{\mathfrak{q}}), v_{\mathfrak{m}_r^*}, v_{V_r}) \\ &= D_{\mathfrak{m}_r^*}(h \circ \pi)(g, \rho, v)(v_{\mathfrak{m}_r^*}) + D_{V_r}(h \circ \pi)(g, \rho, v)(v_{V_r}), \end{aligned}$$

where $D_{\mathfrak{m}_r^*}(h \circ \pi)(g, \rho, v) \in \mathfrak{m}$ and $D_{V_r}(h \circ \pi)(g, \rho, v) \in V^*$ are the partial derivatives of $h \circ \pi$ with respect to the $\mathfrak{m}_r^*$ and V_r variables, respectively. Using this remark and (7.2.2), the expression (7.7.2) can be rewritten as

$$\begin{aligned}
&D_{\mathfrak{m}_r^*}(h\circ\pi)(g,\,\rho,\,v)\cdot(v_{\mathfrak{m}_r^*})+D_{V_r}(h\circ\pi)(g,\,\rho,\,v)(v_{V_r})\\
&\quad=\langle v_{\mathfrak{m}_r^*},\,X_{\mathfrak{m}}\rangle+\langle T_v\mathbf{J}_V(v_{V_r}),\,X_{\mathfrak{g}_m}\rangle\\
&\qquad-\langle X_{\mathfrak{m}_r^*},\,v_{\mathfrak{m}}\rangle-\langle T_v\mathbf{J}_V(X_{V_r}),\,v_{\mathfrak{g}_m}\rangle\\
&\qquad+\langle\rho+\mathbf{J}_V(v),\,[X_{\mathfrak{g}_m}+X_{\mathfrak{m}}+X_{\mathfrak{q}},\,v_{\mathfrak{g}_m}+v_{\mathfrak{m}}+v_{\mathfrak{q}}]\rangle\\
&\qquad+\Psi(m)(X_{\mathfrak{g}_m}+X_{\mathfrak{m}}+X_{\mathfrak{q}},\,v_{\mathfrak{g}_m}+v_{\mathfrak{m}}+v_{\mathfrak{q}})\\
&\qquad+\omega_V(X_{V_r},\,v_{V_r}).
\end{aligned}\tag{7.7.3}$$

This expression admits several simplifications. First, using the explicit expression of $\mathbf{J}_V$ in terms of ω_V, we compute $T_v\mathbf{J}_V(v_{V_r})$. For any $\xi\in\mathfrak{g}_m$,

$$\begin{aligned}
\langle T_v\mathbf{J}_V(v_{V_r}),\,\xi\rangle&=\left.\frac{d}{dt}\right|_{t=0}\langle\mathbf{J}_V(v+tv_{V_r}),\,\xi\rangle\\
&=\left.\frac{d}{dt}\right|_{t=0}\frac{1}{2}\omega_V(\xi_{V_r}(v+tv_{V_r}),\,v+tv_{V_r})\\
&=\frac{1}{2}\big(\omega_V(\xi_{V_r}(v_{V_r}),\,v)+\omega_V(\xi_{V_r}(v),\,v_{V_r})\big)\\
&=\omega_V(\xi_{V_r}(v),\,v_{V_r}),
\end{aligned}$$

since the $\mathfrak{g}_m$-representation on V_r is symplectic. In particular, if we take $X_{\mathfrak{g}_m}\in\mathfrak{g}_m$, we obtain

$$\langle T_v\mathbf{J}_V(v_{V_r}),\,X_{\mathfrak{g}_m}\rangle=\omega_V((X_{\mathfrak{g}_m})_{V_r}(v),\,v_{V_r}).$$

Second, as $X_{\mathfrak{g}_m}+X_{\mathfrak{m}},\,v_{\mathfrak{g}_m}+v_{\mathfrak{m}}\in\mathfrak{k}$ we have

$$\begin{aligned}
&\Psi(m)(X_{\mathfrak{g}_m}+X_{\mathfrak{m}}+X_{\mathfrak{q}},\,v_{\mathfrak{g}_m}+v_{\mathfrak{m}}+v_{\mathfrak{q}})\\
&\qquad\qquad=\Psi(m)(X_{\mathfrak{q}},\,v_{\mathfrak{q}})=\omega(m)((X_{\mathfrak{q}})_M(m),\,(v_{\mathfrak{q}})_M(m)).
\end{aligned}$$

Moreover, the bilinear pairing $\langle\cdot,\cdot\rangle_{\mathfrak{q}}$ in $\mathfrak{q}$ defined using the Chu map by

$$\langle\xi,\eta\rangle_{\mathfrak{q}}:=\Psi(m)(\xi,\eta)=\omega(m)\left(\xi_M(m),\eta_M(m)\right),\qquad\xi,\eta\in\mathfrak{k}$$

is nondegenerate. These two points substituted into (7.7.3) yield

$$\begin{aligned}
&D_{\mathfrak{m}_r^*}(h\circ\pi)(g,\,\rho,\,v)(v_{\mathfrak{m}_r^*})+D_{V_r}(h\circ\pi)(g,\,\rho,\,v)(v_{V_r})\\
&\quad=\langle v_{\mathfrak{m}_r^*},\,X_{\mathfrak{m}}\rangle+\omega_V((X_{\mathfrak{g}_m})_{V_r}(v),\,v_{V_r})\\
&\qquad-\langle X_{\mathfrak{m}_r^*},\,v_{\mathfrak{m}}\rangle-\langle T_v\mathbf{J}_V(X_{V_r}),\,v_{\mathfrak{g}_m}\rangle\\
&\qquad+\langle\rho+\mathbf{J}_V(v),\,[X_{\mathfrak{g}_m}+X_{\mathfrak{m}}+X_{\mathfrak{q}},\,v_{\mathfrak{g}_m}+v_{\mathfrak{m}}+v_{\mathfrak{q}}]\rangle\\
&\qquad+\langle X_{\mathfrak{q}},\,v_{\mathfrak{q}}\rangle_{\mathfrak{q}}+\omega_V(X_{V_r},\,v_{V_r}).
\end{aligned}\tag{7.7.4}$$

Since this equality holds for any $v_{\mathfrak{g}_m}\in\mathfrak{g}_m$, $v_{\mathfrak{m}}\in\mathfrak{m}$, $v_{\mathfrak{q}}\in\mathfrak{q}$, $v_{\mathfrak{m}_r^*}\in\mathfrak{m}^*$, and $v_{V_r}\in V$, (7.7.4) separates into the following five identities all evaluated at an arbitrary point

$[g, \rho, v]$ for $g \in G$, $\rho \in \mathfrak{m}^*$, and $v \in V_r$

$$X_{\mathfrak{m}} = D_{\mathfrak{m}_r^*}(h \circ \pi) \tag{7.7.5}$$

$$\mathbf{i}_{X_{V_r}+(X_{\mathfrak{g}_m})_{V_r}}\omega_V = D_{V_r}(h \circ \pi) \tag{7.7.6}$$

$$\mathbb{P}_{\mathfrak{q}^*}\left(\mathrm{ad}^*_{(X_{\mathfrak{m}}+X_{\mathfrak{g}_m}+X_{\mathfrak{q}})}(\rho + \mathbf{J}_V(v))\right) + \langle X_{\mathfrak{q}}, \cdot\rangle_{\mathfrak{q}} = 0 \tag{7.7.7}$$

$$\mathbb{P}_{\mathfrak{m}^*}\left(\mathrm{ad}^*_{(X_{\mathfrak{m}}+X_{\mathfrak{g}_m}+X_{\mathfrak{q}})}(\rho + \mathbf{J}_V(v))\right) = X_{\mathfrak{m}_r^*} \tag{7.7.8}$$

$$\mathbb{P}_{\mathfrak{g}_m^*}\left(\mathrm{ad}^*_{(X_{\mathfrak{m}}+X_{\mathfrak{g}_m}+X_{\mathfrak{q}})}(\rho + \mathbf{J}_V(v))\right) = T_v\mathbf{J}_V(X_{V_r}), \tag{7.7.9}$$

where $\mathbb{P}_{\mathfrak{g}_m^*}$, $\mathbb{P}_{\mathfrak{m}^*}$, and $\mathbb{P}_{\mathfrak{q}^*}$ denote the projections from $\mathfrak{g}^*$ onto $\mathfrak{g}_m^*$, $\mathfrak{m}^*$, and $\mathfrak{q}^*$, respectively, according to the $\mathrm{Ad}^*_{G_m}$-invariant splitting

$$\mathfrak{g}^* = \mathfrak{g}_m^* \oplus \mathfrak{m}^* \oplus \mathfrak{q}^*.$$

7.7.3 Solution of the equations (7.7.5)–(7.7.9). We now solve for the unknowns $X_{\mathfrak{g}_m}$, $X_{\mathfrak{m}}$, $X_{\mathfrak{q}}$, $X_{\mathfrak{m}_r^*}$, and X_{V_r} in the equations (7.7.5)–(7.7.9) and we therefore determine the Hamiltonian vector field associated to the G-invariant Hamiltonian $h \in C^\infty(Y_r)^G$.

We first note that since $h \in C^\infty(Y_r)^G$ is G-invariant, the corresponding Hamiltonian vector field is going to be G-equivariant; therefore it suffices to consider the preceding equations at the point $[e, \rho, v]$ which will give us the value of the vector field $X_h([e, \rho, v])$ and then, using the equivariance, we obtain $X_h([g, \rho, v])$ by writing

$$X_h([g, \rho, v]) = T_{[e,\rho,v]}\Phi_g(X_h([e, \rho, v])),$$

where $\Phi : G \times Y_r \to Y_r$ denotes the canonical left G-action on (Y_r, ω_{Y_r}). An additional simplification consists of assuming that $X_{\mathfrak{g}_m} = 0$ and to verify later that, within the framework of this assumption, we are able to solve for the unknowns $X_{\mathfrak{m}}$, $X_{\mathfrak{q}}$, $X_{\mathfrak{m}_r^*}$, and X_{V_r} satisfying the equations (7.7.5)–(7.7.9).

We start solving the equations. First, note that (7.7.5) and (7.7.6), together with the assumption $X_{\mathfrak{g}_m} = 0$, uniquely determine $X_{\mathfrak{m}}$ and X_{V_r}. In particular, we can write

$$X_{\mathfrak{m}} = D_{\mathfrak{m}_r^*}(h \circ \pi), \qquad X_{V_r} = B_V^\sharp(D_{V_r}(h \circ \pi)), \tag{7.7.10}$$

where $B_V^\sharp$ is the isomorphism associated to the Poisson tensor B_V given by the symplectic structure ω_V of V. We now study (7.7.7) at the point $[e, \rho, v]$ with the assumption $X_{\mathfrak{g}_m} = 0$, that is,

$$\mathbb{P}_{\mathfrak{q}^*}\left(\mathrm{ad}^*_{(X_{\mathfrak{m}}+X_{\mathfrak{q}})}(\rho + \mathbf{J}_V(v))\right) + \langle X_{\mathfrak{q}}, \cdot\rangle_{\mathfrak{q}} = 0. \tag{7.7.11}$$

We start by defining

$$\begin{aligned} F : \mathfrak{k} \times \mathfrak{k}^* \times \mathfrak{q} &\longrightarrow \mathfrak{q}^* \\ (\xi, \lambda, \tau) &\longmapsto \mathbb{P}_{\mathfrak{q}^*}\left(\mathrm{ad}^*_{(\xi+\tau)}\lambda\right) + \langle \tau, \cdot\rangle_{\mathfrak{q}}. \end{aligned} \tag{7.7.12}$$

The nondegeneracy of the pairing $\langle\cdot,\cdot\rangle_{\mathfrak{q}}$ implies that the mapping F is such that $F(0,0,0) = 0$ and that its derivative $D_{\mathfrak{q}}F(0,0,0) : \mathfrak{q}^* \to \mathfrak{q}^*$ is a linear isomorphism. The Implicit Function Theorem implies the existence of a locally defined function $\tau : \mathfrak{k}\times\mathfrak{k}^* \to \mathfrak{q}$ around the origin such that $\tau(0,0)=0$, and $\mathbb{P}_{\mathfrak{q}^*}\left(\mathrm{ad}^*_{(\xi+\tau(\xi,\lambda))}\lambda\right) + \langle\tau(\xi,\lambda),\cdot\rangle_{\mathfrak{q}} = 0$. Let now $\psi : \mathfrak{m}^*\times V_r \to \mathfrak{q}$ be the locally defined function given by

$$\psi(\rho,v) := \tau\left(D_{\mathfrak{m}^*}(h\circ\pi)(\rho,v), \rho + \mathbf{J}_V(v)\right).$$

We now note that it suffices to take

$$X_{\mathfrak{q}} = \psi \tag{7.7.13}$$

to solve equation (7.7.11). Additionally, (7.7.8) is automatically solved if we take

$$\begin{aligned} X_{\mathfrak{m}_r^*}(g,\rho,v) = \mathbb{P}_{\mathfrak{m}^*}&\left(\mathrm{ad}^*_{D_{\mathfrak{m}^*}(h\circ\pi)}\rho\right) \\ &+ \mathrm{ad}^*_{D_{\mathfrak{m}^*}(h\circ\pi)}\mathbf{J}_V(v) + \mathbb{P}_{\mathfrak{m}^*}\left(\mathrm{ad}^*_{\psi(\rho,v)}(\rho+\mathbf{J}_V(v))\right), \end{aligned} \tag{7.7.14}$$

where we used the fact that $\mathrm{ad}^*_{D_{\mathfrak{m}^*}(h\circ\pi)}\mathbf{J}_V(v) \in \mathfrak{m}^*$ which is true because, as $D_{\mathfrak{m}^*}(h\circ\pi)\in\mathfrak{m}^*\subset\mathfrak{k}^*$, $\mathbf{J}_V(v)\in\mathfrak{h}^*\subset\mathfrak{k}^*$, and $\mathfrak{k}$ is a Lie algebra, we have that $\mathrm{ad}^*_{D_{\mathfrak{m}^*}(h\circ\pi)}\mathbf{J}_V(v) \in \mathfrak{k}^*$. Moreover, for any $\eta\in\mathfrak{g}_m$ we can write

$$\langle \mathrm{ad}^*_{D_{\mathfrak{m}_r^*}(h\circ\pi)}\mathbf{J}_V(v), \eta\rangle = \langle \mathbf{J}_V(v), \mathrm{ad}_{X_{\mathfrak{m}}}\eta\rangle = \langle \mathbf{J}_V(v), -\mathrm{ad}_\eta X_{\mathfrak{m}}\rangle = 0,$$

since $\mathbf{J}_V(v)\in\mathfrak{g}_m^*$ and $-\mathrm{ad}_\eta X_{\mathfrak{m}}\in\mathfrak{m}$ by the Ad_{G_m}-invariance of $\mathfrak{m}$. This implies that $\mathrm{ad}^*_{D_{\mathfrak{m}_r^*}(h\circ\pi)}\mathbf{J}_V(v)\in\mathfrak{m}^*$ and hence

$$\mathbb{P}_{\mathfrak{m}^*}\left(\mathrm{ad}^*_{D_{\mathfrak{m}_r^*}(h\circ\pi)}\mathbf{J}_V(v)\right) = \mathrm{ad}^*_{D_{\mathfrak{m}_r^*}(h\circ\pi)}\mathbf{J}_V(v).$$

To conclude we have to make sure that the expressions that we proposed in (7.7.10), (7.7.13), and (7.7.14), together with $X_{\mathfrak{g}_m} = 0$, satisfy the equations (7.7.5)–(7.7.9). By construction this comes down to verifying that when we insert those expressions into (7.7.9) we obtain an automatic identity, that is,

$$\mathbb{P}_{\mathfrak{g}_m^*}\left(\mathrm{ad}^*_{\left(D_{\mathfrak{m}_r^*}(h\circ\pi)+\psi(\rho,v)\right)}(\rho+\mathbf{J}_V(v))\right) = T_v\mathbf{J}_V(X_{V_r}). \tag{7.7.15}$$

We first study the right-hand side. All the expressions that follow are implicitly assumed to be written at the point $[e,\rho,v]\in Y_r$. Let $\xi\in\mathfrak{g}_m$ be an arbitrary element;

then

$$
\begin{aligned}
\langle T_v \mathbf{J}_V(X_{V_r}), \xi\rangle &= \omega_V(\xi_{V_r}(v), X_{V_r}) \\
&= -D_{V_r}(h\circ\pi)(\xi_{V_r}(v)) \qquad \text{(by (7.7.10))} \\
&= -\left.\frac{d}{dt}\right|_{t=0} (h\circ\pi)(\rho, \exp t\xi\cdot v) \\
&= -\left.\frac{d}{dt}\right|_{t=0} (h\circ\pi)(\exp(-t\xi)\cdot\rho, v) \quad (G_m\text{-invariance of } h\circ\pi) \\
&= D_{\mathfrak{m}_r^*}(h\circ\pi)(\xi_{\mathfrak{m}_r^*}(\rho)) \\
&= -D_{\mathfrak{m}_r^*}(h\circ\pi)(\mathrm{ad}_\xi^*\rho) \\
&= -\langle \mathrm{ad}_\xi^*\rho, X_{\mathfrak{m}}\rangle \qquad \text{(by (7.7.10))}.
\end{aligned} \tag{7.7.16}
$$

At the same time, we can write the left-hand side paired with $\xi \in \mathfrak{g}_m$ as

$$
\begin{aligned}
&\langle \mathbb{P}_{\mathfrak{g}_m^*}(\mathrm{ad}^*_{\left(D_{\mathfrak{m}_r^*}(h\circ\pi)+\psi(\rho,v)\right)}(\rho + \mathbf{J}_V(v))), \xi\rangle \\
&\qquad = \left\langle \mathrm{ad}^*_{\left(D_{\mathfrak{m}_r^*}(h\circ\pi)+\psi(\rho,v)\right)}(\rho + \mathbf{J}_V(v)), \xi \right\rangle \\
&\qquad = \langle \rho + \mathbf{J}_V(v), -\mathrm{ad}_\xi\left(D_{\mathfrak{m}_r^*}(h\circ\pi) + \psi(\rho, v)\right)\rangle.
\end{aligned}
$$

Note that since $\mathfrak{m}$ and $\mathfrak{q}$ are Ad_{G_m}-invariant and $\xi \in \mathfrak{g}_m$, we have that

$$
\mathrm{ad}_\xi\left(D_{\mathfrak{m}_r^*}(h\circ\pi) + \psi(\rho, v)\right) = \mathrm{ad}_\xi\left(D_{\mathfrak{m}_r^*}(h\circ\pi)\right) + \mathrm{ad}_\xi\left(\psi(\rho, v)\right),
$$

with $\mathrm{ad}_\xi\left(D_{\mathfrak{m}_r^*}(h\circ\pi)\right) \in \mathfrak{m}$ and $\mathrm{ad}_\xi\left(\psi(\rho, v)\right) \in \mathfrak{q}$. Also, as $\rho \in \mathfrak{m}_r^*$ and $\mathbf{J}_V(v) \in \mathfrak{g}_m^*$, we have that

$$
\langle \rho + \mathbf{J}_V(v), -\mathrm{ad}_\xi\left(D_{\mathfrak{m}_r^*}(h\circ\pi) + \psi(\rho, v)\right)\rangle = \langle \rho, -\mathrm{ad}_\xi X_{\mathfrak{m}}\rangle = -\langle \mathrm{ad}_\xi^*\rho, X_{\mathfrak{m}}\rangle,
$$

which coincides with (7.7.16), as required.

7.7.4 The reconstruction equations. The previous arguments allow us to conclude that X_h is given by

$$
X_h([g, \rho, v]) = T_{(g,\rho,v)}\pi(X_G(g, \rho, v), X_{\mathfrak{m}_r^*}(g, \rho, v), X_{V_m}(g, \rho, v)),
$$

where

$$
X_G(g, \rho, v) = T_e L_g\left(\psi(\rho, v) + D_{\mathfrak{m}_r^*}(h\circ\pi)(\rho, v)\right) \tag{7.7.17}
$$

$$
X_{\mathfrak{m}_r^*}(g, \rho, v) = \mathbb{P}_{\mathfrak{m}^*}\left(\mathrm{ad}^*_{D_{\mathfrak{m}_r^*}(h\circ\pi)}\rho\right) + \mathrm{ad}^*_{D_{\mathfrak{m}_r^*}(h\circ\pi)}\mathbf{J}_V(v) + \mathbb{P}_{\mathfrak{m}^*}\left(\mathrm{ad}^*_{\psi(\rho,v)}(\rho + \mathbf{J}_V(v))\right) \tag{7.7.18}
$$

$$
X_{V_r}(g, \rho, v) = B_V^\sharp(D_{V_r}(h\circ\pi)(\rho, v))). \tag{7.7.19}
$$

7.7.5 The reconstruction equations in the presence of a standard momentum map. A particular case in which the previous equations present special relevance in terms of applications has to do with the presence of a momentum map. Hence assume that the G-action on the symplectic manifold (M, ω) has an associated coadjoint equivariant momentum map $\mathbf{J} : M \to \mathfrak{g}^*$ such that $\mathbf{J}(m) = \mu \in \mathfrak{g}^*$. Notice that in this situation the Lie algebra $\mathfrak{k}$ coincides with the Lie algebra $\mathfrak{g}_\mu$ of the coadjoint isotropy subgroup G_μ of μ. In order to keep the exposition simple we will assume that the momentum value μ is ***split*** (see GUILLEMIN et al. (1996)). This means that the complement $\mathfrak{q}$ to $\mathfrak{g}_\mu$ in $\mathfrak{g}$ can be chosen to be Ad_{G_μ}-invariant.

The main simplification that takes place in the reconstruction equations under these hypotheses is due to the fact that $X_\mathfrak{q} = 0$, as will be shown below. We start by noting that the defining property of the momentum map and its equivariance implies that in this situation the equation (7.7.7) takes the form

$$\mathbb{P}_{\mathfrak{q}^*}\left(\mathrm{ad}^*_{(X_\mathfrak{m}+X_\mathfrak{q})}(\rho + \mathbf{J}_V(v) + \mu)\right) = 0.$$

Notice that, as $X_\mathfrak{m} \in \mathfrak{m} \subset \mathfrak{g}_\mu$, $\rho + \mathbf{J}_V(v) \in \mathfrak{g}_\mu^*$, and $\mathfrak{q}$ is by hypothesis Ad_{G_μ}-invariant, we have that $\mathrm{ad}^*_{X_\mathfrak{m}}(\rho+\mathbf{J}_V(v)) \in \mathfrak{g}_\mu^*$ and $\mathrm{ad}^*_{X_\mathfrak{m}}\mu = 0$. Therefore, the previous expression can be rewritten as

$$\mathbb{P}_{\mathfrak{q}^*}\left(\mathrm{ad}^*_{X_\mathfrak{q}}(\rho + \mathbf{J}_V(v) + \mu)\right) = 0. \tag{7.7.20}$$

We define the linear map

$$\begin{aligned} L([e,\ \rho,\ v]) : \mathfrak{q} &\longrightarrow \mathfrak{q}^* \\ \zeta &\longmapsto \mathbb{P}_{\mathfrak{q}^*}\left(\mathrm{ad}^*_\zeta(\rho + \mathbf{J}_V(v) + \mu)\right). \end{aligned}$$

We shall prove that $L([e,\ 0,\ 0])$ is an isomorphism. Indeed, if there is an element $\zeta \in \mathfrak{q}$ such that $L([e,\ 0,\ 0])(\zeta) = \mathbb{P}_{\mathfrak{q}^*}\left(\mathrm{ad}^*_\zeta\mu\right) = 0$, it follows that $\mathrm{ad}^*_\zeta\mu \in \mathfrak{m}^* \oplus \mathfrak{g}_m^*$. At the same time, $\mathrm{ad}^*_\zeta\mu$ belongs to $(\mathfrak{g}_\mu)^\circ$, the annihilator of $\mathfrak{g}_\mu$ in $\mathfrak{g}^*$. Since by the construction of the splittings we have $(\mathfrak{g}_\mu)^\circ = \mathfrak{q}^*$, this implies $\mathrm{ad}^*_\zeta\mu = 0$ and therefore $\zeta \in \mathfrak{g}_\mu$. Given that $\zeta \in \mathfrak{q}$ and $\mathfrak{q} \cap \mathfrak{g}_\mu = \{0\}$, it follows that $\zeta = 0$, which implies that $L([e,\ 0,\ 0])$ is injective. Since $\dim \mathfrak{q} = \dim \mathfrak{q}^*$, $L([e,\ 0,\ 0])$ is an isomorphism.

The continuity of the momentum map $\mathbf{J}_V$ guarantees that the symplectic manifold Y_r can be chosen (by shrinking r if necessary) so that $L([e,\ \rho,\ v])$ is an isomorphism for any element in Y_r of the form $[e,\ \rho,\ v]$.

We now go back to studying (7.7.20), which can be written as

$$L([e,\ \rho,\ v])(X_\mathfrak{q}) = 0.$$

If we assume that $[e,\ \rho,\ v] \in Y_r$, we have that

$$X_\mathfrak{q} = 0. \tag{7.7.21}$$

Consequently, in this situation X_h is given by

$$X_h([g, \rho, v]) = T_{(g,\rho,v)}\pi(X_\mathfrak{m}(g, \rho, v), X_{\mathfrak{m}_r^*}(g, \rho, v), X_{V_m}(g, \rho, v)),$$

where

$$X_{\mathfrak{m}}(g,\rho,v) = T_eL_g(D_{\mathfrak{m}_r^*}(h\circ\pi)(\rho,v)) \tag{7.7.22}$$

$$X_{V_r} = B_V^{\sharp}(D_{V_r}(h\circ\pi)(\rho,v)) \tag{7.7.23}$$

$$X_{\mathfrak{m}_r^*} = \mathbb{P}_{\mathfrak{m}^*}\left(\mathrm{ad}^*_{D_{\mathfrak{m}_r^*}(h\circ\pi)}\rho\right) + \mathrm{ad}^*_{D_{\mathfrak{m}_r^*}(h\circ\pi)}\mathbf{J}_V(v). \tag{7.7.24}$$

An important conclusion that can be drawn just by looking at the reconstruction equations is that, under these hypotheses, the motion in the group part of the symplectic tube coordinates, around the point $m \simeq [e,\ 0,\ 0]$, takes place only in the subgroup G_μ. Said differently, since $X_{\mathfrak{q}} = 0$, the *group drift* part of the motion only happens in the G_μ-direction. This is consistent with the studies on the stability of relative equilibria in symmetric Hamiltonian systems, where, using different approaches, one concludes that around a stable relative equilibrium with split momentum value the group drift takes place only in the G_μ-directions (see PATRICK (1992); LERMAN AND SINGER (1998); ORTEGA AND RATIU (1999a)).

There exist a variety of Lie algebraic hypotheses capable of producing different simplifications of the reconstruction equations. An exhaustive study along those lines has been carried out in ROBERTS et al. (2002) along with a generalization of these ideas to the framework of time reversible symmetries.

Chapter 8

Singular Reduction and the Stratification Theorem

This chapter studies the structure of the symplectic reduced spaces introduced in Chapter 6 when the hypothesis on the freeness of the canonical group action is dropped. In this new scenario, standard momentum maps are not submersions anymore and consequently, the reduced spaces are not necessarily smooth manifolds, but just quotient topological spaces. The main result proved here shows that these quotients are symplectic Whitney stratified spaces in the sense that the strata are symplectic manifolds in a very natural way; moreover, the local properties of this Whitney stratification make it into a cone space in the sense of Definition 1.7.3. This statement is referred to as the ***Symplectic Stratification Theorem***. This symplectic stratification is well adapted to the study of G-invariant dynamics since the flows of Hamiltonian vector fields associated to G-invariant Hamiltonian functions naturally reduce to Hamiltonian systems on these strata.

The study of these reduced spaces from the stratifications point of view was started by SJAMAAR (1990) and SJAMAAR AND LERMAN (1991), who considered reduction at the zero value of the momentum map for compact Lie group actions. BATES AND LERMAN (1997) extended some of these results to proper group actions in the context of orbit reduction, provided that the coadjoint orbits are embedded submanifolds of the dual of the Lie algebra. In this chapter we will present a theory of singular reduction for proper group actions that covers both the point and orbit reduction approaches without any hypothesis on the coadjoint orbits. This degree of generality requires great care with various technical details that are thoroughly presented in the sections below.

As we did in Chapter 6, we will restrict our attention to the general case of symmetric symplectic manifolds without any additional structure. Singular reduction for Kähler manifolds has been carried out by HUEBSCHMANN (2003). The case of symmetric cotangent bundles is still a research topic; the first steps in the study of this problem can be found in SCHMAH (2001) and RODRÍGUEZ-OLMOS et al. (2003).

Throughout this chapter (M, ω) will be a connected symplectic manifold. As already explained in §6.1.2, this assumption is entirely motivated by the need to have a well-defined non-equivariance cocycle $\sigma : M \to \mathfrak{g}^*$ that later is used to define an affine action on $\mathfrak{g}^*$ with respect to which the momentum map $\mathbf{J} : M \to \mathfrak{g}^*$ is equivariant. Consequently, if the momentum map in question is already equivariant, the connectedness hypothesis on M can be dropped in all the results that follow.

Also, the notations and conventions of the previous chapter remain in force. Specifically, given $m \in M$ denote by $\mu = \mathbf{J}(m) \in \mathfrak{g}^*$ and fix an Ad_{G_m}-invariant inner product $\langle \cdot, \cdot \rangle$ on $\mathfrak{g}$, always available by the compactness of G_m. Recall the direct sum splittings (7.1.2)

$$\mathfrak{g}_\mu = \mathfrak{g}_m \oplus \mathfrak{m} \quad \text{and} \quad \mathfrak{g} = \mathfrak{g}_m \oplus \mathfrak{m} \oplus \mathfrak{q},$$

where $\mathfrak{g}_\mu$ is the isotropy algebra at μ of the affine action of G on $\mathfrak{g}^*$ defined by the non-equivariance one-cocycle of $\mathbf{J}$, $\mathfrak{m}$ is the $\langle \cdot, \cdot \rangle$-orthogonal complement of $\mathfrak{g}_m$ in $\mathfrak{g}_\mu$, and $\mathfrak{q}$ is the $\langle \cdot, \cdot \rangle$-orthogonal complement of $\mathfrak{g}_\mu$ in $\mathfrak{g}$. These splittings induce similar ones on the duals (see (7.1.3))

$$\mathfrak{g}_\mu^* = \mathfrak{g}_m^* \oplus \mathfrak{m}^* \quad \text{and} \quad \mathfrak{g}^* = \mathfrak{g}_m^* \oplus \mathfrak{m}^* \oplus \mathfrak{q}^*,$$

where the dual of each subspace is understood as the annihilator of the complement, that is, $\mathfrak{q}^* = \{\langle \xi, \cdot \rangle \mid \xi \in \mathfrak{q}\} = \mathfrak{g}_\mu^\circ$, $\mathfrak{g}_\mu^* = \mathfrak{q}^\circ$, $\mathfrak{m}^* = (\mathfrak{g}_m \oplus \mathfrak{q})^\circ = \mathfrak{g}_m^\circ \cap \mathfrak{q}^\circ$, $\mathfrak{g}_m^* = (\mathfrak{m} \oplus \mathfrak{q})^\circ = \mathfrak{m}^\circ \cap \mathfrak{q}^\circ$.

Recall also from the Witt–Artin Decomposition Theorem 7.1.1 the definition of the symplectic normal space V at $m \in M$. Choose a G_m-invariant inner product $\langle\!\langle \cdot, \cdot \rangle\!\rangle$ in $T_m M$, available by Lemma 2.3.30. The subspace $V \subset T_m M$ is defined to be the orthogonal complement to $\mathfrak{g} \cdot m \cap (\mathfrak{g} \cdot m)^{\omega(m)} = \mathfrak{g}_\mu \cdot m$ in $(\mathfrak{g} \cdot m)^{\omega(m)}$ with respect to $\langle\!\langle \cdot, \cdot \rangle\!\rangle$, that is:

$$(\mathfrak{g} \cdot m)^{\omega(m)} = \mathfrak{g} \cdot m \cap (\mathfrak{g} \cdot m)^{\omega(m)} \oplus V = \mathfrak{g}_\mu \cdot m \oplus V.$$

The subspace V is a symplectic G_m-invariant subspace of $(T_m M, \omega(m))$ such that $V \cap \mathfrak{q} \cdot m = \{0\}$. The linear and canonical G_m-action on $(V, \omega(m)|_V)$ admits (by Example 4.5.7) a standard momentum map $\mathbf{J}_V : V \to \mathfrak{g}_m^*$.

8.1 The symplectic strata

We start by describing the symplectic manifolds that later will be shown to constitute a stratification of the Marsden–Weinstein quotients in the singular context. The stratification of these quotients is strongly inspired by the stratification of G-manifolds by their orbit types, already discussed in Chapter 2. The first work where the stratification by orbit types appears in singular reduction is OTTO (1987), who works in the context of Kähler manifolds. The use of the Symplectic Slice Theorem allowed SJAMAAR AND LERMAN (1991) and BATES AND LERMAN (1997) to extend these ideas to the context of general symplectic manifolds. LERMAN AND WILLETT (2001) have obtained similar results in the context of contact geometry. We will start with the singular analog of symplectic point reduction and later we will deal with the orbit approach.

8.1.1 Theorem (Singular symplectic point strata) *Let (M, ω) be a connected symplectic manifold acted canonically and properly upon by a Lie group G. Suppose that this action has an associated standard momentum map $\mathbf{J} : M \to \mathfrak{g}^*$ with non-equivariance one-cocycle $\sigma : G \to \mathfrak{g}^*$. Let $\mu \in \mathfrak{g}^*$ be a value of $\mathbf{J}$, G_μ the isotropy* *subgroup of* μ *with respect to the affine action* $\Theta : G \times \mathfrak{g}^* \to \mathfrak{g}^*$ *determined by* σ*, and let* $H \subset G$ *be an isotropy subgroup of the* G*-action on* M*. Let* M_H^z *be the connected component of the* H*-isotropy type manifold that contains a given element* $z \in M$ *such that* $\mathbf{J}(z) = \mu$ *and let* $G_\mu M_H^z$ *be its* G_μ*-saturation. Then the following hold:*

(i) *The set* $\mathbf{J}^{-1}(\mu) \cap G_\mu M_H^z$ *is a submanifold of* M*.*

(ii) *The set* $M_\mu^{(H)} := [\mathbf{J}^{-1}(\mu) \cap G_\mu M_H^z]/G_\mu$ *has a unique quotient differentiable structure such that the canonical projection*

$$\pi_\mu^{(H)} : \mathbf{J}^{-1}(\mu) \cap G_\mu M_H^z \longrightarrow M_\mu^{(H)}$$

is a surjective submersion.

(iii) *There is a unique symplectic structure* $\omega_\mu^{(H)}$ *on* $M_\mu^{(H)}$ *characterized by*

$$i_\mu^{(H)*}\omega = \pi_\mu^{(H)*}\omega_\mu^{(H)},$$

where $i_\mu^{(H)} : \mathbf{J}^{-1}(\mu) \cap G_\mu M_H^z \hookrightarrow M$ *is the natural inclusion. The pairs* $(M_\mu^{(H)}, \omega_\mu^{(H)})$ *will be called* ***singular symplectic point strata****.*

(iv) *Let* $h \in C^\infty(M)^G$ *be a* G*-invariant Hamiltonian. Then the flow* F_t *of* X_h *leaves the connected components of* $\mathbf{J}^{-1}(\mu) \cap G_\mu M_H^z$ *invariant and commutes with the* G_μ*-action, so it induces a flow* F_t^μ *on* $M_\mu^{(H)}$ *that is characterized by*

$$\pi_\mu^{(H)} \circ F_t \circ i_\mu^{(H)} = F_t^\mu \circ \pi_\mu^{(H)}.$$

(v) *The flow* F_t^μ *is Hamiltonian on* $M_\mu^{(H)}$*, with* ***reduced Hamiltonian*** *function* $h_\mu^{(H)} : M_\mu^{(H)} \to \mathbb{R}$ *defined by*

$$h_\mu^{(H)} \circ \pi_\mu^{(H)} = h \circ i_\mu^{(H)}.$$

The vector fields X_h *and* $X_{h_\mu^{(H)}}$ *are* $\pi_\mu^{(H)}$*-related.*

(vi) *Let* $k : M \to \mathbb{R}$ *be another* G*-invariant function. Then* $\{h, k\}$ *is also* G*-invariant and*

$$\{h, k\}_\mu^{(H)} = \{h_\mu^{(H)}, k_\mu^{(H)}\}_{M_\mu^{(H)}}$$

where $\{\ ,\ \}_{M_\mu^{(H)}}$ *denotes the Poisson bracket induced by the symplectic structure on* $M_\mu^{(H)}$*.*

The smoothness of the just-introduced strata will be proved using the Symplectic Slice Theorem or, more specifically, a very important corollary of it formulated in the following proposition which is a strengthening of a result in Bates and Lerman (1997).

8.1.2 Proposition *Let (M, ω) be a connected symplectic manifold acted canonically and properly upon by a Lie group G. Suppose that this action has an associated standard momentum map $\mathbf{J} : M \to \mathfrak{g}^*$ with non-equivariance one-cocycle $\sigma : G \to \mathfrak{g}^*$. Let $m \in M$, $\mu = \mathbf{J}(m)$, and $(Y_r = G \times_{G_m} (\mathfrak{m}_r^* \times V_r), \omega_{Y_r})$ be a symplectic tube around that point. Then there is a G_μ-invariant open neighborhood $Y_0 \subset Y_r$ of the orbit $G_\mu \cdot [e, 0, 0]$ such that*

$$\mathbf{J}_{Y_r}^{-1}(\mu) \cap Y_0 = \left\{[g, \rho, v] \in Y_0 \mid g \in G_\mu,\ \rho = 0 \text{ and } \mathbf{J}_V(v) = 0\right\}.$$

The symbol G_μ denotes the isotropy subgroup of the point $\mu \in \mathfrak{g}^*$ with respect to the affine action $\Theta : G \times \mathfrak{g}^* \to \mathfrak{g}^*$ determined by σ.

Proof. Express the momentum map $\mathbf{J}_{Y_r}$ as $\mathbf{J}_{Y_r} = \varepsilon \circ b$, where

$$\begin{aligned} b : G \times_{G_m} (\mathfrak{m}_r^* \times V_r) &\longrightarrow G \times_{G_m} \mathfrak{g}_\mu^* \\ [g, \rho, v] &\longmapsto [g, \rho + \mathbf{J}_V(v)], \end{aligned}$$

$$\begin{aligned} \varepsilon : \quad G \times_{G_m} \mathfrak{g}_\mu^* &\longrightarrow \mathfrak{g}^* \\ [g, \nu] &\longmapsto \Theta_g(\mu + \nu) = \mathrm{Ad}^*_{g^{-1}}(\mu + \nu) + \sigma(g). \end{aligned}$$

We now show that the mapping ε has surjective derivative at the point $[e, 0] \in G \times_{G_m} \mathfrak{g}_\mu^*$. To do this, recall that $\mathfrak{g} = \mathfrak{g}_\mu \oplus \mathfrak{q}$, the direct sum being orthogonal relative to an Ad_{G_m}-invariant inner product on $\mathfrak{g}$. This allows us to write the dual splitting $\mathfrak{g}^* = \mathfrak{g}_\mu^* \oplus \mathfrak{q}^*$ in terms of annihilators $\mathfrak{g}^* = \mathfrak{g}_\mu^\circ \oplus \mathfrak{q}^\circ$, that is, $\mathfrak{g}_\mu^\circ = \mathfrak{q}^*$ and $\mathfrak{q}^\circ = \mathfrak{g}_\mu^*$. Now, an arbitrary vector $v_{[e, 0]} \in T_{[e, 0]}(G \times_{G_m} \mathfrak{g}_\mu^*)$ can be written as $v_{[e, 0]} = \frac{d}{dt}\big|_{t=0} [\exp t\xi, t\nu]$, with $\xi \in \mathfrak{g}$ and $\nu \in \mathfrak{q}^\circ = \mathfrak{g}_\mu^*$. Using the Leibniz rule we compute

$$T_{[e, 0]}\varepsilon \left(v_{[e, 0]}\right) = \left.\frac{d}{dt}\right|_{t=0} \Theta_{\exp t\xi}(\mu + t\nu) = \xi_{\mathfrak{g}^*}(\mu) + \nu,$$

where we used the notation introduced in (4.5.24). Since $\xi_{\mathfrak{g}^*}(\mu) \in T_\mu \mathcal{O}_\mu$ and $\nu \in \mathfrak{q}^\circ$ are arbitrary and since $\mathfrak{g}_\mu^\circ = T_\mu \mathcal{O}_\mu$ by (4.5.26), it is clear that $T_{[e, 0]}\varepsilon$ maps onto $\mathfrak{g}_\mu^\circ \oplus \mathfrak{q}^\circ = \mathfrak{g}^*$ and consequently $T_{[e, 0]}\varepsilon$ is surjective.

Let $\phi : G \times (G \times_{G_m} \mathfrak{g}_\mu^*) \to (G \times_{G_m} \mathfrak{g}_\mu^*)$ be the left action defined by $\phi(g, [h, \nu]) := [gh, \nu]$, $g \in G$, $[h, \nu] \in G \times_{G_m} \mathfrak{g}_\mu^*$. By its very definition the map ε is equivariant, that is,

$$\Theta_g \circ \varepsilon = \varepsilon \circ \phi_g, \quad \text{for any } g \in G. \tag{8.1.1}$$

Therefore, for any $[g, \nu] \in G \times_{G_m} \mathfrak{g}_\mu^*$ we have $T_{[g,\nu]}\varepsilon = T_{\mu+\nu}\Theta_g \circ T_{[e,\nu]}\varepsilon \circ T_{[g,\nu]}\phi_{g^{-1}}$ and hence, as $T_{\mu+\nu}\Theta_g$ and $T_{[g,\nu]}\phi_{g^{-1}}$ are isomorphisms and $T_{[e,\nu]}\varepsilon$ is surjective, it follows that $T_{[g,\nu]}\varepsilon$ is also surjective. Thus ε is a submersion at all points of the form $[g, 0]$, for any $g \in G$. Moreover, as submersivity is an open condition, there is an open neighborhood $U_{[e,0]}$ of the point $[e, 0]$ such that $T_{[g,\nu]}\varepsilon$ is onto, for any $[g, \nu] \in U_{[e,0]}$. The equivariance property (8.1.1) of ε implies that the open G-invariant neighborhood $U := G \cdot U_{[e,0]} = \bigcup_{g \in G} \phi_g(U_{[e,0]})$ of the orbit $G \cdot [e, 0]$ has also the same property.

By the Submersion Theorem, $\varepsilon^{-1}(\mu) \cap U$ is a submanifold of $G \times_{G_m} \mathfrak{g}_\mu^*$ of dimension $\dim G_\mu - \dim G_m$. However, the submanifold $G_\mu \times_{G_m} \{0\}$ of $G \times_{G_m} \mathfrak{g}_\mu^*$ is included in $\varepsilon^{-1}(\mu) \cap U$ and has also dimension $\dim G_\mu - \dim G_m$. Therefore $G_\mu \times_{G_m} \{0\}$ is an open submanifold of $\varepsilon^{-1}(\mu) \cap U$.

We are now going to prove that there exists an open G_μ-invariant subset V of U such that

$$G_\mu \times_{G_m} \{0\} = \varepsilon^{-1}(\mu) \cap V. \tag{8.1.2}$$

We start by noting that as $\varepsilon^{-1}(\mu) \cap U$ is an embedded submanifold of $G \times_{G_m} \mathfrak{g}_\mu^*$, the sets $\varepsilon^{-1}(\mu) \cap U \cap W$, with W open in $G \times_{G_m} \mathfrak{g}_\mu^*$, form a basis of its topology. The same can be said about the family $\mathcal{B} := \{\varepsilon^{-1}(\mu) \cap U_i \mid i \in I, U_i \subset U, U_i \text{ open in } G \times_{G_m} \mathfrak{g}_\mu^*\}$. Consequently, as $G_\mu \times_{G_m} \{0\}$ is an open subset of $\varepsilon^{-1}(\mu) \cap U$, there exists a subset $J \subset I$ such that

$$G_\mu \times_{G_m} \{0\} = \bigcup_{j \in J} \varepsilon^{-1}(\mu) \cap U_j. \tag{8.1.3}$$

Since U is G-invariant, the saturations $G_\mu \cdot U_j$ are open subsets of U. Let us show that

$$G_\mu \times_{G_m} \{0\} = \bigcup_{j \in J} \varepsilon^{-1}(\mu) \cap G_\mu \cdot U_j. \tag{8.1.4}$$

Indeed, the inclusion $G_\mu \times_{G_m} \{0\} \subset \bigcup_{j \in J} \varepsilon^{-1}(\mu) \cap G_\mu \cdot U_j$ is obvious because of (8.1.3). Conversely, given a point $h \cdot z \in \varepsilon^{-1}(\mu) \cap G_\mu \cdot U_j$ with $h \in G_\mu$ and $z \in U_j$ and such that $\varepsilon(h \cdot z) = \mu$, the equivariance property (8.1.1) of ε implies that $\Theta_h(\varepsilon(z)) = \mu$, and hence $\varepsilon(z) = \Theta_{h^{-1}}(\mu) = \mu$ because $h \in G_\mu$. Thus, $z \in \varepsilon^{-1}(\mu) \cap U_j$ and by (8.1.3) it can be written as $z = [g, 0]$ with $g \in G_\mu$. Finally, $h \cdot z = [hg, 0] \in G_\mu \times_{G_m} \{0\}$, as required.

Expression (8.1.4) implies that $G_\mu \times_{G_m} \{0\} = \bigcup_{j \in J} \varepsilon^{-1}(\mu) \cap G_\mu \cdot U_j = \varepsilon^{-1}(\mu) \bigcap \left(\bigcup_{j \in J} G_\mu \cdot U_j\right)$ and therefore (8.1.2) holds by taking $V = \bigcup_{j \in J} G_\mu \cdot U_j$ which is obviously G_μ-invariant and contains $G_\mu \cdot [e, 0]$.

Define $Y_0 := b^{-1}(V) \supset b^{-1}(G_\mu \cdot [e, 0]) \supset G_\mu \cdot [e, 0, 0]$. The set Y_0 is open in Y_r by smoothness of b and is G_μ-invariant by G-equivariance of b (the G-action on the source and target is on the first factor only). Moreover,

$$\begin{aligned} \mathbf{J}_{Y_r}^{-1}(\mu) \cap Y_0 &= b^{-1}(\varepsilon^{-1}(\mu)) \cap b^{-1}(V) = b^{-1}(G_\mu \times_{G_m} \{0\}) \\ &= \{[g, \rho, v] \in Y_0 \mid g \in G_\mu,\ \rho = 0 \text{ and } \mathbf{J}_V(v) = 0\} \end{aligned}$$

and the claim of the proposition is satisfied. ■

Proof of Theorem 8.1.1. Points **(i)** and **(ii)** can be easily proved using Proposition 8.1.2. Let $m \in \mathbf{J}^{-1}(\mu) \cap G_\mu M_H^z$. Assume, without loss of generality, that $G_m = H$ and let $Y_r = G \times_H (\mathfrak{m}_r^* \times V_r)$ be a symplectic tube around the orbit $G \cdot m$. The Symplectic Slice Theorem guarantees the existence of a G-equivariant symplectomorphism $G \times_H (\mathfrak{m}_r^* \times V_r) \to U$ onto some open neighborhood U of $G \cdot m$. Let $\varphi : Y_0 \to W := \varphi(Y_0)$ be the restriction of that symplectomorphism to the open G_μ-invariant neighborhood Y_0 of $G_\mu \cdot m$ provided by Proposition 8.1.2. Then by the construction of Y_0 and by Proposition 2.4.6**(i)** we have that

$$\begin{aligned}\varphi^{-1}(\mathbf{J}^{-1}(\mu)\cap G_\mu M_H^z\cap W) &= \varphi^{-1}(\mathbf{J}^{-1}(\mu))\cap\varphi^{-1}\left(G_\mu M_H^z\right)\cap\varphi^{-1}(W)\\ &= \mathbf{J}_{Y_r}^{-1}(\mu)\cap\varphi^{-1}\left(G_\mu M_H^z\right)\cap Y_0\\ &= \left(G_\mu\times_H\left(\{0\}\times(\mathbf{J}_V^{-1}(0))^H\right)\right)\cap Y_0\\ &= \left(G_\mu\times_H\left(\{0\}\times V_r^H\right)\right)\cap Y_0, \qquad (8.1.5)\end{aligned}$$

where the last equality follows from the relation $V_r^H=(\mathbf{J}_V^{-1}(0))^H$ which in turn is a consequence of the definition of $\mathbf{J}_V$. Indeed, as $\mathbf{J}_V^{-1}(0)\subset V_r$ we have that $(\mathbf{J}_V^{-1}(0))^H\subset V_r^H$. Conversely, if $v\in V_r^H$, then for any $\eta\in\mathfrak{h}$ we have that $\langle\mathbf{J}_V(v),\eta\rangle=\frac{1}{2}\omega_V(\eta\cdot v,v)=0$, since $\eta\cdot v=0$. This implies that $V_r^H\subset(\mathbf{J}_V^{-1}(0))^H$ and consequently the relation $V_r^H=(\mathbf{J}_V^{-1}(0))^H$.

Now, as $\varphi^{-1}(\mathbf{J}^{-1}(\mu)\cap G_\mu M_H^z\cap W)=G_\mu\times_H\left(\{0\}\times V_r^H\right)\cap Y_0$ is a closed G_μ-invariant embedded submanifold of Y_r and we can repeat this argument at any point in $\mathbf{J}^{-1}(\mu)\cap G_\mu M_H^z$ we can conclude that this subset is a locally closed embedded submanifold of M which proves **(i)**.

Due to the G_μ-invariance of Y_0, this argument can be dropped to the quotient by G_μ. Indeed, the G_μ-equivariant symplectomorphism $\varphi: Y_0\to W:=\varphi(Y_0)$ drops to a diffeomorphism $\psi: Y_0/G_\mu\to W/G_\mu$ that allows us to locally see the quotient $\mathbf{J}^{-1}(\mu)\cap G_\mu M_H^z/G_\mu$ as the smooth manifold $G_\mu\times_H\left(\{0\}\times V_r^H\right)/G_\mu\simeq V_r^H/H=V_r^H$ and the map onto the orbit space $\pi_\mu^{(H)}:\mathbf{J}^{-1}(\mu)\cap G_\mu M_H^z\to\mathbf{J}^{-1}(\mu)\cap G_\mu M_H^z/G_\mu$ as the projection $G_\mu/H\times V_r^H\to V_r^H$, which is trivially a surjective submersion. This proves **(ii)**.

Part **(ii)** can also be proved by using facts from the theory of proper group actions. Indeed, given that by Part **(i)** the set $\mathbf{J}^{-1}(\mu)\cap G_\mu M_H^z$ is an embedded (in particular initial) G_μ-invariant submanifold of M, the G-action on M restricts to a proper smooth G_μ-action on $\mathbf{J}^{-1}(\mu)\cap G_\mu M_H^z$ that exhibits the feature of having all its isotropy subgroups conjugate to H. Part **(iv)** in Proposition 2.3.8 concludes the argument.

(iii) We start by proving that for any $m\in\mathbf{J}^{-1}(\mu)\cap G_\mu M_H^z$ we have

$$T_m\left(\mathbf{J}^{-1}(\mu)\cap G_\mu M_H^z\right)=\mathfrak{g}_\mu\cdot m+A_G'(m), \qquad (8.1.6)$$

where A_G' is the distribution polar to the group of diffeomorphisms A_G induced by the G-action. Since (8.1.6) is a local statement we can use the diffeomorphism $\varphi: Y_0\to W$ introduced in the previous point. Without loss of generality, assume that $m=\varphi([e,0,0])$. By (8.1.5) we have that

$$\begin{aligned}T_m(\mathbf{J}^{-1}(\mu)\cap G_\mu M_H^z) &= T_{[e,0,0]}\varphi(G_\mu\times_H(\{0\}\times V_r^H))\\ &= T_{[e,0,0]}\varphi\left(G_\mu\cdot(N_{G_\mu}(H)\times_H(\{0\}\times V_r^H))\right)\\ &= T_{[e,0,0]}\varphi\left(G_\mu\cdot[e,0,0]\right)+T_{[e,0,0]}\varphi\left(N_{G_\mu}(H)\times_H(\{0\}\times V_r^H)\right)\\ &= \mathfrak{g}_\mu\cdot m+T_m\left(\mathbf{J}^{-1}(\mu)\cap M_H^{l_m}\right)=\mathfrak{g}_\mu\cdot m+A_G'(m),\end{aligned}$$

where the last equality is a consequence of (5.5.5). We recall that $N_{G_\mu}(H) := N(H) \cap G_\mu$ (see §2.1.12).

We now use (8.1.6) to define a two-form $\omega_\mu^{(H)}$ on $M_\mu^{(H)}$. Let $v, w \in T_m(\mathbf{J}^{-1}(\mu) \cap G_\mu M_H^z)$ be arbitrary and let $f, g \in C^\infty(M)^G$, $\xi, \eta \in \mathfrak{g}_\mu$ be such that $v = X_f(m) + \xi_M(m)$ and $w = X_g(m) + \eta_M(m)$. We define

$$\omega_\mu^{(H)}\left([m]_\mu^{(H)}\right)\left(T_m\pi_\mu^{(H)} \cdot v,\ T_m\pi_\mu^{(H)} \cdot w\right) := \omega(m)(v, w) = \{f, g\}(m), \qquad (8.1.7)$$

where $\{\cdot, \cdot\}$ is the Poisson bracket associated to the symplectic form ω on M. We first check that the expression (8.1.7) is a good definition for the two-form $\omega_\mu^{(H)}$ in the quotient $M_\mu^{(H)}$. Let $m, m' \in \mathbf{J}^{-1}(\mu) \cap G_\mu M_H^z$ be such that $\pi_\mu^{(H)}(m) = \pi_\mu^{(H)}(m')$ and $v', w' \in T_{m'}\left(\mathbf{J}^{-1}(\mu) \cap G_\mu M_H^z\right)$ be such that

$$T_m\pi_\mu^{(H)}(v) = T_{m'}\pi_\mu^{(H)}(v'), T_m\pi_\mu^{(H)}(w) = T_{m'}\pi_\mu^{(H)}(w').$$

Let $f', g' \in C^\infty(M)^G$ and $\xi', \eta' \in \mathfrak{g}_\mu$ be such that $v' = X_{f'}(m') + \xi'_M(m')$, $w' = X_{g'}(m') + \eta'_M(m')$. The condition $\pi_\mu^{(H)}(m) = \pi_\mu^{(H)}(m')$ implies the existence of an element $k \in G_\mu$ such that $m' = \Phi_k(m)$. We also have that $T_m\pi_\mu^{(H)} = T_{m'}\pi_\mu^{(H)} \circ T_m\Phi_k$. Analogously, because of the equalities $T_m\pi_\mu^{(H)}(v) = T_{m'}\pi_\mu^{(H)}(v')$, $T_m\pi_\mu^{(H)}(w) = T_{m'}\pi_\mu^{(H)}(w')$ there exist two elements $\xi^1, \xi^2 \in \mathfrak{g}_\mu$ such that

$$X_{f'+\mathbf{J}^{\xi'}}(m') - T_m\Phi_k(X_{f+\mathbf{J}^\xi}(m)) = \xi^1_{\mathbf{J}^{-1}(\mu)\cap G_\mu M_H^z}(m') = X_{\mathbf{J}^{\xi^1}}(m'),$$

and

$$X_{g'+\mathbf{J}^{\eta'}}(m') - T_m\Phi_k(X_{g+\mathbf{J}^\eta}(m)) = \xi^2_{\mathbf{J}^{-1}(\mu)\cap G_\mu M_H^z}(m') = X_{\mathbf{J}^{\xi^2}}(m'),$$

whence

$$X_{f'+\mathbf{J}^{\xi'}}(m') = X_{\mathbf{J}^{\xi^1}+f\circ\Phi_{k^{-1}}+\mathbf{J}^\xi\circ\Phi_{k^{-1}}}(m') = X_{\mathbf{J}^{\xi^1}+f+\mathbf{J}^{\mathrm{Ad}_k\xi}}(m'),$$

and $X_{g'+\mathbf{J}^{\eta'}}(m') = X_{\mathbf{J}^{\xi^2}+g+\mathbf{J}^{\mathrm{Ad}_k\eta}}(m')$. In these expressions we have used the identity $\mathbf{J}^\xi \circ \Phi_k = \mathbf{J}^{\mathrm{Ad}_{k^{-1}}\xi} + \langle\sigma(k), \xi\rangle$ which holds for any $\xi \in \mathfrak{g}$ and any $k \in G$; this shows that the functions $\mathbf{J}^\xi \circ \Phi_k$ and $\mathbf{J}^{\mathrm{Ad}_{k^{-1}}\xi}$ differ by a constant which in turn implies that their associated Hamiltonian vector fields coincide. Now, if we use Noether's Theorem 4.5.11, we can write

$$\begin{aligned}
\omega_\mu^{(H)}(\pi_\mu^{(H)}(m'))(v', w') &= \{f' + \mathbf{J}^{\xi'}, g' + \mathbf{J}^{\eta'}\}(m') \\
&= \{\mathbf{J}^{\xi^1} + f + \mathbf{J}^{\mathrm{Ad}_k\xi}, \mathbf{J}^{\xi^2} + g + \mathbf{J}^{\mathrm{Ad}_k\eta}\}(m') \\
&= \{\mathbf{J}^{\xi^1} + f + \mathbf{J}^{\mathrm{Ad}_k\xi}, \mathbf{J}^{\xi^2} + g + \mathbf{J}^{\mathrm{Ad}_k\eta}\}(\Phi_k(m)) \\
&= \{\mathbf{J}^{\mathrm{Ad}_{k^{-1}}\xi^1} + f + \mathbf{J}^\xi, \mathbf{J}^{\mathrm{Ad}_{k^{-1}}\xi^2} + g + \mathbf{J}^\eta\}(m) \\
&= \{f, g\}(m) = \omega_\mu^{(H)}(\pi_\mu^{(H)}(m))(v, w),
\end{aligned}$$

where we used the fact that the Lie algebra $\mathfrak{g}_\mu$ is invariant by the adjoint action of the group G_μ. Consequently, $\omega_\mu^{(H)}$ is a well-defined two-form on the quotient $M_\mu^{(H)}$. Since

$\pi_\mu^{(H)}$ is a smooth surjective submersion, the form $\omega_\mu^{(H)}$ is clearly smooth. The Jacobi identity for the bracket $\{\cdot, \cdot\}$ on M and its antisymmetry imply that $\omega_\mu^{(H)}$ is closed and antisymmetric.

Finally, we show that $\omega_\mu^{(H)}$ is nondegenerate. Indeed, let $m \in \mathbf{J}^{-1}(\mu) \cap G_\mu M_H^z$ and $v \in T_m\left(\mathbf{J}^{-1}(\mu) \cap G_\mu M_H^z\right)$ be such that

$$\omega_\mu^{(H)}([m]_\mu)(T_m\pi_\mu^{(H)}(v),\ T_m\pi_\mu^{(H)}(w)) = 0 \tag{8.1.8}$$

for all $w \in T_m\left(\mathbf{J}^{-1}(\mu) \cap G_\mu M_H^z\right)$. If we write $v = X_f(m) + \xi_M(m)$, for some $f \in C^\infty(M)^G$ and $\xi \in \mathfrak{g}_\mu$, then (8.1.7) and (8.1.8) imply that

$$\left(\pi_\mu^{(H)*}\omega_\mu^{(H)}\right)(m)(X_f(m), X_h(m)) = \{f, h\}(m) = 0, \tag{8.1.9}$$

for all $h \in C^\infty(M)^G$. In order to prove that $\omega_\mu^{(H)}$ is nondegenerate we have to show that $X_f(m) \in \mathfrak{g}_\mu \cdot m$. We will do so by using the semi global coordinates around the point m provided by the Slice Theorem. First, G-invariance of f implies that $X_f(m) \in T_m M_{G_m}^{l_m}$ (see Proposition 3.3.4). Second, in the slice coordinates the local isotropy type manifold $M_{G_m}^{l_m}$ can be represented as $N(G_m) \times_{G_m} B^{G_m}$ (see Proposition 2.4.6 **(i)**) where B is an open G_m-invariant neighborhood of 0 in a vector space V that is G_m-equivariantly isomorphic to $T_m M/\mathfrak{g}\cdot m$. As in this picture the point m is given by $[e, 0]$ we have that $X_f(m)$ is represented by the vector $T_{(e,0)}\pi(\zeta, u)$, where $\pi : G \times B \to G \times_{G_m} B$ is the natural projection, $\zeta \in \mathrm{Lie}(N(G_m))$, and $u \in V^{G_m}$.

In this representation the condition in (8.1.9), that is,

$$\{f, h\}(m) = -\mathbf{d}h(m) \cdot X_f(m) = 0, \qquad \text{for all} \quad h \in C^\infty(M)^G,$$

amounts to saying that $\mathbf{d}g(0)(u) = 0$ for all the functions $g \in C^\infty(B)^{G_m}$. In other words, $u \in \left(\{\mathbf{d}g(0) \mid g \in C^\infty(B)^{G_m}\}\right)^\circ$. Theorem 2.5.10 implies that $u \in ((V^*)^{G_m})^\circ$. Consequently, $u \in V^{G_m} \cap ((V^*)^{G_m})^\circ$. We now show that this intersection is trivial and that therefore $u = 0$.

We start by recalling (see Lemma 2.5.11) that the restriction to $(V^*)^{G_m}$ of the dual map associated to the inclusion $i_{V^{G_m}} : V^{G_m} \hookrightarrow V$ is a G_m-equivariant isomorphism from $(V^*)^{G_m}$ to $(V^{G_m})^*$. Now, as $u \in V^{G_m} \cap ((V^*)^{G_m})^\circ$ we have that $\langle \alpha, u\rangle_V = 0$ for every $\alpha \in ((V)^*)^{G_m}$. The symbol $\langle \cdot, \cdot\rangle_V$ denotes the natural pairing of V with its dual. We can rewrite this condition as

$$0 = \langle \alpha, u\rangle_V = \langle \alpha, i_{V^{G_m}}(u)\rangle_V = \langle i_{V^{G_m}}^*(\alpha), u\rangle_{V^{G_m}}.$$

As the restriction $i_{V^{G_m}}^*|_{(V^*)^{G_m}}$ is an isomorphism, the previous identity is equivalent to $\langle \beta, u\rangle_{V^{G_m}} = 0$ for all $\beta \in (V^{G_m})^*$. Consequently, $u = 0$, as required.

We conclude our argument by noting that as $X_f(m) = \zeta_M(m)$, we have by Noether's Theorem 4.5.11 and the Reduction Lemma 4.5.34 that

$$X_f(m) \in \mathfrak{g} \cdot m \cap \ker T_m\mathbf{J} = \mathfrak{g}_\mu \cdot m, \tag{8.1.10}$$

which proves that $v = X_f(m) + \xi_M(m) \in \mathfrak{g}_\mu \cdot m$. This shows that $\omega_\mu^{(H)}$ is nondegenerate.

The remaining points in the theorem are proved by mimicking the proofs of points **(ii)** through **(iv)** in Theorem 6.1.1. The only claim where there is a difference that deserves some attention is the invariance of $\mathbf{J}^{-1}(\mu) \cap G_\mu M_H^z$ with respect to the Hamiltonian flow F_t associated to an invariant function. The set $\mathbf{J}^{-1}(\mu)$ is invariant by Noether's Theorem and so is $G_\mu M_H^z$ because, as F_t is equivariant, it preserves the connected components M_H^z of the isotropy type manifolds and hence also $G_\mu M_H^z$. ■

8.2 A structure theorem and Sjamaar's principle

The singular symplectic point strata introduced in Theorem 8.1.1 admit a Structure Theorem similar to the one that was formulated for G-manifolds in Theorem 2.4.12. However, in this case such theorem has additional relevance because the base of the bundle to which the singular symplectic point strata are symplectomorphic is given by a symplectic reduced quotient with respect to a free and proper action. In other words, the singular strata can be obtained via regular symplectic reduction. This statement is known as ***Sjamaar's Principle*** since it was introduced in SJAMAAR (1990) for compact group actions at the zero value of the momentum map. Different improvements and generalizations of this idea have been carried out in ORTEGA (1998), where the important role of non-equivariant momentum maps is identified, and in CUSHMAN AND SNIATYCKI (2001).

8.2.1 Theorem (Structure Theorem for the singular point strata) *Let (M, ω) be a connected symplectic manifold acted canonically and properly upon by a Lie group G. Suppose that this action has an associated standard momentum map $\mathbf{J} : M \to \mathfrak{g}^*$ with non-equivariance one-cocycle $\sigma : G \to \mathfrak{g}^*$. Let $\mu \in \mathfrak{g}^*$ be a value of $\mathbf{J}$, G_μ the isotropy subgroup of μ with respect to the affine action $\Theta : G \times \mathfrak{g}^* \to \mathfrak{g}^*$ determined by σ, and $H \subset G$ an isotropy subgroup of the G-action on M. Let M_H^z be the connected component of the H-isotropy type manifold M_H that contains the point $z \in M$ satisfying $\mathbf{J}(z) = \mu$ and let $G_\mu M_H^z$ be its G_μ-saturation. Then the following hold:*

(i) *The canonical projection $\pi_\mu^{(H)} : \mathbf{J}^{-1}(\mu) \cap G_\mu M_H^z \longrightarrow M_\mu^{(H)} := \mathbf{J}^{-1}(\mu) \cap G_\mu M_H^z / G_\mu$ defines a smooth fiber bundle with fiber G_μ / H and structure group $N_{G_\mu}(H)^z / H$. We recall that $N_{G_\mu}(H)^z = N(H)^z \cap G_\mu$, with $N(H)^z$ the closed subgroup of $N(H)$ that leaves M_H^z invariant.*

(ii) *Consider the free, proper, and canonical action of $L^z := N(H)^z / H$ on M_H^z and let $\mathbf{J}_{L^z} : M_H^z \to \mathfrak{l}^*$ be the associated momentum map given by $\mathbf{J}_{L^z}(m) = \Lambda(\mathbf{J}|_{M_H^z}(m) - \mu)$, for any $m \in M_H^z$ (see Theorem 4.6.1). Then the Marsden–Weinstein reduced space $(M_H^z)_0$ at the zero value of this momentum map is given by*

$$(M_H^z)_0 = \mathbf{J}_{L^z}^{-1}(0) / L_0^z = [\mathbf{J}^{-1}(\mu) \cap M_H^z] / (N_{G_\mu}(H)^z / H).$$

(iii) *The projection $\pi_0 : \mathbf{J}_{L^z}^{-1}(0) \to (M_H^z)_0$ defines a smooth principal L_0^z-bundle. Regarding G_μ / H as a right $(N_{G_\mu}(H)^z / H)$-space and $\mathbf{J}^{-1}(\mu) \cap M_H^z$ as a left*

$(N_{G_\mu}(H)^z/H)$-space, consider the bundle with fiber G_μ/H and structure group G_μ associated with π_0, that is,

$$G_\mu/H \times_{N_{G_\mu}(H)^z/H} \left(\mathbf{J}^{-1}(\mu) \cap M_H^z\right) \longrightarrow \mathbf{J}^{-1}(\mu) \cap M_H^z/(N_{G_\mu}(H)^z/H).$$

This bundle is G_μ-symplectomorphic to $\pi_\mu^{(H)} : \mathbf{J}^{-1}(\mu) \cap G_\mu M_H^z \longrightarrow M_\mu^{(H)}$, that is, $G_\mu/H \times_{N_{G_\mu}(H)^z/H} \left(\mathbf{J}^{-1}(\mu) \cap M_H^z\right)$ is G_μ-diffeomorphic to $\mathbf{J}^{-1}(\mu) \cap G_\mu M_H^z$ and $(M_H^z)_0 = \mathbf{J}_{L^z}^{-1}(0)/L_0^z = \mathbf{J}^{-1}(\mu) \cap M_H^z/(N_{G_\mu}(H)^z/H)$ is symplectomorphic to $M_\mu^{(H)}$. We will say that $(M_H^z)_0$ is a ***regularization*** *of the point reduced space $M_\mu^{(H)}$.*

Proof. **(i)** is a straightforward consequence of part **(ii)** in Theorem 8.1.1.
(ii) Given the expression that defines the map $\mathbf{J}_{L^z} : M_H^z \to \mathfrak{l}^*$, it is clear that a point m belongs to $\mathbf{J}_{L^z}^{-1}(0)$ if and only if $\mathbf{J}|_{M_H^z}(m) = \mu$. Consequently, $\mathbf{J}_{L^z}^{-1}(0) = \mathbf{J}^{-1}(\mu) \cap M_H^z$. Additionally, since the affine L^z-action $\Psi : L^z \times \mathfrak{l}^* \to \mathfrak{l}^*$ on $\mathfrak{l}^*$ defined by the non-equivariance cocycle of $\mathbf{J}_{L^z}$ is given by $\Psi(l, \lambda) = \mathrm{Ad}_{l^{-1}}^* \lambda + \Lambda(\sigma(n) + n \cdot \mu - \mu)$, for any $l = nH \in L^z$ and $\lambda \in \mathfrak{l}^*$, we have that $l = nH \in L_0^z$ if and only if $\Lambda(\sigma(n) + n \cdot \mu - \mu) = 0$. This is equivalent to $\sigma(n) + n \cdot \mu = \Theta(n, \mu) = \mu$, which in turn amounts to $n \in N_{G_\mu}(H)^z$. Hence we have that $L_0^z = N_{G_\mu}(H)^z/H$ and consequently

$$\mathbf{J}_{L^z}^{-1}(0)/L_0^z = \mathbf{J}^{-1}(\mu) \cap M_H^z/(N_{G_\mu}(H)^z/H).$$

(iii) The bundle isomorphism in the statement is provided by the maps

$$f : G_\mu/H \times_{N_{G_\mu}(H)^z/H} \left(\mathbf{J}^{-1}(\mu) \cap M_H^z\right) \to \mathbf{J}^{-1}(\mu) \cap G_\mu M_H^z$$

and

$$F : \mathbf{J}^{-1}(\mu) \cap M_H^z/(N_{G_\mu}(H)^z/H) \to \mathbf{J}^{-1}(\mu) \cap G_\mu M_H^z/G_\mu$$

defined by $f([gH, m]) := g \cdot m$, for any $[gH, m] \in G_\mu/H \times_{N_{G_\mu}(H)^z/H} \left(\mathbf{J}^{-1}(\mu) \cap M_H^z\right)$ and by $F(\pi_0(m)) := \pi_\mu^{(H)}(m)$, for $m \in \mathbf{J}^{-1}(\mu) \cap M_H^z$.

The mapping f is clearly well defined and G_μ-equivariant. It is surjective because given any element $g \cdot m \in G_\mu(\mathbf{J}^{-1}(\mu) \cap M_H^z) = \mathbf{J}^{-1}(\mu) \cap G_\mu M_H^z$ with $g \in G_\mu$ and $m \in \mathbf{J}^{-1}(\mu) \cap M_H^z$, we have that $f([gH, m]) = g \cdot m$. The map f is also injective because if $[gH, m], [g'H, m'] \in G_\mu/H \times_{N_{G_\mu}(H)^z/H} \left(\mathbf{J}^{-1}(\mu) \cap M_H^z\right)$ are such that $g \cdot m = g' \cdot m'$, then $m' = (g')^{-1} g \cdot m$. Since $m, m' \in \mathbf{J}^{-1}(\mu) \cap M_H^z$ we have that $(g')^{-1} g = n \in N_{G_\mu}(H)$. Moreover, by Corollary 3.5.2, $n \in N_{G_\mu}(H)^z$. This allows us to write

$$[g', m'] = [g'nH, n^{-1}H \cdot m'] = [g'(g')^{-1}g, g^{-1}g' \cdot m'] = [g, m],$$

which proves the injectivity of f. By Proposition 2.4.5, the map f is an immersion. Additionally, using Proposition 8.1.2 it follows that for any point $m \in \mathbf{J}^{-1}(\mu) \cap M_H^z$ there exists an open neighborhood U_m of m in M such that the submanifolds $U_m \cap (\mathbf{J}^{-1}(\mu) \cap M_H^z)$ and $U_m \cap (\mathbf{J}^{-1}(\mu) \cap G_\mu M_H^z)$ are diffeomorphic to $N_{G_\mu}(H)^z \times_H$

$(\{0\} \times V_r^H)$ and $G_\mu \times_H (\{0\} \times V_r^H)$, respectively. This proves that the domain and the range of f have the same dimension equal to $\dim G_\mu - \dim H + \dim V^H$ and that, consequently, this map is a local diffeomorphism and hence a diffeomorphism. The map F is also a diffeomorphism since it is the projection of the G_μ-equivariant diffeomorphism f onto the quotient spaces by the G_μ-action of its domain and range.

It only remains to be shown that F is a symplectomorphism between $(M_H^z)_0$ and $M_\mu^{(H)}$. Consider first the following commutative diagram

$$\begin{array}{ccc} \mathbf{J}^{-1}(\mu) \cap M_H^z & \xrightarrow{\ i\ } & \mathbf{J}^{-1}(\mu) \cap G_\mu M_H^z \\ \pi_0 \big\downarrow & & \big\downarrow \pi_\mu^{(H)} \\ \mathbf{J}^{-1}(\mu) \cap M_H^z/(N_{G_\mu}(H)^z/H) & \xrightarrow{\ F\ } & \mathbf{J}^{-1}(\mu) \cap G_\mu M_H^z/G_\mu \end{array}$$

where the map $i : \mathbf{J}^{-1}(\mu) \cap M_H^z \to \mathbf{J}^{-1}(\mu) \cap G_\mu M_H^z$ is the inclusion. Additionally, consider the inclusions $i_H : M_H^z \hookrightarrow M$ and $i_0 : \mathbf{J}^{-1}(\mu) \cap M_H^z \hookrightarrow M_H^z$. Then,

$$\begin{aligned} \pi_0^* F^* \omega_\mu^{(H)} &= (F \circ \pi_0)^* \omega_\mu^{(H)} = (\pi_\mu^{(H)} \circ i)^* \omega_\mu^{(H)} = i^* \pi_\mu^{(H)*} \omega_\mu^{(H)} = i^* i_\mu^{(H)*} \omega \\ &= (i_\mu^{(H)} \circ i)^* \omega = (i_H \circ i_0)^* \omega = i_0^* i_H^* \omega = i_0^* \omega_{M_H^z} = \pi_0^* (\omega_{M_H^z})_0. \end{aligned}$$

Since π_0 is a surjective submersion, this proves that $F^* \omega_\mu^{(H)} = (\omega_{M_H^z})_0$ and consequently that F is a symplectomorphism. ■

8.2.2 Remark The structure theorem just proved differs from the corresponding result for G-manifolds (Theorem 2.4.12) in the use of the spaces that intervene in its formulation. More specifically, for G-manifolds we used local isotropy type manifolds $M_H^{l_z}$ while in Theorem 8.2.1 we used the connected components of the isotropy type manifolds M_H^z. In the symplectic symmetric context we are forced to work with the spaces M_H^z since the local isotropy type manifolds are in general not connected and, as we saw in Proposition 4.5.21, this condition is extremely important at the time of defining the non-equivariance one-cocycle of the momentum map associated to the natural action that we have in that space and that we use later for reduction.

8.3 The Symplectic Stratification Theorem

The main goal of this section consists of proving that the symplectic strata $M_\mu^{(H)} = \mathbf{J}^{-1}(\mu) \cap G_\mu M_H^z/G_\mu$ that we introduced in Theorem 8.1.1 constitute a Whitney stratification of the quotient topological space $M_\mu := \mathbf{J}^{-1}(\mu)/G_\mu$ and that, moreover, M_μ is a cone space. We start by describing a Whitney stratification of the fiber $\mathbf{J}^{-1}(\mu)$.

8.3.1 Theorem *Let (M, ω) be a connected symplectic manifold acted canonically and properly upon by a Lie group G. Suppose that this action has an associated standard momentum map $\mathbf{J} : M \to \mathfrak{g}^*$ with non-equivariance one-cocycle $\sigma : G \to \mathfrak{g}^*$. Let $\mu \in \mathfrak{g}^*$ be a value of $\mathbf{J}$ and G_μ the isotropy subgroup of μ with respect to the affine action $\Theta : G \times \mathfrak{g}^* \to \mathfrak{g}^*$ determined by σ. Consider the closed subset $\mathbf{J}^{-1}(\mu) \subset M$ as a topological subspace of M. Then the submanifolds of the type $\mathbf{J}^{-1}(\mu) \cap G_\mu M_H^z$, with M_H^z the connected component of the H-isotropy type submanifold that contains a point z such that $\mathbf{J}(z) = \mu$, form a Whitney stratification of $\mathbf{J}^{-1}(\mu)$.*

Proof. We proceed in two steps.

1). $\mathbf{J}^{-1}(\mu)$ is a decomposed space. We first show that the submanifolds of the type $\mathbf{J}^{-1}(\mu) \cap G_\mu M_H^z$ are the pieces of a decomposition of $\mathbf{J}^{-1}(\mu)$ in the sense of §1.5.1. In turn, this consists of showing that the partition contains locally closed elements, that it is locally finite, and that it satisfies the frontier condition:

a). The submanifolds $\mathbf{J}^{-1}(\mu) \cap G_\mu M_H^z$ are locally closed subsets of $\mathbf{J}^{-1}(\mu)$. This can be easily shown by using Proposition 8.1.2 since for any $m \in \mathbf{J}^{-1}(\mu) \cap G_\mu M_H^z$, the Symplectic Slice Theorem guarantees the existence of a G-equivariant symplectomorphism $G \times_H (\mathfrak{m}_r^* \times V_r) \to U$ onto some open neighborhood U of $G \cdot m$. Let

$$\varphi : Y_0 \to T := \varphi(Y_0) \tag{8.3.1}$$

be the restriction of that symplectomorphism to the open G_μ-invariant neighborhood Y_0 of $G \cdot m$ provided by Proposition 8.1.2. Then, as we saw in (8.1.5), the construction of Y_0 and Proposition 2.4.6 imply that

$$\mathbf{J}^{-1}(\mu) \cap G_\mu M_H^z \cap T = \varphi\left(G_\mu \times_H \left(\{0\} \times V_r^H\right)\right) \cap T. \tag{8.3.2}$$

Given that

$$\mathbf{J}^{-1}(\mu) \cap T = \varphi\left(G_\mu \times_H \left(\{0\} \times \mathbf{J}_V^{-1}(0)\right)\right) \cap T$$

and that $G_\mu \times_H \left(\{0\} \times V_r^H\right)$ is closed in $G_\mu \times_H \left(\{0\} \times \mathbf{J}_V^{-1}(0)\right)$, the local closedness of $\mathbf{J}^{-1}(\mu) \cap G_\mu M_H^z$ in $\mathbf{J}^{-1}(\mu)$ follows.

b). The partition of the space $\mathbf{J}^{-1}(\mu)$ into the submanifolds $\mathbf{J}^{-1}(\mu) \cap G_\mu M_H^z$ is locally finite. In order to prove this we first study the form that the elements of this partition take in the tube given by the symplectomorphism $\varphi : Y_0 \to T := \varphi(Y_0)$ introduced in (8.3.1). Indeed, we will now show that for any isotropy subgroup K of the G-action on M such that $(\mathbf{J}^{-1}(\mu) \cap G_\mu M_K^s) \cap T \neq \emptyset$ for some $s \in M$, we have that

$$(\mathbf{J}^{-1}(\mu) \cap G_\mu M_K^s) \cap T = \varphi\left(G_\mu \times_H \left(\{0\} \times (\mathbf{J}_V^{-1}(0))_{(K)_H^{G_\mu}}\right)\right) \cap T \tag{8.3.3}$$

$$= \varphi\left(\left(G_\mu \times_H \left(\{0\} \times \mathbf{J}_V^{-1}(0)\right)\right)_{(K)^{G_\mu}}\right) \cap T, \tag{8.3.4}$$

where the symbols $(K)_H^{G_\mu}$ and $(K)^{G_\mu}$ are consistent with the notation introduced in Definition 2.4.2. In order to establish these equations, note first that if $\mathbf{J}^{-1}(\mu) \cap G_\mu M_K^s \neq \emptyset$, then $K \subset G_\mu$ because if $z \in \mathbf{J}^{-1}(\mu) \cap G_\mu M_K^s \neq \emptyset$, then there exist elements $g \in G_\mu$ and $z' \in M_K^s$ such that $z = g \cdot z'$. Consequently, $\mathbf{J}(z') = \mathbf{J}(g^{-1} \cdot z) = g^{-1} \cdot \mathbf{J}(z) = g^{-1} \cdot \mu = \mu$. Hence, for any $k \in K$ we have that $k \cdot \mu = k \cdot \mathbf{J}(z') = \mathbf{J}(k \cdot z') = \mathbf{J}(z') = \mu$ and therefore $k \in G_\mu$. The inclusion $K \subset G_\mu$ that we proved and part **(iv)** of Proposition (2.4.6) give us the equality $\varphi\left(G_\mu \times_H \left(\{0\} \times (\mathbf{J}_V^{-1}(0))_{(K)_H^{G_\mu}}\right)\right) \cap T = \varphi\left(\left(G_\mu \times_H \left(\{0\} \times \mathbf{J}_V^{-1}(0)\right)\right)_{(K)^{G_\mu}}\right) \cap T$, that is, (8.3.4).

We now prove the identity (8.3.3) by double inclusion. Let $g \cdot z \in (\mathbf{J}^{-1}(\mu) \cap G_\mu M_K^s) \cap T$, with $g \in G_\mu$, $z \in \mathbf{J}^{-1}(\mu) \cap M_K^s$. By the choice of $T := \varphi(Y_0)$, the point z is such that $z = \varphi([h, 0, v])$, with $h \in G_\mu$, $v \in \mathbf{J}_V^{-1}(0)$, and $K = G_{[h,0,v]} = hH_v h^{-1}$. The last equality implies that $H_v = h^{-1} K h \in (K)_H^{G_\mu}$. Consequently, $g \cdot z = \varphi([gh, 0, v])$ and as $H_v \in (K)_H^{G_\mu}$ and $\mathbf{J}_V(v) = 0$, the inclusion

$$(\mathbf{J}^{-1}(\mu) \cap G_\mu M_K^s) \cap T \subset \varphi\left(G_\mu \times_H \left(\{0\} \times (\mathbf{J}_V^{-1}(0))_{(K)_H^{G_\mu}}\right)\right) \cap T$$

follows. Conversely, take $\varphi([g, 0, v])$ with $g \in G_\mu$, $H_v \in (K)_H^{G_\mu}$, and $\mathbf{J}_V(v) = 0$. The condition $H_v \in (K)_H^{G_\mu}$ guarantees the existence of an element $h \in G_\mu$ such that $H_v = hKh^{-1}$. Then $\varphi([g, 0, v]) = \varphi(gh \cdot [h^{-1}, 0, v]) = gh \cdot \varphi([h^{-1}, 0, v])$, which belongs to $\mathbf{J}^{-1}(\mu) \cap G_\mu M_K^s$ because $G_{\varphi([h^{-1},0,v])} = G_{[h^{-1},0,v]} = h^{-1} H_v h = K$.

Now, the equality

$$\mathbf{J}^{-1}(\mu) \cap G_\mu M_K^s \cap T = \varphi\left(\left(G_\mu \times_H \left(\{0\} \times \mathbf{J}_V^{-1}(0)\right)\right)_{(K)^{G_\mu}}\right) \cap T$$

implies that the decomposition in the statement of our theorem coincides locally with the decomposition of the topological G_μ-space $G_\mu \times_H (\{0\} \times \mathbf{J}_V^{-1}(0))$ into its G_μ-orbit types. Additionally, $G_\mu \times_H (\{0\} \times \mathbf{J}_V^{-1}(0))$ is a closed topological subspace of the manifold $G_\mu \times_H (\{0\} \times V_r)$ where G_μ acts smoothly and properly. Therefore, since by §2.4.14 the decomposition of $G_\mu \times_H (\{0\} \times V_r)$ into its orbit types is locally finite (it is actually finite; see PFLAUM (2001a, Lemma 4.3.6)), so are the decompositions of $G_\mu \times_H (\{0\} \times \mathbf{J}_V^{-1}(0))$ and of $\mathbf{J}^{-1}(\mu)$.

c). The partition satisfies the frontier condition (DS) in §1.5.1. We will prove this fact by showing that any $m \in \mathbf{J}^{-1}(\mu) \cap G_\mu M_H^z$ has an open neighborhood T around it such that $\mathbf{J}^{-1}(\mu) \cap G_\mu M_H^z \cap T$ lies in the closure of all the elements $\mathbf{J}^{-1}(\mu) \cap G_\mu M_K^s$ of our partition that intersect T. If we take $T := \varphi(Y_0)$, with $\varphi : Y_0 \to T$ the symplectomorphism in (8.3.1), this statement reduces to proving that

$$G_\mu \times_H (\{0\} \times V_r^H) \subset \overline{G_\mu \times_H \left(\{0\} \times (\mathbf{J}_V^{-1}(0))_{(K)_H^{G_\mu}}\right)}, \tag{8.3.5}$$

for any $K \subset G_\mu$ and any $s \in M$ such that $\mathbf{J}^{-1}(\mu) \cap G_\mu M_K^s \cap T \neq \emptyset$. In order to prove this inclusion we consider the splitting

$$V = V^H \oplus (V^H)^{\omega_V}. \tag{8.3.6}$$

Notice that the symplectic character of V^H implies that $(V^H)^{\omega_V}$ is also symplectic. Additionally, $(V^H)^{\omega_V}$ is also H-invariant because for any $w \in (V^H)^{\omega_V}$, $v \in V^H$, and $h \in H$ we have that $\omega(h \cdot w, v) = \omega(w, h^{-1} \cdot v) = 0$, which implies that $h \cdot w \in (V^H)^{\omega_V}$.

Denote $W := (V^H)^{\omega_V}$ and let $\mathbf{J}_W : (V^H)^{\omega_V} \to \mathfrak{h}^*$ be the momentum map associated to the H-action on W induced by $\mathbf{J}_V$. We now show that

$$\mathbf{J}_V^{-1}(0) = V^H \oplus \mathbf{J}_W^{-1}(0). \tag{8.3.7}$$

Let $u \in \mathbf{J}_V^{-1}(0)$. By (8.3.6) there exist vectors $v \in V^H$ and $w \in W$ such that $u = v+w$. Using the expression of $\mathbf{J}_W$ we can write, for any $\xi \in \mathfrak{h}$,

$$\begin{aligned}\langle \mathbf{J}_W(w), \xi\rangle &= \langle \mathbf{J}_W(u-v), \xi\rangle = \frac{1}{2}\omega_V(\xi \cdot (u-v), u-v)\\ &= \frac{1}{2}\omega_V(\xi\cdot u, u) - \frac{1}{2}\omega_V(\xi\cdot u, v) = \langle \mathbf{J}_V(u), \xi\rangle + \frac{1}{2}\omega_V(u, \xi\cdot v) = 0,\end{aligned}$$

since $\xi \cdot v = 0$ and $\mathbf{J}_V(u) = 0$. This implies that $w \in \mathbf{J}_W^{-1}(0)$. Conversely, let $u = v+w$, with $v \in V^H$ and $w \in \mathbf{J}_W^{-1}(0)$. For any $\xi \in \mathfrak{h}$ we have that

$$\begin{aligned}\langle \mathbf{J}_V(u), \xi\rangle = \frac{1}{2}\omega_V(\xi\cdot u, u) &= \frac{1}{2}\omega_V(\xi\cdot(v+w), v+w)\\ &= \frac{1}{2}\omega_V(\xi\cdot w, w) = \langle \mathbf{J}_W(w), \xi\rangle = 0.\end{aligned}$$

The expression we want to obtain is the following corollary of (8.3.7), namely,

$$\left(\mathbf{J}_V^{-1}(0)\right)_{(K)_H^{G_\mu}} = V^H \oplus \left(\mathbf{J}_W^{-1}(0)\right)_{(K)_H^{G_\mu}}. \tag{8.3.8}$$

This equality is proved in the following manner. Let $u \in \left(\mathbf{J}_V^{-1}(0)\right)_{(K)_H^{G_\mu}}$. By (8.3.7) there exist vectors $v \in V^H$ and $w \in \mathbf{J}_W^{-1}(0)$ such that $u = v+w$. Since $v \in V^H$ we have that

$$H_w = H_{u-v} = \{h \in H \mid hu - hv = u - v\} = \{h \in H \mid hu = u\} = H_u.$$

Consequently, $H_u \in (K)_H^{G_\mu}$ if and only if $H_w \in (K)_H^{G_\mu}$, which establishes (8.3.8).

Finally, we prove (8.3.5). Indeed,

$$\begin{aligned}G_\mu \times_H (\{0\} \times V_r^H) &\subset G_\mu \times_H \left(\{0\} \times \left(V_r^H \oplus \overline{\left(\mathbf{J}_W^{-1}(0)\right)_{(K)_H^{G_\mu}}}\right)\right)\\ &= G_\mu \times_H \left(\{0\} \times \overline{\left(V_r^H \oplus \left(\mathbf{J}_W^{-1}(0)\right)_{(K)_H^{G_\mu}}\right)}\right)\\ &= G_\mu \times_H \left(\{0\} \times \overline{\left(\mathbf{J}_V^{-1}(0)\right)_{(K)_H^{G_\mu}}}\right)\\ &\subset \overline{G_\mu \times_H \left(\{0\} \times (\mathbf{J}_V^{-1}(0))_{(K)_H^{G_\mu}}\right)},\end{aligned}$$

as required.

2). $\mathbf{J}^{-1}(\mu)$ is a Whitney stratified space. Let $m \in \mathbf{J}^{-1}(\mu) \cap G_\mu M_H^z$ for some $z \in M$ and $H \subset G_\mu$. Suppose that the piece $\mathbf{J}^{-1}(\mu) \cap G_\mu M_H^z$ is incident to another piece $\mathbf{J}^{-1}(\mu) \cap G_\mu M_K^s$ for some $s \in M$ and $K \subset G_\mu$. Let $\{x_n\}_{n\in\mathbb{N}} \subset \mathbf{J}^{-1}(\mu) \cap G_\mu M_K^s$ and $\{y_n\}_{n\in\mathbb{N}} \subset \mathbf{J}^{-1}(\mu) \cap G_\mu M_H^z$ be two sequences such that $x_n \neq y_n$, for all $n \in \mathbb{N}$, and $\lim_{n\to\infty} x_n = \lim_{n\to\infty} y_n = m$. Let $\varphi : Y_0 \to T$ be the tube around the point $m \in M$

introduced in (8.3.1) and that satisfies $\varphi([e,0,0]) = 0$. Recall that in this tube we have that

$$\mathbf{J}^{-1}(\mu) \cap G_\mu M_H^z \cap T = \varphi\left(G_\mu \times_H \left(\{0\} \times V_r^H\right)\right) \cap T \quad \text{and} \tag{8.3.9}$$

$$\mathbf{J}^{-1}(\mu) \cap G_\mu M_K^s \cap T = \varphi\left(G_\mu \times_H \left(\{0\} \times (\mathbf{J}_V^{-1}(0))_{(K)_H^{G_\mu}}\right)\right) \cap T. \tag{8.3.10}$$

Now, given that by (8.3.6) $V = V^H \oplus (V^H)^{\omega_V}$ and that

$$\begin{aligned} T_{[e,0,0]}(G \times_H (\mathfrak{m}_r^* \times V_r)) &\simeq \mathfrak{g}/\mathfrak{h} \times \mathfrak{m}^* \times V \\ &\simeq \mathfrak{m} \times \mathfrak{m}^* \times V = \mathfrak{m} \times \mathfrak{m}^* \times V^H \times (V^H)^{\omega_V}, \end{aligned}$$

we have a chart $\psi : U \to \mathfrak{m} \times \mathfrak{m}^* \times V^H \times (V^H)^{\omega_V}$ of M around the point m defined by the equality

$$\psi^{-1}(\xi, \eta, v, w) = \varphi([\exp \xi, \eta, v + w]),$$

for $\xi \in \mathfrak{m}$, $\eta \in \mathfrak{m}^*$, $v \in V^H$, and $w \in (V^H)^{\omega_V}$, all of them close enough to the origin of their respective spaces. Let now $\{l_n\}_{n\in\mathbb{N}} := \{\overline{\psi(x_n)\psi(y_n)}\}_{n\in\mathbb{N}}$ be the sequence of lines in $\psi(U)$ determined by $\{\psi(x_n)\}_{n\in\mathbb{N}}$ and $\{\psi(y_n)\}_{n\in\mathbb{N}}$ that by hypothesis converges to a line l (if necessary, take subsequences of $\{x_n\}_{n\in\mathbb{N}}$ and $\{y_n\}_{n\in\mathbb{N}}$ so that they are in the domain of ψ). By (8.3.9) and (8.3.10) we have that $\psi(y_n) = (\xi_n, 0, v_n, 0)$ for all $n \in \mathbb{N}$, with $\{\xi_n\}_{n\in\mathbb{N}} \subset \mathfrak{m}$ and $\{v_n\}_{n\in\mathbb{N}} \subset V^H$ sequences such that $\xi_n \to 0$ and $v_n \to 0$. Analogously, $\psi(x_n) = (\zeta_n, 0, u_n, w_n)$ for all $n \in \mathbb{N}$, with $\{\zeta_n\}_{n\in\mathbb{N}} \subset \mathfrak{m}$, $\{u_n\}_{n\in\mathbb{N}} \subset V^H$, and $\{w_n\} \subset (\mathbf{J}_W^{-1}(0))_{(K)_H^{G_\mu}}$ (see (8.3.8)) sequences such that $\zeta_n \to 0$, $u_n \to 0$, and $w_n \to 0$. The limit l in projective space is given by the limit point on the sphere

$$l = \lim_{n\to\infty} \frac{\psi(x_n) - \psi(y_n)}{\|\psi(x_n) - \psi(y_n)\|} = (\xi, 0, v, w) \in \mathfrak{m} \times \{0\} \times V^H \times \overline{(\mathbf{J}_W^{-1}(0))_{(K)_H^{G_\mu}}}.$$

Since for all $n \in \mathbb{N}$, $\varphi(y_n)$ has always zero as its $(V^H)^{\omega_V}$ component, we can choose a subsequence of $\{w_n\}_{n\in\mathbb{N}}$ such that

$$w = \|w\| \lim_{n\to\infty} \frac{w_n}{\|w_n\|},$$

that is, w can be taken to be the limit in projective space of a sequence of lines in $(\mathbf{J}_W^{-1}(0))_{(K)_H^{G_\mu}}$ (notice that $(\mathbf{J}_W^{-1}(0))_{(K)_H^{G_\mu}}$ is invariant with respect to multiplication by nonzero scalars). Consequently, $w \in \lim_{n\to\infty} T_{w_n}(\mathbf{J}_W^{-1}(0))_{(K)_H^{G_\mu}}$, which in turn implies that $l \subset \lim_{n\to\infty} T_{\psi(x_n)}\left(\mathfrak{m} \times \{0\} \times V^H \times (\mathbf{J}_W^{-1}(0))_{(K)_H^{G_\mu}}\right)$ and hence

$$T_{(0,0,0,0)}\psi^{-1} \cdot l \subset \lim_{n\to\infty} T_{x_n}(\mathbf{J}^{-1}(\mu) \cap G_\mu M_K^s),$$

as required. ■

8.3.2 Theorem (Stratification Theorem) *Let (M, ω) be a connected symplectic manifold acted canonically and properly upon by a Lie group G. Suppose that this action has an associated standard momentum map $\mathbf{J} : M \to \mathfrak{g}^*$ with non-equivariance one-cocycle $\sigma : G \to \mathfrak{g}^*$. Let $\mu \in \mathfrak{g}^*$ be a value of $\mathbf{J}$ and let G_μ be the isotropy subgroup of μ with respect to the affine action $\Theta : G \times \mathfrak{g}^* \to \mathfrak{g}^*$ determined by σ. Then the symplectic strata $M_\mu^{(H)}$ introduced in Theorem* 8.1.1 *form a symplectic Whitney stratification of the quotient $M_\mu := \mathbf{J}^{-1}(\mu)/G_\mu$. The quotient M_μ is a cone space when considered as a stratified space with strata $M_\mu^{(H)}$.*

Proof. The symplectic character of the elements of the partition is a consequence of Theorem 8.1.1. In order to prove that this partition provides a decomposition of the quotient topological space M_μ in the sense of §1.5.1 we consider $\pi_\mu : \mathbf{J}^{-1}(\mu) \to \mathbf{J}^{-1}(\mu)/G_\mu$ the projection onto the orbit space and the point $[m]_\mu := \pi_\mu(m) \in \mathbf{J}^{-1}(\mu)/G_\mu$, with $m \in \mathbf{J}^{-1}(\mu)$ such that $G_m =: H \subset G_\mu$. Let $\varphi : Y_0 \to T$ be one of the G_μ-invariant tubes around m provided by Proposition 8.1.2, that is, T is an open neighborhood around m, $\varphi([e, 0, 0]) = m$, Y_0 and T are G_μ-invariant, and φ is G_μ-equivariant. These features imply the existence of a homeomorphism $\phi : Y_0/G_\mu \to T/G_\mu =: \mathcal{T}$. The set $M_\mu \cap \mathcal{T} = (\mathbf{J}^{-1}(\mu) \cap T)/G_\mu$ is an open neighborhood of $[m]_\mu \in \mathbf{J}^{-1}(\mu)/G_\mu$ such that

$$\begin{aligned}\phi^{-1}(M_\mu \cap \mathcal{T}) &= \phi^{-1}((\mathbf{J}^{-1}(\mu) \cap T)/G_\mu) \\ &= (G_\mu \times_H (\{0\} \times \mathbf{J}_V^{-1}(0))) \cap Y_0/G_\mu. \end{aligned} \tag{8.3.11}$$

Also, using (8.3.2) we obtain that

$$\begin{aligned}\phi^{-1}(M_\mu^{(H)} \cap \mathcal{T}) &= \phi^{-1}\left(\mathbf{J}^{-1}(\mu) \cap G_\mu M_H^z \cap T/G_\mu\right) \\ &= (G_\mu \times_H (\{0\} \times V_r^H)) \cap Y_0/G_\mu. \end{aligned} \tag{8.3.12}$$

This expression together with (8.3.11) shows that the symplectic strata $M_\mu^{(H)}$ are locally closed subspaces of M_μ, since $[(G_\mu \times_H (\{0\} \times V_r^H)) \cap Y_0]/G_\mu$ is clearly closed in $[(G_\mu \times_H (\{0\} \times \mathbf{J}_V^{-1}(0))) \cap Y_0]/G_\mu$.

Additionally, for any neighboring symplectic stratum $M_\mu^{(K)} = \mathbf{J}^{-1}(\mu) \cap G_\mu M_K^s/G_\mu$ we have that

$$\phi^{-1}(M_\mu^{(K)} \cap \mathcal{T}) = \left(G_\mu \times_H \left(\{0\} \times \left(\mathbf{J}_V^{-1}(0)\right)_{(K)_H^{G_\mu}}\right) \cap Y_0\right)/G_\mu \tag{8.3.13}$$

$$= \left(G_\mu \times_H \left(\{0\} \times \mathbf{J}_V^{-1}(0)\right)_{(K)^{G_\mu}} \cap Y_0\right)/G_\mu. \tag{8.3.14}$$

The finiteness of the partition of $G_\mu \times_H (\{0\} \times \mathbf{J}_V^{-1}(0))$ in terms of its G_μ-orbit types (see the proof of Theorem 8.3.1) implies, in view of (8.3.12) and (8.3.14) the local finiteness of the partition of M_μ into its symplectic strata.

We now show that this partition satisfies the frontier condition by proving that $M_\mu^{(H)} \cap \mathcal{T}$ lies in the closure of all the symplectic strata $M_\mu^{(K)}$ that intersect $\mathcal{T}$. Indeed, the frontier condition for the stratification of $\mathbf{J}^{-1}(\mu)$ in Theorem 8.3.1 implies that

$\mathbf{J}^{-1}(\mu) \cap G_\mu M_H^z \subset \overline{\mathbf{J}^{-1}(\mu) \cap G_\mu M_K^s}$ for any piece $\mathbf{J}^{-1}(\mu) \cap G_\mu M_K^s$ that intersects T and hence

$$M_\mu^{(H)} = \pi_\mu \left(\mathbf{J}^{-1}(\mu) \cap G_\mu M_H^z \right) \subset \pi_\mu \left(\overline{\mathbf{J}^{-1}(\mu) \cap G_\mu M_K^s} \right)$$
$$\subset \overline{\pi_\mu \left(\mathbf{J}^{-1}(\mu) \cap G_\mu M_K^s \right)} = \overline{M_\mu^{(K)}}.$$

We conclude by showing that our decomposition of M_μ satisfies the Whitney condition (B). Since this is a local property, and in view of (8.3.13) it is clear that we will be done if we prove that the manifolds $(\mathbf{J}_V^{-1}(0))_{(K)_H^{G_\mu}}/H$ form a Whitney (B) stratification of the quotient $\mathbf{J}_V^{-1}(0)/H$. We start by endowing the quotient topological space with a smooth structure. This can be achieved by following the prescription described in §2.5.4 and this consists in using a minimal Hilbert basis $\mathcal{B} = \{\sigma_1, \ldots, \sigma_k\}$ of homogeneous H-invariant polynomials for the H-action on V and in defining

$$\begin{aligned} \sigma : \mathbf{J}_V^{-1}(0)/H &\longrightarrow \mathbb{R}^k \\ [v] &\longmapsto (\sigma_1(v), \ldots, \sigma_k(v)). \end{aligned}$$

The map σ is a global stratified chart for $\mathbf{J}_V^{-1}(0)/H$ compatible with its partition into the manifolds of the type $(\mathbf{J}_V^{-1}(0))_{(K)_H^{G_\mu}}/H$. The minimality of the basis $\mathcal{B}$ implies that $\sigma(V^H)$ is a linear subspace of $\mathbb{R}^k$ and that (see (8.3.7))

$$\sigma(\mathbf{J}_V^{-1}(0)/H) = \sigma(\mathbf{J}_W^{-1}(0)/H) \times \sigma(V^H). \tag{8.3.15}$$

Let now $\{x_n\}_{n\in\mathbb{N}} \subset (\mathbf{J}_V^{-1}(0))_{(K)_H^{G_\mu}}/H$ and $\{y_n\}_{n\in\mathbb{N}} \subset V^H/H$ be two sequences such that $x_n \neq y_n$, for all $n \in \mathbb{N}$, and $x_n, y_n \to [0]$. By (8.3.15) we have that $\sigma(y_n) = (0, u_n)$ and $\sigma(x_n) = (w_n, v_n)$, with $\{u_n\}_{n\in\mathbb{N}}, \{v_n\}_{n\in\mathbb{N}} \subset \sigma(V^H)$ and $\{w_n\}_{n\in\mathbb{N}} \subset \sigma\left((\mathbf{J}_W^{-1}(0))_{(K)_H^{G_\mu}}/H\right)$ such that $u_n, v_n, w_n \to 0$.

Let now $\{l_n\}_{n\in\mathbb{N}} := \{\overline{\sigma(x_n)\sigma(y_n)}\}_{n\in\mathbb{N}}$ be the sequence of lines in $\sigma(\mathbf{J}_V^{-1}(0)/H)$ determined by $\{\sigma(x_n)\}_{n\in\mathbb{N}}$ and $\{\sigma(y_n)\}_{n\in\mathbb{N}}$, which by hypothesis converges to a line l. Notice that these lines are indeed in $\sigma(\mathbf{J}_V^{-1}(0)/H)$ since in projective space they can be seen as the elements on the sphere given by

$$\frac{\sigma(x_n) - \sigma(y_n)}{\|\sigma(x_n) - \sigma(y_n)\|} = \frac{(w_n, v_n - u_n)}{\|\sigma(x_n) - \sigma(y_n)\|} \in \sigma(\mathbf{J}_W^{-1}(0)/H) \times \sigma(V^H)$$
$$= \sigma(\mathbf{J}_V^{-1}(0)/H),$$

(we recall that $\sigma(V^H)$ is a linear subspace of $\mathbb{R}^k$). Let now $w \in \overline{\sigma\left((\mathbf{J}_W^{-1}(0))_{(K)_H^{G_\mu}}/H\right)}$, $v \in \sigma(V^H)$, be such that

$$l = \lim_{n\to\infty} \frac{\sigma(x_n) - \sigma(y_n)}{\|\sigma(x_n) - \sigma(y_n)\|} = (w, v).$$

Since $w = \|w\| \lim_{n\to\infty} (w_n/\|w\|)$, w can be seen as the limit in projective space of a sequence of lines in $\sigma\left((\mathbf{J}_W^{-1}(0))_{(K)_H^{G_\mu}}/H\right)$ and consequently

$$w \subset \lim_{n\to\infty} T_{w_n}\sigma\left((\mathbf{J}_W^{-1}(0))_{(K)_H^{G_\mu}}/H\right)$$

which in turn implies that

$$l \subset \lim_{n\to\infty} T_{\sigma(x_n)}\Big(\sigma\big((\mathbf{J}_W^{-1}(0))_{(K)_H^{G_\mu}}/H\big)\times\sigma(V^H)\Big),$$

or equivalently, that

$$\begin{aligned} T_0\sigma^{-1}\cdot l &\subset \lim_{n\to\infty} T_{x_n}\Big((\mathbf{J}_W^{-1}(0))_{(K)_H^{G_\mu}}/H\oplus V^H\Big)\\ &= \lim_{n\to\infty} T_{x_n}\Big((\mathbf{J}_V^{-1}(0))_{(K)_H^{G_\mu}}/H\Big), \end{aligned}$$

as required.

We proved that $\sigma(\mathbf{J}_V^{-1}(0)/H)$ is a Whitney stratified subset of $\mathbb{R}^k$. By Mather's theory of control data (see MATHER (1970) and Section 1.7.5) $\sigma(\mathbf{J}_V^{-1}(0)/H)$ is a cone space and consequently so are $\mathbf{J}_V^{-1}(0)/H$ and M_μ.

We conclude by showing the construction of the link of $\mathbf{J}_V^{-1}(0)/H$ considered as a stratified space by the manifolds of the form $(\mathbf{J}_V^{-1}(0))_{(K)_H^{G_\mu}}/H$. Our approach follows a slight generalization of the one taken on page 400 of SJAMAAR AND LERMAN (1991).

Take an H-invariant positive definite inner product $\langle\cdot,\cdot\rangle$ on W (always available by the compactness of H) and let $S^{2n-1} := \{w\in W \mid \langle w,w\rangle = 1\}$ be the unit sphere in W constructed using $\langle\cdot,\cdot\rangle$. We now show that $\mathbf{J}_W^{-1}(0)$ and S^{2n-1} intersect transversely. If $\mathbf{J}_W^{-1}(0)\cap S^{2n-1}\neq\emptyset$, let $v\in\mathbf{J}_W^{-1}(0)\cap S^{2n-1}$ and $w := \frac{d}{dt}\big|_{t=0}(v+tv)\in T_vW$. As $v\in\mathbf{J}_W^{-1}(0)$, a straightforward computation shows that $w\in\ker T_v\mathbf{J}_W$. Since $w\notin T_vS^{2n-1}$ and S^{2n-1} has codimension one in W we have that

$$T_vW = \ker T_w\mathbf{J}_W + T_vS^{2n-1}.$$

Since $v\in\mathbf{J}_W^{-1}(0)\cap S^{2n-1}$ is arbitrary the claim follows and hence

$$L := \mathbf{J}_W^{-1}(0)\cap S^{2n-1}$$

is an H-invariant submanifold of W that admits a decomposition into submanifolds of the form $L_{(K)_H^{G_\mu}}$. Additionally, the map $\rho : L\times(0,\infty)\to\mathbf{J}_W^{-1}(0)\setminus\{0\}$ defined by $\rho(z,t) := \sqrt{t}z$, $(z,t)\in L\times(0,\infty)$ is a decomposition preserving an H-equivariant homeomorphism that drops to a homeomorphism of the reduced spaces $\big(\mathbf{J}_W^{-1}(0)\setminus\{0\}\big)/H$ and $(L/H)\times(0,\infty)$. In turn, this homeomorphism extends to a diffeomorphism $C(L/H)\longrightarrow\mathbf{J}_W^{-1}(0)/H$ of stratified spaces with smooth structure. Using the decomposition (8.3.7) this map can be immediately used to construct a cone map

$$U\longrightarrow(U\cap V_H)\times C(L/H)$$

for some open neighborhood U of the origin in $\mathbf{J}_V^{-1}(0)/H$. ■

8.3.3 The minimality of the symplectic stratification. Unlike the orbit type stratification of any orbit space of a proper group action on a manifold, the symplectic stratification described in Theorem 8.3.2 is, in general, *not* minimal among all the

Whitney stratifications of the quotient $\mathbf{J}^{-1}(\mu)/G_\mu$ when the value $\mu \in \mathfrak{g}^*$ is not zero. Indeed, in the proof of the theorem we showed that the symplectic stratification of $\mathbf{J}^{-1}(\mu)/G_\mu$ can be locally identified with the stratification of the quotient $\mathbf{J}_V^{-1}(0)/H$ by the manifolds $\mathbf{J}_V^{-1}(0)_{(K)_H^{G_\mu}}/H$. When the value $\mu \neq 0$, this stratification is in general strictly finer than the stratification by the standard orbit types $\mathbf{J}_V^{-1}(0)_{(K)}/H$ which we know *is* minimal among all the Whitney stratifications of $\mathbf{J}_V^{-1}(0)/H$ (see §2.4.13). When $\mu = 0$, the symplectic stratification coincides with the orbit type stratification.

8.3.4 As a corollary of M_μ being a cone space it can be proved (see Theorem 5.9 in SJAMAAR AND LERMAN (1991)) that each connected component of M_μ contains a unique open stratum that is connected, open, and dense in the connected component of M_μ containing it.

8.3.5 An example that has attracted much attention in connection with the symplectically stratified reduced quotients in Theorem 8.3.2 is the moduli spaces of flat connections on a Riemann surface introduced in ATIYAH AND BOTT (1982). The reader interested in the relation of this construction with the theorem that we proved is encouraged to check with GOLDMAN (1984); GURUPRASAD et al. (1997); HUEBSCHMANN (1995a, 1996, 1998); KARSHON (1997), and references therein.

8.3.6 Other properties of the quotient $\mathbf{J}^{-1}(\mu)/G_\mu$. KIRWAN (1984a, 1988) has studied the cohomology of the quotient $\mathbf{J}^{-1}(\mu)/G_\mu$ when the symplectic manifold and the symmetry group in question are compact, μ is a regular value of the momentum map $\mathbf{J}$, and the G_μ-action on $\mathbf{J}^{-1}(\mu)$ is locally free, that is, its isotropy subgroups are finite. The local freeness assumption reduces the study of the cohomology of the quotient $\mathbf{J}^{-1}(\mu)/G_\mu$ to the G_μ-equivariant cohomology of the numerator $\mathbf{J}^{-1}(\mu)$ (with rational coefficients). This can be done in much detail using the properties of the critical points of the function $\|\mathbf{J} - \mu\|^2$ ($\|\cdot\|$ is an Ad_G-invariant inner product). The reader is encouraged to check with the original papers for explicit formulas of the Betti numbers of $\mathbf{J}^{-1}(\mu)/G_\mu$.

8.3.7 The Duistermaat–Heckman Theorem. Let $\mathbb{T}$ be a torus acting canonically on the symplectic manifold (M, ω). Assume that this action is locally free (the isotropy subgroups are finite) and that it has an associated proper invariant momentum map $\mathbf{J} : M \to \mathfrak{t}^*$. The properness hypothesis on $\mathbf{J}$ implies that the subset $\mathfrak{t}^*_{reg}$ of $\mathfrak{t}^*$ consisting of the regular values of $\mathbf{J}$ is open. Sard's Theorem implies that $\mathfrak{t}^*_{reg}$ is also dense in $\mathfrak{t}^*$. The local freeness of the action implies that the quotients $\mathbf{J}^{-1}(\mu)/\mathbb{T}$, with $\mu \in \mathfrak{t}^*_{reg}$ are ***orbifolds*** or ***V-manifolds*** in the sense of SATAKE (1956) or, more specifically, symplectic orbifolds in the sense of WEINSTEIN (1977b).

The orbifolds are a generalization of the notion of manifold where most of the notions of differential and integral calculus can be appropriately defined (see SATAKE (1957)). For instance, the concepts of differential form and of de Rham cohomology are well defined. The introduction of the orbifolds and the identification of the quotients $\mathbf{J}^{-1}(\mu)/\mathbb{T}$ as such objects allow the statement of a straightforward generalization of the Marsden–Weinstein Theorem that says that the symplectic form ω induces a natural symplectic structure ω_μ on the orbifold $\mathbf{J}^{-1}(\mu)/\mathbb{T}$.

The first theorem in DUISTERMAAT AND HECKMAN (1982) describes how the cohomology class $[\omega_\mu]$ of the reduced symplectic form varies when we move μ in the subset $\mathfrak{t}^*_{reg}$. To be more specific, fix some $\mu_0 \in \mathbf{J}(M)$ and let U denote the connected component of $\mathfrak{t}^*_{reg}$ containing μ_0. Then:

(i) For each $\mu \in U$ there is a canonical isomorphism

$$H^2(\mathbf{J}^{-1}(\mu)/\mathbb{T}; \mathbb{R}) \to H^2(\mathbf{J}^{-1}(\mu_0)/\mathbb{T}; \mathbb{R}).$$

Denote by $\widehat{\sigma}$ the image by this isomorphism of the element $\sigma \in H^2(\mathbf{J}^{-1}(\mu)/\mathbb{T}; \mathbb{R})$.

(ii) For any $\mu \in U$ we have that

$$\widehat{[\omega_\mu]} = [\omega_{\mu_0}] - \langle \mu - \mu_0, \overline{c} \rangle,$$

where $c \in H^2(\mathbf{J}^{-1}(\mu_0); \mathfrak{g})$ is the Chern class (see §5.1.4) of the principal $\mathbb{T}$-bundle $\pi_{\mu_0} : \mathbf{J}^{-1}(\mu_0) \to \mathbf{J}^{-1}(\mu_0)/\mathbb{T}$ and $\overline{c} \in H^2(\mathbf{J}^{-1}(\mu_0)/\mathbb{T}; \mathfrak{g})$ is its projection to the quotient (which can be done by the Abelian character of $\mathbb{T}$).

The ***Duistermaat–Heckman function*** $\varphi : \mathfrak{t}^*_{reg} \to \mathbb{R}$ is defined by

$$\varphi(\mu) := \text{volume}\left(M_\mu\right) = \int_{M_\mu} \frac{\omega_\mu^d}{d!},$$

where $d = \frac{1}{2} \dim M - \dim \mathbb{T}$ is half the dimension of the reduced space M_μ. The Duistermaat–Heckman Theorem guarantees that the Duistermaat–Heckman function is a polynomial in μ of degree at most d on each connected component of $\mathfrak{t}^*_{reg}$.

8.4 Singular orbit reduction

So far in this chapter we have adopted a point reduction approach even though the entire reduction program can be carried out in the orbit reduction context under the same set of hypotheses. In particular, the Stratification Theorem 8.3.2, which provides a symplectic stratification of the quotient $\mathbf{J}^{-1}(\mu)/G_\mu$, has an analog describing a symplectic stratification of $\mathbf{J}^{-1}(\mathcal{O}_\mu)/G$; these two quotients are isomorphic as symplectic stratified spaces.

For the study of the set $\mathbf{J}^{-1}(\mathcal{O}_\mu)/G$ a very important technical point is the choice of its topology. In the point reduction approach we thought of $\mathbf{J}^{-1}(\mu)$ as a topological subspace of M and of $\mathbf{J}^{-1}(\mu)/G_\mu$ as the resulting topological quotient. This is *not* the right way to proceed when dealing with orbit reduction; in this situation $\mathbf{J}^{-1}(\mathcal{O}_\mu)$ needs to be endowed not with the relative topology but with the initial topology induced by the map

$$\begin{aligned} \mathbf{J}_{\mathcal{O}_\mu} : \mathbf{J}^{-1}(\mathcal{O}_\mu) &\longrightarrow \mathcal{O}_\mu \\ z &\longmapsto \mathbf{J}(z), \end{aligned}$$

where the orbit $\mathcal{O}_\mu$ comes with the orbit smooth structure introduced in Proposition 2.3.12. We will refer to this choice as the ***initial topology*** of $\mathbf{J}^{-1}(\mathcal{O}_\mu)$. Recall that the initial topology induced by the map $\mathbf{J}_{\mathcal{O}_\mu} : \mathbf{J}^{-1}(\mathcal{O}_\mu) \longrightarrow \mathcal{O}_\mu$ is characterized by the fact that for any topological space Z and any map $\varphi : Z \to \mathbf{J}^{-1}(\mathcal{O}_\mu)$, we have that φ is continuous if and only if $\mathbf{J}_{\mathcal{O}_\mu} \circ \varphi$ is continuous. Additionally, the set $\mathcal{B} = \{\mathbf{J}_{\mathcal{O}_\mu}^{-1}(U) \mid U \text{ open in } \mathcal{O}_\mu\}$ is a subbasis of this topology. In particular, this implies that $\mathbf{J}^{-1}(\mathcal{O}_\mu)$ is first countable.

The following proposition shows that the initial topology of $\mathbf{J}^{-1}(\mathcal{O}_\mu)$ generalizes to the singular case the smooth structure for this set considered in the regular situation (see Proposition 6.3.2).

8.4.1 Proposition *Let (M, ω) be a connected symplectic manifold acted canonically and properly upon by a Lie group G. Suppose that this action has an associated standard momentum map $\mathbf{J} : M \to \mathfrak{g}^*$ with non-equivariance one-cocycle $\sigma : G \to \mathfrak{g}^*$. Let $\mathcal{O}_\mu := G \cdot \mu \subset \mathfrak{g}^*$ be the G-orbit of the point $\mu \in \mathfrak{g}^*$ with respect to the affine action of G on $\mathfrak{g}^*$ associated to σ. Then, if we consider $\mathbf{J}^{-1}(\mathcal{O}_\mu)$ with its initial topology, the map*

$$\begin{aligned} f : G \times_{G_\mu} \mathbf{J}^{-1}(\mu) &\longrightarrow \mathbf{J}^{-1}(\mathcal{O}_\mu) \\ [g, z] &\longmapsto g \cdot z, \end{aligned}$$

is a homeomorphism.

Proof. Since $\mathbf{J}^{-1}(\mathcal{O}_\mu) = G \cdot \mathbf{J}^{-1}(\mathcal{O}_\mu)$, the map f is clearly a bijection whose inverse is given by

$$\begin{aligned} f^{-1} : \mathbf{J}^{-1}(\mathcal{O}_\mu) &\longrightarrow G \times_{G_\mu} \mathbf{J}^{-1}(\mu) \\ z = g \cdot t &\longmapsto [g, t], \end{aligned}$$

where $g \in G$ and $t \in \mathbf{J}^{-1}(\mu)$. We now show that both f and f^{-1} are continuous.

The choice of the initial topology for $\mathbf{J}^{-1}(\mathcal{O}_\mu)$ implies that f is continuous if and only if the map

$$\begin{aligned} G \times_{G_\mu} \mathbf{J}^{-1}(\mu) &\longrightarrow \mathcal{O}_\mu \\ [g, z] &\longmapsto g \cdot \mu \end{aligned}$$

is continuous. By Proposition 2.3.12, this is equivalent (the projection $G \times \mathbf{J}^{-1}(\mu) \to G \times_{G_\mu} \mathbf{J}^{-1}(\mu)$ is a submersion) to the map $G \times \mathbf{J}^{-1}(\mu) \to G/G_\mu$ defined by $(g, z) \in G \times \mathbf{J}^{-1}(\mu) \mapsto gG_\mu$ being continuous, which is true.

We now show that the inverse $f^{-1} : \mathbf{J}^{-1}(\mathcal{O}_\mu) \to G \times_{G_\mu} \mathbf{J}^{-1}(\mu)$ of f given by $g \cdot z \mapsto [g, z]$ is continuous. Let $j : \mathcal{O}_\mu \to G/G_\mu$ be the standard diffeomorphism and $\sigma : U_{gG_\mu} \subset G/G_\mu \to G$ a local smooth section of the submersion $G \to G/G_\mu$ in a neighborhood U_{gG_μ} of $gG_\mu \in G/G_\mu$. Let $V := \mathbf{J}_{\mathcal{O}_\mu}^{-1}(j^{-1}(U_{gG_\mu}))$. V is an open neighborhood of z in $\mathbf{J}^{-1}(\mathcal{O}_\mu)$ because $(j \circ \mathbf{J}_{\mathcal{O}_\mu})(z) = j(g \cdot \mu) = gG_\mu \in U_{gG_\mu}$. We now note that for any $m \in V$ we have

$$f^{-1}(m) = \left[(\sigma \circ j \circ \mathbf{J}_{\mathcal{O}_\mu})(m), \left((\sigma \circ j \circ \mathbf{J}_{\mathcal{O}_\mu})(m)\right)^{-1} \cdot m\right].$$

The continuity of all the maps involved in this expression guarantees the continuity of f^{-1}. ■

We now introduce the manifolds that constitute the symplectic strata of the orbit reduced space $\mathbf{J}^{-1}(\mathcal{O}_\mu)/G$.

8.4.2 Theorem (Singular symplectic orbit strata) *Let (M, ω) be a connected symplectic manifold acted canonically and properly upon by a Lie group G. Suppose that this action has an associated standard momentum map $\mathbf{J} : M \to \mathfrak{g}^*$ with non-equivariance one-cocycle $\sigma : G \to \mathfrak{g}^*$. Let $\mu \in \mathfrak{g}^*$ be a value of $\mathbf{J}$, $\mathcal{O}_\mu := G \cdot \mu$ the orbit of this element with respect to the affine action $\Theta : G \times \mathfrak{g}^* \to \mathfrak{g}^*$ determined by σ, and let $H \subset G$ be an isotropy subgroup of the G-action on M. Let M_H^z be the connected component of the H-isotropy type manifold that contains a given element $z \in M$. Then the following hold:*

(i) *The set $G \cdot (\mathbf{J}^{-1}(\mu) \cap M_H^z)$ is an initial submanifold of M whose tangent space is given by*

$$T_m\left(G \cdot (\mathbf{J}^{-1}(\mu) \cap M_H^z)\right) = \operatorname{span}\{\xi_M(m) + X_f(m) \mid \xi \in \mathfrak{g},\ f \in C^\infty(M)^G\}$$
$$= \mathfrak{g} \cdot m + A'_G(m), \tag{8.4.1}$$

with A'_G the polar distribution associated to the G-action on M.

(ii) *The set $M_{\mathcal{O}_\mu}^{(H)} := G \cdot (\mathbf{J}^{-1}(\mu) \cap M_H^z)/G$ has a unique quotient differentiable structure such that the canonical projection*

$$\pi_{\mathcal{O}_\mu}^{(H)} : G \cdot (\mathbf{J}^{-1}(\mu) \cap M_H^z) \longrightarrow M_{\mathcal{O}_\mu}^{(H)}$$

is a surjective submersion.

(iii) *There is a unique symplectic structure $\omega_{\mathcal{O}_\mu}^{(H)}$ on $M_{\mathcal{O}_\mu}^{(H)}$ characterized by*

$$i_{\mathcal{O}_\mu}^{(H)*}\omega = \pi_{\mathcal{O}_\mu}^{(H)*}\omega_{\mathcal{O}_\mu}^{(H)} + \mathbf{J}_{\mathcal{O}_\mu}^{(H)*}\omega_{\mathcal{O}_\mu}^{+}, \tag{8.4.2}$$

where $i_{\mathcal{O}_\mu}^{(H)} : G \cdot (\mathbf{J}^{-1}(\mu) \cap M_H^z) \hookrightarrow M$ is the inclusion, $\mathbf{J}_{\mathcal{O}_\mu}^{(H)} : G \cdot (\mathbf{J}^{-1}(\mu) \cap M_H^z) \to \mathcal{O}_\mu$ is obtained by restriction of the momentum map $\mathbf{J}$, and $\omega_{\mathcal{O}_\mu}^{+}$ is the $+$-symplectic form on $\mathcal{O}_\mu$ introduced in Theorem 4.5.31. The pairs of the form $(M_{\mathcal{O}_\mu}^{(H)}, \omega_{\mathcal{O}_\mu}^{(H)})$ will be referred to as the ***singular symplectic orbit strata****.*

(iv) *Let $h \in C^\infty(M)^G$ be a G-invariant Hamiltonian. Then the flow F_t of X_h leaves the connected components of $G \cdot (\mathbf{J}^{-1}(\mu) \cap M_H^z)$ invariant and commutes with the G-action, so it induces a flow $F_t^{\mathcal{O}_\mu}$ on $M_{\mathcal{O}_\mu}^{(H)}$ that is characterized by*

$$\pi_{\mathcal{O}_\mu}^{(H)} \circ F_t \circ i_{\mathcal{O}_\mu}^{(H)} = F_t^{\mathcal{O}_\mu} \circ \pi_{\mathcal{O}_\mu}^{(H)}.$$

(v) *The flow $F_t^{\mathcal{O}_\mu}$ is Hamiltonian on $(M_{\mathcal{O}_\mu}^{(H)}, \omega_{\mathcal{O}_\mu}^{(H)})$, with Hamiltonian function $h_{\mathcal{O}_\mu}^{(H)} : M_{\mathcal{O}_\mu}^{(H)} \to \mathbb{R}$ defined by*

$$h_{\mathcal{O}_\mu}^{(H)} \circ \pi_{\mathcal{O}_\mu}^{(H)} = h \circ i_{\mathcal{O}_\mu}^{(H)}.$$

The vector fields X_h and $X_{h^{(H)}_{\mathcal{O}_\mu}}$ are $\pi^{(H)}_{\mathcal{O}_\mu}$-related. We will call $h^{(H)}_{\mathcal{O}_\mu}$ the ***reduced Hamiltonian****.*

(vi) *Let $k : M \to \mathbb{R}$ be another G-invariant function. Then $\{h, k\}$ is also G-invariant and*

$$\{h, k\}^{(H)}_{\mathcal{O}_\mu} = \{h^{(H)}_{\mathcal{O}_\mu}, k^{(H)}_{\mathcal{O}_\mu}\}_{M^{(H)}_{\mathcal{O}_\mu}}$$

where $\{\cdot, \cdot\}_{M^{(H)}_{\mathcal{O}_\mu}}$ denotes the Poisson bracket induced by the symplectic structure in $M^{(H)}_{\mathcal{O}_\mu}$.

Proof. **(i)** We start by recalling that by the properness of the G-action on M, the polar distribution A'_G is spanned by the vector fields associated to the functions in $C^\infty(M)^G$ (see (5.5.2)). Its saturating distribution D (see §3.4.5) defined by $D(m) = \mathfrak{g} \cdot m + A'_G(m)$, for any $m \in M$, is integrable by Proposition 3.4.6 and its maximal integral submanifold going through an arbitrary point $m \in M$ is the initial submanifold $G^0 \cdot (\mathbf{J}^{-1}(\mu) \cap M^m_{G_m})^m$ (see Corollary 5.5.18), where G^0 is the connected component of G containing the identity, $\mu = \mathbf{J}(m)$, and $(\mathbf{J}^{-1}(\mu) \cap M^z_H)^m$ is the connected component of $\mathbf{J}^{-1}(\mu) \cap M^z_H$ that contains a given point $m \in M$.

Recall now that by Corollary 4.6.2 the set $\mathbf{J}^{-1}(\mu) \cap M^z_H$ is an embedded submanifold of M whose connected components coincide with the orbits of the polar pseudogroup A'_G, that is, $(\mathbf{J}^{-1}(\mu) \cap M^z_H)^m = A'_G \cdot m$ (see Corollary 5.5.18). Additionally, as we saw in the proof of Corollary 4.6.2, $\mathbf{J}^{-1}(\mu) \cap M^z_H = \mathbf{J}^{-1}_{L^z}(0)$, where $L^z := N(H)^z/H$ and $\mathbf{J}_{L^z} : M^z_H \to \mathfrak{l}^* := (\operatorname{Lie}(L^z))^*$ is the momentum map associated to the L^z-action on M^z_H by Theorem 4.6.1. Also, in the proof of Theorem 8.2.1 we noticed that $\mathbf{J}^{-1}(\mu) \cap M^z_H = \mathbf{J}^{-1}_{L^z}(0)$ is invariant with respect to the free action of the group

$$L^z_0 := N_{G_\mu}(H)^z/H. \tag{8.4.3}$$

We can use these remarks to construct the twisted product

$$G \times_{\left(N_{G_\mu}(H)^z/H\right)} \left(\mathbf{J}^{-1}(\mu) \cap M^z_H\right)$$

which, due to the freeness of the $\left(N_{G_\mu}(H)^z/H\right)$-action on $\mathbf{J}^{-1}(\mu) \cap M^z_H$ is endowed with a natural smooth quotient manifold structure. It will be proved below that the mapping

$$\begin{array}{rcl} i : G \times_{\left(N_{G_\mu}(H)^z/H\right)} (\mathbf{J}^{-1}(\mu) \cap M^z_H) & \longrightarrow & M \\ {[g, m]} & \longmapsto & g \cdot m \end{array}$$

is an injective immersion onto $G \cdot \left(\mathbf{J}^{-1}(\mu) \cap M^z_H\right)$ such that, for any

$$[g, m] \in G \times_{\left(N_{G_\mu}(H)^z/H\right)} \left(\mathbf{J}^{-1}(\mu) \cap M^z_H\right),$$

$$T_{[g,m]}i\left(T_{[g,m]}\left(G\times_{\left(N_{G_\mu}(H)^z/H\right)}\left(\mathbf{J}^{-1}(\mu)\cap M_H^z\right)\right)\right)=D(g\cdot m). \tag{8.4.4}$$

This implies that $G\cdot\left(\mathbf{J}^{-1}(\mu)\cap M_H^z\right)$ is an integral submanifold of the saturating distribution D when endowed with the smooth structure that makes the bijection

$$G\times_{\left(N_{G_\mu}(H)^z/H\right)}\left(\mathbf{J}^{-1}(\mu)\cap M_H^z\right)\to G\cdot\left(\mathbf{J}^{-1}(\mu)\cap M_H^z\right)$$

into a diffeomorphism.

Injectivity of i is a direct consequence of its definition. In order to prove that i is an immersion, it suffices to reproduce the proof of Proposition 3.4.8 **(ii)** taking A'_G as the invariant distribution D_F, $\mathbf{J}^{-1}(\mu)\cap M_H^z$ as $\mathcal{J}_F^{-1}(\rho)$, and $N_{G_\mu}(H)^z/H$ as G_ρ. The immersive character of i in this case follows from the relation

$$\mathfrak{g}\cdot m\cap A'_G(m)=T_m\left(\left(N_{G_\mu}(H)^z/H\right)\cdot m\right),$$

which is a consequence of the following chain of equalities:

$$\begin{aligned}\mathfrak{g}\cdot m\cap A'_G(m)&=\mathfrak{g}\cdot m\cap\ker T_m\mathbf{J}_{L^z}\quad\text{(by (5.5.7) and (5.5.6))}\\&=\mathfrak{g}\cdot m\cap\ker T_m\mathbf{J}_{L^z}\cap T_mM_H^z\\&=T_m(N(H)\cdot m)\cap\ker T_m\mathbf{J}_{L^z}\quad\text{(by (2.4.5) and (2.4.1))}\\&=T_m\left(\left(N(H)^z/H\right)\cdot m\right)\cap\ker T_m\mathbf{J}_{L^z}\quad\text{(by (2.4.9))}\\&=T_m(L_0^z\cdot m)\quad\text{(by Lemma 4.5.34)}\\&=T_m\left(\left(N_{G_\mu}(H)^z/H\right)\cdot m\right)\quad\text{(by (8.4.3)).}\end{aligned}$$

Finally, (8.4.4) is a direct consequence of the formula for i and the defining relation of the saturating distribution D.

We now show that the connected components of $G\cdot(\mathbf{J}^{-1}(\mu)\cap M_H^z)$ endowed with the smooth structure that makes the bijection $G\times_{\left(N_{G_\mu}(H)^z/H\right)}(\mathbf{J}^{-1}(\mu)\cap M_H^z)\to G\cdot\left(\mathbf{J}^{-1}(\mu)\cap M_H^z\right)$ into a diffeomorphism are given by the sets of the form $gG^0\cdot(\mathbf{J}^{-1}(\mu)\cap M_H^z)^m$, with $g\in G$ and $(\mathbf{J}^{-1}(\mu)\cap M_H^z)^m$ the connected component of $\mathbf{J}^{-1}(\mu)\cap M_H^z$ that contains a given point m. Notice first that by Proposition 3.4.6 and Corollary 5.5.18 the sets $gG^0\cdot(\mathbf{J}^{-1}(\mu)\cap M_H^z)^m$ are the maximal integral manifolds of the distribution D and therefore they are connected initial submanifolds of M such that

$$T_s\left(gG^0\cdot(\mathbf{J}^{-1}(\mu)\cap M_H^z)^m\right)=D(s), \tag{8.4.5}$$

for all $s\in gG^0\cdot(\mathbf{J}^{-1}(\mu)\cap M_H^z)^m$. Let now A be the connected component of $G\cdot\left(\mathbf{J}^{-1}(\mu)\cap M_H^z\right)$ that contains $gG^0\cdot(\mathbf{J}^{-1}(\mu)\cap M_H^z)^m$, that is,

$$gG^0\cdot(\mathbf{J}^{-1}(\mu)\cap M_H^z)^m\subset A\subset G\cdot(\mathbf{J}^{-1}(\mu)\cap M_H^z).$$

As $G\cdot(\mathbf{J}^{-1}(\mu)\cap M_H^z)$ is a manifold, it is locally connected and therefore its connected components are open and closed. In particular, since A is an open and connected subset of $G\cdot(\mathbf{J}^{-1}(\mu)\cap M_H^z)$ and the choice of smooth structure for this set guarantees through (8.4.4) that it is an integral submanifold of D, A is consequently a connected integral submanifold of D. The maximality of $gG^0 \cdot (\mathbf{J}^{-1}(\mu) \cap M_H^z)^m$ as an integral submanifold of D implies that $gG^0 \cdot (\mathbf{J}^{-1}(\mu) \cap M_H^z)^m = A$, as required. Hence, as the initial submanifolds of the form $gG^0 \cdot (\mathbf{J}^{-1}(\mu) \cap M_H^z)^m$ are the connected components of $G \cdot (\mathbf{J}^{-1}(\mu) \cap M_H^z)$, Lemma 1.1.11 **(iii)**, guarantees that $G \cdot (\mathbf{J}^{-1}(\mu) \cap M_H^z)$ is an initial submanifold of M that, by (8.4.5), has the tangent space given in the statement of the theorem.

(ii) The initial character of $G \cdot \left(\mathbf{J}^{-1}(\mu) \cap M_H^z\right)$ and its G-invariance implies by §2.3.7 that the G-action on M restricts to a smooth proper G-action on $G \cdot (\mathbf{J}^{-1}(\mu) \cap M_H^z)$. It is also clear that the isotropy subgroups corresponding to the elements in $G \cdot \left(\mathbf{J}^{-1}(\mu) \cap M_H^z\right)$ are all conjugate to H. Proposition 2.3.8 **(iv)** proves the statement.

(iii) The proof of this part follows very closely the proof of the similar statement in the regular case presented in Theorem 6.3.1. The main reason that allows us to reproduce the strategy of that proof is that the tangent space of $\mathbf{J}^{-1}(\mathcal{O}_\mu)$ in the regular situation can be expressed in terms of the same elements as the tangent space of $G \cdot (\mathbf{J}^{-1}(\mu) \cap M_H^z)$ in the singular case (compare (6.3.6) with (8.4.1)) and $\omega_{\mathcal{O}_\mu}^{(H)}$ can be written in terms of only those elements. Indeed, let $m \in G \cdot (\mathbf{J}^{-1}(\mu) \cap M_H^z)$, $[m]_{\mathcal{O}_\mu}^{(H)} = \pi_{\mathcal{O}_\mu}^{(H)}(m)$, and $v, w \in T_m \left(G \cdot (\mathbf{J}^{-1}(\mu) \cap M_H^z)\right)$. By (8.4.1) we have that

$$v = \xi_M(m) + X_f(m), \quad \text{with } \xi \in \mathfrak{g} \text{ and } f \in C^\infty(M)^G \tag{8.4.6}$$

$$w = \eta_M(m) + X_g(m), \quad \text{with } \eta \in \mathfrak{g} \text{ and } g \in C^\infty(M)^G. \tag{8.4.7}$$

We define a two-form $\omega_{\mathcal{O}_\mu}^{(H)}\left([m]_{\mathcal{O}_\mu}^{(H)}\right)$ on $T_{[m]_{\mathcal{O}_\mu}^{(H)}} M_{\mathcal{O}_\mu}^{(H)}$ by

$$\omega_{\mathcal{O}_\mu}^{(H)}\left([m]_{\mathcal{O}_\mu}^{(H)}\right)\left(T_m\pi_{\mathcal{O}_\mu}^{(H)}(v),\, T_m\pi_{\mathcal{O}_\mu}^{(H)}(w)\right) := \{f, g\}(m). \tag{8.4.8}$$

The proof of the fact that (8.4.8) is a good definition for $\omega_{\mathcal{O}_\mu}^{(H)}$ follows line by line the proof that we gave for the regular case changing $\pi_{\mathcal{O}_\mu}^{(H)}$ by $\pi_{\mathcal{O}_\mu}$, $\omega_{\mathcal{O}_\mu}^{(H)}$ by $\omega_{\mathcal{O}_\mu}$, and $[m]_{\mathcal{O}_\mu}^{(H)}$ by $[m]_{\mathcal{O}_\mu}$. In order to show that $\omega_{\mathcal{O}_\mu}^{(H)}$ is nondegenerate we write the arbitrary vectors $v, w \in T_m \left(G \cdot (\mathbf{J}^{-1}(\mu) \cap M_H^z)\right)$ as in (8.4.6) and (8.4.7) and suppose that

$$\omega_{\mathcal{O}_\mu}^{(H)}\left([m]_{\mathcal{O}_\mu}^{(H)}\right)\left(T_m\pi_{\mathcal{O}_\mu}^{(H)}(v),\, T_m\pi_{\mathcal{O}_\mu}^{(H)}(w)\right) = 0$$

for any $w \in T_m \left(G \cdot (\mathbf{J}^{-1}(\mu) \cap M_H^z)\right)$. By the expression (8.4.8) this amounts to having that the function $f \in C^\infty(M)^G$ is such that $\{f, g\}(m) = 0$ for any $g \in C^\infty(M)^G$ which coincides with the condition that we encountered in (8.1.9). As we saw in (8.1.10) this implies that $X_f(m) \in \mathfrak{g}_\mu \cdot m$ and hence that $T_m\pi_{\mathcal{O}_\mu}^{(H)}(v) = 0$, as required. The proof that $\omega_{\mathcal{O}_\mu}^{(H)}$ is determined by (8.4.2) also follows line by line the proof of the regular case.

The remaining points in the statement of the theorem are a diagram chasing exercise that uses Noether's Theorem and the law of conservation of isotropy, according to the scheme presented in the proof of the analogous statements in Theorem 6.1.1. ■

8.4.3 Theorem (Structure Theorem for the singular orbit strata) *Let (M, ω) be a connected symplectic manifold acted canonically and properly upon by a Lie group G. Suppose that this action has an associated standard momentum map $\mathbf{J} : M \to \mathfrak{g}^*$ with non-equivariance one-cocycle $\sigma : G \to \mathfrak{g}^*$. Let $\mu \in \mathfrak{g}^*$ be a value of $\mathbf{J}$, G_μ the isotropy subgroup of μ with respect to the affine action $\Theta : G \times \mathfrak{g}^* \to \mathfrak{g}^*$ determined by σ, and $H \subset G$ an isotropy subgroup of the G-action on M. Let M_H^z be the connected component of the H-isotropy type manifold M_H that contains a given element $z \in M$ such that $\mathbf{J}(z) = \mu$. Then the following hold:*

(i) *The canonical projection $\pi_{\mathcal{O}_\mu}^{(H)} : G(\mathbf{J}^{-1}(\mu) \cap M_H^z) \longrightarrow M_{\mathcal{O}_\mu}^{(H)} = G(\mathbf{J}^{-1}(\mu) \cap M_H^z)/G$ defines a smooth fiber bundle with fiber G/H and structure group $N(H)^z/H$. We recall that $N(H)^z$ is the closed subgroup of $N(H)$ that leaves M_H^z invariant.*

(ii) *Consider the free, proper, and canonical action of $L^z := N(H)^z/H$ on M_H^z and let $\mathbf{J}_{L^z} : M_H^z \to \mathfrak{l}^*$ be the associated momentum map given by $\mathbf{J}_{L^z}(m) = \Lambda(\mathbf{J}|_{M_H^z}(m) - \mu)$, for any $m \in M_H^z$ (see Theorem 4.6.1). Then the regular orbit reduced space $(M_H^z)_{\mathcal{O}_0}$ in the affine orbit corresponding to $0 \in \mathfrak{l}^*$ is given by*

$$(M_H^z)_{\mathcal{O}_0} = \mathbf{J}_{L^z}^{-1}(\mathcal{O}_0)/L^z = \mathbf{J}^{-1}(N(H)^z \cdot \mu) \cap M_H^z/(N(H)^z/H) \qquad (8.4.9)$$

(iii) *The projection $\pi_{\mathcal{O}_0} : \mathbf{J}_{L^z}^{-1}(\mathcal{O}_0) \to (M_H^z)_{\mathcal{O}_0}$ defines a smooth principal L^z-bundle. Regarding G/H as a right $(N(H)^z/H)$-space and $\mathbf{J}^{-1}(N(H)^z \cdot \mu) \cap M_H^z$ as a left $(N(H)^z/H)$-space, consider the bundle with fiber G/H and structure group G associated to $\pi_{\mathcal{O}_0}$, that is,*

$$G/H \times_{N(H)^z/H} \left(\mathbf{J}^{-1}(N(H)^z \cdot \mu) \cap M_H^z\right) \\ \longrightarrow \left[\mathbf{J}^{-1}(N(H)^z \cdot \mu) \cap M_H^z/(N(H)^z\right]/H).$$

This bundle is G-symplectomorphic to $\pi_{\mathcal{O}_\mu}^{(H)} : G(\mathbf{J}^{-1}(\mu) \cap M_H^z) \longrightarrow M_{\mathcal{O}_\mu}^{(H)}$, that is, $G/H \times_{N(H)^z/H} (\mathbf{J}^{-1}(N(H)^z \cdot \mu) \cap M_H^z)$ is G-diffeomorphic to $G(\mathbf{J}^{-1}(\mu) \cap M_H^z)$ and $(M_H^z)_{\mathcal{O}_0} = \mathbf{J}_{L^z}^{-1}(\mathcal{O}_0)/L^z = \mathbf{J}^{-1}(N(H)^z \cdot \mu) \cap M_H^z/(N(H)^z/H)$ is symplectomorphic to $M_{\mathcal{O}_\mu}^{(H)}$. We say that $(M_H^z)_{\mathcal{O}_0}$ is a ***regularization*** *of the orbit reduced space $M_{\mathcal{O}_\mu}^{(H)}$.*

Proof. The arguments follow very closely those for the similar statement in the context of point reduction (Theorem 8.2.1). We will focus on the differences with respect to that proof. We start by establishing (8.4.9) which amounts to showing that $\mathbf{J}_{L^z}^{-1}(\mathcal{O}_0) = \mathbf{J}^{-1}(N(H)^z \cdot \mu) \cap M_H^z$. Indeed, using the expression of the affine action induced by the non-equivariance cocycle in (4.6.3) we have that a point $m \in \mathbf{J}_{L^z}^{-1}(\mathcal{O}_0)$ if and only if there exists an element $l = nH \in N(H)^z/H$ such that $\mathbf{J}_{L^z}(m) = l \cdot 0$. Using the expression of $\mathbf{J}_{L^z}$ this amounts to $\Lambda(\mathbf{J}|_{M_H^z}(m) - \mu) = \Lambda(\sigma(n) + n \cdot \mu - \mu)$ which in turn is equivalent to $\mathbf{J}|_{M_H^z}(m) = n \cdot \mu + \sigma(n)$ and hence $\mathbf{J}_{L^z}^{-1}(\mathcal{O}_0) = \mathbf{J}^{-1}(N(H)^z \cdot \mu) \cap M_H^z$, as required.

Regarding the point **(iii)**, in this case the bundle isomorphism is implemented by the following two maps. First consider the smooth map

$$\begin{aligned} G/H \times_{N(H)^z/H} \left(\mathbf{J}^{-1}(N(H)^z \cdot \mu) \cap M_H^z\right) &\longrightarrow G \cdot (\mathbf{J}^{-1}(\mu) \cap M_H^z) \\ [gH, n \cdot m] &\longmapsto gn \cdot m, \end{aligned} \tag{8.4.10}$$

with $n \in N(H)^z$, $m \in \mathbf{J}^{-1}(\mu) \cap M_H^z$, and where we used the fact that $\mathbf{J}^{-1}(N(H)^z \cdot \mu) \cap M_H^z = N(H)^z \cdot (\mathbf{J}^{-1}(\mu) \cap M_H^z)$. This map is clearly G-invariant and well defined. Corollary 3.5.2 and Proposition 2.4.5 allow one to prove that it is also an injective immersion. Surjectivity is obvious and an argument that mimics the one used in the proof of part **(iii)** of Theorem 8.2.1 shows that this map is a local diffeomorphism and hence a diffeomorphism. Second, the G-equivariance of (8.4.10) allows us to drop this map to a diffeomorphism F between

$$\begin{aligned} (M_H^z)_{\mathcal{O}_0} &= \mathbf{J}^{-1}(N(H)^z \cdot \mu) \cap M_H^z / \left(N(H)^z/H\right) \\ &\simeq \left(G/H \times_{N(H)^z/H} \left(\mathbf{J}^{-1}(N(H)^z \cdot \mu) \cap M_H^z\right)\right)/G \end{aligned} \tag{8.4.11}$$

and $G \cdot (\mathbf{J}^{-1}(\mu) \cap M_H^z)/G = M_{\mathcal{O}_\mu}^{(H)}$. The expression (8.4.11) follows from (8.4.9). To be more specific, the diffeomorphism F is the unique map that makes the following diagram commutative

$$\begin{array}{ccc} \mathbf{J}^{-1}(N(H)^z \cdot \mu) \cap M_H^z & \xrightarrow{\ i\ } & G \cdot (\mathbf{J}^{-1}(\mu) \cap M_H^z) \\ {\scriptstyle \pi_{\mathcal{O}_0}}\big\downarrow & & \big\downarrow{\scriptstyle \pi_{\mathcal{O}_\mu}^{(H)}} \\ \mathbf{J}^{-1}(N(H)^z \cdot \mu) \cap M_H^z/(N(H)^z/H) & \xrightarrow{\ F\ } & G \cdot (\mathbf{J}^{-1}(\mu) \cap M_H^z)/G = M_{\mathcal{O}_\mu}^{(H)}, \end{array} \tag{8.4.12}$$

where i is the inclusion.

We now show that F is a symplectomorphism. Following the diagram (8.4.12) and the expression (8.4.2) we obtain that

$$\begin{aligned} \pi_{\mathcal{O}_0}^* F^* \omega_{\mathcal{O}_\mu}^{(H)} &= (F \circ \pi_{\mathcal{O}_0})^* \omega_{\mathcal{O}_\mu}^{(H)} = (\pi_{\mathcal{O}_\mu}^{(H)} \circ i)^* \omega_{\mathcal{O}_\mu}^{(H)} = i^* \pi_{\mathcal{O}_\mu}^{(H)*} \omega_{\mathcal{O}_\mu}^{(H)} \\ &= i^* \left(i_{\mathcal{O}_\mu}^{(H)*} \omega - \mathbf{J}_{\mathcal{O}_\mu}^{(H)*} \omega_{\mathcal{O}_\mu}^+\right). \end{aligned} \tag{8.4.13}$$

On the other hand and according to (6.3.1), the form $\omega_{\mathcal{O}_0}$ is defined by the relation

$$\pi_{\mathcal{O}_0}^* \omega_{\mathcal{O}_0} = i_{\mathcal{O}_0}^* \omega_{M_H^z} - (\mathbf{J}_{L^z})_{\mathcal{O}_0}^* \omega_{\mathcal{O}_0}^+. \tag{8.4.14}$$

Let now $i_H : M_H^z \hookrightarrow M$ be the inclusion. It is clear that

$$i_{\mathcal{O}_\mu}^{(H)} \circ i = i_H \circ i_{\mathcal{O}_0}$$

and hence

$$i^* i_{\mathcal{O}_\mu}^{(H)*} \omega = i_{\mathcal{O}_0}^* i_H^* \omega = i_{\mathcal{O}_0}^* \omega_{M_H^z}. \tag{8.4.15}$$

Let now $m \in \mathbf{J}_{L^z}^{-1}(\mathcal{O}_0) = \mathbf{J}^{-1}(N(H)^z \cdot \mu) \cap M_H^z$ and consider $v, w \in T_m\big(\mathbf{J}_{L^z}^{-1}(\mathcal{O}_0)\big)$. By (6.3.6) there exist elements $\xi, \eta \in \mathfrak{l} := \mathrm{Lie}(L^z)$ and functions $f, g \in C^\infty(M_H^z)^{L^z}$ such that $v = \xi_M(m) + X_f(m)$ and $w = \eta_M(m) + X_g(m)$. Hence if $\xi_1, \eta_1 \in \mathrm{Lie}(N(H)^z)$ are such that $\xi = \frac{d}{dt}\big|_{t=0} \exp t\xi_1 H$ and $\eta = \frac{d}{dt}\big|_{t=0} \exp t\eta_1 H$, τ is the non-equivariance cocycle of $\mathbf{J}_{L^z}$, and Σ_τ is the associated infinitesimal two-cocycle, we have

$$
\begin{aligned}
(\mathbf{J}_{L^z})^*_{\mathcal{O}_0}\omega^+_{\mathcal{O}_0}(m)(v, w) &= \omega^+_{\mathcal{O}_0}(\mathbf{J}_{L^z}(m))\left(T_m\mathbf{J}_{L^z}(\xi_M(m)), T_m\mathbf{J}_{L^z}(\eta_M(m))\right)\\
&= \omega^+_{\mathcal{O}_0}(\mathbf{J}_{L^z}(m))\big(-\mathrm{ad}^*_\xi \mathbf{J}_{L^z}(m) + \Sigma_\tau(\xi, \cdot), -\mathrm{ad}^*_\eta \mathbf{J}_{L^z}(m) + \Sigma_\tau(\eta, \cdot)\big)\\
&= \langle \mathbf{J}_{L^z}(m), [\xi, \eta]\rangle - \Sigma_\tau(\xi, \eta)\\
&= \langle \Lambda(\mathbf{J}(m) - \mu), [\xi, \eta]\rangle - \Sigma_\sigma(\xi_1, \eta_1) + \langle \mu, [\xi_1, \eta_1]\rangle\\
&= \langle \mathbf{J}(m) - \mu, [\xi_1, \eta_1]\rangle - \Sigma_\sigma(\xi_1, \eta_1) + \langle \mu, [\xi_1, \eta_1]\rangle\\
&= \langle \mathbf{J}(m), [\xi_1, \eta_1]\rangle - \Sigma_\sigma(\xi_1, \eta_1)\\
&= (\mathbf{J}^{(H)*}_{\mathcal{O}_\mu}\omega^+_{\mathcal{O}_\mu})(m)((\xi_1)_M(m), (\eta_1)_M(m)).
\end{aligned} \tag{8.4.16}
$$

In this chain of equalities we used the expressions (4.6.1), (4.6.2), and (4.6.4). Now, by Propositions 2.5.9 and 4.2.9 **(i)** there exist functions $F, G \in C^\infty(M)^G$ such that

$$
T_m(i^{(H)}_{\mathcal{O}_\mu} \circ i)\big(X_f(m)\big) = X_F(m) \quad \text{and} \quad T_m(i^{(H)}_{\mathcal{O}_\mu} \circ i)\big(X_g(m)\big) = X_G(m).
$$

Consequently, we have that

$$
\begin{aligned}
&(i^*\mathbf{J}^{(H)*}_{\mathcal{O}_\mu}\omega^+_{\mathcal{O}_\mu})(m)(v, w)\\
&= \omega^+_{\mathcal{O}_\mu}(\mathbf{J}(m))(T_m(\mathbf{J} \circ i^{(H)}_{\mathcal{O}_\mu} \circ i)(X_f(m) + \xi_M(m)),\\
&\qquad\qquad T_m(\mathbf{J} \circ i^{(H)}_{\mathcal{O}_\mu} \circ i)(X_g(m) + \eta_M(m)))\\
&= \omega^+_{\mathcal{O}_\mu}(\mathbf{J}(m))(T_m\mathbf{J}(X_F(m)) + T_m(\mathbf{J} \circ i_{\mathcal{O}_\mu} \circ i)(\xi_M(m)),\\
&\qquad\qquad T_m\mathbf{J}(X_G(m)) + T_m(\mathbf{J} \circ i_{\mathcal{O}_\mu} \circ i)(\eta_M(m)))\\
&= \left(\mathbf{J}^{(H)*}_{\mathcal{O}_\mu}\omega^+_{\mathcal{O}_\mu}\right)(m)\,((\xi_1)_M(m), (\eta_1)_M(m))\\
&= \left((\mathbf{J}_{L^z})^*_{\mathcal{O}_0}\,\omega^+_{\mathcal{O}_0}\right)(m)(v, w) \quad \text{(by (8.4.16))}.
\end{aligned}
$$

Since in this chain of equalities the point $m \in \mathbf{J}_{L^z}^{-1}(\mathcal{O}_0) = \mathbf{J}^{-1}(N(H)^z \cdot \mu) \cap M_H^z$ and the vectors $v, w \in T_m\left(\mathbf{J}_{L^z}^{-1}(\mathcal{O}_0)\right)$ are arbitrary we have that

$$
i^*\mathbf{J}^{(H)*}_{\mathcal{O}_\mu}\omega^+_{\mathcal{O}_\mu} = (\mathbf{J}_{L^z})^*_{\mathcal{O}_0}\,\omega^+_{\mathcal{O}_0}. \tag{8.4.17}
$$

Now, the expressions (8.4.15) and (8.4.17) guarantee that (8.4.13) and (8.4.14) coincide and consequently

$$
\pi^*_{\mathcal{O}_0}F^*\omega^{(H)}_{\mathcal{O}_\mu} = \pi^*_{\mathcal{O}_0}\omega_{\mathcal{O}_0}.
$$

Since $\pi_{\mathcal{O}_0}$ is a surjective submersion we necessarily have that

$$
F^*\omega^{(H)}_{\mathcal{O}_\mu} = \omega_{\mathcal{O}_0},
$$

as required. ■

8.4.4 Theorem (Orbit reduction stratification theorem and the singular reduction diagram) *Let (M, ω) be a connected symplectic manifold acted canonically and properly upon by a Lie group G. Suppose that this action has an associated standard momentum map $\mathbf{J} : M \to \mathfrak{g}^*$ with non-equivariance one-cocycle $\sigma : G \to \mathfrak{g}^*$. Let $\mu \in \mathfrak{g}^*$ be a value of $\mathbf{J}$ and $\mathcal{O}_\mu := G \cdot \mu$ the orbit of this element with respect to the affine action $\Theta : G \times \mathfrak{g}^* \to \mathfrak{g}^*$ determined by σ. Let $l_\mu : \mathbf{J}^{-1}(\mu) \hookrightarrow \mathbf{J}^{-1}(\mathcal{O}_\mu)$ be the inclusion and $L_\mu : \mathbf{J}^{-1}(\mu)/G_\mu \to \mathbf{J}^{-1}(\mathcal{O}_\mu)/G$ the map defined by the commutative diagram*

$$\begin{array}{ccc}
\mathbf{J}^{-1}(\mu) & \xrightarrow{\ l_\mu\ } & \mathbf{J}^{-1}(\mathcal{O}_\mu) \\
{\scriptstyle \pi_\mu}\big\downarrow & & \big\downarrow{\scriptstyle \pi_{\mathcal{O}_\mu}} \\
\mathbf{J}^{-1}(\mu)/G_\mu & \xrightarrow{\ L_\mu\ } & \mathbf{J}^{-1}(\mathcal{O}_\mu)/G.
\end{array}$$

Consider $\mathbf{J}^{-1}(\mu)/G_\mu$ as a symplectically stratified topological space with the stratification introduced in Theorem 8.3.2. *Then:*

(i) *The submanifolds in Theorem* 8.4.2 *induce a symplectic Whitney stratification on $\mathbf{J}^{-1}(\mathcal{O}_\mu)/G$.*

(ii) *The map L_μ is a homeomorphism of symplectically Whitney stratified spaces.*

Proof. The map L_μ is a homeomorphism because, by Proposition 8.4.1, the quotient $\mathbf{J}^{-1}(\mathcal{O}_\mu)/G$ is homeomorphic to $(G \times_{G_\mu} \mathbf{J}^{-1}(\mu))/G$ which, by Proposition 2.3.19, is homeomorphic to $\mathbf{J}^{-1}(\mu)/G_\mu$. The composition of these two homeomorphisms equals L_μ^{-1} and hence L_μ is a homeomorphism.

A simple diagram-chasing exercise shows that the restriction $L_\mu^{(H)}$ of the map L_μ to the symplectic stratum $M_\mu^{(H)} = \mathbf{J}^{-1}(\mu) \cap G_\mu M_H^z / G_\mu$ is a symplectomorphism between $M_\mu^{(H)}$ and $M_{\mathcal{O}_\mu}^{(H)}$. Since this statement is true for any stratum, **(i)** and **(ii)** follow. ■

The Structure Theorem, Sjamaar's Principle, and the singular reduction diagram are illustrated in the following array of maps:

$$
\begin{array}{ccc}
\mathbf{J}^{-1}(\mu) & \xrightarrow{\ l_\mu\ } & \mathbf{J}^{-1}(\mathcal{O}_\mu) \\
\Big\downarrow \pi_\mu & & \Big\downarrow \pi_{\mathcal{O}_\mu} \\
\mathbf{J}^{-1}(\mu)/G_\mu & \xrightarrow{\ L_\mu\ } & \mathbf{J}^{-1}(\mathcal{O}_\mu)/G \\
\Big\uparrow & & \Big\uparrow \\
\mathbf{J}^{-1}(\mu) \cap G_\mu M_H^z/G_\mu & \xrightarrow{\ L_\mu^{(H)}\ } & G \cdot (\mathbf{J}^{-1}(\mu) \cap M_H^z)/G \\
\Big\uparrow f_\mu^{(H)} & & \Big\uparrow f_{\mathcal{O}_\mu}^{(H)} \\
\mathbf{J}_{L^z}^{-1}(0)/L_0^z & \xrightarrow{\ L_0\ } & \mathbf{J}_{L^z}^{-1}(\mathcal{O}_0)/L^z \\
\| & & \| \\
\mathbf{J}^{-1}(\mu) \cap M_H^z/\left(N_{G_\mu}(H)^z/H\right) & & \mathbf{J}^{-1}(N(H)^z \cdot \mu) \cap M_H^z/\left(N(H)^z/H\right)
\end{array}
$$

In this diagram, the symplectomorphisms $f_\mu^{(H)}$ and $f_{\mathcal{O}_\mu}^{(H)}$ are the explicit implementation of Sjamaar's Principle (see theorems 8.2.1 and 8.4.3). We recall that L_0 and $L_\mu^{(H)}$ are also symplectomorphisms and that L_μ is a homeomorphism of symplectically Whitney stratified spaces.

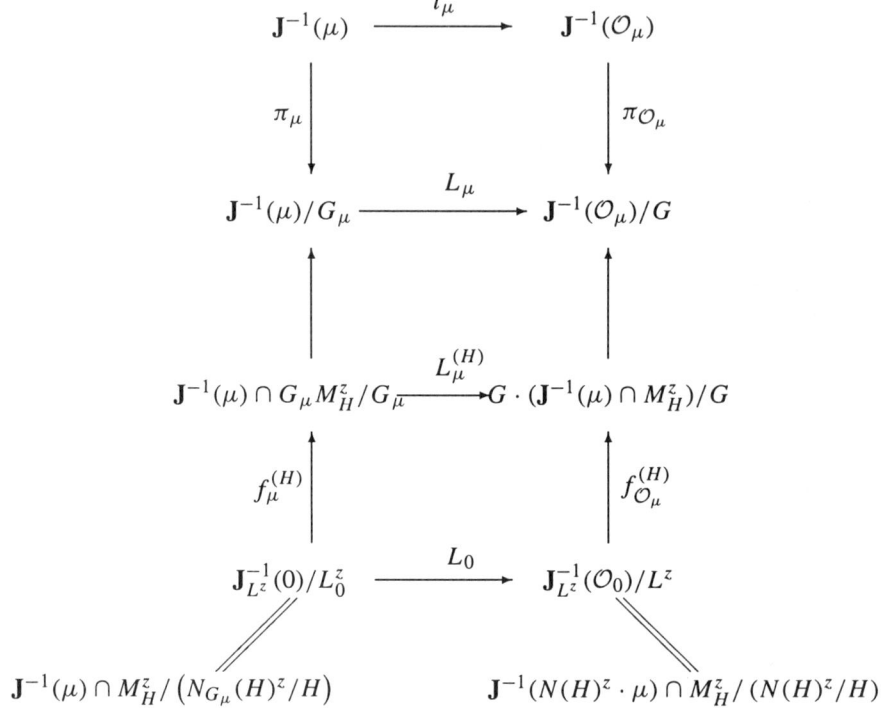

Chapter 9

Optimal Reduction

In this chapter we use the optimal momentum map introduced in Chapter 5 to carry out symplectic reduction. The reduced spaces that we will obtain generalize the symplectic strata presented in Chapter 8 to the categories in which the optimal momentum map is well defined. More specifically, the optimal momentum map approach allows the construction of symplectic point and orbit reduced spaces purely within the Poisson category under hypotheses that do not necessarily imply the existence of a standard momentum map.

Another subject at the core of this chapter is the notion of ***polar reduction***. This procedure shows that the orbits in the momentum space of the optimal momentum map (we call them polar reduced spaces) admit a presymplectic structure that generalizes the Kostant–Kirillov–Souriau symplectic structure (4.5.25) of the coadjoint orbits in the dual of a Lie algebra. Using this presymplectic structure, the optimal orbit reduced spaces are symplectic with a form that satisfies a relation identical to the classical one presented in (6.3.1). In the Poisson case we will provide a sufficient condition for the polar reduced spaces to be symplectic and in the symplectic case we will see that the polar reduced spaces are symplectic if and only if a certain relation between the tangent space to the orbit and its symplectic orthogonal with the tangent space to the isotropy type submanifolds is satisfied. In general, the presymplectic polar reduced spaces are foliated by symplectic submanifolds that are obtained through a generalization to the optimal context of the Sjamaar Principle.

The results presented in this chapter are contained in ORTEGA (2002a, 2003c).

9.1 Optimal point reduction

We reproduce the program that we carried out when performing reduction with the standard momentum map. We start with the point reduction approach.

9.1.1 Theorem (Optimal point reduction by Poisson actions) *Let* $(M, \{\cdot, \cdot\})$ *be a smooth Poisson manifold and* G *a Lie group acting canonically and properly on* M. *Let* $\mathcal{J} : M \to M/A'_G$ *be the optimal momentum map associated to this action. Then, for any* $\rho \in M/A'_G$ *whose isotropy subgroup* G_ρ *acts properly on* $\mathcal{J}^{-1}(\rho)$, *we have that*

(i) *The orbit space* $M_\rho := \mathcal{J}^{-1}(\rho)/G_\rho$ *is a smooth symplectic regular quotient manifold with symplectic form* ω_ρ *defined by*

$$\pi_\rho^*\omega_\rho(m)(X_f(m), X_h(m)) = \{f, h\}(m), \tag{9.1.1}$$

for any $m \in \mathcal{J}^{-1}(\rho)$ *and any* $f, h \in C^\infty(M)^G$. *We will refer to the pair* (M_ρ, ω_ρ) *as the* **(optimal) point reduced space** *of* $(M, \{\cdot,\cdot\})$ *at* ρ.

(ii) *Let* $h \in C^\infty(M)^G$. *The flow* F_t *of* X_h *leaves* $\mathcal{J}^{-1}(\rho)$ *invariant, commutes with the* G*-action, and therefore induces a flow* F_t^ρ *on* M_ρ *uniquely determined by the relation* $\pi_\rho \circ F_t \circ i_\rho = F_t^\rho \circ \pi_\rho$, *where* $i_\rho : \mathcal{J}^{-1}(\rho) \hookrightarrow M$ *is the inclusion.*

(iii) *The flow* F_t^ρ *in* (M_ρ, ω_ρ) *is Hamiltonian with the Hamiltonian function* $h_\rho \in C^\infty(M_\rho)$ *given by the equality* $h_\rho \circ \pi_\rho = h \circ i_\rho$.

(iv) *Let* $k \in C^\infty(M)^G$ *be another* G*-invariant function on* M *and* $\{\cdot,\cdot\}_\rho$ *the Poisson bracket associated to the symplectic form* ω_ρ *on* M_ρ. *Then* $\{h, k\}_\rho = \{h_\rho, k_\rho\}_\rho$.

Proof. **(i)** Since by hypothesis the G_ρ-action on $\mathcal{J}^{-1}(\rho)$ is proper and by Proposition 3.4.4 (iii) it has fixed isotropies, the quotient $\mathcal{J}^{-1}(\rho)/G_\rho$ is by Proposition 2.3.8 a regular quotient manifold and hence the projection $\pi_\rho : \mathcal{J}^{-1}(\rho) \to \mathcal{J}^{-1}(\rho)/G_\rho$ is a smooth surjective submersion.

We start the proof of the symplectic nature of M_ρ by showing that (9.1.1) is a good definition for the form ω_ρ on the quotient M_ρ. Let $m, m' \in \mathcal{J}^{-1}(\rho)$ be such that $\pi_\rho(m) = \pi_\rho(m')$, and let $v, w \in T_m\mathcal{J}^{-1}(\rho)$, $v', w' \in T_{m'}\mathcal{J}^{-1}(\rho)$ be such that $T_m\pi_\rho(v) = T_{m'}\pi_\rho(v')$, $T_m\pi_\rho(w) = T_{m'}\pi_\rho(w')$. Let $f, f', g, g' \in C^\infty(M)^G$ be such that $v = X_f(m)$, $v' = X_{f'}(m')$, $w = X_g(m)$, $w' = X_{g'}(m')$. The condition $\pi_\rho(m) = \pi_\rho(m')$ implies the existence of an element $k \in G_\rho$ such that $m' = \Phi_k^\rho(m)$. We also have that $T_m\pi_\rho = T_{m'}\pi_\rho \circ T_m\Phi_k^\rho$. Analogously, because of the equalities $T_m\pi_\rho(v) = T_{m'}\pi_\rho(v')$, $T_m\pi_\rho(w) = T_{m'}\pi_\rho(w')$, there exist G-invariant functions $h^1, h^2 \in C^\infty(M)^G$ and elements $\xi^1, \xi^2 \in \mathfrak{g}_\rho$ such that

$$\begin{aligned}
X_{f'}(m') - T_m\Phi_k^\rho\left(X_f(m)\right) &= \xi^1_{\mathcal{J}^{-1}(\rho)}(m') = X_{h^1}(m'),\\
X_{g'}(m') - T_m\Phi_k^\rho\left(X_g(m)\right) &= \xi^2_{\mathcal{J}^{-1}(\rho)}(m') = X_{h^2}(m'),
\end{aligned}$$

or, analogously,

$$\begin{aligned}
X_{f'}(m') &= X_{h^1+f\circ\Phi_{k^{-1}}}(m') = X_{h^1+f}(m'),\\
X_{g'}(m') &= X_{h^2+g\circ\Phi_{k^{-1}}}(m') = X_{h^2+g}(m').
\end{aligned}$$

Hence, we can write

$$\begin{aligned}
\omega_\rho(\pi_\rho(m'))(v', w') &= \{f', g'\}(m') = \{h^1 + f, h^2 + g\}(m')\\
&= \{h^1 + f, h^2 + g\}(m)\\
&= \{f, g\}(m) + \{f, h^2\}(m) + \{h^1, g\}(m) + \{h^1, h^2\}(m)\\
&= \{f, g\}(m) + \mathbf{d}f(m)\left(\xi^2_{\mathcal{J}^{-1}(\rho)}(m)\right) - \mathbf{d}(g + h^2)(m)\left(\xi^1_{\mathcal{J}^{-1}(\rho)}(m)\right)\\
&= \{f, g\}(m) = \omega_\rho(\pi_\rho(m))(v, w).
\end{aligned}$$

Consequently, ω_ρ is a well-defined two-form on the quotient M_ρ. Since π_ρ is a smooth surjective submersion, the form ω_ρ is clearly smooth. The Jacobi identity for the bracket $\{\cdot, \cdot\}$ on M implies that ω_ρ is closed. These two features of the form ω_ρ can also be immediately read from Expression (9.1.3), whose equivalence with (9.1.1) is straightforward.

It only remains to be shown that ω_ρ is nondegenerate. We start our argument with some notation and remarks. Let $H \subset G$ be the isotropy subgroup of all the elements in $\mathcal{J}^{-1}(\rho)$ with respect to the smooth G_ρ-action on this manifold. Recall that by Proposition 3.4.4 this isotropy subgroup coincides with an isotropy H of the G-action on M. Since by hypothesis the G-action on M is proper, the subgroup $H \subset G_\rho$ is necessarily compact. Moreover, the Slice Theorem guarantees that for any point $m \in \mathcal{J}^{-1}(\rho)$, there is a G-invariant neighborhood U of m in M that is G-equivariantly diffeomorphic to the twisted product $G \times_H V_r$, where V_r is a ball of radius r around the origin in some vector space V on which H acts linearly.

Let $m \in \mathcal{J}^{-1}(\rho)$. Suppose that the vector $X_f(m)$, $f \in C^\infty(M)^G$, is such that

$$\pi_\rho^*\omega_\rho(m)(X_f(m), X_h(m)) = \{f, h\}(m) = 0, \ \text{ for all } h \in C^\infty(M)^G. \qquad (9.1.2)$$

In order to prove that ω_ρ is nondegenerate we have to show that $X_f(m) \in T_m(G_\rho \cdot m)$. We will do so by using the local coordinates around the point m provided by the Slice Theorem. First of all, as f is G-invariant, it follows that $X_f(m) \in T_m M_H^{l_m}$. Since in the slice model $M_H^{l_m} \simeq N(H) \times_H V_r^H$, we have that $X_f(m) = T_{(e,0)}\pi(\zeta, v)$, where $\pi : G \times V_r \to G \times_H V_r$ is the natural projection, $\zeta \in \operatorname{Lie}(N(H))$, and $v \in V^H$. We recall that V^H denotes the H-fixed point subspace in V.

We now rephrase the condition in (9.1.2) in this slice representation. Indeed, the fact that

$$\pi_\rho^*\omega_\rho(m)(X_f(m), X_h(m)) = \{f, h\}(m) = -\mathbf{d}h(m) \cdot X_f(m) = 0,$$

for all $h \in C^\infty(M)^G$ amounts to saying that $\mathbf{d}g(0)(v) = 0$ for all functions $g \in C^\infty(V_r)^H$. In other words, $v \in \left(\{\mathbf{d}g(0) \mid g \in C^\infty(V_r)^H\}\right)^\circ$. By Theorem 2.5.10 this implies that $v \in ((V^*)^H)^\circ$. Consequently, $v \in V^H \cap ((V^*)^H)^\circ$. We now show that this intersection is trivial and that therefore $v = 0$.

We start by recalling (see Lemma 2.5.11) that the restriction to $(V^*)^H$ of the dual map associated to the inclusion $i_{V^H} : V^H \hookrightarrow V$ is an H-equivariant isomorphism from $(V^*)^H$ to $(V^H)^*$. Now, as $v \in V^H \cap ((V^*)^H)^\circ$ we have that $\langle \alpha, v \rangle_V = 0$ for every $\alpha \in ((V)^*)^H$. The symbol $\langle \cdot, \cdot \rangle_V$ denotes the natural pairing of V with its dual. We can rewrite this condition as

$$0 = \langle \alpha, v \rangle_V = \langle \alpha, i_{V^H}(v) \rangle_V = \langle i_{V^H}^*(\alpha), v \rangle_{V^H}.$$

As the restriction $i_{V^H}^*|_{(V^*)^H}$ is an isomorphism, the previous identity is equivalent to $\langle \beta, v \rangle_{V^H} = 0$ for all $\beta \in (V^H)^*$. Consequently, $v = 0$, as required.

We conclude our argument by noting that since $X_f(m) = T_{(e,0)}\pi(\zeta, 0)$, we have that $X_f(m) \in T_m(G \cdot m) \cap A'_G(m) = T_m(G_\rho \cdot m)$ (see Proposition 3.4.4) which proves the nondegeneracy of ω_ρ.

The proof of the remaining points is a simple commutative diagram exercise that is left to the reader. ■

9.1.2 The very definition of the polar distribution implies that for any $\rho \in M/A'_G$ there is a unique symplectic leaf $\mathcal{L}_\rho$ of the Poisson manifold $(M, \{\cdot, \cdot\})$ such that $\mathcal{J}^{-1}(\rho) \subset \mathcal{L}_\rho$. Let $i_{\mathcal{L}_\rho} : \mathcal{J}^{-1}(\rho) \hookrightarrow \mathcal{L}_\rho$ be the inclusion of $\mathcal{J}^{-1}(\rho)$ into the symplectic leaf $(\mathcal{L}_\rho, \omega_{\mathcal{L}_\rho})$ of $(M, \{\cdot, \cdot\})$ in which it is sitting. As $\mathcal{L}_\rho$ is an initial submanifold of M, the injection $i_{\mathcal{L}_\rho}$ is a smooth map. The form ω_ρ can also be written in terms of the symplectic structure of the leaf $\mathcal{L}_\rho$ as

$$\pi_\rho^* \omega_\rho = i_{\mathcal{L}_\rho}^* \omega_{\mathcal{L}_\rho}. \tag{9.1.3}$$

The reader should be warned that this statement does *not* imply that the previous theorem could be obtained by just performing symplectic optimal reduction in the symplectic leaves of the Poisson manifold, basically because those leaves are not G-manifolds. As we already emphasized several times, the fact that the G-action on M is Poisson does not imply that it preserves its symplectic leaves.

In view of this remark we can obtain the standard Symplectic Foliation Theorem 4.1.28 of Poisson manifolds as a straightforward corollary of Theorem 9.1.1 by taking the group $G = \{e\}$. In that case the distribution A'_G coincides with the characteristic distribution of the Poisson manifold and the level sets of the optimal momentum map, and thereby the symplectic quotients M_ρ, are exactly the symplectic leaves. We explicitly point this out in our next statement.

9.1.3 Corollary (Symplectic Foliation Theorem) *Let $(M, \{\cdot, \cdot\})$ be a smooth Poisson manifold. Then M is the disjoint union of the maximal integral leaves of the integrable generalized distribution E given by*

$$E(m) := \{X_f(m) \mid f \in C^\infty(M)\}, \quad m \in M.$$

These leaves are symplectic initial submanifolds of M.

9.1.4 On the properness of the G_ρ-action. The only extra hypothesis in the statement of Theorem 9.1.1 with respect to the hypotheses used in the reduction theorems, which use standard momentum maps in the symplectic context, is the properness of the G_ρ-action on $\mathcal{J}^{-1}(\rho)$. This is a real hypothesis in the sense that the properness of the G_ρ-action is not automatically inherited from the properness of the G-action on M, as it used to be the case in the presence of a standard momentum map. From this reduction point of view, we can think of the presence of a standard momentum map as an extra integrability feature of the polar distribution that makes its integrable leaves imbedded (and not just initial) submanifolds of M and their isotropy subgroups automatically closed.

In order to illustrate this comment we now present a situation in which the G_ρ-action on $\mathcal{J}^{-1}(\rho)$ is not proper, while the G-action on M satisfies this condition. Let $M := \mathbb{T}^2 \times \mathbb{T}^2$ be the product of two two-tori whose elements will be denoted by the four-tuples $(e^{i\theta_1}, e^{i\theta_2}, e^{i\psi_1}, e^{i\psi_2})$. We endow M with the symplectic structure ω defined by $\omega := \mathbf{d}\theta_1 \wedge \mathbf{d}\theta_2 + \sqrt{2}\, \mathbf{d}\psi_1 \wedge \mathbf{d}\psi_2$. We now consider the canonical two-torus action given by $(e^{i\phi_1}, e^{i\phi_2}) \cdot (e^{i\theta_1}, e^{i\theta_2}, e^{i\psi_1}, e^{i\psi_2}) := (e^{i(\theta_1+\phi_1)}, e^{i(\theta_2+\phi_2)}, e^{i(\psi_1+\phi_1)}, e^{i(\psi_2+\phi_2)})$. First, notice that since the two-torus is compact this action is necessarily proper. Moreover, as $\mathbb{T}^2$ acts freely, the corresponding orbit space $M/A_{\mathbb{T}^2}$ is a smooth manifold such that the projection $\pi_{A_{\mathbb{T}^2}} : M \to M/A_{\mathbb{T}^2}$ is a surjective submersion. The polar distribution $A'_{\mathbb{T}^2}$ does not have that property. Indeed, $C^\infty(M)^{\mathbb{T}^2}$ comprises all the functions f of the form $f \equiv f(e^{i(\theta_1-\psi_1)}, e^{i(\theta_2-\psi_2)})$. An inspection of the Hamiltonian flows associated to such functions readily shows that the leaves of $A'_{\mathbb{T}^2}$, that is, the level sets of the optimal momentum map $\mathcal{J}$, are the products of two leaves of an irrational foliation in a two-torus. Moreover, it can be checked that for any $\rho \in M/A'_{\mathbb{T}^2}$, the isotropy subgroup $\mathbb{T}^2_\rho$ is the product of two discrete subgroups of S^1, each of which fill densely the circle. We can use this density property to show that the $\mathbb{T}_\rho$-action on $\mathcal{J}^{-1}(\rho)$ is not proper. Let $\{(e^{i\tau_n}, e^{i\sigma_n})\}_{n\in\mathbb{N}}$ be a sequence of elements in $\mathbb{T}^2_\rho$ that converges to (e, e) in $\mathbb{T}^2$. The sequences $\{\tau_n\}_{n\in\mathbb{N}}, \{\sigma_n\}_{n\in\mathbb{N}} \subset \mathbb{R}$ are chosen such that $\tau_n, \sigma_n \to 0$ and $0 < \tau_{n+1} < \tau_n$, $0 < \sigma_{n+1} < \sigma_n$, for any $n \in \mathbb{N}$. Then for any sequence $\{z_n\}_{n\in\mathbb{N}} \subset \mathcal{J}^{-1}(\rho)$ such that $z_n \to z \in \mathcal{J}^{-1}(\rho)$ in $\mathcal{J}^{-1}(\rho)$ we have that $(e^{i\tau_n}, e^{i\sigma_n}) \cdot z_n \to z$ in $\mathcal{J}^{-1}(\rho)$. However, since $\mathbb{T}^2_\rho$ is endowed with the discrete topology and $\{\tau_n\}_{n\in\mathbb{N}}, \{\sigma_n\}_{n\in\mathbb{N}}$ are chosen to be strictly monotone, the sequence $\{(e^{i\tau_n}, e^{i\sigma_n})\}_{n\in\mathbb{N}}$ has no convergent subsequences, which implies that G_ρ does not act properly on $\mathcal{J}^{-1}(\rho)$.

9.1.5 Example A simplified version of the previous example provides a situation in which the hypotheses of Theorem 9.1.1 are satisfied while all the standard reduction theorems that we presented in Chapters 6 and 8 do not hold.

Let $M := \mathbb{T}^2 \times \mathbb{T}^2$ with the symplectic structure of the previous example. We now consider the canonical circle action given by $e^{i\phi} \cdot (e^{i\theta_1}, e^{i\theta_2}, e^{i\psi_1}, e^{i\psi_2}) := (e^{i(\theta_1+\phi)}, e^{i\theta_2}, e^{i(\psi_1+\phi)}, e^{i\psi_2})$. In this case, $C^\infty(M)^{S^1}$ consists of all the functions f of the form $f(e^{i\theta_1}, e^{i\theta_2}, e^{i\psi_1}, e^{i\psi_2}) \equiv g(e^{i\theta_2}, e^{i\psi_2}, e^{i(\theta_1-\psi_1)})$, for some function $g \in C^\infty(\mathbb{T}^3)$. An inspection of the Hamiltonian flows associated to such functions readily shows that the leaves of A'_{S^1}, that is, the level sets $\mathcal{J}^{-1}(\rho)$ of the optimal momentum map $\mathcal{J}$ are the product of a two-torus with a leaf of an irrational foliation (Kronecker submanifold) of another two-torus. Obviously this is not compatible with the existence of a ($\mathbb{R}^2$ or $\mathbb{T}^2$-valued) momentum map. Nevertheless, the isotropies S^1_ρ coincide with the circle S^1, whose compactness guarantees that its action on $\mathcal{J}^{-1}(\rho)$ is proper. Theorem 9.1.1 automatically guarantees that the quotients of the form

$$M_\rho := \mathcal{J}^{-1}(\rho)/S^1_\rho \simeq \left(S^1 \times_{S^1} S^1\right) \times \{\text{Kronecker submanifold of } \mathbb{T}^2\}.$$

are symplectic.

9.1.6 A Poisson example. We now use Theorem 9.1.1 to carry out the symplectic reduction of a Poisson symmetric manifold that was already used in Chapter 5 to illustrate the construction of the optimal momentum map. Let $(\mathbb{R}^3, \{\cdot, \cdot\})$ be the Poisson manifold formed by the Euclidean three-dimensional space $\mathbb{R}^3$ together with the Poisson structure induced by the Poisson tensor B that in Euclidean coordinates takes the form:

$$B = \begin{pmatrix} 0 & 1 & 0 \\ -1 & 0 & 1 \\ 0 & -1 & 0 \end{pmatrix}.$$

Consider the action of the additive group $(\mathbb{R}, +)$ on $\mathbb{R}^3$ given by $\lambda \cdot (x, y, z) := (x + \lambda, y, z)$, for any $\lambda \in \mathbb{R}$ and any $(x, y, z) \in \mathbb{R}^3$. This action is proper and does not have an associated standard momentum map. Nevertheless, it is a Poisson action and it has an optimal momentum map $\mathcal{J}$ associated to it given by the expression

$$\begin{array}{rccc} \mathcal{J}: & \mathbb{R}^3 & \longrightarrow & \mathbb{R} \\ & (x, y, z) & \longmapsto & x + z. \end{array}$$

The level sets $\mathcal{J}^{-1}(c)$ of the optimal momentum map are the planes given by the equation $x + z = c$ and the isotropy subgroups $\mathbb{R}_c$ are always trivial. Therefore, by Theorem 9.1.1, the planes of the form $x + z = c$ are symplectic submanifolds of the Poisson manifold $(\mathbb{R}^3, \{\cdot, \cdot\})$. Actually, it is easy to verify that these planes constitute its symplectic leaves.

9.1.7 What if the action admits a standard momentum map? Suppose that (M, ω) is a symplectic manifold and that there is a canonical and proper G-action on it with an associated standard momentum map $\mathbf{J} : M \to \mathfrak{g}^*$. A natural question to ask is how the optimal symplectic reduced spaces introduced in Theorem 9.1.1 compare with the usual point reduced spaces obtained by using standard momentum maps. The answer comes from Corollary 5.5.18, which says that for any $m \in M$ such that $\mathbf{J}(m) = \mu$ and $\mathcal{J}(m) = \rho$, we have that

$$\mathcal{J}^{-1}(\rho) = (\mathbf{J}^{-1}(\mu) \cap M^m_{G_m})^m,$$

where $(\mathbf{J}^{-1}(\mu) \cap M^m_{G_m})^m$ denotes the connected component of $\mathbf{J}^{-1}(\mu) \cap M^m_{G_m}$, which contains the point m. Additionally, $G_\rho = N_{G_\mu}(G_m)^{c(m)}$, where $N_{G_\mu}(G_m)^{c(m)}$ is the closed subgroup of $N_{G_\mu}(G_m)$ that leaves the connected component $(\mathbf{J}^{-1}(\mu) \cap M^m_{G_m})^m$ of $\mathbf{J}^{-1}(\mu) \cap M^m_{G_m}$ invariant. Consequently,

$$\mathcal{J}^{-1}(\rho)/G_\rho = (\mathbf{J}^{-1}(\mu) \cap M^m_{G_m})^m / N_{G_\mu}(G_m)^{c(m)}.$$

A particularly interesting situation takes place when $\mathbf{J}^{-1}(\mu) \cap M^m_{G_m}$ is connected. In that case $\mathcal{J}^{-1}(\rho) = \mathbf{J}^{-1}(\mu) \cap M^m_{G_m}$ and $N_{G_\mu}(G_m)^{c(m)} = N_{G_\mu}(G_m)^m := G_\mu \cap N(G_m)^m$ and hence, by Theorem 8.2.1,

$$\mathcal{J}^{-1}(\rho)/G_\rho = \mathbf{J}^{-1}(\mu) \cap M^m_{G_m} / N_{G_\mu}(G_m)^m = \mathbf{J}^{-1}_{L^m}(0)/L^m_0$$

which is naturally symplectomorphic to the symplectic stratum $M_\mu^{(G_m)}$ via Sjamaar's Principle. The conclusion of this computation is that in the symplectic case and in the presence of a standard momentum map, the optimal reduced spaces are naturally symplectomorphic, up to connected components, to the symplectic point reduced strata.

9.1.8 The symplectic case and Sjamaar's Principle. When M is a symplectic manifold with form ω, the optimal point reduction by the G-action on M produces the same result as the reduction of the isotropy type submanifold in which our momentum level set is sitting by the relevant remaining group action on it. This idea translates in the optimal momentum map context the Sjamaar Principle that we discussed in the framework of reduction with a standard momentum map. There is one difference: in the standard momentum map situation there is a natural symplectomorphism between the reduced spaces using the original manifold and the isotropy type submanifolds. In the optimal momentum map context this symplectomorphism is an identity.

Let $\mathcal{J} : M \to M/A'_G$ be the optimal momentum map corresponding to the proper G-action on (M, ω). Fix $\rho \in M/A'_G$ a momentum value of $\mathcal{J}$ and let $H \subset G$ be the unique G-isotropy subgroup such that $\mathcal{J}^{-1}(\rho) \subset M_H$ and $G_\rho \subset H$. Recall that the normalizer $N(H)$ of H in G acts naturally on M_H. This action induces a free action of the quotient group $L := N(H)/H$ on M_H. Let M_H^ρ be the unique connected component of M_H containing $\mathcal{J}^{-1}(\rho)$ and let L^ρ be the closed subgroup of L that leaves it invariant. Obviously, L^ρ can be written as $L^\rho = N(H)^\rho/H$ for some closed subgroup $N(H)^\rho$ of $N(H)$.

The subset M_H^ρ is a symplectic embedded submanifold of M on which the group L^ρ acts freely and canonically. We will denote by $\mathcal{J}_{L^\rho} : M_H^\rho \to M_H^\rho/A'_{L^\rho}$ the associated optimal momentum map. The following proposition explains the interest of this construction.

9.1.9 Proposition (Optimal Sjamaar's Principle) *Let G be a Lie group that acts properly and canonically on the symplectic manifold (M, ω), with associated optimal momentum map $\mathcal{J} : M \to M/A'_G$. Let $\rho \in M/A'_G$ and let $H \subset G$ be the unique G-isotropy subgroup such that $\mathcal{J}^{-1}(\rho) \subset M_H$ and $G_\rho \subset H$. Then, with the notation introduced in the previous sections, we have the following statements:*

(i) *Let $i_H^\rho : M_H^\rho \hookrightarrow M$ be the inclusion. For any $z \in M_H^\rho$ we have the equality $T_z i_H^\rho \left(A'_{L^\rho}(z)\right) = A'_G(z)$.*

(ii) *Let $z \in \mathcal{J}^{-1}(\rho)$ be such that $\mathcal{J}_{L^\rho}(z) =: \sigma \in M_H^\rho/A'_{L^\rho}$. Then $\mathcal{J}^{-1}(\rho) = \mathcal{J}_{L^\rho}^{-1}(\sigma)$.*

(iii) *$L_\sigma^\rho = G_\rho/H$.*

(iv) *$(M_H^\rho)_\sigma = \mathcal{J}_{L^\rho}^{-1}(\sigma)/L_\sigma^\rho = \mathcal{J}^{-1}(\rho)/(G_\rho/H) = \mathcal{J}^{-1}(\rho)/G_\rho = M_\rho$. Moreover, if G_ρ acts properly on $\mathcal{J}^{-1}(\rho)$ this equality is true when we consider M_ρ and $(M_H^\rho)_\sigma$ as symplectic spaces, that is,*

$$(M_\rho, \omega_\rho) = ((M_H^\rho)_\sigma, (\omega|_{M_H^\rho})_\sigma).$$

Proof. It is a straightforward consequence of Proposition 2.5.9. ■

9.1.10 Definition *Suppose that the hypotheses of the previous proposition hold. The symplectic reduced space $((M_H^\rho)_\sigma, (\omega|_{M_H^\rho})_\sigma)$ is called the **regularization** of the point reduced space (M_ρ, ω_ρ).*

9.2 Optimal orbit reduction

9.2.1 Theorem (Optimal orbit reduction by Poisson actions) *Let $(M, \{\cdot, \cdot\})$ be a smooth Poisson manifold and G a Lie group acting canonically and properly on M. Let $\mathcal{J} : M \to M/A'_G$ be the optimal momentum map associated to this action and $\rho \in M/A'_G$. Suppose that G_ρ acts properly on $\mathcal{J}^{-1}(\rho)$. If we denote $\mathcal{O}_\rho := G \cdot \rho$, then:*

(i) *There is a unique smooth structure on $\mathcal{J}^{-1}(\mathcal{O}_\rho)$ that makes it into an initial submanifold of M.*

(ii) *The G-action on $\mathcal{J}^{-1}(\mathcal{O}_\rho)$ by restriction of the G-action on M is smooth and proper and all its isotropy subgroups are conjugate to a given compact isotropy subgroup of the G-action on M.*

(iii) *The quotient $M_{\mathcal{O}_\rho} := \mathcal{J}^{-1}(\mathcal{O}_\rho)/G$ admits a unique smooth structure that makes the projection $\pi_{\mathcal{O}_\rho} : \mathcal{J}^{-1}(\mathcal{O}_\rho) \to \mathcal{J}^{-1}(\mathcal{O}_\rho)/G$ a surjective submersion.*

(iv) *The quotient $M_{\mathcal{O}_\rho} := \mathcal{J}^{-1}(\mathcal{O}_\rho)/G$ admits a unique symplectic structure $\omega_{\mathcal{O}_\rho}$ that makes it symplectomorphic to the point reduced space M_ρ. We will refer to the pair $(M_{\mathcal{O}_\rho}, \omega_{\mathcal{O}_\rho})$ as the* **(optimal) orbit reduced space** *of $(M, \{\cdot, \cdot\})$ at $\mathcal{O}_\rho$.*

(v) *Let $h \in C^\infty(M)^G$. The flow F_t of X_h leaves $\mathcal{J}^{-1}(\mathcal{O}_\rho)$ invariant, commutes with the G-action, and therefore induces a flow $F_t^{\mathcal{O}_\rho}$ on $M_{\mathcal{O}_\rho}$ uniquely determined by the relation $\pi_{\mathcal{O}_\rho} \circ F_t \circ i_{\mathcal{O}_\rho} = F_t^{\mathcal{O}_\rho} \circ \pi_{\mathcal{O}_\rho}$, where $i_{\mathcal{O}_\rho} : \mathcal{J}^{-1}(\mathcal{O}_\rho) \hookrightarrow M$ is the inclusion.*

(vi) *The flow $F_t^{\mathcal{O}_\rho}$ in $(M_{\mathcal{O}_\rho}, \omega_{\mathcal{O}_\rho})$ is Hamiltonian with the Hamiltonian function $h_{\mathcal{O}_\rho} \in C^\infty(M_{\mathcal{O}_\rho})$ given by the equality $h_{\mathcal{O}_\rho} \circ \pi_{\mathcal{O}_\rho} = h \circ i_{\mathcal{O}_\rho}$.*

(vii) *Let $k \in C^\infty(M)^G$ be another G-invariant function on M and $\{\cdot, \cdot\}_{\mathcal{O}_\rho}$ the Poisson bracket associated to the symplectic form $\omega_{\mathcal{O}_\rho}$ on $M_{\mathcal{O}_\rho}$. Then $\{h, k\}_{\mathcal{O}_\rho} = \{h_{\mathcal{O}_\rho}, k_{\mathcal{O}_\rho}\}_{\mathcal{O}_\rho}$.*

Proof. **(i)** It is a corollary of Theorem 3.4.10. **(ii)** By Proposition 3.4.4 there exists a unique isotropy subgroup H of the G-action on M such that $\mathcal{J}^{-1}(\rho) \subset M_H$. Thus $\mathcal{J}^{-1}(\mathcal{O}_\rho) = G \cdot \mathcal{J}^{-1}(\rho) \subset G \cdot M_H = M_{(H)}$. **(iii)** Since $\mathcal{J}^{-1}(\mathcal{O}_\rho)$ is an initial G-invariant submanifold of M, the G-action on $\mathcal{J}^{-1}(\mathcal{O}_\rho)$ is smooth and by §2.3.7 it is also proper. The claim is then a consequence of Proposition 2.3.8. **(iv)** Recall that by Theorem 3.4.10, the smooth structure on $\mathcal{J}^{-1}(\mathcal{O}_\rho)$, which makes it into an initial submanifold of M, coincides with the smooth structure that makes it diffeomorphic to $G \times_{G_\rho} \mathcal{J}^{-1}(\rho)$. Consequently, the quotient manifold $\mathcal{J}^{-1}(\mathcal{O}_\rho)/G$ is naturally diffeomorphic to the symplectic point reduced space. Indeed,

$$\mathcal{J}^{-1}(\mathcal{O}_\rho)/G \simeq G \times_{G_\rho} \mathcal{J}^{-1}(\rho)/G \simeq \mathcal{J}^{-1}(\rho)/G_\rho.$$

This diffeomorphism can be explicitly implemented as follows. Let $f_\rho : \mathcal{J}^{-1}(\rho) \to \mathcal{J}^{-1}(\mathcal{O}_\rho)$ be the inclusion. Since the inclusion $\mathcal{J}^{-1}(\rho) \hookrightarrow M$ is smooth and $\mathcal{J}^{-1}(\mathcal{O}_\rho)$ is initial, the map f_ρ is smooth. Also, since f_ρ is (G_ρ, G)-equivariant, it drops to a unique smooth map $F_\rho : \mathcal{J}^{-1}(\rho)/G_\rho \to \mathcal{J}^{-1}(\mathcal{O}_\rho)/G$ that makes the following diagram

$$\begin{array}{ccc} \mathcal{J}^{-1}(\rho) & \xrightarrow{f_\rho} & \mathcal{J}^{-1}(\mathcal{O}_\rho) \\ {\scriptstyle \pi_\rho}\downarrow & & \downarrow{\scriptstyle \pi_{\mathcal{O}_\rho}} \\ \mathcal{J}^{-1}(\rho)/G_\rho & \xrightarrow{F_\rho} & \mathcal{J}^{-1}(\mathcal{O}_\rho)/G \end{array} \tag{9.2.1}$$

commutative. F_ρ is a smooth bijection. In order to show that its inverse is also smooth we will think of $\mathcal{J}^{-1}(\mathcal{O}_\rho)$ as $G \times_{G_\rho} \mathcal{J}^{-1}(\rho)$. First of all notice that the projection $G \times \mathcal{J}^{-1}(\rho) \to \mathcal{J}^{-1}(\rho)$ is G_ρ-(anti)equivariant and therefore induces a smooth map $G \times_{G_\rho} \mathcal{J}^{-1}(\rho) \to \mathcal{J}^{-1}(\rho)/G_\rho$ given by $[g, z] \mapsto [z]$, $[g, z] \in G \times_{G_\rho} \mathcal{J}^{-1}(\rho)$. This map is G-invariant and therefore drops to another smooth mapping $G \times_{G_\rho} \mathcal{J}^{-1}(\rho)/G \to \mathcal{J}^{-1}(\rho)/G_\rho$ that coincides with F_ρ^{-1}, the inverse of F_ρ, which is consequently a diffeomorphism.

The orbit reduced space $\mathcal{J}^{-1}(\mathcal{O}_\rho)/G$ can be therefore trivially endowed with a symplectic structure $\omega_{\mathcal{O}_\rho}$ by defining $\omega_{\mathcal{O}_\rho} := (F_\rho^{-1})^*\omega_\rho$.

That the proof of the remaining points is implied by the commutativity of a diagram is left to the reader. ■

9.2.2 The regularization of the orbit reduced spaces. We conclude this section with a description of the orbit version of the regularized reduced spaces introduced in Definition 9.1.10 for the symplectic case.

9.2.3 Proposition *Let G be a Lie group that acts properly and canonically on the symplectic manifold (M, ω), with associated optimal momentum map $\mathcal{J} : M \to M/A'_G$. Let $\rho \in M/A'_G$, $\mathcal{O}_\rho := G \cdot \rho$, and let $H \subset G$ be the unique G-isotropy subgroup such that $\mathcal{J}^{-1}(\rho) \subset M_H$ and $G_\rho \subset H$. Assume that G_ρ acts properly on $\mathcal{J}^{-1}(\rho)$. Let M_H^ρ be the connected component of M_H containing $\mathcal{J}^{-1}(\rho)$, $N(H)^\rho$ the closed subgroup of $N(H)$ that leaves it invariant, and $L^\rho := N(H)^\rho/H$.*

(i) *Let $z \in \mathcal{J}^{-1}(\rho)$ be such that $\mathcal{J}_{L^\rho}(z) =: \sigma \in M_H^\rho/A'_{L^\rho}$ and $\mathcal{N}_\rho := N(H)^\rho \cdot \rho \subset M/A'_G$. The set $\mathcal{J}_{L^\rho}^{-1}(L^\rho \cdot \sigma) = \mathcal{J}^{-1}(\mathcal{N}_\rho)$ is an embedded submanifold of $\mathcal{J}^{-1}(\mathcal{O}_\rho)$.*

(ii) *The initial submanifold $\mathcal{J}^{-1}(\mathcal{O}_\rho)$ can be written as a disjoint union of its embedded submanifolds:*

$$\mathcal{J}^{-1}(\mathcal{O}_\rho) = \coprod_{[g] \in G/N(H)^\rho} \mathcal{J}^{-1}(\mathcal{N}_{g \cdot \rho}). \tag{9.2.2}$$

(iii) *The symplectic quotient $(\mathcal{J}_{L^\rho}^{-1}(L^\rho \cdot \sigma)/L^\rho, (\omega|_{M_H^\rho})_{L^\rho \cdot \sigma})$ is naturally symplectomorphic to the orbit reduced space $(\mathcal{J}^{-1}(\mathcal{O}_\rho)/G, \omega_{\mathcal{O}_\rho})$. We will say that $(\mathcal{J}_{L^\rho}^{-1}(L^\rho \cdot \sigma)/L^\rho, (\omega|_{M_H^\rho})_{L^\rho \cdot \sigma})$ is **an orbit regularization** of $(\mathcal{J}^{-1}(\mathcal{O}_\rho)/G, \omega_{\mathcal{O}_\rho})$.*

Proof. **(i)** If we use the statements in Proposition 9.1.9 it is easy to see that $\mathcal{J}_{L^\rho}^{-1}(L^\rho \cdot \sigma) = L^\rho \cdot \mathcal{J}_{L^\rho}^{-1}(\sigma) = N(H)^\rho \cdot \mathcal{J}^{-1}(\rho) = \mathcal{J}^{-1}(\mathcal{N}_\rho)$, with $\mathcal{N}_\rho := N(H)^\rho \cdot \rho \subset M/A'_G$. The set $\mathcal{J}_{L^\rho}^{-1}(L^\rho \cdot \sigma) = \mathcal{J}^{-1}(\mathcal{N}_\rho)$ is an embedded submanifold of $\mathcal{J}^{-1}(\mathcal{O}_\rho)$ since $N(H)^\rho \times_{G_\rho} \mathcal{J}^{-1}(\rho) \simeq \mathcal{J}^{-1}(\mathcal{N}_\rho)$ is embedded in $G \times_{G_\rho} \mathcal{J}^{-1}(\rho) \simeq \mathcal{J}^{-1}(\mathcal{O}_\rho)$.
(ii) We now prove (9.2.2). This equality is a straightforward consequence of the fact that for any $g \in G$,

$$\begin{aligned} M_{gHg^{-1}}^{g\rho} &= \Phi_g(M_H^\rho), \\ N(gHg^{-1})^{g\rho} &= gN(H)^\rho g^{-1}, \\ \mathcal{J}^{-1}(\mathcal{N}_{g \cdot \rho}) &= gN(H)^\rho \mathcal{J}^{-1}(\rho). \end{aligned}$$

The last relation implies that if $g, g' \in G$ are such that $[g] = [g'] \in G/N(H)^\rho$, then $\mathcal{J}^{-1}(\mathcal{N}_{g\cdot\rho}) = \mathcal{J}^{-1}(\mathcal{N}_{g'\cdot\rho})$. We now show that the union in (9.2.2) is indeed disjoint. Let $gn \cdot z \in \mathcal{J}^{-1}(\mathcal{N}_{g\cdot\rho})$ and $g'n' \cdot z' \in \mathcal{J}^{-1}(\mathcal{N}_{g'\cdot\rho})$ be such that $gn \cdot z = g'n' \cdot z'$, with $g, g' \in G$, $n, n' \in N(H)^\rho$, and $z, z' \in \mathcal{J}^{-1}(\rho)$. Since $gn \cdot z = g'n' \cdot z'$, we necessarily have that $G_{gn\cdot z} = G_{g'n'\cdot z'}$ which implies that $gHg^{-1} = g'H(g')^{-1}$, and hence $g^{-1}g' \in N(H)$. We now recall that M_H^ρ is by Theorem 3.5.1 the accessible set going through z or z' of the integrable generalized distribution B'_G defined by

$$B'_G := \operatorname{span}\{X \in \mathfrak{X}(U)^G \mid U \text{ open } G\text{-invariant set in } M\},$$

where the symbol $\mathfrak{X}(U)^G$ denotes the set of G-equivariant vector fields defined on U. Let $\mathcal{B}'_G$ be the pseudogroup of transformations of M consisting of the G-equivariant flows of the vector fields that span B'_G. Now, as the points $n \cdot z, n' \cdot z' \in M_H^\rho$, there exists $\mathcal{F}_T \in \mathcal{B}'_G$ such that $n' \cdot z' = \mathcal{F}_T(n \cdot z)$, hence $(g')^{-1}gn \cdot z = \mathcal{F}_T(n \cdot z)$. Moreover, as any element in M_H^ρ can be written as $\mathcal{G}_T(n \cdot z)$ with $\mathcal{G}_T \in \mathcal{B}'_G$, we have that

$$(g')^{-1}g \cdot \mathcal{G}_T(n \cdot z) = \mathcal{G}_T((g')^{-1}gn \cdot z) = \mathcal{G}_T(\mathcal{F}_T(n \cdot z)) \in M_H^\rho,$$

which implies that $(g')^{-1}g \in N(H)^\rho$ and therefore $[g] = [g'] \in G/N(H)^\rho$, as required.

(iii) It is straightforward. ■

9.3 Polar reduction

In Theorem 9.2.1 we showed that the optimal orbit reduced spaces $\mathcal{J}^{-1}(\mathcal{O}_\rho)/G$ are symplectic manifolds. Nevertheless, one of the most interesting parts of the theory of orbit reduction in the context of standard momentum maps is so far missing, namely, formula (6.3.1) that provides the explicit expression of the symplectic form of the orbit reduced space in terms of the symplectic structure of the original manifold and of the Kostant–Kirillov–Souriau symplectic form of the affine orbit in the dual of the Lie algebra.

The goal of this section is the construction of a replacement for the affine orbits in the optimal momentum map context that allows us to state a formula similar to (6.3.1). These "generalized affine orbits" will be referred to as ***polar reduced spaces***. The justification for this denomination, which will be more apparent in Chapter 11 in the context of our study of dual pairs, is based on the fact that these reduced spaces are constructed using the invariance properties of $\mathcal{J}^{-1}(\mathcal{O}_\rho)$ with respect to the polar distribution. More specifically, the initial submanifold $\mathcal{J}^{-1}(\mathcal{O}_\rho)$ is invariant with respect to the action of both the group of diffeomorphisms A_G as well as of its polar A'_G. The quotient $\mathcal{J}^{-1}(\mathcal{O}_\rho)/A_G$ gives us the orbit reduced space and the leaf space $\mathcal{J}^{-1}(\mathcal{O}_\rho)/A'_G$ is the polar reduced space that serves as a replacement of the affine orbits in $\mathfrak{g}^*$.

In this section we use a stronger hypothesis on G_ρ with respect to the one that we used in the previous sections, namely, we will assume that G_ρ is closed in G which, as we pointed out in Proposition 3.4.4, implies that the G_ρ action on $\mathcal{J}^{-1}(\rho)$ is proper.

9.3.1 Proposition *Let $(M, \{\cdot,\cdot\})$ be a smooth Poisson manifold and G a Lie group acting canonically and properly on M. Let $\mathcal{J} : M \to M/A'_G$ be the optimal momentum map associated to this action and $\rho \in M/A'_G$. Suppose that G_ρ is closed in G.*

Then the polar distribution A'_G restricts to a smooth integrable regular distribution on $\mathcal{J}^{-1}(\mathcal{O}_\rho)$, which we will also denote by A'_G. The leaf space $M'_{\mathcal{O}_\rho} := \mathcal{J}^{-1}(\mathcal{O}_\rho)/A'_G$ admits a unique smooth structure that makes it into a regular quotient manifold and diffeomorphic to the homogeneous manifold G/G_ρ. With this smooth structure the projection $\mathcal{J}_{\mathcal{O}_\rho} : \mathcal{J}^{-1}(\mathcal{O}_\rho) \to \mathcal{J}^{-1}(\mathcal{O}_\rho)/A'_G$ is a smooth surjective submersion. We will refer to $M'_{\mathcal{O}_\rho}$ as the ***polar reduced space****.*

Proof. Let $m \in \mathcal{J}^{-1}(\mathcal{O}_\rho)$. By Theorem 3.4.10 the initial submanifold $\mathcal{J}^{-1}(\mathcal{O}_\rho)$ is an integral submanifold of the saturating distribution of A'_G, and hence for any $m \in \mathcal{J}^{-1}(\mathcal{O}_\rho)$ we have that $T_m\mathcal{J}^{-1}(\mathcal{O}_\rho) = D(m) = \mathfrak{g}\cdot m + A'_G(m)$, which implies that the restriction of A'_G to $\mathcal{J}^{-1}(\mathcal{O}_\rho)$ is tangent to it. Consequently, as $\mathcal{J}^{-1}(\mathcal{O}_\rho)$ is an immersed submanifold of M, there exists for each Hamiltonian vector field $X_f \in \mathfrak{X}(M)$, $f \in C^\infty(M)^G$, a vector field $X'_f \in \mathfrak{X}(\mathcal{J}^{-1}(\mathcal{O}_\rho))$ such that $Ti_{\mathcal{O}_\rho} \circ X'_f = X_f \circ i_{\mathcal{O}_\rho}$, with $i_{\mathcal{O}_\rho} : \mathcal{J}^{-1}(\mathcal{O}_\rho) \hookrightarrow M$ the injection (see §1.2.4). The restriction $A'_G|_{\mathcal{J}^{-1}(\mathcal{O}_\rho)}$ of A'_G to $\mathcal{J}^{-1}(\mathcal{O}_\rho)$ is generated by the vector fields of the form X'_f and it is therefore smooth. It is also integrable since for any point $m = g\cdot z \in \mathcal{J}^{-1}(\mathcal{O}_\rho)$, $z \in \mathcal{J}^{-1}(\rho)$, the embedded submanifold $\mathcal{J}^{-1}(g\cdot\rho)$ of $\mathcal{J}^{-1}(\mathcal{O}_\rho)$ is the maximal integral submanifold of $A'_G|_{\mathcal{J}^{-1}(\mathcal{O}_\rho)}$. This is so because the flows F_t and F'_t of X_f and X'_f, respectively, satisfy the identity $i_{\mathcal{O}_\rho} \circ F'_t = F_t \circ i_{\mathcal{O}_\rho}$. It is then clear that $A'_G|_{\mathcal{J}^{-1}(\mathcal{O}_\rho)}$ has constant rank since $\dim A'_G|_{\mathcal{J}^{-1}(\mathcal{O}_\rho)} = \dim \mathcal{J}^{-1}(\rho)$. This all shows that the leaf space $\mathcal{J}^{-1}(\mathcal{O}_\rho)/A'_G$ is well defined.

In order to show that the leaf space $\mathcal{J}^{-1}(\mathcal{O}_\rho)/A'_G$ is a regular quotient manifold we first note that

$$\mathcal{J}^{-1}(\mathcal{O}_\rho)/A'_G \simeq (G \times_{G_\rho} \mathcal{J}^{-1}(\rho))/A'_G$$

is in bijection with the quotient G/G_ρ that, by the hypothesis on the closedness of G_ρ, is a smooth homogeneous manifold. Take on $M'_{\mathcal{O}_\rho} := \mathcal{J}^{-1}(\mathcal{O}_\rho)/A'_G$ the smooth structure that makes the bijection with G/G_ρ a diffeomorphism. It turns out that this smooth structure is the unique one that makes $M'_{\mathcal{O}_\rho}$ into a regular quotient manifold since it can be readily verified that the map

$$\begin{array}{rccc} \mathcal{J}_{\mathcal{O}_\rho} : \mathcal{J}^{-1}(\mathcal{O}_\rho) \simeq G \times_{G_\rho} \mathcal{J}^{-1}(\rho) & \longrightarrow & \mathcal{J}^{-1}(\mathcal{O}_\rho)/A'_G \simeq G/G_\rho \\ [g, z] & \longmapsto & gG_\rho \end{array}$$

is a surjective submersion. ■

9.3.2 Regularization of the polar reduced spaces. We now introduce the regularized polar reduced subspaces of $M'_{\mathcal{O}_\rho}$, available when M is symplectic. We assume the ideas and notation introduced in Proposition 9.2.3. Let $(\mathcal{J}_{L^\rho}^{-1}(L^\rho\cdot\sigma)/L^\rho, (\omega|_{M_H^\rho})_{L^\rho\cdot\sigma})$ be an orbit regularization of $(\mathcal{J}^{-1}(\mathcal{O}_\rho)/G, \omega_{\mathcal{O}_\rho})$. A straightforward application of Proposition 9.1.9 implies that the reduced space polar to $(\mathcal{J}_{L^\rho}^{-1}(L^\rho\cdot\sigma)/L^\rho, (\omega|_{M_H^\rho})_{L^\rho\cdot\sigma})$ equals

$$\mathcal{J}_{L^\rho}^{-1}(L^\rho\cdot\sigma)/A'_{L^\rho} = \mathcal{J}^{-1}(\mathcal{N}_\rho)/A'_G$$

which is naturally diffeomorphic to $N(H)^\rho/G_\rho$. We will say that $\mathcal{J}^{-1}(\mathcal{N}_\rho)/A'_G$ is ***a regularized polar reduced subspace*** of $M'_{\mathcal{O}_\rho}$. We will write $M'_{\mathcal{N}_\rho} := \mathcal{J}^{-1}(\mathcal{N}_\rho)/A'_G$ and denote by $\mathcal{J}_{\mathcal{N}_\rho} : \mathcal{J}^{-1}(\mathcal{N}_\rho) \to \mathcal{J}^{-1}(\mathcal{N}_\rho)/A'_G$ the canonical projection. Note that the spaces $M'_{\mathcal{N}_\rho}$ are embedded submanifolds of $M'_{\mathcal{O}_\rho}$. Finally, the decomposition (9.2.2) implies that the polar reduced space can be written as the following disjoint union of regularized polar reduced subspaces:

$$M'_{\mathcal{O}_\rho} = \mathcal{J}^{-1}(\mathcal{O}_\rho)/A'_G = \coprod_{[g]\in G/N(H)^\rho} \mathcal{J}^{-1}(\mathcal{N}_{g\cdot\rho})/A'_G = \coprod_{[g]\in G/N(H)^\rho} M'_{\mathcal{N}_{g\cdot\rho}}. \tag{9.3.1}$$

Equivalently, we have that

$$G/G_\rho = \coprod_{[g]\in G/N(H)^\rho} gN(H)^\rho/G_\rho, \tag{9.3.2}$$

where the quotient $gN(H)^\rho/G_\rho$ denotes the orbit space of the free and proper action of G_ρ on $gN(H)^\rho$ by $h \cdot gn := gnh$, $h \in G_\rho$, $n \in N(H)^\rho$.

9.3.3 The W-spanning condition. Before we state our next result we need some terminology. We will denote by $C^\infty\left(\mathcal{J}^{-1}(\mathcal{O}_\rho)/A'_G\right)$ the set of smooth real valued functions on $M'_{\mathcal{O}_\rho}$ with the smooth structure introduced in Proposition 9.3.1. Recall now that, as we pointed out in §1.1.19, there is a notion of presheaf C^∞_{M/A'_G} of smooth functions on the quotient M/A'_G whose global sections are given by the set $C^\infty_{M/A'_G}(M/A'_G) := \{f \text{ function on } M/A'_G \mid f \circ \mathcal{J} \in C^\infty(M)^{A'_G}\}$. Analogously, for each open A'_G-invariant subset U of M we have $C^\infty_{M/A'_G}(U/A'_G) :=$ function on $U/A'_G \mid f \circ \mathcal{J}|_U \in C^\infty(U)^{A'_G}\}$. Consider now the set of functions defined by

$$\begin{aligned} W^\infty\left(\mathcal{J}^{-1}(\mathcal{O}_\rho)/A'_G\right) := \{ & f \text{ real function on } M'_{\mathcal{O}_\rho} \mid \\ & f = F|_{M'_{\mathcal{O}_\rho}}, \text{ with } F \in C^\infty_{M/A'_G}(U/A'_G), M'_{\mathcal{O}_\rho} \subset U/A'_G\}. \end{aligned}$$

In the previous expression U/A'_G is an open subset of M/A'_G. The definitions and the fact that $\mathcal{J}_{\mathcal{O}_\rho}$ is a submersion imply that

$$W^\infty\left(\mathcal{J}^{-1}(\mathcal{O}_\rho)/A'_G\right) \subset C^\infty\left(\mathcal{J}^{-1}(\mathcal{O}_\rho)/A'_G\right).$$

Indeed, let $f \in W^\infty\left(\mathcal{J}^{-1}(\mathcal{O}_\rho)/A'_G\right)$ be arbitrary. By definition, there exists an open set U/A'_G containing $\mathcal{J}^{-1}(\mathcal{O}_\rho)/A'_G$ and a function $F \in C^\infty_{M/A'_G}(U/A'_G)$ such that $f = F|_{M'_{\mathcal{O}_\rho}}$. As $F \in C^\infty_{M/A'_G}(U/A'_G)$ we have that $F \circ \mathcal{J} \in C^\infty(U)$. Also, as $\mathcal{J}^{-1}(\mathcal{O}_\rho)$ is an immersed initial submanifold of M, the injection $i_{\mathcal{O}_\rho} : \mathcal{J}^{-1}(\mathcal{O}_\rho) \hookrightarrow U$ is smooth and therefore so is $F \circ \mathcal{J} \circ i_{\mathcal{O}_\rho} = F \circ \mathcal{J}_{\mathcal{O}_\rho}$. Consequently, $f \circ \mathcal{J}_{\mathcal{O}_\rho} = F \circ \mathcal{J}_{\mathcal{O}_\rho}$ is smooth. As $\mathcal{J}_{\mathcal{O}_\rho}$ is a submersion f is necessarily smooth, that is, $f \in C^\infty\left(\mathcal{J}^{-1}(\mathcal{O}_\rho)/A'_G\right)$, as required.

9.3.4 Definition *We say that $M'_{\mathcal{O}_\rho}$ is* ***W-spanned*** *when the differentials of the elements in $W^\infty(M'_{\mathcal{O}_\rho})$ span its cotangent bundle, that is,*

$$\operatorname{span}\{\mathbf{d}f(\sigma) \mid f \in W^\infty(M'_{\mathcal{O}_\rho})\} = T^*_\sigma M'_{\mathcal{O}_\rho}, \quad \textit{for all} \quad \sigma \in M'_{\mathcal{O}_\rho}.$$

A sufficient (but not necessary!) condition for $M'_{\mathcal{O}_\rho}$ to be W-spanned is that $W^\infty(M'_{\mathcal{O}_\rho}) = C^\infty(M'_{\mathcal{O}_\rho})$.

Consider now the double polar distribution defined by the group of diffeomorphisms A_G, that is,

$$A''_G := (A'_G)' = \{X_f \mid f \in C^\infty(U)^{A'_G}, \text{ with } U \text{ open and } A'_G\text{-invariant}\}.$$

9.3.5 Definition *We say that the group of diffeomorphisms A_G is* ***weakly von Neumann*** *when for all $z \in M$ we have that*

$$\mathfrak{g} \cdot z = T_z(A_G(z)) \subset A''_G(z).$$

This condition, as well as the situations under which it holds, will be studied in Chapter 11 in the context of the theory of dual pairs.

9.3.6 Theorem (Polar reduction of a Poisson manifold) *Let $(M, \{\cdot,\cdot\})$ be a smooth Poisson manifold and G a Lie group acting canonically and properly on M. Let $\mathcal{J} : M \to M/A'_G$ be the optimal momentum map associated to this action and $\rho \in M/A'_G$ such that G_ρ is closed in G. If A_G is weakly von Neumann, then for each point $z \in \mathcal{J}^{-1}(\mathcal{O}_\rho)$ and vectors $v, w \in T_z\mathcal{J}^{-1}(\mathcal{O}_\rho)$, there exists an open A'_G-invariant neighborhood U of z and two smooth functions $f, g \in C^\infty(U)$ such that $v = X_f(z)$ and $w = X_g(z)$. Moreover, there is a unique presymplectic form $\omega'_{\mathcal{O}_\rho}$ on the polar reduced space $M'_{\mathcal{O}_\rho}$ that satisfies*

$$\{f, g\}|_U(z) = \pi^*_{\mathcal{O}_\rho}\omega_{\mathcal{O}_\rho}(z)(v, w) + \mathcal{J}^*_{\mathcal{O}_\rho}\omega'_{\mathcal{O}_\rho}(z)(v, w) \tag{9.3.3}$$

If $M'_{\mathcal{O}_\rho}$ is W-spanned then the form $\omega'_{\mathcal{O}_\rho}$ is symplectic.

When the Poisson manifold $(M, \{\cdot,\cdot\})$ is actually a symplectic manifold with symplectic form ω the von Neumann condition in the previous result is no longer needed. Moreover, the conditions under which the form $\omega'_{\mathcal{O}_\rho}$ is symplectic can be completely characterized and the regularized polar subspaces appear as symplectic submanifolds of the polar space containing them.

9.3.7 Theorem (Polar reduction of a symplectic manifold) *Let (M, ω) be a smooth symplectic manifold and G a Lie group acting canonically and properly on M. Let $\mathcal{J} : M \to M/A'_G$ be the optimal momentum map associated to this action and let $\rho \in M/A'_G$ be such that G_ρ is closed in G.*

(i) *There is a unique presymplectic form $\omega'_{\mathcal{O}_\rho}$ on the polar reduced space $M'_{\mathcal{O}_\rho} \simeq G/G_\rho$ that satisfies*

$$i^*_{\mathcal{O}_\rho}\omega = \pi^*_{\mathcal{O}_\rho}\omega_{\mathcal{O}_\rho} + \mathcal{J}^*_{\mathcal{O}_\rho}\omega'_{\mathcal{O}_\rho}. \tag{9.3.4}$$

The form $\omega'_{\mathcal{O}_\rho}$ *is symplectic if and only if for one point* $z \in \mathcal{J}^{-1}(\mathcal{O}_\rho)$ *(and hence for all) we have that*

$$\mathfrak{g} \cdot z \cap (\mathfrak{g} \cdot z)^\omega \subset T_z M^{l_z}_{G_z}. \tag{9.3.5}$$

(ii) *Let* $M'_{\mathcal{N}_\rho} = \mathcal{J}^{-1}(\mathcal{N}_\rho)/A'_G \simeq N(H)^\rho/G_\rho$ *be a regularized polar reduced subspace of* $M'_{\mathcal{O}_\rho}$. *Let* $j_{\mathcal{N}_\rho} : \mathcal{J}^{-1}(\mathcal{N}_\rho)/A'_G \hookrightarrow \mathcal{J}^{-1}(\mathcal{O}_\rho)/A'_G$ *be the inclusion and* $\omega'_{\mathcal{O}_\rho}$ *the presymplectic form defined in* **(i)**. *Then the form*

$$\omega'_{\mathcal{N}_\rho} := j^*_{\mathcal{N}_\rho} \omega'_{\mathcal{O}_\rho} \tag{9.3.6}$$

is symplectic, that is, the regularized polar subspaces are symplectic submanifolds of the polar space containing them.

9.3.8 The condition for symplecticity in the presence of a standard momentum map. The characterization (9.3.5) of the symplecticity of $\omega'_{\mathcal{O}_\rho}$ admits a particularly convenient formulation when the G-action on the symplectic manifold (M, ω) admits a standard momentum map $\mathbf{J} : M \to \mathfrak{g}^*$. Indeed, assume that M is connected and let $z \in M$ be such that $\mathbf{J}(z) = \mu \in \mathfrak{g}^*$ and $G_z = H$. Then, if the symbol G_μ denotes the isotropy subgroup of μ with respect to the affine G-action on $\mathfrak{g}^*$ defined with the non-equivariance cocycle of $\mathbf{J}$, the Reduction Lemma 4.5.34 implies that (9.3.5) is equivalent to

$$\mathfrak{g} \cdot z \cap (\mathfrak{g} \cdot z)^\omega = \mathfrak{g}_\mu \cdot z \subset T_z M^{l_z}_H,$$

which in turn, by Proposition 2.4.5, amounts to $\mathfrak{g}_\mu \cdot z \subset \mathfrak{g}_\mu \cdot z \cap T_z M^{l_z}_H = \mathrm{Lie}(N(H) \cap G_\mu) \cdot z$. Let $N_{G_\mu}(H) := N(H) \cap G_\mu$. With this notation, the condition can be rewritten as $\mathfrak{g}_\mu + \mathfrak{h} \subset \mathrm{Lie}(N_{G_\mu}(H)) + \mathfrak{h} \subset \mathfrak{g}_\mu$ or, equivalently, as

$$\mathfrak{g}_\mu = \mathrm{Lie}(N_{G_\mu}(H)). \tag{9.3.7}$$

Proof of Theorem 9.3.6. Since A_G is weakly von Neumann we have that $\mathfrak{g} \cdot z \subset A''_G(z)$, for any $z \in M$, or, equivalently, that for any $z \in M$ and any $\xi \in \mathfrak{g}$, there is an A'_G-invariant neighborhood U of z and a function $F \in C^\infty\left(U/A'_G\right)$ such that $\xi_M(z) = X_{F \circ \mathcal{J}}(z)$. Consequently, for any vector $v \in T_z\mathcal{J}^{-1}(\mathcal{O}_\rho)$ there exists $f \in C^\infty(M)^G$ and $F \in C^\infty\left(U/A'_G\right)$ (shrink U if necessary) such that $v = X_f(z) + X_{F \circ \mathcal{J}}(z) = X_{f|_U + F \circ \mathcal{J}}(z)$. Let $w \in T_z\mathcal{J}^{-1}(\mathcal{O}_\rho)$, $l \in C^\infty(M)^G$, and let $L \in C^\infty\left(U/A'_G\right)$ be such that $w = X_l(z) + X_{L \circ \mathcal{J}}(z) = X_{l|_U + L \circ \mathcal{J}}(z)$. Expression (9.3.3) can then be rewritten as

$$\begin{aligned}
\mathcal{J}^*_{\mathcal{O}_\rho} \omega'_{\mathcal{O}_\rho}(z)(v, w) &= \mathcal{J}^*_{\mathcal{O}_\rho} \omega'_{\mathcal{O}_\rho}(z)(X_{f|_U + F \circ \mathcal{J}}(z), X_{l|_U + L \circ \mathcal{J}}(z)) \\
&= \{f + F \circ \mathcal{J}, l + L \circ \mathcal{J}\}|_U(z) \\
&\quad - \pi^*_{\mathcal{O}_\rho} \omega_{\mathcal{O}_\rho}(z)(X_{f|_U + F \circ \mathcal{J}}(z), X_{l|_U + L \circ \mathcal{J}}(z)) \\
&= \{F \circ \mathcal{J}, L \circ \mathcal{J}\}|_U(z).
\end{aligned} \tag{9.3.8}$$

We now show that $\omega'_{\mathcal{O}_\rho}$ is well defined. Indeed, let $z' \in \mathcal{J}^{-1}(\mathcal{O}_\rho)$ and $v', w' \in T_{z'}\mathcal{J}^{-1}(\mathcal{O}_\rho)$ be such that $T_z\mathcal{J}_{\mathcal{O}_\rho}(v) = T_{z'}\mathcal{J}_{\mathcal{O}_\rho}(v')$ and $T_z\mathcal{J}_{\mathcal{O}_\rho}(w) = T_{z'}\mathcal{J}_{\mathcal{O}_\rho}(w')$. First of all these equalities imply the existence of an element $\mathcal{F}_T$ in the polar pseudogroup of A_G such that $z' = \mathcal{F}_T(z)$. As $\mathcal{F}_T$ is a local diffeomorphism such that $\mathcal{J}_{\mathcal{O}_\rho} \circ \mathcal{F}_T = \mathcal{J}_{\mathcal{O}_\rho}$, we have that $T_z\mathcal{J}_{\mathcal{O}_\rho} = T_{z'}\mathcal{J}_{\mathcal{O}_\rho} \circ T_z\mathcal{F}_T$. Now, we can rewrite the conditions $T_z\mathcal{J}_{\mathcal{O}_\rho}(v) = T_{z'}\mathcal{J}_{\mathcal{O}_\rho}(v')$ and $T_z\mathcal{J}_{\mathcal{O}_\rho}(w) = T_{z'}\mathcal{J}_{\mathcal{O}_\rho}(w')$ as $T_{z'}\mathcal{J}_{\mathcal{O}_\rho}(T_z\mathcal{F}_T(v)) = T_{z'}\mathcal{J}_{\mathcal{O}_\rho}(v')$ and $T_{z'}\mathcal{J}_{\mathcal{O}_\rho}(T_z\mathcal{F}_T(w)) = T_{z'}\mathcal{J}_{\mathcal{O}_\rho}(w')$, respectively, which implies the existence of two functions $f', l' \in C^\infty(M)^G$ such that

$$
\begin{aligned}
v' &= T_z\mathcal{F}_T(X_f(z) + X_{F\circ\mathcal{J}}(z)) + X_{f'}(\mathcal{F}_T(z))\\
w' &= T_z\mathcal{F}_T(X_l(z) + X_{L\circ\mathcal{J}}(z)) + X_{l'}(\mathcal{F}_T(z))
\end{aligned}
$$

or, equivalently:

$$
\begin{aligned}
v' &= X_{f\circ\mathcal{F}_{-T}}(\mathcal{F}_T(z)) + X_{F\circ\mathcal{J}}(\mathcal{F}_T(z)) + X_{f'}(\mathcal{F}_T(z))\\
w' &= X_{l\circ\mathcal{F}_{-T}}(\mathcal{F}_T(z)) + X_{L\circ\mathcal{J}}(\mathcal{F}_T(z)) + X_{l'}(\mathcal{F}_T(z)).
\end{aligned}
$$

Therefore, using (9.3.3), we have that

$$
\begin{aligned}
&\mathcal{J}^*_{\mathcal{O}_\rho}\omega'_{\mathcal{O}_\rho}(z')(v', w')\\
&\quad = \{f \circ \mathcal{F}_{-T} + F \circ \mathcal{J} + f', l \circ \mathcal{F}_{-T} + L \circ \mathcal{J} + l'\}|_V(\mathcal{F}_T(z))\\
&\quad\quad - \pi^*_{\mathcal{O}_\rho}\omega_{\mathcal{O}_\rho}(\mathcal{F}_T(z))(X_{f\circ\mathcal{F}_{-T}|_V + F\circ\mathcal{J} + f'}(z), X_{l\circ\mathcal{F}_{-T}|_V + L\circ\mathcal{J} + l'}(z))\\
&\quad = \{F \circ \mathcal{J}, L \circ \mathcal{J}\}|_V(\mathcal{F}_T(z)) = \{F \circ \mathcal{J}, L \circ \mathcal{J}\}|_U(z)\\
&\quad = \mathcal{J}^*_{\mathcal{O}_\rho}\omega'_{\mathcal{O}_\rho}(z)(v, w),
\end{aligned}
$$

where $V = U \cap \mathcal{F}_T(\mathrm{Dom}(\mathcal{F}_T)) = \mathcal{F}_T(U \cap \mathrm{Dom}(\mathcal{F}_T))$. Hence, the form $\omega'_{\mathcal{O}_\rho}$ is well defined.

The closedness and skew-symmetric character of $\omega'_{\mathcal{O}_\rho}$ is obtained as a consequence of $\mathcal{J}_{\mathcal{O}_\rho}$ being a surjective submersion, $\omega_{\mathcal{O}_\rho}$ being closed and skew-symmetric, and $\{\cdot, \cdot\}$ being a Poisson bracket. An equivalent fashion to realize this is by writing $\omega'_{\mathcal{O}_\rho}$ in terms of the symplectic structure of the leaves of M. Indeed, as A_G is weakly von Neumann, each connected component of $\mathcal{J}^{-1}(\mathcal{O}_\rho)$ lies in a single symplectic leaf of $(M, \{\cdot, \cdot\})$. In order to simplify the exposition suppose that $\mathcal{J}^{-1}(\mathcal{O}_\rho)$ is connected and let $\mathcal{L}_{\mathcal{O}_\rho}$ be the unique symplectic leaf of M containing it (otherwise one has just to proceed connected component by connected component). Let $i_{\mathcal{L}_{\mathcal{O}_\rho}} : \mathcal{J}^{-1}(\mathcal{O}_\rho) \to \mathcal{L}_{\mathcal{O}_\rho}$ be the natural injection. Given that $i_{\mathcal{O}_\rho} : \mathcal{J}^{-1}(\mathcal{O}_\rho) \to M$ is smooth and $\mathcal{L}_{\mathcal{O}_\rho}$ is an initial submanifold of M, the map $i_{\mathcal{L}_{\mathcal{O}_\rho}}$ is therefore smooth. If we denote by $\omega_{\mathcal{L}_{\mathcal{O}_\rho}}$ the symplectic form of the leaf $\mathcal{L}_{\mathcal{O}_\rho}$, the expression (9.3.3) can be rewritten as

$$
i^*_{\mathcal{L}_{\mathcal{O}_\rho}}\omega_{\mathcal{L}_{\mathcal{O}_\rho}} = \pi^*_{\mathcal{O}_\rho}\omega_{\mathcal{O}_\rho} + \mathcal{J}^*_{\mathcal{O}_\rho}\omega'_{\mathcal{O}_\rho}. \tag{9.3.9}
$$

The antisymmetry and closedness of $\omega'_{\mathcal{O}_\rho}$ appears then as a consequence of the antisymmetry and closedness of $\omega_{\mathcal{O}_\rho}$ and $\omega_{\mathcal{L}_{\mathcal{O}_\rho}}$.

It remains to be shown that if $M'_{\mathcal{O}_\rho}$ is W-spanned, then the form $\omega'_{\mathcal{O}_\rho}$ is nondegenerate. Let $z \in \mathcal{J}^{-1}(\mathcal{O}_\rho)$ and $v \in T_z\mathcal{J}^{-1}(\mathcal{O}_\rho)$ be such that

$$\omega'_{\mathcal{O}_\rho}(\mathcal{J}_{\mathcal{O}_\rho}(z))(T_z\mathcal{J}_{\mathcal{O}_\rho}(v), T_z\mathcal{J}_{\mathcal{O}_\rho}(w)) = 0, \tag{9.3.10}$$

for all $w \in T_z\mathcal{J}^{-1}(\mathcal{O}_\rho)$. Now take $f \in C^\infty(M)^G$ and $F \in C^\infty\left(U/A'_G\right)$ such that $v = X_f(z) + X_{F\circ\mathcal{J}}(z)$. Condition (9.3.10) is equivalent to

$$\omega'_{\mathcal{O}_\rho}(\mathcal{J}_{\mathcal{O}_\rho}(z))\left(T_z\mathcal{J}_{\mathcal{O}_\rho}(X_{F\circ\mathcal{J}}(z)), T_z\mathcal{J}_{\mathcal{O}_\rho}(X_{L\circ\mathcal{J}}(z))\right) = 0, \tag{9.3.11}$$

for all $L \in C^\infty\left(V/A'_G\right)$ and all open A'_G-invariant neighborhoods V of z. By (9.3.8) we can rewrite (9.3.11) as

$$\{F \circ \mathcal{J}, L \circ \mathcal{J}\}|_{U\cap V}(z) = 0. \tag{9.3.12}$$

Now notice that for any $h \in W^\infty(M'_{\mathcal{O}_\rho})$ there exists a function $H \in C^\infty(U/A'_G)$ with U open and A'_G-invariant such that $\mathcal{J}^{-1}(\mathcal{O}_\rho)/A'_G \subset U/A'_G$ and $H|_{M'_{\mathcal{O}_\rho}} = h$. Moreover, by (9.3.12) we have that

$$\begin{aligned}
&\mathbf{d}h(\mathcal{J}_{\mathcal{O}_\rho}(z))\left(T_z\mathcal{J}_{\mathcal{O}_\rho}\left(X_{F\circ\mathcal{J}}(z)\right)\right)\\
&\qquad = \mathbf{d}(h \circ \mathcal{J}_{\mathcal{O}_\rho})(z)\left(X_{F\circ\mathcal{J}}(z)\right) = \mathbf{d}(H \circ \mathcal{J})(z)\left(X_{F\circ\mathcal{J}}(z)\right) = 0.
\end{aligned}$$

Since the previous equality holds for any $h \in W^\infty(M'_{\mathcal{O}_\rho})$ and $M'_{\mathcal{O}_\rho}$ is W-spanned, we conclude that $T_z\mathcal{J}_{\mathcal{O}_\rho}\left(X_{F\circ\mathcal{J}}(z)\right) = T_z\mathcal{J}_{\mathcal{O}_\rho}(v) = 0$. ■

Proof of Theorem 9.3.7. **(i)** The well-definedness and presymplectic character of $\omega'_{\mathcal{O}_\rho}$ in this case can be obtained as a consequence of Theorem 9.3.6. This is particularly evident when we think of $\omega'_{\mathcal{O}_\rho}$ as the form characterized by equality (9.3.9) and we recall that in the symplectic case $\omega_{\mathcal{L}_{\mathcal{O}_\rho}} = \omega$.

It just remains to be shown that the form $\omega'_{\mathcal{O}_\rho}$ is nondegenerate if and only if condition (9.3.5) holds. We proceed by showing first that if condition (9.3.5) holds for the point $z \in \mathcal{J}^{-1}(\mathcal{O}_\rho)$ then it holds for all points of $\mathcal{J}^{-1}(\mathcal{O}_\rho)$. We will then prove that (9.3.5) at the point z is equivalent to the nondegeneracy of $\omega'_{\mathcal{O}_\rho}$ at $\mathcal{J}_{\mathcal{O}_\rho}(z)$.

Suppose first that the point $z \in \mathcal{J}^{-1}(\mathcal{O}_\rho)$ is such that $\mathfrak{g} \cdot z \cap (\mathfrak{g} \cdot z)^\omega \subset T_zM^{l_z}_{G_z}$. Notice now that any element in $\mathcal{J}^{-1}(\mathcal{O}_\rho)$ can be written as $\Phi_g(\mathcal{F}_T(z))$ with $g \in G$ and $\mathcal{F}_T$ in the polar pseudogroup of A_G. It is easy to show that the relation

$$\mathfrak{g} \cdot (\Phi_g(\mathcal{F}_T(z))) \cap (\mathfrak{g} \cdot (\Phi_g(\mathcal{F}_T(z))))^\omega \subset T_{\Phi_g(\mathcal{F}_T(z))}M^{l_{\Phi_g(\mathcal{F}_T(z))}}_{G_{\Phi_g(\mathcal{F}_T(z))}}$$

is equivalent to $T_z(\Phi \circ \mathcal{F}_T)(\mathfrak{g} \cdot z \cap (\mathfrak{g} \cdot z)^\omega) \subset T_z(\Phi \circ \mathcal{F}_T)M^{l_z}_{G_z}$ and therefore to $\mathfrak{g} \cdot z \cap (\mathfrak{g} \cdot z)^\omega \subset T_zM^{l_z}_{G_z}$.

Now let $v \in T_z\mathcal{J}^{-1}(\mathcal{O}_\rho)$ be such that

$$\omega'_{\mathcal{O}_\rho}(\mathcal{J}_{\mathcal{O}_\rho}(z))(T_z\mathcal{J}_{\mathcal{O}_\rho}(v), T_z\mathcal{J}_{\mathcal{O}_\rho}(w)) = 0, \tag{9.3.13}$$

for all $w \in T_z\mathcal{J}^{-1}(\mathcal{O}_\rho)$. Now take $f \in C^\infty(M)^G$ and $\xi \in \mathfrak{g}$ such that $v = X_f(z) + \xi_M(z)$. Condition (9.3.13) is equivalent to

$$\omega'_{\mathcal{O}_\rho}(\mathcal{J}_{\mathcal{O}_\rho}(z))(T_z\mathcal{J}_{\mathcal{O}_\rho}(\xi_M(z)), T_z\mathcal{J}_{\mathcal{O}_\rho}(\eta_M(z))) = 0, \quad \text{for all } \eta \in \mathfrak{g}, \tag{9.3.14}$$

which, by (9.3.4), can be rewritten as

$$\omega(z)(\xi_M(z), \eta_M(z)) = 0, \qquad \text{for all } \eta \in \mathfrak{g}, \tag{9.3.15}$$

and thereby amounts to having that $\xi_M(z) \in \mathfrak{g} \cdot z \cap (\mathfrak{g} \cdot z)^\omega$. Hence, $\omega'_{\mathcal{O}_\rho}(\mathcal{J}_{\mathcal{O}_\rho}(z))$ is nondegenerate if and only if $\xi_M(z) \in \ker T_z\mathcal{J}_{\mathcal{O}_\rho} = A'_G(z)$.

Suppose now that condition (9.3.5) holds; then, as $\xi_M(z) \in \mathfrak{g} \cdot z \cap (\mathfrak{g} \cdot z)^\omega$ we have that $\xi_M(z) \in T_z M_{G_z}^{l_z}$. Theorem 5.5.17 allows us to conclude that $\xi_M(z) \in A'_G(z)$, as required. Conversely, suppose that $\omega'_{\mathcal{O}_\rho}$ is symplectic. The previous equalities immediately imply that $\mathfrak{g} \cdot z \cap (\mathfrak{g} \cdot z)^\omega \subset A'_G(z) \subset T_z M_{G_z}^{l_z}$, as required.

(ii) The form $\omega'_{\mathcal{N}_\rho}$ is clearly closed and antisymmetric. We now show that it is nondegenerate. Recall first that the tangent space to $T_z\mathcal{J}^{-1}(\mathcal{N}_\rho)$ at a given point $z \in \mathcal{J}^{-1}(\mathcal{N}_\rho)$ is given by the vectors of the form $v = X_f(z) + \xi_M(z)$, with $f \in C^\infty(M)^G$ and $\xi \in \operatorname{Lie}(N(H)^\rho)$. Let $v = X_f(z) + \xi_M(z) \in T_z\mathcal{J}^{-1}(\mathcal{N}_\rho)$ be such that

$$\mathcal{J}^*_{\mathcal{N}_\rho}(j^*_{\mathcal{N}_\rho}\omega'_{\mathcal{O}_\rho})(z)(X_f(z) + \xi_M(z), X_g(z) + \eta_M(z)) = 0,$$

for all $\eta \in \operatorname{Lie}(N(H)^\rho)$ and $g \in C^\infty(M)^G$. If we plug into the previous expression the definition of the form $\omega'_{\mathcal{O}_\rho}$ we obtain that

$$\omega(z)(\xi_M(z), \eta_M(z)) = 0, \qquad \text{for all } \eta \in \operatorname{Lie}(N(H)^\rho),$$

that is, $\xi_M(z) \in (\operatorname{Lie}(N(H)^\rho) \cdot z) \cap (\operatorname{Lie}(N(H)^\rho) \cdot z)^\omega = (\operatorname{Lie}(N(H)^\rho) \cdot z) \cap (\operatorname{Lie}(N(H)^\rho) \cdot z)^{\omega|_{M_H^\rho}} = (\operatorname{Lie}(N(H)^\rho/H) \cdot z) \cap A'_{N(H)^\rho/H}(z)$, where the last equality follows from Theorem 5.5.17 and the freeness of the natural $N(H)^\rho/H$-action on M_H^ρ. We now recall (see Proposition 2.5.9) that any $N(H)^\rho/H$-invariant function on M_H^ρ admits a local extension to a G-invariant function on M, hence $\xi_M(z) \in (\operatorname{Lie}(N(H)^\rho/H) \cdot z) \cap A'_G(z)$ and, consequently, $T_z\mathcal{J}_{\mathcal{O}_\rho}(\xi_M(z)) = T_z\mathcal{J}_{\mathcal{O}_\rho}(v) = T_z\mathcal{J}_{\mathcal{N}_\rho}(v) = 0$. ■

9.3.9 The coadjoint orbits as polar reduced spaces. Let G be a Lie group, $\mathfrak{g}$ its Lie algebra, and $\mathfrak{g}^*$ its dual considered as a Lie–Poisson space. In this elementary example we show how the coadjoint orbits appear as the polar reduced spaces of the coadjoint G-action on $\mathfrak{g}^*$.

A straightforward computation shows that the coadjoint action of G on the Lie–Poisson space $\mathfrak{g}^*$ is canonical. Moreover, the polar distribution $A'_G(\mu) = 0$ for all $\mu \in \mathfrak{g}^*$ and therefore the optimal momentum map $\mathcal{J} : \mathfrak{g}^* \to \mathfrak{g}^*$ is the identity map on $\mathfrak{g}^*$. This immediately implies that any open set $U \subset \mathfrak{g}^*$ is A'_G-invariant, that $C^\infty(U)^{A'_G} = C^\infty(U)$, and that therefore $\mathfrak{g} \cdot \mu \subset A''_G(\mu)$, for any $\mu \in \mathfrak{g}^*$. The coadjoint action on $\mathfrak{g}^*$ is therefore weakly von Neumann.

We now look at the corresponding reduced spaces. On the one hand the orbit reduced spaces $\mathcal{J}^{-1}(\mathcal{O}_\rho)/G$ are the quotients $G \cdot \mu/G$ and therefore amount to points. At the same time, we have that $\mathcal{J}^{-1}(\mathcal{O}_\rho)/A'_G = \mathcal{O}_\mu/A'_G = \mathcal{O}_\mu$, that is, the polar reduced spaces are the coadjoint orbits which, by Theorem 9.3.6, are symplectic. Indeed, the W-spanning condition necessary for the application of this result is satisfied since in this case $\operatorname{span}\{\mathbf{d}f(\mu) \mid f \in W^\infty(M'_{\mathcal{O}_\rho})\} = \operatorname{span}\{\mathbf{d}h|_{\mathcal{O}_\mu}(\mu) \mid h \in C^\infty(\mathfrak{g}^*)\} = T^*_\mu \mathcal{O}_\mu$. Note that the last equality is a consequence of the immersed character of the coadjoint orbits $\mathcal{O}_\mu$ as submanifolds of $\mathfrak{g}^*$ (the equality is easily proved using immersion charts around the point μ).

9.3.10 Symplectic decomposition of presymplectic homogeneous manifolds. Let G be a Lie group, $\mathfrak{g}$ its Lie algebra, and $\mathfrak{g}^*$ its dual. Let $\mathcal{O}_{\mu_1}$ and $\mathcal{O}_{\mu_2}$ be two coadjoint orbits of $\mathfrak{g}^*$ that we will consider as symplectic manifolds endowed with the orbit symplectic forms $\omega_{\mathcal{O}_{\mu_1}}$ and $\omega_{\mathcal{O}_{\mu_2}}$, respectively. Consider now the symplectic sum $\mathcal{O}_{\mu_1} \oplus \mathcal{O}_{\mu_2} := (\mathcal{O}_{\mu_1} \times \mathcal{O}_{\mu_2}, \omega_{\mathcal{O}_{\mu_1}} \oplus \omega_{\mathcal{O}_{\mu_2}})$. The diagonal action of G on $\mathcal{O}_{\mu_1} \oplus \mathcal{O}_{\mu_2}$ is canonical with respect to this symplectic structure and, moreover, it has an associated standard equivariant momentum map $\mathbf{J} : \mathcal{O}_{\mu_1} \oplus \mathcal{O}_{\mu_2} \to \mathfrak{g}^*$ given by $\mathbf{J}(\nu, \eta) = \nu + \eta$. We now suppose that this action is proper and we will study, in this particular case, the orbit and polar reduced spaces introduced in the previous sections.

We start by looking at the level sets of the optimal momentum map $\mathcal{J} : \mathcal{O}_{\mu_1} \times \mathcal{O}_{\mu_2} \to (\mathcal{O}_{\mu_1} \times \mathcal{O}_{\mu_2})/A'_G$. By Corollary 5.5.18 we know that in the presence of a standard momentum map the fibers of the optimal momentum map coincide with the connected components of the intersections of the level sets of the momentum map with the isotropy type submanifolds. Hence, in our case, if $\rho = \mathcal{J}(\mu_1, \mu_2)$, we have that

$$\mathcal{J}^{-1}(\rho) = (\mathbf{J}^{-1}(\mu_1 + \mu_2) \cap (\mathcal{O}_{\mu_1} \times \mathcal{O}_{\mu_2})_{G_{(\mu_1,\mu_2)}})_c, \tag{9.3.16}$$

where the subscript c in the previous expression stands for the connected component of $\mathbf{J}^{-1}(\mu_1 + \mu_2) \cap (\mathcal{O}_{\mu_1} \times \mathcal{O}_{\mu_2})_{G_{(\mu_1,\mu_2)}}$ containing $\mathcal{J}^{-1}(\rho)$. Since the isotropy $G_{(\mu_1,\mu_2)} = G_{\mu_1} \cap G_{\mu_2}$, with G_{μ_1} and G_{μ_2} the coadjoint isotropies of μ_1 and μ_2, respectively, the expression (9.3.16) can be rewritten as

$$\begin{aligned}\mathcal{J}^{-1}(\rho) = (\{(\mathrm{Ad}^*_{g^{-1}}\mu_1, \mathrm{Ad}^*_{h^{-1}}\mu_2) \mid g, h \in G,\ \text{such that}\\ \mathrm{Ad}^*_{g^{-1}}\mu_1 + \mathrm{Ad}^*_{h^{-1}}\mu_2 = \mu_1 + \mu_2,\ gG_{\mu_1}g^{-1} \cap hG_{\mu_2}h^{-1} = G_{\mu_1} \cap G_{\mu_2}\})_c.\end{aligned}$$

It is easy to show that in this case

$$G_\rho = N_{G_{\mu_1+\mu_2}}(G_{\mu_1} \cap G_{\mu_2})^c, \tag{9.3.17}$$

where the superscript c denotes the closed subgroup of $N_{G_{\mu_1+\mu_2}}(G_{\mu_1} \cap G_{\mu_2}) := N(G_{\mu_1} \cap G_{\mu_2}) \cap G_{\mu_1+\mu_2}$ that leaves $\mathcal{J}^{-1}(\rho)$ invariant. Theorems 9.1.1 and 9.2.1 guarantee that the quotients $\mathcal{J}^{-1}(\rho)/G_\rho \simeq \mathcal{J}^{-1}(\mathcal{O}_\rho)/G$ are symplectic. Nevertheless, we will focus our attention on the corresponding polar reduced spaces.

According to Theorem 9.3.7 and to (9.3.17), the polar reduced space corresponding to $\mathcal{J}^{-1}(\mathcal{O}_\rho)/G$ is the homogeneous presymplectic manifold

$$G/N_{G_{\mu_1+\mu_2}}(G_{\mu_1} \cap G_{\mu_2})^c. \tag{9.3.18}$$

Expression (9.3.7) states that $G/N_{G_{\mu_1+\mu_2}}(G_{\mu_1} \cap G_{\mu_2})^c$ is symplectic if and only if

$$\mathfrak{g}_{\mu_1+\mu_2} = \mathrm{Lie}(N_{G_{\mu_1+\mu_2}}(G_{\mu_1} \cap G_{\mu_2})),$$

which is obviously true when, for instance, $G_{\mu_1} \cap G_{\mu_2}$ is a normal subgroup of $G_{\mu_1+\mu_2}$. In any case, using (9.3.2) we can write the polar reduced space (9.3.18) as a disjoint union of its regularized symplectic reduced subspaces that, in this case, are of the form $gN(G_{\mu_1} \cap G_{\mu_2})^\rho / N_{G_{\mu_1+\mu_2}}(G_{\mu_1} \cap G_{\mu_2})^c$ with $g \in G$ and where the superscript ρ denotes the closed subgroup of $N(G_{\mu_1} \cap G_{\mu_2})$ that leaves invariant the connected component of $(\mathcal{O}_{\mu_1} \times \mathcal{O}_{\mu_2})_{G_{\mu_1} \cap G_{\mu_2}}$ containing $\mathcal{J}^{-1}(\rho)$. More explicitly, we can write the following symplectic decomposition of the polar reduced space:

$$\begin{aligned} G/N_{G_{\mu_1+\mu_2}}&(G_{\mu_1} \cap G_{\mu_2})^c \\ &= \coprod_{[g] \in G/N(G_{\mu_1} \cap G_{\mu_2})^\rho} gN(G_{\mu_1} \cap G_{\mu_2})^\rho / N_{G_{\mu_1+\mu_2}}(G_{\mu_1} \cap G_{\mu_2})^c. \end{aligned}$$

What we just did in the previous sections for two coadjoint orbits can be inductively generalized to n orbits. We collect the results of that construction under the form of a proposition.

9.3.11 Proposition *Let G be a Lie group, $\mathfrak{g}$ its Lie algebra, and $\mathfrak{g}^*$ its dual. Let $\mu_1, \ldots, \mu_n \in \mathfrak{g}^*$. Then the homogeneous manifold*

$$G/N_{G_{\mu_1+\cdots+\mu_n}}(G_{\mu_1} \cap \ldots \cap G_{\mu_n})^c \tag{9.3.19}$$

has a natural presymplectic structure that is nondegenerate if and only if

$$\mathfrak{g}_{\mu_1+\cdots+\mu_n} = \operatorname{Lie}(N_{G_{\mu_1+\cdots+\mu_n}}(G_{\mu_1} \cap \ldots \cap G_{\mu_n})).$$

Moreover, (9.3.19) *can be written as the following disjoint union of symplectic submanifolds:*

$$\begin{aligned} G/N_{G_{\mu_1+\cdots+\mu_n}}&(G_{\mu_1} \cap \ldots \cap G_{\mu_n})^c \\ &= \coprod_{[g] \in G/N(G_{\mu_1} \cap \ldots \cap G_{\mu_n})^\rho} gN(G_{\mu_1} \cap \ldots \cap G_{\mu_n})^\rho / N_{G_{\mu_1+\cdots+\mu_n}}(G_{\mu_1} \cap \ldots \cap G_{\mu_n})^c. \end{aligned}$$

9.4 Optimal reduction by stages

As we already described in Section 6.7 for the regular case and in the presence of a standard momentum map, the reduction by stages procedure consists of carrying out symplectic reduction in two steps using a normal subgroup of the symmetry group. To be more specific, suppose that we are in the same setup as Theorem 9.1.1 and that the symmetry group G has a closed normal subgroup N. In this section we will spell out the conditions under which reduction by G renders the same result as reduction in the following two stages: we first reduce by N; the resulting space inherits symmetry properties coming from the quotient Lie group G/N that can be used to reduce one more time.

The results in this section extend those presented in Section 6.7 for the optimal setup. As a corollary of this study, we will obtain a generalization to the singular case (nonfree actions) of the Reduction by Stages Theorem 6.7.10 that we will present in the following section.

9.4.1 The polar distribution associated to a normal subgroup. Throughout this section we will work on a Poisson manifold $(M, \{\cdot, \cdot\})$ acted properly and canonically upon by a Lie group G. We will assume that G has a closed normal subgroup that we will denote by N. The closedness of N implies that the N-action on M by restriction is still proper and that G/N is a Lie group when considered as a homogenous manifold. We will denote by A'_G and A'_N the polar distributions associated to the G and N-actions, respectively, and by $\mathcal{J}_G : M \to M/A'_G$ and $\mathcal{J}_N : M \to M/A'_N$ the corresponding optimal momentum maps.

The following proposition provides a characterization of the conditions under which the polar distribution A'_H associated to a closed subgroup H of G is invariant under the lifted action of G to the tangent bundle TM.

9.4.2 Proposition *Let $(M, \{\cdot, \cdot\})$ be a Poisson manifold acted properly and canonically upon by a Lie group G via the map $\Phi : G \times M \to M$. Let H be a closed Lie subgroup of G. Then:*

(i) *The lifted action of G to the tangent bundle TM leaves the H-polar distribution A'_H invariant if and only if $f \circ \Phi_{g^{-1}} \in C^\infty(M)^H$, for any $f \in C^\infty(M)^H$ and any $g \in G$. This condition holds if and only if for all $g \in G$, $h \in H$, and $m \in M$, there exists an element $h' \in H$ such that*

$$gh \cdot m = h'g \cdot m. \tag{9.4.1}$$

(ii) *If the lifted G-action on TM leaves the H-polar distribution A'_H invariant, then it naturally acts on the corresponding momentum space M/A'_H and, relative to this action, the H-optimal momentum map $\mathcal{J}_H : M \to M/A'_H$ is G-equivariant.*

(iii) *For any $m \in M$ we have $A'_G(m) \subset A'_H(m)$. Consequently there is a natural projection $\pi_H : M/A'_G \to M/A'_H$ such that*

$$\mathcal{J}_H = \pi_H \circ \mathcal{J}_G. \tag{9.4.2}$$

Moreover, if the lifted G-action on TM leaves the H-polar distribution A'_H invariant and consequently M/A'_H, then the map π_H is G-equivariant.

Proof. (i) Since H is closed in G, its action on M by restriction of the G-action is still proper. Therefore, $A'_H = \{X_f \mid f \in C^\infty(M)^H\}$. Given that for any $f \in C^\infty(M)^H$ and any $g \in G$ we have that $T\Phi_g \circ X_f = X_{f \circ \Phi_{g^{-1}}} \circ \Phi_g$, we can conclude that the polar distribution A'_H is G-invariant iff $f \circ \Phi_{g^{-1}} \in C^\infty(M)^H$, for any $f \in C^\infty(M)^H$ and any $g \in G$. We now check that this condition is equivalent to (9.4.1).

Suppose that $f \circ \Phi_g \in C^\infty(M)^H$, for all $f \in C^\infty(M)^H$ and $g \in G$. Consequently, for $m \in M$ and $h \in H$ arbitrary, we have $f(gh \cdot m) = f(g \cdot m)$. Since the H-action on M is proper, the set $C^\infty(M)^H$ of H-invariant functions on M separates the H-orbits. Therefore, the points $gh \cdot m$ and $g \cdot m$ are in the same H-orbit and hence there exists an element $h' \in H$ such that $gh \cdot m = h'g \cdot m$.

Conversely, suppose that for all $g \in G$, $h \in H$, and $m \in M$, there exists an element $h' \in H$ such that $gh \cdot m = h'g \cdot m$. Then, if $f \in C^\infty(M)^H$ we have that

$$(f \circ \Phi_g)(h \cdot m) = f(gh \cdot m) = f(h'g \cdot m) = f(g \cdot m) = (f \circ \Phi_g)(m).$$

Consequently, $f \circ \Phi_g \in C^\infty(M)^H$, as required.
(ii) Suppose that the lifted action of G to the tangent bundle TM leaves the H-polar distribution A'_H invariant. We define the action $G \times M/A'_H \to M/A'_H$ by $g \cdot \mathcal{J}_H(m) := \mathcal{J}_H(g \cdot m)$. If we show that it is well defined, then it is clearly a left action. To prove that it is well defined, let $m' \in M$ be such that $m' = \mathcal{F}_T(m)$, with $\mathcal{F}_T \in G_{\mathcal{A}'_H}$. For the sake of simplicity in the exposition suppose that $\mathcal{F}_T = F_T$ with F_T the Hamiltonian flow associated to $f \in C^\infty(M)^H$. Then, for any $g \in G$ we have that

$$\begin{aligned} g \cdot \mathcal{J}_H(m') &= \mathcal{J}_H(g \cdot F_T(m)) \\ &= \mathcal{J}_H\left(G_T^{f\circ\Phi_{g^{-1}}}(g \cdot m)\right) = \mathcal{J}_H(g \cdot m) = g \cdot \mathcal{J}_H(m), \end{aligned}$$

where $G_T^{f\circ\Phi_{g^{-1}}}$ is the Hamiltonian flow associated to the function $f \circ \Phi_{g^{-1}}$ that, by the hypothesis on the G-invariance of A'_H, is H-invariant and guarantees the equality $\mathcal{J}_H\left(G_T^{f\circ\Phi_{g^{-1}}}(g \cdot m)\right) = \mathcal{J}_H(g \cdot m)$.
(iii) The inclusion $A'_G(m) \subset A'_H(m)$ is a direct consequence of the definition of the polar distribution and it implies that each maximal integral leaf of A'_G is included in a single maximal integral leaf of A'_H. This feature constitutes the definition of π_H that assigns to each leaf in M/A'_G the unique leaf in M/A'_H in which it is sitting. With this definition, it is obvious that $\mathcal{J}_H = \pi_H \circ \mathcal{J}_G$. Now, if the lifted G-action on TM leaves the H-polar distribution A'_H invariant, the map $\mathcal{J}_H$ is G-equivariant by part **(ii)**. The G-equivariance of $\mathcal{J}_G$ plus the relation $\mathcal{J}_H = \pi_H \circ \mathcal{J}_G$ implies that π_H is G-invariant. ■

9.4.3 If H is normal in G then, condition (9.4.1) is trivially satisfied and therefore G acts on A'_H. Conversely, if G acts on A'_H and the identity element is an isotropy subgroup of the G-action on M, then H is necessarily normal in G. Indeed, in that case for any $m \in M$, $g \in G$, and $h \in H$, there exists an element $h' \in H$ such that $gh \cdot m = h'g \cdot m$. In particular, if we take an element $m \in M_{\{e\}}$ we have that $gh = h'g$ or, equivalently, that $gHg^{-1} \subset H$, for all $g \in G$, which implies that H is normal in G.

For future reference we state in the following corollary the claims of Proposition 9.4.2 in the particular case in which H is a normal subgroup of G.

9.4.4 Corollary *Let $(M, \{\cdot, \cdot\})$ be a Poisson manifold acted properly and canonically upon by a Lie group G. Let N be a closed normal Lie subgroup of G. Then:*

(i) *The group G acts on A'_N and on the corresponding momentum space M/A'_N with a natural action that makes the N-optimal momentum map $\mathcal{J}_N : M \to M/A'_N$ G-equivariant.*

(ii) *There is a natural G-equivariant projection $\pi_N : M/A'_G \to M/A'_N$ such that $\mathcal{J}_N = \pi_N \circ \mathcal{J}_G$.*

9.4.5 Isotropy subgroups and quotient groups. In this section we introduce the relevant groups and spaces for optimal reduction in two stages.

9.4.6 Lemma *Let $(M, \{\cdot, \cdot\})$ be a Poisson manifold acted properly and canonically upon by a Lie group G. Let N be a closed normal Lie subgroup of G. Let $\rho \in M/A'_G$ and $\nu := \pi_N(\rho) \in M/A'_N$.*

(i) *Let G_ρ and G_ν be the isotropy subgroups of $\rho \in M/A'_G$ and $\nu := \pi_N(\rho) \in M/A'_N$ with respect to the G-actions on M/A'_G and M/A'_N, respectively. Then, $G_\rho \subset G_\nu$.*

(ii) *Let N_ν be the N-isotropy subgroup of $\nu \in M/A'_N$. Then $N_\nu = N \cap G_\nu$ and N_ν is normal in G_ν.*

(iii) *Endow N_ν and G_ν with the unique smooth structures that make them into initial Lie subgroups of G. Then N_ν is closed in G_ν and therefore the quotient $H_\nu := G_\nu/N_\nu$ is a Lie group.*

Proof. **(i)** This is a consequence of the G-equivariance of the projection $\pi_N : M/A'_G \to M/A'_N$. **(ii)** This is straightforward. **(iii)** Let A and B two subsets of a smooth manifold M such that $A \subset B \subset M$. It can be checked by simply using the definition of initial submanifold that if A and B are initial submanifolds of M, then A is an initial submanifold of B. In our setup, this fact implies that N_ν is an initial Lie subgroup of G_ν. We will check that it is a closed Lie subgroup of G_ν. Indeed, let $g \in G_\nu$ be an element in the closure of N_ν in G_ν. Let $\{g_n\}_{n\in\mathbb{N}} \subset N_\nu$ be a sequence of elements in N_ν that converges to g in the topology of G_ν. As G_ν is initial in G, we have that $g_n \to g$ also in the topology of G. Now, as $\{g_n\}_{n\in\mathbb{N}} \subset N$ and N is closed in G, $g \in N$ necessarily. Hence $g \in N \cap G_\nu = N_\nu$, as required. ■

Suppose now that the value $\nu \in M/A'_N$ is such that the action of N_ν on the level set $\mathcal{J}_N^{-1}(\nu)$ is proper. We emphasize that this property is not automatically inherited from the properness of the N-action on M. Theorem 9.1.1 guarantees in that situation that the orbit space $M_\nu := \mathcal{J}_N^{-1}(\nu)/N_\nu$ is a smooth symplectic regular quotient manifold with symplectic form ω_ν defined by

$$\pi_\nu^*\omega_\nu(m)(X_f(m), X_h(m)) = \{f, h\}(m),$$

for any $m \in \mathcal{J}_N^{-1}(\nu)$ and any $f, h \in C^\infty(M)^N$. As customary $\pi_\nu : \mathcal{J}_N^{-1}(\nu) \to \mathcal{J}_N^{-1}(\nu)/N_\nu$ denotes the canonical projection and $i_\nu : \mathcal{J}_N^{-1}(\nu) \hookrightarrow M$ the inclusion. We will refer to the pair (M_ν, ω_ν) as the ***first stage reduced space***.

9.4.7 Proposition *Let $(M, \{\cdot,\cdot\})$ be a Poisson manifold acted properly and canonically upon by a Lie group G via the map $\Phi : G \times M \to M$. Let N be a closed normal Lie subgroup of G. Let $\rho = \mathcal{J}_G(m) \in M/A'_G$, for some $m \in M$, and $\nu := \pi_N(\rho) = \mathcal{J}_N(m) \in M/A'_N$.*

(i) *If the Lie group N_ν acts properly on the level set $\mathcal{J}_N^{-1}(\nu)$, then the Lie group $H_\nu := G_\nu/N_\nu$ acts smoothly and canonically on the first stage reduced space $(\mathcal{J}_N^{-1}(\nu)/N_\nu, \omega_\nu)$ via the map*

$$gN_\nu \cdot \pi_\nu(m) := \pi_\nu(g \cdot m), \tag{9.4.3}$$

for all $gN_\nu \in H_\nu$ and $m \in \mathcal{J}_N^{-1}(\nu)$.

(ii) *Suppose that N_ν and H_ν act properly on $\mathcal{J}_N^{-1}(\nu)$ and M_ν, respectively. Let $\mathcal{J}_{H_\nu} : M_\nu \to M_\nu/A'_{H_\nu}$ be the optimal momentum map associated to the H_ν-action on*

$M_\nu = \mathcal{J}_N^{-1}(\nu)/N_\nu$ *and let* $\sigma =: \mathcal{J}_{H_\nu}(\pi_\nu(m))$. *Then*

$$\mathcal{J}_{H_\nu}\left(\pi_\nu(\mathcal{J}_G^{-1}(\rho))\right) = \sigma. \tag{9.4.4}$$

Proof. (i) We first show that the action given by expression (9.4.3) is well defined and is smooth. The action $\varphi^\nu : G_\nu \times \mathcal{J}_N^{-1}(\nu) \to \mathcal{J}_N^{-1}(\nu)$ obtained by restriction of the domain and range of Φ is smooth since G_ν and $\mathcal{J}_N^{-1}(\nu)$ are initial submanifolds of G and M, respectively. Also, this map is compatible with the action of $N_\nu \times N_\nu$ on $G_\nu \times \mathcal{J}_N^{-1}(\nu)$ defined by $(n, n') \cdot (g, z) := (gn^{-1}, n' \cdot z)$, and the N_ν-action on $\mathcal{J}_N^{-1}(\nu)$. Indeed, for any $(n, n') \in N_\nu \times N_\nu$ and any $(g, z) \in G_\nu \times \mathcal{J}_N^{-1}(\nu)$, the point $(gn^{-1}, n' \cdot z)$ gets sent by this map to $gn^{-1}n' \cdot z$. As N_ν is normal in G_ν, there exists some $n'' \in N_\nu$ such that $gn^{-1}n' \cdot z = n''g \cdot z$ which is in the same N_ν-orbit as $g \cdot z$. Consequently, the map $\varphi^\nu : G_\nu \times \mathcal{J}_N^{-1}(\nu) \to \mathcal{J}_N^{-1}(\nu)$ drops to a smooth map $\phi^\nu : G_\nu/N_\nu \times \mathcal{J}_N^{-1}(\nu)/N_\nu \to \mathcal{J}_N^{-1}(\nu)/N_\nu$ that coincides with (9.4.3) and therefore $\phi^\nu_{kN_\nu} \circ \pi_\nu = \pi_\nu \circ \varphi^\nu_k$, for any $kN_\nu \in H_\nu$.

We now show that the action given by the map ϕ^ν is canonical. Let $kN_\nu \in H_\nu$, $m \in \mathcal{J}_N^{-1}(\nu)$, and $f, h \in C^\infty(M)^N$ be arbitrary. Then, taking into account that $\phi^\nu_{kN_\nu} \circ \pi_\nu = \pi_\nu \circ \varphi^\nu_k$ and that by Proposition 9.4.2**(i)** the functions $f \circ \Phi_{k^{-1}}$ and $h \circ \Phi_{k^{-1}}$ are N-invariant, we can write:

$$\begin{aligned}
\pi_\nu^*((\phi^\nu_{kN_\nu})^*\omega_\nu)(m)(X_f(m), X_h(m)) &= ((\phi^\nu_{kN_\nu} \circ \pi_\nu)^*\omega_\nu)(m)(X_f(m), X_h(m)) \\
&= ((\pi_\nu \circ \varphi^\nu_k)^*\omega_\nu)(m)(X_f(m), X_h(m)) \\
&= (\varphi^\nu_k)^*(\pi_\nu^*\omega_\nu)(m)(X_f(m), X_h(m)) \\
&= (\pi_\nu^*\omega_\nu)(k \cdot m)(T_m\Phi_k(X_f(m)), T_m\Phi_k(X_h(m))) \\
&= (\pi_\nu^*\omega_\nu)(k \cdot m)(X_{f\circ\Phi_{k^{-1}}}(k \cdot m), X_{h\circ\Phi_{k^{-1}}}(k \cdot m)) \\
&= \{f \circ \Phi_{k^{-1}}, h \circ \Phi_{k^{-1}}\}(k \cdot m) = \{f, h\}(m) \\
&= (\pi_\nu^*\omega_\nu)(m)(X_f(m), X_h(m)).
\end{aligned}$$

Since the map π_ν is a surjective submersion, this chain of equalities implies that $(\phi^\nu_{kN_\nu})^*\omega_\nu = \omega_\nu$, as required.

(ii) Let $m' \in \mathcal{J}_G^{-1}(\rho)$ be such that $m' \neq m$. Then, there exists $\mathcal{F}_T \in G_{\mathcal{A}'_G}$ such that $m' = \mathcal{F}_T(m)$. For simplicity in the exposition take $\mathcal{F}_T = F_T$, with F_T the Hamiltonian flow associated to the G-invariant function $f \in C^\infty(M)^G$. Let now $f_\nu \in C^\infty(M_\nu)^{H_\nu}$ be the H_ν-invariant function on M_ν uniquely determined by the relation $f_\nu \circ \pi_\nu = f \circ i_\nu$. The Hamiltonian flow F^ν_T associated to f_ν is related to F_T by the relation $F^\nu_T \circ \pi_\nu = \pi_\nu \circ F_T \circ i_\nu$. Therefore, by Noether's Theorem applied to $\mathcal{J}_{H_\nu}$, we have that

$$\mathcal{J}_{H_\nu}(\pi_\nu(m')) = \mathcal{J}_{H_\nu}(\pi_\nu(F_T(m))) = \mathcal{J}_{H_\nu}(F^\nu_T(\pi_\nu(m))) = \mathcal{J}_{H_\nu}(\pi_\nu(m)) = \sigma,$$

as required. ■

9.4.8 The Optimal Reduction by Stages Theorem. Let $m \in M$ be such that $\rho = \mathcal{J}_G(m)$. Also, let $\nu = \mathcal{J}_N(m)$ and $\sigma = \mathcal{J}_{H_\nu}(\pi_\nu(m))$. The second part of Proposition 9.4.7 guarantees that the restriction of π_ν to $\mathcal{J}_G^{-1}(\rho)$ gives a well-defined map

$$\pi_\nu|_{\mathcal{J}_G^{-1}(\rho)} : \mathcal{J}_G^{-1}(\rho) \longrightarrow \mathcal{J}_{H_\nu}^{-1}(\sigma).$$

This map is smooth because $\mathcal{J}_{H_\nu}^{-1}(\sigma)$ is an initial submanifold of M_ν and also because $\mathcal{J}_G^{-1}(\rho)$ and $\mathcal{J}_N^{-1}(\nu)$ are initial submanifolds of M such that $\mathcal{J}_G^{-1}(\rho) \subset \mathcal{J}_N^{-1}(\nu) \subset M$; this implies that $\mathcal{J}_G^{-1}(\rho)$ is an initial submanifold of $\mathcal{J}_N^{-1}(\nu)$ (this argument is a straightforward consequence of the definition of an initial submanifold). Denote by $i_{\rho,\nu} : \mathcal{J}_G^{-1}(\rho) \hookrightarrow \mathcal{J}_N^{-1}(\nu)$ the corresponding smooth injection. Let $(H_\nu)_\sigma$ be the H_ν-isotropy subgroup of the element $\sigma \in M_\nu/A'_{H_\nu}$. Then the map $\pi_\nu|_{\mathcal{J}_G^{-1}(\rho)} = \pi_\nu \circ i_{\rho,\nu} : \mathcal{J}_G^{-1}(\rho) \to \mathcal{J}_{H_\nu}^{-1}(\sigma)$ is smooth and $(G_\rho, (H_\nu)_\sigma)$-equivariant. Indeed, let $g \in G_\rho$ and $m \in \mathcal{J}_G^{-1}(\rho)$ be arbitrary. Lemma 9.4.6 and the inclusion $G_\rho \subset G_\nu$ imply that $g \in G_\nu$ and $gN_\nu \in G_\nu/N_\nu$. Using Definition 9.4.3, we have that $\pi_\nu(g \cdot m) = gN_\nu \cdot \pi_\nu(m)$. Additionally, by (9.4.4) we have that

$$\mathcal{J}_{H_\nu}(gN_\nu \cdot \pi_\nu(m)) = \mathcal{J}_{H_\nu}(\pi_\nu(g \cdot m)) = \sigma,$$

because $g \cdot m \in \mathcal{J}_G^{-1}(\rho)$, which shows that $gN_\nu \in (H_\nu)_\sigma$ and therefore guarantees the $(G_\rho, (H_\nu)_\sigma)$-equivariance of $\pi_\nu|_{\mathcal{J}_G^{-1}(\rho)}$. Consequently, the map $\pi_\nu|_{\mathcal{J}_G^{-1}(\rho)}$ drops to a well-defined map F that makes the following diagram

$$\begin{array}{ccc} \mathcal{J}_G^{-1}(\rho) & \xrightarrow{\pi_\nu|_{\mathcal{J}_G^{-1}(\rho)}} & \mathcal{J}_{H_\nu}^{-1}(\sigma) \\ {\scriptstyle \pi_\rho}\downarrow & & \downarrow{\scriptstyle \pi_\sigma} \\ \mathcal{J}_G^{-1}(\rho)/G_\rho & \xrightarrow{F} & \mathcal{J}_{H_\nu}^{-1}(\sigma)/(H_\nu)_\sigma . \end{array}$$

commutative. We remind the reader once more that the G_ρ and $(H_\nu)_\sigma$-actions on $\mathcal{J}_G^{-1}(\rho)$ and $\mathcal{J}_{H_\nu}^{-1}(\sigma)$, respectively, are not automatically proper as a consequence of the properness of the G-action on M. If this happens to be the case, the map F is smooth. Moreover, in this situation, Theorem 9.1.1 guarantees that the quotients $M_\rho := \mathcal{J}_G^{-1}(\rho)/G_\rho$ and $(M_\nu)_\sigma := \mathcal{J}_{H_\nu}^{-1}(\sigma)/(H_\nu)_\sigma$ are symplectic manifolds. We will refer to the symplectic manifold $(\mathcal{J}_{H_\nu}^{-1}(\sigma)/(H_\nu)_\sigma, \omega_\sigma)$ as the ***second stage reduced space***. Recall that the symplectic form ω_σ is uniquely determined by the equality $\pi_\sigma^*\omega_\sigma = i_\sigma^*\omega_\nu$, where $i_\sigma : \mathcal{J}_{H_\nu}^{-1}(\sigma) \hookrightarrow \mathcal{J}_N^{-1}(\nu)/N_\nu$ is the injection and $\pi_\sigma : \mathcal{J}_{H_\nu}^{-1}(\sigma) \to \mathcal{J}_{H_\nu}^{-1}(\sigma)/(H_\nu)_\sigma$ the projection.

Our goal in this section will consist in proving a theorem that, under certain hypotheses, states that the map F is a symplectomorphism between the one-step reduced space $(\mathcal{J}_G^{-1}(\rho)/G_\rho, \omega_\rho)$ and the reduced space in two steps $(\mathcal{J}_{H_\nu}^{-1}(\sigma)/(H_\nu)_\sigma, \omega_\sigma)$.

Because the properness assumptions appear profusely we will simplify the exposition by grouping them all in the following definition.

9.4.9 Definition *Let $(M, \{\cdot,\cdot\})$ be a Poisson manifold acted properly and canonically upon by a Lie group G via the map $\Phi : G \times M \to M$. Let N be a closed normal Lie subgroup of G. Let $\rho \in M/A'_G$, $\nu := \pi_N(\rho)$, $H_\nu := G_\nu/N_\nu$, and $\sigma = \mathcal{J}_{H_\nu}\left(\pi_\nu(\mathcal{J}_G^{-1}(\rho))\right) \in M_\nu/A'_{H_\nu}$. We will say that we have **proper actions at** ρ whenever G_ρ acts properly on $\mathcal{J}_G^{-1}(\rho)$, N_ν acts properly on $\mathcal{J}_N^{-1}(\nu)$, H_ν acts properly on $\mathcal{J}_N^{-1}(\nu)/N_\nu$, and $(H_\nu)_\sigma$ acts properly on $\mathcal{J}_{H_\nu}^{-1}(\sigma)$.*

Let $(G_\nu)_\sigma \subset G_\nu$ be the unique subgroup of G_ν such that $(H_\nu)_\sigma = (G_\nu)_\sigma / N_\nu$. We say that the element $\rho \in M/A'_G$ satisfies the **stages hypothesis** *when for any other element $\rho' \in M/A'_G$ such that*

$$\pi_N(\rho) = \pi_N(\rho') =: \nu \quad \text{and} \quad \mathcal{J}_{H_\nu}(\pi_\nu(\mathcal{J}_G^{-1}(\rho))) = \mathcal{J}_{H_\nu}(\pi_\nu(\mathcal{J}_G^{-1}(\rho'))) = \sigma$$

there exists an element $h \in (G_\nu)_\sigma$ such that $\rho' = h \cdot \rho$.

We say that the element $\nu \in M/A'_N$ has the **local extension property** *when any function $f \in C^\infty(\mathcal{J}_N^{-1}(\nu))^{G_\nu}$ is such that for any $m \in \mathcal{J}_N^{-1}(\nu)$ there is an open N-invariant neighborhood U of m and a function $F \in C^\infty(M)^G$ such that $F|_U = f|_U$.*

9.4.10 Theorem (Optimal Reduction by Stages) *Let $(M, \{\cdot, \cdot\})$ be a Poisson manifold acted properly and canonically upon by a Lie group G via the map $\Phi : G \times M \to M$. Let N be a closed normal Lie subgroup of G. Let $\rho \in M/A'_G$, $\nu := \pi_N(\rho)$, $H_\nu := G_\nu / N_\nu$, and $\sigma = \mathcal{J}_{H_\nu}\left(\pi_\nu(\mathcal{J}_G^{-1}(\rho))\right) \in M_\nu / A'_{H_\nu}$. Then, if ρ satisfies the stages hypothesis, the actions are proper at ρ, and the quotient manifold $\mathcal{J}_G^{-1}(\rho)/G_\rho$ is either Lindelöf or paracompact, the map*

$$\begin{aligned} F : \left(\mathcal{J}_G^{-1}(\rho)/G_\rho, \omega_\rho\right) &\longrightarrow \left(\mathcal{J}_{H_\nu}^{-1}(\sigma)/(H_\nu)_\sigma, \omega_\sigma\right) \\ \pi_\rho(m) &\longmapsto \pi_\sigma(\pi_\nu(m)) \end{aligned}$$

is a symplectomorphism between the one-step reduced space $(\mathcal{J}_G^{-1}(\rho)/G_\rho, \omega_\rho)$ and $(\mathcal{J}_{H_\nu}^{-1}(\sigma)/(H_\nu)_\sigma, \omega_\sigma)$ that is obtained by reduction in two steps.

Proof. F **is injective**: Let $\pi_\rho(m)$ and $\pi_\rho(m') \in \mathcal{J}_G^{-1}(\rho)/G_\rho$ be such that $F(\pi_\rho(m)) = F(\pi_\rho(m'))$. By the definition of F this implies that $\pi_\sigma(\pi_\nu(m)) = \pi_\sigma(\pi_\nu(m'))$. Hence, there exists an element $gN_\nu \in (H_\nu)_\sigma$ such that $\pi_\nu(m') = gN_\nu \cdot \pi_\nu(m)$ which, by the definition (9.4.3), is equivalent to $\pi_\nu(m') = \pi_\nu(g \cdot m)$. Therefore, there exists $n \in N_\nu$ such that $m' = ng \cdot m$. However, since both m and m' are in $\mathcal{J}_G^{-1}(\rho)$ we have that $ng \in G_\rho$ and, consequently, $\pi_\rho(m) = \pi_\rho(m')$, as required.

F **is surjective**: Let $\pi_\sigma(\bar{z}) \in (M_\nu)_\sigma = \mathcal{J}_{H_\nu}^{-1}(\sigma)/(H_\nu)_\sigma$. Take any $z \in \mathcal{J}_N^{-1}(\nu)$ such that $\pi_\nu(z) = \bar{z}$ and let $\rho' := \mathcal{J}_G(z)$. It is clear that $\pi_N(\rho') = \pi_N(\mathcal{J}_G(z)) = \mathcal{J}_N(z) = \nu = \pi_N(\rho)$ and also, as $\mathcal{J}_{H_\nu}(\pi_\nu(z)) = \sigma$, Lemma 9.4.7 guarantees that $\mathcal{J}_{H_\nu}(\pi_\nu(\mathcal{J}_G^{-1}(\rho'))) = \sigma = \mathcal{J}_{H_\nu}(\pi_\nu(\mathcal{J}_G^{-1}(\rho)))$. By the stages hypothesis, there exists $h \in (G_\nu)_\sigma$ such that $\rho' = h \cdot \rho$. Now, we have that

$$F(\pi_\rho(h^{-1} \cdot z)) = \pi_\sigma(\pi_\nu(h^{-1} \cdot z)) = \pi_\sigma(h^{-1}N_\nu \cdot \pi_\nu(z)) = \pi_\sigma(\pi_\nu(z)) = \pi_\sigma(\bar{z}),$$

which proves the surjectivity of F.

F **is a symplectic map**: we will show that $F^*\omega_\sigma = \omega_\rho$. Let $m \in \mathcal{J}_G^{-1}(\rho)$ and $f, g \in$

$C^\infty(M)^G$ arbitrary. Then,

$$\begin{aligned}
\pi_\rho^*(F^*\omega_\sigma)(m)(X_f(m), X_g(m)) &= (F\circ\pi_\rho)^*\omega_\sigma(m)(X_f(m), X_g(m))\\
&= \left(\pi_\sigma\circ\pi_\nu|_{\mathcal{J}_G^{-1}(\rho)}\right)^*\omega_\sigma(m)(X_f(m), X_g(m))\\
&= (\pi_\sigma\circ\pi_\nu\circ i_{\rho,\nu})^*\omega_\sigma(m)(X_f(m), X_g(m))\\
&= \left((\pi_\nu\circ i_{\rho,\nu})^*(\pi_\sigma^*\omega_\sigma)\right)(m)(X_f(m), X_g(m))\\
&= \left((\pi_\nu\circ i_{\rho,\nu})^*(i_\sigma^*\omega_\nu)\right)(m)(X_f(m), X_g(m))\\
&= (i_\sigma\circ\pi_\nu\circ i_{\rho,\nu})^*\omega_\nu(m)(X_f(m), X_g(m))\\
&= \pi_\nu^*\omega_\nu(m)(X_f(m), X_g(m)) = \{f, g\}(m)\\
&= \pi_\rho^*\omega_\rho(m)(X_f(m), X_g(m)).
\end{aligned}$$

This chain of equalities guarantees that $\pi_\rho^*(F^*\omega_\sigma) = \pi_\rho^*\omega_\rho$. Since the map π_ρ is a surjective submersion we have that $F^*\omega_\sigma = \omega_\rho$, and consequently F is a symplectic map.

F **is a symplectomorphism**: Since F is a bijective symplectic map, it is necessarily an immersion. Since by hypothesis the space $\mathcal{J}_G^{-1}(\rho)/G_\rho$ is either Lindelöf or paracompact, Lemma 1.1.17 guarantees that F is actually a diffeomorphism. ■

9.4.11 Proposition *Let $(M, \{\cdot,\cdot\})$ be a Poisson manifold acted properly and canonically upon by a Lie group G via the map $\Phi : G\times M \to M$. Let N be a closed normal Lie subgroup of G. Let $\rho \in M/A'_G$, $\nu := \pi_N(\rho)$, $H_\nu := G_\nu/N_\nu$, and $\sigma = \mathcal{J}_{H_\nu}\left(\pi_\nu(\mathcal{J}_G^{-1}(\rho))\right) \in M_\nu/A'_{H_\nu}$. If ν has the local extension property and N_ν acts properly on $\mathcal{J}_N^{-1}(\nu)$, then $\pi_\nu(\mathcal{J}_G^{-1}(\rho)) = \mathcal{J}_{H_\nu}^{-1}(\sigma)$ and ρ satisfies the stages hypothesis.*

Proof. The inclusion $\pi_\nu(\mathcal{J}_G^{-1}(\rho)) \subset \mathcal{J}_{H_\nu}^{-1}(\sigma)$ is guaranteed by (9.4.4). In order to prove the equality take $\pi_\nu(m) \in \pi_\nu(\mathcal{J}_G^{-1}(\rho))$ and $f \in C^\infty(M_\nu)^{H_\nu}$ to be arbitrary, such that the Hamiltonian vector field X_f on M_ν has flow F_t. Let $\bar{f} \in C^\infty(\mathcal{J}_N^{-1}(\nu))^{G_\nu}$ be the function defined by $\bar{f} := f\circ\pi_\nu$. The H_ν-invariance of f implies that $\bar{f}$ is G_ν-invariant. In principle, the point $F_T(\pi_\nu(m))$ lies somewhere in $\mathcal{J}_{H_\nu}^{-1}(\sigma)$. However, we will show that it actually stays in $\pi_\nu(\mathcal{J}_G^{-1}(\rho))$, which will prove the desired equality. Indeed, as the curve $\{F_t(\pi_\nu(m))\}_{t\in[0,T]}$ is compact it can be covered by a finite number of open sets $\{U_1, \ldots, U_n\}$. Suppose that we have chosen the neighborhoods U_i such that $\pi_\nu(m) \in U_1$, $F_T(\pi_\nu(m)) \in U_n$, $U_i \cap U_j \neq \emptyset$ iff $|j-i| = 1$, and for each open N-invariant set $\pi_\nu^{-1}(U_i)$, there is a $g_i \in C^\infty(M)^G$ such that $\bar{f}|_{\pi_\nu^{-1}(U_i)} = g_i|_{\pi_\nu^{-1}(U_i)}$, where the function $\bar{f}$ admits local extensions to G-invariant functions on M. Denote by G_t^i the flow of the Hamiltonian vector field X_{g_i} on M associated to $g_i \in C^\infty(M)^G$. The flows G_t^i and F_t are related by the equality $F_t\circ\pi_\nu|_{\mathcal{J}_N^{-1}(\nu)\cap\pi_\nu^{-1}(U_i)} = \pi_\nu\circ G_t^i\circ i_\nu|_{\mathcal{J}_N^{-1}(\nu)\cap\pi_\nu^{-1}(U_i)}$. Due to the G-invariance of g we have that $\mathcal{J}_G\circ G_t^i = \mathcal{J}_G$ and thus $\{F_t(\pi_\nu(m))\}_{t\in[0,T]} \subset \pi_\nu(\mathcal{J}_G^{-1}(\rho))$, as required. This proves that $\pi_\nu(\mathcal{J}_G^{-1}(\rho)) = \mathcal{J}_{H_\nu}^{-1}(\sigma)$.

We conclude by showing that this equality implies that ρ satisfies the stages hypothesis. Indeed, if $\rho' \in M/A'_G$ is such that $\mathcal{J}_{H_\nu}(\pi_\nu(\mathcal{J}_G^{-1}(\rho'))) = \sigma$, then

$$\pi_\nu(\mathcal{J}_G^{-1}(\rho')) \subset \mathcal{J}_{H_\nu}^{-1}(\sigma) = \pi_\nu(\mathcal{J}_G^{-1}(\rho)).$$

Consequently, for any $\pi_\nu(z') \in \pi_\nu(\mathcal{J}_G^{-1}(\rho'))$, $z' \in \mathcal{J}_G^{-1}(\rho')$, there exists an element $z \in \mathcal{J}_G^{-1}(\rho)$ such that $\pi_\nu(z') = \pi_\nu(z)$. Hence, there is an element $n \in N_\nu \subset (G_\nu)_\sigma$ such that $z' = n \cdot z$ which, by applying the map $\mathcal{J}_G$ to both sides of this equality, implies that $\rho' = n \cdot \rho$. ■

9.5 Singular reduction by stages in the presence of a standard momentum map

In this section we will assume that M is a symplectic manifold with symplectic form ω and that the G-action is proper and canonical. We will also assume that this action admits a standard coadjoint equivariant momentum map $\mathbf{J}_G : M \to \mathfrak{g}^*$ and that G contains a closed normal subgroup N. Recall that the inclusion $N \subset G$ and the normal character of N in G implies that $\mathfrak{n}$ is an ideal in $\mathfrak{g}$. Let $i : \mathfrak{n} \hookrightarrow \mathfrak{g}$ be the inclusion. As we saw in Section 6.7, the N-action on M is also globally Hamiltonian with a G-equivariant momentum map $\mathbf{J}_N : M \to \mathfrak{n}^*$ given by $\mathbf{J}_N = i^*\mathbf{J}_G$.

We shall see below how our understanding of the optimal reduction by stages procedure allows us to generalize the results in Section 6.7 to the nonfree actions case. More specifically, we will see that the reduced spaces and subgroups involved in the Optimal Reduction by Stages Theorem 9.4.10 admit in this case a very precise characterization in terms of level sets of the standard momentum maps present in the problem and of various subgroups of G obtained as a byproduct of isotropy subgroups related to the G and N-actions on M and the coadjoint actions on $\mathfrak{g}^*$ and $\mathfrak{n}^*$.

We start our study by looking in this setup at the level sets of the G and N-optimal momentum maps. The characterization provided by Corollary 5.5.18 establishes the following facts. Let $m \in M$ be such that $\mathcal{J}_G(m) = \rho$, $\mathbf{J}_G(m) = \mu$, and $G_m =: H$. Then $\mathcal{J}_G^{-1}(\rho)$ equals the unique connected component of the submanifold $\mathbf{J}_G^{-1}(\mu) \cap M_H^m$ containing m. Analogously, if $\mathcal{J}_N(m) = \nu$, $\mathbf{J}_N(m) = \eta$, and $N_m = H \cap N$, then $\mathcal{J}_N^{-1}(\nu)$ equals the unique connected component of the submanifold $\mathbf{J}_N^{-1}(\eta) \cap M_{H\cap N}^m$ containing m. Throughout this section we will assume the following:

Connectedness hypothesis: The submanifolds $\mathbf{J}_G^{-1}(\mu) \cap M_H$ and $\mathbf{J}_N^{-1}(\eta) \cap M_{H\cap N}$ are connected and coincide with $\mathbf{J}_G^{-1}(\mu) \cap M_H^m$ and $\mathbf{J}_N^{-1}(\eta) \cap M_{H\cap N}^m$, respectively.

This hypothesis is *not* realistic. However, it will make the presentation that follows much more clear and accessible. The statement and proof of the reduction by stages theorem presented below is independent, qualitatively speaking, of the above connectedness hypothesis. However, the notational additions necessary to accommodate the most general case would make the following pages very difficult to read. In order to adapt the results discussed in this section to the general situation one would need to work with the connected components of $\mathbf{J}_G^{-1}(\mu) \cap M_H$ and $\mathbf{J}_N^{-1}(\eta) \cap M_{H\cap N}$. Additionally, one should replace in the following presentation each group that leaves these two sets invariant by their relevant closed subgroups that leave the previously chosen connected components invariant. If this is carried out explicitly, the notation becomes immediately very convoluted but the ideas involved in the proof are the same.

We continue our characterization of the ingredients for reduction by stages in the following proposition.

9.5.1 Proposition *Let (M, ω) be a symplectic manifold acted properly and canonically upon by a Lie group G and suppose that this action admits a standard equivariant momentum map $\mathbf{J}_G : M \to \mathfrak{g}^*$. Let $N \subset G$ be a closed normal subgroup of G. For $m \in M$, denote $\mathcal{J}_G(m) =: \rho$, $\mathbf{J}_G(m) =: \mu$, and $G_m =: H$. Then:*

(i) $\mathcal{J}_G^{-1}(\rho) = \mathbf{J}_G^{-1}(\mu) \cap M_H$,

(ii) $\mathcal{J}_N(m) = \pi_N(\rho) =: \nu$, $\mathbf{J}_N(m) = i^*\mu =: \eta$, *and* $\mathcal{J}_N^{-1}(\nu) = \mathbf{J}_N^{-1}(\eta) \cap M_{N_\eta \cap H}$,

(iii) $G_\rho = N_{G_\mu}(H)$, $N_\nu = N_{N_\eta}(N_\eta \cap H)$, *and* $G_\nu = N_{G_\eta}(N_\eta \cap H)$. *Recall that $N_{G_\mu}(H) := N(H) \cap G_\mu$, where $N(H)$ denotes the normalizer of H in G. The subgroup $N_{G_\mu}(H)$ is the normalizer of H in G_μ.*

Proof. The proof of the equalities $\mathcal{J}_G^{-1}(\rho) = \mathbf{J}_G^{-1}(\mu) \cap M_H$ and $G_\rho = N_{G_\mu}(H)$ is a consequence of Corollary 5.5.18 and of the connectedness hypothesis.

We now show that $\mathcal{J}_N^{-1}(\nu) = \mathbf{J}_N^{-1}(\eta) \cap M_{N_\eta \cap H}$. It suffices to show that $N_m = N_\eta \cap H$. Indeed, as the G-equivariance of $\mathbf{J}_N$ implies that $H = G_m \subset G_\eta$, we have that $N_m = H \cap N = H \cap G_\eta \cap N = N_\eta \cap H$. Consequently, the same result that gave us $G_\rho = N_{G_\mu}(H)$ can be applied to the N-action on M to obtain $N_\nu = N_{N_\eta}(N_\eta \cap H)$.

Finally, we prove the identity $G_\nu = N_{G_\eta}(N_\eta \cap H)$ by double inclusion. First let $g \in G_\nu$. The equality $g \cdot \nu = \nu$ implies that $g \cdot m = \mathcal{F}_T(m)$, with $\mathcal{F}_T \in A'_N$. For simplicity suppose that $\mathcal{F}_T = F_t$, with F_t the Hamiltonian flow associated to an N-invariant function on M. The standard Noether Theorem implies that $g \cdot m = F_t(m) \in \mathbf{J}_N^{-1}(\eta)$ and therefore $g \in G_\eta$. Also, as the flow F_t is N-equivariant, we have that

$$N_\eta \cap H = N_m = N_{F_t(m)} = N_{g \cdot m} = g N_m g^{-1} = g(N_\eta \cap H)g^{-1},$$

and consequently $g \in N_{G_\eta}(N_\eta \cap H)$. The reverse inclusion is trivial. ■

9.5.2 A major consequence of the previous proposition is the fact that the subgroups G_ν and N_ν, and those that will derive from them, are automatically closed subgroups. This circumstance implies that the proper actions hypothesis given in Definition 9.4.9, which is necessary for reduction by stages, is automatically satisfied in this setup.

9.5.3 The previous proposition allows us to explicitly write down in our setup the full reduced space:

$$M_\rho := \mathcal{J}_G^{-1}(\rho)/G_\rho = \mathbf{J}_G^{-1}(\mu) \cap M_H / N_{G_\mu}(H), \tag{9.5.1}$$

as well as the first stage reduced space:

$$M_\nu := \mathcal{J}_N^{-1}(\nu)/N_\nu = \mathbf{J}_N^{-1}(\eta) \cap M_{N_\eta \cap H} / N_{N_\eta}(N_\eta \cap H).$$

9.5.4 We now proceed with the construction of the second stage reduced space. As it was already the case in the general optimal setup, the quotient group

$$\mathcal{H}_\nu := G_\nu/N_\nu = \frac{N_{G_\eta}(N_\eta \cap H)}{N_{N_\eta}(N_\eta \cap H)}$$

acts canonically on the quotient M_ν with associated optimal momentum map $\mathcal{J}_{\mathcal{H}_\nu} : M_\nu \to M_\nu/A'_{\mathcal{H}_\nu}$. In this setup we can say more. Indeed, in this case the $\mathcal{H}_\nu$-action on M_ν is automatically proper and has an associated standard momentum map $\mathbf{J}_{\mathcal{H}_\nu} : M_\nu \to \mathrm{Lie}(\mathcal{H}_\nu)^*$, where the symbol $\mathrm{Lie}(\mathcal{H}_\nu)$ denotes the Lie algebra of the group $\mathcal{H}_\nu$. An explicit expression for $\mathbf{J}_{\mathcal{H}_\nu}$ can be obtained by mimicking the computations made in Section 6.7 for the free case. In order to write it down we introduce the following maps: let $\pi_{G_\nu} : G_\nu \to G_\nu/N_\nu$ be the projection, $r_\nu = T_e\pi_{G_\nu} : \mathfrak{g}_\nu \to \mathrm{Lie}(\mathcal{H}_\nu) \simeq \mathfrak{g}_\nu/\mathfrak{n}_\nu$ its derivative at the identity, and $r_\nu^* : \mathrm{Lie}(\mathcal{H}_\nu)^* \to \mathfrak{g}_\nu^*$ the corresponding dual map. Then, if N_ν is connected, we have that for any $\pi_\nu(z) \in M_\nu$ and any $r_\nu(\xi) \in \mathrm{Lie}(\mathcal{H}_\nu)$, the momentum map $\mathbf{J}_{\mathcal{H}_\nu}$ is given by the expression

$$\langle \mathbf{J}_{\mathcal{H}_\nu}(\pi_\nu(z)), r_\nu(\xi)\rangle = \langle \mathbf{J}_G(z), \xi\rangle - \langle \bar{\eta}, \xi\rangle, \tag{9.5.2}$$

where $\bar{\eta} \in \mathfrak{g}_\nu^*$ is some chosen extension of the restriction $\eta|_{\mathfrak{n}_\nu}$ to a linear functional on $\mathfrak{g}_\nu$. This momentum map is not equivariant. Indeed, if M_ν is connected, its non-equivariance cocycle $\bar{\omega}$ is given by the expression

$$r_\nu^*(\bar{\omega}(\pi_{G_\nu}(h))) = \mathrm{Ad}^*_{h^{-1}}\bar{\eta} - \bar{\eta},$$

for any $\pi_{G_\nu}(h) \in G_\nu/N_\nu$. The map $\mathbf{J}_{\mathcal{H}_\nu}$ becomes equivariant if we replace the coadjoint action of $\mathcal{H}_\nu$ on the dual of its Lie algebra by the affine action defined by

$$\pi_{G_\nu}(h) \cdot \lambda := \mathrm{Ad}^*_{(\pi_{G_\nu}(h))^{-1}}\lambda + \bar{\omega}(\pi_{G_\nu}(h)), \tag{9.5.3}$$

for any $\pi_{G_\nu}(h) \in \mathcal{H}_\nu$ and any $\lambda \in \mathrm{Lie}(\mathcal{H}_\nu)^*$. Now let $\tau \in \mathrm{Lie}(\mathcal{H}_\nu)^*$ be the element defined by

$$\langle \tau, r_\nu(\xi)\rangle = \langle \mu, \xi\rangle - \langle \bar{\nu}, \xi\rangle, \tag{9.5.4}$$

for any $r_\nu(\xi) \in \mathrm{Lie}(\mathcal{H}_\nu)$. A straightforward calculation, similar to the one in Section 6.7, shows that the isotropy subgroup $(\mathcal{H}_\nu)_\tau$ of τ with respect to the affine action (9.5.3) of $\mathcal{H}_\nu$ on the dual of its Lie algebra is given by

$$(\mathcal{H}_\nu)_\tau = \pi_{G_\nu}\left((G_\nu)_{\mu|_{\mathfrak{g}_\nu}}\right) = \frac{\left(\left(N_{G_\eta}(N_\eta \cap H)\right)_{\mu|_{\mathrm{Lie}\left(N_{G_\eta}(N_\eta\cap H)\right)}}\right)}{N_{N_\eta}(N_\eta \cap H)}. \tag{9.5.5}$$

Now, for any $m \in \mathcal{J}_G^{-1}(\rho)$, the choice of $\tau \in \mathrm{Lie}(\mathcal{H}_\nu)^*$ in (9.5.4) guarantees that $\mathbf{J}_{\mathcal{H}_\nu}(\pi_\nu(m)) = \tau$ and, moreover, if $\mathcal{J}_{\mathcal{H}_\nu}(\pi_\nu(m)) = \sigma \in M/A'_{\mathcal{H}_\nu}$, then

$$\mathcal{J}_{\mathcal{H}_\nu}^{-1}(\sigma) = \mathbf{J}_{\mathcal{H}_\nu}^{-1}(\tau) \cap (M_\nu)_{(\mathcal{H}_\nu)_{\pi_\nu(m)}} \tag{9.5.6}$$

since, by extension of the connectedness hypothesis, we will suppose that $\mathbf{J}^{-1}_{\mathcal{H}_\nu}(\tau) \cap (M_\nu)_{(\mathcal{H}_\nu)_{\pi_\nu(m)}}$ is also connected.

We compute the isotropy subgroup $(\mathcal{H}_\nu)_{\pi_\nu(m)}$ in terms of the groups that already appeared in our study. We will now show that

$$(\mathcal{H}_\nu)_{\pi_\nu(m)} = \frac{N_{N_\eta}(H \cap N_\eta)H}{N_{N_\eta}(N_\eta \cap H)}. \tag{9.5.7}$$

First take an element $\pi_{G_\nu}(g) \in \mathcal{H}_\nu$ such that $\pi_{G_\nu}(g) \cdot \pi_\nu(m) = \pi_\nu(m)$ or, equivalently, $\pi_\nu(g \cdot m) = \pi_\nu(m)$. Hence, there exists a group element $n \in N_\nu = N_{N_\eta}(N_\eta \cap H)$ such that $g \cdot m = n \cdot m$. Since $G_m = H$ it follows that $n^{-1}g \in H$ and hence $g \in N_{N_\eta}(N_\eta \cap H)H$ and $\pi_{G_\nu}(g) \in N_{N_\eta}(N_\eta \cap H)H/N_{N_\eta}(N_\eta \cap H)$. Conversely, if $\pi_{G_\nu}(g) \in N_{N_\eta}(N_\eta \cap H)H/N_{N_\eta}(N_\eta \cap H)$, we can write $g = nh$, with $n \in N_{N_\eta}(N_\eta \cap H)$ and $h \in H$; therefore $\pi_{G_\nu}(g) \cdot \pi_\nu(m) = \pi_\nu(nh \cdot m) = \pi_\nu(n \cdot m) = \pi_\nu(m)$, as required.

In order to write down the second stage reduced space we have to compute the isotropy subgroup $(\mathcal{H}_\nu)_\sigma$. From (9.5.5), (9.5.7), and Proposition 9.5.1 adapted to the optimal momentum map $\mathcal{J}_{\mathcal{H}_\nu}$ it follows that

$$(\mathcal{H}_\nu)_\sigma = N_{(\mathcal{H}_\nu)_\tau}\left(\frac{N_{N_\eta}(H \cap N_\eta)H}{N_{N_\eta}(N_\eta \cap H)}\right), \tag{9.5.8}$$

where the group $(\mathcal{H}_\nu)_\tau$ is given by (9.5.5). We now recall a standard result about normalizers that says that if $A \subset B \subset C$ are groups such that A is normal in both B and C, then

$$N_{C/A}(B/A) = N_C(B)/A.$$

If we apply this equality to (9.5.8), we obtain

$$\begin{aligned}(\mathcal{H}_\nu)_\sigma &= \frac{N_{(G_\nu)_{\mu|\mathfrak{g}_\nu}}(N_{N_\eta}(H \cap N_\eta)H)}{N_{N_\eta}(N_\eta \cap H)}\\ &= \frac{N_{\left(\left(N_{G_\eta}(N_\eta \cap H)\right)_{\mu|_{\mathrm{Lie}\left(N_{G_\eta}(N_\eta\cap H)\right)}}\right)}(N_{N_\eta}(H \cap N_\eta)H)}{N_{N_\eta}(N_\eta \cap H)}\end{aligned} \tag{9.5.9}$$

All the computations just carried out allow us to explicitly write down the second stage reduced space. Namely, combining expressions (9.5.6), (9.5.7), and (9.5.9), we obtain

$$(M_\nu)_\sigma = \mathcal{J}^{-1}_{\mathcal{H}_\nu}(\sigma)/(\mathcal{H}_\nu)_\sigma = \frac{\mathbf{J}^{-1}_{\mathcal{H}_\nu}(\tau) \cap (M_\nu)_{\frac{N_{N_\eta}(H\cap N_\eta)H}{N_{N_\eta}(N_\eta\cap H)}}}{\frac{N_{(G_\nu)_{\mu|\mathfrak{g}_\nu}}(N_{N_\eta}(H\cap N_\eta)H)}{N_{N_\eta}(N_\eta\cap H)}}, \tag{9.5.10}$$

where $(G_\nu)_{\mu|\mathfrak{g}_\nu} = \left(N_{G_\eta}(N_\eta \cap H)\right)_{\mu|_{\mathrm{Lie}\left(N_{G_\eta}(N_\eta\cap H)\right)}}$.

9.5.5 The Singular Stages Hypothesis. The Optimal Reduction by Stages Theorem 9.4.10 guarantees that the second stage reduced space (9.5.10) is symplectomorphic to the one-step reduced space (9.5.1) if the Stages Hypothesis introduced in Definition 9.4.9 holds. In this setup, that hypothesis can be completely reformulated in terms of relations between Lie algebraic elements and isotropy subgroups. More specifically, in the globally Hamiltonian framework, the Stages Hypothesis is equivalent to the following condition:

Hamiltonian Stages Hypothesis: Let $\mu \in \mathfrak{g}^*$ and $H \subset G$. We say that the pair (μ, H) satisfies the Hamiltonian Stages Hypothesis whenever for any other similar pair (μ', H') such that

$$\begin{cases} i^*\mu = i^*\mu' =: \eta \in \mathfrak{n}^* \\ N_\eta \cap H = N_\eta \cap H' =: K \end{cases} \quad \text{and}$$

$$\begin{cases} \mu|_{\mathrm{Lie}\left(N_{G_\eta}(K)\right)} = \mu'|_{\mathrm{Lie}\left(N_{G_\eta}(K)\right)} =: \zeta \in \mathrm{Lie}\left(N_{G_\eta}(K)\right)^* \\ N_{N_\eta}(K)H = N_{N_\eta}(K)H' =: L, \end{cases}$$

there exists an element $n \in N_{(N_{G_\eta}(K))_\zeta}(L)$ such that

$$\mu' = \mathrm{Ad}^*_{n^{-1}}\mu \quad \text{and} \quad H' = nHn^{-1}.$$

A quick inspection shows that when the G-action is free, that is, when all the isotropy subgroups $H = \{e\}$, the previous condition collapses to the Stages Hypothesis introduced in Section 6.7.

We recall that, in the same fashion in which the proper actions hypothesis introduced in Definition 9.4.9 is automatically satisfied in this setup, so is the Lindelöf hypothesis on the one-step reduced space M_ρ, if one assumes that M is Lindelöf. This is so because closed subsets and continuous images of Lindelöf spaces are always Lindelöf.

The Optimal Reduction by Stages Theorem together with the ideas that we just introduced implies in this setup the following highly nontrivial symplectomorphism, which we state in the form of a theorem. The following statement is consistent with previously introduced notations.

9.5.6 Theorem (Hamiltonian Reduction by Stages) *Let (M, ω) be a symplectic manifold acted properly and canonically upon by a Lie group G that has a closed normal subgroup N. Suppose that this action has an associated equivariant momentum map $\mathbf{J}_G : M \to \mathfrak{g}^*$. Let $\mu \in \mathfrak{g}^*$ be a value of $\mathbf{J}_G$ and let H be an isotropy subgroup of the G action on M. Suppose that the manifold M is Lindelöf, that the pair (μ, H) satisfies the Hamiltonian Stages Hypothesis, and that $M_\nu = \frac{\mathbf{J}_N^{-1}(\eta)\cap M_{N_\eta \cap H}}{N_{N_\eta}(N_\eta \cap H)}$ and $N_\nu = N_{N_\eta}(N_\eta \cap H)$ are connected, where $\eta = i^*\mu$. Then the symplectic reduced spaces*

$$\frac{\mathbf{J}_G^{-1}(\mu) \cap M_H}{N_{G_\mu}(H)} \quad \textit{and} \quad \frac{\mathbf{J}_{\mathcal{H}_\nu}^{-1}(\tau) \cap (M_\nu)_{\frac{N_{N_\eta}(H\cap N_\eta)H}{N_{N_\eta}(N_\eta\cap H)}}}{\frac{N_{(G_\nu)_{\mu|\mathfrak{g}_\nu}}(N_{N_\eta}(H\cap N_\eta)H)}{N_{N_\eta}(N_\eta\cap H)}}$$

are symplectomorphic. In this expression,

$$\mathcal{H}_\nu = \frac{N_{G_\eta}(N_\eta \cap H)}{N_{N_\eta}(N_\eta \cap H)}, \qquad (G_\nu)_{\mu|_{\mathfrak{g}_\nu}} = \left(N_{G_\eta}(N_\eta \cap H)\right)_{\mu|_{\mathrm{Lie}\left(N_{G_\eta}(N_\eta \cap H)\right)}},$$

$\mathbf{J}_{\mathcal{H}_\nu} : M_\nu \to \mathrm{Lie}(\mathcal{H}_\nu)^*$ *is the momentum map associated to the* $\mathcal{H}_\nu$*-action on* M_ν *defined in* (9.5.2), *and* $\tau \in \mathrm{Lie}(\mathcal{H}_\nu)^*$ *is the element defined in* (9.5.4).

9.5.7 When the G-action is free, the previous theorem coincides with the Reduction by Stages Theorem presented in Section 6.7.

9.5.8 Discrete reduction by stages. A special but very important particular case of Theorem 9.5.6 takes place when the group G is discrete ($\mathfrak{g} = \{0\}$). In this situation, all the standard momentum maps in the construction vanish and the theorem gives us a highly nontrivial relation between quotients of isotropy type submanifolds. We start by reformulating the Hamiltonian Stages Hypothesis in this particular case.

Discrete Reduction by Stages Hypothesis: Let G be a discrete group, N a normal subgroup, and H a subgroup. We say that H satisfies the Discrete Reduction by Stages Hypothesis with respect to N if for any other subgroup H' such that

$$N \cap H = N \cap H' =: K \quad \text{and} \quad N_N(K)H = N_N(K)H' =: L,$$

there exists an element $n \in N_{N_G(K)}(L)$ such that $H' = nHn^{-1}$.

9.5.9 Theorem (Discrete Reduction by Stages) *Let* (M, ω) *be a symplectic manifold acted properly and canonically upon by a discrete Lie group* G *that has a closed normal subgroup* N*. Let* H *be an isotropy subgroup of the* G *action on* M *and* $K := N \cap H$*. If the manifold* M *is Lindelöf and* H *satisfies the Discrete Reduction by Stages Hypothesis with respect to* N*, then the symplectic reduced spaces*

$$\frac{M_H}{N_G(H)} \quad \textit{and} \quad \frac{\left(\frac{M_K}{N_N(K)}\right)_{\frac{N_N(K)H}{N_N(K)}}}{\frac{N_{N_G(K)}(N_N(K)H)}{N_N(K)}} \tag{9.5.11}$$

are symplectomorphic.

9.5.10 When the G-action on M is free, the Discrete Reduction by Stages Hypothesis is trivially satisfied and Theorem 9.5.9 produces the obvious symplectomorphism

$$M/G \simeq (M/N)/(G/N).$$

Hence, it is in the presence of singularities that the relation established in (9.5.11) is really visible and nontrivial.

Chapter 10

Poisson Reduction

The preceding chapters presented various procedures that led to the construction of new symplectic manifolds from Poisson or symplectic manifolds acted canonically upon by a Lie group. The present goal is the description of a general method to obtain new Poisson manifolds, or at least Poisson algebras, by reducing symmetric Poisson manifolds. In parallel, we will show the relationship between these reduced spaces and the symplectic reduced spaces that were previously introduced. References related to this topic are MARSDEN AND RATIU (1986), VAISMAN (1996b), and ORTEGA AND RATIU (1998).

Regular Poisson reduction has proved to be very useful in the theory of integrable systems, applied both directly or by using of the method of compatible Poisson brackets (PEDRONI (1995); VANHAECKE (1996); MAGRI (1997)). The singular case has not found yet its way into the bifurcation analysis of the Liouville tori.

The Poisson reduction results in this chapter admit a generalization to the category of ***Leibniz manifolds***; these are manifolds endowed with a bilinear bracket that is only required to satisfy the Leibniz rule on each of its entries. The reader interested in those structures which naturally appear in applications is encouraged to check with ORTEGA AND PLANAS (2003b).

10.1 Regular Poisson reduction

The hypotheses of the following theorem are the most stringent ones that ensure a minimal amount of technical complication. Unfortunately, these assumptions are rarely fulfilled in concrete applications. Nevertheless, the statement below will serve as a model for the rest of the chapter, where similar results are obtained under more realistic conditions.

Let $(M, \{\cdot, \cdot\})$ be a Poisson manifold and let G be a Lie group acting canonically on M. If the G-action is free and proper, Proposition 2.3.8 guarantees that the orbit space M/G is a smooth manifold and that the canonical projection $\pi : M \to M/G$ is a smooth surjective submersion. The following result shows that the Poisson structure on M can be naturally projected, or ***reduced***, to the quotient, along with any dynamics generated on $(M, \{\cdot, \cdot\})$ by any G-invariant Hamiltonian.

10.1.1 Theorem (Regular Poisson reduction) *Let $(M, \{\cdot, \cdot\})$ be a Poisson manifold and let G be a Lie group acting canonically, freely, and properly on M via the map $\Phi : G \times M \to M$. Let $\mathcal{J} : M \to M/A'_G$ be the corresponding optimal momentum map. Then:*

(i) *The orbit space M/G is a Poisson manifold with the Poisson bracket $\{\cdot, \cdot\}^{M/G}$, uniquely characterized by the relation*

$$\{f, g\}^{M/G}(\pi(m)) = \{f \circ \pi, g \circ \pi\}(m), \tag{10.1.1}$$

for any $m \in M$ and where $f, g : M/G \to \mathbb{R}$ are two arbitrary smooth functions.

(ii) *The Poisson structure induced by the bracket $\{\cdot, \cdot\}^{M/G}$ on M/G is the only one for which the projection $\pi : (M, \{\cdot, \cdot\}) \to (M/G, \{\cdot, \cdot\}^{M/G})$ is a Poisson map.*

(iii) *Let $h \in C^\infty(M)^G$ be a G-invariant smooth function on M. The Hamiltonian flow F_t of X_h commutes with the G-action, so it induces a flow $F_t^{M/G}$ on M/G characterized by*

$$\pi \circ F_t = F_t^{M/G} \circ \pi.$$

The flow $F_t^{M/G}$ is Hamiltonian on $(M/G, \{\cdot, \cdot\}^{M/G})$, for the **reduced Hamiltonian** *function $[h] \in C^\infty(M/G)$ defined by*

$$[h] \circ \pi = h.$$

The vector fields X_h and $X_{[h]}$ are π-related.

(iv) *The symplectic leaves of $(M/G, \{\cdot, \cdot\}^{M/G})$ are given by the optimal orbit reduced spaces $\left(\mathcal{J}^{-1}(\mathcal{O}_\rho)/G, \omega_{\mathcal{O}_\rho}\right)$, $\rho \in M/A'_G$, introduced in Theorem 9.2.1.*

(v) *If the Poisson manifold $(M, \{\cdot, \cdot\})$ is actually symplectic with form ω and the G-action has an associated standard momentum map $\mathbf{J} : M \to \mathfrak{g}^*$, then the symplectic leaves of $(M/G, \{\cdot, \cdot\}^{M/G})$ are given by the spaces $\left(M^c_{\mathcal{O}_\mu} := G \cdot \mathbf{J}^{-1}(\mu)^c/G, \omega^c_{\mathcal{O}_\mu}\right)$, where $\mathbf{J}^{-1}(\mu)^c$ is a connected component of the fiber $\mathbf{J}^{-1}(\mu)$ and $\omega^c_{\mathcal{O}_\mu}$ is the restriction to $M^c_{\mathcal{O}_\mu}$ of the symplectic form $\omega_{\mathcal{O}_\mu}$ of the orbit reduced space $M_{\mathcal{O}_\mu}$ introduced in Theorem 6.3.1. If, additionally, G is compact, M is connected, and the momentum map $\mathbf{J}$ is proper, then $M^c_{\mathcal{O}_\mu} = M_{\mathcal{O}_\mu}$.*

Proof. (i) We start by checking that the expression (10.1.1) is well defined. Let $m, m' \in M$ be such that $\pi(m) = \pi(m')$. This equality implies that there exists a group element $h \in G$ such that $m' = \Phi_h(m)$. Consequently, as the G-action is canonical,

$$\begin{aligned}\{f, g\}^{M/G}(\pi(m')) &= \{f \circ \pi, g \circ \pi\}(m') = \{f \circ \pi, g \circ \pi\}(\Phi_h(m))\\ &= \{f \circ \pi \circ \Phi_h, g \circ \pi \circ \Phi_h\}(m) = \{f \circ \pi, g \circ \pi\}(m)\\ &= \{f, g\}^{M/G}(\pi(m)),\end{aligned}$$

as required. The bracket $\{\cdot,\cdot\}^{M/G}$ is clearly bilinear and antisymmetric. The rest of the defining properties of a Poisson bracket are inherited from the Poisson character of the bracket $\{\cdot,\cdot\}$. Indeed, for any $f, g, h \in C^\infty(M/G)$ we have

$$\begin{aligned}\{f, gh\}^{M/G}(\pi(m)) &= \{f\circ\pi, (g\circ\pi)(h\circ\pi)\}(m)\\ &= (g\circ\pi)(m)\{f\circ\pi, h\circ\pi\}(m) + (h\circ\pi)(m)\{f\circ\pi, g\circ\pi\}(m)\\ &= g(\pi(m))\{f,h\}^{M/G}(\pi(m)) + h(\pi(m))\{f,g\}^{M/G}(\pi(m)),\end{aligned}$$

which establishes the Leibniz rule. Additionally,

$$\begin{aligned}&\{f,\{g,h\}^{M/G}\}^{M/G}(\pi(m)) + \{g,\{h,f\}^{M/G}\}^{M/G}(\pi(m))\\ &\quad + \{h,\{f,g\}^{M/G}\}^{M/G}(\pi(m))\\ &\quad = \{f\circ\pi, \{g,h\}^{M/G}\circ\pi\}(m) + \{g\circ\pi, \{h,f\}^{M/G}\circ\pi\}(m)\\ &\quad + \{h\circ\pi, \{f,g\}^{M/G}\circ\pi\}(m)\\ &\quad = \{f\circ\pi, \{g\circ\pi, h\circ\pi\}\}(m) + \{g\circ\pi, \{h\circ\pi, f\circ\pi\}\}(m)\\ &\quad + \{h\circ\pi, \{f\circ\pi, g\circ\pi\}\}(m) = 0,\end{aligned}$$

which, as the map π is surjective, proves that $\{\cdot,\cdot\}_{M/G}$ satisfies the Jacobi identity. The bracket $\{\cdot,\cdot\}^{M/G}$ is the unique Poisson bracket satisfying (10.1.1) since π is surjective.

(ii) It is a straightforward consequence of (10.1.1) and the surjective character of π.

(iii) The existence of $F_t^{M/G}$ is a direct consequence of part **(iii)** in Proposition 5.5.4. The rest is a straightforward verification that uses part **(ii)** and Proposition 4.1.19.

(iv) Let $m \in M$ be a point such that $\mathcal{J}(m) = \rho \in M/A'_G$. The symplectic leaf $\mathcal{L}_{\pi(m)}$ in $(M/G, \{\cdot,\cdot\})$ containing the point $\pi(m) \in M/G$ is by definition the accessible set from $\pi(m)$ via a finite composition of flows F_t associated to functions $h \in C^\infty(M/G)$. Since by **(iii)** the flow F'_t associated to the function $h\circ\pi \in C^\infty(M)^G$ satisfies $\pi\circ F'_t = F_t\circ\pi$, we can conclude that

$$\begin{aligned}\mathcal{L}_{\pi(m)} = \pi(\mathcal{J}^{-1}(\rho)) &= \pi(G\cdot\mathcal{J}^{-1}(\rho))\\ &= \pi(\mathcal{J}^{-1}(\mathcal{O}_\rho)) = \mathcal{J}^{-1}(\mathcal{O}_\rho)/G. \qquad (10.1.2)\end{aligned}$$

It only remains to be shown that the symplectic forms $\omega_{\mathcal{L}_{\pi(m)}}$ and $\omega_{\mathcal{O}_\rho}$ on $\mathcal{J}^{-1}(\mathcal{O}_\rho)/G$ coincide. Let $v_{\pi(m)}, w_{\pi(m)} \in T_{\pi(m)}\mathcal{L}$ be arbitrary. Since the symplectic leaf $\mathcal{L}_{\pi(m)}$ is an integral manifold of the characteristic distribution associated to the Poisson structure $\{\cdot,\cdot\}^{M/G}$, there exist functions $f, g \in C^\infty(M/G)$ such that $v_{\pi(m)} = X_f(\pi(m))$ and $w_{\pi(m)} = X_g(\pi(m))$. Now, by (4.1.18) we have that

$$\omega_{\mathcal{L}(\pi(m))}(v_{\pi(m)}, w_{\pi(m)}) = \{f,g\}^{M/G}(\pi(m)) = \{f\circ\pi, g\circ\pi\}(m). \qquad (10.1.3)$$

We now check that this coincides with $\omega_{\mathcal{O}_\rho}(v_{\pi(m)}, w_{\pi(m)})$. Using the commutativity of the diagram (9.2.1) we obtain

$$\begin{aligned}T_{[m]_\rho}L_\rho\left(T_m\pi_\rho\left(X_{f\circ\pi}(m)\right)\right) &= T_{[m]_{\mathcal{O}_\rho}}\pi_{\mathcal{O}_\rho}\left(X_{f\circ\pi}(m)\right)\\ T_{[m]_\rho}L_\rho\left(T_m\pi_\rho\left(X_{g\circ\pi}(m)\right)\right) &= T_{[m]_{\mathcal{O}_\rho}}\pi_{\mathcal{O}_\rho}\left(X_{g\circ\pi}(m)\right), \qquad (10.1.4)\end{aligned}$$

where $[m]_\rho := \pi_\rho(m)$ and $[m]_{\mathcal{O}_\rho} := \pi_{\mathcal{O}_\rho}(m)$. Additionally, since the point m has been chosen without loss of generality to satisfy $\mathcal{J}(m) = \rho$, we conclude

$$L_\rho^{-1}(\pi_{\mathcal{O}_\rho}(m)) = \pi_\rho(m). \tag{10.1.5}$$

Consequently, $\omega_{\mathcal{O}_\rho} := (L_\rho^{-1})^* \omega_\rho$, (10.1.4), and (10.1.5) imply

$$\begin{aligned}
\omega_{\mathcal{O}_\rho}(\pi(m))(v_{\pi(m)}, w_{\pi(m)}) &= \omega_{\mathcal{O}_\rho}(\pi_{\mathcal{O}_\rho}(m))(X_f(\pi(m)), X_g(\pi(m))) \\
&= \omega_{\mathcal{O}_\rho}(\pi_{\mathcal{O}_\rho}(m))(T_m\pi_{\mathcal{O}_\rho}(X_{f\circ\pi}(m)), T_m\pi_{\mathcal{O}_\rho}(X_{g\circ\pi}(m))) \\
&= ((L_\rho^{-1})^*\omega_\rho)(\pi_{\mathcal{O}_\rho}(m))(T_m\pi_{\mathcal{O}_\rho}(X_{f\circ\pi}(m)), T_m\pi_{\mathcal{O}_\rho}(X_{g\circ\pi}(m))) \\
&= \omega_\rho(\pi_\rho(m))(T_m\pi_\rho(X_{f\circ\pi}(m)), T_m\pi_\rho(X_{g\circ\pi}(m))) \\
&= (\pi_\rho^*\omega_\rho)(m)(X_{f\circ\pi}(m), X_{g\circ\pi}(m)) \\
&= \{f\circ\pi, g\circ\pi\}(m) \quad \text{(by (9.1.1))}.
\end{aligned}$$

This equality, together with (10.1.3), proves the statement.

(v) In the symplectic case and in the presence of a standard momentum map, Corollary 5.5.18 implies that (10.1.2) can be rewritten as

$$\mathcal{L}_{\pi(m)} = \pi(\mathcal{J}^{-1}(\rho)) = \pi(\mathbf{J}^{-1}(\mu)^m) = \pi(G\cdot\mathbf{J}^{-1}(\mu)^m) = G\cdot\mathbf{J}^{-1}(\mu)^m/G,$$

where $\mu := \mathbf{J}(m)$ and $\mathbf{J}^{-1}(\mu)^m$ is the connected component of $\mathbf{J}^{-1}(\mu)$ that contains m.

Finally, if G is compact, M is connected, and $\mathbf{J}$ is a proper map, then $\mathbf{J}^{-1}(\mu)^m = \mathbf{J}^{-1}(\mu)$ necessarily (see Section 4.7) and hence the last claim follows. ■

10.2 The reduction of a presheaf of Poisson algebras

The reduction theorem presented in the previous section contains extremely strong regularity hypotheses that ensure the smoothness of the orbit space onto which the Poisson bracket and the corresponding equivariant dynamics can be projected. When these hypotheses are not present, the orbit space is no longer smooth but nevertheless, the Poisson algebra, or more specifically, the ***presheaf of Poisson algebras*** associated to the bracket admits, under certain circumstances, a projection to the quotient.

10.2.1 Definition *Let M be a topological space with a presheaf $\mathcal{F}$ of smooth functions. A **presheaf of Poisson algebras** on $(M, \mathcal{F})$ is a map $\{\cdot,\cdot\}$ that assigns to each open set $U \subset M$ a bilinear operation $\{\cdot,\cdot\}_U : \mathcal{F}(U)\times\mathcal{F}(U) \to \mathcal{F}(U)$ such that the pair $(\mathcal{F}(U), \{\cdot,\cdot\}_U)$ is a Poisson algebra. A presheaf of Poisson algebras will be usually denoted as a triple $(M, \mathcal{F}, \{\cdot,\cdot\})$.*

*We say that the presheaf of Poisson algebras $(M, \mathcal{F}, \{\cdot,\cdot\})$ is **nondegenerate** if the following condition holds: if $f \in \mathcal{F}(U)$ is such that $\{f, g\}_{U\cap V} = 0$, for any $g \in \mathcal{F}(V)$ and any open set of V, then f is constant on the connected components of U.*

10.2.2 Example Any Poisson manifold $(M, \{\cdot,\cdot\})$ has a natural presheaf of Poisson algebras on its presheaf of smooth functions that associates to any open subset U of M the restriction $\{\cdot,\cdot\}|_U$ of $\{\cdot,\cdot\}$ to $C^\infty(U)\times C^\infty(U)$.

The main goal of this section is to prove a result that fully characterizes the situations in which the presheaf of Poisson algebras in Example 10.2.2 behaves properly under restrictions to subsets and projections to the orbit spaces of pseudogroups of local Poisson diffeomorphisms of $(M, \{\cdot, \cdot\})$.

We start by elaborating on some concepts introduced in §2.2.11. Let M be a smooth manifold and $A \subset \mathrm{Diff}_L(M)$ a pseudogroup of local diffeomorphisms of M. Let $S \subset M$ be a subset of M endowed with a topology $\mathcal{T}$ that, in general, does not coincide with the relative or subspace topology. The presheaf C^∞_M of smooth functions on M induces a quotient presheaf $C^\infty_{M/A}$ on the orbit space M/A. Consider now the subset

$$A_S := \{a \in A \mid a(s) \in S \text{ for any } s \in S \cap \mathrm{Dom}(a)\}.$$

Throughout this chapter we will assume that A_S is a subpseudogroup of A. This hypothesis will allow us to construct the quotients S/A_S and M/A_S. Given that the quotient S/A_S can be seen as a subset of M/A_S, there is a well-defined presheaf of Whitney smooth functions $C^\infty_{S/A_S, M/A_S}$ on S/A_S induced by C^∞_{M/A_S}. Since the projection $M \to M/A_S$ is an open map, Proposition 1.1.20 guarantees that

$$C^\infty_{S/A_S, M/A_S} = C^\infty_{S/A_S}(\cdot)^{A_S},$$

where $C^\infty_{S/A_S}(\cdot)^{A_S}$ is the quotient presheaf on S/A_S associated to the presheaf $C^\infty_{S,M}(\cdot)^{A_S}$ of Whitney A_S-invariant functions on S induced by $C^\infty_M(\cdot)^{A_S}$.

Notation. In order to simplify notation, define

$$W^\infty_{S/A_S} := C^\infty_{S/A_S, M/A_S} = C^\infty_{S/A_S}(\cdot)^{A_S}.$$

We recall that for any open set $V \subset S/A_S$, the elements $f \in W^\infty_{S/A_S}(V)$ are characterized by the fact that if $\pi_S : S \to S/A_S$ is the projection onto orbit space, then for any $m \in \pi_S^{-1}(V)$ there exists an open A_S-invariant neighborhood of m in M and $F \in C^\infty_M(U_m)^{A_S}$ such that

$$f \circ \pi_S|_{\pi_S^{-1}(V) \cap U_m} = F|_{\pi_S^{-1}(V) \cap U_m}. \tag{10.2.1}$$

We will say that F is a ***local extension*** of $f \circ \pi_S$ at the point m.

10.2.3 Definition *Let M be a smooth manifold, $A \subset \mathrm{Diff}_L(M)$ a pseudogroup of local diffeomorphisms of M, and S a subset of M endowed with a topology $\mathcal{T}$ that is stronger than or equal to the relative topology. We say that the presheaf W^∞_{S/A_S} has the (A, A_S)-**local extension property** when A_S is a subpseudogroup of A and for any $f \in W^\infty_{S/A_S}(V)$ and $m \in \pi_S^{-1}(V)$ there exists an open A-invariant neighborhood U_m of m in M and $F \in C^\infty_M(U_m)^A$ such that*

$$f \circ \pi_S|_{\pi_S^{-1}(V) \cap U_m} = F|_{\pi_S^{-1}(V) \cap U_m}.$$

*We will say that F is an A-**invariant local extension** of $f \circ \pi_S$ at m.*

10.2.4 Definition *Let $(M, \{\cdot, \cdot\})$ be a smooth Poisson manifold, $A \subset \mathcal{P}_L(M)$ a pseudogroup of local Poisson diffeomorphisms of M, and $S \subset M$ a subset of M such that*

W^{∞}_{S/A_S} *has the* (A, A_S)*-local extension property. We say that* $(M, \{\cdot,\cdot\}, A, S)$ *is* ***Poisson reducible*** *when* $\left(S/A_S, W^{\infty}_{S/A_S}, \{\cdot,\cdot\}^{S/A_S}\right)$ *is a well-defined presheaf of Poisson algebras where, for any open set* $V \subset S/A_S$*, the bracket* $\{\cdot,\cdot\}_V^{S/A_S} : W^{\infty}_{S/A_S}(V) \times W^{\infty}_{S/A_S}(V) \to W^{\infty}_{S/A_S}(V)$ *is given by*

$$\{f, g\}_V^{S/A_S}(\pi_S(m)) = \{F, G\}(m) \tag{10.2.2}$$

for any $m \in \pi_S^{-1}(V)$ *and where* F, G *are* A*-invariant local extensions at* m *of* $f \circ \pi_S$ *and* $g \circ \pi_S$*, respectively.*

10.2.5 Theorem (ORTEGA AND RATIU **(1998)**) *Let* $(M, \{\cdot,\cdot\})$ *be a smooth Poisson manifold,* $A \subset \mathcal{P}_L(M)$ *a pseudogroup of local Poisson diffeomorphisms of* M*, and* $S \subset M$ *a subset of* M *such that* W^{∞}_{S/A_S} *has the* (A, A_S)*-local extension property. Let* $B^{\sharp} : T^*M \to TM$ *be the bundle map associated to the Poisson tensor of* $(M, \{\cdot,\cdot\})$*. Then* $(M, \{\cdot,\cdot\}, A, S)$ *is Poisson reducible if and only if for any* $m \in S$ *we have*

$$B^{\sharp}(\Delta_m) \subset \left[\Delta_m^S\right]^{\circ}, \tag{10.2.3}$$

where $\Delta_m := \{\mathbf{d}F(m) \mid F \in C^{\infty}_M(U_m)^A$*, for any open* A*-invariant neighborhood* U_m *of* m *in* $M\}$*, and where* $\Delta_m^S := \{\mathbf{d}F(m) \in \Delta_m \mid F|_{U_m \cap V_m}$ *is constant, for an open* A*-invariant neighborhood* U_m *of* m *in* M *and an open* A_S*-invariant neighborhood* V_m *of* m *in* $S\}$.

10.2.6 Remark If S has the relative topology, then

$$\Delta_m^S = \{\mathbf{d}F(m) \in \Delta_m \mid F|_{U_m \cap S} \text{ is constant}\},$$

for an open A-invariant neighborhood U_m of m in M.

10.2.7 Remark Even though in Theorem 10.2.5 we require the pseudogroup A to consist of local Poisson diffeomorphisms of $(M, \{\cdot,\cdot\})$, a weaker hypothesis on A is sufficient for this result to hold. Indeed, a careful inspection of the proof of this result presented below shows that it is enough to assume that the presheaf $C^{\infty}_M(\cdot)^A$ of A-invariant functions on M is closed under the bracket $\{\cdot,\cdot\}$. Obviously, if the elements in A are local Poisson diffeomorphisms this condition is satisfied.

10.2.8 Remark Even though in the previous theorem only the subpseudogroup A_S intervenes in the construction of the quotient space S/A_S, the full pseudogroup A is used in the definition of the Poisson bracket on this quotient when $(M, \{\cdot,\cdot\}, A, S)$ is Poisson reducible. Actually, in spite of the fact that the reduction of $(M, \{\cdot,\cdot\}, A, S)$ and $(M, \{\cdot,\cdot\}, A_S, S)$ renders the same quotient manifold S/A_S it does *not* yield the same Poisson brackets on this quotient since different sets of functions are involved. Moreover, as we will see in §10.4.13 there are instances in which $(M, \{\cdot,\cdot\}, A, S)$ is Poisson reducible whereas $(M, \{\cdot,\cdot\}, A_S, S)$ is not.

10.2.9 Lemma *Let* M *be a smooth manifold,* $A \subset \mathrm{Diff}_L(M)$ *a pseudogroup of local transformations of* M*, and* $S \subset M$ *a subset whose topology is stronger than the relative* *topology and such that A_S is a subpseudogroup of A. If $\pi_S : S \rightarrow S/A_S$ is the projection, $U \subset M$ is an open A-invariant subset of M, $F \in C^\infty_M(U)^A$, and $V := \pi_S(U \cap S)$, then there exists a unique function $f \in W^\infty_{S/A_S}(V)$ such that*

$$f \circ \pi_S|_{U\cap S} = f \circ \pi_S|_{\pi_S^{-1}(V)\cap U} = F|_{\pi_S^{-1}(V)\cap U}. \tag{10.2.4}$$

Proof. Since by hypothesis the topology of S is stronger than the relative topology, for any open A-invariant subset U of M, the intersection $U \cap S$ is an open A_S-invariant subset of S. As the projection π_S is an open map, the set $V := \pi_S(U \cap S)$ is open in S/A_S. Also, the A_S-invariance of $U \cap S$ implies that $U \cap S = \pi_S^{-1}(V)$ and hence

$$\pi_S^{-1}(V) \cap U = U \cap S \cap U = U \cap S, \tag{10.2.5}$$

which proves the first equality in (10.2.4).

Now, the invariance properties of F and S imply the existence of a unique map f defined on V such that $f \circ \pi_S|_{U\cap S} = F|_{U\cap S}$ or equivalently, by (10.2.5), $f \circ \pi_S|_{\pi_S^{-1}(V)\cap U} = F|_{\pi_S^{-1}(V)\cap U}$. Since $\pi_S^{-1}(V) \subset U$ by construction, for any $m \in \pi_S^{-1}(V)$, the map f satisfies (10.2.1) by taking in that characterization U (instead of U_m) and F. This implies that $f \in W^\infty_{S/A_S}(V)$. ■

Proof of Theorem 10.2.5. We first show that if $(M, \{\cdot,\cdot\}, A, S)$ is Poisson reducible, then $\Delta_m^S \subset \left[B^\sharp(\Delta_m)\right]^\circ$, for all $m \in S$. Let $\alpha_m \in \Delta_m^S$; by definition, there exists an open A-invariant neighborhood U_m of m in M and a function $K \in C^\infty_M(U_m)^A$ such that $\alpha_m = \mathbf{d}K(m)$ and $K|_{V_m\cap U_m}$ is constant for an open A_S-invariant neighborhood of m in S. Notice now that, by definition, any element in $B^\sharp(\Delta_m)$ can be written as $X_F(m)$ with $F \in C^\infty(W_m)^A$, W_m an open A-invariant neighborhood of m in M. By Lemma 10.2.9 there exist functions $k \in W^\infty_{S/A_S}(\pi_S(U_m \cap S))$ and $f \in W^\infty_{S/A_S}(\pi_S(W_m \cap S))$ such that

$$k \circ \pi_S|_{U_m\cap S} = K|_{U_m\cap S}, \qquad f \circ \pi_S|_{W_m\cap S} = F|_{W_m\cap S}.$$

Hence, by the Poisson reducibility of $(M, \{\cdot,\cdot\}, A, S)$ we have that

$$\langle \alpha_m, X_F(m)\rangle = \{K, F\}(m) = \{k|_W, f|_W\}_W^{S/A_S}(\pi_S(m)),$$

where $W = \pi_S(U_m \cap S) \cap \pi_S(W_m \cap S)$. However, given that the function C on M that is constant and equal to $K(m)$ is also an A-invariant local extension of $k \circ \pi_S$ at m, we have that

$$\{k|_W, f|_W\}_W^{S/A_S}(\pi_S(m)) = \{C, F\}(m) = 0,$$

which implies that $\langle \alpha_m, X_F(m)\rangle = 0$. Since $X_F(m) \in B^\sharp(\Delta_m)$ is arbitrary it follows that $\alpha_m \in \left[B^\sharp(\Delta_m)\right]^\circ$.

Suppose now that the inclusion (10.2.3) holds. We will prove the reducibility of $(M, \{\cdot,\cdot\}, A, S)$. Let $f, g \in W^\infty_{S/A_S}(V)$ and $F, G \in C^\infty_M(U_m)^A$ be local A-invariant extensions of $f \circ \pi_S$ and $g \circ \pi_S$, respectively, at a point $m \in \pi_S^{-1}(V)$. We now show that the equality

$$\{f, g\}_V^{S/A_S}(\pi_S(m)) = \{F, G\}_{U_m}(m) \tag{10.2.6}$$

provides a well-defined presheaf of Poisson algebras. The only point that requires a proof is that the expression (10.2.6) does not depend on the local extensions utilized in the definition. The fact that $\{\cdot, \cdot\}^{S/A_S}$ determines a presheaf of Poisson algebras is inherited from the properties of the bracket $\{\cdot, \cdot\}$ on M. Let $G' \in C^\infty_M(U_m)^A$ be another local extension of $g \circ \pi_S$ at m. This implies that $G - G'|_{\pi_S^{-1}(V)\cap U_m} = 0$ and hence $\mathbf{d}(G - G')(m) \in \Delta_m^S \subset [B^\sharp(\Delta_m)]^\circ$. Consequently,

$$0 = \langle \mathbf{d}(G - G')(m), X_F(m) \rangle = \{G - G', F\}_{U_m}(m),$$

which implies that $\{G, F\}_{U_m}(m) = \{G', F\}_{U_m}(m)$ and hence guarantees the independence of (10.2.6) with respect to the choice of local extension for $g \circ \pi_S$. This argument and the antisymmetry of $\{\cdot, \cdot\}$ guarantee that this definition is also independent of the choice of extension for $f \circ \pi_S$. Therefore, the expression (10.2.6) defines a function $\{f, g\}_V^{S/A_S}$ on V that actually belongs to $W^\infty_{S/A_S}(V)$ because if F and G are local extensions of $f \circ \pi_S$ and $g \circ \pi_S$, respectively, at any point $m \in \pi_S^{-1}(V)$, then so is the function $\{F, G\}$ with respect to $\{f, g\}_V^{S/A} \circ \pi_S$ (recall that the canonical character of the action implies that if F and G are A-invariant, then so is the function $\{F, G\}$). ■

10.2.10 Corollary *Let S be an embedded submanifold of the Poisson manifold $(M, \{\cdot, \cdot\})$. The triple $(M, \{\cdot, \cdot\}, S)$ is Poisson reducible if and only if*

$$B^\sharp(T_m^*M) \subset T_m S, \quad \textit{for any } m \in S, \tag{10.2.7}$$

or, equivalently, whenever

$$T_m \mathcal{L}_m \subset T_m S, \quad \textit{for any } m \in S, \tag{10.2.8}$$

where $\mathcal{L}_m$ is the symplectic leaf of $(M, \{\cdot, \cdot\})$ containing the point $m \in S$; this condition amounts to S being a quasi-Poisson submanifold of M (see (4.1.21)*). If S is only an immersed submanifold of M, then the conditions* (10.2.7) *or* (10.2.8) *are sufficient but, in general, not necessary conditions for the Poisson reducibility of $(M, \{\cdot, \cdot\}, S)$. In both cases, the Poisson reducibility of $(M, \{\cdot, \cdot\}, S)$ implies that $(S, \{\cdot, \cdot\}|_S)$ is a Poisson manifold.*

Proof. The equivalence between (10.2.7) and (10.2.8) is obvious once we recall that $B^\sharp(T^*M)$ is the characteristic distribution associated to the Poisson manifold $(M, \{\cdot, \cdot\})$ (see §4.1.14) whose maximal integral submanifolds are its symplectic leaves (see §4.1.27). We now show that in the hypotheses of the statement, the reducibility inclusion (10.2.3) coincides with (10.2.7). First, we obviously have that in this case

$$\Delta_m = T_m^*M, \tag{10.2.9}$$

for any $m \in S$. Second, by definition of Δ_m^S it is also obvious that

$$T_m S \subset [\Delta_m^S]^\circ. \tag{10.2.10}$$

Third, fix $m \in S$ and take a chart (U, φ) of M around m with the submanifold property with respect to S. More specifically, we choose $\varphi : U \to V \times W$ with V and W

vector spaces such that $\varphi(m) = (0, 0)$ and for which there exists another chart (U', ψ) of S around $m \in S$ such that $\psi : U' \to V$ and the inclusion $i : S \hookrightarrow M$ is such that $i(U') \subset U$ and $(\varphi \circ i \circ \psi^{-1})(v) = (v, 0)$, for any $v \in \psi(U')$. Using these coordinates we have that $\varphi(U \cap S) = \varphi(U) \cap (V \times \{0\})$ and hence the submanifold S is represented by the points of the form $(v, 0) \in V \times W$. Analogously, the elements in $(T_m S)^\circ$ can be expressed as $\langle (0, w_0), \cdot \rangle$ for any $w_0 \in W$ and where $\langle \cdot, \cdot \rangle$ is the Euclidean pairing. Since for any $w_0 \in W$ we can write $\langle (0, w_0), \cdot \rangle = \mathbf{d}F(0, 0)$ with $F(u, v) := \langle (0, w_0), (u, v) \rangle = \langle w_0, v \rangle$, for any $(u, v) \in V \times W$, and the just defined function F is constant on $\varphi(U \times S)$, it follows that $[T_m S]^\circ \subset \Delta_m^S$ or, equivalently, that

$$[\Delta_m^S]^\circ \subset T_m S. \tag{10.2.11}$$

This inclusion, together with expressions (10.2.9) and (10.2.10), implies that (10.2.3) coincides with (10.2.7).

If S is just an immersed submanifold of M and not embedded, the inclusion (10.2.11) does not necessarily hold and hence (10.2.9) and (10.2.10) guarantee that, in this case, (10.2.7) implies (10.2.3). Finally, part **(ii)** in Proposition 1.1.24 guarantees the last claim. ■

10.2.11 Corollary *Let $(M, \{\cdot, \cdot\})$ be a smooth Poisson manifold, $B \in \Lambda^2(T^*M)$ the associated Poisson tensor, and D a smooth, integrable, and regular distribution on M generated by a family of local infinitesimal Poisson automorphisms of M. Then there is a unique Poisson bracket $\{\cdot, \cdot\}^{M/D}$ on the quotient manifold M/D for which the projection $\pi_D : M \to M/D$ is a Poisson map.*

If M is symplectic with form ω, then the rank of the Poisson structure $\{\cdot, \cdot\}^{M/D}$ at the point $\pi_D(m)$ is

$$\begin{aligned}\operatorname{rank}&\left(B^\sharp_{M/D}(\pi_D(m))\right)\\ &= \dim M - \dim D(m) - \dim\left[(D(m))^\omega \cap D(m)\right],\end{aligned} \tag{10.2.12}$$

where $B_{M/D} \in \Lambda^2(T^*(M/D))$ is the Poisson tensor associated to the bracket $\{\cdot,\cdot\}^{M/D}$ on M/D.

Proof. Let F be a generating family for D containing locally defined infinitesimal Poisson automorphisms of $(M, \{\cdot, \cdot\})$. By definition, the pseudogroup A_F of local diffeomorphisms constructed using the flows of the elements in F is included in $\mathcal{P}_L(M)$. Given that in this case $W^\infty_{M/D} = W^\infty_{M/A_F} = C^\infty_{M/D}$ (see (1.1.7) and use the fact that π_D is a submersion) the claim in the statement reduces to showing that $(M, \{\cdot, \cdot\}, A_F)$ is reducible in the sense of Theorem 10.2.5 or, equivalently, that (10.2.3) holds. This is actually the case because for any $m \in M$ we have

$$\begin{aligned}\Delta_m &= \{\mathbf{d}F(m) \mid F \in C^\infty(U_m)^D,\ U_m \text{ open } D\text{-invariant}\}\\ &= \{\mathbf{d}F(m) \mid F \in C^\infty(M)^D\} \quad \text{(by Proposition 3.2.9)}\\ &= \{\mathbf{d}(f \circ \pi_D)(m) \mid f \in C^\infty(M/D)\} \quad \text{(since } \pi_D \text{ is a submersion)}\\ &= (\ker T_m\pi_D)^\circ \quad \text{(by (1.1.5))}\\ &= (D(m))^\circ.\end{aligned}$$

However, in this case we also have $[\Delta_m^S]^\circ = T_m^* M$, for any $m \in M$, by the definition of Δ_m^S (see the statement of Theorem 10.2.5) and hence (10.2.3) trivially holds.

We now prove expression (10.2.12). Since π_D is a Poisson map, by (4.1.9), we can write

$$B_{M/D}^\sharp(\pi_D(m)) = T_m\pi_D \circ B^\sharp(m) \circ T_m^*\pi_D$$

and hence

$$\ker\left(B_{M/D}^\sharp(\pi_D(m))\right) = \left(T_m^*\pi_D\right)^{-1}\left(B^\sharp(m)\right)^{-1}(\ker T_m\pi_D)\,.$$

By Proposition 4.1.32

$$\left(B^\sharp(m)\right)^{-1}(\ker T_m\pi_D) = \left((\ker T_m\pi_D)^\omega\right)^\circ$$

and thus

$$\ker\left(B_{M/D}^\sharp(\pi_D(m))\right) = \left(T_m^*\pi_D\right)^{-1}\left(\left((\ker T_m\pi_D)^\omega\right)^\circ\right).$$

Since $T_m^*\pi_D : T_{\pi_D(m)}^*(M/D) \to T_m^* M$ is injective,

$$\begin{aligned}\dim\left(\left(T_m^*\pi_D\right)^{-1}\left(\left((\ker T_m\pi_D)^\omega\right)^\circ\right)\right) &= \dim\left[\left((\ker T_m\pi_D)^\omega\right)^\circ \cap \operatorname{range}\left(T_m^*\pi_D\right)\right]\\ &= \dim\left[\left((\ker T_m\pi_D)^\omega\right)^\circ \cap (\ker T_m\pi_D)^\circ\right],\end{aligned}$$

because $\operatorname{range}\left(T_m^*\pi_D\right) = (\ker T_m\pi_D)^\circ$. Consequently, by Lemma 4.1.31

$$\begin{aligned}\dim\left(\ker\left(B_{M/D}^\sharp(\pi_D(m))\right)\right) &= \dim\left[\left((\ker T_m\pi_D)^\omega + \ker T_m\pi_D\right)^\circ\right]\\ &= \dim M - \dim(D(m))^\omega - \dim D(m) + \dim\left[(D(m))^\omega \cap D(m)\right]\\ &= \dim\left[(D(m))^\omega \cap D(m)\right],\end{aligned}$$

by §4.1.5. Finally,

$$\begin{aligned}\operatorname{rank}\left(B_{M/D}^\sharp(\pi_D(m))\right) &= \dim M/D - \dim\left(\ker\left(B_{M/D}^\sharp(\pi_D(m))\right)\right)\\ &= \dim M - \dim D(m) - \dim\left[(D(m))^\omega \cap D(m)\right]. \quad \blacksquare\end{aligned}$$

10.2.12 Corollary *Let G be a Lie group acting freely, properly, and canonically on the Poisson manifold $(M, \{\cdot,\cdot\})$ via the map $\Phi : G \times M \to M$. Let $A := A_G = \{\Phi_g \mid g \in G\} \subset \mathcal{P}(M)$ and let S be an embedded G-invariant submanifold of M. Then $(M, \{\cdot,\cdot\}, A, S)$ is Poisson reducible if and only if*

$$B^\sharp((\mathfrak{g}\cdot m)^\circ) \subset T_m S, \quad \textit{for any} \quad m \in S. \tag{10.2.13}$$

If the G-action on M is not free, the inclusion

$$B^\sharp\left(\left((\mathfrak{g}\cdot m)^\circ\right)^{G_m}\right) \subset T_m S, \quad \textit{for any} \quad m \in S, \tag{10.2.14}$$

implies that $(M, \{\cdot,\cdot\}, A, S)$ is Poisson reducible.

10.2.13 Remark When the G-action is free and the submanifold S paracompact, Corollary 2.5.8 and part **(v)** of Proposition 2.3.8 guarantee that the reducibility of $(M, \{\cdot,\cdot\}, A, S)$ amounts to $\left(S/G, C^{\infty}_{S/G}, \{\cdot,\cdot\}^{S/G}\right)$ being a sheaf of Poisson algebras. In particular, Theorem 10.1.1 **(i)** follows from the previous corollary in a straightforward manner since (10.2.13) is always satisfied in this setup.

Proof of Corollary 10.2.12. Assume first that the G-action is free. We will show that in this situation the inclusions (10.2.13) and (10.2.3) coincide. First of all, Theorem 2.5.10 guarantees that in this case $\Delta_m = (\mathfrak{g}\cdot m)^\circ$, for any $m \in S$. We now prove by double inclusion that $T_m S = [\Delta^S_m]^\circ$, for any $m \in S$. Let $\alpha_m \in \Delta^S_m$; by definition, $\alpha_m = \mathbf{d}F(m)$, with $F \in C^\infty_M(U_m)^G$ and where U_m is a G-invariant open neighborhood of m in M such that the restriction $F|_{U_m\cap S}$ is a constant function. Consequently, for any $v_m \in T_m S$ we have that $\langle \alpha_m, v_m\rangle = \mathbf{d}F(m)(v_m) = 0$, which proves that $T_m S \subset [\Delta^S_m]^\circ$. We will now show that $[T_m S]^\circ \subset \Delta^S_m$; given the local character of the claim and the properness of the action, the Slice Theorem 2.3.31 and part **(vii)** in Proposition 2.4.6 guarantee that it suffices to prove this inclusion when the manifold M is given by $G \times B_1 \times B_2$ and S equals $G \times B_1 \times \{0\}$, where B_1 and B_2 are two neighborhoods of zero in two vector spaces V_1 and V_2, respectively, and we take $(e, 0, 0)$ as the point m. Let $\langle\cdot,\cdot\rangle_{\mathfrak{g}}$, $\langle\cdot,\cdot\rangle_1$, and $\langle\cdot,\cdot\rangle_2$ be nondegenerate inner products on $\mathfrak{g}$, V_1, and V_2, respectively, and let $\langle\cdot,\cdot\rangle$ be the nondegenerate inner product on $\mathfrak{g} \times V_1 \times V_2$ given by

$$\langle(\xi_1, v_1, u_1), (\xi_2, v_2, u_2)\rangle = \langle\xi_1, \xi_2\rangle_{\mathfrak{g}} + \langle v_1, v_2\rangle_1 + \langle u_1, u_2\rangle_2,$$

for any $\xi_1, \xi_2 \in \mathfrak{g}$, $v_1, v_2 \in V_1$, and $u_1, u_2 \in V_2$. Using this inner product we can write

$$(T_m S)^\circ = \{\langle(0,0,u),\cdot\rangle \mid u \in V_2\}.$$

Since for any arbitrary element $u_0 \in V_2$, the map $F \in C^\infty(M)^G$ defined by $F(g, v, u) := \langle u_0, u\rangle_2$, $(g, v, u) \in G \times B_1 \times B_2$, is constant on S and $\mathbf{d}F(e, 0, 0) = \langle(0, 0, u_0),\cdot\rangle$, the inclusion $[T_m S]^\circ \subset \Delta^S_m$ follows.

When the G-action on M is not free, Theorem 2.5.10 implies $\Delta_m = ((\mathfrak{g}\cdot m)^\circ)^{G_m}$, for any $m \in S$. Since we obviously have $T_m S \subset \left[\Delta^S_m\right]^\circ$, the inclusion (10.2.14) implies that (10.2.3) in Theorem 10.2.5 holds. Thus the reducibility of $(M, \{\cdot,\cdot\}, G, S)$ is guaranteed. ■

10.3 Applications of the Poisson Reduction Theorem 10.2.5

The following sections will apply Theorem 10.2.5 to obtain some of the reduction methods found in the literature.

10.3.1 Symplectic orbit reduction. Let (M, ω) be a connected symplectic manifold acted freely, properly, and canonically upon by a Lie group G. Let $\mathbf{J} : M \to \mathfrak{g}^*$ be a standard momentum map associated to this action. Let $\mathcal{O}_\mu \subset \mathfrak{g}^*$ be the orbit of the point $\mu \in \mathfrak{g}^*$ with respect to the affine G-action on $\mathfrak{g}^*$ constructed using the non-equivariance cocycle of $\mathbf{J}$. As we showed in Theorem 6.3.1, the preimage $\mathbf{J}^{-1}(\mathcal{O}_\mu)$ is an initial G-invariant submanifold of M and the orbit space $\mathbf{J}^{-1}(\mathcal{O}_\mu)/G$

is a regular quotient manifold. When $\mathbf{J}^{-1}(\mathcal{O}_\mu)$ is actually an embedded paracompact submanifold of M, then Proposition 1.1.22 and the condition (10.2.13) implies that $\left(\mathbf{J}^{-1}(\mathcal{O}_\mu)/G, C^\infty_{\mathbf{J}^{-1}(\mathcal{O}_\mu)/G}, \{\cdot,\cdot\}^{\mathbf{J}^{-1}(\mathcal{O}_\mu)/G}\right)$ is a sheaf of Poisson algebras. Indeed, in this case, this inclusion reduces to the condition $(\mathfrak{g}\cdot m)^\omega = \ker T_m\mathbf{J} \subset T_m\mathbf{J}^{-1}(\mathcal{O}_\mu)$, for any $m \in \mathbf{J}^{-1}(\mathcal{O}_\mu)$, which is obviously true. This sheaf of Poisson algebras coincides with the one that we would obtain by using the Poisson bracket $\{\cdot,\cdot\}_{M_{\mathcal{O}_\mu}}$ induced by the symplectic reduced structure $\omega_{\mathcal{O}_\mu}$ on $\mathbf{J}^{-1}(\mathcal{O}_\mu)/G$ introduced in Theorem 6.3.1.

If we drop the hypothesis on $\mathbf{J}^{-1}(\mathcal{O}_\mu)$ being an embedded paracompact submanifold of M, the system $(M, \{\cdot,\cdot\}, G, \mathbf{J}^{-1}(\mathcal{O}_\mu))$ is still Poisson reducible and hence $\left(\mathbf{J}^{-1}(\mathcal{O}_\mu)/G, W^\infty_{\mathbf{J}^{-1}(\mathcal{O}_\mu)/G}, \{\cdot,\cdot\}^{\mathbf{J}^{-1}(\mathcal{O}_\mu)/G}\right)$ is a presheaf of Poisson algebras. Nevertheless, in this case the presheaves of smooth functions $W^\infty_{\mathbf{J}^{-1}(\mathcal{O}_\mu)/G}$ and $C^\infty_{\mathbf{J}^{-1}(\mathcal{O}_\mu)/G}$ do *not* coincide.

10.3.2 Universal reduction. Let (M,ω) be a connected symplectic manifold acted properly and canonically upon by a Lie group G. Let $\mathbf{J} : M \to \mathfrak{g}^*$ be a standard momentum map associated to this action, $\mu \in \mathbf{J}(M)$ one of its values, and G_μ the isotropy subgroup of this element with respect to the affine action of G on $\mathfrak{g}^*$ constructed using the non-equivariance cocycle of $\mathbf{J}$. In Chapter 8 we showed that the quotient $\mathbf{J}^{-1}(\mu)/G_\mu$ is a symplectically stratified Hausdorff topological space. A predecessor of this result is the so-called ***universal reduction procedure*** of ARMS et al. (1991), which constructed a natural presheaf of Poisson algebras on $\mathbf{J}^{-1}(\mu)/G_\mu$ out of the Poisson bracket on M associated to its symplectic form. As we show in the following sections, this construction can be obtained as a corollary to Theorem 10.2.5 by showing that the system $(M, \{\cdot,\cdot\}, G, \mathbf{J}^{-1}(\mu))$ is Poisson reducible in the sense of Definition 10.2.4, where $\{\cdot,\cdot\}$ is the Poisson bracket associated to the symplectic form ω and $\mathbf{J}^{-1}(\mu)$ will be considered as a topological subspace of M. We proceed in two steps:

(i) $W^\infty_{\mathbf{J}^{-1}(\mu)/G_\mu}$ **satisfies the** (G, G_μ)**-local extension property.** This statement can be obtained as a corollary to Proposition 8.1.2. In what follows we use the notation introduced in the statement of that result. Let $m \in \mathbf{J}^{-1}(\mu)$ and $f \in W^\infty_{\mathbf{J}^{-1}(\mu)/G_\mu}(V)$, for some open neighborhood V of $\pi_\mu(m)$ in $\mathbf{J}^{-1}(\mu)/G_\mu$. We recall that $\pi_\mu : \mathbf{J}^{-1}(\mu) \to \mathbf{J}^{-1}(\mu)/G_\mu$ is the projection onto the orbit space. By definition, there exists an open G_μ-invariant neighborhood U_m of m in M and a function $F \in C^\infty_M(U_m)^G$ such that

$$f \circ \pi_\mu|_{\pi_\mu^{-1}(V)\cap U_m} = F|_{\pi_\mu^{-1}(V)\cap U_m}.$$

We can assume, without loss of generality, that $U_m = Y_0$, with Y_0 the G_μ-invariant neighborhood of m in Proposition 8.1.2. Now let $U'_m = G \cdot U_m$ and $F' \in C^\infty_M(U'_m)^G$ be the function defined by $F'([g,\eta,v]) = F([e,\eta,v])$, for any $[g,\eta,v] \in U'_m$. The G_μ-invariance of F implies that F' is well defined. Also, by construction, F' is G-invariant and

$$f \circ \pi_\mu|_{\pi_\mu^{-1}(V)\cap U'_m} = F'|_{\pi_\mu^{-1}(V)\cap U'_m},$$

which shows that F' is a G-invariant local extension of $f \circ \pi_\mu$ at m.

(ii) Condition (10.2.3) holds and hence $\left(W^{\infty}_{\mathbf{J}^{-1}(\mu)/G_\mu}, \{\cdot,\cdot\}^{\mathbf{J}^{-1}(\mu)/G_\mu}\right)$ is a presheaf of Poisson algebras, where the bracket $\{\cdot,\cdot\}^{\mathbf{J}^{-1}(\mu)/G_\mu}$ is given by the expression (10.2.2). Indeed, for any $m \in \mathbf{J}^{-1}(\mu)$ and $\mathbf{d}F(m) \in \Delta_m$, where $F \in C^\infty(U_m)^G$, for some open G-invariant neighborhood U_m of m in M, we have that $X_F(m) = B^\sharp(m)(\mathbf{d}F(m))$ is such that $\mathbf{d}G(m)\,(X_F(m)) = 0$, for any $\mathbf{d}G(m) \in \Delta^S_m$. This is so because, by Noether's Theorem 4.5.11, the integral curve $F_t(m)$ of $X_F(m)$ going through m stays in $\mathbf{J}^{-1}(\mu)$ since the function G is constant on a neighborhood of m in $\mathbf{J}^{-1}(\mu)$. Consequently, (10.2.3) holds.

10.3.3 Casimir Reduction. (Ortega and Planas (2003a)) The Casimir functions of a Poisson manifold $(M, \{\cdot,\cdot\})$ provide a natural way to obtain submanifolds S that satisfy the condition (10.2.7). Indeed, let $\{C_1, \ldots, C_n\} \subset C^\infty(M)$ be Casimir functions of the Poisson manifold $(M, \{\cdot,\cdot\})$ and let $F : M \to \mathbb{R}^n$ be the associated smooth function defined by $m \longmapsto (C_1(m), \ldots, C_n(m))$, $m \in M$. Let $v \in \mathbb{R}^n$ be a regular value of F and let $S := F^{-1}(v)$ be the corresponding fiber. By the Submersion Theorem (see §1.1.13), S is a closed embedded submanifold of M such that $T_zS = \ker T_zF$, for all $z \in S$. Additionally, let $f \in C^\infty(U)$, for some open neighborhood U of a given point $z \in S$. Then

$$\begin{aligned} T_zF\left(X_f(z)\right) &= \left.\frac{d}{dt}\right|_{t=0} (C_1(F_t(z)), \ldots, C_n(F_t(z))) \\ &= (\{C_1, f\}(z), \ldots, \{C_2, f\}(z)) = 0, \end{aligned}$$

where F_t is the flow of X_f. This equality implies that $X_f(z) \in T_zS$, for any $z \in S$. Since $X_f(z) \in B^\sharp(z)(T_z^*M)$ is arbitrary, condition (10.2.7) holds and hence $(S, \{\cdot,\cdot\}|_S)$ is a Poisson manifold.

If the Poisson manifold $(S, \{\cdot,\cdot\}|_S)$ has Casimir functions, we can repeat this reduction process to obtain another Poisson manifold $(S_2, \{\cdot,\cdot\}|_{S_2})$ satisfying $S_2 \subset S \subset M$. The iteration of this procedure is limited by the fact that at each step there is a drop in dimensions and also because there exists a symplectic leaf $\mathcal{L}$ of $(M, \{\cdot,\cdot\})$ such that

$$\mathcal{L} \subset S_k \subset S_{k-1} \subset \ldots \subset S_2 \subset S \subset M.$$

The Casimir functions of $(S_i, \{\cdot,\cdot\}|_{S_i})$ are usually referred to as the ith-***level pseudo-Casimir functions*** of $(M, \{\cdot,\cdot\})$.

10.3.4 The Hilbert–Weyl Theorem and reduced Poisson structures. An important problem that we have not yet addressed in our discussion is the explicit construction of the quotients that constitute the Poisson or symplectic reduced spaces by a group action. When the symmetry group is compact, a theorem of Gotay and Tuynman (1991) suggests an approach based on the Theory of Invariants, which we briefly reviewed §2.5.1. This theorem shows that every canonical action of a compact Lie group G on a symplectic manifold can be modeled by a symplectic linear representation of G on a certain finite dimensional symplectic vector space $V \simeq \mathbb{R}^{2n}$. Once we are in this framework, the array of tools introduced in §2.5.1 are fully applicable. We now present two simple examples with compact symmetry group that illustrate how to implement this idea. For more examples the reader is encouraged to check with the book by CUSHMAN AND BATES (1997) and with the article devoted to examples written by LERMAN, MONTGOMERY, AND SJAMAAR (1993).

(i) A canonical circle action. We reproduce here a part of the reduction treatment presented in ARMS et al. (1991) and CUSHMAN AND BATES (1997). We will start by considering a symplectic action of the circle S^1 on $(T\mathbb{R}^3, \omega)$. Let $(\mathbf{x}, \mathbf{y})$ denote points in $T\mathbb{R}^3$, with Cartesian coordinates $\mathbf{x} = (x_1, x_2, x_3)$, $\mathbf{y} = (y_1, y_2, y_3)$. The canonical symplectic form is

$$\omega = \sum_{i=1}^{3} dx_i \wedge dy_i. \tag{10.3.1}$$

In these coordinates, we consider the S^1-action given by

$$\begin{array}{rccc} \Phi: & S^1 \times T\mathbb{R}^3 & \longrightarrow & T\mathbb{R}^3 \\ & (e^{i\phi}, (\mathbf{x}, \mathbf{y})) & \longmapsto & (R_\phi \mathbf{x}, R_\phi \mathbf{y}), \end{array}$$

where

$$R_\phi = \begin{pmatrix} \cos\phi & -\sin\phi & 0 \\ \sin\phi & \cos\phi & 0 \\ 0 & 0 & 1 \end{pmatrix}.$$

It is straightforward to verify that this action is linear and canonical and hence, by Example 4.5.7, it has an associated equivariant standard momentum map $\mathbf{J} : T\mathbb{R}^3 \to \mathbb{R}$ (the usual angular momentum) given by

$$\mathbf{J}(\mathbf{x}, \mathbf{y}) = x_1 y_2 - x_2 y_1.$$

We now use the invariant polynomials of this action. A Hilbert basis of the algebra of S^1-invariant polynomials is given by

$$\begin{array}{lll} \sigma_1 = x_3 & \sigma_3 = y_1^2 + y_2^2 + y_3^2 & \sigma_5 = x_1^2 + x_2^2 \\ \sigma_2 = y_3 & \sigma_4 = x_1 y_1 + x_2 y_2 & \sigma_6 = x_1 y_2 - x_2 y_1. \end{array} \tag{10.3.2}$$

The elements of this basis satisfy the semialgebraic relations

$$\sigma_4^2 + \sigma_6^2 = \sigma_5(\sigma_3 - \sigma_2^2), \qquad \sigma_3 \geq 0, \qquad \sigma_5 \geq 0. \tag{10.3.3}$$

In this case the Hilbert map is given by

$$\begin{array}{rccc} \sigma: & T\mathbb{R}^3 & \longrightarrow & \mathbb{R}^6 \\ & (\mathbf{x}, \mathbf{y}) & \longmapsto & (\sigma_1(\mathbf{x}, \mathbf{y}), \ldots, \sigma_6(\mathbf{x}, \mathbf{y})). \end{array}$$

Hence, using the arguments provided in §2.5.1, the S^1-orbit space $T\mathbb{R}^3/S^1$ can be identified with the semialgebraic variety $\sigma(T\mathbb{R}^3) \subset \mathbb{R}^6$, defined by the relations (10.3.3).

Consider now the tangent space TS^2 to the 2-sphere S^2 as a submanifold of $T\mathbb{R}^3$. In this sense, we can write

$$TS^2 = \{(\mathbf{x}, \mathbf{y}) \in T\mathbb{R}^3 \mid x_1^2 + x_2^2 + x_3^2 = 1 \text{ and } x_1 y_1 + x_2 y_2 + x_3 y_3 = 0\}.$$

Note that TS^2 is obviously S^1-invariant and constitutes a symplectic submanifold of $T\mathbb{R}^3$. Using the previously introduced invariants and their associated Hilbert map we can characterize the quotient TS^2/S^1 as the semialgebraic variety $\sigma(TS^2)$ defined by the relations (10.3.3) and

$$\sigma_5 + \sigma_1^2 = 1 \qquad \sigma_4 + \sigma_1\sigma_2 = 0,$$

which allow us to solve for σ_4 and σ_5, yielding

$$\begin{aligned} TS^2/S^1 &= \sigma(TS^2) = \{(\sigma_1, \sigma_2, \sigma_3, \sigma_6) \in \mathbb{R}^4 \mid \\ &\qquad \sigma_1^2\sigma_2^2 + \sigma_6^2 = (1-\sigma_1^2)(\sigma_3 - \sigma_2^2),\ |\sigma_1| \leq 1,\ \sigma_3 \geq 0\}. \end{aligned} \tag{10.3.4}$$

Since the S^1-momentum map $\mathbf{J} : T\mathbb{R}^3 \to \mathbb{R}$ coincides with the invariant σ_6, it is very easy to write down the symplectic point reduced spaces associated to the S^1-action on TS^2 using the Hilbert basis of invariant polynomials. Indeed, for any $\mu \in \mathbb{R} \simeq \operatorname{Lie}(S^1)$

$$(TS^2)_\mu = \sigma(\mathbf{J}|_{TS^2}^{-1}(\mu)) = \sigma(\mathbf{J}^{-1}(\mu) \cap TS^2),$$

which can be obtained from (10.3.4) by fixing $\sigma_6 = \mu$. Therefore, $(TS^2)_\mu$ is the semialgebraic subset of $\mathbb{R}^3$ defined by

$$(1-\sigma_1^2)\sigma_3 = \sigma_2^2 + \mu^2, \quad |\sigma_1| \leq 1, \quad \sigma_3 \geq 0. \tag{10.3.5}$$

Notice that the nonzero values of the momentum map $\mathbf{J}|_{TS^2} : TS^2 \to \mathbb{R}$ are regular and hence the corresponding fibers are closed submanifolds acted freely and properly upon by S^1. Consequently, for those values $\mu \in \mathbb{R}$, the reduced spaces $(TS^2)_\mu := \mathbf{J}|_{TS^2}^{-1}(\mu)/S^1$ are smooth symplectic reduced spaces. Using our approach based on the use of a Hilbert basis, this is reflected in the fact that $(TS^2)_\mu$ appears as the graph of the smooth function

$$\sigma_3 = \frac{\sigma_2^2 + \mu^2}{1-\sigma_1^2}, \qquad |\sigma_1| < 1.$$

The case $\mu = 0$ is singular and $(TS^2)_0$ is not a smooth manifold. In the model provided by the Hilbert basis, this is translated into the fact that the semialgebraic subset defined by (10.3.5) is no longer the smooth graph of a function. However, as we proved in §10.3.2, there exists a presheaf of Poisson algebras on the functions on that quotient which is inherited from the Poisson structure associated to the symplectic form in (10.3.1). As we will now see, this Poisson algebra admits a very explicit expression in terms of the elements of the Hilbert basis of polynomials associated to our action.

Firstly, the symplectic structure on $T\mathbb{R}^3$ induces, via the Hilbert map obtained from the invariants (10.3.2), a Poisson bracket on $\mathbb{R}^6$ whose structure matrix is given in the following table:

$\{\cdot,\cdot\}_{\mathbb{R}^6}$	σ_1	σ_2	σ_3	σ_4	σ_5	σ_6
σ_1	0	1	$2\sigma_2$	0	0	0
σ_2	-1	0	0	0	0	0
σ_3	$-2\sigma_2$	0	0	$-2(\sigma_3 - \sigma_2^2)$	$-4\sigma_4$	0
σ_4	0	0	$2(\sigma_3 - \sigma_2^2)$	0	$-2\sigma_5$	0
σ_5	0	0	$4\sigma_4$	$2\sigma_5$	0	0
σ_6	0	0	0	0	0	0

It is easy to see that $C_1 = \sigma_4^2 + \sigma_6^2 - \sigma_5(\sigma_3 - \sigma_2^2)$ and $C_2 = \sigma_6$ are Casimir functions of this Poisson structure. If we understand the Poisson variety TS^2/S^1 as $\sigma(TS^2) \subset \mathbb{R}^6$, the Poisson bracket $\{\cdot, \cdot\}_{\mathbb{R}^6}$ induces a Poisson bracket $\{\cdot, \cdot\}^{TS^2/S^1}$ on TS^2/S^1, whose structure matrix is given in the following table:

$\{\cdot, \cdot\}^{TS^2/S^1}$	σ_1	σ_2	σ_3	σ_6
σ_1	0	$1 - \sigma_1^2$	$2\sigma_2$	0
σ_2	$-(1 - \sigma_1^2)$	0	$-2\sigma_1\sigma_3$	0
σ_3	$-2\sigma_2$	$2\sigma_1\sigma_3$	0	0
σ_6	0	0	0	0

Since the reduced spaces $(TS^2)_\mu \subset \sigma(TS^2)$ are defined by $C_2 = \mu$, the Poisson bracket $\{\cdot, \cdot\}^{(TS^2)_\mu}$ on $(TS^2)_\mu$ has the same structure matrix as the one given in the previous table with the last row and column deleted.

Using this table it is easy to verify that for any $f_\mu, g_\mu \in C^\infty((TS^2)_\mu)$

$$\{f_\mu, g_\mu\}^{(TS^2)_\mu} = (\nabla f_\mu \times \nabla g_\mu) \cdot \nabla \psi,$$

where

$$\psi(\sigma_1, \sigma_2, \sigma_3) = \sigma_3(1 - \sigma_1^2) - \sigma_2^2 - \mu^2,$$

and ∇ denotes the gradient with respect to the $(\sigma_1, \sigma_2, \sigma_3)$-coordinates.

(ii) A problem with spherical symmetry. Consider now the tangent bundle $T\mathbb{R}^3$ with the symplectic form ω given by the expression

$$\omega = \sum_{i=1}^{3} dx_i \wedge dy_i,$$

where $(\mathbf{x}, \mathbf{y}) \in T\mathbb{R}^3 \simeq \mathbb{R} \times \mathbb{R}^3$ and $\mathbf{x} = (x_1, x_2, x_3)$, $\mathbf{y} = (y_1, y_2, y_3)$ are the Cartesian coordinates of $\mathbf{x}$ and $\mathbf{y}$, respectively. Consider also the canonical action of the group $SO(3)$ on $(T\mathbb{R}^3, \omega)$ given by

$$\begin{aligned} \Phi : SO(3) \times T\mathbb{R}^3 &\longrightarrow \quad T\mathbb{R}^3 \\ (A, (\mathbf{x}, \mathbf{y})) &\longmapsto (A\mathbf{x}, A\mathbf{y}). \end{aligned}$$

It is straightforward to verify that this action is linear and canonical; hence, by Example 4.5.7, it has an associated equivariant standard momentum map $\mathbf{J} : T\mathbb{R}^3 \to \mathfrak{so}(3)^*$ (the usual spatial angular momentum) given by

$$\mathbf{J}(\mathbf{x}, \mathbf{y}) = \mathbf{x} \times \mathbf{y}, \qquad (\mathbf{x}, \mathbf{y}) \in T\mathbb{R}^3.$$

A Hilbert basis of the algebra of $SO(3)$-invariant polynomials of this action is given by (see WEYL (1946); BELTRAME et al. (1997))

$$\pi_1 = x_1^2 + x_2^2 + x_3^2 \qquad \pi_2 = y_1^2 + y_2^2 + y_3^2 \qquad \pi_3 = x_1 y_1 + x_2 y_2 + x_3 y_3, \tag{10.3.6}$$

where these polynomials are subjected to the following semialgebraic relations

$$\pi_1\pi_2 - \pi_3^2 \geq 0, \qquad \pi_1 \geq 0, \qquad \pi_2 \geq 0. \tag{10.3.7}$$

The corresponding Hilbert map is given by

$$\begin{aligned} \sigma : T\mathbb{R}^3 &\longrightarrow \mathbb{R}^3 \\ (\mathbf{x},\, \mathbf{y}) &\longmapsto (\pi_1(\mathbf{x},\, \mathbf{y}),\, \pi_2(\mathbf{x},\, \mathbf{y}),\, \pi_3(\mathbf{x},\, \mathbf{y})). \end{aligned}$$

Hence, the $SO(3)$-orbit space $T\mathbb{R}^3/SO(3)$ can be identified with the semialgebraic variety $\sigma(T\mathbb{R}^3) \subset \mathbb{R}^3$, defined by the relations (10.3.7). The symplectic structure in $T\mathbb{R}^3$ via the Hilbert map obtained from the invariants (10.3.6), induces a Poisson algebra structure on the quotient presheaf of smooth functions on $T\mathbb{R}^3/SO(3)$ whose structure matrix is given in the following table:

$\{\cdot, \cdot\}^{T\mathbb{R}^3/SO(3)}$	π_1	π_2	π_3
π_1	0	$4\pi_3$	$2\pi_1$
π_2	$-4\pi_3$	0	$-2\pi_2$
π_3	$-2\pi_1$	$2\pi_2$	0

Notice that the function

$$C := \pi_1\pi_2 - \pi_3^2 \tag{10.3.8}$$

is a Casimir for this Poisson structure that will play an important role later..

We now compute the orbit reduced spaces $(T\mathbb{R}^3)_{\mathcal{O}_\mu}$. First, recall that if we make the usual identification between $\mathbb{R}^3$ and the Lie algebra $\mathfrak{so}(3)$ given by the map

$$(a,\, b,\, c) \longmapsto \begin{pmatrix} 0 & -c & b \\ c & 0 & -a \\ -b & a & 0 \end{pmatrix},$$

the coadjoint orbit $\mathcal{O}_\mu$ of the point $\mu \in \mathbb{R}^3$ is the sphere in $\mathbb{R}^3$ of radius $\|\mu\| \geq 0$, hence

$$\mathbf{J}^{-1}(\mathcal{O}_\mu) = \{(\mathbf{x},\, \mathbf{y}) \in T\mathbb{R}^3 \mid \|\mathbf{x} \times \mathbf{y}\| = \|\mu\|\}.$$

Given that

$$\|\mathbf{x} \times \mathbf{y}\|^2 = \|\mathbf{x}\|^2\|\mathbf{y}\|^2(1 - \frac{\langle \mathbf{x},\, \mathbf{y}\rangle^2}{\|\mathbf{x}\|^2\|\mathbf{y}\|^2}) = \pi_1\pi_2 - \pi_3^2,$$

we can conclude that $(T\mathbb{R}^3)_{\mathcal{O}_\mu}$ is the subset of $T\mathbb{R}^3/SO(3)$ defined by the algebraic constraint

$$\pi_1\pi_2 - \pi_3^2 = \|\mu\|^2$$

or, in other words, the $\|\mu\|^2$-level set of the Casimir (10.3.8), which guarantees that the Poisson bracket $\{\cdot,\cdot\}^{(T\mathbb{R}^3)_{\mathcal{O}_\mu}}$ has the same structure matrix as $T\mathbb{R}^3/SO(3)$. More explicitly, if $\psi := 2\pi_3^2 - 2\pi_1\pi_2 = -2C$, then

$$\{\pi_i, \pi_j\}^{(T\mathbb{R}^3)_{\mathcal{O}_\mu}} = \sum_k \epsilon_{ijk}\frac{\partial\psi}{\partial\pi_k},$$

where ϵ_{ijk} denotes the completely antisymmetric symbol. This implies that for any $f_{\mathcal{O}_\mu}, g_{\mathcal{O}_\mu} \in C^\infty((T\mathbb{R}^3)_{\mathcal{O}_\mu})$

$$\{f_{\mathcal{O}_\mu}, g_{\mathcal{O}_\mu}\}^{(T\mathbb{R}^3)_{\mathcal{O}_\mu}} = (\nabla f_{\mathcal{O}_\mu} \times \nabla g_{\mathcal{O}_\mu}) \cdot \nabla\psi.$$

10.4 Poisson reduction by distributions

The Poisson reduction theorem presented in Section 10.2 requires the presence of a pseudogroup of transformations defined on the entire manifold. As we will see in the examples at the end of this section, sometimes one may wish to reduce with respect to an invariance property defined only on a subset of the manifold in question. The study of this situation is the main goal of the following pages. We start by introducing the setup.

10.4.1 Definition *Let M be a differentiable manifold and $S \subset M$ a decomposed subset of M. Let $\{S_i\}_{i\in I}$ be the pieces of this decomposition. The topology of S is not necessarily the relative topology as a subset of M. We say that $D \subset TM|_S$ is a* ***smooth distribution on S adapted to the decomposition*** *$\{S_i\}_{i\in I}$, if $D \cap TS_i$ is a smooth distribution on S_i for all $i \in I$. The distribution D is said to be* ***integrable*** *if $D \cap TS_i$ is integrable for each $i \in I$.*

In the situation described by the previous definition and if D is integrable, the integrability of the distributions $D_{S_i} := D \cap TS_i$ on S_i allows us to partition each S_i into the corresponding maximal integral manifolds. Thus, there is an equivalence relation on S_i whose equivalence classes are precisely these maximal integral manifolds. Doing this on each S_i, we obtain an equivalence relation D_S on the whole set S by taking the union of the different equivalence classes corresponding to all the D_{S_i}. We define the quotient space S/D_S as

$$S/D_S := \bigcup_{i\in I} S_i/D_{S_i}.$$

We will denote by $\pi_{D_S} : S \to S/D_S$ the natural projection.

10.4.2 Definition *Let $(M, \{\cdot,\cdot\})$ be a Poisson manifold and $D \subset TM$ a smooth distribution on M. The distribution D is called* ***Poisson*** *or* ***canonical*** *if the condition $\mathbf{d}f|_D = \mathbf{d}g|_D = 0$, for any $f, g \in C^\infty_M(U)$ and any open subset $U \subset P$, implies that $\mathbf{d}\{f, g\}|_D = 0$.*

10.4.3 Concerning the previous definition, note that if D is spanned by a family of infinitesimal Poisson automorphisms, then D is a Poisson distribution by (4.1.15). The converse is not necessarily true as we show in the following counterexample.

Consider $\mathbb{R}^3$ as a Poisson manifold with the Poisson tensor given by

$$B(x, y, z) = \begin{pmatrix} 0 & x & y \\ -x & 0 & z \\ -y & -z & 0 \end{pmatrix}$$

and the additive group $(\mathbb{R}^2, +)$ acting on it by the map $\Phi : \mathbb{R}^2 \times \mathbb{R}^3 \to \mathbb{R}^3$ defined by

$$\Phi((a, b), (x, y, z)) := (x, e^a y, e^b z), \quad \text{for any } (a, b) \in \mathbb{R}^2, (x, y, z) \in \mathbb{R}^3.$$

This action is not canonical. Indeed, since $\{y, z\} = z$, we have $\Phi^*_{(a,a)}\{y, z\} = e^a z$, whereas $\{\Phi^*_{(a,a)} y, \Phi^*_{(a,a)} z\} = e^{2a} z$, for any $a \in \mathbb{R}$. Moreover, the infinitesimal generators are not infinitesimal Poisson automorphisms but, as we shall see below, the distribution that they span is canonical in the sense of Definition 10.4.2. First, the infinitesimal generators are

$$(1, 0)_{\mathbb{R}^3}(x, y, z) = (0, y, 0), \ (0, 1)_{\mathbb{R}^3}(x, y, z) = (0, 0, z), \ \text{for any } (x, y, z) \in \mathbb{R}^3.$$

The infinitesimal generators are not infinitesimal Poisson automorphisms. For example, $(1, 0)_{\mathbb{R}^3}[\{y, z\}] = 0$, whereas $\{(1, 0)_{\mathbb{R}^3}[y], z\} + \{y, (1, 0)_{\mathbb{R}^3}[z]\} = z$ and hence (4.1.15) is violated. Second, let D be the distribution spanned by $(1, 0)_{\mathbb{R}^3}$ and $(0, 1)_{\mathbb{R}^3}$. We now verify that D is canonical according to Definition 10.4.2. Since for any function $f \in C^\infty(\mathbb{R}^3)$

$$(1, 0)_{\mathbb{R}^3}[f](x, y, z) = y\frac{\partial f}{\partial y} \quad \text{and} \quad (0, 1)_{\mathbb{R}^3}[f](x, y, z) = z\frac{\partial f}{\partial z}$$

it follows that $\mathbf{d}f|_D = 0$ if and only if f is a function only of the first component. Since the bracket of two such functions is zero, the canonical character of D follows.

10.4.4 Example. Let $\mathfrak{g}$ be a Lie algebra acting canonically on the Poisson manifold $(M, \{\cdot, \cdot\})$ and let D be the distribution spanned by all vector fields $\xi_M \in \mathfrak{X}(M), \xi \in \mathfrak{g}$. This distribution is canonical since it is spanned by a family of infinitesimal Poisson automorphisms (see §10.4.3 above).

10.4.5 The presheaf of smooth functions on S/D_S. In this section we will consider a presheaf of smooth functions on S/D_S that requires fewer invariance properties in its definition than those appearing in the context of quotients by pseudogroups of transformations. We define the presheaf of smooth functions C^∞_{S/D_S} on S/D_S as the map that associates to any open subset V of S/D_S the set of functions $C^\infty_{S/D_S}(V)$ characterized by the following property: $f \in C^\infty_{S/D_S}(V)$ if and only if for any $z \in V$ there exists $m \in \pi^{-1}_{D_S}(V)$, U_m an open neighborhood of m in M, and $F \in C^\infty_M(U_m)$ such that

$$f \circ \pi_{D_S}|_{\pi^{-1}_{D_S}(V) \cap U_m} = F|_{\pi^{-1}_{D_S}(V) \cap U_m}. \tag{10.4.1}$$

We say that F is a ***local extension*** of $f \circ \pi_{D_S}$ at the point $m \in \pi^{-1}_{D_S}(V)$.

The presheaf C^∞_{S/D_S} is said to have the (D, D_S)-***local extension property*** when the topology of S is stronger than the relative topology and, at the same time, the local extensions of $f \circ \pi_{D_S}$ defined in (10.4.1) can always be chosen to satisfy

$$\mathbf{d}F(n)|_{D(n)} = 0, \quad \text{for any} \quad n \in \pi^{-1}_{D_S}(V) \cap U_m.$$

We say that F is a ***local D-invariant extension*** of $f \circ \pi_{D_S}$ at the point $m \in \pi_{D_S}^{-1}(V)$.

10.4.6 Proposition *Suppose that S is a smooth embedded submanifold of M and that D_S is a smooth, integrable, and regular distribution on S. Then the presheaf C^{∞}_{S/D_S} coincides with the presheaf of smooth functions on S/D_S when considered as a regular quotient manifold.*

Proof. In order to distinguish both presheaves of functions, denote by T^{∞}_{S/D_S} the presheaf of smooth functions on S/D_S when considered as a regular quotient manifold. We will now show that for any open subset V of S/D_S

$$T^{\infty}_{S/D_S}(V) = C^{\infty}_{S/D_S}(V).$$

Let $f \in T^{\infty}_{S/D_S}(V)$. The regularity of the distribution D_S implies that the projection π_{D_S} is a surjective submersion and hence $f \circ \pi_{D_S} \in C^{\infty}_S(\pi_{D_S}^{-1}(V))$. Now let m be an arbitrary element of $\pi_{D_S}^{-1}(V)$ and U_m a submanifold chart for S in M in a neighborhood of m such that $U_m \cap S \subset \pi_{D_S}^{-1}(V)$. The neighborhood U_m can be thought of as a product $W_1 \times W_2$ of two open neighborhoods of the origin in two vector spaces such that $U_m \cap S = W_1 \times \{0\}$. Define now $F \in C^{\infty}_M(U_m)$ by $F(w_1, w_2) := (f \circ \pi_{D_S})(w_1)$, for any $(w_1, w_2) \in W_1 \times W_2 = U_m$. By construction, we have that

$$f \circ \pi_{D_S}|_{\pi_{D_S}^{-1}(V) \cap U_m} = F|_{\pi_{D_S}^{-1}(V) \cap U_m}.$$

Since this can be done for any $m \in \pi_{D_S}^{-1}(V)$ we have proved that $f \in C^{\infty}_{S/D_S}(V)$.

Conversely, let $f \in C^{\infty}_{S/D_S}(V)$. By definition, for any $z \in V$ there exists $m \in \pi_{D_S}^{-1}(V)$, U_m open neighborhood of m in M, and $F \in C^{\infty}_M(U_m)$ such that

$$f \circ \pi_{D_S}|_{\pi_{D_S}^{-1}(V) \cap U_m} = F|_{\pi_{D_S}^{-1}(V) \cap U_m}.$$

Since π_{D_S} is a submersion and $\pi_{D_S}^{-1}(V) \cap U_m$ is an open neighborhood of m in S, it follows that $N_z := \pi_{D_S}(\pi_{D_S}^{-1}(V) \cap U_m)$ is an open neighborhood of z in S/D_S and that $f|_{N_z}$ is smooth. Since in this argument $z \in V$ was arbitrary, we have proved that $f \in T^{\infty}_{S/D_S}(V)$. ■

10.4.7 Definition *Let $(M, \{\cdot,\cdot\})$ be a Poisson manifold, S a decomposed subset of M, and $D \subset TM|_S$ a Poisson integrable generalized distribution adapted to the decomposition of S. Assume that C^{∞}_{S/D_S} has the (D, D_S)-local extension property. We say that $(M, \{\cdot,\cdot\}, D, S)$ is* **Poisson reducible** *when $(S/D_S, C^{\infty}_{S/D_S}, \{\cdot,\cdot\}^{S/D_S})$ is a well-defined presheaf of Poisson algebras where, for any open set $V \subset S/D_S$, the bracket $\{\cdot,\cdot\}^{S/D_S}_V : C^{\infty}_{S/D_S}(V) \times C^{\infty}_{S/D_S}(V) \to C^{\infty}_{S/D_S}(V)$ is given by*

$$\{f, g\}^{S/D_S}_V(\pi_{D_S}(m)) := \{F, G\}(m),$$

for any $m \in \pi_{D_S}^{-1}(V)$. The maps F, G are local D-invariant extensions at m of $f \circ \pi_{D_S}$ and $g \circ \pi_{D_S}$, respectively.

The proof of the following theorem imitates the corresponding implication in Theorem 10.2.5.

10.4.8 Theorem *Let $(M, \{\cdot, \cdot\})$ be a Poisson manifold with associated Poisson tensor $B \in \Lambda^2(T^*M)$, S a decomposed space, and $D \subset TM|_S$ a Poisson integrable generalized distribution adapted to the decomposition of S in the Definitions* 10.4.1 *and* 10.4.2. *Assume that C^∞_{S/D_S} has the (D, D_S)-local extension property. Then $(M, \{\cdot, \cdot\}, D, S)$ is Poisson reducible if for any $m \in S$*

$$B^\sharp(\Delta_m) \subset \left[\Delta_m^S\right]^\circ \tag{10.4.2}$$

where $\Delta_m := \{\mathbf{d}F(m) \mid F \in C^\infty_M(U_m), \mathbf{d}F(z)|_{D(z)} = 0$, for all $z \in U_m \cap S$, and for any open neighborhood U_m of m in $M\}$ and $\Delta_m^S := \{\mathbf{d}F(m) \in \Delta_m \mid F|_{U_m \cap V_m}$ is constant for an open neighborhood U_m of m in M and an open neighborhood V_m of m in S $\}$.

10.4.9 Remark If S is endowed with the relative topology, then $\Delta_m^S := \{\mathbf{d}F(m) \in \Delta_m \mid F|_{U_m \cap V_m}$ is constant for an open neighborhood U_m of m in $M\}$.

10.4.10 Remark As opposed to the situation in Theorem 10.2.5, the condition (10.4.2) is sufficient for Poisson reducibility but, in general, is not necessary. The reason fort this is that the functions defining the spaces Δ_m and Δ_m^S are not defined on saturated open sets which prevents the formulation of a result similar to Lemma 10.2.9. As we will see in Theorem 10.4.12, an alternative hypothesis that makes this condition necessary and sufficient is, roughly speaking, the regularity of the distribution $D_S := D \cap TS$.

10.4.11 Reduction by regular canonical distributions. Let $(M, \{\cdot, \cdot\})$ be a Poisson manifold and S an embedded submanifold of M. Let $D \subset TM|_S$ be a subbundle of the tangent bundle of M restricted to S such that $D_S := D \cap TS$ is a smooth, integrable, regular distribution on S and D is canonical. In this situation we can prove the following theorem.

10.4.12 Theorem (Marsden and Ratiu **(1986)**) *Let $(M, \{\cdot, \cdot\})$ be a Poisson manifold with associated Poisson tensor $B \in \Lambda^2(T^*M)$ and S an embedded smooth submanifold of M. Let $D \subset TM|_S$ be a canonical subbundle of the tangent bundle of M restricted to S such that $D_S := D \cap TS$ is a smooth, integrable, and regular distribution on S. Then $(M, \{\cdot, \cdot\}, D, S)$ is Poisson reducible if and only if*

$$B^\sharp(D^\circ) \subset TS + D. \tag{10.4.3}$$

10.4.13 Remark Even though in the previous theorem only the distribution D_S intervenes in the construction of the quotient manifold S/D_S, the full distribution D is used in the definition of the Poisson bracket on this quotient when $(M, \{\cdot, \cdot\}, D, S)$ is Poisson reducible. Actually, in spite of the fact that the reduction of $(M, \{\cdot, \cdot\}, D, S)$ and $(M, \{\cdot, \cdot\}, D_S, S)$ renders the same quotient manifold S/D_S it does *not* yield the same Poisson brackets on this quotient since different sets of functions are involved. This is particularly evident in the following example in which we show, using Theorem 10.4.12, that $(M, \{\cdot, \cdot\}, D, S)$ is reducible whereas $(M, \{\cdot, \cdot\}, D_S, S)$ is not.

Let (M, ω) be a connected symplectic manifold acted freely and canonically upon by a connected and compact Lie group G. Let $\mathbf{J} : M \to \mathfrak{g}^*$ be a coadjoint equivariant standard momentum map associated to this action, $\mu \in \mathfrak{g}^*$ one of its values, and $G_\mu \subset G$ its coadjoint isotropy subgroup. Let D be the distribution on M given by the tangent spaces to the orbits of the G-action and $S := \mathbf{J}^{-1}(\mu)$, which is a smooth closed submanifold of M because of the freeness of the action. The Reduction Lemma 4.5.34 implies that D_S equals, in this case, the distribution given by the tangent spaces to the orbits of the G_μ-action. The compactness and connectedness of G implies that G_μ is connected (see Theorem 3.3.1 in DUISTERMAAT AND KOLK (1999)) and hence $S/D_S = \mathbf{J}^{-1}(\mu)/G_\mu$.

The quadruple $(M, \omega, D, \mathbf{J}^{-1}(\mu))$ satisfies (10.4.3) and it is hence Poisson reducible. Indeed, in this case the expression (10.4.3) reads $(\mathfrak{g} \cdot m)^\omega \subset \ker T_m\mathbf{J} + \mathfrak{g} \cdot m$, for any $m \in \mathbf{J}^{-1}(\mu)$, which by the Proposition 4.5.14**(ii)** amounts to $\ker T_m\mathbf{J} \subset \ker T_m\mathbf{J} + \mathfrak{g} \cdot m$, which is an obvious inclusion.

On the other hand, the quadruple $(M, \omega, D_S, \mathbf{J}^{-1}(\mu))$ is *not* Poisson reducible even though the corresponding quotient manifold is the same as for $(M, \omega, D, \mathbf{J}^{-1}(\mu))$. Indeed, condition (10.4.3) reads in this case $(\mathfrak{g}_\mu \cdot m)^\omega \subset \ker T_m\mathbf{J} + \mathfrak{g}_\mu \cdot m = \ker T_m\mathbf{J}$, for any $m \in \mathbf{J}^{-1}(\mu)$. However, by the Reduction Lemma 4.5.34 and Proposition 4.5.14**(ii)** we have $(\mathfrak{g}_\mu \cdot m)^\omega = (\ker T_m\mathbf{J} \cap \mathfrak{g} \cdot m)^\omega = (\ker T_m\mathbf{J})^\omega + (\mathfrak{g} \cdot m)^\omega = \mathfrak{g} \cdot m + \ker T_m\mathbf{J}$ which is, in general, not a subset of $\ker T_m\mathbf{J}$.

Before we start with the proof of Theorem 10.4.12 we need the following technical result.

10.4.14 Lemma *Let M be a smooth manifold and S an embedded submanifold of M. Let $D \subset TM|_S$ be a subbundle of the tangent bundle of M restricted to S such that $D_S := D \cap TS$ is a smooth, integrable, regular distribution on S. Then the presheaf C^∞_{S/D_S} has the (D, D_S)-local extension property.*

Proof. Since the statement that we want to prove is of a local nature, we can assume without loss of generality that $M = U \times V$ and $S = U$, where U and V are open neighborhoods of the origin in two vector spaces $\mathbf{F}_1$ and $\mathbf{F}_2$, respectively. Analogously, we can write $TM := U \times V \times (\mathbf{F}_1 \oplus \mathbf{F}_2)$, $TS = U \times \mathbf{F}_1$, and $D = U \times \mathbf{E}$, where $\mathbf{E}$ is a vector subspace $\mathbf{E} \subset \mathbf{F}_1 \oplus \mathbf{F}_2$. The last statement is a consequence of the fact that D is a subbundle of the tangent bundle of M restricted to S. Moreover, $D_S := D \cap TN = U \times \mathbf{E}_1$ with $\mathbf{E}_1 := \mathbf{E} \cap (\mathbf{F}_1 \oplus \{0\}))$. The regularity of the distribution D_S on S implies that $S/D_S = W$, where W is an open neighborhood of 0 in a vector space $\mathbf{E}_2$ such that

$$\mathbf{F}_1 = \mathbf{E}_1 \oplus \mathbf{E}_2 \quad \text{and} \quad \mathbf{E} \cap \mathbf{E}_2 = \{0\}. \tag{10.4.4}$$

Let $T \subset \mathbf{E}_1$ be such that $U = T \times W$. In these local coordinates, the projection $\pi_{D_S} : S = U = T \times W \to S/D_S = W$ is just the projection onto the second factor.

Now, given any smooth function $f \in C^\infty(S/D_S) = C^\infty(W)$, we define the function $F \in C^\infty(M)$ by

$$F(x, y, z) := (f \circ \pi_{D_S})(A(x, y, z)), \tag{10.4.5}$$

for any $(x, y, z) \in M = T \times W \times V \subset \mathbf{E}_1 \oplus \mathbf{E}_2 \oplus \mathbf{F}_2$ and where $A : \mathbf{E}_1 \oplus \mathbf{E}_2 \oplus \mathbf{F}_2 \to \mathbf{E}_1 \oplus \mathbf{E}_2$ denotes a linear map that constructed as follows: Let $\mathbf{E}' := (\mathbf{E}_2 \oplus \mathbf{F}_2) \cap \mathbf{E}$

and $\pi_{\mathbf{F}_2} : \mathbf{E}' \to \mathbf{F}_2$, $\pi_{\mathbf{E}_2} : \mathbf{E}' \to \mathbf{E}_2$ be the natural projections. Define $L : \pi_{\mathbf{F}_2}(\mathbf{E}') \to \pi_{\mathbf{E}_2}(\mathbf{E}')$ by $L(w) = -v$, where $w \in \pi_{\mathbf{F}_2}(\mathbf{E}')$ and $v \in \pi_{\mathbf{E}_2}(\mathbf{E}')$ is an element such that $v+w \in \mathbf{E}'$. The map L is well defined because if $v' \in \pi_{\mathbf{E}_2}(\mathbf{E}')$ is such that $v'+w \in \mathbf{E}'$, then $v - v' \in \mathbf{E}' \cap \mathbf{E}_2 \subset \mathbf{E} \cap \mathbf{E}_2 = \{0\}$ by (10.4.4), which shows that $v = v'$. The map L is clearly linear and can be extended to a linear map $\overline{A} : \mathbf{F}_2 \to \mathbf{E}_2$. Define $A : \mathbf{E}_1 \oplus \mathbf{E}_2 \oplus \mathbf{F}_2 \to \mathbf{E}_1 \oplus \mathbf{E}_2$ by $A(x, y, z) := (0, y + \overline{A}z)$, for any $(x, y, z) \in \mathbf{E}_1 \oplus \mathbf{E}_2 \oplus \mathbf{F}_2$.

We conclude by verifying that the map $F \in C^\infty(M)$ introduced in (10.4.5) satisfies the following two conditions:

(i) $F|_S = f \circ \pi_{D_S}$.
Let $(x, y, 0) \in S$ be arbitrary. By definition:

$$\begin{aligned} F(x, y, 0) &= (f \circ \pi_{D_S})(A(x, y, 0)) = (f \circ \pi_{D_S})(0, y) \\ &= (f \circ \pi_{D_S})(x, y), \end{aligned} \tag{10.4.6}$$

where the last equality is a consequence of the fact that $\pi_{D_S}(x, y) = \pi_{D_S}(0, y) = y$.

(ii) $\mathbf{d}F(x, y, 0)(u, v, w) = 0$ for any $(x, y, 0) \in S$ and any $(u, v, w) \in \mathbf{E}$.
This is so because by the definition (10.4.5) and by (10.4.6)

$$\begin{aligned} \mathbf{d}F(x, y, 0)(u, v, w) &= \mathbf{d}f(y) \cdot (\pi_{D_S}(A(u, v, w))) \\ &= \mathbf{d}f(y) \cdot (v + \overline{A}w). \end{aligned} \tag{10.4.7}$$

Since $(u, v, w) \in \mathbf{E}$ and $u \in \mathbf{E}_1 \subset \mathbf{E}$ we can conclude that $v+w \in \mathbf{E} \cap (\mathbf{E}_2 \oplus \mathbf{F}_2) = \mathbf{E}'$ and hence $w \in \pi_{\mathbf{F}_2}(\mathbf{E}')$. Consequently, $\overline{A}(w) = L(w) = -v$ and hence by (10.4.7) we get $\mathbf{d}F(x, y, 0)(u, v, w) = 0$, as required. ■

Proof of Theorem 10.4.12. We first prove that the condition (10.4.3) implies the Poisson reducibility of $(M, \{\cdot, \cdot\}, D, S)$. This implication can be obtained as a corollary of Theorem 10.4.8. Indeed, since Lemma 10.4.14 guarantees that in the hypotheses on the theorem the presheaf C^∞_{S/D_S} has the (D, D_S)-local extension property, it suffices to show that in this situation

$$\Delta_m = D(m)^\circ, \tag{10.4.8}$$

$$\left[\Delta_m^S\right]^\circ = T_m S + D(m). \tag{10.4.9}$$

In order to prove (10.4.8) notice first that, by definition, $\Delta_m \subset D(m)^\circ$. To prove the converse inclusion, take $\alpha_m \in D(m)^\circ tobe$ arbitrary. As we did in the proof of Lemma 10.4.14 we take a submanifold chart U_m of S around m that we can think of as $U \times V \subset \mathbf{F}_1 \oplus \mathbf{F}_2$ with the property that $U_m \cap S = U$. Analogously, we can locally take $D = U \times \mathbf{E}$, with $\mathbf{E}$ a vector subspace of $\mathbf{F}_1 \oplus \mathbf{F}_2$, $TU_m = U \times V \times (\mathbf{F}_1 \oplus \mathbf{F}_2)$, and $T^*U_m = U \times V \times (\mathbf{F}_1 \oplus \mathbf{F}_2)^*$. In these coordinates $m \equiv (0, 0)$ and $\alpha_m \equiv (0, 0, \alpha)$ with $\alpha \in \mathbf{E}^\circ$. Define $F : U \times V \to \mathbb{R}$ by $F(u, v) = \langle \alpha, (u, v) \rangle$. Note that $\mathbf{d}F(m) \equiv \mathbf{d}F(0, 0) = \alpha \equiv \alpha_m$ and that for any $(u, 0, w) \in D(u, 0)$, $u \in U$, $w \in \mathbf{E}$, we have that $\mathbf{d}F(u, 0)(w) = \langle \alpha, w \rangle = 0$, which implies that $\mathbf{d}F(z)|_{D(z)} = 0$, for any $z \in U_m \cap S$, as required.

We now prove (10.4.9). By definition $T_m S + D(m) \subset [\Delta_m^S]^\circ$. Conversely, the inclusion $[\Delta_m^S]^\circ \subset T_m S + D(m)$ holds if and only if $D(m)^\circ \cap T_m S^\circ \subset \Delta_m^S$ which, by (10.4.8), amounts to $\Delta_m \cap T_m S^\circ \subset \Delta_m^S$. We prove this inclusion by using again the same adapted local submanifold coordinates around the point m. Let $\alpha_m \in \Delta_m \cap T_m S^\circ$ be arbitrary. As we saw above, there exists $\alpha \in \mathbf{E}^\circ$ such that $\alpha_m = \mathbf{d}F(0,0)$, with $F : U \times V \to \mathbb{R}$ given by $F(u,v) := \langle \alpha, (u,v) \rangle$, $(u,v) \in U \times V$. Since $\alpha_m \in (T_m S)^\circ$ we have $\mathbf{d}F(0,0)(u,0) = \langle \alpha, (u,0) \rangle = 0$, for any $u \in \mathbf{F}_1$. This equality implies that $F(u,v) = \langle \alpha, (u,0) \rangle + \langle \alpha, (0,v) \rangle = \langle \alpha, (0,v) \rangle$, for any $(u,v) \in U \times V$, and hence F is constant on $U = U_m \cap S$ which shows that $\alpha_m \in \Delta_m^S$, as required.

We now show that the reducibility of $(M, \{\cdot,\cdot\}, D, S)$ implies (10.4.3) or, equivalently, that for any $m \in S$

$$T_m S^\circ \cap D(m)^\circ \subset \left(B^\sharp(m)(D(m)^\circ) \right)^\circ . \tag{10.4.10}$$

We proceed again by using the same local coordinates. On this occasion we consider a nondegenerate inner product $\langle \cdot,\cdot \rangle_{\mathbf{F}_1 \oplus \mathbf{F}_2}$ on $\mathbf{F}_1 \oplus \mathbf{F}_2$ defined by $\langle (u_1,u_2),(v_1,v_2) \rangle_{\mathbf{F}_1 \oplus \mathbf{F}_2} = \langle u_1, v_1 \rangle_{\mathbf{F}_1} + \langle u_2, v_2 \rangle_{\mathbf{F}_2}$, with $\langle \cdot,\cdot \rangle_{\mathbf{F}_1}$ and $\langle \cdot,\cdot \rangle_{\mathbf{F}_2}$ nondegenerate inner products in $\mathbf{F}_1$ and $\mathbf{F}_2$, respectively, and $u_1, v_1 \in \mathbf{F}_1$, $u_2, v_2 \in \mathbf{F}_2$. Given that $U_m \cap S = U \subset \mathbf{F}_1$, any element $\alpha_m \in T_m S^\circ$ can be written as $\alpha_m = \langle (0,u_0), \cdot \rangle_{\mathbf{F}_1 \oplus \mathbf{F}_2}$, for some $u_0 \in \mathbf{F}_2$ or, analogously, as $\alpha_m = \mathbf{d}K(m)$, with $K \in C^\infty(U_m)$ defined by

$$K(u,v) := \langle (0,u_0),(u,v) \rangle_{\mathbf{F}_1 \oplus \mathbf{F}_2} = \langle u_0, v \rangle_{\mathbf{F}_2}. \tag{10.4.11}$$

Moreover, if $\alpha_m \in T_m S^\circ \cap D(m)^\circ$ then, as D in these coordinates looks like $D = U \times \mathbf{E}$ for some vector subspace $\mathbf{E} \subset \mathbf{F}_1 \oplus \mathbf{F}_2$, we have that the function K defined in (10.4.11) is such that

$$K|_{U_m \cap S} = 0 \text{ and } \mathbf{d}K(z)|_{D(z)} = 0, \quad \text{for any } z \in U_m \cap S.$$

We have thus proved that any $\alpha_m \in T_m S^\circ \cap D(m)^\circ$ can be written as $\alpha_m = \mathbf{d}K(m)$ with K a local D-invariant extension of the zero function in S at the point $m \in S$.

Now let $\beta_m \in D(m)^\circ$. Due to the nondegeneracy of the inner product $\langle \cdot,\cdot \rangle_{\mathbf{F}_1 \oplus \mathbf{F}_2}$ there exists $w_0 \in \mathbf{F}_1 \oplus \mathbf{F}_2$ such that $\beta_m = \mathbf{d}F(m)$ with $F(u) := \langle w_0, u \rangle_{\mathbf{F}_1 \oplus \mathbf{F}_2}$, $u \in U_m$, and such that $\langle w_0, w \rangle_{\mathbf{F}_1 \oplus \mathbf{F}_2} = 0$ for any $w \in \mathbf{E}$. The regularity of the distribution D_S implies via a result of Godement (see Lemma 3.5.26 in ABRAHAM et al. (1988)) that the neighborhood U_m can be shrunk so that there exists a smooth submanifold T of $U_m \cap S$ (called a ***local slice*** of D_S) and a smooth map $s : U_m \cap S \to T$ such that $s|_T$ is the identity map on T and the integral leaf $\mathcal{L}_u$ of D_S that contains any arbitrary point $u \in U_m \cap S$ is such that $\mathcal{L}_u \cap T = \{s(u)\}$. Notice now that since $\mathbf{d}F|_{U_m \cap S}\left(D_S|_{U_m \cap S}\right) = 0$, we can use the slice and the map $F|_{U_m \cap S}$ to define another map $f \in C^\infty_{S/D_S}(\pi_{D_S}(U_m \cap S))$ as the unique map that satisfies

$$(f \circ \pi_{D_S})(z) = F(z) = F(s(z)), \quad \text{for any } z \in U_m \cap S.$$

Using the constructions in the last two sections we can now write for any $\alpha_m \in T_m S^\circ \cap D(m)^\circ$ and any $\beta_m \in D(m)^\circ$

$$\begin{aligned} &\langle \alpha_m, B^\sharp(m)(\beta_m) \rangle \\ &\quad = \{K, F\}(m) = \{0, f\}^{S/D_S}_{\pi_{D_S}(U_m \cap S)}(\pi_{D_S}(m)) = \{0, F\}(m) = 0, \end{aligned} \tag{10.4.12}$$

where in the last equality we used the Poisson reducibility of $(M, \{\cdot, \cdot\}, D, S)$ to write $\{0, f\}^{S/D_S}_{\pi_{D_S}(U_m \cap S)}(\pi_{D_S}(m)) = \{0, F\}(m)$ since the zero function on M is also an extension of the zero function on S that can be used instead of K in the definition of the bracket. The expression (10.4.12) establishes (10.4.10). ■

10.4.15 Point reduction of Poisson manifolds. Let $(M, \{\cdot, \cdot\})$ be a Poisson manifold acted canonically upon by a Lie group G admitting an infinitesimally equivariant standard momentum map $\mathbf{J} : M \to \mathfrak{g}^*$. Let $\mu \in \mathfrak{g}^*$ be a regular value of $\mathbf{J}$ and assume that the coadjoint isotropy group G_μ is connected. Let D be the generalized distribution on M given by $D(m) = \mathfrak{g} \cdot m$, for any $m \in M$, and let $D_{\mathbf{J}^{-1}(\mu)}(z) := D(z) \cap T_z(\mathbf{J}^{-1}(\mu)) = \mathfrak{g}_\mu \cdot z$, for any $z \in \mathbf{J}^{-1}(\mu)$ (see (4.5.28) for the last equality). If $C^\infty_{\mathbf{J}^{-1}(\mu)/G_\mu}$ has the $(D, D_{\mathbf{J}^{-1}(\mu)})$-local extension property, then $(M, \{\cdot, \cdot\}, D, \mathbf{J}^{-1}(\mu))$ is Poisson reducible. Indeed, notice first that the canonical character of the G-action on $(M, \{\cdot, \cdot\})$ guarantees that D is Poisson in the sense of Definition 10.4.2. Second, since $D_{\mathbf{J}^{-1}(\mu)}$ is clearly smooth and integrable it only remains to be shown that the inclusion (10.4.2) holds; this is so since for any $m \in \mathbf{J}^{-1}(\mu)$ we have $\Delta_m = (\mathfrak{g} \cdot m)^\circ$ and hence by Proposition 4.5.14 **(iii)** $B^\sharp(m)(\Delta_m) \subset \ker T_m\mathbf{J}$, which implies (10.4.2) for $\ker T_m\mathbf{J}$ is obviously included in $[\Delta_m^S]^\circ$.

Recall that if the G-action on M is free and the G_μ-action on $\mathbf{J}^{-1}(\mu)$ is proper, then Lemma 10.4.14 ensures that $C^\infty_{\mathbf{J}^{-1}(\mu)/G_\mu}$ has the $(D, D_{\mathbf{J}^{-1}(\mu)})$-local extension property.

10.4.16 Poisson foliation reduction. In the following sections we use Theorems 10.4.8 and 10.4.12 in order to introduce a few constructions that generalize to the Poisson context the ideas presented in §6.1.5.

10.4.17 Proposition *Let $(M, \{\cdot, \cdot\})$ be a Poisson manifold with associated Poisson tensor $B \in \Lambda^2(T^*M)$. Let S be an embedded submanifold of M and $D := B^\sharp((TS)^\circ) \subset TM|_S$. Assume that the* ***characteristic distribution*** *$D_S := D \cap TS$ of S relative to the Poisson bracket $\{\cdot, \cdot\}$ is a smooth and integrable generalized distribution on S such that C^∞_{S/D_S} has the (D, D_S)-local extension property. Then $(M, \{\cdot, \cdot\}, D, S)$ is Poisson reducible.*

Proof. According to Theorem 10.4.8 we need to check that the distribution D is Poisson and that the inclusion (10.4.2) holds. In order to prove that D is Poisson consider two functions $f, g \in C^\infty(M)$ such that $\mathbf{d}f|_D = \mathbf{d}g|_D = 0$. The condition $\mathbf{d}f|_D = 0$ is equivalent to $\mathbf{d}f(s) \in \left(B^\sharp(s)\left((T_sS)^\circ\right)\right)^\circ$, for any point $s \in S$. Hence by Proposition 4.1.33 we have that $X_f(s) \in T_sS \cap T_s\mathcal{L}_s$, where $\mathcal{L}_s$ is the symplectic leaf that contains $s \in S$. Analogously, $X_g(s) \in T_sS \cap T_s\mathcal{L}_s$, for any $s \in S$. Take now $v \in D(s)$. By Lemma 1.1.9 there exists a function $h \in C^\infty(M)$ such that $h|_S \equiv 0$ and $v = X_h(s)$. Let F_t, G_s be the flows of the vector fields X_f and X_g, respectively. Then

$$\mathbf{d}\{f, g\}(s) \cdot X_h(s) = \{\{f, g\}, h\}(s) = -\{\{g, h\}, f\}(s) - \{\{h, f\}, g\}(s).$$

We shall prove that each term in this sum vanishes separately. Indeed,

$$\begin{aligned}\{\{g,h\},f\}(s) &= \mathbf{d}\{g,h\}(s)\cdot X_f(s) = \left.\frac{d}{dt}\right|_{t=0}\{g,h\}(F_t(s))\\ &= -\left.\frac{d}{dt}\right|_{t=0}\mathbf{d}h(F_t(s))\cdot X_g(F_t(s))\\ &= -\left.\frac{d}{dr}\right|_{r=0}\left.\frac{d}{dt}\right|_{t=0} h(G_r(F_t(s)))=0\end{aligned}$$

because G_r and F_t leave S invariant and $h|_S \equiv 0$. Similarly, one proves that $\big\{\{h,f\}, g\big\}(s)=0$, for any $s \in S$.

Finally, we verify that the inclusion (10.4.2) holds. In this case $\Delta_s = D(s)^\circ$. Therefore, by Proposition 4.1.33

$$B^\sharp(s)(\Delta_s) = B^\sharp(s)\left(\left(B^\sharp(s)\left((T_sS)^\circ\right)\right)^\circ\right) = T_sS \cap T_s\mathcal{L}_s,$$

where $\mathcal{L}_s$ is the symplectic leaf containing the point $s \in S$. Since $T_sS \subset \left[\Delta_s^S\right]^\circ$, the proposition follows. ■

10.4.18 Reduction of coisotropic submanifolds. Let $(M, \{\cdot,\cdot\})$ be a Poisson manifold with associated Poisson tensor $B \in \Lambda^2(T^*M)$ and S an immersed smooth submanifold of M. Denote by $(TS)^\circ := \{\alpha_s \in T_s^*M \mid \langle\alpha_s, v_s\rangle = 0, \text{ for all } s \in S, v_s \in T_sS\} \subset T^*M$ the ***conormal bundle*** of the manifold S; it is a vector subbundle of $T^*M|_S$. The manifold S is called ***coisotropic*** if $B^\sharp\left((TS)^\circ\right) \subset TS$.

Note that this is the straightforward generalization of the definition of a coisotropic submanifold of a symplectic manifold (see §4.1.5). Indeed, if the Poisson bracket on M is defined by a symplectic form $\omega \in \Omega^2(M)$, Proposition 4.1.32 states that $B^\sharp\left((TS)^\circ\right) = (TS)^\omega$ and the condition given above becomes $(TS)^\omega \subset TS$, that is, S is coisotropic in (M,ω).

The notion of coisotropic submanifold was introduced in the Poisson category by LIBERMANN AND MARLE (1987) and SÁNCHEZ DE ALVAREZ (1986, 1989). In the symplectic case, coisotropic submanifolds appear sometimes in the physics literature under the name ***first class constraints*** (see for instance BOJOWALD AND STROBL (2002) and references therein).

10.4.19 Proposition *Let $(M, \{\cdot,\cdot\})$ be a Poisson manifold with associated Poisson tensor $B \in \Lambda^2(T^*M)$ and S an embedded smooth submanifold of M. The following are equivalent:*

(i) *S is coisotropic;*

(ii) *if $f \in C^\infty(M)$ satisfies $f|_S \equiv 0$, then $X_f|_S \in \mathfrak{X}(S)$;*

(iii) *for any $s \in S$, any open neighborhood U_s of s in M, and any function $g \in C^\infty(U_s)$ such that $X_g(s) \in T_sS$, if $f \in C^\infty(U_s)$ satisfies $\{f,g\}(s) = 0$, it follows that $X_f(s) \in T_sS$;*

(iv) *the subalgebra $\{f \in C^\infty(M) \mid f|_S \equiv 0\}$ is a Poisson subalgebra of $(C^\infty(M), \{\cdot,\cdot\})$.*

Proof. Assume that S is coisotropic and $f \in C^\infty(M)$ satisfies $f|_S \equiv 0$. Then $\mathbf{d}f(s) \in (T_sS)^\circ$, for any $s \in S$, and hence $X_f(s) \in B^\sharp(s)((T_sS)^\circ) \subset TS$ which shows that **(i)** implies **(ii)**. Conversely, assume that **(ii)** holds. By Lemma 1.1.9 any element $\alpha_s \in (T_sS)^\circ$ can be written as $\alpha_s = \mathbf{d}f(s)$, with $f \in C^\infty(M)$ such that $f|_S \equiv 0$. Hence $B^\sharp(s)(\alpha_s) = B^\sharp(s)(\mathbf{d}f(s)) = X_f(s) \in T_sS$, by hypothesis. This implies that **(i)** and **(ii)** are equivalent.

Assume that S is coisotropic. Let $s \in S$ be arbitrary and U_s an arbitrary open neighborhood of s in M. Let $f \in C^\infty(U_s)$ be such that $0 = \{f, g\}(s) = \mathbf{d}f(s) \cdot X_g(s)$ for any function $g \in C^\infty(U_s)$ satisfying $X_g(s) \in T_sS$. Denoting by $\mathcal{L}_s$ the symplectic leaf containing s, this relation states that $\mathbf{d}f(s) \in (T_s\mathcal{L}_s \cap T_sS)^\circ = (T_s\mathcal{L}_s)^\circ + (T_sS)^\circ$ by Lemma 4.1.31. Therefore,

$$X_f(s) = B^\sharp(\mathbf{d}f(s)) \in B^\sharp\left((T_s\mathcal{L}_s)^\circ + (T_sS)^\circ\right) = B^\sharp\left((T_sS)^\circ\right) \subset T_sS,$$

since $B^\sharp\left((T_s\mathcal{L}_s)^\circ\right) = \{0\}$. This shows that **(i)** implies **(iii)**.

Assume now that **(iii)** holds and let $\alpha_s \in (T_sS)^\circ$ be arbitrary. Then there is an open neighborhood U_s of s in M and a function $f \in C^\infty(U_s)$ such that $\mathbf{d}f(s) = \alpha_s$. Let $g \in C^\infty(U_s)$ be any functions such that $X_g(s) \in T_sS$. Since $\mathbf{d}f(s) \in (T_sS)^\circ$, we have $\{f, g\}(s) = \mathbf{d}f(s) \cdot X_g(s) = 0$, that is, the hypotheses in **(iii)** hold for the function f and hence $B^\sharp(\alpha_s) = B^\sharp(\mathbf{d}f(s)) = X_f(s) \in T_sS$, which shows that **(i)** holds. Thus **(i)** and **(iii)** are equivalent.

Assume that S is coisotropic and let $f, g \in C^\infty(M)$ be such that $f|_S \equiv 0$ and $g|_S \equiv 0$. Thus, for any $s \in S$, we have $\mathbf{d}f(s), \mathbf{d}g(s) \in (T_sS)^\circ$. Since S is coisotropic, it follows that $X_f(s) = B^\sharp(\mathbf{d}f(s)) \in B^\sharp((T_sS)^\circ) \subset T_sS$. Thus, $\{f, g\}(s) = -\mathbf{d}g(s) \cdot X_f(s) = 0$. Since this argument is valid at any point $s \in S$, it follows that $\{f, g\}|_S \equiv 0$ which shows that **(i)** implies **(iv)**.

Conversely, assume that **(iv)** holds and let $\alpha_s \in (T_sS)^\circ$. Now let $\alpha_s \in (T_sS)^\circ$ be arbitrary. By Lemma 1.1.9 there is a function $f \in C^\infty(M)$ such that $f|_S \equiv 0$ and $\mathbf{d}f(s) = \alpha_s$. In addition, any element of $(T_sS)^\circ$ is of the form $\mathbf{d}g(s)$ for some $g \in C^\infty(M)$ satisfying $g|_S \equiv 0$. Thus, by **(iv)**,

$$\langle \mathbf{d}g(s), B^\sharp(\alpha_s)\rangle = \langle \mathbf{d}g(s), B^\sharp(\mathbf{d}f(s))\rangle = \{g, f\}(s) = 0,$$

which says that $B^\sharp(\alpha_s) \in ((T_sS)^\circ)^\circ = T_sS$. Since $s \in S$ and $\alpha_s \in (T_sS)^\circ$ were arbitrary, this shows that $B^\sharp((T_sS)^\circ) \subset T_sS$ for any $s \in S$, that is, S is a coisotropic submanifold of M. ■

10.4.20 Proposition *Let $(M, \{\cdot, \cdot\})$ be a Poisson manifold with associated Poisson tensor $B \in \Lambda^2(T^*M)$. Let S be an embedded coisotropic submanifold of M and $D := B^\sharp((TS)^\circ)$. Then*

(i) *$D = D \cap TS = D_S$ is a smooth generalized distribution on S.*

(ii) *D is integrable.*

(iii) *If C^∞_{S/D_S} has the (D, D_S)-local extension property, then $(M, \{\cdot, \cdot\}, D, S)$ is Poisson reducible.*

Proof. **(i)** Since S is coisotropic $D \subset TS$, which implies that $D = D \cap TS = D_S$. At the same time we can write, by (1.1.3), $D(z) = D_S(z) = \{X_f(z) \mid f \in C^\infty(M),\ f|_S \equiv 0\}$, for any $z \in S$, which ensures the smoothness of D.

(ii) Let $f, g \in C^\infty(M)$ be such that $f|_S = g|_S \equiv 0$ and F_t the flow of X_f. The coisotropy of S implies that X_f is tangent to S and hence by §1.2.4 F_t leaves S invariant. Then for any $s \in \mathrm{Dom}(F_t)$ we have $T_sF_t \cdot X_g(s) = X_{g\circ F_{-t}}(F_t(s))$. Since $g \circ F_{-t}|_{S\cap\mathrm{Dom}(F_{-t})} \equiv 0$ we can conclude that $X_{g\circ F_{-t}}(F_t(s)) \in D(F_t(s))$ which guarantees the integrability of D by the Stefan–Sussmann Theorem 3.2.1.

(iii) It is a corollary of parts **(i)** and **(ii)** together with Proposition 10.4.17.

10.4.21 Let $(M, \{\cdot,\cdot\})$ be a Poisson manifold and $B \in \Lambda^2(T^*M)$ the corresponding Poisson tensor. Let S be an embedded submanifold such that the characteristic distribution $D_S := B^\sharp((TS)^\circ) \cap TS$ is a smooth, integrable, Poisson, and regular distribution on S. Even though the quotient manifolds associated to the quadruples $(M, \{\cdot,\cdot\}, D, S)$ and $(M, \{\cdot,\cdot\}, D_S, S)$ are the same and $(M, \{\cdot,\cdot\}, D, S)$ is reducible by Proposition 10.4.17, the quadruplet $(M, \{\cdot,\cdot\}, D_S, S)$ is, in general, *not* reducible. Actually, its reducibility is, by Theorem 10.4.12, equivalent to S being a coisotropic submanifold of M. Indeed, by Proposition 10.4.20 and Lemma 10.4.14 if S is coisotropic, then $(M, \{\cdot,\cdot\}, D_S, S)$ is reducible. Conversely, if $(M, \{\cdot,\cdot\}, D_S, S)$ is reducible, then by Theorem 10.4.12 we have $B^\sharp(D_S^\circ) \subset TS + D_S = TS$ since $D_S \subset TS$. Additionally, note that by Lemma 4.1.31, $B^\sharp((TS)^\circ) \subset B^\sharp((TS)^\circ) + B^\sharp([B^\sharp((TS)^\circ)]^\circ) = B^\sharp((TS)^\circ + [B^\sharp((TS)^\circ)]^\circ) = B^\sharp([B^\sharp((TS)^\circ) \cap TS]^\circ) = B^\sharp(D_S^\circ)$. Thus *$(M, \{\cdot,\cdot\}, D_S, S)$ is Poisson reducible relative to the characteristic distribution if and only if S is coisotropic.*

The difference in terms of reducibility between the quadruples $(M, \{\cdot,\cdot\}, D, S)$ and $(M, \{\cdot,\cdot\}, D_S, S)$ is the same as that between the systems $(M, \omega, D, \mathbf{J}^{-1}(\mu))$ and $(M, \omega, D_S, \mathbf{J}^{-1}(\mu))$ considered in Remark 10.4.13.

10.4.22 Coisotropic submanifolds and integrals in involution. Let $(M, \{\cdot,\cdot\})$ be a Poisson manifold with Poisson tensor B and let $f_1, \ldots, f_k \in C^\infty(M)$ be k-smooth functions in ***involution***, that is,

$$\{f_i, f_j\} = 0, \qquad \text{for any} \quad i, j \in \{1, \ldots, k\}.$$

Assume that $\mathbf{0} \in \mathbb{R}^k$ is a regular value of the function $F := (f_1, \ldots, f_k) : M \to \mathbb{R}^k$ and let $S := F^{-1}(0)$. Since for any $s \in S$, $\mathrm{span}\{\mathbf{d}f_1(s), \ldots, \mathbf{d}f_k(s)\} \subset (T_sS)^\circ$ and the dimensions of both sides of this inclusion are equal, we get

$$\mathrm{span}\{\mathbf{d}f_1(s), \ldots, \mathbf{d}f_k(s)\} = (T_sS)^\circ .$$

Hence,

$$B^\sharp(s)((T_sS)^\circ) = \mathrm{span}\left\{X_{f_1}(s), \ldots, X_{f_k}(s)\right\},$$

and $B^\sharp(s)((T_sS)^\circ) \subset T_sS$ by the involutivity of the components of F. Consequently, S is a coisotropic submanifold of $(M, \{\cdot,\cdot\})$ and Proposition 10.4.20 can be applied to it.

10.4.23 Example: Inverse images of coadjoint orbits are coisotropic. Let $(M, \{\cdot, \cdot\})$ be a Poisson manifold with associated Poisson tensor $B \in \Lambda^2(T^*M)$. Let G be a Lie group acting freely and canonically on $(M, \{\cdot, \cdot\})$ with an associated coadjoint equivariant standard momentum map $\mathbf{J} : M \to \mathfrak{g}^*$. Let $\mu \in \mathfrak{g}^*$ be a value of the momentum map and $\mathcal{O}_\mu \subset \mathfrak{g}^*$ its coadjoint orbit. We will now show that the inverse image $\mathbf{J}^{-1}(\mathcal{O}_\mu)$ is an initial coisotropic submanifold of M. Indeed, as we saw in the proof of Theorem 6.3.1, $\mathbf{J}^{-1}(\mathcal{O}_\mu)$ is an initial submanifold of M such that for any $m \in \mathbf{J}^{-1}(\mathcal{O}_\mu)$

$$T_m\left(\mathbf{J}^{-1}(\mathcal{O}_\mu)\right) = \ker T_m\mathbf{J} + \mathfrak{g} \cdot m. \tag{10.4.13}$$

Hence if $m \in \mathbf{J}^{-1}(\mathcal{O}_\mu)$ and $\mathcal{L}$ is the symplectic leaf of $(M, \{\cdot, \cdot\})$ containing m, we can write

$$\begin{aligned} B^\sharp(m)\left(\left(T_m(\mathbf{J}^{-1}(\mathcal{O}_\mu))\right)^\circ\right) &= B^\sharp(m)\left((\ker T_m\mathbf{J} + \mathfrak{g} \cdot m)^\circ\right) \qquad \text{(by (10.4.13))} \\ &= ((\ker T_m\mathbf{J} + \mathfrak{g} \cdot m) \cap T_m\mathcal{L})^{\omega_{\mathcal{L}}(m)} \\ &= ((\ker T_m\mathbf{J} \cap T_m\mathcal{L}) + \mathfrak{g} \cdot m)^{\omega_{\mathcal{L}}(m)}, \end{aligned}$$

where we used that, by Proposition 4.5.15**(i)**, $\mathfrak{g} \cdot m \subset T_m\mathcal{L}$. Additionally, by part **(ii)** of the same result, if we write $\mathbf{J}_{\mathcal{L}} := \mathbf{J}|_{\mathcal{L}} : \mathcal{L} \to \mathfrak{g}^*$, then

$$\begin{aligned} B^\sharp(m)\left(\left(T_m(\mathbf{J}^{-1}(\mathcal{O}_\mu))\right)^\circ\right) &= (\ker T_m\mathbf{J}_{\mathcal{L}} + \mathfrak{g} \cdot m)^{\omega_{\mathcal{L}}(m)} \\ &= (\ker T_m\mathbf{J}_{\mathcal{L}} + \mathfrak{g} \cdot m)^{\omega_{\mathcal{L}}(m)} \\ &= (\ker T_m\mathbf{J}_{\mathcal{L}})^{\omega_{\mathcal{L}}(m)} \cap (\mathfrak{g} \cdot m)^{\omega_{\mathcal{L}}(m)} \\ &= \mathfrak{g} \cdot m \cap T_m\mathbf{J}_{\mathcal{L}} \qquad \text{(by Proposition 4.5.14)} \\ &= \mathfrak{g}_\mu \cdot m \qquad \text{(by Lemma 4.5.34)} \\ &\subset \ker T_m\mathbf{J} + \mathfrak{g} \cdot m = T_m\left(\mathbf{J}^{-1}(\mathcal{O}_\mu)\right). \end{aligned}$$

It should be noted that the level set $\mathbf{J}^{-1}(\mu)$ is in general not coisotropic when $\mu \neq 0$.

10.5 Cosymplectic submanifolds and Dirac's formula

In this section we will show that the Poisson Reduction Theorem 10.4.12 allows us to define Poisson structures on certain embedded submanifolds that are not Poisson submanifolds.

10.5.1 Definition *Let $(M, \{\cdot, \cdot\})$ be a Poisson manifold and let $B \in \Lambda^2(T^*M)$ be the corresponding Poisson tensor. An embedded submanifold $S \subset M$ is called* ***cosymplectic*** *if*

(i) $B^\sharp((TS)^\circ) \cap TS = \{0\}$.

(ii) $T_sS + T_s\mathcal{L}_s = T_sM$,

for any $s \in S$ and $\mathcal{L}_s$ the symplectic leaf of $(M, \{\cdot, \cdot\})$ containing $s \in S$.

The cosymplectic submanifolds of a symplectic manifold (M, ω) are its symplectic submanifolds. Indeed, condition **(ii)** is always satisfied in this context since $T_s\mathcal{L}_s = T_sM$ and, by Proposition 4.1.32, condition **(i)** becomes $(TS)^\omega \cap TS = \{0\}$. Also in the symplectic case, coisotropic submanifolds appear sometimes in the physics literature under the name ***second class constraints*** (see for instance BOJOWALD AND STROBL (2002) and references therein).

10.5.2 Proposition *Let $(M, \{\cdot,\cdot\})$ be a Poisson manifold, $B \in \Lambda^2(T^*M)$ the corresponding Poisson tensor, and S a cosymplectic submanifold of M. Then for any $s \in S$*

(i) $T_s\mathcal{L}_s = (T_sS \cap T_s\mathcal{L}_s) \oplus B^\sharp(s)((T_sS)^\circ)$, *where $\mathcal{L}_s$ is the symplectic leaf of $(M, \{\cdot,\cdot\})$ that contains $s \in S$.*

(ii) $(T_sS)^\circ \cap \ker B^\sharp(s) = \{0\}$.

(iii) $T_sM = B^\sharp(s)((T_sS)^\circ) \oplus T_sS$.

(iv) *$B^\sharp((TS)^\circ)$ is a subbundle of $TM|_S$ and hence $TM|_S = B^\sharp((TS)^\circ) \oplus TS$.*

(v) *The symplectic leaves of $(M, \{\cdot,\cdot\})$ intersect S transversely and hence $S \cap \mathcal{L}$ is an initial submanifold of S, for any symplectic leaf $\mathcal{L}$ of $(M, \{\cdot,\cdot\})$.*

Proof. **(i)** Recall that by Proposition 4.1.33, $B^\sharp(s)\left(\left(B^\sharp(s)((T_sS)^\circ)\right)^\circ\right) = T_sS \cap T_s\mathcal{L}_s$ and hence by Lemma 4.1.31 and the cosymplectic hypothesis **(i)** we get

$$
\begin{aligned}
(T_sS \cap T_s\mathcal{L}_s) + B^\sharp(s)((T_sS)^\circ) &= B^\sharp(s)\left(\left(B^\sharp(s)((T_sS)^\circ)\right)^\circ + (T_sS)^\circ\right) \\
&= B^\sharp(s)\left(\left(B^\sharp(s)((T_sS)^\circ) \cap T_sS\right)^\circ\right) \\
&= B^\sharp(s)(T_s^*M) = T_s\mathcal{L}_s.
\end{aligned}
$$

Finally, the cosymplectic hypothesis **(i)** implies that $(T_sS \cap T_s\mathcal{L}_s) \cap B^\sharp(s)((T_sS)^\circ) = \{0\}$ and hence the equality follows.

(ii) It follows from Lemma 4.1.31 and the equality $(T_s\mathcal{L}_s)^\circ = \ker B^\sharp(s)$.

(iii) Using part **(i)** and the cosymplectic hypothesis **(ii)** we get

$$
\begin{aligned}
T_sM = T_sS + T_s\mathcal{L}_s &= T_sS + \left[(T_sS \cap T_s\mathcal{L}_s) \oplus B^\sharp(s)((T_sS)^\circ)\right] \\
&= T_sS + B^\sharp(s)((T_sS)^\circ).
\end{aligned}
$$

The cosymplectic hypothesis **(i)** guarantees that the sum is direct.

(iv) From **(iii)** it follows that $\dim B^\sharp(s)((T_sS)^\circ) = \dim M - \dim S$, for any $s \in S$. Therefore, $B^\sharp((TS)^\circ)$ is a vector subbundle of $TM|_S$ since it is the image of the subbundle $(TS)^\circ \subset T^*M|_S$ by the constant rank vector bundle map $B^\sharp|_S : T^*M|_S \to TM|_S$ (see for instance Proposition 3.4.18 in ABRAHAM et al. (1988)).

(v) The cosymplectic hypothesis **(ii)** is equivalent to saying that the inclusion $S \hookrightarrow M$ is transversal to the leaf $\mathcal{L}_s$. Since $\mathcal{L}_s$ is an initial submanifold of M by Theorem 4.1.28, the intersection $S \cap \mathcal{L}_s$ is an initial submanifold of S by Theorem 1.1.15. ■

The following theorem is due to WEINSTEIN (1983a).

10.5.3 Theorem (The Poisson structure of a cosymplectic submanifold) *Let $(M, \{\cdot,\cdot\})$ be a Poisson manifold, $B \in \Lambda^2(T^*M)$ the corresponding Poisson tensor, and S a cosymplectic submanifold of M. Let $D := B^\sharp((TS)^\circ) \subset TM|_S$. Then*

(i) *$(M, \{\cdot,\cdot\}, D, S)$ is Poisson reducible.*

(ii) *The corresponding quotient manifold equals S and the reduced bracket $\{\cdot,\cdot\}^S$ is given by*

$$\{f, g\}^S(s) = \{F, G\}(s), \tag{10.5.1}$$

where $f, g \in C^\infty_{S,M}(V)$ are arbitrary and $F, G \in C^\infty_M(U)$ are local D-invariant extensions of f and g around $s \in S$, respectively.

(iii) *The Hamiltonian vector field X_f of an arbitrary function $f \in C^\infty_{S,M}(V)$ is given by*

$$Ti \circ X_f = X_F \circ i, \tag{10.5.2}$$

where $F \in C^\infty_M(U)$ is a local D-invariant extension of f and $i : S \hookrightarrow M$ is the inclusion.

(iv) *The Hamiltonian vector field X_f of an arbitrary function $f \in C^\infty_{S,M}(V)$ can be written as*

$$Ti \circ X_f = \pi_S \circ X_F \circ i, \tag{10.5.3}$$

where $F \in C^\infty_M(U)$ is an arbitrary local extension of f and $\pi_S : TM|_S \to TS$ is the projection induced by the Whitney sum decomposition $TM|_S = B^\sharp((TS)^\circ) \oplus TS$ of $TM|_S$.

(v) *The symplectic leaves of $(S, \{\cdot,\cdot\}^S)$ are the connected components of the intersections $S \cap \mathcal{L}$, with $\mathcal{L}$ a symplectic leaf of $(M, \{\cdot,\cdot\})$. Any symplectic leaf of $(S, \{\cdot,\cdot\}^S)$ is a symplectic submanifold of the symplectic leaf of $(M, \{\cdot,\cdot\})$ that contains it.*

(vi) *Let $\mathcal{L}_s$ and $\mathcal{L}^S_s$ be the symplectic leaves of $(M, \{\cdot,\cdot\})$ and $(S, \{\cdot,\cdot\}^S)$, respectively, that contain the point $s \in S$. Let $\omega_{\mathcal{L}_s}$ and $\omega_{\mathcal{L}^S_s}$ be the corresponding symplectic forms. Then $B^\sharp(s)((T_sS)^\circ)$ is a symplectic subspace of $T_s\mathcal{L}_s$ and*

$$B^\sharp(s)((T_sS)^\circ) = \left(T_s\mathcal{L}^S_s\right)^{\omega_{\mathcal{L}_s}(s)}. \tag{10.5.4}$$

(vii) *Let $B_S \in \Lambda^2(T^*S)$ be the Poisson tensor associated to $(S, \{\cdot,\cdot\}^S)$. Then*

$$B^\sharp_S = \pi_S \circ B^\sharp|_S \circ \pi^*_S, \tag{10.5.5}$$

where $\pi^*_S : T^*S \to T^*M|_S$ is the dual of $\pi_S : TM|_S \to TS$.

10.5.4 Remark Note that this theorem provides a presheaf of Poisson algebras on $(S, C^\infty_{S,M})$ and by (1.1.11) also on (S, C^∞_S). When S is paracompact both presheaves coincide.

10.5.5 Remark There exists a generalization of the notion of cosymplectic submanifold for which the existence of a natural Poisson structure similar to the one presented in the previous theorem is guaranteed (see XU (2001)). These generalized cosymplectic submanifolds S are referred to as ***Dirac submanifolds*** and are defined by the existence of a Whitney sum decomposition of the form

$$TM|_S = TS \oplus V_S,$$

where $V_S \subset TM$ is a coisotropic submanifold of TM, when endowed with its natural Poisson structure (see SÁNCHEZ DE ALVAREZ (1993)).

Proof of Theorem 10.5.3 . **(i)** and **(ii)** can be obtained as a consequence of Proposition 10.4.17. Indeed, in this case, $D_S := D \cap TS = \{0\}$ by the cosymplectic hypothesis **(i)**. Hence all the conditions on D_S in the statement of Proposition 10.4.17 are trivially satisfied. Moreover, $S/D_S = S$ and using the definitions in §10.4.5 it is easy to see that $C^\infty_{S/D_S} = C^\infty_{S,M}$. Finally, by part **(iv)** in Proposition 10.5.2 D is a subbundle of $TM|_S$ and hence by Lemma 10.4.14 the presheaf $C^\infty_{S,M}$ has the (D, D_S)-local extension property, which proves the statement.

(iii) Let $f \in C^\infty_{S,M}(V)$ and $F \in C^\infty_M(U)$ be a local D-invariant extension of f. The D-invariance of F implies that $\mathbf{d}F(s) \in \left(B^\sharp(s)((T_sS)^\circ)\right)^\circ$, for any $s \in V$, and hence by Proposition 4.1.33

$$X_F(s) \in B^\sharp(s)\left(\left(B^\sharp(s)((T_sS)^\circ)\right)^\circ\right) = T_sS \cap T_s\mathcal{L}_s, \tag{10.5.6}$$

with $\mathcal{L}_s$ the symplectic leaf of $(M, \{\cdot, \cdot\})$ containing $s \in S$. Now, by (10.5.1), we have for any $g \in C^\infty_{S,M}(V)$ and any local D-invariant extension $G \in C^\infty_M(U)$ of g

$$\mathbf{d}g(s) \cdot X_f(s) = \{g, f\}^S(s) = \{G, F\}(s) = \mathbf{d}G(s) \cdot X_F(s). \tag{10.5.7}$$

By (10.5.6) we can write

$$\begin{aligned} \mathbf{d}G(s) \cdot X_F(s) &= \mathbf{d}G(i(s)) \cdot T_si(X_F(i(s))) = i^*(\mathbf{d}G)(s) \cdot X_F(i(s)) \\ &= \mathbf{d}(i^*G)(s) \cdot X_F(i(s)) = \mathbf{d}g(s) \cdot X_F(i(s)). \end{aligned} \tag{10.5.8}$$

Since $g \in C^\infty_{S,M}(V)$ is arbitrary, expressions (10.5.7) and (10.5.8) guarantee that (10.5.2) holds true.

(iv) Let $f \in C^\infty_{S,M}(V)$ and $F, F_D \in C^\infty_M(U)$ be two local extensions of f, with F_D D-invariant and F arbitrary. Since $(F - F_D)|_V \equiv 0$ we can write that $X_{F-F_D}(s) \in B^\sharp(s)((T_sS)^\circ)$, for any $s \in V$. Additionally, by part **(iii)**, $X_{F_D}(s) \in T_sS \cap T_s\mathcal{L}_s$, for any $s \in V$. Therefore, by (10.5.2)

$$X_F(s) = X_{F_D}(s) + X_{F-F_D}(s) = X_f(s) + X_{F-F_D}(s) \in T_sS \oplus B^\sharp(s)((T_sS)^\circ)$$

and hence $\pi_S(X_F(s)) = X_f(s)$, as required.

(v) It suffices to show that

$$T_s\mathcal{L}^S_s = T_s\mathcal{L}_s \cap T_sS,$$

for any $s \in S$, where $\mathcal{L}_s$ and $\mathcal{L}_s^S$ are the symplectic leaves containing $s \in S$ and corresponding to the Poisson structures $(M, \{\cdot, \cdot\})$ and $(S, \{\cdot, \cdot\}^S)$, respectively. Let $X_f(s) \in T_s\mathcal{L}_s^S$, for some $f \in C^\infty_{S,M}(V)$. Since by part **(iii)** $X_f(s) = X_F(s)$, for some D-invariant local extension $F \in C^\infty_M(U)$ of f, it follows that $X_f(s) \in T_s\mathcal{L}_s \cap T_sS$. Conversely, let $X_F(s) \in T_s\mathcal{L}_s \cap T_sS$. By (10.5.3) we can write

$$X_F(s) = \pi_S(X_F(s)) = X_{F|_S}(s) \in T_s\mathcal{L}_s^S.$$

We now show that if $i_{\mathcal{L}_s^S} : \mathcal{L}_s^S \hookrightarrow \mathcal{L}_s$ is the inclusion and $\omega_{\mathcal{L}_s^S}, \omega_{\mathcal{L}_s}$ are the symplectic forms on $\mathcal{L}_s^S$ and $\mathcal{L}_s$, respectively, then

$$i^*_{\mathcal{L}_s^S}\omega_{\mathcal{L}_s} = \omega_{\mathcal{L}_s^S}.$$

Let $f, g \in C^\infty_{S,M}(V)$ be arbitrary and let $F, G \in C^\infty_M(U)$ be two local D-invariant extensions of f and g, respectively. Then by (10.5.2) and (10.5.1) we get

$$\begin{aligned} i^*_{\mathcal{L}_s^S}\omega_{\mathcal{L}_s}(s)(X_f(s), X_g(s)) &= \omega_{\mathcal{L}_s}(s)(T_s i_{\mathcal{L}_s^S} \cdot X_f(s), T_s i_{\mathcal{L}_s^S} \cdot X_g(s)) \\ &= \omega_{\mathcal{L}_s}(s)(X_F(s), X_G(s)) = \{F, G\}(s) = \{f, g\}^S(s) \\ &= \omega_{\mathcal{L}_s^S}(s)(X_f(s), X_g(s)). \end{aligned}$$

(vi) We shall prove that $\omega_{\mathcal{L}_s}(s)$ is nondegenerate on $B^\sharp(s)((T_sS)^\circ)$. Let $X_F(s) \in B^\sharp(s)((T_sS)^\circ)$ and assume that

$$\omega_{\mathcal{L}_s}(s)(X_F(s), X_G(s)) = 0, \tag{10.5.9}$$

for any $X_G(s) \in B^\sharp(s)((T_sS)^\circ)$. By part **(i)** of Proposition 10.5.2, any vector $w \in T_s\mathcal{L}_s$ can be written as $w = X_G(s) + v$ with $X_G(s) \in B^\sharp(s)((T_sS)^\circ)$ and $v \in T_sS \cap T_s\mathcal{L}_s$. Hence by (10.5.9)

$$\begin{aligned} \omega_{\mathcal{L}_s}(s)(X_F(s), w) &= \omega_{\mathcal{L}_s}(s)(X_F(s), X_G(s) + v) \\ &= \omega_{\mathcal{L}_s}(s)(X_F(s), v) = \mathbf{d}F(s) \cdot v = 0, \end{aligned}$$

since $\mathbf{d}F(s) \in (T_sS)^\circ$. Finally, as $w \in T_s\mathcal{L}_s$ is arbitrary and $\mathcal{L}_s$ is symplectic, $X_F(s) = 0$ necessarily.

Expression (10.5.4) is a consequence of part **(v)**, Proposition 10.5.2**(v)**, and Proposition 4.1.32.

(vii) The decomposition $T_sM = B^\sharp(s)((T_sS)^\circ) \oplus T_sS$ induces the direct sum decomposition

$$T_s^*M = (T_sS)^\circ \oplus \left(B^\sharp(s)((T_sS)^\circ)\right)^\circ. \tag{10.5.10}$$

Let $f \in C^\infty_{S,M}(V)$ be arbitrary and let $F \in C^\infty_M(U)$ be a local extension of f (not necessarily D-invariant). Let $\mathbf{d}F(s) = \alpha_1 + \alpha_2$ be the decomposition of $\mathbf{d}F(s)$ according to (10.5.10). We shall compute α_2. By definition $\alpha_2 = \mathbf{d}F(s)|_{T_sS}$. Let $v \in T_sS$ be arbitrary. Then

$$\begin{aligned} \langle \alpha_2, v \rangle &= \mathbf{d}F(s) \cdot v = \mathbf{d}F(i(s)) \cdot T_s i(v) = (i^*\mathbf{d}F)(s) \cdot v = \mathbf{d}(i^*F)(s) \cdot v \\ &= \mathbf{d}f(s) \cdot v = \mathbf{d}f(s) \cdot \pi_S(v) = \langle \pi_S^*\mathbf{d}f(s), v \rangle. \end{aligned} \tag{10.5.11}$$

If $u \in B^\sharp(s)((T_sS)^\circ)$, we have that $\langle \alpha_2, u\rangle = 0$ and

$$\langle \pi_S^* \mathbf{d} f(s), u\rangle = \langle \mathbf{d} f(s), \pi_S(u)\rangle = 0$$

which together with (10.5.11) proves that for any $w \in T_sM$, $\langle \alpha_2, w\rangle = \langle \pi_S^* \mathbf{d} f(s), w\rangle$. This proves that

$$\alpha_2 = \pi_S^* \mathbf{d} f(s). \tag{10.5.12}$$

We now prove (10.5.5). Using prior notation, (10.5.3), and (10.5.12) we get

$$\begin{aligned} B_S^\sharp(s)(\mathbf{d} f(s)) &= X_f(s) = \pi_S(X_F(s)) = \pi_S(B^\sharp(s)(\mathbf{d}F(s))) \\ &= \pi_S\left(B^\sharp(s)(\alpha_1 + \alpha_2)\right) = \pi_S\left(B^\sharp(s)(\alpha_1)\right) + \pi_S\left(B^\sharp(s)(\alpha_2)\right) \\ &= \pi_S\left(B^\sharp(s)(\pi_S^* \mathbf{d} f(s))\right), \end{aligned}$$

since by (10.5.10) there exists a function $G \in C^\infty(M)$ such that $\mathbf{d}G(s)|_{T_sS} = 0$ and $\mathbf{d}G(s) = \alpha_1$. Thus

$$\pi_S\left(B^\sharp(s)(\alpha_1)\right) = \pi_S\left(B^\sharp(s)(\mathbf{d}G(s))\right) \in \pi_S\left(B^\sharp(s)((T_sS)^\circ)\right) = \{0\}. \qquad \blacksquare$$

10.5.6 Corollary *Let $(M, \{\cdot, \cdot\})$ be a Poisson manifold and $S \subset M$ an embedded submanifold. Then S is a cosymplectic submanifold of $(M, \{\cdot, \cdot\})$ if and only if it satisfies the following two properties:*

(i) *$T_sS \cap T_s\mathcal{L}_s$ is a symplectic subspace of $(T_s\mathcal{L}_s, \omega_{\mathcal{L}_s}(s))$, for any $s \in S$, where $\mathcal{L}_s$ is the symplectic leaf of $(M, \{\cdot, \cdot\})$ that contains $s \in S$;*

(ii) *$T_sS + T_s\mathcal{L}_s = T_sM$, for any $s \in S$.*

Proof. If S is a cosymplectic submanifold of M, Theorem 10.5.3**(v)** shows that S satisfies **(i)**. Conversely, if **(i)** and **(ii)** hold, then by Proposition 4.1.32 we get, for any $s \in S$,

$$\begin{aligned} B^\sharp(s)((T_sS)^\circ) \cap T_sS &= B^\sharp(s)((T_sS)^\circ) \cap T_sS \cap T_s\mathcal{L}_s \\ &= (T_sS \cap T_s\mathcal{L}_s)^{\omega_{\mathcal{L}_s}} \cap (T_sS \cap T_s\mathcal{L}_s) = \{0\}. \qquad \blacksquare \end{aligned}$$

10.5.7 The Dirac constraints formula. In this section we show that the classical formula in DIRAC (1950) for constrained brackets generalizes to the Poisson context if the constraint is a cosymplectic submanifold. Let $(M, \{\cdot, \cdot\})$ be an n-dimensional Poisson manifold and let S be a k-dimensional cosymplectic submanifold of M. Let z_0 be an arbitrary point in S and $(U, \overline{\kappa})$ a submanifold chart around z_0 such that $\overline{\kappa} = (\overline{\varphi}, \overline{\psi}) : U \to V_1 \times V_2$. V_1 and V_2 are two open neighborhoods of the origin in two Euclidean spaces such that $\overline{\kappa}(z_0) = \left(\overline{\varphi}(z_0), \overline{\psi}(z_0)\right) = (0, 0)$ and

$$\overline{\kappa}(U \cap S) = V_1 \times \{0\}. \tag{10.5.13}$$

Let $\overline{\varphi} =: (\overline{\varphi}^1, \ldots, \overline{\varphi}^k)$ be the components of $\overline{\varphi}$ and define $\widehat{\varphi}^1 := \overline{\varphi}^1|_{U\cap S}, \ldots, \widehat{\varphi}^k := \overline{\varphi}^k|_{U\cap S}$. Use now Lemma 10.4.14 to extend $\widehat{\varphi}^1, \ldots, \widehat{\varphi}^k$ to D-invariant functions φ^1, $\ldots, \varphi^k$ on U. Since the differentials $\mathbf{d}\widehat{\varphi}^1(s), \ldots, \mathbf{d}\widehat{\varphi}^k(s)$ are linearly independent for any $s \in U \cap S$, we can assume (by shrinking U if necessary) that $\mathbf{d}\varphi^1(z), \ldots, \mathbf{d}\varphi^k(z)$ are also linearly independent for any $z \in U$. Consequently, (U, κ) with $\kappa := (\varphi^1, \ldots, \varphi^k, \psi^1, \ldots, \psi^{n-k})$ is a submanifold chart for M around z_0 with respect to S such that, by construction,

$$\mathbf{d}\varphi^1(s)|_{B^\sharp(s)((T_sS)^\circ)} = \cdots = \mathbf{d}\varphi^k(s)|_{B^\sharp(s)((T_sS)^\circ)} = 0,$$

for any $s \in U \cap S$. This implies that for any $i \in \{1, \ldots, k\}$, $j \in \{1, \ldots, n-k\}$, and $s \in S$

$$\{\varphi^i, \psi^j\}(s) = \mathbf{d}\varphi^i(s) \cdot X_{\psi^j}(s) = 0$$

since $\mathbf{d}\psi^j(s) \in (T_sS)^\circ$ by (10.5.13) and hence

$$X_{\psi^j}(s) \in B^\sharp(s)((T_sS)^\circ). \tag{10.5.14}$$

Additionally, since the functions $\varphi^1, \ldots, \varphi^k$ are D-invariant we have, by (10.5.2), that

$$X_{\varphi^1}(s) = X_{\widehat{\varphi}^1}(s) \in T_sS, \ldots, X_{\varphi^k}(s) = X_{\widehat{\varphi}^k}(s) \in T_sS,$$

for any $s \in S$. Consequently, $\{X_{\varphi^1}(s), \ldots, X_{\varphi^k}(s), X_{\psi^1}(s), \ldots, X_{\psi^{n-k}}(s)\}$ spans $T_s\mathcal{L}_s$ with

$$\{X_{\varphi^1}(s), \ldots, X_{\varphi^k}(s)\} \subset T_sS \cap T_s\mathcal{L}_s$$

and

$$\{X_{\psi^1}(s), \ldots, X_{\psi^{n-k}}(s)\} \subset B^\sharp(s)((T_sS)^\circ).$$

By Proposition 10.5.2**(i)**,

$$\operatorname{span}\{X_{\varphi^1}(s), \ldots, X_{\varphi^k}(s)\} = T_sS \cap T_s\mathcal{L}_s$$

and

$$\operatorname{span}\{X_{\psi^1}(s), \ldots, X_{\psi^{n-k}}(s)\} = B^\sharp(s)((T_sS)^\circ).$$

Since $\dim\left(B^\sharp(s)((T_sS)^\circ)\right) = n-k$ by Proposition 10.5.2 **(iii)**, it follows that $\{X_{\psi^1}(s), \ldots, X_{\psi^{n-k}}(s)\}$ is a basis of $B^\sharp(s)((T_sS)^\circ)$.

Given that by Theorem 10.5.3**(vi)** $B^\sharp(s)((T_sS)^\circ)$ is a symplectic subspace of $T_s\mathcal{L}_s$, there exists some $r \in \mathbb{N}$ such that $n-k = 2r$ and, additionally, the matrix $C(s)$ with entries

$$C^{ij}(s) := \{\psi^i, \psi^j\}(s), \qquad i, j \in \{1, \ldots, n-k\}$$

is invertible. Therefore, in the coordinates $(\varphi^1, \ldots, \varphi^k, \psi^1, \ldots, \psi^{n-k})$ the matrix associated to the Poisson tensor $B(s)$ is

$$\begin{pmatrix} B_S & 0 \\ 0 & C \end{pmatrix}.$$

Let $C_{ij}(s)$ be the entries of the matrix $C^{-1}(s)$.

10.5.8 Proposition (Dirac formulas) *In the coordinate neighborhood* $(\varphi^1, \ldots, \varphi^k, \psi^1, \ldots, \psi^{n-k})$ *constructed above and for* $s \in S$ *we have, for any* $f, g \in C^\infty_{S,M}(V)$

$$X_f(s) = X_F(s) - \sum_{i,j=1}^{n-k} \{F, \psi^i\}(s) C_{ij}(s) X_{\psi^j}(s) \tag{10.5.15}$$

and

$$\{f, g\}^S(s) = \{F, G\}(s) - \sum_{i,j=1}^{n-k} \{F, \psi^i\}(s) C_{ij}(s) \{\psi^j, G\}(s), \tag{10.5.16}$$

where $F, G \in C^\infty_M(U)$ *are arbitrary local extensions of* f *and* g*, respectively, around* $s \in S$.

Proof. By Theorem 10.5.3**(iv)**, $X_f(s) = \pi_S(X_F(s))$. Therefore, the equality (10.5.15) is equivalent to

$$(\mathrm{Id} - \pi_S)\, X_F(s) = \sum_{i,j=1}^{n-k} \{F, \psi^i\}(s) C_{i,j}(s) X_{\psi^j}(s). \tag{10.5.17}$$

By Proposition 10.5.2**(ii)** this amounts to showing that the right-hand side of (10.5.17) is the projection of $X_F(s)$ onto $B^\sharp(s)((T_sS)^\circ)$. We will do this by proving that this term

(i) is an element of $B^\sharp(s)((T_sS)^\circ)$;

(ii) equals $X_F(s)$ if $X_F(s) \in B^\sharp(s)((T_sS)^\circ)$;

(iii) equals 0 if $X_F(s) \in T_sS$.

Part **(i)** follows from (10.5.14). To prove **(ii)** assume that $X_F(s) \in B^\sharp(s)((T_sS)^\circ)$. Since $\{X_{\psi^1}(s), \ldots, X_{\psi^{n-k}}(s)\}$ is a basis of $B^\sharp(s)((T_sS)^\circ)$, there exist constants $\{a_1, \ldots, a_k\}$ such that

$$X_F(s) = \sum_{l=1}^{n-k} a_l X_{\psi^l}(s).$$

We shall prove that

$$\sum_{i,j=1}^{n-k} \{F, \psi^i\}(s) C_{ij}(s) X_{\psi^j}(s) = X_F(s).$$

Indeed,

$$
\begin{aligned}
\sum_{i,j=1}^{n-k} \{F, \psi^i\}(s) C_{ij}(s) X_{\psi^j}(s) &= - \sum_{i,j=1}^{n-k} \mathbf{d}\psi^i(s) \cdot X_F(s) C_{ij}(s) X_{\psi^j}(s) \\
&= - \sum_{i,j,l=1}^{n-k} a_l \mathbf{d}\psi^i(s) \cdot X_{\psi^l}(s) C_{ij}(s) X_{\psi^j}(s) \\
&= \sum_{i,j,l=1}^{n-k} a_l \{\psi^l, \psi^i\}(s) C_{ij}(s) X_{\psi^j}(s) \\
&= \sum_{i,j,l=1}^{n-k} a_l C^{li}(s) C_{ij}(s) X_{\psi^j}(s) = \sum_{j,l=1}^{n-k} a_l \delta^l_j X_{\psi^j}(s) \\
&= \sum_{l=1}^{n-k} a_l X_{\psi^l}(s) = X_F(s).
\end{aligned}
$$

Finally, to show **(iii)** let $X_F(s) \in T_s S$. Since by construction $\mathbf{d}\psi^i(s) \in (T_s S)^\circ$, for any $i \in \{1, \ldots, n-k\}$, we get $\{F, \psi^i\}(s) = -\mathbf{d}\psi^i(s) \cdot X_F(s) = 0$. ■

10.5.9 Remark Dirac's formula (10.5.16) provides an explicit local expression for the transverse Poisson structure of a Poisson manifold $(M, \{\cdot, \cdot\})$ at any of its points since by Theorem 4.1.16 the local transverse slice given by the points of the form $(\mathbf{0}, \mathbf{0}, \mathbf{z})$ is a local cosymplectic submanifold of M.

If one applies this formula to the Lie–Poisson structure on $\mathfrak{g}^*$ at a point μ satisfying the condition $\mathfrak{g} = \mathfrak{g}_\mu \oplus \mathfrak{k}$, with $\mathfrak{k}$ a linear subspace such that $[\mathfrak{g}_\mu, \mathfrak{k}] \subset \mathfrak{k}$, then the transverse Poisson structure is that of $\mathfrak{g}^*_\mu$ (WEINSTEIN (1983a); MOLINO (1984), Givental). If $\mathfrak{g}_\mu$ has a complement that is a Lie subalgebra, then the transverse structure as expressed by the Dirac formula is at most quadratic, a result due to OH (1986).

Chapter 11

Dual Pairs

The notion of polarity, as well as its use in the context of optimal reduction, can be seen as a particular case of a more general construction known generically as ***dual pair***. This chapter reviews some of the definitions of the concept of dual pair encountered in the literature, the relations between them, and some of their applications in the context of momentum maps and reduction theory.

11.1 Regular dual pairs

11.1.1 Definition *Let (M, ω) be a symplectic manifold, $(P_1, \{\cdot, \cdot\}_{P_1})$ and $(P_2, \{\cdot, \cdot\}_{P_2})$ two Poisson manifolds, and $\pi_1 : M \to P_1$ and $\pi_2 : M \to P_2$ two Poisson surjective submersions. The diagram*

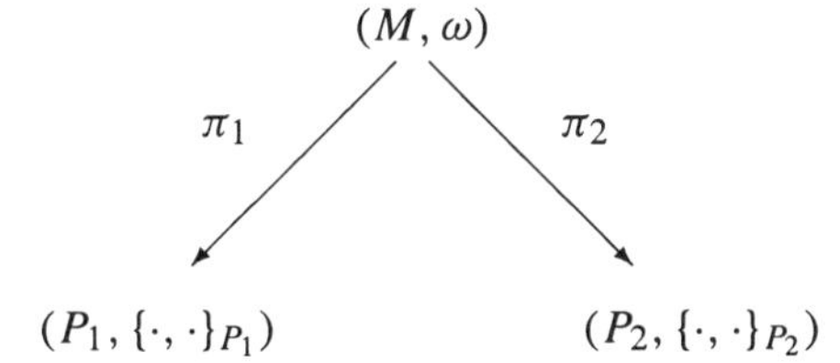

*is called a **Howe dual pair** if the Poisson subalgebras $\pi_1^* C^\infty(P_1)$ and $\pi_2^* C^\infty(P_2)$ centralize each other with respect to the Poisson structure $\{\cdot, \cdot\}$ in M associated to the symplectic form ω, that is,*

$$\pi_1^* C^\infty(P_1) = (\pi_2^* C^\infty(P_2))^c \text{ and } \pi_2^* C^\infty(P_2) = (\pi_1^* C^\infty(P_1))^c, \tag{11.1.1}$$

where the superscript c denotes the centralizer with respect to the bracket $\{\cdot, \cdot\}$. The expressions in (11.1.1) *are sometimes referred to as **double commutant relations**.*

*Analogously, we say that this diagram forms a **Lie–Weinstein dual pair** when* $\ker T\pi_1$ *and* $\ker T\pi_2$ *are symplectically orthogonal distributions, that is, for any $m \in M$ the following relation holds:*

$$(\ker T_m \pi_1)^\omega = \ker T_m \pi_2. \tag{11.1.2}$$

The notion of Howe dual pair has its origins in the study of group representations arising in quantum mechanics (see for instance HOWE (1989), KASHIWARA AND VERGNE (1978), STERNBERG AND WOLF (1978), JAKOBSEN AND VERGNE (1979), and references therein) and it appears for the first time in the context of Poisson geometry in WEINSTEIN (1983a). The definition of Lie–Weinstein dual pair can be traced back to LIE (1890) and, in its modern formulation, is due to WEINSTEIN (1983a). Examples of dual pairs arising in classical mechanics can be found in MARSDEN et al. (1984a); MARSDEN AND WEINSTEIN (1983), and references therein.

11.1.2 Example. Let D be an integrable regular distribution on the symplectic manifold (M, ω) that is the span of local infinitesimal Poisson automorphisms. Under these circumstances, by Corollary 10.2.11, the space of leaves M/D is a Poisson manifold whose bracket $\{\cdot, \cdot\}^{M/D}$ is uniquely determined by the condition that the canonical projection $\pi_D : M \to M/D$ is a Poisson surjective submersion. Let D^ω be the ***symplectic orthogonal distribution*** to D defined by

$$D^\omega(m) := \{v \in T_m M \mid \omega(m)(v, w) = 0 \quad \text{for all} \quad w \in D(m)\}.$$

We now show that D^ω is smooth and integrable.

(i) D^ω is a smooth constant rank distribution spanned by a family of infinitesimal Poisson automorphisms: Since, by hypothesis, the projection π_D is a surjective submersion by (1.1.5) we conclude that $(\ker T_m \pi_D)^\circ = \operatorname{span}\{\mathbf{d}(f \circ \pi_D)(m) \mid f \in C^\infty(M/D)\}$ and consequently

$$\begin{aligned} D^\omega(m) = [D(m)]^\omega = (\ker T_m\pi_D)^\omega = B^\sharp(m)\left((\ker T_m\pi_D)^\circ\right) \\ = \{X_{f\circ\pi_D}(m) \mid f \in C^\infty(M/D)\}, \end{aligned} \tag{11.1.3}$$

which shows that D^ω is generated by the family of Hamiltonian vector fields $\{X_{f\circ\pi_D} \mid f \in C^\infty(M/D)\}$ and hence proves the smoothness of D^ω. The regularity of D implies that it has constant rank. Since $\dim D^\omega(m) = \dim M - \dim D(m)$, for any $m \in M$, D^ω has also constant rank.

(ii) D^ω coincides with the polar distribution in the sense of Definition 5.5.2: Given that π_D is a surjective submersion

$$\{X_{f\circ\pi_D}(m) \mid f \in C^\infty(M/D)\} = \{X_F(m) \mid F \in C^\infty(M)^D\}.$$

By Propositions 3.2.9 and 5.5.3 this is a polar family for the pseudogroup of local Poisson diffeomorphisms generated by the flows of the vector fields that span D, that is,

$$D^\omega = \{X_F \mid F \in C^\infty(M)^D\} = D'. \tag{11.1.4}$$

We emphasize that in order to prove this equality we used in particular the regularity hypothesis on D. Indeed, in the absence of this hypothesis, only the inclusion $D' \subset D^\omega$ is valid.

(iii) D^ω is integrable: This is a consequence of (11.1.4) and Part **(i)** of Proposition 5.5.4.

If we assume that the space of leaves M/D^ω corresponding to D^ω is a regular quotient manifold, then the above facts and Corollary 10.2.11, imply that the space of leaves M/D^ω is a Poisson manifold whose bracket $\{\cdot,\cdot\}^{M/D^\omega}$ is uniquely determined by the condition that the canonical projection $\pi_{D^\omega} : M \to M/D^\omega$ is a Poisson surjective submersion. We emphasize that the regularity hypothesis on D^ω is genuine since the regularity of D does not imply, in general, that of its symplectic orthogonal distribution D^ω. The following counterexample is due to J. Montaldi and T. Tokieda. Let $M := \mathbb{T}^2 \times \mathbb{T}^2$ be the product of two-tori whose elements will be denoted by the four-tuples $(e^{i\theta_1}, e^{i\theta_2}, e^{i\psi_1}, e^{i\psi_2})$. Endow M with the symplectic structure ω defined by $\omega := \mathbf{d}\theta_1 \wedge \mathbf{d}\theta_2 + \sqrt{2}\,\mathbf{d}\psi_1 \wedge \mathbf{d}\psi_2$. Consider the regular distribution D on M spanned by the tangent spaces to the orbits of the canonical circle action given by $e^{i\phi} \cdot (e^{i\theta_1}, e^{i\theta_2}, e^{i\psi_1}, e^{i\psi_2}) := (e^{i(\theta_1+\phi)}, e^{i\theta_2}, e^{i(\psi_1+\phi)}, e^{i\psi_2})$. A quick inspection shows that the symplectic orthogonal distribution D^ω of D is smooth and integrable but not regular.

The importance of this construction resides in the fact that when M/D^ω is a regular quotient manifold the diagram $(M/D, \{\cdot,\cdot\}^{M/D}) \stackrel{\pi_D}{\longleftarrow} M \stackrel{\pi_{D^\omega}}{\longrightarrow} (M/D^\omega, \{\cdot,\cdot\}^{M/D^\omega})$ is a Lie–Weinstein dual pair.

The following proposition shows that Howe and Lie–Weinstein dual pairs are very closely related and that the previous construction provides a Howe dual pair.

11.1.3 Proposition *Let (M, ω) be a symplectic manifold, and let $(P_1, \{\cdot,\cdot\}_{P_1})$ and $(P_2, \{\cdot,\cdot\}_{P_2})$ be two Poisson manifolds, and $\pi_1 : M \to P_1$ and $\pi_2 : M \to P_2$ two Poisson surjective submersions.*

(i) *If the diagram $(P_1, \{\cdot,\cdot\}_{P_1}) \stackrel{\pi_1}{\leftarrow} (M, \omega) \stackrel{\pi_2}{\rightarrow} (P_2, \{\cdot,\cdot\}_{P_2})$ forms a Lie–Weinstein dual pair and if the submersions π_1 and π_2 have connected fibers, then it is also a Howe dual pair.*

(ii) *If the diagram $(P_1, \{\cdot,\cdot\}_{P_1}) \stackrel{\pi_1}{\leftarrow} (M, \omega) \stackrel{\pi_2}{\rightarrow} (P_2, \{\cdot,\cdot\}_{P_2})$ forms a Howe dual pair and if* $\dim M = \dim P_1 + \dim P_2$, *then it is also a Lie–Weinstein dual pair.*

Proof. **(i)** We start by noting that since π_1 and π_2 are surjective submersions, the equality (1.1.5) implies that

$$(\ker T_m\pi_1)^\circ = \operatorname{span}\{\mathbf{d}\,(f \circ \pi_1)\,(m) \mid f \in C^\infty(P_1)\}, \tag{11.1.5}$$
$$(\ker T_m\pi_2)^\circ = \operatorname{span}\{\mathbf{d}\,(f \circ \pi_2)\,(m) \mid f \in C^\infty(P_2)\}. \tag{11.1.6}$$

We now prove by double inclusion the identity $\pi_2^* C^\infty(P_2) = (\pi_1^* C^\infty(P_1))^c$. Let $f \in C^\infty(P_2)$ and denote by B the Poisson tensor associated to the symplectic form ω. Since $\mathbf{d}(f \circ \pi_2)(m) \in (\ker T_m\pi_2)^\circ$ and, by hypothesis $(\ker T_m\pi_1)^\omega = \ker T_m\pi_2$, for any $m \in M$, we have that

$$\begin{aligned} X_{f\circ\pi_2}(m) \in B^\sharp(m)\left((\ker T_m\pi_2)^\circ\right) &= (\ker T_m\pi_2)^\omega = \ker T_m\pi_1 \\ &= \operatorname{span}\{\mathbf{d}\,(g \circ \pi_1)\,(m) \mid g \in C^\infty(P_1)\}, \end{aligned}$$

where the last equality follows from (11.1.5). Consequently, for an arbitrary $g \in C^\infty(P_1)$, we conclude that

$$\{g \circ \pi_1, f \circ \pi_2\} = \mathbf{d}(g \circ \pi_1)\left(X_{f\circ\pi_2}\right) = 0,$$

which implies that $f \circ \pi_2 \in (\pi_1^* C^\infty(P_1))^c$.

Conversely, let $f \in (\pi_1^* C^\infty(P_1))^c$. In order to prove that $f \in \pi_2^* C^\infty(P_2)$ we start by showing that this map is constant on the fibers of π_2. Indeed, since π_2 is a surjective submersion and the diagram $(P_1, \{\cdot,\cdot\}_{P_1}) \overset{\pi_1}{\leftarrow} (M, \omega) \overset{\pi_2}{\rightarrow} (P_2, \{\cdot,\cdot\}_{P_2})$ forms a Lie–Weinstein dual pair, for any $m \in M$ we have

$$T_m\left(\pi_2^{-1}(\pi_2(m))\right) = \ker T_m\pi_2 = (\ker T_m\pi_1)^\omega = B^\sharp(m)\left((\ker T_m\pi_1)^\circ\right).$$

This equality, together with (11.1.5), guarantees that any vector $v \in T_m\left(\pi_2^{-1}(\pi_2(m))\right)$ tangent to the fiber $\pi_2^{-1}(\pi_2(m))$ can be written as $v = X_{g\circ\pi_1}(m)$, for some $g \in C^\infty(P_1)$. Hence,

$$\mathbf{d}f(m)(v) = \mathbf{d}f(m)\left(X_{g\circ\pi_1}(m)\right) = \{f, g\circ\pi_1\}(m) = 0.$$

Since $v \in T_m\left(\pi_2^{-1}(\pi_2(m))\right)$ and $m \in M$ are arbitrary and, by hypothesis, the fibers of π_2 are connected, this equality guarantees that the function $f \in (\pi_1^* C^\infty(P_1))^c$ is constant on the fibers of π_2. This implies that there exists a unique function $\overline{f} : P_2 \to \mathbb{R}$ that satisfies the equality $f = \overline{f} \circ \pi_2$. Since f is smooth and π_2 is a submersion, it follows that $\overline{f}$ is smooth and hence $f \in \pi_2^* C^\infty(P_2)$.

The identity $\pi_1^* C^\infty(P_1) = (\pi_2^* C^\infty(P_2))^c$ is proved analogously.

(ii) Let $m \in M$ be an arbitrary point. We will now show that under the hypotheses in the statement we have that $(\ker T_m\pi_2)^\omega = \ker T_m\pi_1$. Indeed, the expression (11.1.6) and the double commutant relations, imply that

$$\begin{aligned}(\ker T_m\pi_2)^\omega &= B^\sharp(m)(\{\mathbf{d}\,(f\circ\pi_2)\,(m) \mid f \in C^\infty(P_2)\}) \\ &= \{X_{f\circ\pi_2}(m) \mid f \in C^\infty(P_2)\} \\ &= \{X_g(m) \mid g \in (\pi_1^* C^\infty(P_1))^c\}.\end{aligned}$$

By definition, this set can be rewritten as

$$\{X_g(m) \mid \text{for all } g \text{ such that } \{g, h\circ\pi_1\} = 0,\ h \in C^\infty(P_1)\},$$

or, equivalently as $\{X_g(m) \mid \text{for all } g \text{ such that } \mathbf{d}(h\circ\pi_1)(z)\left(X_g(z)\right) = 0, \text{ for all } h \in C^\infty(P_1) \text{ and } z \in M\}$. This subspace is clearly included in $(\{\mathbf{d}\,(f\circ\pi_1)\,(m) \mid f \in C^\infty(P_1)\})^\circ = \ker T_m\pi_1$ and hence $(\ker T_m\pi_2)^\omega \subset \ker T_m\pi_1$. The hypothesis on the dimensions and the submersivity of π_1 and π_2 imply that

$$(\ker T_m\pi_2)^\omega = \dim P_2 = \dim M - \dim P_1 = \ker T_m\pi_1,$$

and hence $(\ker T_m\pi_2)^\omega = \ker T_m\pi_1$, as required. ■

The dimension hypothesis in part **(ii)** of Proposition 11.1.3 can be dropped if we assume a localized version of the Howe dual pair condition in Definition 11.1.1. The following definition and the next two propositions have been introduced in Montaldi et al. (2003).

11.1.4 Definition *Let (M, ω) be a symplectic manifold, $(P_1, \{\cdot, \cdot\}_{P_1})$ and $(P_2, \{\cdot, \cdot\}_{P_2})$ two Poisson manifolds, and $\pi_1 : M \to P_1$ and $\pi_2 : M \to P_2$ two open Poisson surjective maps. The diagram $(P_1, \{\cdot, \cdot\}_{P_1}) \stackrel{\pi_1}{\leftarrow} (M, \omega) \stackrel{\pi_2}{\to} (P_2, \{\cdot, \cdot\}_{P_2})$ is called a* ***localized Howe dual pair*** *if for any $m \in M$ and any open neighborhood U_m of m there exists an open subset $U \subset U_m$, $m \in U$, such that*

$$\pi_1^{U\,*} C^\infty_{P_1}(\pi_1(U)) = (\pi_2^{U\,*} C^\infty_{P_2}(\pi_2(U)))^c$$
$$\pi_2^{U\,*} C^\infty_{P_2}(\pi_2(U)) = (\pi_1^{U\,*} C^\infty_{P_1}(\pi_1(U)))^c,$$

where $\pi_1^U : U \to \pi_1(U)$ and $\pi_2^U : U \to \pi_2(U)$ are the restricted maps.

11.1.5 Proposition *Let (M, ω) be a symplectic manifold, and let $(P_1, \{\cdot, \cdot\}_{P_1})$ and $(P_2, \{\cdot, \cdot\}_{P_2})$ be two Poisson manifolds, and $\pi_1 : M \to P_1$ and $\pi_2 : M \to P_2$ two Poisson surjective submersions.*

(i) *If the diagram $(P_1, \{\cdot, \cdot\}_{P_1}) \stackrel{\pi_1}{\leftarrow} (M, \omega) \stackrel{\pi_2}{\to} (P_2, \{\cdot, \cdot\}_{P_2})$ forms a Lie–Weinstein dual pair, then it is also a localized dual pair.*

(ii) *If the diagram $(P_1, \{\cdot, \cdot\}_{P_1}) \stackrel{\pi_1}{\leftarrow} (M, \omega) \stackrel{\pi_2}{\to} (P_2, \{\cdot, \cdot\}_{P_2})$ forms a localized Howe dual pair, then it is also a Lie–Weinstein dual pair.*

Proof. **(i)** In order to prove this statement it suffices to show that for any $m \in M$ there exists an open neighborhood U of m such that the level sets of both π_1^U and π_2^U are connected; if this is the case, an argument mimicking the proof of part **(i)** in Proposition 11.1.3 gives us the result.

This claim is proved by taking a smooth function $h : M \to \mathbf{R}$ with a nondegenerate minimum at m and such that $h(m) = 0$. Let B_ϵ be the connected component containing m of the infralevel set $\{x \in M \mid h(x) < \epsilon\}$. Then we claim there is an $\epsilon > 0$ such that for each $c_j \in \pi_j(B_\epsilon)$, $j \in \{1, 2\}$, the fibers $E_{c,\epsilon} := \pi_j^{-1}(c_j) \cap B_\epsilon$ are connected. Indeed, let π be either π_1 or π_2. Choose submersion coordinates near m and $\pi(m)$ such that we can locally write $\pi(x, y) = y$. Since the restriction of h to the submanifold consisting of the points of the form $(0, y)$ has a nondegenerate minimum at $(0, 0)$, the Splitting Lemma (see for instance page 125 of BRÖCKER AND LANDER (1975)) guarantees the existence of a change of coordinates of the form $X = X(x, y)$ and $Y = Y(y)$ such that $h(X, Y) = Q(X) + g(Y)$, where Q is a positive definite quadratic form, and g is a smooth function. In these coordinates, it is clear that each $E_{y,\epsilon}$ is diffeomorphic to an open ball and so it is connected.

(ii) Let $m \in M$ be a given point in M. We start by identifying a connected open neighborhood U of m such that $\pi_1(U)$ and $\pi_2(U)$ are the domains of two Darboux–Weinstein charts (see Theorem 4.1.16) around $\pi_1(m)$ and $\pi_2(m)$, respectively. This can be easily done by taking two open neighborhoods $U_1 \subset P_1$, $U_2 \subset P_2$ of $\pi_1(m)$ and $\pi_2(m)$, respectively, that are the domains of two Darboux–Weinstein charts. Define U as the connected component of $\pi_1^{-1}(U_1) \cap \pi_2^{-1}(U_2)$ containing m. Since $\pi_1(U)$ and $\pi_2(U)$ are connected and $\pi_1(U) \subset U_1$ and $\pi_2(U) \subset_2$, U is the neighborhood that we need.

Now let $(\mathbf{q}, \mathbf{p}, \mathbf{z})$ and $(\mathbf{q}', \mathbf{p}', \mathbf{z}')$ be the local coordinates in $\pi_1(U)$ and $\pi_2(U)$, respectively, where $\pi_1(m) \equiv (\mathbf{0}, \mathbf{0}, \mathbf{0})$ and $\pi_2(m) \equiv (\mathbf{0}, \mathbf{0}, \mathbf{0})$. The localized Howe dual pair condition implies that the functions $(\mathbf{q} \circ \pi_1^U, \mathbf{q}' \circ \pi_2^U, \mathbf{p} \circ \pi_1^U, \mathbf{p}' \circ \pi_2^U)$ satisfy the canonical commutation relations and hence by Theorem 4.1.9 we can find further functions $\mathbf{q}''$ and $\mathbf{p}''$ on U such that $(\mathbf{q} \circ \pi_1^U, \mathbf{q}' \circ \pi_2^U, \mathbf{q}'', \mathbf{p} \circ \pi_1^U, \mathbf{p}' \circ \pi_2^U, \mathbf{p}'')$ is a canonical coordinate system on U and $m \equiv (\mathbf{0}, \mathbf{0}, \mathbf{0}, \mathbf{0}, \mathbf{0}, \mathbf{0})$. Additionally, the component functions $z_r \circ \pi_1^U$ of $\mathbf{z} \circ \pi_1^U$ commute with the components $q^i \circ \pi_1^U$ and $p_j \circ \pi_1^U$ since π_1 is a Poisson map. The localized Howe dual pair condition implies that they also commute with $(q')^i \circ \pi_2^U$ and $p_j' \circ \pi_2^U$. This implies that $z_r \circ \pi_1^U$ is only a function of $(\mathbf{q}'', \mathbf{p}'')$. Analogously, one proves that $z_r' \circ \pi_2^U$ is only a function of $(\mathbf{q}'', \mathbf{p}'')$. This argument shows that the mappings π_1^U and π_2^U can be written as $\pi_1^U(\mathbf{q} \circ \pi_1^U, \mathbf{q}' \circ \pi_2^U, \mathbf{q}'', \mathbf{p} \circ \pi_1^U, \mathbf{p}' \circ \pi_2^U, \mathbf{p}'') = (\mathbf{q}, \mathbf{p}, \mathbf{z}(\mathbf{q}'', \mathbf{p}''))$ and $\pi_2^U(\mathbf{q} \circ \pi_1^U, \mathbf{q}' \circ \pi_2^U, \mathbf{q}'', \mathbf{p} \circ \pi_1^U, \mathbf{p}' \circ \pi_2^U, \mathbf{p}'') = (\mathbf{q}', \mathbf{p}', \mathbf{z}'(\mathbf{q}'', \mathbf{p}''))$. Consequently, it suffices to show that the diagram

$$\begin{array}{ccc} & V := \{(\mathbf{q}'', \mathbf{p}'')\} & \\ \pi_1 \swarrow & & \searrow \pi_2 \\ \pi_1(V) = \{\mathbf{z}(\mathbf{q}'', \mathbf{p}'')\} & & \{\mathbf{z}'(\mathbf{q}'', \mathbf{p}'')\} = \pi_2(V) \end{array}$$

forms a Lie–Weinstein dual pair at $(\mathbf{q}'', \mathbf{p}'') = (\mathbf{0}, \mathbf{0})$.

Notice first that since the Poisson structures on $\pi_1(V)$ and $\pi_2(V)$ at $\pi_1(\mathbf{0}, \mathbf{0})$ and $\pi_2(\mathbf{0}, \mathbf{0})$ have zero rank we have

$$\{z_i, z_j\}_{\pi_1(V)}(\mathbf{0}) = \{z_i', z_j'\}_{\pi_2(V)}(\mathbf{0}) = 0.$$

Hence, since π_1 is a Poisson map

$$\begin{aligned} 0 = \{z_i, z_j\}_{\pi_1(V)}(\mathbf{0}) &= \mathbf{d}z_i(\mathbf{0}) \cdot X_{z_j}(\mathbf{0}) = \mathbf{d}z_i(\pi_1(\mathbf{0}, \mathbf{0})) \cdot X_{z_j}(\pi_1(\mathbf{0}, \mathbf{0})) \\ &= \mathbf{d}z_i(\pi_1(\mathbf{0}, \mathbf{0})) \cdot T_{(\mathbf{0},\mathbf{0})}\pi_1 \left(X_{z_j \circ \pi_1}\right)(\mathbf{0}, \mathbf{0}) = \mathbf{d}(z_i \circ \pi_1)(\mathbf{0}, \mathbf{0}) \cdot X_{z_j \circ \pi_1}(\mathbf{0}, \mathbf{0}), \end{aligned}$$

which shows that

$$\mathcal{B} := \{X_{z_1 \circ \pi_1}(\mathbf{0}, \mathbf{0}), \ldots, X_{z_k \circ \pi_1}(\mathbf{0}, \mathbf{0})\} \subset \ker T_{(\mathbf{0},\mathbf{0})}\pi_1.$$

Note: the elements in the family $\mathcal{B}$ are linearly independent since $\mathbf{d}(z_i \circ \pi_1)(\mathbf{0}, \mathbf{0}) = T^*_{(\mathbf{0},\mathbf{0})}\pi_1(\mathbf{d}z_i(\mathbf{0}))$, the covectors in the family $\{\mathbf{d}z_1(\mathbf{0}), \ldots, \mathbf{d}z_k(\mathbf{0})\}$ are linearly independent, and $T^*_{(\mathbf{0},\mathbf{0})}\pi_1$ is injective because π_1 is by hypothesis a submersion. Furthermore, since $X_{z_i \circ \pi_1}(\mathbf{0}, \mathbf{0}) \in \left(\ker T_{(\mathbf{0},\mathbf{0})}\pi_1\right)^\omega$, for any $i \in \{1, \ldots, k\}$, and

$$\dim \left(\ker T_{(\mathbf{0},\mathbf{0})}\pi_1\right)^\omega = \dim V - \dim(\ker T_{(\mathbf{0},\mathbf{0})}\pi_1) = \dim\left(\pi_1(V)\right) = k,$$

we can conclude that the family $\mathcal{B}$ is a basis of $\left(\ker T_{(\mathbf{0},\mathbf{0})}\pi_1\right)^\omega$. Analogously, the family $\mathcal{B}' := \{X_{z_1 \circ \pi_2}(\mathbf{0}, \mathbf{0}), \ldots, X_{z_{k'} \circ \pi_2}(\mathbf{0}, \mathbf{0})\}$ is a basis for $\left(\ker T_{(\mathbf{0},\mathbf{0})}\pi_2\right)^\omega$. In particular, we have shown that

$$\left(\ker T_{(\mathbf{0},\mathbf{0})}\pi_1\right)^\omega \subset \ker T_{(\mathbf{0},\mathbf{0})}\pi_1, \tag{11.1.7}$$

$$\left(\ker T_{(\mathbf{0},\mathbf{0})}\pi_2\right)^\omega \subset \ker T_{(\mathbf{0},\mathbf{0})}\pi_2. \tag{11.1.8}$$

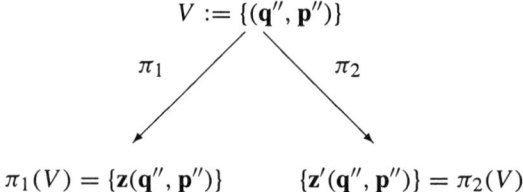

We will now prove that

$$\ker T_{(\mathbf{0},\mathbf{0})}\pi_2 \subset \left(\ker T_{(\mathbf{0},\mathbf{0})}\pi_1\right)^{\omega}. \tag{11.1.9}$$

Let $v \in \ker T_{(\mathbf{0},\mathbf{0})}\pi_2$. The symplectic nature of V implies that $v = X_f(\mathbf{0},\mathbf{0})$, for some $f \in C^{\infty}(V)$ such that $0 = \mathbf{d}(z'_j \circ \pi_2)(\mathbf{0},\mathbf{0}) \cdot X_f(\mathbf{0},\mathbf{0}) = -\mathbf{d}f(\mathbf{0},\mathbf{0}) \cdot X_{z'_j \circ \pi_2}(\mathbf{0},\mathbf{0})$, for any $j \in \{1,\ldots,k'\}$, which shows that $\mathbf{d}f(\mathbf{0},\mathbf{0}) \in \left(\left(\ker T_{(\mathbf{0},\mathbf{0})}\pi_2\right)^{\omega}\right)^{\circ}$. Since $\left(\left(\ker T_{(\mathbf{0},\mathbf{0})}\pi_2\right)^{\omega}\right)^{\circ}$ is a subbundle of T^*V, the function f can be chosen so that $\mathbf{d}f(\mathbf{q}'',\mathbf{p}'') \in \left(\left(\ker T_{(\mathbf{q}'',\mathbf{p}'')}\pi_2\right)^{\omega}\right)^{\circ}$, for any $(\mathbf{q}'',\mathbf{p}'') \in V$, which implies that

$$0 = \langle \mathbf{d}(z'_j \circ \pi_2), X_f\rangle = \{z'_j \circ \pi_2, f\}$$

on V and hence $f \in (\pi_2^{V*} C^{\infty}_{P_2}(\pi_2(V)))^c = \pi_1^{V*} C^{\infty}_{P_1}(\pi_1(V))$. Consequently, there exists a map $g \in C^{\infty}_{P_1}(\pi_1(V))$ such that $f = g \circ \pi_1|_V$ and therefore

$$v = X_f(\mathbf{0},\mathbf{0}) = X_{g\circ\pi_1|_V}(\mathbf{0},\mathbf{0}) \in \left(\ker T_{(\mathbf{0},\mathbf{0})}\pi_1\right)^{\omega},$$

which proves the inclusion (11.1.9).

Finally, (11.1.7), (11.1.8), (11.1.9), and the symplectic orthogonal version of (11.1.9) guarantee that

$$\left(\ker T_{(\mathbf{0},\mathbf{0})}\pi_2\right)^{\omega} \subset \ker T_{(\mathbf{0},\mathbf{0})}\pi_2 \subset \left(\ker T_{(\mathbf{0},\mathbf{0})}\pi_1\right)^{\omega} \subset \ker T_{(\mathbf{0},\mathbf{0})}\pi_1 \subset \left(\ker T_{(\mathbf{0},\mathbf{0})}\pi_2\right)^{\omega}.$$

Hence,

$$\left(\ker T_{(\mathbf{0},\mathbf{0})}\pi_2\right)^{\omega} = \ker T_{(\mathbf{0},\mathbf{0})}\pi_1,$$

as required. ■

11.1.6 The relation between local and global dual pairs. A common misconception about Howe pairs is the belief that the localized and global versions of these objects are equivalent provided that the projections π_1 and π_2 are submersions and have connected fibers. This mistake appears most of the times in the literature under the form of the equivalence between the Lie–Weinstein and the global Howe conditions in the presence of projections π_1 and π_2 with connected fibers. While, as we will see in Proposition 11.1.7, localized Howe pairs are Howe pairs, MONTALDI et al. (2003) have found a counterexample for the reverse implication, which we reproduce below.

Let $M := \mathbb{T}^3 \times \mathbb{R}$ and $\lambda_1, \lambda_2, \lambda_3 \in \mathbb{R}$ be three rationally independent constants such that $\lambda_i/\lambda_j \in \mathbb{R}\backslash\mathbb{Q}$, $i, j \in \{1,2,3\}$, $i \neq j$. Consider in M the symplectic structure ω whose Poisson tensor $B \in \Lambda^2(T^*M)$ is such that

$$B^{\sharp} = \begin{pmatrix} 0 & 1 & 0 & \lambda_1 \\ -1 & 0 & 0 & \lambda_2 \\ 0 & 0 & 0 & \lambda_3 \\ -\lambda_1 & -\lambda_2 & -\lambda_3 & 0 \end{pmatrix}.$$

Let $\pi : \mathbb{T}^3 \times \mathbb{R} \to \mathbb{R}$ be the projection onto the $\mathbb{R}$ factor. Then the diagram $\mathbb{R} \xleftarrow{\pi} (M,\omega) \xrightarrow{\pi} \mathbb{R}$ is a Howe pair but clearly not a Lie–Weinstein dual pair and hence not a localized Howe dual pair. In order to see that it is a Howe dual pair let $g \in (\pi^* C^{\infty}(\mathbb{R}))^c$. The trajectories of the Hamiltonian vector field X_{π} on M are irrational windings which are dense in the fibers of π. Since g is invariant under this Hamiltonian flow it must be constant on the fibers of π and hence $g \in \pi^* C^{\infty}(\mathbb{R})$, as required.

11.1.7 Proposition *Let (M,ω) be a symplectic manifold and let $(P_1,\{\cdot,\cdot\}_{P_1})$ and $(P_2,\{\cdot,\cdot\}_{P_2})$ be two Poisson manifolds. Let $\pi_1 : M \to P_1$ and $\pi_2 : M \to P_2$ be two open Poisson surjective maps with connected fibers. If the diagram $(P_1,\{\cdot,\cdot\}_{P_1}) \stackrel{\pi_1}{\leftarrow} (M,\omega) \stackrel{\pi_2}{\rightarrow} (P_2,\{\cdot,\cdot\}_{P_2})$ is a localized dual pair, then the global double commutant relations* (11.1.1) *hold.*

Proof. Using the symmetry of the statement with respect to the exchange of π_1 by π_2 it suffices to show that, for instance, $(\pi_1^* C^\infty(P_1))^c = \pi_2^* C^\infty(P_2)$.

First, given $f \in C^\infty(P_1)$ and $g \in C^\infty(P_2)$ it is easy to show that $\{f \circ \pi_1, g \circ \pi_2\} = 0$. Indeed, let $m \in M$ and U_m be the open neighborhood of m given by the localized Howe hypothesis. Then, $f \circ \pi_1|_{U_m} \in \pi_1^{U_m\,*} C^\infty_{P_1}(\pi_1(U_m))$ and $g \circ \pi_2|_{U_m} \in \pi_2^{U_m\,*} C^\infty_{P_2}(\pi_2(U_m))$ and hence

$$\{f \circ \pi_1, g \circ \pi_2\}(m) = \{f \circ \pi_1|_{U_m}, g \circ \pi_2|_{U_m}\}_{U_m}(m) = 0.$$

Since $m \in M$, $f \in C^\infty(P_1)$, and $g \in C^\infty(P_2)$ are arbitrary, it follows that

$$\pi_2^* C^\infty(P_2) \subset (\pi_1^* C^\infty(P_1))^c.$$

Second, let $g \in (\pi_1^* C^\infty(P_1))^c$; one needs to show that $g \in \pi_2^* C^\infty(P_2)$. Let $m \in M$ be arbitrary and U_m the open neighborhood of m on which the localized Howe condition is satisfied. We claim that $g|_{U_m} \in \left(\pi_1^{U_m\,*} C^\infty_{P_1}(\pi_1(U_m))\right)^c$. It then follows by hypothesis that $g|_{U_m} \in \pi_2^{U_m\,*} C^\infty_{P_2}(\pi_2(U_m))$, which allows us to write

$$g|_{U_m} = \overline{g}_m \circ \pi_2|_{U_m} \tag{11.1.10}$$

for some $\overline{g}_m \in C^\infty_{P_2}(\pi_2(U_m))$. Our second claim is that there is a function $\overline{g} \in C^\infty(P_2)$ such that $\overline{g}_m = \overline{g}|_{\pi_2(U_m)}$. The result then follows as $g = \overline{g} \circ \pi_2$. We now establish the two claims.

The first claim proceeds by contradiction. Suppose that there exists a function $f \in C^\infty_{P_1}(\pi_1(U_m))$ and $x \in U_m$ such that $\{g|_{U_m}, f \circ \pi_1|_{U_m}\}^{U_m}(x) \neq 0$. Let V_x be an open neighborhood of x such that $\overline{V_x} \subset \pi_1(U_m)$. Then there is an extension $F \in C^\infty(P_1)$ of $f|_{V_x}$. Since $g \in (\pi_1^* C^\infty(P_1))^c$ it follows that

$$0 = \{g, F \circ \pi_1\}(x) = \{g|_{U_m}, f \circ \pi_1|_{U_m}\}_{U_m}(x) \neq 0.$$

As to the second claim, notice that (11.1.10) implies that g is locally constant along the fibers of π_2. Since by hypothesis these fibers are connected, g is constant on the fibers of π_2 and $\overline{g}$ is therefore well defined. Moreover, on the open sets of the form $\pi_2(U_m)$ $\overline{g}$ coincides with $\overline{g}_m$ and so it is smooth. ■

11.1.8 Remark The connectedness hypothesis in the previous proposition and in part **(i)** of Proposition 11.1.3 are necessary, as we show in the following example provided to us by A. Giacobbe. Let $\mathbb{T}^2 = \left\{\left(e^{i\theta_1}, e^{i\theta_2}\right) \mid (\theta_1,\theta_2) \in \mathbb{R}^2\right\}$ be the two-torus considered as a symplectic manifold with the area form $\omega := \mathbf{d}\theta_1 \wedge \mathbf{d}\theta_2$. Consider the diagram $S^1 \stackrel{\pi_1}{\leftarrow} (\mathbb{T}^2,\omega) \stackrel{\pi_2}{\rightarrow} S^1$ with $\pi_j(e^{i\theta_1}, e^{i\theta_2}) := e^{ij\theta_1}$, $j \in \{1,2\}$. The fibers of π_2 have two connected components. It is easy to see that $S^1 \stackrel{\pi_1}{\leftarrow} (\mathbb{T}^2,\omega) \stackrel{\pi_2}{\rightarrow} S^1$ forms a localized Howe dual pair and a Lie–Weinstein dual pair. However, the function $\cos(\theta_1)$ belongs to $\left(\pi_1^* C^\infty(S^1)\right)^c$ but not to $\pi_2^* C^\infty(S^1)$.

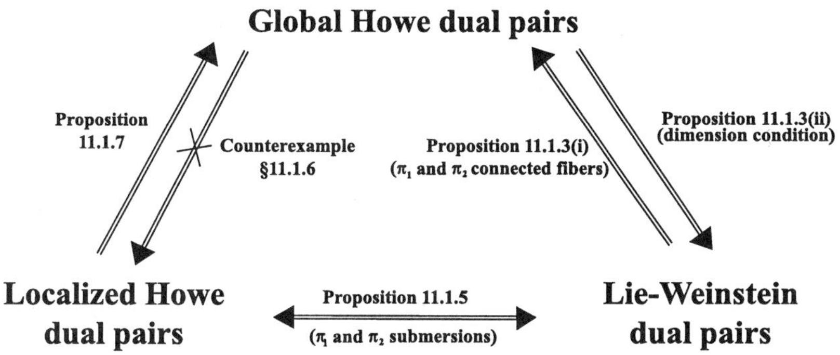

The following theorem shows that the Poisson manifolds occurring in a Lie–Weinstein dual pair are very closely related. This result is due to KAZHDAN et al. (1978); GUILLEMIN AND STERNBERG (1980), WEINSTEIN (1983a), and BLAOM (2001).

11.1.9 Theorem (Symplectic Leaf Correspondence) *Let (M, ω) be a symplectic manifold, $(P_1, \{\cdot, \cdot\}_{P_1})$ and $(P_2, \{\cdot, \cdot\}_{P_2})$ two Poisson manifolds, and $\pi_1 : M \to P_1$ and $\pi_2 : M \to P_2$ two Poisson surjective submersions with connected fibers. Suppose that the diagram $(P_1, \{\cdot, \cdot\}_{P_1}) \overset{\pi_1}{\leftarrow} (M, \omega) \overset{\pi_2}{\rightarrow} (P_2, \{\cdot, \cdot\}_{P_2})$ forms a Lie–Weinstein dual pair. Then:*

(i) *The generalized distribution D on M defined by $D(m) := \ker T_m\pi_1 + \ker T_m\pi_2$, $m \in M$, is smooth and integrable*

(ii) *Let M/D be the leaf space of the distribution D and let $P_1/\{\cdot, \cdot\}_{P_1}$ and $P_2/\{\cdot, \cdot\}_{P_2}$ be the space of symplectic leaves of the Poisson manifolds $(P_1, \{\cdot, \cdot\}_{P_1})$ and $(P_2, \{\cdot, \cdot\}_{P_2})$, respectively. The maps*

$$\begin{array}{ccccc} P_1/\{\cdot, \cdot\}_{P_1} & \longleftarrow & M/D & \longrightarrow & P_2/\{\cdot, \cdot\}_{P_2} \\ \pi_1(\mathcal{K}) & \longleftarrow \mathcal{K} & & \mathcal{K} \longrightarrow & \pi_2(\mathcal{K}) \end{array}$$

are bijections. In particular, the map $P_1/\{\cdot, \cdot\}_{P_1} \to P_2/\{\cdot, \cdot\}_{P_2}$ given by $\mathcal{L} \longmapsto \pi_2\left(\pi_1^{-1}(\mathcal{L})\right)$, with $\mathcal{L}$ an arbitrary symplectic leaf of P_1, defines a bijective correspondence between the symplectic leaves of P_1 and those of P_2. The inverse of this map is given by $\mathcal{L} \longmapsto \pi_1\left(\pi_2^{-1}(\mathcal{L})\right)$, for any symplectic leaf $\mathcal{L}$ of P_2.

(iii) *Let $(\mathcal{L}_1, \omega_{\mathcal{L}_1})$ and $(\mathcal{L}_2, \omega_{\mathcal{L}_2})$ be two symplectic leaves of P_1 and P_2 in correspondence, that is, $\mathcal{L}_2 = \pi_2(\pi_1^{-1}(\mathcal{L}_1))$. Let $\mathcal{K} = \pi_1^{-1}(\mathcal{L}_1) = \pi_2^{-1}(\mathcal{L}_2)$ be the corresponding integral leaf of D. If $i_{\mathcal{K}} : \mathcal{K} \hookrightarrow M$ is the inclusion, then*

$$i_{\mathcal{K}}^*\omega = \pi_1|_{\mathcal{K}}^*\omega_{\mathcal{L}_1} + \pi_2|_{\mathcal{K}}^*\omega_{\mathcal{L}_2}. \tag{11.1.11}$$

11.1.10 A result in WEINSTEIN (1983a) that we will not prove here shows that when the diagram $(P_1, \{\cdot, \cdot\}_{P_1}) \overset{\pi_1}{\leftarrow} (M, \omega) \overset{\pi_2}{\rightarrow} (P_2, \{\cdot, \cdot\}_{P_2})$ forms a localized Howe dual pair and the maps π_1 and π_2 are submersions then, for any $m \in M$, the transverse Poisson structures on P_1 and P_2 at $\pi_1(m)$ and $\pi_2(m)$, respectively, are anti-isomorphic. We should warn the reader that even though the hypothesis that one finds in the literature for this result and for the Symplectic Leaf Correspondence is Howe's condition these results do require the localized Howe condition to hold. A counterexample that justifies this claim can be found in MONTALDI et al. (2003).

Proof of Theorem 11.1.9. **(i)** We start by recalling again that since π_1 and π_2 are surjective submersions, the expression (1.1.5) implies that

$$\ker T_m\pi_1 = \left(\operatorname{span}\{\mathbf{d}\,(f \circ \pi_1)\,(m) \mid f \in C^\infty(P_1)\}\right)^\circ, \tag{11.1.12}$$

$$\ker T_m\pi_2 = \left(\operatorname{span}\{\mathbf{d}\,(f \circ \pi_2)\,(m) \mid f \in C^\infty(P_2)\}\right)^\circ. \tag{11.1.13}$$

Hence the Lie–Weinstein hypothesis on the diagram $(P_1, \{\cdot,\cdot\}_{P_1}) \overset{\pi_1}{\leftarrow} (M,\omega) \overset{\pi_2}{\rightarrow} (P_2, \{\cdot,\cdot\}_{P_2})$, implies that

$$\begin{aligned}\ker T\pi_1 &= (\ker T\pi_2)^\omega \\ &= B^\sharp\left((\ker T\pi_2)^\circ\right) = \operatorname{span}\{X_{f\circ\pi_2} \mid f \in C^\infty(P_2)\},\end{aligned} \tag{11.1.14}$$

where B is the Poisson tensor associated to (M,ω). Analogously, we have

$$\ker T\pi_2 = \operatorname{span}\{X_{f\circ\pi_1} \mid f \in C^\infty(P_1)\}. \tag{11.1.15}$$

Consequently, we can think of $\ker T\pi_1$ and $\ker T\pi_2$ as constant rank smooth distributions on M spanned by the families of Hamiltonian vector fields

$$F_1 := \{X_{f\circ\pi_2} \mid f \in C^\infty(P_2)\} \quad \text{and} \quad F_2 := \{X_{f\circ\pi_1} \mid f \in C^\infty(P_1)\}.$$

These distributions are integrable and, by the connectedness hypothesis on the fibers of π_1 and π_2, these fibers constitute the maximal integral leaves of $\ker T\pi_1$ and $\ker T\pi_2$, respectively.

The distribution $D := \ker T\pi_1 + \ker \pi_2$ on M is smooth since it is spanned by the family of vector fields $F := F_1 \cup F_2$. Let $A_F, A_{F_1}, A_{F_2} \subset \mathcal{P}_L(M)$ be the pseudogroups of local canonical transformations generated by the flows of the elements in F, F_1, and F_2, respectively. We will now show that D is integrable by using the Stefan–Sussmann Theorem 3.2.1. More specifically, we will show that for any $m \in M$, $X \in F$, and $\mathcal{F}_T \in A_F$, we have $T_m\mathcal{F}_T\,(X(m)) \in D(\mathcal{F}_T(m))$. Since $\ker T\pi_1$ and $\ker T\pi_2$ are already integrable, it suffices to prove that for any two arbitrary functions $f \in C^\infty(P_1)$ and $g \in C^\infty(P_2)$ such that F_t and G_t are the flows of $X_{f\circ\pi_1}$ and $X_{g\circ\pi_2}$, respectively, we have

$$T_m F_t\left(X_{g\circ\pi_2}(m)\right) \in D(F_t(m)) \quad \text{and} \quad T_m G_t\left(X_{f\circ\pi_1}(m)\right) \in D(G_t(m)).$$

We will only prove the first relation since the second is obtained similarly. Our main tool will be the Lie derivative formula (1.2.8) which, in this particular case, guarantees that

$$\frac{d}{dt} F_t^* X_{g\circ\pi_2} = F_t^*[X_{f\circ\pi_1}, X_{g\circ\pi_2}]. \tag{11.1.16}$$

Since $[X_{f\circ\pi_1}, X_{g\circ\pi_2}] = X_{\{g\circ\pi_2, f\circ\pi_1\}}$, $\{g\circ\pi_2, f\circ\pi_1\} = \mathbf{d}(g\circ\pi_2)\left(X_{f\circ\pi_1}\right) = 0$, and $X_{f\circ\pi_1} \in \ker T\pi_2$, we have

$$[X_{f\circ\pi_1}, X_{g\circ\pi_2}] = 0, \tag{11.1.17}$$

which substituted in (11.1.16) implies that

$$\frac{d}{dt}F_t^* X_{g\circ\pi_2} = 0$$

and hence $F_t^* X_{g\circ\pi_2} = F_0^* X_{g\circ\pi_2} = X_{g\circ\pi_2}$. The evaluation of this identity at the point $m \in M$ yields

$$T_m F_t\left(X_{g\circ\pi_2}(m)\right) = X_{g\circ\pi_2}(F_t(m)) \in D(F_t(m)),$$

as required.

(ii) We know from Theorem 4.1.28 that the symplectic leaves of P_1 and P_2 are the maximal integral leaves of the distributions spanned by the families of vector fields

$$H_1 := \{X_g \mid g \in C^\infty(P_2)\} \quad \text{and} \quad H_2 := \{X_g \mid g \in C^\infty(P_1)\},$$

respectively. We will denote by $A_{H_1} \subset \mathcal{P}_L(P_2)$ and $A_{H_2} \subset \mathcal{P}_L(P_1)$ the pseudogroups of local canonical transformations generated by the flows of the elements in H_1 and H_2, respectively.

Now let $\mathcal{K} \subset M$ be the integral manifold of D containing a given point $m \in M$. By the Stefan–Sussmann Theorem, the integrability of D guarantees that $\mathcal{K}$ coincides with the A_F-orbit of $m \in M$, that is, $\mathcal{K} = A_F \cdot m$. Since $F = F_1 \cup F_2$, any element in A_F can be written as the finite composition of elements in A_{F_1} and A_{F_2}. Additionally, since by (11.1.17) the elements in A_{F_1} and A_{F_2} commute, we can write

$$\mathcal{K} = A_F \cdot m = A_{F_1}\left(A_{F_2} \cdot m\right) = A_{F_2}\left(A_{F_1} \cdot m\right). \tag{11.1.18}$$

We now recall that by (11.1.14) and (11.1.15), the elements in A_{F_1} and A_{F_2} preserve the fibers of π_1 and π_2, respectively. Consequently, (11.1.18) implies that

$$\pi_1(\mathcal{K}) = \pi_1(A_F \cdot m) = \pi_1(A_{F_1}(A_{F_2} \cdot m)) = \pi_1(A_{F_2} \cdot m).$$

Any element $\mathcal{F}_T \in A_{F_2}$ is of the form $\mathcal{F}_T = F^1_{t_1} \circ \cdots \circ F^n_{t_n}$, with $F^i_{t_i}$ the flow of a vector field $X_{f_i\circ\pi_1}$, $f_i \in C^\infty(P_1)$, $i \in \{1, \ldots, n\}$. Since π_1 is a canonical map

$$\pi_1(\mathcal{F}_T(m)) = \pi_1((F^1_{t_1} \circ \cdots \circ F^n_{t_n})(m)) = (G^1_{t_1} \circ \cdots \circ G^n_{t_n})(\pi_1(m)),$$

with $G^i_{t_i}$ the flow of X_{f_i}, $i \in \{1, \ldots, n\}$. This equality implies that

$$\pi_1(\mathcal{K}) = \pi_1(A_{F_2} \cdot m) = A_{H_2} \cdot \pi_1(m) = \mathcal{L}_{\pi_1(m)}, \tag{11.1.19}$$

where $\mathcal{L}_{\pi_1(m)}$ denotes the symplectic leaf of $(P_1, \{\cdot,\cdot\}_{P_1})$ containing the point $\pi_1(m)$. Analogously, one shows that $\pi_2(\mathcal{K}) = \mathcal{L}_{\pi_2(m)}$, where $\mathcal{L}_{\pi_2(m)}$ denotes the symplectic leaf of $(P_2, \{\cdot,\cdot\}_{P_2})$ containing the point $\pi_2(m)$. These equalities allow us to define the map

$$\begin{array}{ccc} M/D & \longrightarrow & P_1/\{\cdot,\cdot\}_{P_1} \\ \mathcal{K} = A_F \cdot m & \longmapsto & \pi_1(\mathcal{K}) = A_{H_2} \cdot \pi_1(m). \end{array} \tag{11.1.20}$$

This map is obviously onto. We will now show that the map given by $\pi_1(\mathcal{K}) \longmapsto \pi_1^{-1}(\pi_1(\mathcal{K}))$, for any $\mathcal{K} \in M/D$, is an inverse of the mapping defined in (11.1.20). Indeed, let $\mathcal{K} = A_F \cdot m$, for some $m \in M$; the relation (11.1.19) yields

$$\pi_1^{-1}(\pi_1(\mathcal{K})) = \pi_1^{-1}(\pi_1(A_{F_2} \cdot m)) = \bigcup_{\mathcal{F}_T \in A_{F_2}} \pi_1^{-1}(\pi_1(\mathcal{F}_T(m))). \tag{11.1.21}$$

As already noted, the connectedness hypothesis on the fibers of π_1 implies that these fibers are the maximal integral leaves of the distribution $\ker T\pi_1$ and hence

$$\pi_1^{-1}(\pi_1(\mathcal{F}_T(m))) = A_{F_1} \cdot \mathcal{F}_T(m),$$

for any $\mathcal{F}_T \in A_{F_2}$. Substituting this identity in (11.1.21) gives

$$\pi_1^{-1}(\pi_1(\mathcal{K})) = \bigcup_{\mathcal{F}_T \in A_{F_2}} A_{F_1} \cdot \mathcal{F}_T(m) = A_{F_1} \cdot (A_{F_2} \cdot m) = A_F \cdot m = \mathcal{K},$$

as required.

Analogously, one shows that the map

$$\begin{array}{ccc} M/D & \longrightarrow & P_2/\{\cdot,\cdot\}_{P_2} \\ \mathcal{K} & \longmapsto & \pi_2(\mathcal{K}) \end{array}$$

is a bijection whose inverse is $\pi_2(\mathcal{K}) \longmapsto \pi_2^{-1}(\pi_2(\mathcal{K}))$, for any $\mathcal{K} \in M/D$.

(iii) Let $\mathcal{K} \in M/D$, $\mathcal{L}_1 \in P_1/\{\cdot,\cdot\}_{P_1}$, and $\mathcal{L}_2 \in P_2/\{\cdot,\cdot\}_{P_2}$ be three leaves in correspondence, $m \in \mathcal{K} \subset M$, and $u, v \in T_m\mathcal{K}$. Using the symplectic orthogonality of $\ker T\pi_1$ and $\ker T\pi_2$ we can write $u = u_1 + u_2$, $v = v_1 + v_2$, for some $u_1, v_1 \in \ker T_m\pi_1$, $u_2, v_2 \in \ker T_m\pi_2$, $m \in M$. Therefore, since $T_m\pi_1(u_2) = 0$, $T_m\pi_1(v_2) = 0$, $T_m\pi_2(u_1) = 0$, $T_m\pi_2(v_1) = 0$, we get

$$\begin{aligned} \omega(m)(u, v) &= \omega(m)(u_1, v_1) + \omega(m)(u_2, v_2) \\ &= \omega_{\mathcal{L}_2}(\pi_2(m))(T_m\pi_2(u_2), T_m\pi_2(v_2)) + \omega_{\mathcal{L}_1}(\pi_1(m))(T_m\pi_1(u_1), T_m\pi_1(v_1)) \\ &= \omega_{\mathcal{L}_2}(\pi_2(m))(T_m\pi_2(u), T_m\pi_2(v)) + \omega_{\mathcal{L}_1}(\pi_1(m))(T_m\pi_1(u), T_m\pi_1(v)) \\ &= \pi_2^*\omega_{\mathcal{L}_2}(m)(u, v) + \pi_1^*\omega_{\mathcal{L}_1}(m)(u, v). \end{aligned}$$

Since $u, v \in T_m\mathcal{K}$ and $m \in \mathcal{K}$ are arbitrary, the equality (11.1.11) holds. ■

11.1.11 Example. Let (M, ω) be a symplectic manifold and G a connected Lie group acting freely, canonically, and properly on M. Suppose that this action admits a proper momentum map $\mathbf{J} : M \to \mathfrak{g}^*$ with corresponding infinitesimal non-equivariance two-cocycle $\Sigma \in Z^2(\mathfrak{g}, \mathbb{R})$. The connectedness of G and the results quoted in Section 4.7 imply that both the projection $\pi : M \to M/G$ and $\mathbf{J}$ have connected fibers. If we denote by $\mathfrak{g}_{\mathbf{J}}^*$ the image of M by $\mathbf{J}$, Corollary 4.5.13 and Theorem 4.5.28 guarantee that $\mathbf{J} : (M, \omega) \to (\mathfrak{g}_{\mathbf{J}}^*, \{\cdot, \cdot\}_+^{\Sigma}|_{\mathfrak{g}_{\mathbf{J}}^*})$ is a canonical map. Moreover, by Corollary 10.2.12 there exists a unique Poisson bracket $\{\cdot, \cdot\}^{M/G}$ on M/G with respect to which the projection $\pi : M \to M/G$ is a canonical map. Finally, Proposition 4.5.14 guarantees that the diagram

$$\begin{array}{ccc} & (M, \omega) & \\ {}^{\pi}\swarrow & & \searrow^{\mathbf{J}} \\ (M/G, \{\cdot, \cdot\}^{M/G}) & & (\mathfrak{g}_{\mathbf{J}}^*, \{\cdot, \cdot\}_+^{\Sigma}|_{\mathfrak{g}_{\mathbf{J}}^*}) \end{array}$$

is a Lie–Weinstein dual pair. The connectedness of the fibers of π and $\mathbf{J}$, and Proposition 11.1.3 imply that this diagram is also a Howe pair.

The Symplectic Leaf Correspondence Theorem 11.1.9 applied to this particular example provides a beautiful insight into the Symplectic Orbit Reduction Theorem 6.3.1. Indeed, in the notation of that result, the submanifolds $\mathbf{J}^{-1}(\mathcal{O}_\mu) \subset M$, $\mathbf{J}^{-1}(\mathcal{O}_\mu)/G \subset M/G$, and $\mathcal{O}_\mu \subset \mathfrak{g}_{\mathbf{J}}^*$ are leaves in correspondence, for some $\mu \in \mathfrak{g}^*$. Note that the connectedness hypotheses guarantee that $(\mathcal{O}_\mu^+, \omega_{\mathcal{O}_\mu^+})$ is a symplectic leaf of $(\mathfrak{g}_{\mathbf{J}}^*, \{\cdot, \cdot\}_+^{\Sigma}|_{\mathfrak{g}_{\mathbf{J}}^*})$ by Theorem 4.5.31 and that $\left(\mathbf{J}^{-1}(\mathcal{O}_\mu)/G, \omega_{\mathcal{O}_\mu}\right)$ is a symplectic leaf of $(M/G, \{\cdot, \cdot\}^{M/G})$ by Theorems 10.1.1 (part **(iv)**) and Corollary 5.5.18. In this context, the defining expression (6.3.1) for the symplectic form on the orbit reduced space can be seen as a particular case of the relation (11.1.11) between the symplectic forms associated to the symplectic leaves in correspondence.

Much of the results that we will present in the next sections are motivated by the search for a similar interpretation in situations in which the group action is no longer free or a standard momentum map does not exist.

In the specific case when there is a standard momentum map, much effort has been dedicated to the study of the situations in which the diagram $(M, \{\cdot, \cdot\}^{M/G}) \stackrel{\pi}{\leftarrow} (M, \omega) \stackrel{\mathbf{J}}{\to} (\mathfrak{g}_{\mathbf{J}}^*, \{\cdot, \cdot\}_+^{\Sigma}|_{\mathfrak{g}_{\mathbf{J}}^*})$ is still a Lie–Weinstein or a Howe dual pair, even when M/G and $\mathfrak{g}_{\mathbf{J}}^*$ cease to be manifolds. In this situation we say that we have a Lie–Weinstein dual pair when $(\ker T_m\mathbf{J})^\omega = \mathfrak{g} \cdot m$, $m \in M$, and a Howe dual pair when the algebras $C^\infty(M)^G$ of G-invariant functions on M and $\mathbf{J}^*C^\infty(\mathfrak{g}^*)$ of ***collective functions*** centralize each other. By Proposition 4.5.14, the Lie–Weinstein condition is always satisfied; however, as we shall describe below, this is not the case for the Howe condition.

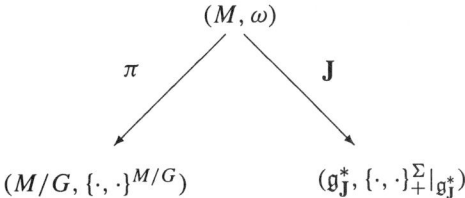

The Howe condition for standard momentum maps has attracted much attention due in part to physical motivations (see for instance GUILLEMIN AND STERNBERG (1980) and references therein). GUILLEMIN AND STERNBERG (1984d) conjectured that the Howe relation holds for the momentum maps associated to any compact group action and they proved it for toral actions. However, LERMAN (1989) gave a counterexample to this conjecture that showed the first indications of the great complexity underlying the relation between Lie–Weinstein and Howe dual pairs in the case of non-free actions. This relation, which may eventually become very sophisticated, has been the subject of a variety of studies (see for instance KARSHON (1993); KARSHON AND LERMAN (1997); KNOP (1990, 1997), and references therein). We reproduce without proof the following result from KARSHON AND LERMAN (1997).

11.1.12 Theorem *Let G be a compact Lie group acting canonically on the symplectic manifold (M, ω) and suppose that this action has an associated coadjoint equivariant proper momentum map $\mathbf{J} : M \to \mathfrak{g}^*$. Then the centralizer of the algebra $C^\infty(M)^G$ of G-invariant functions on M consists of those smooth functions on M that are pull-backs by $\mathbf{J}$ of continuous functions on $\mathfrak{g}^*$, that is,*

$$\left(C^\infty(M)^G\right)^c = C^\infty(M) \cap \mathbf{J}^* C^0(\mathfrak{g}^*).$$

Additionally, if M and G are connected, the algebra $C^\infty(M)^G$ of G-invariant functions on M and the algebra of smooth functions on M that are constant on the level sets of $\mathbf{J}$ centralize each other.

11.1.13 Morita equivalence. An important example of Lie–Weinstein dual pairs is provided by the Poisson manifolds in Morita equivalence MORITA (1958); XU (1991). We say that two Poisson manifolds $(P_1, \{\cdot,\cdot\}_{P_1})$ and $(P_2, \{\cdot,\cdot\}_{P_2})$ are ***Morita equivalent*** when there exists a symplectic manifold (M, ω) such that the diagram

$$\begin{array}{ccc} & (M,\omega) & \\ \pi_1 \swarrow & & \searrow \pi_2 \\ (P_1, -\{\cdot,\cdot\}_{P_1}) & & (P_2, \{\cdot,\cdot\}_{P_2}) \end{array}$$

is a Lie–Weinstein dual pair, the fibers of π_1 and π_2 are connected and simply connected, and π_1 and π_2 are complete maps. We recall that a Poisson map $\varphi : (M, \{\cdot,\cdot\}_M) \to (N, \{\cdot,\cdot\}_N)$ is ***complete*** when for each complete Hamiltonian vector field X_h, $h \in C^\infty(N)$, the vector field $X_{h\circ\varphi}$ is also complete.

Despite this denomination, Morita equivalence is not an equivalence relation since it is not reflexive.

Two symplectic manifolds are Morita equivalent if and only if they have isomorphic fundamental groups (see CANNAS AND WEINSTEIN (1999)). If two Poisson manifolds are Morita equivalent, then they have isomorphic fundamental groups (see GINZBURG AND LU (1992)).

11.2 Bifoliations

In this section we present a generalization of the dual pairs introduced in Definition 11.1.1 that appears in the context of integrable systems. The constructions to be described aim at dealing with situations similar to those presented in Example 11.1.2 in which the regularity hypothesis on the distribution has been dropped as well as the fact that it is generated by local infinitesimal Poisson isomorphisms. The reader interested in these topics is encouraged to check with LIBERMANN (1983); DAZORD AND

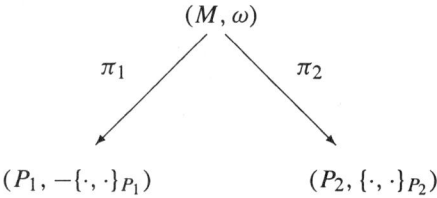

DELZANT (1987); KARASEV AND MASLOV (1993). We will follow very closely the presentation in FASSÒ (1996, 1999).

11.2.1 Definition *Let (M, ω) be a symplectic manifold and $D \subset TM$ a subbundle. Define the* ***symplectic orthogonal distribution*** *$D^\omega \subset TM$ by*

$$D^\omega(m) := \{v \in T_mM \mid \omega(m)(v, w) = 0, \text{ for any } w \in D(m)\}.$$

We recall that a ***local first integral*** of a distribution D on M is a function $f \in C^\infty(U)$, where U is an open subset of M, such that $\mathbf{d}f|_D \equiv 0$.

11.2.2 Proposition *Let (M, ω) be a symplectic manifold and $D \subset TM$ a subbundle of TM.*

(i) *D^ω is a subbundle of TM.*

(ii) $(D^\omega)^\omega = D$.

(iii) *A function $f \in C^\infty(U)$ is a local first integral of D if and only if its Hamiltonian vector field X_f takes values in D^ω.*

(iv) *If D is integrable, then for any $m \in M$ there exists an open neighborhood U of m and k local first integrals of D, $f_1, \ldots, f_k \in C^\infty(U)$, $k = \operatorname{corank}(D)$, such that for any $z \in U$*

$$\begin{aligned} D^\omega(z) &= \operatorname{span}\{X_{f_1}(z), \ldots, X_{f_k}(z)\} && (11.2.1)\\ &= \{X_f(z) \mid f \text{ local first integral of } D\}. && (11.2.2)\end{aligned}$$

Proof. **(i)** If $B \in \Lambda^2(T^*M)$ is the Poisson tensor defined by ω, then $D^\omega = B^\sharp(D^\circ)$ by Proposition 4.1.32. Therefore D^ω is the image of the subbundle $D^\circ \subset T^*M$ by the vector bundle isomorphism $B^\sharp$ and hence a subbundle of TM (see for instance the Proposition 3.4.18 in ABRAHAM et al. (1988)).

(ii) This follows trivially from §4.1.5.

(iii) The function $f \in C^\infty(U)$ is a local first integral of D if and only if $\mathbf{d}f(m) \cdot v = \omega(m)(X_f(m), v) = 0$, for any $m \in M$ and $v \in D(m)$, that is, $X_f(m) \in (D(m))^\omega = D^\omega(m)$.

(iv) The integrability of D and the expression (1.4.1) imply the existence of the k local first integrals $f_1, \ldots, f_k \in C^\infty(U)$ of D such that

$$D(z) = (\operatorname{span}\{\mathbf{d}f_1(z), \ldots, \mathbf{d}f_k(z)\})^\circ,$$

for any $z \in U$. Hence by Proposition 4.1.32

$$\begin{aligned} D^\omega(z) &= (D(z))^\omega = B^\sharp(z)\left((D(z))^\circ\right)\\ &= B^\sharp(z)\left(\operatorname{span}\{\mathbf{d}f_1(z), \ldots, \mathbf{d}f_k(z)\}\right)\\ &= \operatorname{span}\left\{X_{f_1}(z), \ldots, X_{f_k}(z)\right\}.\end{aligned}$$

The use of (1.4.2) yields the second equality in the statement. ■

11.2.3 Definition *Let (M, ω) be a symplectic manifold and $D \subset TM$ an integrable subbundle of TM. We say that D determines a* **bifoliation** *or that D is* **symplectically complete** *if D^{ω} is integrable.*

11.2.4 Proposition *Let (M, ω) be a symplectic manifold and $D \subset TM$ an integrable subbundle. Then D determines a bifoliation if and only if the Poisson bracket of any two local first integrals of D is a local first integral of D.*

Proof. Assume that D determines a bifoliation. Let $f, g \in C^{\infty}(U)$ be two local first integrals of D. By Proposition 11.2.2 **(iii)**, the vector fields X_f and X_g take values in D^{ω}. Since D^{ω} is integrable, $[X_f, X_g] = -X_{\{f,g\}}$ takes values in D^{ω} and hence, again by Proposition 11.2.2 **(iii)**, the Poisson bracket $\{f, g\}$ is a local first integral of D. Conversely, by Proposition 11.2.2 **(iv)**, for any $m \in M$ there exists an open neighborhood U of m and $k := \operatorname{codim}(D)$ local first integrals $f_1, \ldots, f_k \in C^{\infty}(U)$ of D such that for any $z \in U$

$$D^{\omega}(z) = \operatorname{span}\left\{X_{f_1}(z), \ldots, X_{f_k}(z)\right\}.$$

Let $X = \sum_{i=1}^{k} g_i X_{f_i}$, $Y = \sum_{j=1}^{k} h_i X_{f_i}$, for some arbitrary functions $g_i, h_i \in C^{\infty}(U)$, $i \in \{1, \ldots, k\}$. We now note that

$$[X, Y] = \sum_{i,j=1}^{k} g_i\{h_j, f_i\}X_{f_j} - h_j\{g_i, f_j\}X_{f_i} - g_i h_j X_{\{f_i, f_j\}}.$$

Since by hypothesis the Poisson brackets $\{f_i, f_j\}$ are local first integrals of D, Proposition 11.2.2 **(iii)** implies that $X_{\{f_i, f_j\}}$ takes values in D^{ω} and hence so does $[X, Y]$. By Frobenius' Theorem D^{ω} is integrable. ■

11.2.5 Example Let (M, ω) be a symplectic manifold and $D \subset TM$ an integrable subbundle of TM. If the fibers of D are coisotropic, that is, $D(m)^{\omega} \subset D(m)$ for any $m \in M$, then D determines a bifoliation. Indeed, by Proposition 11.2.4 it suffices to show that if $f, g \in C^{\infty}(U)$ are local first integrals of D, then so is their Poisson bracket $\{f, g\}$.

Let Y be an arbitrary local vector field that takes values in D. Since f and g are local first integrals of D we have $\mathbf{d}f \cdot Y = \mathbf{d}g \cdot Y \equiv 0$. Additionally, by the closedness of ω and formula (1.3.13)

$$\begin{aligned} 0 = \mathbf{d}\omega(X_f, X_g, Y) &= X_f[\omega(X_g, Y)] + X_g[\omega(X_f, Y)] + Y[\omega(X_f, X_g)] \\ &\quad - \omega([X_f, X_g], Y) + \omega([X_f, Y], X_g) - \omega([X_g, Y], X_f) \\ &= X_f[\mathbf{d}g \cdot Y] + X_g[\mathbf{d}f \cdot Y] + 2Y[\{f, g\}] + \omega([X_f, Y], X_g) \\ &\quad - \omega([X_g, Y], X_f). \end{aligned} \tag{11.2.3}$$

By Proposition 11.2.2 **(iv)**, X_f and X_g take values in $D^{\omega} \subset D$. The integrability of D implies that $[X_f, Y]$ and $[X_g, Y]$ also take values in D and hence the last two terms in (11.2.3) vanish. As $\mathbf{d}f \cdot Y = \mathbf{d}g \cdot Y \equiv 0$, the expression (11.2.3) guarantees that $\mathbf{d}\{f, g\} \cdot Y \equiv 0$. Since Y is an arbitrary local section of D this implies that $\{f, g\}$ is a local first integral of D.

11.2.6 Proposition *Let (M, ω) be a symplectic manifold and $D \subset TM$ an integrable subbundle that determines a bifoliation. Then*

(i) *for any $m \in M$ there exists a neighborhood U of $m \in M$ and $d :=$ rank D local first integrals $f_1, \ldots, f_d \in C^\infty(U)$ of D^ω such that for any $z \in U$*

$$\begin{aligned} D(z) &= \operatorname{span}\{X_{f_1}(z), \ldots, X_{f_d}(z)\} \\ &= \{X_f(z) \mid f \text{ local first integral of } D^\omega\}. \end{aligned} \tag{11.2.4}$$

(ii) *The function $f \in C^\infty(U)$, $U \subset M$, is a local first integral of D if and only if $\{f, g\}^{U\cap V} = 0$, for any local first integral $g \in C^\infty(V)$ of D^ω.*

Proof. **(i)** This follows from parts **(ii)** and **(iv)** of Proposition 11.2.2. **(ii)** follows from part **(i)** and Proposition 11.2.2**(iv)**. ■

The following proposition shows a very interesting interplay between bifoliations and reducibility of Poisson structures.

11.2.7 Proposition *Let (M, ω) be a symplectic manifold and let $D \subset TM$ be an integrable regular subbundle. Denote by $\pi_D : M \to M/D$ the associated surjective submersion. Then D determines a bifoliation if and only if the triple (M, ω, D) is Poisson reducible, that is, there exists a unique Poisson bracket $\{\cdot, \cdot\}^{M/D}$ on M/D with respect to which π_D is a Poisson map.*

Proof. First note that since $\pi_D : M \to M/D$ is a surjective submersion, (1.1.5) implies that

$$D^\circ = (\ker T\pi_D)^\circ = \{\mathbf{d}(f \circ \pi_D) \mid f \in C^\infty(M/D)\},$$

which combined with Proposition 4.1.32 guarantees that

$$D^\omega = \{X_{f\circ\pi_D} \mid f \in C^\infty(M/D)\}. \tag{11.2.5}$$

Suppose now that D determines a bifoliation. Under this hypothesis, Proposition 11.2.4 implies that D is a canonical distribution in the sense of Definition 10.4.2. Thus Theorem 10.4.12 implies the reducibility of $(M, \{\cdot, \cdot\}, D)$, with $\{\cdot, \cdot\}$ the bracket determined by ω.

Conversely, suppose that there exists a bracket $\{\cdot, \cdot\}^{M/D}$ on M/D such that, for any $f, g \in C^\infty(M/D)$, $\{f, g\}^{M/D} \circ \pi_D = \{f \circ \pi_D, g \circ \pi_D\}$. We will now show that the family of vector fields defining D^ω on the right-hand side of (11.2.5) is closed under the formation of the Lie bracket and then by Frobenius' Theorem 3.2.4 the integrability of D^ω will be guaranteed. Indeed, for any $f, g \in C^\infty(M/D)$,

$$[X_{f\circ\pi_D}, X_{g\circ\pi_D}] = -X_{\{f\circ\pi_D, g\circ\pi_D\}} = -X_{\{f,g\}^{M/D}\circ\pi_D}.$$

Since $\{f, g\}^{M/D} \in C^\infty(M/D)$, the result follows. ■

In Example 11.2.5 we showed that integrable coisotropic distributions determine a bifoliation. This is in general not the case for isotropic integrable distributions. We study this situation in the following proposition.

11.2.8 Proposition *Let (M, ω) be a symplectic manifold and $D \subset TM$ an integrable regular subbundle that determines a bifoliation. Denote by $\pi_D : M \to M/D$ the corresponding submersion onto the leaf space. Then the following four conditions are equivalent:*

(i) *D is isotropic, that is, $D(m) \subset D^\omega(m)$, for any $m \in M$.*

(ii) *The reduced Poisson structure $\left(M/D, \{\cdot,\cdot\}^{M/D}\right)$ has constant rank which equals* $\dim M - 2\operatorname{rank} D$.

(iii) *For any $m \in M$ there exists an open neighborhood U of m in M and $d :=$ rank D independent local Casimir functions (see §4.1.17) $c_1, \ldots, c_d \in C^\infty(\pi_D(U))$ of $\{\cdot,\cdot\}^{M/D}_{\pi_D(U)}$ such that for any $z \in U$*

$$D(z) = \operatorname{span}\{X_{c_1\circ\pi_D}(z), \ldots, X_{c_d\circ\pi_D}\}.$$

(iv) *Let $f \in C^\infty(U)$ be a local first integral of D^ω. Then for any $m \in U$ there exists an open neighborhood $V \subset U$ of m and d independent local Casimir functions $c_1, \ldots, c_d \in C^\infty(\pi_D(V))$ of $\{\cdot,\cdot\}^{M/D}_{\pi_D(V)}$ such that*

$$\mathbf{d}f = \sum_{i=1}^{d} h_i \mathbf{d}(c_i \circ \pi_D),$$

for some functions $h_1, \ldots, h_d \in C^\infty(V)$.

Proof. **(i)⇔(ii)** By (10.2.12) we have for any $m \in M$

$$\operatorname{rank}\left(B^\sharp_{M/D}(\pi_D(m))\right) = \dim M - \operatorname{rank} D - \dim\left((D(m))^\omega \cap D(m)\right). \quad (11.2.6)$$

If D is isotropic, then $D(m) \subset D(m)^\omega$ and hence

$$\dim\left(D(m)^\omega \cap D(m)\right) = \dim(D(m)) = \operatorname{rank} D$$

and **(ii)** follows. Conversely, if $\operatorname{rank}\left(B^\sharp_{M/D}(\pi_D(m))\right) = \dim M - 2\operatorname{rank} D$, for any $m \in M$, then (11.2.6) implies that $\dim\left((D(m))^\omega \cap D(m)\right) = \operatorname{rank} D = \dim D(m)$, which implies that $D(m) \subset D(m)^\omega = D^\omega(m)$, for any $m \in M$. Thus **(i)** and **(ii)** are equivalent.

(ii)⇒(iii) The constancy of the rank of $B_{M/D}$ implies by Weinstein's Theorem (see §4.1.17) that for any point $\pi_D(m)$ there exists an open neighborhood V of $\pi_D(m)$ in M/D and d independent local Casimir functions $c_1, \ldots, c_d \in C^\infty(V)$. Hence for any point $\pi_D(z) \in V$, the covectors $\{\mathbf{d}c_1(\pi_D(z)), \ldots, \mathbf{d}c_d(\pi_D(z))\}$ are linearly independent. Since $T_z^*\pi_D$ is injective, the family $\{T_z^*\pi_D \cdot \mathbf{d}c_1(\pi_D(z)), \ldots, T_z^*\pi_D \cdot \mathbf{d}c_d(\pi_D(z))\} = \{\mathbf{d}(c_1 \circ \pi_D)(z), \ldots, \mathbf{d}(c_d \circ \pi_D)(z)\}$ is also composed by linearly independent covectors for any $z \in \pi_D^{-1}(V) =: U$. The nondegeneracy of the symplectic form ω implies that $\{X_{c_1\circ\pi_D}(z), \ldots, X_{c_d\circ\pi_D}(z)\}$ are linearly independent, for any $z \in U$. Since π_D is a Poisson map, Proposition 4.1.19 implies that

$$T_z\pi_D \cdot X_{c_i\circ\pi_D}(z) = X_{c_i}(\pi_D(z)) = 0,$$

for any $i \in \{1, \ldots, d\}$ and $z \in U$. Thus,

$$\operatorname{span}\{X_{c_1 \circ \pi_D}(z), \ldots, X_{c_d \circ \pi_D}(z)\} \subset D(z). \tag{11.2.7}$$

Since $\dim D(z) = d$ we necessarily have equality in (11.2.7).

(iii)⇒(iv) Let $f \in C^\infty(U)$ be a local first integral of D^ω. Since by (11.2.5) D^ω is spanned by the vector fields of the form $X_{f \circ \pi_D}$, $f \in C^\infty(M/D)$, we have $\mathbf{d}f \circ X_{g \circ \pi_D} = \omega(X_f, X_{g \circ \pi_D}) = 0$, for any $g \in C^\infty(M/D)$, which implies that $X_f(m) \in ((D(m))^\omega)^\omega = D(m)$, for any $m \in U$. Consequently, by part **(iii)**, there exists for any $m \in U$ an open neighborhood $V \subset U$ of m, d independent local Casimir functions $c_1, \ldots, c_d \in C^\infty(\pi_D(V))$ of $\{\cdot, \cdot\}^{M/D}$, and d smooth functions $h_1, \ldots, h_d \in C^\infty(V)$ such that

$$X_f(z) = \sum_{i=1}^{d} h_i(z) X_{c_i \circ \pi_D}(z),$$

for any $z \in V$. Using the nondegeneracy of the symplectic form ω we have

$$\mathbf{d}f(z) = \sum_{i=1}^{d} h_i(z) \mathbf{d}(c_i \circ \pi_D)(z).$$

(iv)⇒(i) Let $z \in M$ be given and let $v, w \in D(z)$ be arbitrary. By (11.2.4) there exist $f, g \in C^\infty(U)$ two local first integrals of D^ω, such that $v = X_f(z)$ and $w = X_g(z)$. By **(iv)**, $X_f(z)$ and $X_g(z)$ can be rewritten as $X_f(z) = \sum_{i=1}^{d} \lambda_i X_{c_i \circ \pi_D}(z)$, $X_g(z) = \sum_{i=1}^{d} \mu_i X_{c_i \circ \pi_D}(z)$, for some constants $(\lambda_1, \ldots, \lambda_d), (\mu_1, \ldots, \mu_d) \in \mathbb{R}^d$, and $c_1, \ldots, c_d \in C^\infty(U)$ local Casimir functions of $\{\cdot, \cdot\}^{M/D}$. Now,

$$\begin{aligned}
\omega(z)(v, w) &= \sum_{i,j=1}^{d} \lambda_i \mu_j \omega(z) \left(X_{c_i \circ \pi_D}(z), X_{c_j \circ \pi_D}(z)\right) \\
&= \sum_{i,j=1}^{d} \lambda_i \mu_j \{c_i \circ \pi_D, c_j \circ \pi_D\}(z) \\
&= \sum_{i,j=1}^{d} \lambda_i \mu_j \{c_i, c_j\}^{M/D} \circ \pi_D(z) = 0,
\end{aligned}$$

which guarantees that $D(z) \subset D(z)^\omega$. ■

Our next proposition shows that in the presence of the isotropy hypothesis, the symplectic leaves of the Poisson manifold $\left(M/D, \{\cdot, \cdot\}^{M/D}\right)$ can be easily found.

11.2.9 Proposition *Let (M, ω) be a symplectic manifold and $D \subset TM$ a regular isotropic subbundle of TM that determines a bifoliation. Additionally, suppose that the projection $\pi_{D^\omega} : M \to M/D^\omega$ is a surjective submersion. Then there exists a unique Poisson surjective submersion $\pi : M/D \to M/D^\omega$ such that the diagram*

$$
\begin{array}{ccc}
 & (M, \omega) & \\
 {}^{\pi_D}\swarrow & & \searrow^{\pi_{D^\omega}} \\
(M/D, \{\cdot,\cdot\}^{M/D}) & \xrightarrow{\ \pi\ } & (M/D^\omega, \{\cdot,\cdot\}^{M/D^\omega})
\end{array}
$$

commutes. The symplectic leaves of $(M/D, \{\cdot,\cdot\}^{M/D})$ *are the connected components of the fibers of* π.

Proof. Since D determines a bifoliation, $\pi_{D^\omega} : M \to M/D^\omega$ is well defined and by hypothesis is a surjective submersion. If we denote by $\{\cdot,\cdot\}$ the Poisson structure associated to ω, then by Proposition 11.2.7 the triple $(M, \{\cdot,\cdot\})$ is Poisson reducible. The isotropy hypothesis $D \subset D^\omega$ implies that the maximal integral leaves of D are contained in those of D^ω and hence the relation $\pi \circ \pi_D = \pi_{D^\omega}$ determines the map π uniquely. Since π_D and π_{D^ω} are smooth surjective submersions so is π (see the comments at the end of §1.1.13). In order to check that π is Poisson, take $f, g \in C^\infty(M/D^\omega)$ to be arbitrary. Then

$$
\begin{aligned}
\{f, g\}^{M/D^\omega} \circ \pi \circ \pi_D = \{f, g\}^{M/D^\omega} \circ \pi_{D^\omega} &= \{f \circ \pi_{D^\omega}, g \circ \pi_{D^\omega}\} \\
&= \{f \circ \pi \circ \pi_D, g \circ \pi \circ \pi_D\} = \{f \circ \pi, g \circ \pi\}^{M/D} \circ \pi_D.
\end{aligned}
$$

The surjectivity of π_D guarantees that

$$
\{f, g\}^{M/D^\omega} \circ \pi = \{f \circ \pi, g \circ \pi\}^{M/D}.
$$

In order to prove the last statement in the theorem it suffices to check that for any $m \in M$

$$
T_{\pi_D(m)}(\pi^{-1}(\pi_{D^\omega}(m))) = \{X_f(\pi_D(m)) \mid f \in C^\infty(M/D)\}.
$$

Indeed, since π is a submersion (11.2.5) implies that

$$
\begin{aligned}
T_{\pi_D(m)}(\pi^{-1}(\pi_{D^\omega}(m))) &= \ker T_{\pi_D(m)}\pi \\
&= \{T_m\pi_D(u) \mid u \in \ker T_m\pi_{D^\omega} = D^\omega(m)\} \\
&= \{T_m\pi_D \cdot X_{f\circ\pi_D}(m) \mid f \in C^\infty(M/D)\} \\
&= \{X_f(\pi_D(m)) \mid f \in C^\infty(M/D)\}. \qquad \blacksquare
\end{aligned}
$$

11.2.10 Bifoliations and noncommutative integrability. The ideas introduced in the preceding propositions allow the formulation in geometric terms of a generalization of the ***Liouville–Mineur–Arnold theorem*** for integrable systems to more degenerate situations. This generalization is due, at various levels of generality, to NEKHOROSHEV (1972); MISHCHENKO AND FOMENKO (1978); DAZORD AND DELZANT (1987); KARASEV AND MASLOV (1993). We follow the presentation in FASSÒ (1996, 1999), where the proof of the following result is described in detail.

11.2.11 Theorem *Let* (M, ω) *be a* $2n$*-dimensional symplectic manifold and let* $\pi_1 : M \to B$ *be a fibration with compact, connected, and isotropic fibers of dimension* $d \leq n$. *Assume that the subbundle* $D \subset TM$ *given by* $D := \ker T\pi_1$ *determines a bifoliation. Then the following hold:*

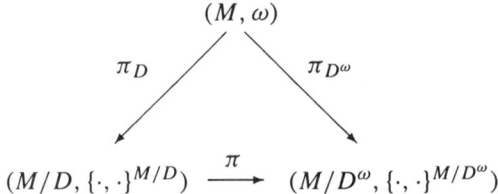

(i) *The fibers of π_1 are diffeomorphic to d-tori $\mathbb{T}^d$.*

(ii) *Every fiber of π_1 has a neighborhood U that can be endowed with* ***generalized action-angle coordinates****, that is, there exist an open neighborhood $\mathcal{B} \subset \mathbb{R}^{2n-d}$ and a diffeomorphism $b \times \alpha : U \to \mathcal{B} \times \mathbb{T}^d$ such that the fibers of π_1 coincide with the sets of the form $(b \times \alpha)^{-1}(\{c\} \times \mathbb{T}^d)$, $c \in \mathcal{B}$. Additionally, if $(a_1, \ldots, a_d, q^1, \ldots, q^{n-d}, p_1, \ldots, p_{n-d})$ are coordinates for $\mathcal{B}$ and $(\alpha_1, \ldots, \alpha_d)$ are coordinates for $\mathbb{T}^d$,, then*

$$\left((b \times \alpha)^{-1}\right)^* \omega = \sum_{i=1}^{n-d} \mathbf{d}q^i \wedge \mathbf{d}p_i + \sum_{i=1}^{d} \mathbf{d}a_i \wedge \mathbf{d}\alpha_i.$$

(iii) *If the leaf space $A := M/D^\omega$ is a regular quotient manifold and the projection $\pi_2 : M \to A$ is a fibration, then A is called the* ***action space*** *and it has an integer affine structure. This means that if we have two different sets $\triangleright \mathbf{a}, \mathbf{q}, \mathbf{p}, \boldsymbol{\alpha} \triangleleft$ and $\triangleright \mathbf{a}', \mathbf{q}', \mathbf{p}', \boldsymbol{\alpha}' \triangleleft$ of generalized action-angle variables, then they are related, on each connected component of the intersection of their domains, by equations of the form*

$$\begin{aligned} \mathbf{a}' &= Z\mathbf{a} + \mathbf{z} \\ (\mathbf{q}', \mathbf{p}') &= F(\mathbf{a}, \mathbf{q}, \mathbf{p}) \\ \boldsymbol{\alpha}' &= (Z^{-1})^T \boldsymbol{\alpha} + L(\mathbf{a}, \mathbf{q}, \mathbf{p}) \pmod{2\pi}, \end{aligned}$$

for some matrix $Z \in \mathrm{SL}_{\pm}(\mathbb{Z}; d)$, some vector $\mathbf{z} \in \mathbb{R}^d$, and some smooth mappings F and L.

Motivated by this theorem, we say that a Hamiltonian dynamical system (M, ω, h) is ***integrable in the noncommutative sense*** of ***super integrable*** when there is a fibration $\pi_1 : M \to B$ of M with compact and connected fibers such that the tangent space to these fibers induces an isotropic foliation of M such that

(i) X_h is tangent to the leaves of this foliation and

(ii) this foliation is symplectically complete.

If we are in the hypotheses of parts **(i)** through **(iii)** of Theorem 11.2.11, Propositions 11.2.8 and 11.2.9 guarantee the existence of a unique Poisson surjective submersion $\pi : B \to A$ such that the diagram

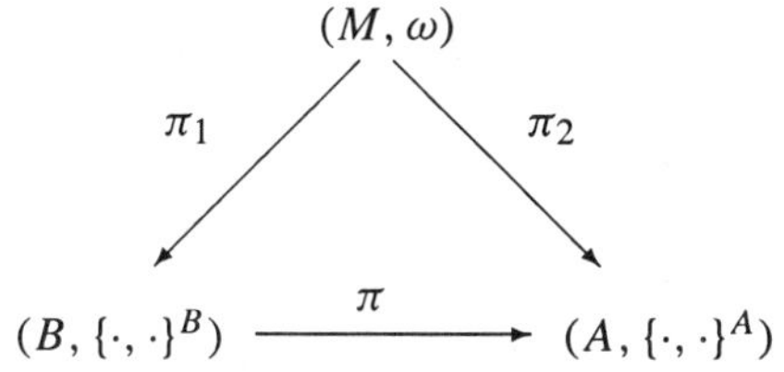

commutes. We recall that the symplectic leaves of $(B, \{\cdot, \cdot\}^B)$ are the connected components of the fibers of π.

Theorem 11.2.11 guarantees that if the system (M, ω, h) is integrable in the noncommutative sense, then the motion leaves invariant the tori given by the fibers of π_1 and that it is quasi-periodic on these tori. Moreover, the motions in all the tori contained in a fiber of π_2 are quasi-periodic with the same frequency. Equivalently, the motions on all the tori of the form $\pi_1^{-1}(b)$ with b in a fixed symplectic leaf of $(B, \{\cdot, \cdot\}^B)$ are quasi-periodic with the same frequency. Many classical mechanical systems are integrable in the noncommutative sense; for instance the rigid body, the two-dimensional isotropic harmonic oscillator, and the Kepler system belong to this class.

The reader interested in generalizations of these ideas to more singular situations is encouraged to check VEY (1978); ELIASSON (1990); ITO (1991); ZUNG (1996, 2003), and references therein.

11.3 Singular dual pairs

In the rest of the chapter we will study a generalization of the dual pairs introduced in Definition 11.1.1, which incorporates less regular situations. The main ingredient in the concepts that we will consider in the sections that follow is the notion of polarity in Definition 5.5.2. Most of the results presented are included in ORTEGA (2003b).

11.3.1 Definition *Let $(M, \{\cdot, \cdot\})$ be a Poisson manifold and $A, B \subset \mathcal{P}_L(M)$ two pseudosubgroups of local Poisson diffeomorphisms. The diagram*

$$\begin{array}{ccc} & (M, \{\cdot, \cdot\}) & \\ \quad\swarrow^{\pi_A} & & \searrow^{\pi_B}\quad \\ (M/A, \{\cdot, \cdot\}^{M/A}) & & (M/B, \{\cdot, \cdot\}^{M/B}) \end{array}$$

is called a **(singular) dual pair** *on $(M, \{\cdot, \cdot\})$ when the polar distributions A' and B' of A and B, respectively, are integrable and they satisfy*

$$M/A' = M/B \text{ and } M/B' = M/A. \tag{11.3.1}$$

This diagram is called a **(singular) Howe pair** *on $(M, \{\cdot, \cdot\})$ if the following two conditions hold:*

$$\left(\pi_A^* C^\infty(M/A)\right)^c = \pi_B^* C^\infty(M/B) \tag{11.3.2}$$

$$\left(\pi_B^* C^\infty(M/B)\right)^c = \pi_A^* C^\infty(M/A). \tag{11.3.3}$$

11.3.2 Remark The identity $M/A' = M/B$ in (11.3.1) means that the partition of M into B-orbits coincides with that of M into A'-leaves. In general, this condition can hold without B being equal to A' as pseudogroups; only the orbit spaces are required to coincide. Notice also that two pseudogroups A and B in duality are necessarily integrable.

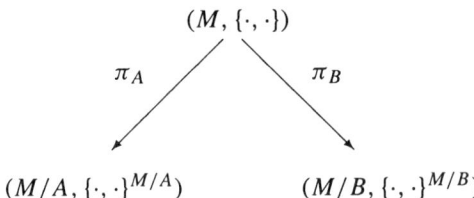

Additionally, the brackets $\{\cdot,\cdot\}^{M/A}$ and $\{\cdot,\cdot\}^{M/B}$ denote the bilinear operations on the quotient presheaves $C^\infty_{M/A}$ and $C^\infty_{M/B}$, respectively, defined by (10.2.2) from the obvious Poisson reducibility of $(M,\{\cdot,\cdot\},A)$ and $(M,\{\cdot,\cdot\},B)$.

11.3.3 Example: The dual pair associated to a regular distribution. The regular dual pairs introduced in Example 11.1.2 are also an example of dual pairs in the sense of the previous definition. Indeed, let D be an integrable regular distribution on the symplectic manifold (M,ω) that is the span of local infinitesimal Poisson automorphisms, and consider the Poisson structure $\{\cdot,\cdot\}^{M/D}$ on the space of leaves M/D for which the canonical projection $\pi_D : M \to M/D$ is a Poisson surjective submersion. Consider now the symplectic orthogonal distribution D^ω to D, defined by

$$D^\omega(m) := \{v \in T_mM \mid \omega(m)(v,w) = 0 \quad \text{for all} \quad w \in D(m)\}.$$

We showed that D^ω is smooth, integrable, and that it coincides with the polar distribution in the sense of Definition 5.5.2, that is,

$$D^\omega = \{X_F \mid F \in C^\infty(M)^D\} = D'. \tag{11.3.4}$$

If M/D^ω is a regular quotient manifold, then the diagram $(M/D,\{\cdot,\cdot\}^{M/D}) \stackrel{\pi_D}{\longleftarrow} M \stackrel{\pi_{D^\omega}}{\longrightarrow} (M/D^\omega,\{\cdot,\cdot\}^{M/D^\omega})$ is a dual pair in the sense of Definition 11.3.1. Indeed, (11.3.4) implies that $M/D' = M/D^\omega$. Additionally, applying (11.3.4) to the distribution D^ω yields $(D^\omega)' = (D^\omega)^\omega = D$ and hence $M/(D^\omega)' = M/D$, as required in (11.3.1).

11.3.4 Example: Lie–Weinstein dual pairs with connected fibers are dual pairs in the sense of Definition 11.3.1. Let $(P_1,\{\cdot,\cdot\}_{P_1}) \stackrel{\pi_1}{\leftarrow} (M,\omega) \stackrel{\pi_2}{\rightarrow} (P_2,\{\cdot,\cdot\}_{P_2})$ be a Lie–Weinstein dual pair such that π_1 and π_2 have connected fibers. We will now see that this connectedness hypothesis allows us to realize the diagram $(P_1,\{\cdot,\cdot\}_{P_1}) \stackrel{\pi_1}{\leftarrow} (M,\omega) \stackrel{\pi_2}{\rightarrow} (P_2,\{\cdot,\cdot\}_{P_2})$ as a dual pair in the sense of Definition 11.3.1. Indeed, as we saw in the proof of Theorem 11.1.9, we can think of $\ker T\pi_1$ and $\ker T\pi_2$ as constant rank smooth distributions on M spanned by the families of Hamiltonian vector fields

$$F_1 := \{X_{f\circ\pi_2} \mid f \in C^\infty(P_2)\} \quad \text{and} \quad F_2 := \{X_{f\circ\pi_1} \mid f \in C^\infty(P_1)\}.$$

These distributions are integrable and, by the connectedness hypothesis on the fibers of π_1 and π_2, these fibers constitute the maximal integral leaves of $\ker T\pi_1$ and $\ker T\pi_2$, respectively. In view of this we can make the natural identifications:

$$P_1 \simeq M/\ker T\pi_1 = M/A_{F_1} \quad \text{and} \quad P_2 \simeq M/\ker T\pi_2 = M/A_{F_2}.$$

Using these identifications we can rewrite the Lie–Weinstein dual pair $(P_1,\{\cdot,\cdot\}_{P_1}) \stackrel{\pi_1}{\leftarrow} (M,\omega) \stackrel{\pi_2}{\rightarrow} (P_2,\{\cdot,\cdot\}_{P_2})$ as

$$\left(M/A_{F_1},\{\cdot,\cdot\}^{M/A_{F_1}}\right) \stackrel{\pi_{A_{F_1}}}{\leftarrow} (M,\omega) \stackrel{\pi_{A_{F_2}}}{\rightarrow} \left(M/A_{F_2},\{\cdot,\cdot\}^{M/A_{F_2}}\right)$$

which, as a corollary of the previous example, is a dual pair in the sense of Definition 11.3.1. Indeed, since $A_{F_1} = \ker T\pi_1$ is a regular integrable distribution, (11.3.4) gives $A'_{F_1} = A^\omega_{F_1} = (\ker T\pi_1)^\omega = \ker T\pi_2 = A_{F_2}$, which implies that $M/A'_{F_1} = M/A_{F_2}$. Analogously, it can be shown that $M/A'_{F_2} = M/A_{F_1}$.

11.3.5 Definition *Let* $(M, \{\cdot,\cdot\})$ *be a smooth Poisson manifold and* $A \subset \mathcal{P}_L(M)$ *a pseudogroup of local Poisson diffeomorphisms. We say that A is* ***von Neumann*** *when the diagram* $(M/A, \{\cdot,\cdot\}^{M/A}) \xleftarrow{\pi_A} (M, \{\cdot,\cdot\}) \xrightarrow{\pi_{A'}} (M/A', \{\cdot,\cdot\}^{M/A'})$ *is a dual pair or, equivalently, when the distributions* A' *and* $(A')'$ *are integrable and*

$$M/(A')' = M/A. \tag{11.3.5}$$

If this condition holds, the dual pair

$$(M/A, \{\cdot,\cdot\}^{M/A}) \xleftarrow{\pi_A} (M, \{\cdot,\cdot\}) \xrightarrow{\pi_{A'}} (M/A', \{\cdot,\cdot\}^{M/A'})$$

is called the ***von Neumann pair*** *associated to* $A \subset \mathcal{P}_L(M)$. *In what follows we will write* $A'' := (A')'$.

11.3.6 Remark. If in the previous definition A is actually a subgroup of $\mathcal{P}(M)$, part **(i)** in Proposition 5.5.4 automatically guarantees the integrability of A'.

11.3.7 Remark. Von Neumann groups have connected and path connected orbits since relation (11.3.5) implies that for any point $m \in M$, the orbit $A \cdot m$ coincides with $A'' \cdot m$ which is a connected and path connected set.

11.3.8 Remark. The terminology in the previous definition has been chosen due to the similarity of condition (11.3.5) with the von Neumann or double commutant relation for $*$-algebras of bounded operators on a Hilbert space.

11.3.9 Example: Canonical Lie group actions and the optimal momentum map. Let G be a connected Lie group acting on the symplectic manifold (M, ω) in a free, proper, and canonical fashion via the map $\Phi : G \times M \to M$. Let $A_G := \{\Phi_g : M \to M \mid g \in G\}$ and $\pi : M \to M/G$ be the surjective submersion obtained by projecting M onto the orbit space $M/G = M/A_G$. Let $\mathcal{J} : M \to M/A'_G$ be the corresponding optimal momentum map and assume that this action admits a standard momentum map $\mathbf{J} : M \to \mathfrak{g}^*$ with infinitesimal non-equivariance two-cocycle $\Sigma \in Z^2(\mathfrak{g}, \mathbb{R})$. We will suppose that the level sets of $\mathbf{J}$ are connected. In the presence of these hypotheses, Corollary 5.5.18 guarantees that $A'_G(m) = \ker T_m\mathbf{J}$ for all $m \in M$. Consequently, in this particular situation, the diagram

$$(M/A_G, \{\cdot,\cdot\}^{M/A_G}) \xleftarrow{\pi} (M, \omega) \xrightarrow{\mathcal{J}} (M/A'_G, \{\cdot,\cdot\}^{M/A'_G})$$

can be identified with

$$(M/G, \{\cdot,\cdot\}^{M/G}) \xleftarrow{\pi} (M, \omega) \xrightarrow{\mathbf{J}} (\mathfrak{g}^*_{\mathbf{J}}, \{\cdot,\cdot\}^{\Sigma}_{+}|_{\mathfrak{g}^*_{\mathbf{J}}}).$$

By Proposition 4.5.14, $(\ker T_m\mathbf{J})^{\omega} = \mathfrak{g} \cdot m$ for any $m \in M$, which shows that this diagram is a Lie–Weinstein dual pair with connected fibers. By Example 11.3.4, it is also a dual pair in the sense of Definition 11.3.1. We have therefore shown that the subgroup A_G associated to a free canonical connected Lie group action admitting a momentum map with connected fibers is von Neumann.

One of the main goals of Section 11.6 will be the study of A_G and A'_G in more realistic situations, namely when the G-action is no longer free, as well as the search for cases in which the diagram

$$(M/A_G, \{\cdot,\cdot\}^{M/A_G}) \xleftarrow{\pi} (M, \{\cdot,\cdot\}) \xrightarrow{\mathcal{J}} (M/A'_G, \{\cdot,\cdot\}^{M/A'_G})$$

is a dual pair.

11.3.10 Proposition *Let $(M, \{\cdot,\cdot\})$ be a smooth Poisson manifold, $A \subset \mathcal{P}_L(M)$ a pseudogroup, and A' its polar. If A has the extension property and A' is integrable,*

$$\left(\pi_A^* C^\infty(M/A)\right)^c = \pi_{A'}^* C^\infty(M/A') \tag{11.3.6}$$

$$\left(\pi_{A'}^* C^\infty(M/A')\right)^c \supset \pi_A^* C^\infty(M/A). \tag{11.3.7}$$

Moreover, if A is von Neumann and A' has the extension property, then the diagram $(M/A, \{\cdot,\cdot\}^{M/A}) \xleftarrow{\pi_A} (M, \{\cdot,\cdot\}) \xrightarrow{\pi_{A'}} (M/A', \{\cdot,\cdot\}^{M/A'})$ is a Howe pair.

Proof. We first establish (11.3.6), which is equivalent to proving that $\left(C^\infty(M)^A\right)^c = C^\infty(M)^{A'}$. Let $f \in C^\infty(M)$ be arbitrary and let $g \in C^\infty(M)^A$ be an A-invariant function with associated Hamiltonian flow F_t. Then, for any $m \in M$

$$\frac{d}{dt} f(F_t(m)) = \mathbf{d}f(F_t(m))\left(X_g(F_t(m))\right) = \{f, g\}(F_t(m)). \tag{11.3.8}$$

Now, if $f \in \left(C^\infty(M)^A\right)^c$, then $\{f, g\} = 0$ in (11.3.8) and therefore $(f \circ F_t)(m) = f(m)$. Since the A-invariant function g and the point m are arbitrary and A has the extension property, we can conclude that $f \in C^\infty(M)^{A'}$. Conversely, if $f \in C^\infty(M)^{A'}$, then $f \circ F_t = f$ and therefore (11.3.8) implies that $f \in \left(C^\infty(M)^A\right)^c$.

Expression (11.3.7) can be obtained by taking centralizers on both sides of (11.3.6).

Suppose now that A is von Neumann. In order to conclude that we have a Howe pair we just need to show that $\left(\pi_{A'}^* C^\infty(M/A')\right)^c \subset \pi_A^* C^\infty(M/A)$ or, equivalently, that $\left(C^\infty(M)^{A'}\right)^c \subset C^\infty(M)^A$. Let $f \in \left(C^\infty(M)^{A'}\right)^c$. Since A' has the extension property, any element $\mathcal{F}_T \in A''$ can be written as the finite composition of locally defined flows F_t associated to the Hamiltonian vector fields of A'-invariant globally defined functions $h \in C^\infty(M)^{A'}$. Since for any of those functions we have $\{f, h\} = 0$, it is clear that $f \circ F_t = f|_{\mathrm{Dom}(F_t)}$. Now, the von Neumann character of A implies that for any $\phi \in A$ and $m \in M$, there exists $\mathcal{F}_T \in A''$ such that $(f \circ \phi)(m) = (f \circ \mathcal{F}_T)(m) = f(m)$, which guarantees the A-invariance of f. ■

11.3.11 Corollary *Let $(M, \{\cdot,\cdot\})$ be a smooth Poisson manifold and $A, B \subset \mathcal{P}_L(M)$ two pseudosubgroups of its local Poisson diffeomorphism pseudogroup $\mathcal{P}_L(M)$ that have the extension property. If the diagram*

$$(M/A, \{\cdot,\cdot\}^{M/A}) \overset{\pi_A}{\leftarrow} (M, \omega) \overset{\pi_B}{\rightarrow} (M/B, \{\cdot,\cdot\}^{M/B})$$

is a dual pair, then it is also a Howe pair.

As a corollary to the previous result we can easily recover Part **(i)** of Proposition 11.1.3.

11.3.12 Corollary *If the diagram* $(P_1, \{\cdot,\cdot\}_{P_1}) \stackrel{\pi_1}{\leftarrow} (M, \omega) \stackrel{\pi_2}{\rightarrow} (P_2, \{\cdot,\cdot\}_{P_2})$ *is a Lie–Weinstein dual pair with connected fibers, then it is a Howe pair.*

Proof. In Example 11.3.4 we saw that any Lie–Weinstein dual pair with connected fibers can be understood as a dual pair in the sense of Definition 11.3.1 with respect to two pseudogroups of $\mathcal{P}_L(M)$ that automatically have the extension property by Proposition 3.2.9. Consequently, the hypotheses of the previous corollary are satisfied and hence the claim follows. ■

Even though Corollary 11.3.11 shows that in the presence of the extension property any dual pair is a Howe pair, the following example demonstrates that the converse is, in general, not true.

11.3.13 Example: A Howe pair that is not a dual pair. Let $(\mathbb{T}^2, \omega)$ be the two-torus thought of as a symplectic manifold with the form ω given by the standard area form. Consider a Poisson action of the additive group $(\mathbb{R}, +)$ on $\mathbb{T}^2$ via an irrational flow. It is straightforward to check that the Poisson diffeomorphisms group $A_{\mathbb{R}} \subset \mathcal{P}(\mathbb{T}^2)$ associated to this action generates a Howe pair

$$\mathbb{T}^2/A_{\mathbb{R}} \stackrel{\pi_{A_{\mathbb{R}}}}{\longleftarrow} \mathbb{T}^2 \stackrel{\pi_{A'_{\mathbb{R}}}}{\longrightarrow} \mathbb{T}^2/A'_{\mathbb{R}}$$

that is not a dual pair. Indeed, notice first that the only $A_{\mathbb{R}}$-invariant open subset of $\mathbb{T}^2$ is $\mathbb{T}^2$ itself. Since $C^\infty(\mathbb{T}^2)^{A_{\mathbb{R}}}$ consists of constant functions, the dual distribution $A'_{\mathbb{R}}$ is trivial and hence $C^\infty(\mathbb{T}^2)^{A'_{\mathbb{R}}} = C^\infty(\mathbb{T}^2)$. It is clear that in these circumstances $(C^\infty(\mathbb{T}^2)^{A_{\mathbb{R}}})^c = C^\infty(\mathbb{T}^2)^{A'_{\mathbb{R}}}$ and $(C^\infty(\mathbb{T}^2)^{A'_{\mathbb{R}}})^c = C^\infty(\mathbb{T}^2)^{A_{\mathbb{R}}}$. Nevertheless, the orbits of the $A_{\mathbb{R}}$-action are strictly contained in the only leaf of the distribution $A''_{\mathbb{R}}$, which implies that $A_{\mathbb{R}}$ is not von Neumann and thereby does not generate a dual pair.

11.4 Dual pairs and symplectic leaf correspondence

As we saw in Theorem 11.1.9, there is a bijective correspondence between the symplectic leaves of the two Poisson manifolds forming the legs of a Lie–Weinstein dual pair with connected fibers. In this section we will see that the situation is analogous for the two Poisson varieties forming the legs of the dual pairs in the sense of Definition 11.3.1. Nevertheless, since in this context there is no Symplectic Foliation Theorem we need to start by defining what we mean by the symplectic leaves of a quotient Poisson space.

Let $(M, \{\cdot,\cdot\})$ be a smooth Poisson manifold, $A \subset \mathcal{P}(M)$ a subgroup of its Poisson diffeomorphism group, and $(M/A, \{\cdot,\cdot\}^{M/A})$ the associated quotient Poisson space. The bracket $\{\cdot,\cdot\}^{M/A}$ denotes the bilinear operation on the quotient presheaf $C^\infty_{M/A}$ defined by (10.2.2) out of the Poisson reducibility of the triple $(M, \{\cdot,\cdot\}, A)$.

Let $\pi_A : M \to M/A$ be the projection, $V \subset M/A$ an open subset of M/A, and $h \in C^\infty_{M/A}(V)$ a smooth function defined on it. Denoting $U := \pi_A^{-1}(V)$, the vector field $X_{h\circ\pi_A|_U}$ belongs to A' and, by part **(iii)** of Proposition 5.5.4, its flow $(F_t, \mathrm{Dom}(F_t))$ uniquely determines a local Poisson diffeomorphism $(\bar{F}_t, \pi_A(\mathrm{Dom}(F_t)))$ of M/A. We will say that $(\bar{F}_t, \pi_A(\mathrm{Dom}(F_t)))$ is the ***Hamiltonian flow*** associated to h. This terminology is justified by the fact that for any other function $f \in C^\infty_{M/A}(V)$ and any

$[m]_A \in \pi_A(\mathrm{Dom}(F_t)))$ we have

$$\frac{d}{dt} f(\bar{F}_t([m]_A)) = \{f, h\}_V^{M/A}(\bar{F}_t([m]_A)). \tag{11.4.1}$$

Nevertheless, the expression (11.4.1) does not fully characterize, in general, the flow $\bar{F}_t$ since there could be other mappings f for which such an equality holds. This could be rephrased by saying that in the category in which we are working, any function has an associated Hamiltonian flow but, unlike the situation encountered in the smooth Poisson category, its uniqueness is not guaranteed. The next result gives a convenient sufficient condition that guarantees uniqueness.

11.4.1 Proposition (SJAMAAR AND LERMAN (1991)) *Let $(M, \{\cdot, \cdot\})$ be a smooth Poisson manifold, let $A \subset \mathcal{P}(M)$ be a subgroup of its Poisson diffeomorphism group, and let $(M/A, \{\cdot, \cdot\}^{M/A})$ be the associated quotient Poisson space. If the functions in $C^\infty(M/A)$ separate the points of M/A, then the Hamiltonian flow $F_t : \mathrm{Dom}(F_t) \to F_t(\mathrm{Dom}(F_t))$ associated to a given function $h \in C^\infty_{M/A}(V)$ is the only Poisson morphism that satisfies* (11.4.1).

Recall that we say that the functions in $C^\infty(M/A)$ separate the points of M/A when the following condition is satisfied: if $x, y \in M/A$ are such that $f(x) = f(y)$ for all $f \in C^\infty(M/A)$, then $x = y$.

Proof. Let G_t be another Poisson morphism satisfying (11.4.1). Suppose, for simplicity, that $\mathrm{Dom}(F_t) = \mathrm{Dom}(G_t) = V$. The functions in $C^\infty_{M/A}(M/A)$ separate points. In order to show that F_t and G_t coincide we will prove that for any $f \in C^\infty_{M/A}(M/A)$, $\pi_A(m) \in V$, and any time t for which the expression $G_t(F_{-t}(\pi_A(m)))$ is defined, we have

$$f(G_t(F_{-t}(\pi_A(m)))) = f(\pi_A(m)).$$

This identity holds as a consequence of the following computation, in which we use the chain rule, the fact that F_{-t} is a Hamiltonian flow for the function $-h$, and that G_t is a Poisson morphism:

$$\begin{aligned}\frac{d}{dt} f(G_t(F_{-t}(\pi_A(m)))) &= \{f,\ h\}_V^{M/A}(G_t(F_{-t}(\pi_A(m))))\\ &\quad + \{f \circ G_t,\ -h \circ G_t\}_V^{M/A}(F_{-t}(\pi_A(m))) = 0. \quad \blacksquare\end{aligned}$$

We will define the symplectic leaves of M/A as the accessible sets from a point in this quotient via finite compositions of Hamiltonian flows. Since it is not clear how to define these flows by projection of A-equivariant flows when A is a pseudogroup of local transformations in $\mathcal{P}_L(M)$, we will restrict to the case $A \subset \mathcal{P}(M)$ in this section.

11.4.2 Definition *Let $(M, \{\cdot, \cdot\})$ be a smooth Poisson manifold, $A \subset \mathcal{P}(M)$ a subgroup of its Poisson diffeomorphism group and $(M/A, \{\cdot, \cdot\}^{M/A})$ the associated quotient Poisson structure. Given a point $[m]_A \in M/A$, the* ***symplectic leaf*** *$\mathcal{L}_{[m]_A}$ through it is defined as the (path connected) set formed by all the points that can be obtained* *from $[m]_A$ by applying to it a finite number of Hamiltonian flows associated to functions in $C^\infty_{M/A}(V)$, with $V \subset M/A$ any open subset of M/A, that is,*

$$\mathcal{L}_{[m]_A} := \{(F^1_{t_1} \circ F^2_{t_2} \circ \cdots \circ F^k_{t_k})([m]_A) \mid k \in \mathbb{N},\ F_{t_i} \text{ flow of some } X_{h_i},\ h_i \in C^\infty_{M/A}(V),\ V \subset M/A \text{ open}\}.$$

The relation being in the same symplectic leaf *determines an equivalence relation in M/A whose corresponding space of equivalence classes will be denoted by $(M/A)/\{\cdot,\cdot\}^{M/A}$.*

Even though in Definition 11.4.2 we called $\mathcal{L}_{[m]_A}$ a symplectic leaf, there is, in general, no natural way to define on this set a smooth structure and a symplectic form that would make it into a symplectic manifold. Nevertheless, there is still something we can do to justify our terminology. Indeed, consider the set $\mathcal{L}_{[m]_A}$ as a topological subspace of M/A endowed not with the quotient topology but with the strongest topology such that for any $k \in \mathbb{N}$, any $\mathcal{F}_T = F^1_{t_1} \circ \cdots \circ F^k_{t_k}$, with F_{t_i} the flow of some X_{h_i}, $h_i \in C^\infty_{M/A}(V)$, $V \subset M/A$ open, and any $z \in M$, the map

$$T \longmapsto \mathcal{F}_T(z)$$

defined on an open neighborhood of the origin in $\mathbb{R}^k$ and with values in M/A is continuous (see §3.2.2). We will refer to this topology on $\mathcal{L}_{[m]_A}$ as the ***symplectic leaf topology***. This topology is, in general, strictly stronger than the one obtained by considering $\mathcal{L}_{[m]_A}$ as a topological subspace of M/A endowed with the quotient topology.

Denote by $C^\infty_{\mathcal{L}_{[m]_A}}$ the presheaf of Whitney smooth functions on $\mathcal{L}_{[m]_A}$ induced by the quotient presheaf $C^\infty_{M/A}$. We recall that given any open subset $V \subset \mathcal{L}_{[m]_A}$ in the symplectic leaf topology, a function $f \in C^\infty_{\mathcal{L}_{[m]_A}}(V)$ if for any $z \in V$ there is a open neighborhood U_z of z in M/A (with the quotient topology) and a function $F \in C^\infty_{M/A}(U_z)$ such that

$$f|_{U_z \cap V} = F|_{U_z \cap V}.$$

The next result will show that the Poisson bracket $\{\cdot,\cdot\}$ on M naturally induces a bracket $\{\cdot,\cdot\}^{\mathcal{L}_{[m]_A}}$ on $\mathcal{L}_{[m]_A}$ with respect to which the triple $\left(\mathcal{L}_{[m]_A}, C^\infty_{\mathcal{L}_{[m]_A}}, \{\cdot,\cdot\}^{\mathcal{L}_{[m]_A}}\right)$ is a nondegenerate presheaf of Poisson algebras in the sense of Definition 10.2.1.

11.4.3 Theorem *Let $(M, \{\cdot,\cdot\})$ be a smooth Poisson manifold, $A \subset \mathcal{P}(M)$ a subgroup of its Poisson diffeomorphism group, and $(M/A, \{\cdot,\cdot\}^{M/A})$ the associated quotient Poisson structure. Let $[m]_A \in M/A$ and $\mathcal{L}_{[m]_A}$ the symplectic leaf through it. Then the triple $\left(\mathcal{L}_{[m]_A}, C^\infty_{\mathcal{L}_{[m]_A}}, \{\cdot,\cdot\}^{\mathcal{L}_{[m]_A}}\right)$ is a nondegenerate presheaf of Poisson algebras with the bracket defined by*

$$\begin{aligned}\{f,g\}^{\mathcal{L}_{[m]_A}}_V([z]_A) :&= \{F,G\}^{M/A}_{U_{[z]_A}}([z]_A)\\ &= \{F \circ \pi_A, G \circ \pi_A\}_{\pi_A^{-1}(U_{[z]_A})}(z),\end{aligned} \tag{11.4.2}$$

for any open set V in $\mathcal{L}_{[m]_A}$, $[z]_A \in V$, $f, g \in C^\infty_{\mathcal{L}_{[m]_A}}(V)$, and for any $F, G \in C^\infty_{M/A}(U_{[z]_A})$ such that $F|_{V \cap U_{[z]_A}} = f|_{V \cap U_{[z]_A}}$ and $G|_{V \cap U_{[z]_A}} = g|_{V \cap U_{[z]_A}}$, where $U_{[z]_A}$ is an open neighborhood of $[z]_A$ in M/A in the quotient topology.

Proof. In order to establish the first part of the Theorem it suffices to show that the bracket (11.4.2) is well defined or, more explicitly, that its value does not depend on the functions $F, G \in C^\infty_{M/A}(U_{[z]_A})$ used in its definition. Let $G' \in C^\infty_{M/A}(U_{[z]_A})$ be another function such that $G'|_{V\cap U_{[z]_A}} = g|_{V\cap U_{[z]_A}}$ and let $(F_t, \mathrm{Dom}(F_t))$ be the flow of the Hamiltonian vector field $X_{F\circ\pi_A}$. Suppose, for simplicity, that $\mathrm{Dom}(F_t) = \pi_A^{-1}(U_{[z]_A})$ and let $(\bar{F}_t, U_{[z]_A})$ be the Hamiltonian flow of F uniquely determined by the relation $\bar{F}_t \circ \pi_A|_{\pi_A^{-1}(U_{[z]_A})} = \pi_A \circ F_t|_{\pi_A^{-1}(U_{[z]_A})}$. Then,

$$
\begin{aligned}
\{F, G'\}^{M/A}_{U_{[z]_A}}([z]_A) &= \{F\circ\pi_A, G'\circ\pi_A\}_{\pi_A^{-1}(U_{[z]_A})}(z) \\
&= -\mathbf{d}(G'\circ\pi_A)(z)\left(X_{F\circ\pi_A}(z)\right) \\
&= -\left.\frac{d}{dt}\right|_{t=0}(G'\circ\pi_A)(F_t(z)) = \left.\frac{d}{dt}\right|_{t=0}(G'\circ\bar{F}_t)([z]_A) \\
&= -\left.\frac{d}{dt}\right|_{t=0}(G\circ\bar{F}_t)([z]_A) = \{F, G\}^{M/A}_{U_{[z]_A}}([z]_A),
\end{aligned}
$$

as required. Notice that the choice of topology for $\mathcal{L}_{[m]_A}$ allowed us to write for small enough t that $(G'\circ\bar{F}_t)([z]_A) = (G\circ\bar{F}_t)([z]_A)$ since for that range of t we have that $\bar{F}_t([z]_A) \in V\cap U_{[z]_A}$ and hence $(G'\circ\bar{F}_t)([z]_A) = g(\bar{F}_t([z]_A)) = (G\circ\bar{F}_t)([z]_A)$. Analogously, if we take another function $F' \in C^\infty_{M/A}(U_{[z]_A})$ such that $F'|_{V\cap U_{[z]_A}} = f|_{V\cap U_{[z]_A}}$ we have for any $[z]_A \in \mathcal{L}_{[m]_A}$, $\{F', G'\}^{M/A}_{U_{[z]_A}}([z]_A) = \{F, G\}^{M/A}_{U_{[z]_A}}([z]_A)$, which proves that the bracket $\{\cdot,\cdot\}^{\mathcal{L}_{[m]_A}}$ is well defined. The rest of the defining properties of a Poisson bracket for $\{\cdot,\cdot\}^{\mathcal{L}_{[m]_A}}$ are a straightforward verification and are automatically inherited from the bracket $\{\cdot,\cdot\}$.

We now show that the bracket $\{\cdot,\cdot\}^{\mathcal{L}_{[m]_A}}$ is nondegenerate. Let $f \in C^\infty_{\mathcal{L}_{[m]_A}}(V)$ be such that $\{f, g\}^{\mathcal{L}_{[m]_A}}_{U\cap V} = 0$ for all $g \in C^\infty_{\mathcal{L}_{[m]_A}}(U)$. Before we proceed, note that the definition of symplectic leaf implies that $\mathcal{L}_{[m]_A} = A'\cdot[m]_A$, where the symbol $A'\cdot[m]_A$ denotes the orbit of the point $[m]_A$ with respect to the action of A' on M/A explicitly written in Part **(iii)** of Proposition 5.5.4. By definition, any element $\mathcal{F}_T \in A'$ can be written as a finite composition of Hamiltonian flows associated to functions in $C^\infty_M(U)^A$, $U\subset M$ open and A-invariant; for the sake of simplicity we take $\mathcal{F}_T = F_t$, with $(F_t, \mathrm{Dom}(F_t))$ the flow of X_g, $g \in C^\infty_M(U)^A$, and $\mathrm{Dom}(F_t) = U$. Let $G \in C^\infty_{M/A}(\pi_A(U))$ be the function uniquely determined by the equality $g = G\circ\pi_A$, let $(\bar{F}_t, \pi_A(U))$ be the Hamiltonian flow of G defined by the relation $\bar{F}_t\circ\pi_A|_U = \pi_A\circ F_t$, and let $g' = G|_{\mathcal{L}_{[m]_A}} \in C^\infty_{\mathcal{L}_{[m]_A}}(\mathcal{L}_{[m]_A}\cap\pi_A(U))$. The definition of the presheaf of Whitney smooth functions $C^\infty_{\mathcal{L}_{[m]_A}}$ on $\mathcal{L}_{[m]_A}$ guarantees that around any point $\bar{F}_t([m]_A) \in \mathcal{L}_{[m]_A}$ there exists an open neighborhood $U_{\bar{F}_t([m]_A)} \subset M/A$ and a function $F \in C^\infty_{M/A}(U_{\bar{F}_t([m]_A)})$ such that $f|_{V\cap U_{\bar{F}_t([m]_A)}} = F|_{V\cap U_{\bar{F}_t([m]_A)}}$. Using these ingredients we can write

$$\begin{aligned}\frac{d}{dt}(f\circ\bar{F}_t)([m]_A) &= \frac{d}{dt}F(\bar{F}_t([m]_A)) = \frac{d}{dt}(F\circ\pi_A\circ F_t)(m)\\ &= \mathbf{d}(F\circ\pi_A)(F_t(m))\big(X_g(F_t(m))\big)\\ &= \{F\circ\pi_A, g\}_{U\cap\pi_A\left(U_{\bar{F}_t([m]_A)}\right)}(F_t(m))\\ &= \{F\circ\pi_A, G\circ\pi_A\}_{U\cap\pi_A\left(U_{\bar{F}_t([m]_A)}\right)}(F_t(m))\\ &= \{F,G\}^{M/A}_{\pi_A(U)\cap U_{\bar{F}_t([m]_A)}}(\bar{F}_t([m]_A))\\ &= \{f,g'\}^{\mathcal{L}_{[m]_A}}_{V\cap\mathcal{L}_{[m]_A}\cap\pi_A(U)}(\bar{F}_t([m]_A)) = 0,\end{aligned}$$

implying $(f\circ\bar{F}_t)([m]_A) = f([m]_A)$. Since $\bar{F}_t$ is arbitrary, we have $f\,(G_{\mathcal{A}'}\cdot[m]_A\cap V) = f(\mathcal{L}_{[m]_A}\cap V) = f([m]_A)$, which shows that the function $f\in C^\infty_{\mathcal{L}_{[m]_A}}(V)$ is constant, as required. ■

11.4.4 Theorem (Symplectic Leaf Correspondence) *Let $(M,\{\cdot,\cdot\})$ be a smooth Poisson manifold, $A, B\subset\mathcal{P}(M)$ two subgroups of its Poisson diffeomorphism group, and $A', B'\subset\mathcal{P}_L(M)$ the standard polar pseudogroups. Denote by $(M/A)/\{\cdot,\cdot\}^{M/A}$ and $(M/B)/\{\cdot,\cdot\}^{M/B}$ the space of symplectic leaves of the Poisson spaces $(M/A, \{\cdot,\cdot\}^{M/A})$ and $(M/B,\{\cdot,\cdot\}^{M/B})$, respectively. Then:*

(i) *The symplectic leaves of M/A and M/B are given by the orbits of the A' and B'-actions on M/A and M/B, respectively, as defined in Part* **(iii)** *of Proposition 5.5.4. Therefore*

$$(M/A)/\{\cdot,\cdot\}^{M/A} = (M/A)/A' \quad \textit{and} \tag{11.4.3}$$

$$(M/B)/\{\cdot,\cdot\}^{M/B} = (M/B)/B'. \tag{11.4.4}$$

(ii) *If the diagram $(M/A,\{\cdot,\cdot\}^{M/A}) \xleftarrow{\pi_A} (M,\{\cdot,\cdot\}) \xrightarrow{\pi_B} (M/B,\{\cdot,\cdot\}^{M/B})$ is a dual pair, then the map*

$$\begin{array}{ccc}(M/A)/\{\cdot,\cdot\}^{M/A} & \longrightarrow & (M/B)/\{\cdot,\cdot\}^{M/B}\\ \mathcal{L}_{[m]_A} & \longmapsto & \mathcal{L}_{[m]_B}\end{array} \tag{11.4.5}$$

is a bijection. The symbols $\mathcal{L}_{[m]_A}$ and $\mathcal{L}_{[m]_B}$ denote the symplectic leaves in M/A and M/B containing the points $[m]_A$ and $[m]_B$ respectively.

Proof. **(i)** This is a straightforward consequence of the definition of symplectic leaf and of the actions of the polar pseudogroups on the quotients described in Part **(iii)** of Proposition 5.5.4.
(ii) Recall that by Part **(ii)** of Proposition 5.5.4, A and A' (resp. B and B') commute. Thus, using the duality hypothesis, we can write

$$(M/A)/\{\cdot,\cdot\}^{M/A} = (M/A)/A' \simeq (M/A')/A = (M/B)/A, \tag{11.4.6}$$

and the same relation for the subgroup B, that is,

$$(M/B)/\{\cdot,\cdot\}^{M/B} = (M/B)/B' \simeq (M/B')/B = (M/A)/B. \tag{11.4.7}$$

In the previous expressions $(M/B)/A$ and $(M/A)/B$ should be understood as the orbit spaces of the A and B actions on M/B and M/A, respectively, inherited from considering these quotients as M/A' and M/B'. More explicitly, for any $a \in A$ and any $[m]_B \in M/B$ we define $a \cdot [m]_B := a \cdot [m]_{A'} = [a \cdot m]_{A'} = [a \cdot m]_B$. Analogously, for any $b \in B$ and any $[m]_A \in M/A$, we define $b \cdot [m]_A := [b \cdot m]_A$. With these conventions and in view of (11.4.6) and (11.4.7) the bijective character of the map in the statement will be proved if we show that the map

$$\begin{aligned} F : (M/B)/A &\longrightarrow (M/A)/B \\ [[m]_B]_A &\longmapsto [[m]_A]_B \end{aligned}$$

is a well defined bijection. It is indeed so since if $[[m]_B]_A = [[m']_B]_A$, there exist elements $a \in A$ and $b \in B$ such that $a \cdot m = b \cdot m'$ and hence $F([[m]_B]_A) = [[m]_A]_B = [[a \cdot m]_A]_B = [[b \cdot m']_A]_B = [b \cdot [m']_A]_B = [[m']_A]_B = F([[m']_B]_A)$, which shows that the map F is well defined. Analogously one shows that F is one to one and onto, as required. ■

11.4.5 Example. As a consequence of the previous theorem and Example 11.3.4 we can conclude that the symplectic leaves of two Poisson manifolds in the legs of a Lie–Weinstein dual pair $(P_1, \{\cdot,\cdot\}_{P_1}) \stackrel{\pi_1}{\leftarrow} (M,\omega) \stackrel{\pi_2}{\rightarrow} (P_2, \{\cdot,\cdot\}_{P_2})$ in which the projections π_1 and π_2 are complete (see §11.1.13) and have connected fibers are in bijective correspondence. Modulo the completeness hypothesis, this recovers Theorem 11.1.9. In the present context, the completeness condition shows up when formulating the Lie–Weinstein dual pair defining condition by making the identifications $P_1 \simeq M/\mathcal{A}_c$ and $P_2 \simeq M/\mathcal{B}_c$, with

$$\mathcal{A}_c = \operatorname{span}\left\{X_{f\circ\pi_2} \mid f \in C_c^\infty(P_2)\right\} \text{ and } \mathcal{B}_c = \operatorname{span}\left\{X_{f\circ\pi_1} \mid f \in C_c^\infty(P_1)\right\}.$$

The subscript c in $C_c^\infty(P_1)$ and $C_c^\infty(P_2)$ denotes compactly supported functions. The completeness of the projections π_1 and π_2 ensures that the vector fields that span $\mathcal{A}_c$ and $\mathcal{B}_c$ are complete and hence the pseudogroups of transformations associated to their flows are actual subgroups of $\mathcal{P}(M)$, as required in the hypotheses of Theorem 11.4.4. In Theorem 11.1.9 we saw that the completeness hypothesis, that appears very frequently in the literature associated to the Symplectic Leaf Correspondence Theorem, is actually not needed.

11.4.6 The Howe condition and leaf correspondence. Example 11.3.13 shows that Howe's condition is not enough to ensure symplectic leaf correspondence. Indeed, the remarks made in that example indicate that the Howe pair associated to the group $A_\mathbb{R}$ is $\mathbb{T}^2/A_\mathbb{R} \stackrel{\pi_{A_\mathbb{R}}}{\longleftarrow} \mathbb{T}^2 \stackrel{\text{id}}{\longrightarrow} \mathbb{T}^2$ (we use the notation introduced in Example 11.3.13). Now, the right leg of this pair has just one symplectic leaf (the entire two-torus $\mathbb{T}^2$) while, for the left leg, every point in $\mathbb{T}^2/A_\mathbb{R}$ is a symplectic leaf since $C^\infty(\mathbb{T}^2/A_\mathbb{R})$ consists of constant functions.

11.5 Hamiltonian Poisson subgroups

In this section we introduce several families of groups of Poisson automorphisms that provide Howe pairs and dual pairs. The reader should be aware that the terminology introduced in the next definition is not completely standard.

11.5.1 Definition *Let A be a subgroup of the Poisson automorphisms group $\mathcal{P}(M)$ of the Poisson manifold $(M, \{\cdot,\cdot\})$. Denote by $C^\infty(M)^A$ the set of A-invariant smooth functions on M and by $(C^\infty(M)^A)^c$ the centralizer of $C^\infty(M)^A$ with respect to the Poisson algebra induced by the bracket $\{\cdot,\cdot\}$ on $C^\infty(M)$.*

(i) *The subgroup A is* **strongly Hamiltonian** *when every element $g \in A$ can be written as $g = F^1_{t_1} \circ F^2_{t_2} \circ \cdots \circ F^k_{t_k}$, with $F^i_{t_i}$ the flow of a Hamiltonian vector field X_{h_i} associated to a function h_i in the centralizer $\left(C^\infty(M)^A\right)^c$.*

(ii) *The subgroup A is* **weakly Hamiltonian** *when for every element $g \in A$ and any $m \in M$ we can write $g \cdot m = (F^1_{t_1} \circ F^2_{t_2} \circ \cdots \circ F^k_{t_k})(m)$, with $F^i_{t_i}$ the flow of a Hamiltonian vector field X_{h_i} associated to a function $h_i \in \left(C^\infty(M)^A\right)^c$.*

(ii) *The subgroup A is* **tubewise strongly (resp. weakly) Hamiltonian** *if for every element $g \in A$ and any $m \in M$ there is an A-invariant neighborhood U of m such that we can write $g = F^1_{t_1} \circ F^2_{t_2} \circ \cdots \circ F^k_{t_k}$ (resp. $g \cdot m = (F^1_{t_1} \circ F^2_{t_2} \circ \cdots \circ F^k_{t_k})(m)$), with $F^i_{t_i}$ the flow of a Hamiltonian vector field X_{h_i} associated to a function $h_i \in \left(C^\infty(U)^A\right)^c$.*

11.5.2 Example: Connected Lie group actions with a standard momentum map are strongly Hamiltonian. Let G be a connected Lie group acting canonically on the Poisson manifold $(M, \{\cdot,\cdot\})$ via the map $\Phi : G \times M \to M$. Suppose that the G-action admits a standard momentum map $\mathbf{J} : M \to \mathfrak{g}^*$. Let $A_G \subset \mathcal{P}(M)$ be the subgroup of $\mathcal{P}(M)$ defined by $A_G := \{\Phi_g : M \to M \mid g \in G\}$. Then A_G is a Hamiltonian subgroup of $\mathcal{P}(M)$. Indeed, by the connectedness of A, every element $g \in G$ can be written as $g = \exp \xi_1 \cdot \ldots \cdot \exp \xi_n$, with $\xi_i \in \mathfrak{g}$ in the Lie algebra $\mathfrak{g}$ of G. Consequently, $\Phi_g = F^{\xi_1}_1 \circ F^{\xi_2}_1 \circ \cdots \circ F^{\xi_n}_1$, with $F^{\xi_i}_t$ the flow of $X_{\langle \mathbf{J}, \xi_i \rangle}$. But, by Noether's Theorem 4.5.11, $\langle \mathbf{J}, \xi_i \rangle \in \left(C^\infty(M)^G\right)^c$.

11.5.3 Example: A weakly and tubewise Hamiltonian group action that is not Hamiltonian. Let $M = S^1 \times S^1 = \mathbb{T}^2$ be the two-torus with the symplectic form $\omega = d\theta_1 \wedge d\theta_2$ given by its area form. Let $G = S^1$ acting canonically on M by $e^{i\phi} \cdot (e^{i\theta_1}, e^{i\theta_2}) := (e^{i(\phi+\theta_1)}, e^{i\theta_2})$ and let A_{S^1} be the associated subgroup of $\mathcal{P}(\mathbb{T}^2)$. It is easy to see that every S^1-invariant smooth function f can be written as $f(e^{i\theta_1}, e^{i\theta_2}) = g(e^{i\theta_2})$, with $g \in C^\infty(S^1)$. Its associated Hamiltonian vector field is given by $X_f = \frac{\partial g}{\partial \theta_2} \frac{\partial}{\partial \theta_1}$. With these remarks it is easy to see that A_{S^1} is weakly Hamiltonian, tubewise strongly Hamiltonian, but *not* strongly Hamiltonian.

11.5.4 In this section we study the properties of the diagrams $(M/A, \{\cdot,\cdot\}^{M/A}) \xleftarrow{\pi_A} (M, \{\cdot,\cdot\}) \xrightarrow{\pi_{A'}} (M/A', \{\cdot,\cdot\}^{M/A'})$ induced by weakly and strongly Hamiltonian subgroups $A \subset \mathcal{P}(M)$. Since we are dealing with actual subgroups of $\mathcal{P}(M)$, Proposition 5.5.4 guarantees the integrability of the polar distribution A' which we will not need to put as a hypothesis.

In Example 11.3.13 we identified a weakly Hamiltonian subgroup that induced a Howe pair. In our first result in this section, Proposition 11.5.5, we will show that this is not a coincidence since any weakly Hamiltonian subgroup endowed with the extension property always has an associated Howe pair. We also saw in that example that the (weak) Hamiltonian condition is not sufficient to generate a dual pair; in Proposition 11.5.6 we will show that if we add to the Hamiltonian hypothesis the property of separation of A-orbits, then we are guaranteed to obtain a dual pair.

11.5.5 Proposition *Let $(M, \{\cdot, \cdot\})$ be a Poisson manifold and $A \subset \mathcal{P}(M)$ be a weakly Hamiltonian subgroup of its Poisson diffeomorphism group. If A has the extension property, then the diagram $(M/A, \{\cdot, \cdot\}^{M/A}) \xleftarrow{\pi_A} (M, \{\cdot, \cdot\}) \xrightarrow{\pi_{A'}} (M/A', \{\cdot, \cdot\}^{M/A'})$ is a Howe pair.*

Proof. By Proposition 5.5.4 the polar distribution A' is always integrable in this case. The conclusions of Proposition 11.3.10 show that we just need to prove that $\left(\pi_{A'}^* C^\infty(M/A')\right)^c \subset \pi_A^* C^\infty(M/A)$. Hence, let $\phi \in A$ and $m \in M$ be arbitrary. Since, by hypothesis, the group A is weakly Hamiltonian, $\phi(m)$ can be written as $\phi(m) = (F_{t_1}^1 \circ F_{t_2}^2 \circ \cdots \circ F_{t_k}^k)(m)$, with $F_{t_i}^i$ the flow of a Hamiltonian vector field X_{h_i} associated to a function h_i in the centralizer $\left(C^\infty(M)^A\right)^c$. We assume, for the sake of simplicity, that $\phi(m) = F_t(m)$, with F_t the flow of X_h, $h \in \left(C^\infty(M)^A\right)^c$. By (11.3.6), the function h can be written as $h = g \circ \pi_{A'}$, with $g \in C^\infty(M/A')$. Now let $f \in \left(\pi_{A'}^* C^\infty(M/A')\right)^c$. Since $\{f, h\} = \{f, g \circ \pi_{A'}\} = 0$ we can conclude that $(f \circ \phi)(m) = (f \circ F_t)(m) = f(m)$. As we can reproduce this process for any $\phi \in A$ and $m \in M$ we have that $f \in C^\infty(M)^A = \pi_A^* C^\infty(M/A)$, as required. ■

11.5.6 Proposition *Let $(M, \{\cdot, \cdot\})$ be a Poisson manifold and $A \subset \mathcal{P}(M)$ a subgroup of its Poisson diffeomorphism group.*

(i) *If A is strongly (resp. weakly) Hamiltonian and has the extension property, then $A \subset A''$ (resp. $A \cdot m \subset A'' \cdot m$ for any $m \in M$).*

(ii) *If $C^\infty(M)^A$ separates the A-orbits on M, then $A'' \cdot m \subset A \cdot m$ for any $m \in M$.*

(iii) *If A is (strongly or weakly) Hamiltonian, has the extension property, and $C^\infty(M)^A$ separates the A-orbits on M, then it follows that A is von Neumann and the diagram $(M/A, \{\cdot, \cdot\}^{M/A}) \xleftarrow{\pi_A} (M, \{\cdot, \cdot\}) \xrightarrow{\pi_{A'}} (M/A', \{\cdot, \cdot\}^{M/A'})$ is a dual pair. Additionally, if A' has the extension property it is also a Howe pair.*

Proof. (i) Let $\phi \in A$ be arbitrary. Since A is strongly (resp. weakly) Hamiltonian, ϕ (resp. $\phi(m)$ for any $m \in M$) can be written as $\phi = F_{t_1}^1 \circ F_{t_2}^2 \circ \cdots \circ F_{t_k}^k$ (resp. $\phi(m) = (F_{t_1}^1 \circ F_{t_2}^2 \circ \cdots \circ F_{t_k}^k)(m)$), with $F_{t_i}^i$ the flow of a Hamiltonian vector field X_{h_i} associated to a function h_i in the centralizer $\left(C^\infty(M)^A\right)^c$. In order to keep the exposition simple we assume that $\phi = F_t$, with F_t the flow of X_h, $h \in \left(C^\infty(M)^A\right)^c$. By (11.3.6), the function h can be written as $h = l \circ \pi_{A'}$, with $l \in C^\infty(M/A')$. Consequently, $X_h = X_{l \circ \pi_{A'}}$ and hence $F_t = \phi \in A''$ (resp. $F_t(m) = \phi(m) \in A'' \cdot m$), as required.

(ii) Any element $\mathcal{F}_T \in A''$ can be written as a finite composition of Hamiltonian flows F_t associated to functions $f \circ \pi_{A'}|_U$, $f \in C^\infty(U/A')$, U an open A'-invariant set. Then, for any $h \in C^\infty(M)^A$ and any $m \in U$, we have that

$$\begin{aligned}\frac{d}{dt}h\left(F_t(m)\right) &= \{h|_U, f \circ \pi_{A'}|_U\}_U\left(F_t(m)\right)\\ &= -\mathbf{d}(f \circ \pi_{A'}|_U)\left(F_t(m)\right)\left(X_h\left(F_t(m)\right)\right) = 0,\end{aligned}$$

that is, any function $h \in C^\infty(M)^A$ is constant along the Hamiltonian flow of $f \circ \pi_{A'}|_U$. Now, since $C^\infty(M)^A$ separates the A-orbits on M, we can conclude that, for any point $m \in M$, the set $F_t(m)$ is included in a single A-orbit, namely, $F_t(m) \subset A \cdot m$ and therefore $A'' \cdot m \subset A \cdot m$, as required.

(iii) Parts **(i)** and **(ii)** imply in the present hypotheses that for any $m \in M$, $A \cdot m = A'' \cdot m''$ and, consequently, $M/A = M/A''$. This proves that A is von Neumann and therefore that the diagram $(M/A, \{\cdot,\cdot\}^{M/A}) \xleftarrow{\pi_A} (M, \{\cdot,\cdot\}) \xrightarrow{\pi_{A'}} (M/A', \{\cdot,\cdot\}^{M/A'})$ is a dual pair. Corollary 11.3.11 ensures that it is also a Howe pair in the presence of the extension property for A'. ■

11.6 Dual pairs induced by canonical Lie group actions

In this section we will analyze under what circumstances we can construct von Neumann and Howe pairs using the subgroups $A_G := \{\Phi_g \mid g \in G\}$ of the Poisson diffeomorphism group $\mathcal{P}(M)$ associated to the canonical action $\Phi : G \times M \to M$ of a Lie group G on a Poisson manifold $(M, \{\cdot,\cdot\})$. Recall that in this setup the polar distribution A'_G is always integrable (Proposition 5.5.4) and the projection onto the corresponding leaf space $\mathcal{J} : M \to M/A'_G$ is the optimal momentum map.

11.6.1 Theorem *Let G be a Lie group acting canonically and properly on the Poisson manifold $(M, \{\cdot,\cdot\})$ via the map $\Phi : G \times M \to M$. Let $A_G \subset \mathcal{P}(M)$ be the subgroup of $\mathcal{P}(M)$ defined by $A_G := \{\Phi_g : M \to M \mid g \in G\}$ and A'_G its polar. Let $\pi : M \to M/A_G$ be the canonical projection of M onto the quotient M/A_G and $\mathcal{J} : M \to M/A'_G$ the associated optimal momentum map. If A_G is (strongly or weakly) Hamiltonian, then it is von Neumann and therefore the diagram $(M/A_G, \{\cdot,\cdot\}^{M/A_G}) \xleftarrow{\pi} (M, \{\cdot,\cdot\}) \xrightarrow{\mathcal{J}} (M/A'_G, \{\cdot,\cdot\}^{M/A'_G})$ is a dual pair.*

Proof. This is a straightforward consequence of Propositions 11.5.6 and of the fact that by the properness of the G-action, A_G has the extension property and $C^\infty(M)^{A_G} = C^\infty(M)^G$ separates the G-orbits (see Propositions 2.5.6 and 2.3.8). ■

11.6.2 Corollary *In the same setup as in the previous theorem, if A_G is (strongly or weakly) Hamiltonian, then the G-orbits are connected and path connected.*

Proof. The condition on A_G being Hamiltonian implies, by the previous theorem, that A_G is von Neumann and therefore, for any $m \in M$, the orbit $G \cdot m$ equals $A'' \cdot m$ which is connected and path connected. ■

Theorem 11.6.1 shows that properness of a canonical G-action is a condition that, added to the Hamiltonian character, is sufficient to ensure that the corresponding transformation group $A_G \subset \mathcal{P}(M)$ is von Neumann. However, as the following example shows, this condition is not necessary.

11.6.3 The coadjoint action produces von Neumann subgroups of $\mathcal{P}(\mathfrak{g}^*)$. Let G be a connected Lie group, $\mathfrak{g}$ its Lie algebra, and $\mathfrak{g}^*$ its dual. Let $\{\cdot,\cdot\}_+$ be the +-Lie–Poisson bracket that makes $\mathfrak{g}^*$ into a Poisson manifold (see Example 4.1.13). The coadjoint action of G on $\mathfrak{g}^*$ is canonical and the identity is a standard associated momentum map.

We now check that A_G is von Neumann. Let $\mu \in \mathfrak{g}^*$ be arbitrary, $U \subset \mathfrak{g}^*$ be an open G-invariant neighborhood of the coadjoint orbit of the element μ, and $f \in C^\infty(U)^G$. Then for any $\xi \in \mathfrak{g}$ and $\rho \in U$ we have that

$$\begin{aligned}\langle X_f(\rho), \xi\rangle &= -\left\langle \mathrm{ad}^*_{\frac{\delta f}{\delta\rho}}\rho, \xi\right\rangle = -\left\langle \rho, \left[\frac{\delta f}{\delta \rho}, \xi\right]\right\rangle = \frac{d}{dt}\bigg|_{t=0}\left\langle \rho, \mathrm{Ad}_{\exp t\xi}\frac{\delta f}{\delta\rho}\right\rangle \\ &= \frac{d}{dt}\bigg|_{t=0}\left\langle \mathrm{Ad}^*_{\exp t\xi}\rho, \frac{\delta f}{\delta\rho}\right\rangle = \left\langle \mathrm{ad}^*_{\xi}\rho, \frac{\delta f}{\delta\rho}\right\rangle \\ &= \frac{d}{dt}\bigg|_{t=0} f\left(\mathrm{Ad}^*_{\exp t\xi}\rho\right) = 0,\end{aligned}$$

where the last equality follows from the G-invariance of the function f. This computation shows that $A'_G(\mu) = \{0\}$ for all $\mu \in \mathfrak{g}^*$. The connectedness of the group G automatically implies that $A''_G = A_G$ and therefore A_G is von Neumann.

The symplectic leaf correspondence for the legs of the diagram $\mathfrak{g}^*/G \leftarrow \mathfrak{g}^* \rightarrow \mathfrak{g}^*$, guaranteed in this case by Theorem 11.4.4, is a restatement of the fact, proved in Theorem 4.5.31, that the symplectic leaves of $(\mathfrak{g}^*, \{\cdot,\cdot\})$ are the coadjoint orbits.

11.6.4 Definition *Let G be a compact connected Lie group with Lie algebra $\mathfrak{g}$ acting canonically on the symplectic manifold (M,ω) via the map $\Phi : G \times M \rightarrow M$. Let $A_G \subset \mathcal{P}(M)$ be the subgroup of $\mathcal{P}(M)$ defined by $A_G := \{\Phi_g : M \rightarrow M \mid g \in G\}$. Let $\xi \in \mathfrak{g}$ and $T(\xi)$ be the torus defined by $T(\xi) := \overline{\{\exp t\xi \mid t \in \mathbb{R}\}}$. The element ξ is said to have an associated* **coisotropic torus** *if the orbits of the $T(\xi)$-action on M are coisotropic.*

11.6.5 Theorem *Let G be a compact connected Lie group with Lie algebra $\mathfrak{g}$ acting canonically on the symplectic manifold (M,ω) via the map $\Phi : G \times M \rightarrow M$. Let $A_G \subset \mathcal{P}(M)$ be the subgroup of $\mathcal{P}(M)$ defined by $A_G := \{\Phi_g : M \rightarrow M \mid g \in G\}$ and A'_G be its polar. Let $\pi : M \rightarrow M/A_G$ be the canonical projection of M onto the quotient M/A_G and $\mathcal{J} : M \rightarrow M/A'_G$ the associated optimal momentum map. Let $\mathbb{T}$ be a maximal torus of G and suppose that its Lie algebra $\mathfrak{t}$ has a basis $\{\xi_1, \ldots, \xi_k\}$ whose elements have associated coisotropic tori $T(\xi_i)$. Then, A_G is weakly Hamiltonian and von Neumann.*

Proof. Since the action of any compact group is always proper, according to Theorem 11.6.1, it suffices to prove that A_G is weakly Hamiltonian, which will be a consequence of the following lemma.

11.6.6 Lemma *Suppose that we are in the hypotheses Theorem 11.6.5. Then for any* $\xi \in \mathfrak{g}$ *that has a coisotropic torus* $T(\xi)$ *associated and any* $m \in M$ *there is a function* $f \in (C^\infty(M)^{A_G})^c$ *such that* $\exp \xi \cdot m = F_1(m)$, *where* F_t *is the flow of the Hamiltonian vector field* X_f.

Proof. Let $\xi_M \in \mathfrak{X}(M)$ be the infinitesimal generator vector field associated to the element $\xi \in \mathfrak{g}$. Since the action is canonical, we have

$$0 = \mathcal{L}_{\xi_M}\omega = \mathbf{i}_{\xi_M}\mathbf{d}\omega + \mathbf{d}\left(\mathbf{i}_{\xi_M}\omega\right) = \mathbf{d}\left(\mathbf{i}_{\xi_M}\omega\right),$$

that is, the one form $\alpha := \mathbf{i}_{\xi_M}\omega$ is closed. Consider now the subsets of G defined by $K := \{\exp t\xi \mid t \in \mathbb{R}\}$ and define $T(\xi) := \bar{K}$, where the bar over K means closure. As we already pointed out, the subset $T(\xi)$ is a closed connected Abelian subgroup of G and therefore a torus. Notice that for any $m \in M$ we have $T(\xi) \cdot m \subset \overline{K \cdot m}$; indeed, if $t \cdot m \in T(\xi) \cdot m$, there exists a sequence $\{k_n\} \subset K$ of elements in K such that $k_n \to t$, which implies that $k_n \cdot m \to t \cdot m$ and therefore $t \cdot m \in \overline{K \cdot m}$. Hence, since the restriction $\alpha|_{K \cdot m} = 0$ we have that $\alpha|_{\overline{K \cdot m}} = 0$, and therefore $\alpha|_{T(\xi)\cdot m} = 0$. By the Relative Poincaré Lemma 1.3.5 there exists a neighborhood U of $T(\xi) \cdot m$, which by the compactness of $T(\xi)$ can be chosen to be $T(\xi)$-invariant, and a function $h \in C^\infty(U)$ such that $\mathbf{d}h|_U = \alpha|_U$. This statement amounts to saying that the function $h \in C^\infty(U)$ is a momentum map for the canonical action of K on the symplectic manifold $(U, \omega|_U)$.

Now, by shrinking U if necessary and using the hypothesis on the coisotropic character of the torus $T(\xi)$, we can use the Symplectic Slice Theorem 7.4.1 to represent U by a $T(\xi)$-symplectic tube, that is,

$$U \cong T(\xi) \times_{T(\xi)_m} \mathfrak{m}_r^*,$$

where $\mathfrak{m}$ is an $\mathrm{Ad}_{T(\xi)_m}$-invariant complement to the Lie algebra $\mathrm{Lie}\,(T(\xi)_m)$ in $\mathfrak{k} := \{\eta \in \mathrm{Lie}\,(T(\xi)) \mid \eta_M(m) \in (\mathrm{Lie}\,(T(\xi)) \cdot m)^\omega\}$. The point m is represented in these coordinates by $[e, 0]$, and $\mathfrak{m}_r^*$ is a $T(\xi)_m$-equivariant ball of radius $r > 0$, small enough and centered at the origin of $\mathfrak{m}^*$. Let $\phi_r : \mathfrak{m}^* \to \mathbb{R}$ be a smooth, $T(\xi)_m$-invariant, and compactly supported function such that $\phi_r(\eta) = 0$, for any $\eta \in \mathfrak{m}^* \setminus \mathfrak{m}_r^*$, and $\phi_r(W) = 1$ for a $T(\xi)_m$-invariant neighborhood $W \subset \mathfrak{m}_r^*$. Let Φ be the $T(\xi)$-invariant function defined by

$$\begin{array}{rcl} \Phi : U \cong T(\xi) \times_{T(\xi)_m} \mathfrak{m}_r^* & \longrightarrow & \mathbb{R} \\ {}[k, \eta] & \longmapsto & \phi_r(\eta). \end{array}$$

Notice that Φ is zero off the open $T(\xi)$-invariant set U and therefore it can be trivially extended to a $T(\xi)$-invariant function on the entire space; this extended function will also be denoted by $\Phi \in C^\infty(M)^{T(\xi)}$. The reconstruction equations (7.7.17)–(7.7.19) applied to Φ (using the Abelian character of $T(\xi)$) imply that the Hamiltonian vector field X_Φ equals

$$X_\Phi(z) = \begin{cases} (D_{\mathfrak{m}^*}\phi_r)_M\,(m) \text{ if } & m \in U \\ 0 & \text{if } m \in M \setminus U. \end{cases}$$

Let $f = \Phi h$. Since $X_f = X_{\Phi h} = \Phi X_h + h X_\Phi$ and Φ is constant on the $T(\xi)$-invariant neighborhood $N \simeq T(\xi) \times_{T(\xi)_m} W$ around m, we have that $X_f(z) = X_h(z) = \xi_M(z)$

for any $z \in N$. Consequently, if F_t is the flow of the vector field X_f, it is clear that $\exp \xi \cdot m = F_1(m)$.

In order to finish the proof we just need to show that $f \in (C^\infty(M)^{A_G})^c$. This is indeed so because for any G-invariant function $l \in C^\infty(M)^{A_G}$ we have that $\{f, l\}(z) = \mathbf{d}f(z)\,(X_l(z)) = 0$ for any $z \in M \setminus U$. Also, if $z \in U$

$$\begin{aligned}\{f,l\}(z) &= \{\Phi h, l\}(z) = \Phi(z)\{h,l\}(z) + h(z)\{\Phi, l\}(z)\\ &= -\Phi(z)\,(\mathbf{d}l(z)(X_h(z))) - h(z)\,(\mathbf{d}l(z)(X_\Phi(z)))\\ &= -\Phi(z)\,(\mathbf{d}l(z)(\xi_M(z))) - h(z)\,\big(\mathbf{d}l(z)((D_{\mathfrak{m}^*}\phi_r)_M(z))\big) = 0,\end{aligned}$$

due to the G-invariance of l. Hence $\{f, l\} = 0$ for any $l \in C^\infty(M)^{A_G}$ and, consequently, $f \in (\pi^* C^\infty(M/A_G))^c$, as required. ▼

We conclude the proof of the theorem by noting that since the group G is compact and connected, any element $g \in G$ can be written as $g = hlh^{-1}$, with $l \in \mathbb{T}$. As $\mathbb{T}$ is Abelian and connected, there exist real numbers $t_1, \ldots, t_k$ such that $l = \exp t_1\xi_1 \cdots \exp t_k\xi_k$. Hence, for any $m \in M$ we can write

$$\begin{aligned}g \cdot m &= h \exp t_1\xi_1 \cdots \exp t_k\xi_k h^{-1} \cdot m\\ &= h \exp t_1\xi_1 h^{-1} \cdots h \exp t_k\xi_k h^{-1} \cdot m\\ &= \exp t_1(\mathrm{Ad}_h\xi_1) \cdots \exp t_k(\mathrm{Ad}_h\xi_k) \cdot m.\end{aligned}$$

A straightforward computation shows that $T(\mathrm{Ad}_h(\xi_i)) = hT(\xi_i)h^{-1}$ and hence the coisotropy of the torus $T(\xi_i)$ implies that of $T(\mathrm{Ad}_h(\xi_i))$. Therefore, by the previous lemma we have that $g \cdot m = (F_1^1 \circ \cdots \circ F_1^k)(m)$, with each F_t^i the flow of a Hamiltonian vector field X_{f_i} associated to a function $f_i \in (\pi^* C^\infty(M/A_G))^c$. ■

The reader may be wondering if the coisotropy hypothesis in the statement of Theorem 11.6.5 is not just a technical requirement appearing in the proof of Lemma 11.6.6 that could be eliminated by using different techniques. The following example, due to J. Montaldi and T. Tokieda, shows that this is not the case, that is, in the absence of additional hypotheses, compactness and connectedness of the Lie group G acting symplectically do not suffice to ensure that the corresponding transformation group A_G is von Neumann.

11.6.7 Example: A compact and connected canonical group action that is not von Neumann. Let $M := \mathbb{T}^2 \times \mathbb{T}^2$ be the product of two-tori whose elements will be denoted by the four-tuples $(e^{i\theta_1}, e^{i\theta_2}, e^{i\psi_1}, e^{i\psi_2})$. Endow M with the symplectic structure ω defined by $\omega := \mathbf{d}\theta_1 \wedge \mathbf{d}\theta_2 + \sqrt{2}\,\mathbf{d}\psi_1 \wedge \mathbf{d}\psi_2$. Consider the canonical circle action given by $e^{i\phi} \cdot (e^{i\theta_1}, e^{i\theta_2}, e^{i\psi_1}, e^{i\psi_2}) := (e^{i(\theta_1+\phi)}, e^{i\theta_2}, e^{i(\psi_1+\phi)}, e^{i\psi_2})$. This action does not satisfy the coisotropy hypothesis and, as we will now verify, the associated transformation group $A_{S^1} \subset \mathcal{P}(M)$ is not von Neumann. Indeed, the set $C^\infty(M)^{S^1}$ comprises the functions f of the form $f \equiv f(e^{i\theta_2}, e^{i\psi_2}, e^{i(\theta_1-\psi_1)})$. An inspection of the Hamiltonian flows associated to such functions readily shows that the leaves of A'_{S^1} fill densely the manifold M. This implies that $C^\infty(M)^{A'_{S^1}}$ consists only of constant functions and therefore $A''_{S^1}(m) = \{0\}$, for all $m \in M$. Consequently, A_{S^1} is not von Neumann.

Notice that this example shows that the polar of a regular integrable distribution, even though it is integrable, it is not, in general, regular. More specifically, even though the projection $\pi_{A_{S^1}} : M \to M/A_{S^1}$ is a surjective submersion, this is not true for the projection $\pi_{A'_{S^1}} : M \to M/A'_{S^1}$.

11.6.8 Tubewise Hamiltonian actions and dual pairs. In Section 7.5 we carried out an in depth study of the conditions under which a proper canonical action of a connected Lie group G on a symplectic manifold (M, ω) is strongly tubewise Hamiltonian. More specifically, in that section we explained how, in the presence of certain hypotheses, for any point $m \in M$ there is an open G-invariant neighborhood of its orbit such that the restriction of the G-action to this neighborhood admits a standard momentum map, thus implying that the action is strongly tubewise Hamiltonian. The following question arises thus naturally: when is a strongly tubewise Hamiltonian action weakly Hamiltonian and therefore induces a dual pair? The result below provides some partial answers.

11.6.9 Theorem *Let G be a connected Lie group with Lie algebra $\mathfrak{g}$ acting canonically and properly on the symplectic manifold (M, ω) via the map $\Phi : G \times M \to M$. Let $A_G \subset \mathcal{P}(M)$ be the subgroup of $\mathcal{P}(M)$ defined by $A_G := \{\Phi_g : M \to M \mid g \in G\}$ and let A'_G be its polar. Let $\pi : M \to M/A_G$ be the canonical projection of M onto the quotient M/A_G and $\mathcal{J} : M \to M/A'_G$ the associated optimal momentum map. For any $m \in M$ let $\mathfrak{k}_m \subset \mathfrak{g}$ be the Lie subalgebra of $\mathfrak{g}$ defined by $\mathfrak{k}_m = \{\eta \in \mathfrak{g} \mid \eta_M(m) \in (\mathfrak{g} \cdot m)^{\omega}\}$, denote by $K_m \subset G$ the (unique) connected Lie subgroup with Lie algebra $\mathfrak{k}_m$, and let $\gamma_m \in \Omega^1(G; \mathfrak{g}^*)$ be the G-equivariant, $\mathfrak{g}^*$-valued one form defined by*

$$\langle \gamma_m(g) \left(T_e L_g(\eta)\right), \xi \rangle := -\omega(m) \left(\left(\mathrm{Ad}_{g^{-1}} \xi\right)_M (m), \eta_M(m) \right),$$

for any $g \in G$, $\xi, \eta \in \mathfrak{g}$. Suppose that for any $m \in M$, the following conditions hold:

- *the orbit $G \cdot m$ is coisotropic,*
- *there exists an $\mathrm{Ad}(K_m)$-invariant complement to $\mathfrak{k}_m$ in $\mathfrak{g}$,*
- *$\mathfrak{k}_m$ is Abelian, and*
- *γ_m is exact (which happens for instance when $H^1(G) = 0$ or when the orbit $G \cdot m$ is isotropic).*

Then A_G is weakly Hamiltonian and von Neumann.

Proof. We will show that the hypotheses imply a conclusion similar to that of Lemma 11.6.6, that is, we will prove that for any $\xi \in \mathfrak{g}$ and any $m \in M$, there exists a function $f \in (C^\infty(M)^{A_G})^c$ such that $\exp \xi \cdot m = F_1(m)$, where F_t is the flow of the Hamiltonian vector field X_f. Indeed, for a fixed $m \in M$, the exactness of γ_m guarantees, by Proposition 7.5.2, that there exists a G-invariant neighborhood U of the orbit $G \cdot m$ on which the restriction of the G-action admits a standard momentum map. Consequently, for any $\xi \in \mathfrak{g}$ we have $\exp \xi \cdot m = F_1(m)$, with F_t the flow of the Hamiltonian vector field in U associated to the ξ-component of the tubular momentum map on U.

The proof imitates that of Lemma 11.6.6. First, by shrinking U if necessary, we can represent it by a normal form coordinate chart around the point m, that is, we can assume without loss of generality that $U \cong G \times_{G_m} \mathfrak{m}_r^*$, where $\mathfrak{m}$ is an Ad_{G_m}-invariant complement to the Lie algebra $\mathrm{Lie}\,(G_m)$ in $\mathfrak{k}_m$. The point m is represented in these coordinates by $[e, 0]$, and $\mathfrak{m}_r^*$ is an open G_m-equivariant ball of radius $r > 0$, small enough and centered at the origin of $\mathfrak{m}^*$. Let $\phi_r : \mathfrak{m}^* \to \mathbb{R}$ be a smooth G_m-invariant compactly supported function such that $\phi_r(\eta) = 0$, for any $\eta \in \mathfrak{m}^* \setminus \mathfrak{m}_r^*$, and $\phi_r(W) = 1$ on a G_m-invariant neighborhood $W \subset \mathfrak{m}_r^*$. Let Φ be the G-invariant function defined by

$$\begin{aligned} \Phi : U \cong G \times_{G_m} \mathfrak{m}_r^* &\longrightarrow \quad \mathbb{R} \\ [k, \eta] \qquad &\longmapsto \phi_r(\eta). \end{aligned}$$

Notice that Φ is zero off the open G-invariant set U and therefore can be trivially extended to a G-invariant function on the entire space, also denoted by $\Phi \in C^\infty(M)^G$. The reconstruction equations (7.7.17)–(7.7.19) applied to Φ imply that the Hamiltonian vector field X_Φ equals

$$X_\Phi(z) = \begin{cases} (D_{\mathfrak{m}^*}\phi_r)_M\,(m) \text{ if } & m \in U \\ \qquad 0 & \text{if } m \in M \setminus U. \end{cases} \tag{11.6.1}$$

Indeed, the hypothesis on the existence of an $\mathrm{Ad}(K_m)$-invariant complement to $\mathfrak{k}_m$ in $\mathfrak{g}$ implies that the map F in (7.7.12) reduces to $F(\xi, \lambda, \tau) = \mathbb{P}_{\mathfrak{q}^*}\left(\mathrm{ad}_\tau^* \lambda\right) + \langle \tau, \cdot \rangle_{\mathfrak{q}}$, whose unique solution for τ is $\tau \equiv 0$. This implies that the map $\psi \equiv 0$ and therefore the reconstruction equation (7.7.18) vanishes since, by hypothesis, the Lie algebra $\mathfrak{k}_m$ is Abelian. This proves (11.6.1).

Let $f = \Phi h$. Since $X_f = X_{\Phi h} = \Phi X_h + h X_\Phi$ and Φ is constant on the G-invariant neighborhood $N \simeq G \times_{G_m} W$ around m, we have that $X_f(z) = X_h(z) = \xi_M(z)$ for any $z \in N$. Consequently, if F_t is the flow of the vector field X_f, it is clear that $\exp \xi \cdot m = F_1(m)$. In order to finish the proof we just need to show that $f \in (C^\infty(M)^{A_G})^c$. This is indeed so because for any G-invariant function $l \in C^\infty(M)^{A_G}$ we have $\{f, l\}(z) = \mathbf{d}f(z)\,(X_l(z)) = 0$ for any $z \in M \setminus U$. Also, if $z \in U$

$$\begin{aligned} \{f, l\}(z) &= \{\Phi h, l\}(z) = \Phi(z)\{h, l\}(z) + h(z)\{\Phi, l\}(z) \\ &= -\Phi(z)\,(\mathbf{d}l(z)\,(X_h(z))) - h(z)\,(\mathbf{d}l(z)\,(X_\Phi(z))) \\ &= -\Phi(z)\,(\mathbf{d}l(z)\,(\xi_M(z))) - h(z)\left(\mathbf{d}l(z)\left((D_{\mathfrak{m}^*}\phi)_M\,(z)\right)\right) = 0, \end{aligned}$$

due to the G-invariance of l. Hence $\{f, l\} = 0$ for any $l \in C^\infty(M)^{A_G}$ and hence $f \in (\pi^* C^\infty(M/A_G))^c$, as required. ■

11.6.10 Symplectic leaf correspondence in dual pairs induced by canonical Lie group actions. In Corollary 3.3.6 we proved that the generalized distribution spanned by a family of equivariant vector fields with respect to a proper Lie group action is complete. In particular, this result implies that the polar distribution A'_G corresponding to the canonical and proper action of a Lie group G on the Poisson manifold $(M, \{\cdot, \cdot\})$ is complete and therefore the quotient space M/A'_G can be written as the orbit space relative to the action on M of a subgroup of $\mathcal{P}(M)$. This consideration, that in principle seems rather technical, gains importance when we recall the definition of the symplectic leaves (Definition 11.4.2) and the Symplectic Leaf Correspondence Theorem (Theorem 11.4.4) where we saw that all these objects are well behaved when we deal with the quotients of M by *genuine subgroups* of $\mathcal{P}(M)$.

Since the symplectic leaves of M/A_G are obviously given by the orbit reduced spaces $\mathcal{J}^{-1}(\mathcal{O}_\rho)/A_G$ and those of M/A'_G are well defined for proper symplectic actions, the Symplectic Leaf Correspondence Theorem 11.4.4 combined with Theorems 11.6.1, 11.6.5, and 11.6.9 prove that in a variety of situations the symplectic leaves of M/A'_G are given by the polar reduced spaces $\mathcal{J}^{-1}(\mathcal{O}_\rho)/A'_G$. We summarize those situations in the following corollary.

11.6.11 Corollary *Let G be a Lie group that acts canonically and properly on the symplectic manifold (M, ω) and let $A_G \subset \mathcal{P}(M)$ be the associated Poisson diffeomorphisms subgroup. Suppose that at least one of the following conditions is satisfied:*

(i) *A_G is (weakly or strongly) Hamiltonian,*

(ii) *the Lie group G is compact and connected, and the tori of A_G are coisotropic,*

(iii) *for any point $m \in M$, the orbit $G \cdot m$ is coisotropic, there exists an $\mathrm{Ad}(K_m)$-invariant complement to $\mathfrak{k}_m$ in $\mathfrak{g}$, $\mathfrak{k}_m$ is Abelian, and γ_m is exact (see Theorem 11.6.9 for these notations).*

Then the map

$$\begin{array}{ccc} (M/A_G)/\{\cdot,\cdot\}^{M/A_G} & \longrightarrow & (M/A'_G)/\{\cdot,\cdot\}^{M/A'_G} \\ \mathcal{J}^{-1}(\mathcal{O}_\rho)/A_G & \longmapsto & \mathcal{J}^{-1}(\mathcal{O}_\rho)/A'_G \end{array}$$

establishes a bijection between the symplectic leaves of $(M/A_G, \{\cdot,\cdot\}^{M/A_G})$ and those of $(M/A'_G, \{\cdot,\cdot\}^{M/A'_G})$ that, in any of these situations, coincide with the optimal orbit reduced spaces and the polar reduced spaces, respectively.

The following diagram represents all the spaces that we worked with in the context of singular dual pairs induced by canonical Lie group actions and optimal reduction. The part of the diagram dealing with the regularized spaces refers only to the situation in which M is symplectic.

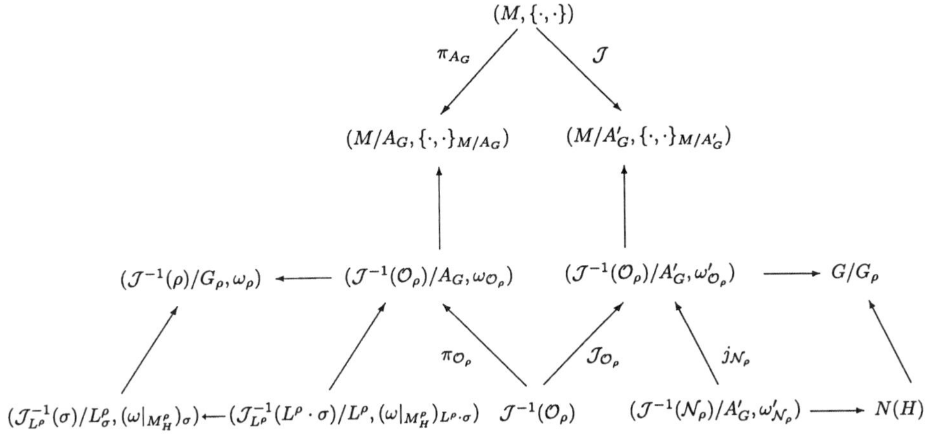
$(M, \{\cdot,\cdot\})$
π_{A_G}
$\mathcal{J}$
$(M/A_G, \{\cdot,\cdot\}_{M/A_G})$
$(M/A'_G, \{\cdot,\cdot\}_{M/A'_G})$
$(\mathcal{J}^{-1}(\rho)/G_\rho, \omega_\rho)$
$(\mathcal{J}^{-1}(\mathcal{O}_\rho)/A_G, \omega_{\mathcal{O}_\rho})$
$(\mathcal{J}^{-1}(\mathcal{O}_\rho)/A'_G, \omega'_{\mathcal{O}_\rho})$
G/G_ρ
$\pi_{\mathcal{O}_\rho}$
$\mathcal{J}_{\mathcal{O}_\rho}$
$j_{\mathcal{N}_\rho}$
$(\mathcal{J}_{L^\rho}^{-1}(\sigma)/L^\rho_\sigma, (\omega|_{M^\rho_H})_\sigma)$
$(\mathcal{J}_{L^\rho}^{-1}(L^\rho \cdot \sigma)/L^\rho, (\omega|_{M^\rho_H})_{L^\rho \cdot \sigma})$
$\mathcal{J}^{-1}(\mathcal{O}_\rho)$
$(\mathcal{J}^{-1}(\mathcal{N}_\rho)/A'_G, \omega'_{\mathcal{N}_\rho})$
$N(H)$

Bibliography

Abraham, R., and Marsden, J.E. [1978] *Foundations of Mechanics*. Second edition, Addison-Wesley.

Abraham, R., Marsden, J.E., and Ratiu, T.S. [1988] *Manifolds, Tensor Analysis, and Applications. Applied Mathematical Sciences*, Vol. 75. Springer-Verlag.

Abraham, R., and Robbin, J. [1967] *Transversal Mappings and Flows*. W. A. Benjamin, Inc.

Abud, M., and Sartori, G. [1981] The geometry of orbit–space and natural minima of Higgs potentials. *Physics Letters*, **104B**(2), 147–152.

Abud, M., and Sartori, G. [1983] The geometry of spontaneous symmetry breaking. *Annals of Physics*, **150**, 307–372.

Aharonov, Y. and Anandan, J. [1987] Phase change during acyclic quantum evolution. *Phys. Rev. Lett.*, **58**, 1593–1596.

Anandan, J. [1987] Geometric angles in quantum and classical physics. *Phys. Lett. A*, **129**, 201–207.

Albert, C. [1988], Théorèmes de réduction pour les variétés cosymplectiques. In: *Action hamiltoniennes de groupes. Troisième théorème de Lie (Lyon, 1986). Travaux en Cours*, Vol. 27. Hermann, Paris, pp. 1–12.

Albert, C. [1989] Le théorème de réduction de Marsden–Weinstein en géométrie cosymplectique et de contact. *J. Geom. Physics*, **6**, 627–649.

Albert, C. and Dazord, P. [1988] Théorie des groupoïdes symplectiques. Chapitre I. Théorie générale des groupoïdes de Lie. *Travaux du Séminaire Sud Rhodanien de Géométrie*, II, 51–105, *Publ. Dép. Math. Nouvelle Sér. B*, **88**(4), Univ. Claude-Bernard, Lyon.

Albert, C. and Dazord, P. [1990] Théorie des groupoïdes symplectiques. Chapitre I. Groupoïdes symplectiques. *Publ. Dép. Math. Nouvelle Sér. B*, 27–99, Univ. Claude-Bernard, Lyon.

Alekseev, A. Y. [1997] On Poisson actions of compact Lie groups on symplectic manifolds. *J. Differential Geom.* **45**, 241–256.

Alekseev, A., Malkin, A., and Meinrenken, E. [1998] Lie group valued momentum maps. *J. Differential Geom.*, **48**, 445–495.

Alekseev, A., Meinrenken, E., and Woodward, C. [2000] Group-valued equivariant localization. *Invent. Math.*, **140**(2), 327–350.

Alekseev, A., Meinrenken, E., and Woodward, C. [2001] The Verlinde formulas as fixed point formulas. *J. Symplectic Geom.*, **1**(1), 1–46.

Ambrose, W. and Singer, I. M. [1953] A theorem on holonomy. *Trans. Amer. Math. Soc.*, **75**, 428–443.

Arms, J.M., Cushman, R., and Gotay, M.J. [1991] A universal reduction procedure for Hamiltonian group actions. In: *The Geometry of Hamiltonian Systems*, (T.S. Ratiu ed.) Springer-Verlag, pp. 33–51.

Arms, J.M., Fischer, A.E., and Marsden, J.E. [1975] Une approche symplectique pour des théorèmes de la décomposition en géométrie ou relativité générale. *C. R. Acad. Sci. Paris Sér. I Math.*, **281**, 517–520.

Arms, J.M., Gotay, M., and Jennings, G. [1990] Geometric and algebraic reduction for singular momentum maps. *Adv. in Math.*, **79**, 43–103.

Arms, J.M., Marsden, J.E., and Moncrief, V. [1981] Symmetry and bifurcations of momentum mappings. *Comm. Math. Phys.*, **78**, 455–478.

Arnold, V. I. [1965] Conditions for nonlinear stability of the stationary plane curvilinear flows of an ideal fluid. *Dokl. Mat. Nauk SSSR*, **162**, 773–777.

Arnold, V.I. [1966a] Sur la géometrie différentielle des groupes de Lie de dimension infinie et ses applications à l'hydrodynamique des fluids parfaits. *Ann. Ins. Fourier, Grenoble*, **16**, 319–361.

Arnold, V. I. [1966b] On an a priori estimate in the theory of hydrodynamical stability. *Izv. Vyssh. Uchebn. Zaved. Mat. Nauk*, **54**, 3–5. English Translation: Amer. Math. Soc. Transl. **79** (1969), 267–269.

Arnold, V.I. [1989] *Mathematical Methods of Classical Mechanics*. Second edition, *Graduate Texts in Mathematics*, Vol, 60, Springer-Verlag.

Arnold, V.I., Kozlov, V.V., and Neishtadt, A.I. [1988] *Mathematical Aspects of Classical and Celestial Mechanics. Encyclopedia of Mathematical Sciences*, Vol. 3. Springer-Verlag.

Artin, E. [1963] *Geometric Algebra*. Interscience, New York.

Atiyah, M. F. [1982] Convexity and commuting hamiltonians. *Bull. London Math. Soc.,* **14**, 1–15.

Atiyah, M. F. and Bott, R. [1982] The Yang–Mills equations over Riemann surfaces. *Phil. Trans. R. Soc. Lond. A*, **308**, 523–615.

Atiyah, M. and N. J. Hitchin [1985] Low-energy scattering of non-Abelian magnetic monopoles. *Phil. Trans. Roy. Soc. Lond. A* **315**, 459–469.

Atiyah, M. and Pressley, A. N. [1982] Convexity and Loop Groups. In: *Arithmetic and geometry, Vol. II*, Prog. Math., **36**, Birkhäuser Boston, pp. 33–63.

Audin, M. [1991] *The Topology of Torus Actions on Symplectic Manifolds*. Prog. Math., **93**, Birkhäuser Boston.

Avron, J., Sadun, L., Segert, J., and Simon, B. [1989] Chern numbers and Berry's phases in Fermi systems. *Comm. Math. Phys.*, **124**, 595–627.

Baer, A. [1929] Zur Einführung des Scharbegriffs. *J. Reine Angew. Math.*, **160**, 199–207.

Balan, R. [1994] A note about the integrability of distributions with singularities. *Bolletino della Unione Matematica Italiana*, **7**(8-A), 335–344.

Barre, R. [1973] De quelques aspects de la théorie des q-variétés différentielles et analytiques. *Ann. Inst. Fourier*, **23**(3), 227–312.

Bartsch, T. [1994] *Topological Methods for Variational Problems with Symmetries*. Lecture Notes in Mathematics, Vol. 1560. Springer.

Bates, L. [1998] Examples of singular nonholonomic reduction. *Rep. Math. Phys.*, 231–247.

Bates, L. [2002] Problems and progress in nonholonomic reduction. *Rep. Math. Phys.* **49**, 143–149. XXXIII Symposium on Mathematical Physics (Torún, 2001).

Bates, L. [1998] Examples of singular nonholonomic reduction. *Rep. Math. Phys.*, **42**(1/2), 231–247.

Bates, L., Graumann, H., and MacDonnell, C. [1996] Examples of gauge conservation laws in nonholonomic systems. *Rep. Math. Phys.*, **37**(3), 295–308.

Bates, L. and Lerman, E. [1997] Proper group actions and symplectic stratified spaces. *Pacific J. Math.*, **181**(2), 201–229.

Bates, L. and Sniatycki, J. [1993] Nonholonomic reduction. *Rep. Math. Phys.*, **32**(1), 99–115.

Battaglia, F. [1999] Circle actions and Morse theory on quaternion-Kähler manifolds. *J. London Math. Soc. (2)* **59**, 345–358.

Bayen, F., Flato, M., Fronsdal, C., Lichnerowicz, A., and Sternheimer, D. [1978] Deformation theory and quantization I. Deformations of symplectic structures. *Ann. Physics*, **111**(1), 61–110.

Beltrame, P., Chossat, P., and Laure, P. [1997] Convection en double diffusion dans une coque sphérique. *C. R. Acad. Sci. Paris Sér. I Math.*, **325**, 1049–1052.

Benenti, S. [1983] The category of symplectic reductions. Proceedings of the international meeting *Geometry and Physics*. M. Modugno (editor). Pitagora Editrice, Bologna, pp. 11–41.

Benenti, S. and Tulczyjew, W. M. [1982] Remarques sur les réductions symplectiques. *C. R. Acad. Sci. Paris*, **294**, 561–564.

Berry, M. [1984] Quantal phase factors accompanying adiabatic changes. *Proc. Royal Soc. London A*. **392**, 43–57.

Berry, M. [1985] Classical adiabatic angles and quantal adiabatic phase. *J. Phys. A: Math. Gen.*. **18**, 15–27.

Berry, M. [1988] The geometric phase. *Scientific American*, December.

Berry, M. and Hannay, J. [1988] Classical non-adiabatic angles. *J. Phys. A: Math. Gen.*, **21**, 325–333.

Besse, A. [1987] *Einstein Manifolds*. Ergebnisse der Mathematik und ihrer Grenzgebiete, Dritte Folge, **3**. Springer-Verlag, Berlin.

Bielawski, R. [1999a] Twistor quotients of hyper-Kähler manifolds. In: *Quaternionic Structures in Mathematics and Physics (Rome, 1999)*, number 2002e: 53070 (electronic). Univ. Studi Roma "La Sapienza", Rome, pp. 7–21.

Bielawski, R. [1999b] Complete hyper-Kähler $4n$-manifolds with a local tri-Hamiltonian $\mathbf{R}^n$-action. *Math. Ann.* **314**, 505–528.

Bielawski, R. [1999c] Complete hyper-Kähler $4n$-manifolds with a local tri-Hamiltonian $\mathbf{R}^n$-action. *Math. Ann.* **314**, 505–528.

Bielawski, R. and Dancer, A.S. [2000] The geometry and topology of toric hyperkähler manifolds. *Comm. Anal. Geom.* **8**, 727–760.

Bierstone, E. [1975] Lifting isotopies from orbit spaces. *Topology*, **14**, 245–252.

Bierstone, E. [1980] *The Structure of Orbit Spaces and the Singularities of Equivariant Mappings*. Monografias de Matemática, Instituto de Matemática Pura e Aplicada, Vol. 35. Rio de Janeiro.

Biquard, O. and Gauduchon, P. [1997] Géométrie hyperkählérienne des espaces hermitiens symétriques complexifiés. In: *Séminaire de Théorie Spectrale et Géométrie, Vol. 16, Année 1997–1998, Sémin. Théor. Spectr. Géom.*, Vol. 16. Univ. Grenoble I, Saint-Martin-d'Hères, pp. 127–173.

Biquard, O. and Gauduchon, P. [1998] Constructions de métriques l.c.K. par quotient symplectique tordu. Unpublished notes.

Birtea, P. [2002] Private communication.

Birtea, P. [2003] *Poisson Geometry, Momentum Maps, and Applications*. Ph.D. thesis, Ecole Polytechnique Fédérale de Lausanne.

Birtea, P., Ortega, J.-P., and Ratiu, T.S. [2003] A unified approach to the convexity of momentum maps. *In preparation.*

Birtea, P., Puta, M., Ratiu, T.S., and Tudoran, R. [2003] Bifurcation of relative equilibria for torus actions. *Preprint.*

Blaom, A. [2000] Reconstruction phases via Poisson reduction. *Differential Geom. Appl.*, **12**(3), 231–252.

Blaom, A. [2001] A geometric setting for Hamiltonian perturbation theory. *Memoirs of the Amer. Math. Soc.*, **153**(727).

Blankenstein, G. [2000] *Implicit Hamiltonian Systems: Symmetry and Interconnection.* Ph.D. Thesis, University of Twente, The Netherlands, November 2000.

Blankenstein, G. and Ratiu, T.S. [2003] Singular reduction of implicit Hamiltonian systems. *Rep. Math. Phys.*, to appear.

Blankenstein, G. and van der Schaft, A.J. [2001] Symmetry and reduction in implicit generalized Hamiltonian systems. *Rep. Math. Phys.*, **47**(1), 57–100.

Bloch, A. [2003] *Nonholonomic Mechanics and Control. Interdisciplinary Applied Mathematics*, Vol. 24. Springer-Verlag.

Bloch, A.M., Brockett, R., and Ratiu, T.S. [1992] Completely integrable gradient flows. *Comm. Math. Phys.*, **147**, 57–74.

Bloch, A.M., Flaschka, H., and Ratiu, T.S. [1993] A Schur–Horn–Kostant convexity theorem for the diffeomorphism group of the annulus. *Invent. Math.*, **113**, 511–529.

Bloch, A. M., Krishnaprasad, P.S., Marsden, J. E., and Murray, R. M. [1996] Nonholonomic mechanical systems with symmetry. *Arch. Rational Mech. Anal.*, **136**, 21–99.

Bloch, A. M., P. S. Krishnaprasad, J. E. Marsden, and T. S. Ratiu [1994] Dissipation induced instabilities. *Ann. Inst. H. Poincaré, Analyse Nonlinéaire* **11**, 37–90.

Bloch, A.M., Krishnaprasad, P.S., Marsden, J.E., and Ratiu, T.S. [1996] The Euler–Poincaré equations and double bracket dissipation. *Comm. Math. Phys.*, **175**, 1–42.

Bobenko, A.I. and Suris, Yu.B. [1999] Discrete time Lagrangian mechanics on Lie groups, with an application to the Lagrange top, *Comm. Math. Phys.*, **204**(1), 147–188.

Bogoyavlenskij, O.I. [1996a] A concept of integrability of dynamical systems. *C. R. Math. Rep. Acad. Sci. Canada*, **18**(4), 163–168.

Bogoyavlenskij, O.I. [1996b] Necessary conditions for existence of non-degenerate Hamiltonian structures. *Comm. Math. Phys.*, **182**(2), 253–289.

Bogoyavlenskij, O.I. [1996c] Theory of tensor invariants of integrable Hamiltonian systems. I. Incompatible Poisson structures. *Comm. Math. Phys.*, **180**(3), 529–586.

Bogoyavlenskij, O.I. [1996d] The A-B-C-cohomologies for dynamical systems. *C. R. Math. Rep. Acad. Sci. Canada*, **18**(5), 199–204.

Bogoyavlenskij, O.I. [1997] Theory of tensor invariants of integrable Hamiltonian systems. II. Theorem on symmetries and its applications. *Comm. Math. Phys.*, **184**(2), 301–365.

Bogoyavlenskij, O.I. [1998a] Extended integrability and bi-Hamiltonian systems. *Comm. Math. Phys.*, **196**(1), 19–51.

Bogoyavlenskij, O.I. [1998b] Necessary and sufficient conditions for two concepts of nondegeneracy in the sense of Poincaré of integrable Hamiltonian systems. *Dokl. Akad. Nauk*, **360**(3), 295–298.

Bogoyavlenskij, O.I. [1998c] Conformal symmetries of dynamical systems and Poincaré 1892 concept of iso-energetic non-degeneracy. *C. R. Acad. Sci. Paris Sér. I Math.*, **326**(2), 213–218.

Bojowald, M. and Strobl, T. [2002] Poisson geometry in constrained systems. *Preprint*. hep-th/0112074.

Bolsinov, A.V. and Fomenko, A.T. [2000] *Integrable Geodesic Flows on Two-Dimensional Surfaces*. Monographs in Contemporary Mathematics. Consultants Bureau, New York.

Bott, R., Tolman, S., and Weitsman, J. [2003] Surjectivity for Hamiltonian loop group spaces. *Preprint*, math.DG/0210036.

Bourbaki, N. [1971] *Variétés différentielles et analytiques. Fascicule de résultats*. Hermann.

Bourbaki, N. [1989a] *General Topology. Chapters 1–4*. Springer-Verlag.

Bourbaki, N. [1989b] *Lie Groups and Lie Algebras. Chapters 1–3*. Springer-Verlag.

Bourguignon, J.P. [1975] Une stratification de l'espace des structures riemanniennes. *Comp. Math.*, **30**, 1–41.

Boyer, C.P. and Galicki, K. [1999] 3-Sasakian manifolds. In: *Surveys in differential geometry: essays on Einstein manifolds*, number 2001m:53076 in Surv. Differ. Geom., VI, Int. Press, Boston, MA, pp. 123–184.

Boyer, C.P. and Galicki, K. [2000] On Sasakian–Einstein geometry. *Internat. J. Math.* **11**, 873–909.

Boyer, C.P., Galicki, K., and Mann, B.M. [1993] Quaternionic reduction and Einstein manifolds. *Comm. Anal. Geom.* **1**, 229–279.

Boyer, C.P., Galicki, K., and Mann, B.M. [1994] The geometry and topology of 3-Sasakian manifolds. *J. Reine Angew. Math.* **455**, 183–220.

Boyer, C.P., Galicki, K., and Mann, B.M. [1998a] A note on smooth toral reductions of spheres. *Manuscripta Math.* **95**, 149–158.

Boyer, C.P., Galicki, K., and Mann, B.M. [1998b] Hypercomplex structures from 3-Sasakian structures. *J. Reine Angew. Math.* **501**, 115–141.

Boyer, C.P., Galicki, K., and Piccinni, P. [2002] 3-Sasakian geometry, nilpotent orbits, and exceptional quotients. *Ann. Global Anal. Geom.* **21**, 85–110.

Brandt, W. [1926] Über eine Verallgemeinerung des Gruppenbegriffes. *Math. Ann.*, **96**, 360–366.

Bredon, G.E. [1972] *Introduction to Compact Transformation Groups*. Academic Press.

Brickell, F. and Clark, R.S. [1970] *Differentiable Manifolds, An Introduction*. Van Nostrand Reinhold Company.

Bröcker, Th. and Lander, L. [1975] *Differentiable Germs and Catastrophes*. *London Mathematical Society Lecture Note Series*, Vol. 17. Cambridge University Press.

Bröcker, Th. and tom Dieck, T. [1985] *Representations of Compact Lie Groups*. *Graduate Text in Mathematics*, Vol. 98. Springer-Verlag.

Burghelea, D., Albu, A., and Ratiu, T. S. [1975] *Compact Lie Group Actions* (In Romanian). *Monografii Matematice*, Vol. 5. Universitatea Timisoara.

Calabi, E. [1970] On the group of automorphisms of a symplectic manifold. In: *Problems in Analysis*, Princeton University Press, 1–26.

Cannas da Silva, A. and Weinstein, A. [1999] *Geometric Models for Noncommutative Algebras*. *Berkeley Mathematics Lecture Note Series*. American Mathematical Society.

Cantrijn, F., de León, M., Marrero, J.C., and de Diego, D.M. [1998] Reduction of nonholonomic mechanical systems with symmetries. *Rep. Math. Phys.*, **42**(1/2), 25–45.

Cartan, É. [1904] Sur la structure des groupes infinis de transformation. *Ann. Ec. Norm. Sup.*, **21**, 153–206.

Cartan, É. [1905] Sur la structure des groupes infinis de transformation. *Ann. Ec. Norm. Sup.*, **22**, 219–308.

Cartan, É. [1908] Les sous-groupes des groupes continus des groupes continus de transformation. *Ann. Ec. Norm. Sup.*, **25**, 57–194.

Cartan, É. [1909] Les groupes continus, infinis, simples. *Ann. Ec. Norm. Sup.*, **26**, 93–161.

Cartan, É. [1922] *Leçons sur les invariants intégraux*. Hermann, Paris.

Castrillón-López, M., García Pérez, P.L., and Ratiu, T.S. [2001] Euler–Poincaré reduction on principal bundles. *Lett. Math. Phys.*, **58**, 167–180.

Castrillón-López, M. and Marsden, J.E. [2003] Some remarks on Lagrangian and Poisson reduction for field theories. *J. Geom. and Physics*, **48**(1), 52–83.

Castrillón-López, M. and Ratiu, T.S. [2003] Reduction in principal bundles: covariant Lagrange–Poincaré equations. *Comm. Math. Phys.*, **236**(2), 223–250.

Castrillón-López, M., Ratiu, T.S., and Shkoller, S. [2000] Reduction in principal fiber bundles: covariant Euler-Poincaré equations, *Proc. Amer. Math. Soc.*, **128**(7), 2155–2164.

Cattaneo, A. and Felder, G. [2001] Poisson sigma models and symplectic groupoids. In: *Quantization of Singular Symplectic Quotients*, N. P. Landsman, M. Pflaum, and M. Schlichenmaier, editors. Prog. Math., **198**. Birkhäuser, Basel, 2001, pp. 61–93.

Cendra, H., Holm, D.D., Marsden, J.E., and Ratiu, T.S. [1998] Lagrangian reduction, the Euler–Poincaré equations, and semidirect products. *Amer. Math. Soc. Translations*, **186**, 1–25.

Cendra, H., Marsden, J.E., and Ratiu, T.S. [2001a] Lagrangian reduction by stages. *Memoirs of the Amer. Math. Soc.*, **152**(722).

Cendra, H., Marsden, J.E., and Ratiu, T.S. [2001b] Geometric mechanics, Lagrangian reduction, and nonholonomic systems. *Mathematics Unlimited – 2001 and Beyond*, B. Engquist and W. Schmid, eds., Springer-Verlag, New York, 221–273.

Chernoff, P.R. and Marsden, J.E. [1974], *Properties of Infinite Dimensional Hamiltonian systems. Lecture Notes in Math.*, **425**. Springer, New York.

Choquet-Bruhat, Y. and DeWitt-Morette, C. [1982] *Analysis, Manifolds, and Physics*. Revised edition. North Holland.

Chossat, P. [1986] Bifurcation secondaire de solutions quasi-périodiques dans un problème de bifurcation invariant par symétrie $O(2)$. *C. R. Acad. Sci. Paris Sér. I Math.*, **302**, 539–541.

Chossat, P. and Dias, F. [1995] The 1:2 resonance with O(2) symmetry and its applications in hydrodynamics. *J. Nonlin. Sci.*, **5**, 105–129.

Chossat, P. and Golubitsky, M. [1988] Iterates of maps with symmetry. *SIAM J. of Math. Anal.*, **19**, 1259–1270.

Chossat, P. and Iooss, G. [1994] *The Couette-Taylor problem. Applied Mathematical Sciences*, Vol. 102. Springer-Verlag.

Chossat, P., Koenig, M., and Montaldi, J. [1995] Bifurcation générique d'ondes d'isotropie maximale. *C. R. Acad. Sci. Paris Sér. I Math.*, **320**, 25–30.

Chossat, P. and Lauterbach, R. [1999] *Methods in equivariant bifurcation and dynamical systems and their applications*. Nonlinear Studies Series, World Scientific, Singapore.

Chossat, P., Lewis, D.K., Ortega, J.-P., and Ratiu, T.S. [2003] Bifurcation of relative equilibria in mechanical systems with symmetry. *Advances in Applied Math.*, **31**, 10–45.

Chossat, P., Ortega, J.-P., and Ratiu, T.S. [2002] Hamiltonian Hopf bifurcation with symmetry. *Arch. Rat. Mech. Anal.*, **163**, 1–33; **167**, 83–84.

Chow, W.L. [1939] Über Systeme von linearen partiellen Differentialgleichungen erster Ordnung. *Math. Ann.*, **117**, 98–105.

Chu, R.Y. [1975] Symplectic homogeneous spaces. *Trans. Amer. Math. Soc.*, **197**, 145–159.

Collins, G. [1974] Quantifier elimination for real closed fields by cylindrical algebraic decomposition. *Lect. Notes in Comp. Sci.*, Vol. 33, 134–183.

Condevaux, M., Dazord, P., and Molino, P. [1988] Géometrie du moment. Travaux du Séminaire Sud-Rhodanien de Géométrie, I, *Publ. Dép. Math. Nouvelle Sér. B* **88-1**, Univ. Claude-Bernard, Lyon, 1988, 131–160.

Costé, A., Dazord, P., and Weinstein, A. [1987] Groupoïdes symplectiques. *Publ. Dép. Math. Nouvelle Sér. A*, **2**, 1–62. Univ. Claude-Bernard, Lyon.

Courant, T. [1990] Dirac manifolds, *Trans. American Math. Soc.*, **319**, 631–661.

Courant, T. and Weinstein, A. [1986] Beyond Poisson structures. In: *Action hamiltoniennes de groupes. Troisième théorème de Lie (Lyon, 1986).* Travaux en Cours, **27**, Hermann, Paris, 1988, pp. 39–49.

Crainic, M. and Loja Fernandes, R. [2000] Integrability of Lie brackets. *Ann. of Math.*, **157**(2), 575–620.

Crainic, M. and Moerdijk, I. [2000] Foliation groupoids and their cyclic homology. *Adv. Math.*, **157**(2), 177–197.

Cushman, R.H. and Bates, L.M. [1997] *Global Aspects of Classical Integrable Systems.* Birkhäuser Verlag.

Cushman, R. and Sniatycki, J. [2001] Differential structure of orbit spaces. *Canad. J. Math.*, **53**(4), 715–755.

Cushman, R. and Sniatycki, J. [2002], Nonholonomic reduction for free and proper actions. *Regul. Chaotic Dyn.* **7**, 61–72.

Dalsmo, M. and van der Schaft, A.J. [2003] On representations and integrability of mathematical structures in energy-conserving physical systems. *SIAM J. Opt. Cont.*, **37**(1), 54–91.

Dancer, A.S. [1999], Hyper-Kähler manifolds. In: *Surveys in differential geometry: essays on Einstein manifolds*, number 2001k:53087 in Surv. Differ. Geom., VI. Int. Press, Boston, MA, pp. 15–38.

Dazord, P. [1985] Feuilletages à singularités. *Nederl. Akad. Wetensch. Indag. Math.*, **47**, 21–39.

Dazord, P. and Delzant, T. [1987] Le problème général des variables action–angle. *J. Diff. Geom.*, **26**, 223–251.

De León, M. and Saralegi, M. [1993] Cosymplectic reduction for singular momentum maps. *J. Phys. A* **26**, 5033–5043.

De León, M. and Tuynman, G.M. [1996] A universal model for cosymplectic manifolds. *J. Geom. Phys.* **20**, 77–86.

Dellnitz, M., Melbourne, I., and Marsden, J.E. [1992] Generic Bifurcation of Hamiltonian vector fields with symmetry. *Nonlinearity*, **5**, 979–996.

Dellnitz, M. and Melbourne, I. [1993] The equivariant Darboux Theorem. *Lectures in Appl. Math.*, **29**, 163–169.

Delzant, T. [1988] Hamiltoniens périodiques et images convexes de l'application moment. *Bull. Soc. Math. France*, **116**, 315–339.

Derks, G., Lewis, D.K., and Ratiu, T.S. [1995] Approximations with curves of relative equilibria in Hamiltonian systems with dissipation. *Nonlinearity*, **8**, 1087–1113.

Dirac, P.A.M. [1950] Generalized Hamiltonian mechanics. *Canad. J. Math.*, **2**, 129–148.

Dirac, P.A.M. [1964] *Lectures on Quantum Mechanics*. Academic Press.

Donaldson, S.K. [1999], Moment maps and diffeomorphisms. *Asian J. Math.* **3**, 1–15.

Dorfman, I. [1993] *Dirac Structures and Integrability of Nonlinear Evolution Equations*. John Wiley, Chichester, 1993.

Dragomir, S. and Ornea, L. [1997] *Locally Conformal Kähler Geometry*. Prog. Math., **155**, Birkhäuser, Boston.

Drăgulete, O., Ornea, L., and Ratiu, T.S. [2003] Reduction of cosphere bundles. *Journ. Sympl. Geom.*, to appear.

Duistermaat, J.J. [1980] On global action angle coodinates. *Comm. Pure Appl. Math.*, **33**, 687–706.

Duistermaat, J.J. [1983] Convexity and tightness for restrictions of Hamiltonian functions to fixed point sets of an antisymplectic involution. *Trans. Amer. Math. Soc.*, **275**(1), 417–429.

Duistermaat, J.J. [1984] On the similarity between the Iwasawa projection and the diagonal part. *Bull. Soc. Math. Fr., 2e série*, Mémoire **15**, 129–138.

Duistermaat, J.J. and Heckman, G.J. [1982] On the variation of the cohomology of the symplectic form of the reduced phase space. *Invent. Math.*, **69**, 259–268. Addendum to: "On the variation in the cohomology of the symplectic form of the reduced phase space". *Invent. Math.*, **72**(1), 153–158.

Duistermaat, J.J. and Kolk, J.A. [1999] *Lie Groups*. Universitext, Springer-Verlag.

Eliasson, L.H. [1990] Normal forms for Hamiltonian systems with Poisson commuting integrals —elliptic case. *Comment. Math. Helv.*, **65**(1), 4–35.

Evens, S. and Lu, J.H. [2001] On the variety of Lagrangian subalgebras. *Ann. Ecol. Norm. Sup. (4)*, **34**(5), 631–668.

Fassò, F. [1996] The Euler–Poinsot top: a non-commutatively integrable system without global action-angle coordinates. *Journal of Applied Mathematics and Physics (ZAMP)*, **47**, 953–976.

Fassò, F. [1999] *Notes on Finite Dimensional Integrable Hamiltonian Systems. Preprint*.

Fassò, F. and Giacobbe, A. [2002] Geometric structure of "broadly integrable" Hamiltonian systems. *J. Geom. Phys.*, **44**(2–3), 156–170.

Fassò, F. and Ratiu, T.S. [1998] Compatibility of symplectic structures adapted to noncommutatively integrable systems. *J. Geom. Phys.*, **27**(3–4), 199–220.

Field, M.J. [1980] Equivariant dynamical systems. *Trans. Amer. Math. Soc.*, **259**(1), 185–205.

Field, M.J. [1982] On the structure of a class of equivariant maps. *Bull. Austral. Math. Soc.*, **26**, 161–180.

Field, M.J. [1983] Isotopy and stability of equivariant diffeomorphisms. *Proc. London Math. Soc.*, **46**, 487–516.

Field, M.J. [1991] Local structure of equivariant dynamics. In: *Singularity Theory and its Applications*. M. Roberts and I. Stewart eds. *Lecture Notes in Mathematics*, Vol. 1463. Springer-Verlag.

Field, M.J. [1994] Blowing-up in equivariant bifurcation theory, Dynamics, Bifurcation, and Symmetry. In: *New Trends and New Tools*, Proceedings of the E.B.T.G. Conference. P. Chossat ed. NATO ASI Series, Series C, nol. 437. Kluwer Academic Publishers, pp. 111–122.

Field, M.J. and Richardson, R.W. [1989] Symmetry breaking and the maximal isotropy subgroup conjecture for reflection groups. *Arch. Rational Mech. Anal.*, **105**, 61–94.

Fischer, A. [1970] A theory of superspace. In: *Relativity*. M. Carmelli et al., eds. Plenum Publishing.

Fischer, A., Marsden, J.E., and Moncrief, V. [1980a] Symmetry breaking in general relativity. In: *Essays in General Relativity*, essay number 7. Academic Press.

Fischer, A., Marsden, J.E., and Moncrief, V. [1980b] The structure of the space of solutions of Einstein's equations. I. One Killing field. *Ann. Inst. Henri Poincaré*, Section A, **33**(2), 147–194.

Flaschka, H. and Ratiu, T.S. [1996] A convexity theorem for Poisson actions of compact Lie groups, *Ann. École Normale Supérieure (4)*, **29**(6), 787–809.

Fomenko, A.T. [1991a] Topological classification of integrable systems. Edited by A.T. Fomenko. Translated from the Russian. *Advances in Soviet Mathematics*, **6**. American Mathematical Society, Providence, RI.

Fomenko, A.T. [1991b] Topological classification of all integrable Hamiltonian differential equations of general type with two degrees of freedom. In: *The Geometry of Hamiltonian Systems* (Berkeley, CA, 1989), Math. Sci. Res. Inst. Publ., **22**, Springer-Verlag, New York, pp. 131–339.

Fomenko, A.T. and Trofimov, V.V. [1988] *Integrable Systems on Lie Algebras and Symmetric Spaces*. Translated from the Russian by A. Karaulov, P. D. Rayfield and A. Weisman. Advanced Studies in Contemporary Mathematics, **2**. Gordon and Breach Science Publishers, New York.

Fong, U. and Meyer, K.R. [1975] Algebras of integrals. *Rev. Colombiana Mat.*, **9**, 75–90.

Foth, P. [2002], Tetraplectic structures, tri-momentum maps, and quaternionic flag manifolds. *J. Geom. Phys.* **41**, 330–343.

Frankel, T. [1959] Fixed points on Kähler manifolds. *Ann. Math.*, **70**, 1–8.

Freeman, M. [1984] Fully integrable Pfaffian systems. *Ann. of Math.*, **119**, 465–510.

Frohlich, C. [1979] Do springboard divers violate angular momentum conservation? *Am. J. Phys.*, **47**, 583–592.

Futaki, A. [1988] *Kähler-Einstein metrics and integral invariants*. Lecture Notes in Mathematics, Vol. 1314. Springer-Verlag, Berlin.

Galicki, K. [1987], A generalization of the momentum mapping construction for quaternionic Kähler manifolds. *Comm. Math. Phys.* **108**, 117–138.

Galicki, K. and Lawson, H.B. [1988], Quaternionic reduction and quaternionic orbifolds. *Math. Ann.* **282**, 1–21.

Geiges, H. [1997] Constructions of contact manifolds. *Math. Proc. Cambridge Philos. Soc.*, **121**, 455–464.

Gibson, C., Wirthmüller, K., Du Plessis, A.A., and Looijenga, E. [1976] Topological stability of smooth mappings. *Springer Lecture Notes in Mathematics*, **553**. Springer-Verlag.

Gini, R., Ornea, L., and Parton, M. [2003] Conformal Kähler reduction. *Preprint*, arXiv: Math.D.G. 0208208

Ginzburg, V. [1992] Some remarks on symplectic actions of compact groups. *Math. Z.*, **210**(4), 625–640.

Ginzburg, V., Guillemin, V., and Karshon, Y. [1999] Assignments and abstract moment maps. *J. Diff. Geometry*, **52**, 259–301.

Ginzburg, V., Guillemin, V., and Karshon, Y. [2002] *Moment Maps, Cobordism, and Hamiltonian Group Actions*. Math. Surv. and Monographs, **98**. Amer. Math. Soc.

Ginzburg, V. and Lu, J.H. [1992] Poisson cohomology of Morita–equivalent Poisson manifolds. *Duke Math. J.*, **68**, A199–A205.

Ginzburg, V. and Weinstein, A. [1992] Lie–Poisson structure on some Poisson Lie groups. *J. Amer. Math. Soc.*, **5**, 445–453.

Gleason A. [1950] Spaces with a compact Lie group of transformations. *Proc. Amer. Math. Soc.*, **1**, 35–43.

Goldman, W.M. [1984] The symplectic nature of the fundamental groups of surfaces. *Adv. Math.*, **54**, 200–225.

Goldstein, H. [1980] *Classical Mechanics*. Second edition, Addison-Wesley.

Golin, S., Knauf, A., and Marmi, S. [1989] The Hannay angles: geometry, adiabaticity, and an example. *Comm. Math. Phys.*, **123**, 95–122.

Golubitsky, M., Marsden, J.E., Stewart, I., and Dellnitz, M. [1995] The constrained Liapunov–Schmidt procedure and periodic orbits. In: *Normal Forms and Homoclinic Chaos*. Langford, W. F. and Nagata, W. eds. Fields Institute Communications, **4**, pp. 81–127

Golubitsky, M. and Schaeffer, D.G. [1985] *Singularities and Groups in Bifurcation Theory: Vol. I*. Vol. 51 in *Applied Mathematical Sciences*. Springer-Verlag.

Golubitsky, M. and Stewart, I. With an appendix by J.E. Marsden. [1987] Generic bifurcation of Hamiltonian systems with symmetry. *Physica D*, **24**. 391–405.

Golubitsky, M. and Stewart, I. [2002] *The Symmetry Perspective. From Equilibrium to Chaos in Phase Space and Physical Space. Prog. Math.*, Vol. 200. Birkhäuser, Basel.

Golubitsky, M., Stewart, I., and Schaeffer, D.G. [1985] *Singularities and Groups in Bifurcation Theory: Vol. II. Applied Mathematical Sciences*, Vol. 69. Springer.

Goresky, M.R. and MacPherson, R. [1988] *Stratified Morse Theory. Ergebnisse der Mathematik und ihrer Grenzgebiete (3)*, Vol. 14. Springer-Verlag, Berlin.

Gotay, M.J. and Tuynman, G.M. [1991] A symplectic analogue of the Mostow–Palais Theorem. *Symplectic geometry, groupoids, and integrable systems* (Berkeley, CA, 1989), *Math. Sci. Res. Inst. Publ.*, **20**. Springer-Verlag, pp. 173–182.

Gozzi, E. and Thacker, W.D. [1987a] Classical adiabatic holonomy in a Grassmannian system. *Phys. Rev. D*, **35**, 2388–2396.

Gozzi, E. and Thacker, W.D. [1987b] Classical adiabatic holonomy and its canonical structure. *Phys. Rev. D*, **35**, 2398–2406.

Greub, W., Halperin, S., and Vanstone, R. [1970a] *Connections, Curvature, and Cohomology: Vol. I.* Pure and applied mathematics, Vol. 47-I. Academic Press.

Greub, W., Halperin, S., and Vanstone, R. [1970b] *Connections, Curvature, and Cohomology: Vol. II.* Pure and applied mathematics, Vol. 47-II. Academic Press.

Grantcharov, G. and Ornea, L. [2001], Reduction of Sasakian manifolds. *J. Math. Phys.* **42**, 3809–3816.

Grantcharov, G., Papadopoulos, G. and Poon, Y.S. [2002], Reduction of HKT-structures. *J. Math. Phys.* **43**, 3766–3782.

Grantcharov, G. and Poon, Y.S. [2000], Geometry of hyper-Kähler connections with torsion. *Comm. Math. Phys.* **213**, 19–37.

Guedira, F. and Lichnerowicz, A. [1984] Géométrie des algèbres de Lie locales de Kirillov. *J. Math. Pures Appl.*, **63**, 407–484.

Guichardet, A. [1984] On notation and vibration motions of molecules. *Ann. Inst. H. Poincaré*, **40**, 329–342.

Guillemin, V., Lerman, E., and Sternberg, S. [1996] *Symplectic Fibrations and Multiplicity Diagrams*. Cambridge University Press.

Guillemin, V. and Sternberg, S. [1980] The moment map and collective motion. *Ann. Physics,* **127**, 220–253.

Guillemin, V. and Sternberg, S. [1982] Convexity properties on the moment mapping. *Invent. Math.*, **67**, 491–513.

Guillemin, V. and Sternberg, S. [1984a] Convexity properties on the moment mapping. II. *Invent. Math.*, **77**, 533–546.

Guillemin, V. and Sternberg, S. [1984b] A normal form for the moment map. In: *Differential Geometric Methods in Mathematical Physics*. S. Sternberg ed. Mathematical Physics Studies, 6. D. Reidel Publishing Company.

Guillemin, V. and Sternberg, S. [1984c] *Symplectic Techniques in Physics*. Cambridge University Press.

Guillemin, V. and Sternberg, S. [1984d] Multiplicity-free spaces. *J. Differential Geometry,* **19**, 31–56.

Guruprasad, K., Huebschmann, J., Jeffrey, L., and Weinstein, A. [1997] Group systems, groupoids, and moduli spaces of parabolic bundles. *Duke Math. J.*, **89**(2), 377–412.

Haller, S. and Rybicki, T. [2001], Reduction for locally conformal symplectic manifolds. *J. Geom. Phys.* **37**, 262–271.

Hannay, J. [1985] Angle variable holonomy in adiabatic excursion of an integrable Hamiltonian. *J. Phys. A: Math. Gen.*, **18**, 221–230.

Hernández, A. and Marsden, J.E. [2003], Regularization of the amended potential and the bifurcation of relative equilibria. *J. of Nonlinear Science (to appear).*

Hilgert, J., Neeb, K.-H., and Plank, W. [1994] Symplectic convexity theorems and coadjoint orbits. *Composition Mathematica*, **94**, 129–180.

Hitchin, N. [2000], L^2-cohomology of hyperkähler quotients. *Comm. Math. Phys.* **211**, 153–165.

Hitchin, N. J., Karlhede, A., Lindström, U., and Roček, M. [1987], Hyper-Kähler metrics and supersymmetry. *Comm. Math. Phys.* **108**, 535–589.

Hochschild, G. [1965] *The structure of Lie Groups*. Holden Day.

Holm, D.D., Marsden, J.E., Ratiu, T.S. and A. Weinstein [1985], Nonlinear stability of fluid and plasma equilibria. *Phys. Rep.* **123**, 1–116.

Holm, D.D., Marsden, J.E. and Ratiu, T.S. [1986] The Hamiltonian structure of continuum mechanics in material, spatial and convective representations. In: *Séminaire de Mathématiques supérieurs*, **100**, Les Presses de L'Univ. de Montréal, Montréal, pp. 11–122.

Holm, D.D., Marsden, J.E., and Ratiu, T.S. [1998] The Euler–Poincaré equations and semidirect products with applications to continuum theories. *Advances in Math.*, **137**, 1–81.

Holm, D.D., Marsden, J.E. and Ratiu, T.S. [2002], The Euler–Poincaré equations in geophysical fluid dynamics. In: Norbury, J. and I. Roulstone, editors, *Large-Scale Atmosphere-Ocean Dynamics II: Geometric Methods and Models* Cambridge Univ. Press, pp. 251–300..

Horn, A. [1954a] Doubly stochastic matrices and the diagonal of a rotation matrix. *Amer. J. Math.* **76**, 620–630.

Horn, A. [1954b] On the eigenvalues of a matrix with prescribed singular values. *Proc. Amer. Math. Soc.*, **5**, 4–7.

Howe, R. [1989] Remarks on classical invariant theory. *Trans. A.M.S.,* **313**, 539–570.

Huebschmann, J. [1995a] Symplectic and Poisson structures of certain moduli spaces. *Duke Math. J.*, **80**, 737–756.

Huebschmann, J. [1995b] Symplectic and Poisson structures of certain moduli spaces II. Projective representations of cocompact planar discrete groups. *Duke Math. J.*, **80**, 757–770.

Huebschmann, J. [1996] Poisson geometry of flat connections for $SU(2)$-bundles on surfaces. *Math. Z.*, **221**, 243–259.

Huebschmann, J. [1998] Smooth structures on certain moduli spaces for bundles on a surface. *J. Pure Appl. Algebra*, **126**, 183–221.

Huebschmann, J. [2001] On the variation of the Poisson structures of certain moduli spaces. *Math. Ann.*, **319**, 267–310.

Huebschmann, J. [2003] Kähler spaces, nilpotent orbits, and singular reduction. *Preprint*. math.DG/0104213.

Huebschmann, J., Guruprasad, K., Jeffrey, L., and Weinstein, A. [1997] Group systems, groupoids, and moduli spaces of parabolic bundles. *Duke Math. J.*, **89**, 377–412.

Huebschmann, J. and Jeffrey, L. [1994] Group cohomology construction of symplectic forms on certain moduli spaces. *Int. Math. Research Notices*, **6**, 245 –249.

Husemoller, D. [1994] *Fibre Bundles. Graduate Texts in Mathematics*, Vol. 20, 3rd Edition. Springer-Verlag.

Ibort, A., de León, M., and Marmo, G. [1997] Reduction of Jacobi manifolds. *J. Phys. A*, **30**(8), 2783–2798.

Ikeda, M. [2000] Moser type theorem for toric hyper-Kähler quotients. *Hokkaido Math. J.* **29**, 585–599.

Isenberg, J. and Marsden, J. [1982] A slice theorem for the space of solutions of Einstein's equations. *Phys. Rep.* **89**(2), 179–222.

Ito, H. [1991] Action–angle coordinates at singularities for analytic integrable systems. *Math. Z.*, **206**, 363–407.

Iwasawa, K. [1949] On some types of topological groups. *Ann. Math.*, **50**, 507–558.

Jacobson, N. [1979] *Lie Algebras*. Dover.

Jakobsen, H.P. and Vergne, M. [1979] Restrictions and expansions of holomorphic representations. *J. Funct. Anal.,* **34**, 29–53.

Jalnapurkar, S.M., Leok, M., Marsden, J.E., and West, M. [2003], Discrete Routh reduction. *Found. Comput. Math. (to appear).*

Jalnapurkar, S.M. and Marsden, J.E. [2000], Reduction of Hamilton's variational principle. *Dynam. Stability Systems* **15**, 287–318.

Jost, R. [1968] Winkel und Wirkungsvariable für allgemeine mechanische Systeme. *Helvetica Physica Acta*, **41**, 965–968.

Joyce, D. [1991], The hypercomplex quotient and the quaternionic quotient. *Math. Ann.* **290**, 323–340.

Karasev, M.V. and Maslov, V.P. [1993] *Nonlinear Poisson Brackets. Geometry and Quantization*. Translations of the Amer. Math. Soc., **119**.

Karshon, Y. [1993] *Hamiltonian Actions of Lie Groups*. Ph.D. Thesis, Harvard University.

Karshon, Y. [1997] An algebraic proof for the symplectic structure of moduli spaces. *Proc. Amer. Math. Soc.*, **275**, 333–343.

Karshon, Y. and Lerman, E. [1997] The centralizer of invariant functions and division properties of the moment map. *Illinois J. Math.* **41**(3), 462–487.

Kashiwara, M. and Vergne, M. [1978] On the Segal–Shale–Weil representations and harmonic polynomials. *Invent. Math.,* **44**, 1–47.

Kawakubo, K. [1991] *The Theory of Transformation Groups*. Oxford University Press.

Kazhdan, D., Kostant, B., and Sternberg, S. [1978] Hamiltonian group actions dynamical systems of Calogero type. *Comm. Pure Appl. Math*, **31**, 481–508.

Kempf, G. [1987] Computing invariants. *Springer Lecture Notes in Mathematics*, **1278**, 62–80. Springer-Verlag.

Kirillov, A.A. [1962] Unitary representations of nilpotent Lie groups. *Russian Math. Surveys*, **17**, 53–104.

Kirillov, A.A. [1976a] Elements of the Theory of Representations. *Grundlehren der mathematischen Wissenschaften*, Vol. 220. Springer-Verlag.

Kirillov, A.A. [1976b] Local Lie algebras. *Russian Math. Surveys*, **31**, 55–75.

Kirwan, F. [1984a] Cohomology of quotients in symplectic and algebraic geometry. *Mathematical Notes*, 31. Princeton University Press.

Kirwan, F. [1984b] Convexity properties of the moment map III. *Invent. Math.*, **77**, 547–552.

Kirwan, F. [1988] The topology of reduced phase spaces of the motion of vortices on a sphere. *Physica D*, **30**, 99–123.

Knapp, A.W. [1996] *Lie Groups Beyond an Introduction. Prog. Math.*, Vol. 140. Birkhäuser.

Knop, F. [1990] Weylgruppe und Momentabbildung. *Invent. Math.,* **99**, 1–23.

Knop, F. [1997] Weyl groups of Hamiltonian manifolds I. *dg-ga 9712010.*

Knop, F. [2002] Convexity of Hamiltonian manifolds. *J. Lie Theory*,**12** (2), 571–582.

Knutson, A. and Tao, T. [1999] The honeycomb model of $GL_n(\mathbb{C})$ tensor products. I. Proof of the saturation conjecture. *J. Amer. Math. Soc.* **12**(4), 1055–1090.

Knutson, A. and Tao, T. [2001] Honeycombs and sums of Hermitian matrices. *Notices Amer. Math. Soc.*, **48**(2), 175–186.

Kobak, P.Z. and Swann, A. [1996] Classical nilpotent orbits as hyper-Kähler quotients. *Internat. J. Math.* **7**, 193–210.

Kobak, P.Z. and Swann, A. [2001] Hyper-Kähler potentials via finite-dimensional quotients. *Geom. Dedicata* **88**, 1–19.

Kobayashi, S. [1995] *Transformation Groups in Differential Geometry. Classics in Mathematics*. Springer-Verlag.

Kobayashi, S. and Nomizu, K. [1963] *Foundations of Differential Geometry*, Vols. 1, 2. Wiley.

Koiller, J. [1992] Reduction of some classical nonholonomic systems with symmetry. *Arch. Rational Mech. Anal.* **118**, 113–148.

Kolář, I., Michor, P.W., and Slovák, J. [1993] *Natural Operations in Differential Geometry*. Springer-Verlag.

Konno, H. [2002] The topology of toric hyper-Kähler manifolds. In: *Minimal surfaces, geometric analysis and symplectic geometry (Baltimore, MD, 1999)*, volume 34 of *Adv. Stud. Pure Math.* Kinokuniya, Tokyo, pp. 173–184.

Koon, W.S. and Marsden, J.E. [1997], The Hamiltonian and Lagrangian approaches to the dynamics of nonholonomic systems. *Reports on Math Phys.* **40**, 21–62.

Koon, W. S. and Marsden, J.E. [1998], The Poisson reduction of nonholonomic mechanical systems. *Reports on Math. Phys.* **42**, 101–134.

Kosmann-Schwarzbach, Y. and Magri, F. [1990] Poisson–Nijenhuis structures. *Ann. Inst. H. Poincaré*, **53**, 35–81.

Kostant, B. [1965] Orbits, symplectic structures and representation theory. *Proc. US–Japan Seminar on Diff. Geom., Kyoto. Nippon Hyronsha, Tokyo*, **77**.

Kostant, B. [1970] Quantization and unitary representations. *Springer Lecture Notes in Mathematics*, Vol. 570. Springer-Verlag.

Kostant, B. [1974] On convexity, the Weyl group and the Iwasawa decomposition. *Ann. Sci. École Norm. Sup.*, **6**(4), 413–455.

Koszul, J.L. [1953] Sur certains groupes de transformation de Lie. In: *Colloque de Géométrie Différentielle, Colloques du CNRS*, Vol. 71. Strasbourg, pp. 137–141.

Krishnaprasad, P.S. [1990], Geometric phases and optimal reconfiguration for multibody systems. In: *Proc. Am. Control Conf.*, pp. 2440–2444.

Krishnaprasad, P.S. and Yang, R. [1991] Geometric phases, anholonomy and optimal movement. *Proceedings of Conference on Robotics and Automation, 1991*, Sacramento, California, 2185–2189.

Kritsis, E. [1987] A topological investigation of the quantum adiabatic phase. *Comm. Math. Phys.*, **111**, 417–437.

Kummer, M. [1981] On the construction of the reduced phase space of a Hamiltonian system with symmetry. *Indiana Univ. Math. Journ.* **30**, 281–291.

Kupershmidt, B.A. and Ratiu, T.S. [1983] Canonical maps between semidirect products with applications to elasticity and superfluids. *Comm. Math. Phys.*, **90**, 235–250.

Krupa, M. [1990] Bifurcations of relative equilibria. *SIAM J. Math. Anal.* **21**(6), 1453–1486.

Kuranishi, M. [1959] On the local theory of continuous infinite pseudogroups I. *Nagoya Math. J.*, **15**, 225–260.

Kuranishi, M. [1961] On the local theory of continuous infinite pseudogroups II. *Nagoya Math. J.*, **19**, 55–91.

Landau, L.D. and Lifshitz, E.M. [1976] *Mechanics. Course of Theoretical Physics*, Vol. 1. Third Edition. Pergamon Press.

Landsman, N.P. [1998] *Mathematical Topics Between Classical and Quantum Mechanics*. Springer Monographs in Mathematics. Springer-Verlag.

Landsman, N.P., Pflaum, M., and Schlichenmaier, M. (editors) [2001] *Quantization of Singular Symplectic Quotients.* Prog. Math., Vol. 198. Birkhäuser Verlag.

Lang, S. [1999] *Fundamentals of Differential Geometry. Graduate Texts in Mathematics*, Vol. 191. Springer-Verlag.

Lang, S. [2002] *Algebra*, Revised third edition. *Graduate Texts in Mathematics*, Vol. 211. Springer-Verlag.

Lerman, E. [1989] On the centralizer of invariant functions on a Hamiltonian G-space. *J. Differential Geometry*, **30**, 805–815.

Lerman, E. [1995] Symplectic cuts. *Mathematical Research Letters*, **2**, 247–258.

Lerman, E. [2001] Contact cuts. *Israel J. Math.* **124**, 77–92.

Lerman, E. [2002] A convexity theorem for torus actions on contact manifolds. *Illinois J. Math.*, **46**(1), 171–184.

Lerman, E., Meinrenken, E., Tolman, S., and Woodward, C. [1998] Nonabelian convexity by symplectic cuts. *Topology*, **37**(2), 245–259.

Lerman, E., Montgomery, R., and Sjamaar, R. [1993] Examples of singular reduction. *Symplectic geometry*, 127–155. London Math. Soc. Lecture Note Ser., **192**. Cambridge Univ. Press, Cambridge.

Lerman, E. and Singer, S.F. [1998] Stability and persistence of relative equilibria at singular values of the moment map. *Nonlinearity*, **11**, 1637-1649.

Lerman, E. and Tokieda, T.F. [1999] On relative normal modes. *C. R. Acad. Sci. Paris Sér. I Math.*, **328**, 413–418.

Lerman, E. and Tolman, S. [1997] Hamiltonian torus actions on symplectic orbifolds and toric varieties. *Trans. Amer. Math. Soc.*, **349**(10), 4201–4230.

Lerman, E. and Willett, C. [2001] The topological structure of contact and symplectic quotients. *Internat. Math. Res. Notices*, **2001**(1), 33–52.

Lew, A., Marsden, J.E., Ortiz, M., and West, M. [2003], Asynchronous variational integrators *Arch. Rat. Mech. An.* **167**, 85–146.

Lewis, D.K. [1993] Bifurcation of liquid drops. *Nonlinearity*, **6**, 491–522.

Lewis, D.K. [1998] Stacked Lagrange tops. *J. Nonlinear Science*, **8**, 63–102.

Lewis, D.K. and Ratiu, T.S. [1996] Rotating n-gon/kn-gon vortex configurations. *Journ. Nonlinear Science* **6**, 385–414.

Lewis, D.K., Ratiu, T.S., Simo, J.C., and Marsden, J.E. [1992] The heavy top: a geometric treatment. *Nonlinearity*, **5**, 1–48.

Libermann, P. [1983] Symplectically regular foliations. In: *Proceedings of the IUTAM–ISIMM Symposium on Modern Developments in Analytical Mechanics. Vol. I, Geometrical Dynamics*, (S. Benenti, M. Francaviglia, A. Lichnerowicz, editors). Pages 239–246. Supplement to vol. 117 of the Acta Academiae Scientiarum Taurinensis.

Libermann, P. and Marle, C.-M. [1987] *Symplectic Geometry and Analytical Mechanics*. Reidel.

Lichnerowicz, A. [1973] Algèbre de Lie des automorphismes infinitésimaux d'une structure de contact. *J. Math. Pures Appl.*, **52**, 473–508.

Lichnerowicz, A. [1977] Les variétés de Poisson et leurs algèbres de Lie associées. *J. Diff. Geom.*, **12**, 253–300.

Lichnerowicz, A. [1979] Les variétés de Jacobi et leurs algèbres de Lie associées. *J. Math. Pures Appl.*, **57**, 453–488.

Lie, S. [1890] *Theorie der Transformationsgruppen. Zweiter Abschnitt.* Teubner.

Lindström, U. and Roček, M. [1983] Scalar tensor duality and $N = 1, 2$ nonlinear σ-models. *Nuclear Phys. B*, **222**(2), 285–308.

Lindström, U., Roček, M., and von Unge, R. [2000] Hyper-Kähler quotients and algebraic curves. *J. High Energy Phys.*, **2000**(1), Paper 22, 22 pages.

Liu, Z.–J., Weinstein, A., and Xu, P. [1998] Dirac structures and Poisson homogeneous spaces, *Comm. Math. Phys.* **192**, 121–144.

Littlejohn, R. [1988] Cyclic evolution in quantum mechanics and the phases of Bohr–Sommerfeld and Maslov. *Phys. Rev. Lett.*, **61**, 2159–2162.

Loose, F. [2000] A remark on the reduction of Cauchy–Riemann manifolds. *Math. Nachr.*, **214**, 39–51.

Loose, F. [2001] Reduction in contact geometry. *J. Lie Theory*, **11**(1), 9–22.

Lu, J.H. [1990] *Multiplicative and Affine Poisson Structures on Lie Groups*, Ph.D. Thesis, University of California, Berkeley.

Lu, J.H. [1991] Momentum mappings and reduction of Poisson actions. In: *Proc. of the Sem. Sud-Rhodanien de Geometrie á Berkeley, 1989*, Springer-Verlag, MSRI Series.

Lu, J.H. and Ratiu T.S. [1991] On the nonlinear convexity theorem of Kostant. *J. Amer. Math. Soc.*, **4**, 349–363.

Lu, J.H. and Weinstein, A. [1990] Poisson Lie groups, dressing transformations, and Bruhat decompositions. *J. of Diff. Geom.*, **31**, 501–526.

Mackenzie, K. [1987] *Lie Groupoids and Lie Algebroids in Differential Geometry.* London Math. Soc. Lecture Notes Series, **124**. Cambridge University Press.

Maddocks, J. [1991] On the stability of relative equilibria. *IMA J. Appl. Math.*, **46**, 71-99.

Magri, F. [1997] Eight lectures on integrable systems. Integrability of nonlinear systems (Pondicherry, 1996), 256–296, *Lecture Notes in Phys.*, **495**, Springer-Verlag, Berlin.

Magri, F. [1978] A simple model of the integrable Hamiltonian equation. *J. Math. Phys.*, **19**(5), 1156–1162.

Marle, C.-M. [1976] Symplectic manifolds, dynamical groups and Hamiltonian mechanics. In: *Differential Geometry and Relativity.* M. Cahen, and M. Flato (eds.). Reidel.

Marle, C.-M. [1984] Le voisinage d'une orbite d'une action hamiltonienne d'un groupe de Lie. *Séminaire Sud–Rhodanien de Géométrie II* (P. Dazord, N. Desolneux-Moulis eds.) pp. 19–35.

Marle, C.-M. [1985] Modéle d'action hamiltonienne d'un groupe the Lie sur une variété symplectique. *Rend. Sem. Mat. Univers. Politecn. Torino*, **43**(2), 227–251.

Marle, C.-M. [1995], Reduction of constrained mechanical systems and stability of relative equilibria. *Comm. Math. Phys.* **174**, 295–318.

Marle, C.-M. [1976] Symplectic manifolds, dynamical groups and Hamiltonian mechanics. In: *Differential Geometry and Relativity*, (M. Cahen and M. Flato, eds.), D. Reidel, Boston, pp. 249–269.

Marle, C.-M. [1998], Various approaches to conservative and nonconservative nonholonomic systems. *Rep. Math. Phys.* **42**, 211–229.

Marrero, J.C., Monterde, J., and Padron, E. [1999] Jacobi–Nijenhuis manifolds and compatible Jacobi structures. *C.R. Acad. Sci. Paris*, Série I, **329**, 797–802.

Marsden, J.E. [1981] *Lectures on Geometric Methods in Mathematica Physics.* CBMS–NSF Regional Conference Series in Applied Mathematics, **37**, SIAM, Philadelphia.

Marsden, J.E. [1992] *Lectures on Mechanics. London Mathematical Society Lecture Notes Series*, **174**. Cambridge University Press.

Marsden, J.E., Misiolek, G., Perlmutter, M., and Ratiu, T.S. [1998] Symplectic reduction for semidirect products and central extensions. *Diff. Geom. and Appl.*, **9**, 173–212.

Marsden, J.E., Misiolek, G., Ortega, J.-P., Perlmutter, M., and Ratiu, T. S. [2002] Symplectic reduction by stages. *Preprint*.

Marsden, J.E., Montgomery, R., Morrison, P.J. and Thompson, W.B. [1986] Covariant Poisson brackets for classical fields. *Annals of Physics*, **169**, 29–48.

Marsden, J.E., Montgomery, R., and Ratiu, T.S. [1989] Cartan–Hannay–Berry phases and symmetry. *Contemp. Math.*, **97**, 279–295.

Marsden, J.E., Montgomery, R., and Ratiu, T.S. [1990] Reduction, symmetry, and phases in mechanics. *Memoirs Amer. Math. Soc*, **88**(436), 1–110.

Marsden, J.E. and Ostrowski, J. [1996] Symmetries in motion: geometric foundations of motion control. *Nonlinear Sci. Today*, 21 pp. (electronic). (http://link.springer-ny.com).

Marsden, J.E., Patrick, G.W., and Shkoller, S. [1998] Multisymplectic geometry, variational integrators and nonlinear PDEs. *Comm. Math. Phys.* **199**, 351–395.

Marsden, J.E., Pekarsky, S., and Shkoller, S. [1999] Discrete Euler–Poincaré and Lie–Poisson equations. *Nonlinearity* **12**, 1647–1662.

Marsden, J.E. and Perlmutter, M. [2000] The orbit bundle picture of cotangent bundle reduction. *C. R. Math. Rep. Acad. Sci. Canada*, **22**, 33–54.

Marsden, J.E. and Ratiu, T.S. [1986] Reduction of Poisson manifolds. *Letters in Mathematical Physics*, **11**, 161–169.

Marsden, J.E. and Ratiu, T.S. [1999] *Introduction to Mechanics and Symmetry*, second edition. First edition [1994]. *Texts in Applied Mathematics*, Vol. 17. Springer-Verlag.

Marsden, J.E. and Ratiu, T.S. [2003] *Mechanics and Symmetry. Reduction Theory. In preparation.*

Marsden, J.E., Ratiu, T.S., and Scheurle, J. [2000] Reduction theory and the Lagrange–Routh equations. *J. Math. Phys.*, **41**, 3379–3429.

Marsden, J.E., Ratiu, T.S., and Weinstein, A. [1984a] Semidirect products and reduction in mechanics. *Trans. Amer. Math. Soc.*, **281**, 147–177.

Marsden, J.E., Ratiu, T.S., and Weinstein, A. [1984b], Reduction and Hamiltonian structures on duals of semidirect product Lie algebras. *Contemp. Math.*, **28**, 55–100. Amer. Math. Soc., Providence, RI.

Marsden, J.E. and Scheurle, J. [1993a] Lagrangian reduction and the double spherical pendulum. *Z. Angew. Math. Phys.*, **44**, 17–43.

Marsden, J.E. and Scheurle, J. [1993b] The reduced Euler–Lagrange equations. *Fields Institute Comm.*, **1**, 139–164.

Marsden, J.E. and Weinstein, A. [1974] Reduction of symplectic manifolds with symmetry. *Rep. Math. Phys.*, **5**(1), 121–130.

Marsden, J.E. and Weinstein, A. [1982] The Hamiltonian structure of the Maxwell-Vlasov equations. *Physica D*, **4**(3), 394–406.

Marsden, J.E. and Weinstein, A. [1983] Coadjoint orbits, vortices, and Clebsch variables for incompressible fluids. *Proc. of the Conference on Order in Chaos,* Los Alamos, 1982. (A. Scott ed.). *Physica D*, **7**(1–3), 305–323.

Marsden, J.E. and Weinstein, A. [2001] Some comments on the history, theory, and applications of symplectic reduction. In: *Quantization of Singular Symplectic Quotients*. Landsman, N. P., Pflaum, M., and Schlichenmaier, M. (editors). Prog. Math., Vol. 198. Birkhäuser, Basel.

Marsden, J.E. and West, M. [2001] Discrete mechanics and variational integrators. *Acta Numerica* **10**, 357–514.

Mather, J.N. [1970] *Notes on Topological Stability*. Mimeographed Lecture Notes. Harvard University.

Mather, J.N. [1973] Stratifications and mappings. In: *Dynamical Systems* (M.M. Peixoto, ed.). Academic Press, pp. 195–232.

Mather, J.N. [1977] Differentiable invariants. *Topology*, **16**, 145–156.

Matsushima, Y. [1972] *Differentiable Manifolds. Pure and Applied Mathematics*, **9**. Marcel Dekker.

McDuff, D. [1988] The moment map for circle actions on symplectic manifolds. *J. Geom. Phys.*, **5**, 149–160.

McDuff, D. and Salamon, D. [1998] *Introduction to Symplectic Topology. Oxford Mathematical Monographs*, Clarendon Press, Oxford. Second Edition.

Mclachlan, R. and Perlmutter, M. [2001] Conformal Hamiltonian systems. *J. Geom. Phys.* **39**, 276–300.

Meinrenken, E. and Woodward, C. [1998] Hamiltonian loop group actions and Verlinde factorization. *J. Differential Geom.*, **50**(3), 417–469.

Meyer, K.R. [1973] Symmetries and integrals in mechanics. In: *Dynamical Systems*, pp. 259–273. M.M. Peixoto, ed. Academic Press.

Michor, P.W. [2001] *Topics in Differential Geometry*. Available at the website http:www.mat.univie.ac.at/~michor/.

Mikami, K. [1989] Reduction of local Lie algebra structures. *Proc. Amer. Math. Soc.*, **105**(3), 686–691.

Mikami, K. and Weinstein, A. [1988] Moments and reduction for symplectic groupoids. *Publ. RIMS, Kyoto Univ.*, **24**, 121–140.

Mineur, H. [1935] Sur les systèmes mécaniques admettant *n* intégrales premières uniformes et l'extension à ces systèmes de la méthode de quantification de Sommerfeld. *C. R. Acad. Sci. Paris*, **200**, 1571–1573.

Mineur, H. [1937] Sur les systèmes mécaniques sur lesquels figurent des paramétres fonctions du temps. Étude des systèmes admettant *n* intégrales premières uniformes en involution. Extension à ces systèmes des conditions de quantification de Sommerfeld. *Journal de l'École Polytechnique, Série III*, 143ème année, 173–191 and 237–270.

Mishchenko, A.S. and Fomenko, A.T. [1978] Generalized Liouville method of integration of Hamiltonian systems. *Funct. Anal. Appl.*, **12**, 113–121.

Moerdijk, I. and Mrčun, J. [2003] *Introduction to Foliations and Lie Groupoids*. Cambridge University Press.

Molino, P. [1984] Structure transverse aux orbites de la représentation coadjointe: le cas des orbites réductives. *Sèminare de Gèometrie de l'Université de Montpellier.*

Montaldi, J.A. [1997a] Persistence and stability of relative equilibria. *Nonlinearity*, **10**, 449–466.

Montaldi, J.A. [1997b] Persistance d'orbites périodiques relatives dans les systèmes hamiltoniens symétriques. *C. R. Acad. Sci. Paris Sér. I Math.*, **324**, 553–558.

Montaldi, J.A., Ortega, J.-P., and Ratiu, T. S. [2003] The relation between local and global dual pairs. *Preprint.*

Montaldi, J.A. and Roberts, R.M. [1999] Relative equilibria of molecules. *J. Nonlin. Sci.*, **9**, 53–88.

Montaldi J.A., Roberts R.M., and Stewart, I.N. [1988] Periodic solutions near equilibria of symmetric Hamiltonian systems. *Phil. Trans. R. Soc. Lond. A*, **325**, 237–293.

Montgomery, D. [1960] Orbits of highest dimension. In: *Seminar on Transfromation Groups*, Ann. of Math. Studies, No. 46, Chapter IX. Princeton University Press.

Montgomery, D. and Yang, C.T. [1957] The existence of a slice. *Ann. of Math.*, **65**, 108–116.

Montgomery, R. [1988] The connection whose holonomy is the classical adiabatic angles of Hannay and Berry and its generalization to the non-integrable case. *Comm. Math. Phys.*, **120**, 269–294.

Montgomery, R. [1990] Isoholonomic problems and some applications. *Comm. Math. Phys.*, **128**(3), 565–592.

Montgomery, R. [1991a] How much does a rigid body rotate? A Berry's phase from the eighteenth century. *Amer. Journ. Physics*, **59**, 394–398.

Montgomery, R. [1991b] Optimal control of deformable bodies and its relation to gauge theory. In: *The Geometry of Hamiltonian Systems*, MSRI publications, **22**, 403–438. Springer-Verlag, 1991.

Montgomery, R. [1996] The geometric phase of the three-body problem. *Nonlinearity*, **9**(5), 1341–1360.

Montgomery, R. [2002] *A Tour of Subriemannian Geometries, Their Geodesics and Applications*. Mathematical Surveys and Monographs, **91**, Amer. Math. Soc., Providence, RI.

Montgomery, R., Marsden, J.E., and Ratiu, T.S. [1984] Gauged Lie Poisson structures. In: *Fluids and Plasmas: Geometry and Dynamics* (J. Marsden, ed.) *Cont. Math.*, **28**, 101–114.

Morita, K. [1958] Duality for modules and its applications to the theory of rings with minimum condition. *Tokyo Kyoiku Diagaku*, **6**, 83–142.

Moser, J. and Veselov, A.P. [1991], Discrete versions of some classical integrable systems and factorization of matrix polynomials. *Comm. Math. Phys.*, **139**, 217–243.

Mostow, G.D. [1957] Equivariant imbeddings in euclidean space. *Ann. of Math.*, **65**, 432–446.

Munkres, J. [1975] *Topology: a First Course*. Prentice Hall.

Nagatomo, Y. [2002], Dimensional reduction and moment maps. *J. Geom. Phys.* **41**, 208–223.

Nekhoroshev, N.N. [1972] Action–angle variables and their generalizations. *Trudy Moskov. Mat. Obsc.*, **26**, 181–198. *Trans. Moskow Math. Soc.*, **26**, 180–198.

Neumann, A. [1999] An infinite-dimensional version of the Schur–Horn convexity theorem. *J. Funct. Anal.*, **161**(2), 418–451.

Neumann, A. [2002] An infinite dimensional version of the Kostant convexity theorem. *J. Funct. Anal.*, **189**(1), 80–131.

Nijmeijer, H. and van der Schaft, A.J. [1990] *Nonlinear Dynamical Control Systems.* Springer-Verlag, New York.

Nitta, T. [1990], Yang–Mills connections on quaternionic Kähler quotients. *Proc. Japan Acad. Ser. A Math. Sci.* **66**, 245–247.

Noether, E. [1918] Invariante Variationsprobleme. *Kgl. Ges. Wiss. Nachr. Göttingen Math. Physik*, **2**, 235–257.

Nunes da Costa, J.M. [1989] Réduction des variétés de Jacobi. *C.R. Acad. Sci. Paris*, **308**, Série I, 101–103.

Nunes da Costa, J.M. [1990] Une généralisation, pour les variétés de Jacobi, du théorème de Marsden–Weinstein. *C.R. Acad. Sci. Paris*, **310**, Série I, 411–414.

Nunes da Costa, J.M. [1997] Reduction of complex Poisson manifolds. *Portugaliae Math.*, **54**, 467–476.

Nunes da Costa, J.M. [1998] Compatible Jacobi manifolds: geometry and reduction. *J. Phys. A*, **31**(3), 1025–1033.

Nunes da Costa, J.M. and Petalidou, F. [2002] Reduction of Jacobi–Nijenhuis manifolds. *J. Geom. Phys.*, **41**(3), 181–195.

Odzijewicz, A. and Ratiu, T.S. [2003] Banach Lie–Poisson spaces and reduction. *Comm. Math. Phys.*, to appear.

Oh, Y.–G. [1986] Some remarks on the transverse Poisson structures of coadjoint orbits. *Lett. Math. Phys.*, **12**(2), 87–91.

Oh, Y.G., Sreenath, N., Krishnaprasad, P.S., and Marsden, J.E. [1989], The dynamics of coupled planar rigid bodies Part 2: bifurcations, periodic solutions and chaos. *Dynamics and Diff. Eq'ns* **1**, 269–298.

Ornea, L. and Piccinni, P. [2001], Cayley 4-frames and a quaternion Kähler reduction related to Spin(7). In: *Global differential geometry: the mathematical legacy of Alfred Gray (Bilbao, 2000)*, Vol. 288 of *Contemp. Math.* Amer. Math. Soc., Providence, RI, pp. 401–405.

Ortega, J.-P. [1998] *Symmetry, Reduction, and Stability in Hamiltonian Systems*. Ph.D. Thesis. University of California, Santa Cruz. June, 1998.

Ortega, J.-P. [2002a] The symplectic reduced spaces of a Poisson action. *C. R. Acad. Sci. Paris Sér. I Math.*, **334**, 999–1004.

Ortega, J.-P. [2002b] Some remarks about the geometry of Hamiltonian conservation laws. In: *Symmetry and Perturbation Theory* Abenda, S., Gaeta, G., and Walcher, S. (editors). World Scientific, pp. 162–170.

Ortega, J.-P. [2003a] Relative normal modes for nonlinear Hamiltonian systems. To appear in *Proc. Roy. Soc. Edinburgh Sect. A*, **133**(3), 665–704.

Ortega, J.-P. [2003b] Singular dual pairs. *Diff. Geom. Appl.*, **19**(1), 61–95.

Ortega, J.-P. [2003c] Optimal reduction. *Preprint.*

Ortega, J.-P. and Planas, V. [2003a] The Casimir reduction method and its applications to the study of stability in Poisson dynamical systems. *In preparation.*

Ortega, J.-P. and Planas, V. [2003b] Dynamics on Leibniz manifolds. *Preprint.*

Ortega, J.-P. and Ratiu, T.S. [1997] Persistence and smoothness of critical relative elements in Hamiltonian systems with symmetry. *C. R. Acad. Sci. Paris Sér. I Math.*, **325**, 1107–1111.

Ortega, J.-P. and Ratiu, T.S. [1998] Singular reduction of Poisson manifolds. *Letters in Mathematical Physics*, **46**, 359–372.

Ortega, J.-P. and Ratiu, T.S. [1999a] Stability of Hamiltonian relative equilibria. *Nonlinearity*, **12**(3), 693–720.

Ortega, J.-P. and Ratiu, T.S. [1999b] A Dirichlet criterion for the stability of periodic and relative periodic orbits in Hamiltonian systems. *J. Geom. Phys.*, **32**, 131–159.

Ortega, J.-P. and Ratiu, T.S. [1999c] Non-linear stability of singular relative periodic orbits in Hamiltonian systems with symmetry. *J. Geom. Phys.*, **32**, 160–188.

Ortega, J.-P. and Ratiu, T.S. [2002a] The optimal momentum map. In *Geometry, Dynamics, and Mechanics: 60th Birthday Volume for J.E. Marsden*. P. Holmes, P. Newton, and A. Weinstein, eds., pages 329–362. Springer-Verlag.

Ortega, J.-P. and Ratiu, T.S. [2002b] A symplectic slice theorem. *Lett. Math. Phys.*, **59**, 81–93.

Ortega, J.-P. and Ratiu, T.S. [2003a] A normal form for the cylinder valued momentum map. *Preprint*.

Ortega, J.-P. and Ratiu, T.S. [2003b] The dynamics around stable and unstable Hamiltonian relative equilibria, *The Royal Society. Proceedings: Math., Phys. and Eng. Sci.*, Series A, to appear.

Ortega, J.-P. and Ratiu, T.S. [2003c] The universal covering and covered spaces of a symplectic Lie algebra action. *Preprint*.

Otto, M. [1987] A reduction scheme for phase spaces with almost Kähler symmetry. Regularity results for momentum level sets. *J. Geom. Phys.*, **4**, 101–118.

Palais, R. [1957] *A Global Formulation of the Lie Theory of Transformation Groups. Mem. Amer. Math. Soc.*, **22**.

Palais, R. [1961] On the existence of slices for actions of non-compact Lie groups. *Ann. Math.*, **73**, 295–323.

Paterson, A.L.T. [1999] *Grupoids, Inverse Semigroups, and their Operator Algebras. Prog. Math.*, Vol. 170. Birkhäuser.

Patrick, G.W. [1992] Relative equilibria in Hamiltonian systems: the dynamic interpretation of nonlinear stability on a reduced phase space. *J. Geom. Phys.*, **9**, 111–119.

Patrick, G.W., Roberts, M., and Wulff, C. [2003] Stability of Poisson equilibria and Hamiltonian relative equilibria by energy methods. *Preprint available at http://arXiv.org/abs/math.DS/0201239.*

Pedroni, M. [1995] Equivalence of the Drinfeld–Sokolov reduction to a bi-Hamiltonian reduction. *Lett. Math. Phys.*, **35**, 291–302.

Pekarsky, S. and Marsden, J.E. [2001] Abstract mechanical connection and abelian reconstruction for almost Kähler manifolds. *Journal of Applied Mathematics* **1**, 1–28.

Perdigão do Carmo, M. [1993] *Riemannian Geometry*. Second printing. Birkhäuser.

Perlmutter, M. and Ratiu, T.S. [2004] Gauged Lie–Poisson structures. *Preprint*.

Petalidou, F. and Nunes da Costa, J.M. [2003] Local structure of Jacobi-Nijenhuis manifolds. *J. Geom. Phys.*, **45**(3-4), 323–367.

Pflaum, M.J. [2001a] *Analytic and Geometric Study of Stratified Spaces. Lecture Notes in Mathematics*, Vol. 510. Springer-Verlag.

Pflaum, M.J. [2001b] Smooth structures on stratified spaces. In: *Quantization of Singular Symplectic Quotients*. Landsman, N. P., Pflaum, M., and Schlichenmaier, M. (editors). *Prog. Math.*, Vol. 198. Birkhäuser Boston.

Poènaru, V. [1976] *Singularités C^∞ en Présence de Symétrie. Lecture Notes in Mathematics*, Vol. 510. Springer-Verlag.

Pradines, J. and Kamga, J.W. [1976] Relations d'équivalence transversalement différentiables. *C. R. Acad. Sci. Paris Sér. A-B*, **283**(2), Ai, A25–A28.

Prato, E. [1994] Convexity properties of the moment map for certain non-compact manifolds. *Comm. Anal. Geom.*, **2**(2), 267–278.

Ratiu, T.S. [1980] *The Euler–Poisson Equations and Integrability*. Ph.D. Thesis, University of California at Berkeley.

Ratiu, T.S. [1981] Euler–Poisson equations on Lie algebras and the N-dimensional heavy rigid body. *Proc. Natl. Acad. Sci., USA*, **78**, 1327–1328.

Ratiu, T. S. [1982] Euler–Poisson equations on Lie algebras and the N-dimensional heavy rigid body. *Amer. J. Math.*, **104**, 1337, 409–448.

Robbin, J. [1973] Relative equilibria in mechanical systems. In: *Dynamical Systems*. M.M. Peixoto, ed. Academic Press, pp. 425–441.

Roberts, M. and de Sousa Dias, M.E.R. [1997] Bifurcations from relative equilibria of Hamiltonian systems. *Nonlinearity*, **10**, 1719–1738.

Roberts, M., Wulff, C., and Lamb, J.S.W. [2002] Hamiltonian systems near relative equilibria. *J. Differential Equations*, **179**(2), 562–604.

Robinson C. [1995] *Dynamical Systems*. CRC Press, Inc.

Rodríguez-Olmos, M., Perlmutter, M., and Sousa Dias, M.E. [2003] Symplectic cotangent bundle reduction. *Preprint*.

Routh, E.J. [1877] *Stability of a given state of motion*. Reprinted in *Stability of motion*. A. T. Fuller (ed.). Halsted Press, New York, 1975.

Routh, E.J. [1884] *Advanced Rigid Dynamics*. MacMillan and Co., London.

Samelson, H. [1952] Topology of Lie groups. *Bull. Amer. Math. Soc.*, **58**, 2–37.

Sánchez de Alvarez, G. [1986] *Geometric Methods of Classical Mechanics Applied to Control theory*, Ph.D. Thesis, UC Berkeley.

Sánchez de Alvarez, G. [1989] Controllability of Poisson control systems with symmetry. *Contemp. Math.*, **97**, 399-412.

Sánchez de Alvarez, G. [1993] Poisson brackets and dynamics. *Dynamical systems* (Santiago, 1990), 230–249, *Pitman Res. Notes Math. Ser.*, **285**, Longman Sci. Tech., Harlow.

Sartori, G. [1983] A theorem on orbit structures (strata) of compact linear Lie groups. *J. Math. Phys.*, **24**(4), 765–768.

Satake, I. [1956] On a generalization of the notion of manifold. *Proc. Nat. Acad. Sci. U.S.A.*, **42**, 359–363.

Satake, I. [1957] The Gauss–Bonnet theorem for V-manifolds. *J. Math. Soc. Japan*, **9**, 464–492.

Satzer, W.J. [1977] Canonical reduction of mechanical systems invariant under Abelian group actions with an application to celestial mechanics. *Indiana Univ. Math. Journ.*, **26**, 951–976.

Scheerer, U. and Wulff, C. [2001] Reduced dynamics for momentum maps with cocycles. *C. R. Acad. Sci. Paris Sér. I Math.*, **333**(11), 999–1004.

Schmah, T. [2001] *Symmetries of Cotangent Bundles*. Ph.D. Thesis. EPFL, Lausanne, Switzerland.

Schmah, T. [2002] A cotangent bundle slice theorem.*Preprint*.

Schur, I. [1923] Über eine Klasse von Mittelbildungen mit Anwendungen auf die Determinantentheorie. *Sitzungsber. der Berliner Math. Gesellschaft*, **22**, 9–20.

Schwarz, G.W. [1974] Smooth functions invariant under the action of a compact Lie group. *Topology*, **14**, 63–68.

Shapere, A. and Wilczek, F. [1987] Self-propulsion at low Reynolds numbers. *Phys. Rev. Lett.*, **58**, 2051–2054.

Shapere, A. and Wilczek, F., eds. [1988] *Geometric Phases in Physics*. World Scientific, Singapore.

Simo, J.C., Lewis, D.K., and Marsden, J.E. [1991] stability of relative equlibria. Part I: The reduced energy–momentum method. *Arch. Rat. Mech. Anal.*, **115**, 15–59.

Simon, B. [1983] Holonomy, the quantum adiabatic theorem, and Berry's phase. *Phys. Rev. Lett.*, **51**, 2167–2170.

Sjamaar, R. [1990] *Singular Orbit Spaces in Riemannian and Symplectic Geometry*. Ph.D. thesis, Rijksuniversiteit te Utrecht.

Sjamaar, R. [1998] Convexity properties of the moment map re-examined. *Adv. Math.*, **138**(1), 46–91.

Sjamaar, R. and Lerman, E. [1991] Stratified symplectic spaces and reduction. *Ann. of Math.*, **134**, 375–422.

Smale, S. [1970] Topology and mechanics. *Invent. Math.*, **10**, 305–331; **11**, 45–64.

Sniatycki, J. [1998] Nonholonomic Noether theorem and reduction of symmetries. *Rep. Math. Phys.*, **42**(1/2), 5–23.

Śniatycki, J. [2001], Almost Poisson spaces and nonholonomic singular reduction. *Rep. Math. Phys.* **48**, 235–248.

Sniatycki, J. and Tulczyjew, W.M. [1972] Generating forms of Lagrangian submanifolds. *Indiana University Math. J.*, **22**, 267–275.

Souriau, J.-M. [1953] Géometrie symplectique différentielle. Applications. Colloque CNRS. *Géométrie différentielle*, Strasbourg. Pages 53–59. Éditions du CNRS. Paris.

Souriau, J.-M. [1965] *Géometrie de l'espace de phases, calcul des variations et mécanique quantique.* Mimeographed notes. Faculté des Sciences de Marseille, 1965.

Souriau, J.-M. [1966] *Quantification géométrique. Comm. Math. Phys.*, **1**, 374–398.

Souriau, J.-M. [1969] *Structure des Systèmes Dynamiques*. Dunod. Paris. English translation by R. H. Cushman and G. M. Tuynman as *Structure of Dynamical Systems. A Symplectic View of Physics.* Vol. 149 of *Prog. Math.*. Birkhäuser, 1997.

Spanier, E.H. [1966] *Algebraic Topology*. Mc-Graw Hill, reprinted by Springer-Verlag.

Spivak, M. [1979] *Differential Geometry*, Vol. I, New printing with corrections. Publish or Perish, Inc. Houston, Texas.

Stefan, P. [1974a] Accessibility and foliations with singularities. *Bull. Amer. Math. Soc.*, **80**, 1142–1145.

Stefan, P. [1974b] Accessible sets, orbits and foliations with singularities. *Proc. Lond. Math. Soc.*, **29**, 699–713.

Sternberg, S. and Wolf, J.A. [1978] Hermitian Lie algebras and metaplectic representations I. *Trans. A.M.S.*, **238**, 1–43.

Suris, Y.B. [2003] *The Problem of Integrable Discretization: Hamiltonian Approach.* Prog. Math., Vol. 219. Birkhäuser Boston.

Sussmann, H. [1973] Orbits of families of vector fields and integrability of distributions. *Trans. Amer. Math. Soc.*, **180**, 171–188.

Swann, A. [1991], Hyper-Kähler and quaternionic Kähler geometry. *Math. Ann.* **289**, 421–450.

Thom, R. [1969] Ensembles et morphismes stratifiés. *Bull. Amer. Math. Soc., New Series*, **75**, 240–284.

tom Dieck, T. [1987] *Transformation Groups*. De Gruyter Studies in Mathematics, number 8. Walter de Gruyter.

Tondeur, P. [1997] *Geometry of Foliations*. Vol. 90 of *Monographs in Mathematics*, Birkhäuser, Basel.

Vaisman, I. [1985] Locally conformal symplectic manifolds. *Int. J. Math. and Math. Sci.*, **8**(3), 521–536.

Vaisman, I. [1996a] *Lectures on the Geometry of Poisson Manifolds*. Prog. Math., Vol. 118. Birkhäuser.

Vaisman, I. [1996b] Reduction of Poisson–Nijenhuis manifolds. *J. Geom. and Physics*, **19**, 90–98.

Vanderbauwhede, A. and van der Meer, J.C. [1995] General reduction method for periodic solutions near equilibria. In: *Normal Forms and Homoclinic Chaos*, Langford, W. F. and Nagata, W. eds. Fields Institute Communications, Vol. 4, pp. 273–294.

van der Meer, J.C. [1986] *The Hamiltonian Hopf Bifurcation. Lecture Notes in Mathematics*, Vol. 1160. Springer-Verlag.

van der Meer, J.C. [1990] Hamiltonian Hopf bifurcation with symmetry. *Nonlinearity*, **3**, 1041–1056.

van der Meer, J.C. [1996] Degenerate Hamiltonian Hopf bifurcations. In: *Conservative Systems and Quantum Chaos*, Bates, L.M. and Rod, D.L., eds. Fields Institute Communications, **8**, pp. 159–176.

van der Schaft, A.J. [1998a] Implicit Hamiltonian systems with symmetry. *Rep. Math. Phys.*, **41**, 203–221.

van der Schaft, A.J. [1998b] *L_2-Gain and Passivity Techniques in Nonlinear Control*. Springer-Verlag, New York; second revised and enlarged edition, 2000.

van der Schaft, A.J. and Maschke, B.M. [1994] On the Hamiltonian formulation of nonholonomic mechanical systems. *Rep. Math. Phys.*, **34**(2), 225–232.

van der Schaft, A.J. and Maschke, B.M. [1995a] The Hamiltonian formulation of energy conserving physical systems with external ports. *Archiv für Elektronik und Übertragungstechnik*, **49**, 362–371.

van der Schaft, A.J. and Maschke, B.M. [1995b] Mathematical modeling of constrained Hamiltonian systems. In: *Preprints of the 3rd Nonlinear Control Systems Design Symposium, NOLCOS*, Tahoe City, CA, 1995.

van der Schaft, A.J. and Maschke, B.M. [1997] Interconnected mechanical systems, part I: geometry of interconnection and implicit Hamiltonian systems. In: A. Astolfi, D.J.N Limebeer, C. Melchiorri, A. Tornambè, and R.B. Vinter, editors, *Modelling and Control of Mechanical Systems*. Imperial College Press, 1997, pp. 1–15.

Vanhaecke, P. [1996] *Integrable Systems in the Realm of Algebraic Geometry. Lecture Notes in Math.*, **1638**, Springer-Verlag, New York.

Vey, J. [1978] Sur certains systèmes dynamiques separables. *Amer. J. Math.*, **100**, 591–614.

Viflyantsev, V.P. [1980] Frobenius theorem for differential systems with singularities. *Vestnik Moscow University*, **3**, 11–14.

Warner, F.W. [1983] *Foundation of Differentiable Manifolds and Lie Groups. Graduate Texts in Mathematics*, Vol. 94. Springer-Verlag.

Wei, J. and Norman, E. [1964] On global representations of the solutions of linear differential equations as a product of exponentials. *Proc. Amer. Math. Soc.*, **15**, 327–334.

Weinstein, A. [1973a] Lagrangian submanifolds and Hamiltonian systems. *Ann. of Math.*, **98**, 377–410.

Weinstein, A. [1973b] Normal modes for nonlinear Hamiltonian systems. *Inventiones Math.*, **20**, 47–57.

Weinstein, A. [1977a] *Lectures on Symplectic Manifolds*. CBMS Conf Series, Amer. Math. Soc. **29**, Providence RI.

Weinstein, A. [1977b] Symplectic V-manifolds, periodic orbits of Hamiltonian systems, and the volume of certain Riemannian manifolds. *Comm. Pure Appl. Math.*, **30**, 265–271.

Weinstein, A. [1983a] The local structure of Poisson manifolds. *J. Differential Geom.*, **18**, 523–557. Errata and addenda: *J. Differential Geom.*, **22**(2), 255 (1985).

Weinstein, A. [1983b] Sophus Lie and symplectic geometry. *Expositio Math.*, **1**, 95–96.

Weinstein, A. [1987a] Symplectic groupoids and Poisson manifolds. *Bull. Amer. Math. Soc.*, **16**, 101–104.

Weinstein, A. [1987b] Poisson geometry of the principal series and nonlinearizable structures. *Differential Geom.*, **25**, 55–73.

Weinstein, A. [1989] Cohomology of symplectomorphism groups and critical values of Hamiltonians. *Math. Zeitschr.*, **201**, 75–82.

Weinstein, A. [1990] Connections of Berry and Hannay type for moving Lagrangian submanifolds. *Adv. Math.*, **82**(2), 133–159.

Weinstein, A. [1996a] Groupoids: unifying internal and external symmetry. A tour through some examples. *Notices Amer. Math. Soc.*, **43**(7), 744–752.

Weinstein, A. [1996b] Lagrangian mechanics and groupoids, *Fields Inst. Commun.*, **7**, 207–231.

Weinstein, A. [1998] Poisson geometry. *Differential Geom. Appl.*, **9**(1–2), 213–238.

Weinstein, A. [2000] Linearization problems for Lie algebroids and Lie groupoids. *Lett. Math. Phys.*, **52**(1), 93–102.

Weinstein, A. [2001] Poisson geometry of discrete series orbits, and momentum convexity for noncompact group actions. *Lett. Math. Phys.*, **56**(1), 17–30.

Weinstein, A. [2002a] The geometry of momentum. *Preprint.*

Weinstein, A. [2002b] Linearization of regular proper groupoids. *J. Inst. Math. Jussieu*, **1**(3), 493–511.

Wendlandt, J.M. and Marsden, J.E. [1997], Mechanical integrators derived from a discrete variational principle, *Physica D* **106**, 223–246.

Weyl, H. [1946] *The Classical Groups*. Second Edition. Princeton University Press.

Weyl, H. [1949] Inequalities between two kinds of eigenvalues of a linear transformation. *Proc. Nat. Acad. Sci. U.S.A.*, **35**, 408–411.

Wilczek, F. and Shapere, A. [1989] Geometry of self-propulsion at low Reynolds number. Efficiencies of self-propulsion at low Reynolds number. *J. Fluid Mech.*, **198**, 557–585, 587–599.

Willett, C. [2002] Contact reduction. *Trans. Amer. Math. Soc.*, **354**, 4245–4260.

Witt, E. [1937] Theorie der quadratischen Formen in beliebigen Körpen. *Journal für die reine und angewandte Mathematik*, **176**, 31–44.

Wulff, C. [2002] Persistence of Hamiltonian relative periodic orbits. *Preprint.*

Wulff, C. [2003] Persistence of relative equilibria in Hamiltonian systems with noncompact symmetry. *Nonlinearity*, **16**(1), 67–91.

Wulff, C. and Roberts, M. [2002] Hamiltonian systems near relative periodic orbits. *SIAM J. Appl. Dyn. Syst.*, **1**, 1–43 (electronic).

Wurzbacher, T. [2001] Fermionic second quantization and the geometry of the restricted Grassmannian. In *Infinite Dimensional Kähler Manifolds (Oberwolfach, 1995)*, A. Huckleberry and T. Wurzbacher eds., pages 287–375, DMV Seminar **31**, Birkhäuser, Basel.

Xu, P. [1991] Morita Equivalence of Poisson manifolds. *Commun. Math. Phys.*, **142**, 493–509.

Xu, P. [2001] Dirac submanifolds and Poisson involutions. *Ann. Sci. École Norm. Sup. (4)*, **4**(3), 403–430.

Yamabe, H. [1950] On an arcwise connected subgroup of a Lie group. *Osaka Math. J.*, **2**, 13–14.

Yang, C.T. [1957] On a problem of Montgomery. *Proc. Amer. Math. Soc.*, **8**, 255–257.

Yang, R. and Krishnaprasad, P.S. [1994], On the geometry and dynamics of floating four–bar linkages. *Dynam. Stability Systems* Vol. 9, 19–45.

Yano, K. and Kon, M. [1984] *Structures on Manifolds*. Series in Pure Mathematics, Vol. 3. World Scientific.

Yoshimura, H. and Marsden, J.E. [2003] Variational principles, Dirac structures and implicit Lagrangian systems. *Preprint.*

Zaalani, N. [1999] Phase space reduction and Poisson structure. *J. Math. Phys.*, **40**(7), 3431–3438.

Zung, N.T. [1996] Symplectic topology of integrable Hamiltonian systems. I: Arnold–Liouville with singularities. *Compositio Math.*, **101**(2), 179–215.

Zung, N.T. [2003] Torus actions and integrable systems. *Preprint.*

Index

D

G

J

K

M

N

U

V

W

Progress in Mathematics

Edited by:

Progress in Mathematics is a series of books intended for professional mathematicians and scientists, encompassing all areas of pure mathematics. This distinguished series, which began in 1979, includes authored monographs and edited collections of papers on important research developments as well as expositions of particular subject areas.

We encourage preparation of manuscripts in some form of TEX for delivery in camera-ready copy which leads to rapid publication, or in electronic form for interfacing with laser printers or typesetters.

Proposals should be sent directly to the editors or to: Birkhäuser Boston, 675 Massachusetts Avenue, Cambridge, MA 02139, USA or to Birkhauser Verlag, 40-44 Viadukstrasse, CH-4051 Basel, Switzerland.

150 SHIOTA. Geometry of Subanalytic and Semialgebraic Sets
151 HUMMEL. Gromov's Compactness Theorem for Pseudo-holomorphic Curves
152 GROMOV. Metric Structures for Riemannian and Non-Riemannian Spaces. Sean Bates (translator)
153 BUESCU. Exotic Attractors: From Liapunov Stability to Riddled Basins
154 BÖTTCHER/KARLOVICH. Carleson Curves, Muckenhoupt Weights, and Toeplitz Operators
155 DRAGOMIR/ORNEA. Locally Conformal Kähler Geometry
156 GUIVARC'H/JI/TAYLOR. Compactifications of Symmetric Spaces
157 MURTY/MURTY. Non-vanishing of L-functions and Applications
158 TIRAO/VOGAN/WOLF (eds). Geometry and Representation Theory of Real and p-adic Groups
159 THANGAVELU. Harmonic Analysis on the Heisenberg Group
160 KASHIWARA/MATSUO/SAITO/SATAKE (eds). Topological Field Theory, Primitive Forms and Related Topics
161 SAGAN/STANLEY (eds). Mathematical Essays in Honor of Gian-Carlo Rota
162 ARNOLD/GREUEL/STEENBRINK. Singularities: The Brieskorn Anniversary Volume
163 BERNDT/SCHMIDT. Elements of the Representaiton Theory of the Jacobi Group
164 ROUSSARIE. Bifurcations of Planar Vector Fields and Hilbert's Sixteenth Problem
165 MIGLIORE. Introduction to Liaison Theory and Deficiency Modules
166 ELIAS/GIRAL/MIRO-ROIG/ZARZUELA (eds). Six Lectures on Commutative Algebra
167 FACCHINI. Module Theory
168 BALOG/KATONA/SZA'SZ/RECSKI (eds). European Congress of Mathematics, Budapest, July 22-26, 1996. Vol. I
169 BALOG/KATONA/SZA'SZ/RECSKI (eds). European Congress of Mathematics, Budapest, July 22-26, 1996. Vol. II
168/169 Sets Vols I, II
170 PATERSON. Groupoids, Inverse Semigroups, and their Operator Algebras
171 REZNIKOV/SCHAPPACHER (eds). Regulators in Analysis, Geometry and Number Theory
172 BRYLINSKI/BRYLINSKI/NISTOR/TSYGAN/XU (eds). Advances in Geometry

173 DRÄXLER/MICHLER/RINGEL (eds). Computational Methods for Representation of Groups and Algebras: Euroconference in Essen
174 GOERSS/JARDINE. Simplicial Homotopy Theory
175 BAÑUELOS/MOORE. Probabilistic Behavior of Harmonic Functions
176 BASS/LUBOTZKY. Tree Lattices
177 BIRKENHAKE/LANGE. Complex Tori
178 PUIG. On the Local Structure of Morita and Rickard Equivalence Between Brauer Blocks
179 MORALES RUIZ. Differential Galois Theory and Non-Integrability of Hamiltonian Systems
180 PATERNAIN. Geodesic Flows
181 HAUSER/LIPMAN/OORT/QUIRÓS (eds). Resolution of Singularities: In Tribute to Oscar Zariski
182 BILLEY/LAKSHMIBAI. Singular Loci of Schubert Varieties
183 KAPOVICH. Hyperbolic Manifolds and Discrete Groups
184 du SAUTOY/SEGAL/SHALEV (eds). New Horizons in pro-p Groups
185 FARAUT/KANEYUKI/KORÁNYI/LU/ROOS. Analysis and Geometry on Complex Homogeneous Domains
186 KLEINERT. Units in Skew Fields
187 BOST/LOESER/RAYNAUD. Courbes Semi-stables et Groupe Fondamental en Géométrie Algébrique
188 DOLBEAULT/IORDAN/HENKIN/SKODA/TRÉPREAU (eds). Complex Analysis and Geometry: International Conference in Honor of Pierre Lelong
189 ANDRÉ/BALDASSARI. DeRham Cohomology of Differential Modules on Algebraic Varieties
190 van den ESSEN. Polynomial Automorphisms and the Jacobian Conjecture.
191 KASHIWARA/MIWA (eds). Physical Combinatorics
192 DEHORNOY. Braids and Self-Distributivity
193 KNUDSON. Homology of Linear Groups
194 JUHL. Cohomological Theory of Dynamical Zeta Functions
195 FABER/van der GEER/OORT (eds). Moduli of Abelian Varieties
196 AGUADÉ/BROTO/CASACUBERTA (eds). Cohomological Methods in Homotopy Theory
197 CHERIX/COWLING/JOLISSAINT/JULG/VALETTE (eds). Groups with the Haagerup property
198 LANDSMAN/PFLAUM/SCHLICHENMAIER (eds). Quantization of Singular Symplectic Quotients
199 PEYRE/TSCHINKEL. Rational Points on Algebraic Varieties
200 GOLUBITSKY/STEWART. The Symmetry Perspective
201 CASACUBERTA/MIRÓ-ROIG/VERDERA/XAMBÓ-DESCAMPS (eds). European Congress of Mathematics, Volume 1
202 CASACUBERTA/MIRÓ-ROIG/VERDERA/XAMBÓ-DESCAMPS (eds). European Congress of Mathematics, Volume 2
203 BLAIR. Riemannian Geometry of Contact and Symplectic Manifolds
204 KUMAR. Kac–Moody Groups, their Flag Varieties, and Representation Theory
205 BOUWKNEGT/WU. Geometric Analysis and Applications to Quantum Field Theory
206 SNAITH. Algebraic K-groups as Galois Modules
207 LONG. Index Theory for Symplectic Paths with Applications
208 TURAEV. Torsions of 3-dimensional Manifolds
209 UNTERBERGER. Automorphic Pseudodifferential Analysis and Higher Level Weyl Calculi
210 JOSEPH/MELNIKOV/RENTSCHLER (eds). Studies in Memory of Issai Schur
211 HERNÁNDEZ-LERMA/LASSERE. Markov Chains and Invariant Probabilities
212 LUBOTZKY/SEGAL. Subgroup Growth
213 DUVAL/GUIEU/OVSIENKO (eds). The Orbit Method in Geometry and Physics. In Honor of A.A. Kirillov
214 DUNGEY/TER ELST/ROBINSON. Analysis on Lie Groups with Polynomial Growth
215 ARONE/HUBBUCK/LEVI/WEISS. Algebraic Topology.
216 BOROVIK/GELFAND/WHITE. Coxeter Matroids.
217 THANGAVELU. An Introduction to the Uncertainty Principle: Hardy's Theorem on Lie Groups.

218 FRESENEL/VAN DER PUT. Rigid Analytical Geometry and its Applications.
219 SURIS. The Problem of Integrable Discretization: Hamiltonian Approach
220 DELORME/VERGNE (eds). Noncommutative Harmonic Analysis: In Honor of Jaques Carmona
221 GRAY. Tubes, 2nd Edition.
222 ORTEGA/RATIU. Momentum Maps and Hamiltonian Reduction
223 ANDREU-VAILLO/CASELLES/MAZÓN. Parabolic Quasilinear Equations Minimizing Linear Growth Functionals
224 CREMONA/LARIO/QUER/RIBET (eds). Modular Curves and Abelian Varieties
225 ANDERSON/PASSARE/SIGURDSSON. Complex Convexity and Analytic Functionals
226 POONEN/TSCHINKEL (eds). Arithmetic of Higher-Dimensional Algebraic Varieties
227 LEPOWSKY/LI. Introduction to Vertex Operator Algebras and Their Representations
228 ANKER/ORSTED (eds). Lie Theory: Lie Algebras and Representations